ASTRONOMY AND ASTROPHYSICS ABSTRACTS

A Publication of the Astronomisches Rechen-Institut Heidelberg
Member of the Abstracting Board of the International
Council of Scientific Unions

Volume 3
Literature 1970, Part 1

Edited by
S. Böhme · W. Fricke · U. Güntzel-Lingner
F. Henn · D. Krahn · G. Zech

Springer-Verlag Berlin Heidelberg GmbH

Astronomisches Rechen-Institut
Heidelberg

Director: Prof. Dr. W. Fricke

Astronomy and Astrophysics Abstracts
Editor-in-Chief: F. Henn

Astronomy and Astrophysics Abstracts
is prepared under the auspices
of the International Astronomical Union

ISBN 978-3-662-00961-1   ISBN 978-3-662-00959-8 (eBook)
DOI 10.1007/978-3-662-00959-8

Preface

Astronomy and Astrophysics Abstracts, which appears in semi-annual volumes, is devoted to the recording, summarizing and indexing of astronomical publications throughout the world. It is prepared under the auspices of the International Astronomical Union (according to a resolution adopted at the 14th General Assembly in 1970).

Astronomy and Astrophysics Abstracts aims to present a comprehensive documentation of literature in all fields of astronomy and astrophysics. Every effort will be made to ensure that the average time interval between the date of receipt of the original literature and publication of the abstracts will not exceed eight months. This time interval is near to that achieved by monthly abstracting journals, compared to which our system of accumulating abstracts for about six months offers the advantage of greater convenience for the user.

Volume 3 contains literature published in 1970 and received before August 15, 1970; some older literature which was received late and which is not recorded in earlier volumes is also included.

The authors of papers who have sent us abstracts on request have effectively contributed to the success of our service. We should like to express our gratitude to them. We acknowledge with thanks contributions to this volume by Dr. J. Bouška, who surveyed journals and publications in the Czech language and supplied us with abstracts in English, by Dr. B. Onderlička, Brno, for providing English abstracts of Russian papers, and by the Commonwealth Scientific and Industrial Research Organization (C.S.I.R.O.), Sydney, for providing titles and abstracts of papers on radio astronomy.

Our warmest thanks also to the ladies of our editorial office, Mrs Monika Betz, and Mrs Utta-Barbara Stegemann, who typed the text of this volume on IBM 72 Composers and compiled the pages from abstract slips in a perfect form for offset reproduction, and to Miss Gisela Nollert, for punching all the material for author index and subject index.

Heidelberg, September 1970

Siegfried Böhme
Walter Fricke
Ulrich Güntzel-Lingner
Frieda Henn
Dietlinde Krahn
Gert Zech

Contents

Stellar Systems

Introduction

Astronomical bibliographies

Astronomy and Astrophysics Abstracts begins documentation and abstracting as from the year 1969. For information on astronomical literature before this date consultation of one of the following bibliographies is suggested:

(1) J. J. de Lalande, Bibliographie Astronomique, Paris 1803 (this work covers the time from 480 B. C. to the year 1803, VIII + 966 pages).

(2) J. C. Houzeau, A. Lancaster, Bibliographie générale de l'astronomie, Volume I (in two parts), Bruxelles 1882, 1887, Volume II, Bruxelles 1889. The complete title of Volume II is "Bibliographie générale de l'astronomie ou catalogue méthodique des ouvrages, des mémoires et des observations astronomiques, publiés depuis l'origine de l'imprimerie jusqu'en 1880". A new edition of these volumes was prepared by D. W. Dewhirst in 1964.

(3) Bibliography of Astronomy, 1881 - 1898. The literature of this period was recorded on standard slips by the Observatoire Royal de Belgique. From the material (some 52.000 items) a microfilm version was produced by University Microfilms Limited, Tylers Green, High Wycombe, Buckinghamshire, England, in 1970.

(4) Astronomischer Jahresbericht, 1899 gegründet von Walter Wislicenus, herausgegeben vom Astronomischen Rechen-Institut in Heidelberg (formerly in Berlin), Verlag W. de Gruyter, Berlin. For the period from 1899 to 1968 sixty-eight volumes were published, each of which, in general, covers the literature of one year.

(5) Bulletin Signalétique – Section Astronomie, Astrophysique, Physique du Globe. Published by Centre de Documentation du Centre National de la Recherche Scientifique, Paris. This publication is a continuation of "Bibliographie Mensuelle de l'Astronomie" founded in 1933 by the Société Astronomique de France. The publication is continued.

(6) Referativnyj Zhurnal. Founded in 1953 and published by Vsesoyuznyj Institut Nauchnoj i Tekhnicheskoj Informatsii, Akademiya Nauk, Moskva. The publication is continued.

Concept of Astronomy and Astrophysics Abstracts

This abstracting service aims to present a comprehensive documentation of the literature in all fields of astronomy and astrophysics. It appears in semi-annual volumes, two of which cover the literature of a calendar year. The half-yearly period of issue is regarded as an optimal period of time for summarizing papers into subject categories and for the presentation of abstracts as quickly as possible after the publication of the original literature. The time limits at which the documentation begins and ends for a volume are not sharply defined, except in the sense that all literature will be covered which was received by the editors within these limits.

Vol. 3 is devoted to the recording, summarizing and indexing of astronomical publications of the year 1970 received from January to August 15, 1970; it also records a number of papers issued before 1970 but received within the given period of time.

The main characteristics of the concept of Astronomy and Astrophysics Abstracts may be summarized briefly.

(1) Titles of papers are given in the language of their authors whenever possible. If they are not in English but supplied with English translations they will be given in English. Abstracts are presented in English, French or German. Titles of papers in Russian are, as a rule, given in English; occasionally, they are given in German.

(2) Authors' abstracts are used whenever possible. In this volume only very few abstracts have been written by persons other than the authors; in such cases the name of the abstractor is given. As a rule, no popular articles were abstracted their titles however given, occasionally with the addition "Popular article".

(3) As a rule, each paper has been classified into one of 108 numbered subject categories and allocated a serial number within the category. In this way each item is numbered by six figures, the first three of which indicate the number of the category. Three further figures indicate the serial number within the category, which was allocated in the order of the receipt of the abstract. Reference to an abstract in Volume 1 is indicated by "01" before the number of the category; for example, 01.074.028, denotes Volume 1, category 074, abstract 028. Vol. 2 is indicated by "02", Vol. 3 by "03". A paper may have been classified into more than one category. Then its abstract has been allocated a number in one of the categories involved, and in the other category (or categories) the paper has been indicated by the title and a reference to the abstract number.

Papers whose authors are not named were treated like those with authors' names, with one exception: reports from correspondents of journals whose names were unknown were not numbered.

(4) There are categories which suggest the presentation of the material in subject groups; these, however, can only be formed immediately before the completion of a volume. For instance, a subject group may be formed by all information received on the same solar eclipse, comet, nova, etc. The unsorted presentation of such material in a subject category would be inconvenient for the user, even if the individual comet, etc. were included in the subject index.

The following subject categories are subdivided into subject groups:

008 Observatories, Institutes. The publications of observatories and astronomical institutes are listed in alphabetical order of the towns of the institutions, each town forming a numbered subject group. For each publication a reference to an abstract number is made.

010 Societies, Associations, Organizations. The publications of each one form a subject group. The groups are presented in alphabetical order.

079 Solar eclipses. All publications related to one solar eclipse form a subject group.

103 Comets, listed objects. All publications related to the same comet form a numbered group.

124 Novae. All publications related to one nova form a subject group.

125 Supernovae. All publications related to one supernova form a subject group.

(5) Border fields of astronomy and astrophysics have been taken into account by presenting titles of papers occasionally without abstracts. The selection of papers for inclusion has been made according to the degree of relevance to astronomical research.

(6) The text of the publication was typed on IBM 72 Composers in the editorial office, and it was given to the printer in a form ready for offset reproduction. The author index and the subject index were compiled and printed by means of electronic computer (Siemens 2002).

(7) While each volume is scheduled to contain an author index and a subject index, the magnetic tapes containing the

index information will be used to produce separate index volumes (authors and subjects) at intervals of a few years.

Transliteration of the Russian Alphabet

The transliteration of the Russian alphabet in use in Astronomy and Astrophysics Abstracts is presented here.

А	а	a	Р	р	r
Б	б	b	С	с	s
В	в	v	Т	т	t
Г	г	g	У	у	u
Д	д	d	Ф	ф	f
Е	е	e	Х	х	kh
Ё	ё	e	Ц	ц	ts
Ж	ж	zh	Ч	ч	ch
З	з	z	Ш	ш	sh
И	и	i	Щ	щ	shch
Й	й	j	Ъ	ъ	''
К	к	k	Ы	ы	y
Л	л	l	Ь	ь	'
М	м	m	Э	э	eh
Н	н	n	Ю	ю	yu
О	о	o	Я	я	ya
П	п	p			

This transliteration was recommended by the Abstracting Board of the International Council of Scientific Unions in 1969. It is essentially the same as the transliteration proposed by the Academy of Sciences, Moscow, and used by the Referativnyj Zhurnal (See Referativnyj Zhurnal, 51. Astronomiya, 1969 No. 1). It may be noted that the letters can be read and printed by usual data processing machines.

In the literature the names of Russian authors can be found transliterated in different ways. We present the names in the form in which they are given in the literature.

Sources of information

The majority of sources of information for this volume are given in section **001 Periodicals** and in section **008 Observatories, Institutes.** The term "periodical" has been used in its widest sense for publications in a sequence of undetermined duration, even if the intervals of appearance are not regular. Section 001 records 265 periodicals with their full titles and with abbreviations which are in use in Astronomy and Astrophysics Abstracts. It may be noted that the titles of the periodicals are given in their original languages, and that Russian titles have been transliterated applying the transliteration given above. Section 008 records 150 periodicals; these are publication series of observatories and astronomical institutes which have not been included in section 001. The abbreviations of the titles of the periodicals have been given so that in most cases they permit recognition of the full title without recourse to the key in section 001. The steadily growing number of periodicals makes it necessary to use more extensive abbreviations and to abandon the use of very condensed ones.

Other abstracting journals have been consulted in order to examine the degree of completeness of our service. Occasionally, in particular in Physics Abstracts, Referativnyj Zhurnal, and Bulletin Signalétique abstracts of papers were found which had not come to our attention. In such cases Astronomy and Astrophysics Abstracts gives the titles with references to the other abstracting service.

Classification into a scheme of subject categories

The subdivision of astronomy and its border fields into subject categories is facilitated by the fact that the astronomical objects appear to be particularly well suited for the formation of categories. Sun, moon, earth, planets, comets, and meteorites, the various kinds of stars, galaxies, radio sources, quasars, and pulsars etc. suggest natural subdivisions. It may be assumed that such subdivisions can be maintained for long periods of time. Experience shows, however, that progress in research may imply changes in the classification scheme, in particular, in fields where the expansion of knowledge is explosive. Probably one of the best examples which shows clearly the reflection of research progress on astronomical documentation is the Crab nebula, which may be classified as a supernova remnant, an emission nebula, a radio source or pulsar. In this volume subject category 134 is devoted to the Crab nebula. Papers related to the Crab nebula will, however, be found in the subject categories 125 (Supernova remnants), 141 (Radio sources, pulsars), 142 (X ray sources), 143 (Cosmic radiation).

A few explanatory remarks may be in order on some of the subject categories. Section 002 includes short news notes whose titles and authors are given, but the authors of the notes have not been included in the author index. In section 003 books on Astronomy and Astrophysics and its border fields are listed which came to our notice from January to August. In cases where books can be classified into one of the subject categories these books are additionally listed under their categories. References to book reviews are given if the reviews appeared quickly.

For completeness of documentation, personal notes (section 006) and obituaries (section 007) are listed. In section 012 (Proceedings of Colloquia, Congresses, Meetings, and Symposia) the proceedings etc. are listed with titles and editors. Whenever the volumes were at hand, the papers were classified into their subject categories and, occasionally, supplied with abstracts.

Author index and subject index

The subject category and the serial number forming six figures for each abstract have been used as a means of reference in the author index and the subject index. These references are more precise than page references. They offer considerable advantages in indexing by means of data processing machines, and they are more convenient for the user.

The author index of this volume contains 5960 names. A complete reference comprises six figures, three for the subject category and three for the serial number within the category. In the case of more than one reference to abstracts in one category, the number of the category is given only once and not repeated in the immediately following references. The total number of papers (some do not give names of authors) recorded in this volume is about 5600.

We consider the subject index as only a first approximation to an optimal index covering all fields of astronomy and astrophysics and its border fields. Several iterative steps appear to be necessary until an index has been compiled for one of the subsequent volumes which may then serve as a kind of standard for the near future. The assigning of one or more key words to a paper is undoubtedly a difficult task. Some journals have started giving key words together with the titles of papers. These key words are chosen by the authors themselves and are in many cases identical with our designations of subject categories with no additional specification. In fact, in some cases it may be more useful to refer to a subject category as a whole than to an item number, in particular, if the total number of abstracts in a category is very small, and if more specific key words do not provide a proper description of the paper.

Abbreviations

AAS	American Astronomical Society	Geogr.	Geography, etc.
AAVSO	American Association of Variable Star Observers	Geophys.	Geophysics, etc.
		Ges.	Gesellschaft
Abh.	Abhandlungen	Glav.	Glavnyj (Main)
Abstr.	Abstract	Gos.	Gosudarstvennyj (State)
Abt.	Abteilung	HRD	Herzsprung-Russell diagram
Acad.	Academy, etc.	Hydrogr.	Hydrography, etc.
Accad.	Accademia	IAF	International Astronautical Federation
Adv.	Advances	IAU	International Astronomical Union
AG	Astronomische Gesellschaft	ICSU	International Council of Scientific Unions
AIAA	American Institute of Aeronautics and Astronautics	IEEE	Institute of Electrical and Electronics Engineers
AJB	Astronomischer Jahresbericht	Industr.	Industry, etc.
Akad.	Akademie	Inform.	Information
An.	Anales, etc.	Inst.	Institute, etc.
Ann.	Annals, etc.	Instn.	Institution
Arch.	Archiv, etc.	Ionosph.	Ionosphere, etc.
Ark.	Arkiv	Issled.	Issledovaniya (Research)
ASA	Astronomical Society of Australia	Ist.	Istituto
Asoc.	Asociación	Izv.	Izvestiya (News)
ASP	Astronomical Society of the Pacific	Jb.	Jahrbuch
Ass.	Association	JO	Journal des Observateurs
ASSA	Astronomical Society of Southern Africa	Journ.	Journal
Astrofis.	Astrofisica, etc.	Kl.	Klasse
Astrofiz.	Astrofizika, etc.	Lab.	Laboratory
Astron.	Astronomy, etc.	Mag.	Magazine
Astronaut.	Astronautics, etc.	Mat.	Matematica, etc.
Astrophys.	Astrophysics, etc.	Math.	Mathematics, etc.
ASV	Astronomical Society of Victoria	Mech.	Mechanics, etc.
ASWA	Astronomical Society of Western Australia	Med.	Mededelingen
Atmosph.	Atmosphere, etc.	Medd.	Meddelande, Meddelser
BA	Bulletin Astronomique	Mekhan.	Mekhanika, etc.
BAA	British Astronomical Association	Mém.	Mémoires
BAN	Bulletin of the Astronomical Institutes of the Netherlands	Mem.	Memoirs, Memorandum, etc.
		Meteorol.	Meteorology, etc.
Ber.	Berichte	MIT	Massachusetts Institute of Technology
BIH	Bureau International de l'Heure (Paris)	Mitt.	Mitteilungen
Bol.	Boletin	MVS Sonneberg	Mitteilungen über Veränderliche Sterne, Sonneberg
Boll.	Bolletino		
Bull.	Bulletin	Nachr.	Nachrichten
Byull.	Byulleten' (Bulletin)	NASA	National Aeronautics and Space Administration
Circ.	Circular		
Cl.	Classe	Nat.	Naturwissenschaftlich, etc.
Coll.	Collection	Naut.	Nautics, etc.
Commun.	Communication	NBS	National Bureau of Standards
Comun.	Comunicazioni	NRAO	National Radio Astronomy Observatory (Green Bank)
Contr.	Contributions, etc.		
COSPAR	Committee on Space Research	NRL	Naval Research Laboratory (Washington)
C.S.I.R.O.	Commonwealth Scientific Industrial Research Organization	Obs.	Observatory, etc.
		OSA	Optical Society of America
Dep.	Department	Oss.	Osservatorio, Osservazioni, etc.
Diss.	Dissertation	Ped.	Pedagogika, etc. (Pedagogics)
Div.	Division	Phil.	Philosophical
Dokl.	Doklady (Reports)	Phys.	Physics, etc.
ESO	European Southern Observatory	Planet.	Planetary
ESRO	European Space Research Organization	Priklad.	Prikladnoj (Applied)
Fis.	Fisica, etc.	Proc.	Proceedings
Fiz.	Fizika, etc.	Progr.	Progress, etc.
Fys.	Fysica, etc.	Pubbl.	Pubblicazioni
Géod.	Géodésie, etc.	Publ.	Publications
Geod.	Geodesy, etc.	Rap.	Raportoj
Geofis.	Geofisica, etc.	RAS	Royal Astronomical Society
Geofiz.	Geofizika, etc.	RAS Canada	Royal Astronomical Society of Canada
Geofys.	Geofysik, etc.	Rech.	Recherches
Geol.	Geology, etc.	Rend.	Rendiconti

3

Abbreviations

Rep.	Report		Techn.	Technics, etc.
Repr.	Reprint		Tekhn.	Tekhnika, etc.
Res.	Research		Teor.	Teoreticheskij
Rev.	Review, etc.		Terr.	Terrestrial, etc.
Ric.	Ricerche		TH	Technische Hochschule
Roy.	Royal, etc.		Theor.	Theoretical
SAF	Société Astronomique de France		Tidssk.	Tidsskrift
SAI	Società Astronomica Italiana		Trans.	Transactions
SAO	Smithsonian Astrophysical Observatory		Trudy	Trudy (Publications)
SAS	Société Astronomique de Suisse		Tsentr.	Tsentral'nyj (Central)
Sci.	Science, etc.		Tsirk.	Tsirkulyar (Circular)
Sect.	Section		TU	Technical University
Ser.	Series, etc.		Uch. Zap.	Uchenye Zapiski (Treatise)
S. I. R.	Service International Rapide des Latitudes		Univ.	University, etc.
Sitz.-Ber.	Sitzungsberichte		URSI	Union Radio Scientifique Internationale
Soc.	Society		Verh.	Verhandlungen
Soobshch.	Soobshcheniya (Communications)		Veröff.	Veröftentlichungen
Sternw.	Sternwarte		Wet.	Wetenschappen
Stud. Cerc.	Studii şi Cercetari		Wiss.	Wissenschaften, etc.
Supl.	Suplemento		Zeitschr.	Zeitschrift
Suppl.	Supplement		ZfA	Zeitschrift für Astrophysik
SuW	Sterne und Weltraum		Zhurn.	Zhurnal (Journal)

Periodicals, Proceedings, Books, Activities

001 Periodicals

Abh. Deutsch. Akad. Wiss. Berlin
Abhandlungen der Deutschen Akademie der Wissenschaften zu Berlin. Klasse Mathematik, Physik und Technik. Publisher: Akademie-Verlag, Berlin.

Acad. Roy. Belgique, Bull. Cl. Sci.
Académie Royale de Belgique, Bulletin de la Classe des Sciences (Koninklijke Academie van België, Mededelingen van de Klasse der Wetenschappen). 5ᵉ Série. Palais des Académies, Bruxelles.

Acta Astron.
Acta Astronomica. Publisher: Polska Akademia Nauk, Warszawa - Kraków.

Acta Phys. Austriaca
Acta Physica Austriaca. Publisher: Springer-Verlag, Wien.

Acta Univ. Carolinae Math. Phys.
Acta Universitatis Carolinae, Mathematica et Physica. Administrace: Matematicko-fyzikálni fakulta University Karlovy, Praha.

Acta Univ. Lundensis
Acta Universitatis Lundensis. Sectio II: Medica, Mathematica, Scientiae Rerum Naturalium. Editor: Royal Physiographic Society of Lund.

Actas Acad. Nacional Cienc. Lima
Actas de la Academia Nacional de Ciencias Exactas, Fisicas y Naturales de Lima. Lima - Peru.

Adv. Astron. Astrophys.
Advances in Astronomy and Astrophysics. Publisher: Academic Press, New York – London.

AIAA Journ.
AIAA Journal. A Publication of the American Institute of Aeronautics and Astronautics, Easton, Pa.

Ann. d'Astrophys.
Annales d'Astrophysique. Revue internationale bimestrielle publiée par le Centre National de la Recherche Scientifique et éditée par son Service d'Astrophysique, Paris. After Vol. 31 replaced by "Astronomy and Astrophysics".

Ann. Françaises Chronométrie Micromécanique
Annales Françaises de Chronométrie et de Micromécanique, publication annuelle de l'Observatoire de Besançon, du Centre Technique de l'Industrie Horlogère et de la Société Française de Chronométrie et de Micromécanique. Rédaction et administration: Observatoire de Besançon. Publiées avec le concours du Centre National de la Recherche Scientifique et des organismes corporatifs.

Ann. Géophys.
Annales de Géophysique. Revue Internationale trimestrielle, publiée par le Centre National de la Recherche Scientifique, Paris.

Ann. Obs. Astron. Météorol. Toulouse
Annales de l'Observatoire Astronomique et Météorologique de Toulouse. Publisher: Gauthier-Villars, Paris.

Ann. Physics
Annals of Physics. Publisher: Academic Press, New York.

Ann. Physik
Annalen der Physik. 7. Folge. Publisher: Johann Ambrosius Barth, Leipzig.

Ann. Physique
Annales de Physique. Publisher: Masson et Cie., Paris.

Ann. Univ.-Sternw. Wien
Annalen der Universitäts-Sternwarte Wien. In Kommission bei Ferd. Dümmlers Verlag, Bonn.

Annual Rev. Astron. Astrophys.
Annual Review of Astronomy and Astrophysics. Publisher: Annual Reviews Inc., Palo Alto, California.

Anzeiger. Österreich. Akad. Wiss. Math.-Nat. Kl.
Anzeiger. Österreichische Akademie der Wissenschaften. Mathematisch-Naturwissenschaftliche Klasse. Publisher: Springer-Verlag, Wien.

Applied Optics
Applied Optics. Published by the Optical Society of America (in Cooperation with the American Institute of Physics), Washington, D.C.

Arch. Sci. Genève
Archives des Sciences, éditées par la Société de Physique et d'Histoire naturelle de Genève. Publisher: Imprimérie Kundig, Genève.

Ann. Soc. Sci. Bruxelles
Annales de la Société Scientifique de Bruxelles. Série I: Sciences Mathématiques, Astronomiques et Physiques. Published by Institut de Physique, Heverlé-Louvain.

Ark. Astron.
Arkiv för Astronomi. Utgivet av Kungliga Svenska Vetenskapsakademien, Stockholm. Printed by Almqvist & Wiksell, Stockholm.

Ark. Fys.
Arkiv för Fysik. Kungliga Svenska Vetenskapsakademien, Stockholm. Printed by Almqvist & Wiksell, Stockholm.

Artificial Satellites
Artificial Satellites. Publication of Polish Scientific Institutions. Polish Academy of Sciences, National Committee of Geophysics and Geodesy, National Committee for Space Research, Warsaw. Publishing Office: Palac Kultury i Nauki, Warszawa.

Asoc. Argentina Astron. Bol.

Asociacion Argentina de Astronomia. Boletin. Editor: Instituto Argentino de Radioastronomia, Provincia de Buenos Aires, Argentina. Printer: Talleres Gráficos "Renovacion", La Plata, República Argentina.

Astrofizika
Astrofizika. Izdatel'stvo Akademii Nauk Armyanskoj SSR, Erevan. [A translation published as "Astrophysics."]

Astron. Astrophys.
Astronomy and Astrophysics. A European Journal. Published by Springer-Verlag, Berlin - Heidelberg - New York.

Astron. in der Schule
Astronomie in der Schule. Zeitschrift für die Hand des Astronomielehrers. Herausgegeben vom Verlag Volk und Wissen, Berlin. Redaktion: Sternwarte Bautzen.

Astron. Journ.
The Astronomical Journal. Published for the American Astronomical Society by the American Institute of Physics, New York.

Astron. Nachr.
Astronomische Nachrichten. Publisher: Akademie-Verlag, Berlin.

Astron. Soc. Pacific Leaflet
Astronomical Society of the Pacific. Leaflet. Edited by the Astronomical Society of the Pacific, San Francisco, California.

Astron. Tidsskr.
Astronomisk Tidsskrift. Edited by Astronomisk Selskab, København; Norsk Astronomisk Selskap, Oslo; Svenska Astronomiska Sällskapet, Stockholm. Printed by John Griegs Boktrykkeri, Bergen.

Astron. Tsirk.
Astronomicheskij Tsirkulyar, izdavaemyj Byuro Astronomicheskikh Soobshchenij Akademii Nauk SSSR. Moskva.

Astron. Vestn.
Astronomicheskij Vestnik. Publishers: Izdatel'stvo "Nauka", Moskva.

Astron. Zhurn. Akad. Nauk SSSR
Astronomicheskij Zhurnal. Akademiya Nauk SSSR. Publishers: Izdatel'stvo "Nauka", Moskva.

Astronaut. Acta
Astronautica Acta. An Archive Journal of the International Academy of Astronautics. Publisher: Pergamon Press, Oxford – New York.

Astronaut. Aeronaut.
Astronautics & Aeronautics. A Publication of the American Institute of Aeronautics and Astronautics. Published monthly by the American Institute of Aeronautics and Astronautics, Easton, Pennsylvania.

Astrophysics
Astrophysics. The Faraday Press cover-to-cover translation of Astrofizika. The Faraday Press, Inc., New York, N. Y.

Astrophys. Journ.
The Astrophysical Journal. Published in collaboration with the American Astronomical Society by the University of Chicago Press, Chicago, Illinois.

Astrophys. Journ. Suppl. Series
The Astrophysical Journal. Supplement Series. Published in collaboration with the American Astronomical Society by the University of Chicago Press, Chicago, Illinois.

Astrophys. Letters
Astrophysical Letters. Published by Gordon and Breach, Science Publisher Ltd., New York - London - Paris.

Astrophys. Norvegica
Astrophysica Norvegica. Edited by The Institute of Theoretical Astrophysics, University of Oslo (Det Norske Videnskaps-Akademi i Oslo). Universitets-forlaget, Oslo.

Astrophys. Space Sci.
Astrophysics and Space Science. An International Journal of Cosmic Physics. Published by D. Reidel Publishing Company, Dordrecht, Holland.

Atti Accad. Nazionale Lincei. Mem.
Atti della Accademia Nazionale dei Lincei. Serie Ottava. Memorie. Classe di Scienze fisiche, matematiche e naturali. Sezione I: Matematica, Meccanica, Astronomia, Geodesia e Geofisica. Published by Accademia Nazionale dei Lincei, Roma.

Atti Accad. Nazionale Lincei Rend.
Atti della Accademia Nazionale dei Lincei. Serie Ottava. Rendiconti. Classe di Scienze fisiche, matematiche e naturali. Published by Accademia Nazionale dei Lincei, Roma.

Australian Journ. Phys.
Australian Journal of Physics. Published by the Commonwealth Scientific and Industrial Research Organization, East Melbourne, Victoria.

Australian Journ. Phys. Astrophys. Suppl.
Australian Journal of Physics, Astrophysical Supplement. Printed by Commenwealth Scientific and Industrial Research Organization, Melbourne, Victoria.

BAV Rundbrief
BAV Rundbrief. Mitteilungsblatt der Berliner Arbeitsgemeinschaft für Veränderliche Sterne. Editor: BAV Berliner Arbeitsgemeinschaft für Veränderliche Sterne eV., Berlin.

Bol. Inst. Mat., Astron., Fis. Univ. Nacional Córdoba
Boletin del Instituto de Matematica, Astronomía y Física, Universidad Nacional de Córdoba (R. A.) Dirección General de Publicaciones, Córdoba (Argentina).

Bol. Liga Latinoamericana Astron.
Boletin de la Liga Latinoamericana de Astronomia. Publicado por la Asociacion Argentina Amigos de la Astronomia, Buenos Aires, Argentina.

Boll. Geod. Sci. Affini
Bolletino di Geodesia e Scienze Affini. Pubblicazione dell' Istituto Geografico Militare, Firenze.

British Astron. Ass. Circ.
British Astronomical Association, Circular. Editorial Office: 97 Hawkswood Drive, Hailsham, Sussex.

Bull. American Astron. Soc.
Bulletin of the American Astronomical Society. Published for the American Astronomical Society by the Ame-

rican Institute of Physics Inc., New York, N. Y.

Bull. Astron. (BA)
Bulletin Astronomique. 3e Série. Publié par le Centre National de la Recherche Scientifique, Paris. After Vol. 3 (1968) replaced by "Astronomy and Astrophysics".

Bull. Astron. Inst. Czechoslovakia (BAC)
Bulletin of the Astronomical Institutes of Czechoslovakia. Published under the auspices of the Czechoslovak Academy of Sciences by Academia, Praha. Editor: Astronomical Institutes of the Czechoslovak Academy of Sciences, Praha.

Bull. Astron. Inst. Netherlands (BAN)
Bulletin of the Astronomical Institutes of the Netherlands. Publisher: North-Holland Publishing Company, Amsterdam. After Vol. 20 replaced by "Astronomy and Astrophysics".

Bull. Astron. Inst. Netherlands, Suppl. Series
Bulletin of the Astronomical Institutes of the Netherlands. Supplement Series. Published by the Astronomical Institutes. Replaced by "Astronomy and Astrophysics", Supplement Series.

Bull. Géod.
Bulletin Géodésique. Nouvelle Série. Publié par le Bureau Central de l'Association Internationale de Géodésie, Paris.

Bull. Geograph. Survey Inst.
Bulletin of the Geographical Survey Institute. Published by the Geographical Survey Institute, Ministry of Construction, Tokyo, Japan.

Bull. Hor.
Bulletin Horaire du Bureau International de l'Heure. Rédaction: BIH, Observatoire de Paris.

Bull. Mesures Ionosph.
Bulletin de Mesures Ionosphériques. Publié par le Centre National d'Etudes des Télécommunications, Issy-les-Moulineaux.

Bull. Obs. Astron. Beograd
Bulletin de l'Observatoire Astronomique de Beograd. Editor: Observatoire Astronomique de Beograd. Printed by Naučna delo, Beograd.

Bull. Sci. Yougoslavie
Bulletin Scientifique. Conseil des Academies des Sciences et des Arts de la RSF de Yougoslavie. Section A: Sciences Naturelles, Techniques et Médicales. Redaction et Administration: Opaticka ul. 18/II, Zagreb (Yougoslavie).

Bull. Signal.
Bulletin Signalétique. Section 120: Astronomie et Astrophysique, Physique du Globe. Centre de Documentation du Centre National de la Recherche Scientifique, Paris.

Bull. Soc. Roy. Sci. Liège
Bulletin de la Société Royale des Sciences de Liège. L'Université, Liège.

Byull. Abastuman. Astrofiz. Obs.
Abastumanskaya Astrofizicheskaya Observatoriya, Gora Kanobili. Byulleten'. Akademiya Nauk Gruzinskoj SSR. Publishers: Izdatel'stvo "Metsniereba", Tbilisi.

Byull. Stantsij Optichesk. Nablyud. Iskusstv. Sputnikov Zemli
Byulleten' Stantsij Opticheskogo Nablyudeniya Iskusstvennykh Sputnikov Zemli. Published by Astronomicheskij Sovet Akademii Nauk SSSR, Moskva.

Canadian Journ. Phys.
Canadian Journal of Physics. Published by the National Research Council of Canada, Ottawa. Printed in Canada by the University of Toronto Press, Toronto,Ont.

Celestial Mechanics
Celestial Mechanics. An International Journal of Space Dynamics. Publishers: D. Reidel Publishing Company, Dordrecht—Holland.

Ciel et Terre
Ciel et Terre. Bulletin de la Société Belge d'Astronomie, de Météorologie et de Physique du Globe. Publié avec le concours du Ministère de l'Education Nationale, par la Société Belge d'Astronomie, Bruxelles.

Circ. d'Information
Circulaire d'Information. Union Astronomique Internationale. Commission des Etoiles Doubles. Address: Observatoire de Meudon, Meudon, France.

Coelum
Coelum. Periodico bimestrale per la Divulgazione dell' Astronomia. Editor: Osservatorio Astronomico Universitario di Bologna.

Comments Astrophys. Space Phys.
Comments on Astrophysics and Space Physics. A Journal of Critical Discussion of the Current Literature. Publishers: Gordon and Breach, Science Publishers, Inc., New York — London.

Comptes Rendus Acad. Sci. Paris
Comptes Rendus hebdomadaires des Séances de l'Académie des Sciences, publié avec le concours du Centre National de la Recherche Scientifique. Imprimérie: Gauthier-Villars, Paris.

Contr. Atmosph. Phys.
Contributions to Atmospheric Physics — Beiträge zur Physik der Atmosphäre. Publisher: Friedrich Vieweg & Sohn, Braunschweig.

Cosmic Electrodynamics
Cosmic Electrodynamics. An International Journal devoted to Geophysical and Astrophysical Plasmas. Printed in The Netherlands by D. Reidel, Publishing Company, Dordrecht, Holland.

COSPAR Inform. Bull.
COSPAR. Information Bulletin. Address: COSPAR Secretariat, Paris.

Deutsche Geod. Kommission Bayer. Akad. Wiss.
Deutsche Geodätische Kommission bei der Bayerischen Akademie der Wissenschaften. Reihe A: Höhere Geodäsie; Reihe B: Angewandte Geodäsie; Reihe C: Dissertationen; Reihe D: Tafelwerke; Reihe E: Geschichte und Entwicklung der Geodäsie. Published by Verlag der Bayerischen Akademie der Wissenschaften, München.

Documentat. Observateurs
Documentation des Observateurs. Rédaction: Station d'Astrophysique de Forcalquier.

Documentat. Observateurs Circ.
Documentation des Observateurs. Circulaire. Rédaction:

Station d'Astrophysique de Forcalquier.

Dokl. Akad. Nauk
Doklady Akademii Nauk SSSR. Seriya Matematika, Fizika. Publishers: Izdatel'stvo "Nauka", Moskva.

Dunsink Obs. Publ.
Dunsik Observatory Publications. The Observatory of the School of Cosmic Physics, Dublin Institute for Advanced Studies, Dublin.

Earth Planet. Sci. Letters
Earth and Planetary Science Letters. A Letter Journal devoted to the Development in Time of the Earth and Planetary System. Publisher: North-Holland Publishing Company, Amsterdam.

El Universo
El Universo. Organo de la Sociedad Astronomica de Mexico, Mexico, D. F.

Endeavour
Eine in vier Sprachen erscheinende Übersicht über Fortschritte der Naturwissenschaft. Published by Imperial Chemical Industries Limited, London.

ESO Bull.
European Southern Observatory, Bulletin. Edited by European Southern Observatory. Office of the Director: Hamburg.

Fortschritte Phys.
Fortschritte der Physik. Publisher: Akademie-Verlag, Berlin.

Gaz. Astron. Mém.
Gazette Astronomique. Mémoires van het Sterrenkundig Genootschap van Antwerpen, (de la Société d'Astronomie d'Anvers), Antwerpen. Printer: «De Voorzorg», A. Van Leuvenhaege, Antwerpen.

Geochim. Cosmochim. Acta
Geochimica et Cosmochimica Acta. Journal of the Geochemical Society. Publishing House: Pergamon Press, Ltd., Oxford.

Geodezja Kartografia
Geodezja i Kartografia. Komitet Geodezji Polskiej Akademii Nauk. Publisher: Państwowe Wydawnictwo Naukowe, Warszawa.

Geomagn. Aeronom.
Geomagnetizm i Aehronomiya. Akademiya Nauk SSSR. Izdatel'stvo "Nauka", Moskva [A translation published as "Geomagnetism and Aeronomy".]

Geophys. Journ.
The Geophysical Journal of the Royal Astronomical Society. Published for the Royal Astronomical Society by Blackwell Scientific Publications, Oxford – Edinburgh.

Gerlands Beiträge Geophys.
Gerlands Beiträge zur Geophysik. Publisher: Akademische Verlagsgesellschaft Geest & Portig K.-G., Leipzig.

Glasnik Mat.
Glasnik Matematički. Published by the Society of Mathematicians and Physicists of the S. R. of Croatia. Publisher: Društvo Matematičara i Fizičara S. R. Hrvats-

ke, Zagreb.

Helvetica Phys. Acta
Helvetica Physica Acta. Publisher: E. Birhäuser, Basel.

Hemel en Dampkring
Maandblad van de Nederlandse Vereniging voor Weer-en Sterrenkunde en van de Vereniging voor Sterrenkunde, Meteorologie, Geophysica en Aanverwante Wetenschappen in Belgie. Publisher: Wolters-Noordhoff N.V., Groningen.

IAU Circ.
International Astronomical Union, Circular. Central Bureau for Astronomical Telegrams, Smithsonian Observatory, Cambridge, Mass.

IBM Journ. Res. Development
IBM Journal of Research and Development. Published bi-monthly by International Business Machines Corporation, Armonk, New York.

ICSU Bull.
ICSU Bulletin. International Council of Scientific Unions. Secretariat: 7, Via Cornelio Celso, Rome, Italy.

Icarus
Icarus. International Journal of the Solar System. Publisher: Academic Press, New York – London.

IEEE Spectrum
IEEE Spectrum. Published monthly by the Institute of Electrical and Electronics Engineers, Inc., New York, N.Y.

Inform. Bull. Southern Hemisphere
Information Bulletin of the Southern Hemisphere. Editorial Office: Observatorio Astronómico, La Plata, Argentina.

Inform. Bull. Variable Stars
Commission 27 of the I.A.U. Information Bulletin on Variable Stars. Konkoly Observatory, Budapest.

Infrared Physics
An International Research Journal. Pergamon Press Ltd., Oxford – London – New York.

Irish Astron. Journ.
The Irish Astronomical Journal. A Quarterly Publication under the auspices of the Observatories of Armagh and Dunsink. Subscription address: Managing Editor, Irish Astronomical Journal, Armagh Observatory, Northern Ireland.

Izv. Akad. Nauk Armyan. SSR
Izvestiya Akademii Nauk Armyanskoj SSR. Fizika Erevan.

Izv. Glav. Astron. Obs. Pulkove
Izvestiya Glavnoj Astronomicheskoj Observatorii v Pulkove. Akademiya Nauk SSSR. Izdanie Glavnoj astronomicheskoj observatorii v Pulkove, Leningrad.

Izv. Komissii Fiz. Planet
Izvestiya Komissii po Fizike Planet. Akademiya Nauk SSSR. Astronomicheskij Sovet. Moskva.

Izv. Krymskoj Astrofiz. Obs.
Izvestiya Krymskoj Astrofizicheskoj Observatorii. Akademiya Nauk SSR. Publishers: Izdatel'stvo "Nauka", Moskva.

JETP Letters
JETP Letters. A translation of JETP Pis'ma v Redaktsiyu of the Academy of Sciences in the USSR. Published semi-monthly by the American Institute of Physics, Lancaster, Pennsylvania.

Journ. Astronaut. Sci.
The Journal of the Astronautical Sciences. Published by the American Astronautical Society Inc., Baltimore, Md.

Journ. Astron. Soc. Victoria
The Journal of the Astronomical Society of Victoria. Printed by D. Buscombe Printers, Glen Waverley, Victoria.

Journ. Astron. Soc. Western Australia
The Journal of the Astronomical Society of Western Australia. Edited by the Astronomical Society of Western Australia, Perth, W. A.

Journ. Atmosph. Terr. Phys.
Journal of Atmospheric and Terrestrial Physics. Publishers: Pergamon Press, Oxford - London - New York.

Journ. British Astron. Ass.
Journal of the British Astronomical Association. Subscription address: Office of the Association, Hounslow West, Middlesex.

Journ. British Interplanet. Soc.
Journal of the British Interplanetary Society. Printed by Unwin Brothers, Ltd., London, and published by the British Interplanetary Society.

Journ. Fluid Mechanics
Journal of Fluid Mechanics. Published by Cambridge University Press, London – New York.

Journ. Geophys. Res.
Journal of Geophysical Research. An International Scientific Publication. Published by the American Geophysical Union, Washington, D. C.

Journ. History Astron.
Journal for the History of Astronomy. Published by Macdonald and Co. (Publishers) Ltd., London. Printed in Great Britain by W. Heffer and Sons Ltd., Cambridge.

Journ. Inst. Navigation
Journal of the Institute of Navigation. Published quarterly by the Institute of Navigation, London.

Journ. Observateurs (JO)
Journal des Observateurs. Publié avec le concours de l'Université d'Aix-Marseille par le Centre National de la Recherche Scientifique, Paris. After Vol. 51 replaced by "Astronomy and Astrophysics".

Journ. Optical Soc. America
Journal of the Optical Society of America. Publisher: American Institute of Physics, New York.

Journ. Phys. A.General Phys.
Journal of Physics A. General Physics. (Proceedings of the Physical Society) Series 2. Published by the Institute of Physics and the Physical Society, London, England, in association with the American Institute of Physics, New York.

Journ. Physique

Journal de Physique. Publication de la Société Française de Physique, Paris.

Journ. Proc. Roy. Soc. New South Wales
Journal and Proceedings of the Royal Society of New South Wales. Published by the Society, Science House, Sydney.

Journ. Quant. Spectrosc. Radiat. Transfer
Journal of Quantitative Spectroscopy & Radiative Transfer. Publisher: Pergamon Press, Oxford – New York.

Journ. Roy. Astron. Soc. Canada
The Journal of the Royal Astronomical Society of Canada, devoted to the Advancement of Astronomy and Allied Sciences. Printed by the University of Toronto Press, Toronto, Ontario.

Kometn. Tsirk. *Kiev*
Kometnyj Tsirkulyar. Gruppa po Issledovaniyu Komet Astrosoveta i Mezhduvedomstvennyj Geofizicheskij Komitet, Akademii Nauk SSSR. Kievskij Universitet im. T. G. Shevchenko.

Komety i Meteory
Komety i Meteory. Akademiya Nauk Tadzhikskoj SSR. Astronomicheskij Sovet Akademii Nauk SSSR. Publishers: Izdatel'stvo "Donish", Dushanbe.

Kosmich. Issled.
Kosmicheskie Issledovaniya. Akademiya Nauk SSSR. Publishers: Izdatel'stvo "Nauka", Moskva.

L'Astronomie
L'Astronomie et Bulletin de la Société Astronomique de France. Revue mensuelle. Rédaction: Société Astronomique de France, Paris.

L'Universo
L'Universo. Rivista dell'Instituto Geografico Militare. Direzione, Redazione e Amministrazione: Istituto Geografico Militare, Firenze.

Magnitnye Polya Solnech. Pyaten
Magnitnye Polya Solnechnykh Pyaten. (Supplements to Solnechnye Dannye. Byulleten' (*Solar Data*)). Publishers: Izdatel'stvo "Nauka", Leningrad.

Math. Rev.
Mathematical Reviews. Published by the American Mathematical Society, Providence, R. I.

Mem. Fac. Sci. Kyoto Univ.
Memoirs of the Faculty of Science, Kyoto University. Series of Physics, Astrophysics, Geophysics, and Chemistry. Printed by Yamashiro Printing Publishing Co. Ltd., Kamigyo, Kyoto.

Mem. Japan Astron. Study Ass.
Memoirs of the Japan Astronomical Study Association. Izumi 59, Yugawara–machi, Kanagawa–ken, Japan.

Mem. Roy. Astron. Soc.
Memoirs of the Royal Astronomical Society. Published for the Royal Astronomical Society by Blackwell Scientific Publications, Oxford – Edinburgh.

Mem. Soc. Astron. Italiana
Memorie della Società Astronomica Italiana. Nuova Serie. Pubblicate sotto gli auspici del Consiglio Nazionale delle Ricerche. Publisher: Tipografia Baccini & Chiappi,

Firenze.

Messtechnik
Messtechnik (Zeitschrift für Instrumentenkunde). Publishers: Verlag Friedrich Vieweg & Sohn GmbH, Braunschweig.

Meteoritics
Meteoritics. The Journal of the Meteoritical Society. Circulation Manager: C. F. Lewis, Center for Meteorite Studies, The Arizona State University, Tempe, Arizona.

Meteoritika
Akademiya Nauk SSSR. Komitet po Meteoritam. Publishers: Izdatel'stvo "Nauka", Moskva.

Mitt. Astron. Ges.
Mitteilungen der Astronomischen Gesellschaft, Hamburg.

Monatsber. Deutsch. Akad. Wiss. Berlin
Monatsberichte der Deutschen Akademie der Wissenschaften zu Berlin. Mitteilungen aus Mathematik, Naturwissenschaft, Medizin und Technik. Publisher: Akademie-Verlag, Berlin.

Monthly Notes Astron. Soc. Southern Africa
Monthly Notes of the Royal Astronomical Society of Southern Africa. Published by the Astronomical Society of Southern Africa, Royal Observatory, Cape Province, South Africa.

Monthly Notices Roy. Astron. Soc.
Monthly Notices of the Royal Astronomical Society. Published for the Royal Astronomical Society by Blackwell Scientific Publications, Oxford – Edinburgh.

MVS Sonneberg
Mitteilungen über Veränderliche Sterne. Edited by Sternwarte Sonneberg (Zentralinstitut für Astrophysik, Bereich Sternphysik) der Deutschen Akademie der Wissenschaften.

Nachr. Akad. Wiss. Göttingen
Nachrichten der Akademie der Wissenschaften in Göttingen. II. Mathematisch-Physikalische Klasse. Vandenhoeck & Ruprecht, Göttingen.

Nachr. Karten-, Vermessungswesen
Nachrichten aus dem Karten- und Vermessungswesen. Editor: Institut für Angewandte Geodäsie (Abt. II des Deutschen Geodätischen Forschungsinstituts). Published by Verlag des Instituts für Angewandte Geodäsie, Frankfurt a. M.

Nature
Nature. A weekly Journal of Science. MacMillan & Co., Ltd., London; St. Martin's Press, Inc., New York.

Naturwissenschaften
Die Naturwissenschaften. Publisher: Springer-Verlag, Berlin – Heidelberg – New York.

Nauchn. Informatsii
Nauchnye Informatsii. Astronomicheskij Sovet Akademii Nauk SSSR, Moskva.

Naučna Misao
Naučna Misao. Drustvo za Unapredivanje i Širenje Nauke, Zagreb. Printer: Novinsko izdavacko poduzeče "Slobodna Dalmacja", Split.

Numerische Math.
Numerische Mathematik. Publisher: Springer-Verlag, Berlin – Heidelberg – New York.

Nuovo Cimento
Il Nuovo Cimento. Rivista Internazionale e Organo della Società Italiana di Fisica, Series A, B. Publisher: Nicola Zanichelli, Editore, Bologna.

Nuovo Cimento Lettere
Lettere al Nuovo Cimento. Rivista internazionale della Società Italiana di Fisica. Serie prima. Editrice Compositori, Bologna.

Nuovo Cimento Rivista
Rivista del Nuovo Cimento a cura della Società Italiana di Fisica. Editrice Compositori, Bologna.

Nuovo Cimento Suppl.
Supplemento al Nuovo Cimento. Nicola Zanichelli, Editore, Bologna.

Observations Artificial Earth Satellites
Observations of Artificial Satellites of the Earth (Nablyudeniya Iskusstvennykh Sputnikov Zemli). Magyar Tudományos Akadémia Csillagvizsgáló Intézete, Budapest.

Observatory
The Observatory. A Review of Astronomy. Publishers: The Editors of "The Observatory", Royal Greenwich Observatory, Herstmonceaux Castle, Hailsham, Sussex, England.

Optik
Optik. Zeitschrift für das gesamte Gebiet der Licht- und Elektronenoptik. Publishers: Wissenschaftliche Verlagsgesellschaft mbH., Stuttgart.

Orion Schaffhausen
Orion. Zeitschrift der Schweizerischen Astronomischen Gesellschaft (SAG). Bulletin de la Société Astronomique de Suisse (SAS). Administration: Generalsekretariat der SAG, Schaffhausen.

Österreich. Zeitschr. Vermessungswesen
Österreichische Zeitschrift für Vermessungswesen. Editor and Publisher: Österreichischer Verein für Vermessungswesen, Wien.

Peremennye Zvezdy, Byull.
Peremennye Zvezdy, Byulleten', izdavaemyj Astronomicheskim Sovetom Akademii Nauk SSSR. Published by Astronomicheskij Sovet Akademii Nauk SSSR, Moskva.

Phil. Mag.
The Philosophical Magazine. A Journal of Theoretical, Experimental and Applied Physics. Eighth Series. Publisher: Taylor & Francis, Ltd., London.

Phil. Trans. Roy. Soc. London
Philosophical Transactions of the Royal Society of London. Series A, Mathematical and Physical Sciences. Published by the Royal Society, London.

Phys. Abstr.
Physics Abstracts. Science Abstracts, Series A. An INSPEC Publication, published by The Institution of Electrical Engineers, London.

Phys. Ber.
Physikalische Berichte. Herausgegeben von der Deutschen Physikalischen Gesellschaft e. V. und von der Deutschen Akademie der Wissenschaften zu Berlin. Friedrich Vieweg & Sohn, Braunschweig.

Phys. Blätter
Physikalische Blätter. Physik-Verlag, Mosbach/Baden.

Phys. Earth Planet. Interiors
Physics of the Earth and Planetary Interiors. A journal devoted to observational and experimental studies of the Earth and Planetary interiors and their theoretical interpretation by the physical Sciences. Publisher: North-Holland Publishing Company, Amsterdam, Netherlands.

Phys. Fluids
The Physics of Fluids. Published by the American Institute of Physics, New York.

Phys. Rev.
The Physical Review. A journal of experimental and theoretical physics. Second Series. Published for The American Physical Society by the American Institute of Physics, Lancaster, Pa., and New York, N. Y.

Phys. Rev. Letters
Physical Review Letters. Published weekly by The American Physical Society, New York, N.Y.

Phys. Today
Physics Today. Published by the American Institute of Physics, New York.

Physica
Physica. Publishers: North-Holland Publishing Company, Amsterdam, The Netherland, on request of the Foundation "Physica", Utrecht.

Planet. Space Sci.
Planetary and Space Science. Pergamon Press, Oxford – London – New York.

Pokroky
Pokroky matematiky, fyziky a astronomie. Vydává Jednota čs. matematiků a fyziků. Publisher: Academia, Praha.

Postępy Astron.
Postępy Astronomii. Czasopismo Poświecone Upowszechnianiu Wiedzy Astronomicznej. Polskie Towarzystwo Astronomiczne, Warszawa. Printed in Poland by Państwowe Wydawnictwo Naukowe, Łódź.

Priroda
Priroda. Publishers: Izdatel'stvo "Nauka", Moskva.

Proc. Astron. Soc. Australia
Proceedings of the Astronomical Society of Australia. Published for the Society by Sydney University Press, Sydney.

Proc. Cambridge Phil. Soc.
Proceedings of the Cambridge Philosophical Society (Mathematical and Physical Sciences). Publishers: Cambridge University Press, London.

Proc. IEEE
Proceedings of the IEEE. Published monthly by the Institute of Electrical and Electronics Engineers, Inc. New York.

Proc. Koninkl. Nederl. Akad. Wet.
Koninklijke Nederlandse Akademie van Wetenschappen. Proceedings. Series B, Physical Sciences. Publishers: North-Holland Publishing Company, Amsterdam.

Proc. National Acad. Sci. U. S. A.
Proceedings of the National Academy of Sciences of the Unites States of America. Published monthly by the National Academy of Sciences, Washington, D. C.

Proc. Roy. Soc.
Proceedings of the Royal Society. Series A, Mathematical and Physical Sciences. Published by the Royal Society, London.

Progr. Theor. Phys. Japan
Progress of Theoretical Physics. Published for the Research Institute for Fundamental Physics and the Physical Society of Japan. Publication Office: Progress of Theoretical Physics, Yukawa Hall, Kyoto University, Kyoto, Japan.

Progr. Theor. Phys. Suppl.
Supplement of the Progress of Theoretical Physics. Published for the Research Institute for Fundamental Physics and The Physical Society of Japan. Publication Office: Progress of Theoretical Physics, Yukawa Hall, Kyoto University, Kyoto, Japan.

PTB Mitt.
PTB Mitteilungen. Amts- und Mitteilungsblatt der Physikalisch-Technischen Bundesanstalt, Braunschweig – Berlin.

Publ. Astron. Soc. Japan
Publications of the Astronomical Society of Japan. Published by the Astronomical Society of Japan. Office of the Society: Tokyo Astronomical Observatory, Mitaka, Tokyo. Agent: Maruzen Co. Ltd. (Export Department), Nihonbashi, Tokyo, Japan.

Publ. Astron. Soc. Pacific
Publications of the Astronomical Society of the Pacific. Published in Provo, Utah, by the Astronomical Society of the Pacific, San Francisco, California. Printed by Brigham Young University Press, Provo, Utah.

Publ. Roy. Obs. Edinburgh
The Royal Observatory, Edinburgh. Publication. Her Majesty's Stationery Office, Edinburgh.

Publ. Tartu Astrofiz. Obs.
W. Struve nimelise, Tartu Astrofüüsika Observatooriumi, Publikatsioonid. Eesti NSV Teaduste Akadeemia, Tartu.

Quarterly Journ. Roy. Astron. Soc.
Quarterly Journal of the Royal Astronomical Society. Published for the Royal Astronomical Society by Blackwell Scientific Publications, Oxford.

Referativ. Zhurn. 51. Astron.
Referativnyj Zhurnal. 51. Astronomiya. Vsesoyuznyj Institut Nauchnoj i Tekhnicheskoj Informatsii. Moskva.

Referativ. Zhurn. 52. Geod. i Aehros"emka.
Referativnyj Zhurnal. 52. Geodeziya i Aehros"emka. Vsesoyuznyj Institut Nauchnoj i Tekhnicheskoj Informatsii. Moskva.

Referativ. Zhurn. 62. Issled. kosm. prostranstv.
Referativnyj Zhurnal. 62. Issledovanie Kosmicheskogo

Prostranstva. Vsesoyuznyj Institut Nauchnoj i Tekhnicheskoj Informatsii. Moskva.

Rep. Progr. Phys.
Reports on Progress in Physics. Institute of Physics and the Physical Society, London.

Rev. Geophys.
Reviews of Geophysics. Published by the American Geophysical Union Washington, D. C.

Revista Astron.
Revista Astronomica. Organo de la Asociación Argentina Amigos de la Astronomia, Buenos Aires.

Rev. Modern Phys.
Reviews of Modern Phys. Published for The American Physical Society by the American Institute of Physics, Lancaster, Pa., and New York, N. Y.

Rev. Sci. Instruments
Reviews of Scientific Instruments. Published by the American Institute of Physics, Lancaster, Pa., and New York, N. Y.

Rezul'taty Nablyud. Sovet. Iskusstv. Sputnikov Zemli
Rezul'taty Nablyudenij Sovetskikh Iskusstvennykh Sputnikov Zemli. Published by Astronomicheskij Sovet Akademii Nauk SSSR, Moskva.

Ric. Sci.
La Ricerca Scientifica. Serie Seconda. Consiglio Nazionale delle Ricerche, Roma.

Říše hvězd
Říše hvězd. Czechoslovak popular astronomical journal. Publisher: Orbis, Praha.

Roy. Astron. Soc. New Zealand Circ.
Royal Astronomical Society of New Zealand, Variable Star Section, Circular. Office: Box 33, Lake Tekapo, New Zealand.

Roy. Astron. Soc. New Zealand Variable Star Sect. Repr.
Royal Astronomical Society of New Zealand. Variable Star Section. Reprint. Address: P. O. Box 33, Lake Tekapo, New Zealand.

Rumanian Sci. Abstr.
Rumanian Scientific Abstracts. Natural Sciences. Publisher: The Scientific Documentation Centre of the Academy of the Socialist Republic of Romania, Bucureşti.

Sci. American
Scientific American. Published monthly by Scientific American, Inc., New York, N. Y.

Sci. Rep. Tôhoku Univ.
The Science Reports of the Tôhoku University. First Series (Physics, Chemistry, Astronomy). Published by the Faculty of Science, Tôhoku University, Sendai, Japan.

Science
Science. American Association for the Advancement of Science, Washington, D. C.

Science Progrès, La Nature
Science Progrès, La Nature. Revue Mensuelle. Publishers: Dunod, Editeur, Paris.

Sitz.-Ber. Bayer. Akad. Wiss.
Bayerische Akademie der Wissenschaften. Mathematisch-Naturwissenschaftliche Klasse. Sitzungsberichte. Publisher: Verlag der Bayerischen Akademie der Wissenschaften, München.

Sitz.-Ber. Deutsch. Akad. Wiss. Berlin
Sitzungsberichte der Deutschen Akademie der Wissenschaften zu Berlin. Klasse für Mathematik, Physik und Technik. Publisher: Akademie-Verlag, Berlin.

Sitz.-Ber. Heidelberger Akad. Wiss.
Sitzungsberichte der Heidelberger Akademie der Wissenschaften. Mathematisch-Naturwissenschaftliche Klasse. Publisher: Springer-Verlag, Heidelberg.

Sitz.-Ber. Österreich. Akad. Wiss.
Sitzungsberichte. Österreichische Akademie der Wissenschaften. Mathematisch-Naturwissenschaftliche Klasse. Abteilung II: Mathematik, Astronomie, Meteorologie und Technik. Publisher: Springer-Verlag, Wien.

Sky Telescope
Sky and Telescope. Published by Sky and Telescope Corporation, Cambridge, Mass.

Smithsonian Contr. Astrophys.
Smithsonian Contributions to Astrophysics. Astrophysical Observatory of the Smithsonian Institution. For sale by the Superintendent of Documents, U. S. Government Printing Office, Washington, D. C.

Smithsonian Year
Smithsonian Year. Annual Report of the Smithsonian Institution, including the financial report of the Executive Committee of the Boards of Regents. Published by the Smithsonian Institution, Washington, D. C.

Solar Physics
Solar Physics. A Journal for Solar Research and the Study of Solar Terrestrial Physics. Publishers: D. Reidel Publishing Company, Dordrecht-Holland.

Solnech. Dannye Byull.
Solnechnye Dannye. Byulleten. *(Solar Data).* Publishers: Izdatel'stvo "Nauka", Leningrad.

Soobshch. Byurakan. Obs.
Soobshcheniya Byurakanskoj Observatorii. Akademiya Nauk Armyanskoj SSR, Erevan.

Soobshch. Gos. Astron. Inst. Shternberg
Soobshcheniya Gosudarstvennogo Astronomicheskogo Instituta im P. K. Shternberga. Publishers: Izdatel'stvo Moskovskogo Universiteta, Moskva.

Southern Stars
Southern Stars. The Journal of the Royal Astronomical Society of New Zealand (Inc.). Address of the Society: P.O. Box 3181, Wellington C1, New Zealand.

Soviet Astron. AJ
Soviet Astronomy AJ. A translation of the Astronomical Journal of the Academy of Sciences of the USSR. Published by the American Institute of Physics, Inc., New York.

Spaceflight
Spaceflight. Published by the British Interplanetary Society, London.

Space Sci. Rev.
Space Science Reviews. Publishers: D. Reidel Publishing Company, Dordrecht-Holland.

Springer Tracts Modern Phys.
Springer Tracts on Modern Physics. (Ergebnisse der exakten Naturwissenschaften). Springer-Verlag, Berlin–Heidelberg–New York.

Sterne
Die Sterne. Zeitschrift für alle Gebiete der Himmelskunde. Johann Ambrosius Barth, Leipzig.

Sternenbote
Sternenbote. Monatsschrift für Österreichs Amateurastronomen. Publisher: Astronomisches Büro, Hermann Mucke, Wien.

Stockholms Obs. Ann.
Stockholms Observatoriums Annaler. Printed by Almqvist & Wiksell, Stockholm.

Strolling Astronomer
The Strolling Astronomer. The Journal of The Association of Lunar and Planetary Observers. Publication Office: The Strolling Astronomer, Box 3AZ, University Park, New Mexico.

Stud. Cerc. Astron.
Studii şi Cercetări de Astronomie. Editura Academiei Republicii Socialiste România. Editorial Office: Observatorul Astronomic, Bucureşti.

Stud. Geophys. Geod.
Studia geophysica et geodaetica. Published for the Geophysical Institute of the Czechoslovak Academy of Sciences by Academia, Praha.

Stud. Univ. Babeş-Bolyai
Studia Universitatis Babeş-Bolyai. Series Mathematica-Physica. Publishers: Intreprinderea Poligrafica, Cluj.

SuW
Sterne und Weltraum. Astronomische Monatsschrift. Verlag Bibliographisches Institut AG, Mannheim.

Tellus
Tellus, a bi-monthly Journal of Geophysics. Svenska Geofysiska Foreningen. Printed in Sweden by Almqvist & Wiksells Boktryckeri AB, Uppsala.

Trans. Astron. Obs. Yale Univ.
Transactions of the Astronomical Observatory of Yale University. Published by the Observatory, New Haven.

Trans. Roy. Soc. Canada
Transactions of the Royal Society of Canada. Published by the Royal Society of Canada, National Research Building, Ottawa.

Trudy Astrofiz. Inst. Alma-Ata
Trudy Astrofizicheskogo Instituta, Alma-Ata. Akademiya Nauk Kazakhskoj SSR. Publishers: Izdatel'stvo "Nauka" Kazakhskoj SSR, Alma Ata.

Trudy Glav. Astron. Obs. Pulkove
Trudy Glavnoj Astronomicheskoj Observatorii v Pulkove. Akademiya Nauk SSSR. Izdanie Glavnoj astronomicheskoj observatorii v Pulkove, Leningrad.

Trudy Inst. Teor. Astron. *Leningrad*

Trudy Instituta Teoreticheskoj Astronomii. Akademiya Nauk SSSR. Publishers: Izdatel'stvo "Nauka", Leningrad.

Trudy Tashkent. Astron. Obs.
Trudy Tashkentskoj Astronomicheskoj Observatorii. Akademiya Nauk Uzbekskoj SSR. Publishers: Izdatel' stvo "FAN" Uzbekskoj SSR, Tashkent.

Tsirk. Astron. Inst. Tashkent
Tsirkulyar Astronomicheskogo Instituta. Akademiya Nauk Uzbekskoj SSR. Izdatel'stvo "FAN" Uzbekskoj SSR, Tashkent.

Tsirk. Astron. Obs. L'vov
Tsirkulyar. Astronomicheskaya Observatoriya. L'vovskij Ordena Lenina Gosudarstvennyj Universitet emeni Ivana Franko. Publisher: Izdatel'stvo L'vovskogo Universiteta, L'vov.

Umschau
Umschau in Wissenschaft und Technik. Umschau-Verlag Frankfurt a. M.

Urania Barcelona
Urania. Revista de Astronomia y Ciencias Afines. Organo de la Sociedad Astronómica de España y América, Barcelona; Unión Nacional de Astronomia y Ciencias Afines, Madrid.

Urania Kraków
Urania. Miesięcznik Polskiego Towarzystwa Milośników Astronomii, Kraków. Publisher: Krakowska Drukarnia Prasowa, Kraków.

Vasiona
Vasiona. Revue d'Astronomie et d'Astronautique. Bulletin de la Société Astronomique "R. Bosković", Beograd.

VdS Nachrichtenblatt
Nachrichtenblatt der Vereinigung der Sternfreunde e.V. After Vol. 18 No. 3 published in combination with "Sterne und Weltraum". Bibliographisches Institut, Mannheim.

Veröff. Astron. Rechen-Inst. Heidelberg
Veröffentlichungen des Astronomischen Rechen-Instituts Heidelberg. Verlag G. Braun, Karlsruhe.

Veröff. Sternw. Sonneberg
Deutsche Akademie der Wissenschaften zu Berlin. Institut für Sternphysik. Veröffentlichungen der Sternwarte in Sonneberg. Publisher: Akademie-Verlag, Berlin.

Vesmír
Vesmír. Přírodovědecky časopis Čs. akademie věd. Publisher: Academia, Praha.

Vestn. Khar'kov. Univ.
Vestnik Khar'kovskogo Universiteta. Seriya Astronomicheskaya. Publishers: Izdatel'stvo Khar'kovskogo Universiteta, Khar'kov.

Vestn. Kiev. Univ.
Vestnik Kievskogo Universiteta. Seriya Astronomii. Publishers: Izdatel'stvo Kievskogo Universiteta, Kiev.

VJS Naturforsch. Ges. Zürich
Vierteljahresschrift der Naturforschenden Gesellschaft in Zürich. Printer and Publisher: Leeman AG, Zürich.

Weltraumfahrt

Weltraumfahrt. Zeitschrift für Astronautik und Raketen-
technik. Umschau-Verlag, Frankfurt a. M.

Wiss. Zeitschr. Humboldt-Univ. Berlin
Wissenschaftliche Zeitschrift der Humboldt-Universität
zu Berlin. Mathematisch-Naturwissenschaftliche Reihe.
Edited by the Rektor der Humboldt-Universität, Berlin.

Yamamoto Circ.
Yamamoto Circular. Published by the Yamamoto Observatory, Kamitanakami – Kiryutyo, Otu, Siga-ken, Japan.

Zeitschr. Angew. Physik
Zeitschrift für Angewandte Physik. Publisher: Springer-
Verlag, Berlin–Heidelberg–New York.

Zeitschr. Astrophys. (ZfA)
Zeitschrift für Astrophysik. Publisher: Springer-Verlag,
Berlin–Heidelberg–New York. After Vol. 69 (1968)
replaced by "Astronomy and Astrophysics".

Zeitschr. Geophys.
Zeitschrift für Geophysik. Publisher: Physica-Verlag,

Würzburg.

Zeitschr. Naturforschung
Zeitschrift für Naturforschung. Verlag der Zeitschrift für
Naturforschung, Tübingen.

Zeitschr. Physik
Zeitschrift für Physik. Publisher: Springer-Verlag, Berlin–
Heidelberg–New York.

Zemlya i Vselennaya
Zemlya i Vselennaya. Nauchno-Populyarnyj Zhurnal
Akademii Nauk SSSR. Publishers: Izdadel'stvo "Nauka",
Moskva.

Zentralblatt Math. Grenzgebiete
Zentralblatt für Mathematik und ihre Grenzgebiete. Publisher: Springer-Verlag, Berlin - Heidelberg - New York.

Zvaigžņota Debess
Latvijas PSR Zinātņu Akadēmijas Radioastrofizikas
Observatorijas Populārzinatnisks Gadalaiku Izdevums.
Izdevnieciba "Zinātne", Riga.

002 Bibliographical Publications

002.001 News notes. G. S. Mumford.
Sky Telescope, Vol. 39, 8, 82 - 83, 159 - 160 (1970).
(1) Widely photographed fireball; Recoating the 200-inch mirror; Compact radio source in M 87. (2) Measuring radial velocities; Rara Astronomica; Bright flares on Mars; Another bright comet (*1969i,Bennett*). (3) Unusual infrared object; Double star measures; A test for Magellanic Cloud members; Hale Observatories; Diameter of Neptune; Comet 1970a appears; Griffith Observatory head.

002.002 Kurzberichte aus der Forschung.
SuW, Vol. 9, 15 - 17, 37 - 39, 69 - 70 (1970).
News notes: (1) Hat der Crabnebel-Pulsar einen Planeten? *(K. Birkle)*; Wasserdampf in Sonnenflecken *(A. Wittmann)*; Satelliten-Radio-Astronomie *(W. Hofmann)*; 3-m-Teleskop im Weltraum vorgeschlagen *(D. L.)*; R CrB und R Y Sgr umgeben von Staubwolken *(G. Fugmann)*; Ein Vergleich zwischen Radialgeschwindigkeiten des neutralen Wasserstoffs (HI) und H II-Gebieten *(T. Neckel)*; Quasistellare Objekte *(T. Schmidt)*. (2) 4-m-Teleskop in Australien *(J. Solf)*; Enden O- und B-Sterne als Supernovae vom Typ II? *(G. Schnur)*; Die Sonne als veränderlicher Stern *(H. Wöhl)*; Bevorzugte Rotverschiebung bei Quasaren entdeckt? *(G. A.)*; Zodiakallichthelligkeit am Pol der Ekliptik *(W. Gabsdil)*; Zur Stabilität von interstellaren Graphitteilchen mit festen H_2-Mänteln *(T. Neckel)*. (3) Aeronomiesatellit DIAL vor dem Start *(C. Leinert)*; Das 6-m-Teleskop fertiggestellt *(J. Classen)*; Organische Farbstoffe in der Jupiter-Atmosphäre *(E. Koch)*.

002.003 News and comments. E. Öpik.
Irish Astron. Journ., Vol. 9, 151 - 162 (1969).
The Apollo 12 lunar landing; The list of orbits of minor planets; Solar granulation; The spatial distribution of stars; The solar wind flux of helium -4; The Hudson Bay Arc as meteor crater? Primordial abundance of helium; Neutrinos and internal structure of the sun; Interstellar meteors; The Alphonsus "eruption"; "Organized elements" in the Orgueil meteorite.

002.004 Mitteilungen aus Wissenschaft und Literatur.
Sterne, 45. Jahrgang, p. 247 - 248 (1969).
Besitzt Barnards Stern ein Planetensystem? *(J. Dorschner)*; Radiosignale der Erde *(J. Classen)*.

002.005 Bibliography of works in the field of astronomy and geodesy. Collected by I. Kurvits, M. Ross.
Astron. Geod. Eston. SSR, Tartu, (see 003.004), p. 127 - 178 (1969). In Russian.

002.006 Chronicle.
Urania Kraków, Vol. 41, 17 - 21, 50 - 57 (1970). In Polish.
News notes: (1) First impressions from Apollo 12 journey *(A. Marks)*; 14 supernovae were discovered at Mt. Palomar in 1968 *(B. Kuchowicz)*; Quarks as catalyzers of thermonuclear reactions in the sun *(B. Kuchowicz)*; Is a new planet forming in the solar system? *(B. Kuchowicz)*. (2) On the doorstep of the satellite bases era *(A. Marks)*; The landing site of Apollo 11 mission *(S. R. Brzostkiewicz)*; How do the stars form? *(B. Kuchowicz)*; A dead quasar in the nucleus of our Galaxy *(B. Kuchowicz)*; Neutron sources in stars investigated in nuclear physics laboratory *(B. Kuchowicz)*; A strange star BL Lac *(B. Kuchowicz)*; Some data about the OSO satellites *(A. Marks)*; New Czechoslovak comet *(W. Sędzielowski)*.

002.007 Nouvelles brèves.
Ciel et Terre, Vol. 86, 72 - 73, 164 - 167 (1970).

(1) La rotation d'un astéroide Troyen; Nouvelle détermination des masses de Jupiter et de Neptune; L'étoile de Barnard possède-t-elle deux compagnons planétaires? La période du pulsar NP 0532. (2) L'instant du maximum des Lyrides; La petite planète Alinda; Eclat exceptionnel de Mira; La distribution galactique des pulsars; Les objets M 104 à M 110; Les objectifs de la mission Viking (1973).

002.008 Forschungen und Publikationen zur Geschichte der Astronomie in der Deutschen Demokratischen Republik. Eine Bibliographie (1949 - 1969).
D. Wattenberg.
Veröff. Archenhold-Sternw. Berlin-Treptow, No. 2, 142 pp. (1969).

002.009 Nouvelles de la Science.
L'Astronomie, 84e année, p. 97 - 99 (1970).
(1) Nouvelles comètes; Sur deux cas de vision à œil nu.

002.010 News notes. G. S. Mumford.
Sky Telescope, Vol. 39, 223 - 224 (1970).
Gunfire in the dome; The rapidly aging earth; 2,304,333 meteors; Nova Serpentis 1970.

002.011 Nouvelles de la Science.
L'Astronomie, 84e année, p. 185 - 186 (1970).
Astrophysique des hautes énergies.

002.012 Mitteilungen aus Wissenschaft und Literatur.
Sterne, 46. Jahrgang, p. 24 - 29 (1970).
Rätselhaftes Mond-Seismometer *(J. Classen)*; Poröse Beschaffenheit des Mondes? *(J. Classen)*; Die Mondproben von Apollo 11 *(J. Classen)*; Photographie der lichtschwachen äusseren Gebiete von Galaxien *(C. Friedemann)*; Die Sterne der Sonnenumgebung näher als 17 Lichtjahre *(H. Lambrecht)*; Interstellare Formaldehyd-Moleküle in Dunkelwolken *(H. Lambrecht)*; Gravitationswellen entdeckt *(J. Gürtler)*.

002.013 Kurzberichte aus der Forschung.
SuW, Vol. 9, 95, 98 - 99, 124 - 125 (1970).
(4) Absenkung des Kontinuums im solaren Spektrum durch Linienabsorption und Balmer-Kontinuum *(H. Wöhl)*; Apollo 11-Sonnenwindexperiment *(W. Gabsdil)*; Solare Häufigkeiten von Calzium und Cadmium *(H. Wöhl)*; Wärmeflußexperiment von Apollo 13; Neues vom Mars *(H. Link)*; Unternehmen "Wiking" 1973; Zur Interpretation von Magnetfeldmessungen in Sonnenflecken *(A. Wittmann)*. (5) Galaxy, ein Meßautomat für photographische Platten; Kein Ammoniak in den Saturnringen *(H. Link)*; Japan startet Satelliten; Gravitationsstrahlung dichter Sternhaufen; Neubestimmung der Solarkonstante *(A. Wittmann)*.

002.014 Science news.
Priroda, No. 1.70, p. 110 - 119; No. 2, p. 106 - 115; No. 3, p. 109 - 118 (1970). In Russian.
(1) Television-pictures of the Martian surface; Antimatter, quasars and the development of galaxies; Can a new planet appear in the solar system? A new variable X-ray source; What is the density of Pluto? The reasons for the variations of the earth's magnetic poles. (2) A new planetary system; Auroras on Jupiter? Concentrations of matter between Saturn's rings; News on the sun's rotation; The radio-frequency radiation of the earth's atmosphere. (3) Krinovit – a new meteoritic material *(A. A. Javnel')*; Unusual increase of solar activity; Carbon and cosmic space; Modeling of the earth's atmosphere.

002.015 Bibliographical references to astrophysical papers

published in the journal "Radiofizika", Vol. 11, Nos. 8 - 12, for the year 1968, and Vol. 12, Nos. 1 - 5 for the year 1969.
Astron. Zhurn. Akad. Nauk SSSR, Vol. 47, 458 - 461 (1970). In Russian. – English translation in Soviet Astron., AJ, Vol. 14, No. 2.

002.016 Recensements des données astrophysiques.
B. Hauck.
Orion Schaffhausen, 28. Jahrgang, p. 15 - 16 (1970).

002.017 Chronicle.
Urania Kraków, Vol. 41, 85 - 87, 113 - 121 (1970). In Polish.
News notes: (3) Flare of the stars sending off X radiation *(B. Kuchowicz);* Photoelectric study of Hektor; Volcans of the bottom of the crater Copernicus? *(S. R. Brzostkiewicz).* (4) First results of the Mariner 6 and 7 flights *(Z. Paprotny);* Continuous link between quasars and pulsars *(B. Kuchowicz);* A nova approaching its maximum of light *(B. Kuchowicz);* Once more theory of tectite origin *(S. R. Brzostkiewicz);* Once more of the fossil traces of particles in meteorites *(B. Kuchowicz);* Infrared emission of Jupiter and Saturn *(B. Kuchowicz);* Flattened bodies better resist the gravitational contraction *(B. Kuchowicz).*

002.018 Chronicle.
Urania Kraków, Vol. 41, 146 - 151, 172 - 180 (1970). In Polish.
(5) Investigations of lunar samples *(A. Marks);* How did the primeval sun lose its angular momentum? *(B. Kuchowicz);* Allende, the largest known carbon chondrite *(B. Lang);* Satellites in orbit *(B. Kuchowicz).* (6) Comet Bennett *(G. Sitarski);* Investigation of lunar samples (2) *(A. Marks);* The wreck of Surveyor 3 *(S. R. Brzostkiewicz);* Maser mechanism in radio sources containing water vapour *(B. Kuchowicz);* Reduction Leonid's activity *(B. Kuchowicz);* New value of solar constant *(S. R. Brzostkiewicz);* "Geological" structure of crater Ciolkowski *(S. R. Brzostkiewicz);* Are there any rivers on the moon? *(S. R. Brzostkiewicz).*

002.019 News notes. G. S. Mumford.
Sky Telescope, Vol. 39, 290, 357 - 358, 364 (1970).
Length of Venus' day. Mariners to test general relativity; Phobos surveyed; Galaxies with bright infrared cores; Scientific meetings; Arecibo's sky enlarged; High-velocity variable star; Lunar orbiters; Corning ships 157-inch mirror blank to Canada; Dutch astronomer honored.

002.020 Nouvelles brèves.
Ciel et Terre, Vol. 86, 259 - 262 (1970).
La petite planète Hidalgo; Un nouvel essaim météorique: Les Lyrides de juin; Les étoiles proches; Les savants ne sont pas d'accord sur l'origine des plaines lunaires.

002.021 Kurzberichte aus der Forschung.
SuW, Vol. 9, 152 - 153 (1970).
Beobachtungen außergewöhnlich hoher Polarisationsbeträge *(G. Schnur);* Kein Bor auf der Sonne? *(H. Wöhl);* Sehr schwacher Stern beobachtet: B = 23.^{m}6 *(H. Krefft);* VRO 42.22.01 = BL Lac, eine ungewöhnliche Strahlungsquelle *(K. Birkle);* Ungewöhnliches Infrarot-Objekt *(C. Thum);* Neubestimmung der Hubble-Konstante *(A. Wittmann);* Lebensdauer von Supergranulen *(H. Wöhl);* Wie entstanden die Mondmeere?

002.022 Nouvelles de la science.
L'Astronomie, 84e année, p. 301 - 302 (1970).
Calcul des perturbations des comètes; Nouvelle des comètes; Nova Serpentis 1970.

002.023 Mitteilungen aus Wissenschaft und Literatur.
Sterne, 46. Jahrgang, p. 88 - 90 (1970).
Emissionslinien im Röntgen-Spektrum? *(S. Görlitz);* Schwere Elemente in 73 Draconis *(S. Görlitz);* Die Struktur des Nebelhaufens in Virgo *(F. Schmeidler);* Journal for the History of Astronomy *(D. B. Herrmann).*

002.024 Bibliography of Astronomy, 1881 - 1898 on Microfilm, with a Guide by J. B. Sykes.
University Microfilms Limited, Tylers Green, High Wycombe, Buckinghamshire, England. 18 reels (two reels for each two-year section). Price $ 180.00 (1970).

002.025 Science news.
Priroda, No. 4.70, p. 108 - 117; No. 5.70, p. 102 - 111; No. 6.70, p. 102 - 112 (1970). In Russian.
(4) News about the radiation belts of the earth *(V. S. Agalakov);* New coronograph in Pulkovo. (5) Variability and nature of quasars; Lepton charge and neutrino astrophysics; Water on Mars; The nature of the magnetic field in the surroundings of Venus; The riddle of the atmosphere of Venus; News on the Tungusic meteorite. (6) Transuranic elements in meteorites? A cosmic factory of diamonds *(B. V. Loginov);* X-ray and infrared radiation in the universe; Of what do the rings of Saturn consist? The iron core of Mars.

002.026 Rassegna delle riviste e notizie brevi. P. Maffei.
Coelum, Vol. 38, 22 - 35, 72 - 87, 124 - 138 (1970).

002.027 Mitteilungen aus Wissenschaft und Literatur.
Sterne, 46. Jahrgang, p. 125 - 127 (1970). – Das Magnetfeld des Mondes *(F. Dorst);* Marsmond Phobos unterliegt keiner Abbremsung *(H. Heuseler);* Die Staubsatelliten der Erde *(E. Krug).*

002.028 Astronomy and Astrophysics Abstracts. Vol. 2, Literature 1969, Part II.
S. Böhme, W. Fricke, U. Güntzel-Lingner, F. Henn, D. Krahn, G. Zech (Editors).
Published for the Astronomisches Rechen-Institut, Heidelberg by Springer-Verlag, Berlin – Heidelberg – New York. 10 + 516 pp. Price DM 72.–, US $ 19.80 respectively [Subscription price per volume DM 57.60; US $ 15.90] (1970).

002.029 Documentation. J. B. Sykes.
Trans. IAU, Vol. 14A, (see 003.028), 11 - 14 (1970). – Report of Commission 5.

002.030 News notes.
Sky Telescope, Vol. 40, 15 - 16 (1970). – New star in the LMC; Optimum sizes for infrared photometric telescopes; RAS commemorative stamp; Another sungrazing comet; Reverberation on the moon.

002.031 News from science and other informations.
Zemlya i Vselennaya, No. 1 (1970). In Russian.
Mars in front of objectives of television cameras *(V. V. Mikhajlov),* p. 17 - 19; Dark regions on Mars – uplifts or lowlands? *(V. V. Golobkov),* p. 19 - 20; Robots in the Orbiting Astronomical Observatory, p. 39; How many natural satellites has the earth? *(Yu. A. Rjabov, V. A. Yurevich),* p. 45 - 46; Earth and Jupiter – sources of radio waves, p. 46; Quasi-stellar objects and clusters of galaxies *(B. V. Komberg),* p. 53 - 54; Once more on extraordinary stars *(Yu. N. Efremov),* p. 54; Interstellar water and formaldehyde, p. 54; Scientific expedition to the moon, p. 55; After the flight of Apollo 11 *(V. V. Mikhajlov),* p. 56; Further stages of the Apollo program, p. 56 - 58; Photography of the flight of the Vilna meteorite, p. 64 - 65; Geological map of the crater Ziolkovsky, p. 65 - 66.

002.032 News from science and other informations.

Zemlya i Vselennaya, No. 2 (1970). In Russian.

Again on extraordinary stars (*Yu. N. Efremov*), p. 10 - 11; The star δ Delphini (*M. S. Frolov*), p. 11 - 12; Start of Intercosmos 2, p. 20; Program of the "large tour", p. 20 - 21; New comet discovered in the USSR (*V. A. Bronshtehn*), p. 57 - 58; Distance between points on the moon and on the earth (*D. Ya. Martynov*), p. 69; Velocity of the earth relative to the relict emission of the universe, p. 69; Standing waves on the moon, p. 69; Scientific biography of I. N. Ul'yanov (*V. V. Radzievskij, G. Danilova*), p. 78, 80.

002.033 News from science and other informations.

Zemlya i Vselennaya, No. 3 (1970). In Russian.

The riddles connected with the moon became larger, p. 11 - 12; The end of Pegasus, p. 14; The Pioneer program, p. 24 - 25; Isotopes of carbon in the interstellar medium, p. 25; Peculiarities of Seyfert galaxies, p. 45; A galaxy with varying brightness, p. 45.

003 Books (Astronomy and Astrophysics)

003.001 **The Proton Flare Project (The July 1966 Event).**
A. C. Stickland. (Editor).
Annals of the IQSY, (International Years of the Quiet Sun),
Vol. 3. The M.I.T. Press, Massachusetts Institute of Technology, Cambridge, Mass., London. 13 + 511 pp. Price £ 10 11s. 0d.
(1969). – Rev. in Solar Physics, Vol. 12, 502, 1970 (*Ž. Svestka*).

003.002 **Principles of Celestial Mechanics.**
P. M. Fitzpatrick.
Academic Press, New York – London. 11 + 405 pp. Price
$ 10.50 (1970). –Contents: 1. Fundamental concepts; 2. The one-body and two-body problems; 3. The determination of position in the orbital plane; 4. Expansions in elliptic motion; 5. Systems of astronomical coordinates; 6. The two-body orbit in space; 7. The Lagrange planetary equations for general perturbing forces; 8. The Lagrange planetary equations for conservative perturbing forces; 9. Lagrange and Hamilton mechanics; 10. Canonical transformations; 11. Hamilton-Jacobi theory with examples; 12. The gravitational potential; 13. Gravity and drag effects on the orbit of an artificial earth satellite; 14. Gravity effects on the rotational motion of an artificial earth satellite.

003.003 **The nature of celestial bodies and their observation.**
N. P. Barabashov.
Khar'kovskij universitet, Khar'kov. 299 pp. Price 1 Rbl. 76 Kop. (1969).   In Russian. – Review in Referativ. Zhurn. 51. Astron., 4.51.48 (1970).

003.004 **Astronomy and geodesy in the Estonian SSR (1940 - 1966).** G. G. Kuzmin (Editor).
Akademiya Nauk Ehstonskoj SSR, Institut Fiziki i Astronomii, Tartu. 179 pp. Price 1 Rbl. 20 Kop. (1969).   In Russian.

003.005 **Nautical astronomy. Textbook for marine schools.**
R. Yu. Titov, G. I. Fajn.
"Transport", Moskva. 384 pp. Price 95 Kop. (1969).
In Russian.

003.006 **Scattering of radiowaves emitted by lunar satellites and interplanetary probes, and by the surfaces of the moon and planets.** S. S. Matyugov, O. I. Yakovlev.
Edited by the Zhurn. Radiotekhn. i ehlektronika AN SSSR, Moskva. 11 pp. (1969). – Review in Referativ. Zhurn. 51. Astron., 4.51.236 (1970).

003.007 **Calculation of the phase dependence of the integral brightness of planets.**
A. A. Rubashevskij, É. G. Yanovitskij.
"Naukova dumka", Kiev. 101 pp. Price 35 Kop. (1969).
In Russian.

003.008 **Minor planets.** F. Yu. Zigel'.
"Nauka", Moskva. 104 pp. Price 19 Kop. (1969).
In Russian.

003.009 **The heights and magnitudes of bright meteors.**
P. B. Babadzhanov.
"Donish", Dushanbe. 44 pp. Price 20 Kop. (1969).
In Russian. – Review in Referativ. Zhurn. 51. Astron., 4.51.327 (1970).

003.010 **The sun and the ionosphere. Short-wave solar radiation and its influence on the ionosphere.**
G. S. Ivanov-Kholodnyj, G. M. Nikolskij.
"Nauka", Moskva. 455 pp. Price 2 Rbl. 79 Kop. (1969).

In Russian. – Review in Referativ. Zhurn. 51. Astron., 4.51.476 (1970).

003.011 **The relation of some events in the earth's troposphere with solar activity.**
A. S. Besprozvannaya, A. I. Ol', (Editors).
Gidrometeoizdat, Leningrad. 160 pp. Price 78 Kop. (1969).
In Russian.

003.012 **Solar-tropospherical connections.**
B. I. Sazonov, V. F. Loginov.
Gidrometeoizdat, Leningrad. 115 pp. Price 47 Kop. (1969).
In Russian.

003.013 **Solar activity and sudden changes in natural processes on the earth.**
I. P. Druzhinin, N. V. Kham'yanova.
"Nauka", Moskva. 224 pp. Price 1 Rbl. 37 Kop. (1969).
In Russian. – Review in Referativ. Zhurn. 51. Astron., 4.51.505 (1970).

003.014 **Introduction to stellar statistics.**   R. Kurth.
Translated from English into Russian. "Mir",
Moskva. 222pp. Price 1 Rbl. 40 Kop. (1969).

003.015 **Perpetual Calendars.**
A. V. Butkevich, M. S. Zelikson.
"Nauka", Moskva. 120 pp. Price 44 Kop. (1969). In Russian.
Review in Referativ. Zhurn. 51. Astron., 5.51.9 (1970).

003.016 **The Analytical Foundations of Celestial Mechanics.**
A. Wintner.
Translated from English into Russian. "Nauka", Moskva.
523 pp. Price 2 Rbl. 32 Kop. (1967). – Review in Referativ.
Zhurn. 51. Astron., 5.51.133 (1970).

003.017 **The Solar Activity.**   Yu. I. Vitinskij.
"Nauka", Moskva. 92 pp. Price 17 Kop. (1969).
In Russian.

003.018 **News on Venus and Mars.**
V. N. Konashenok, K. Ya. Kondrat'ev.
Gidrometeoizdat, Leningrad. 51 pp. Price 20 Kop. (1970).
In Russian. – Review in Referativ. Zhurn. 62. Issled. kosm. prostranstv., 6.62.159 (1970).

003.019 **Investigations on time and frequency measurements.**
Moskva. 205 pp. Price 1 Rbl. 48 Kop. (1969).
In Russian. – Review in Referativ. Zhurn. 51. Astron., 6.51.203 (1970).

003.020 **Physical processes in the upper atmosphere of the earth.**
L. A. Katasev, S. M. Poloskov (Editors).
Gidrometeoizdat, Moskva. 268 pp. Price 1 Rbl. 22 Kop. (1969).   In Russian.

003.021 **The solar atmosphere.** H. Zirin.
Translated from English into Russian.
"Mir", Moskva. 584 pp. Price 3 Rbl. 54 Kop. (1969).

003.022 **Modulation of cosmic rays in interplanetary space.**
G. F. Krymskij.
"Nauka", Moskva. 152 pp. Price 84 Kop. (1969). In Russian.
Review in Referativ. Zhurn. 51. Astron., 6.51.648 (1970).

003.023 **Cosmical Tracking. Radio-Technical Methods and**

Mathematical Reduction of Data.
P. A. Agadzhanova, V. E. Dulevicha, A. A. Korosteleva.
"Sovetskoe radio", Moskva. 498 pp. Price 1 Rbl. 71 Kop.
(1969). – Review in Referativ. Zhurn. 62. Issled. kosm.
prostranstv., 5.62.306 (1970).

003.024 Physics of the sun and stars.
V. I. Voroshilov (Editor).
Respublikanskij Mezhvedomstvennyj Sbornik. Ser. Astrometriya i Astrofizika No. 8, Akademiya Nauk Ukrainskoj SSR, Glav. Astron. Obs. Izdatel'stvo "Nauka Dumka", Kiev. 100 pp. Price 85 Kop. (1969). In Russian. – The papers included are abstracted in their subject categories.

003.025 Isaac Newton. L. Nový, J. Smolka.
Orbis, Praha. 196 pp. Price Kčs 12.– (1969).
In Czech. – Review in Říše hvězd, Vol. 51, 142 (1970).

003.026 Astronautical multilingual dictionary of the International Academy of Astronautics.
R. Pešek (Scientific editor).
Academia, Praha, 936 pp. Price $ 34.75 (1970). – The dictionary contains about forty thousand scientific and technical terms in English, Russian, German, French, Italian, Spanish and Czech and about six thousands space law terms.

003.027 Cosmic rays. Results of researches on the International Geophysical Projects. Articles No. 11.
S. N. Vernov, L. I. Dorman (Editors).
Publishing House "Nauka" Moscow. 207 pp. Price 1 Rbl. 45 Kop. (1969). In Russian.

003.028 Reports on Astronomy.
C. de Jager (Editor).
Transactions of the International Astronomical Union, Vol. 14A. D. Reidel Publishing Company, Dordrecht–Holland. 8 + 566 pp. Price $ 22.50 (1970).

003.029 Interesting cosmonautics.
F. Yu. Zigel'.
"Mashinostroenie", Moskva. 304 pp. Price 60 Kop. (1970). Review in Referativ. Zhurn. 62. Issled. kosm. prostranstv., 7.62.49 (1970).

003.030 Handbook on the physical parameters of the atmosphere. Yu. A. Glagolev.
Gidrometeoizdat, Leningrad. 211 pp. Price 1 Rbl. 9 Kop. (1970). In Russian. – Review in Referativ. Zhurn. 62. Issled. kosm. prostranstv., 7.62.50 (1970).

003.031 Variable stars and methods of their study.
V. P. Tsesevich.
"Pedagogika", Moskva. 239 pp. Price 51 Kop. (1970). In Russian. – Review in Referativ. Zhurn. 51. Astron., 7.51.33 (1970).

003.032 The Stars. Their Structure and Evolution.
R. J. Tayler.
Wykeham, London. 11 + 207 pp. Price 30s (1970). – Reviews in Nature, Vol. 226, 1276; (1970), (*J. Gribbin*); Phys. Abstr., Vol. 73, No. 45257 (1970).

003.033 Atlas for Objective Prism Spectra. (Bonner Spektral Atlas I). W. C. Seitter.
Veröff. Astron. Inst. Bonn. F. Dümmler Verlag, Bonn. 56 pp. and 65 plates. Subscription price DM 180.00 (1970).

003.034 J. C. Poggendorff, Biographisch-Literarisches Handwörterbuch der exakten Naturwissenschaften.
Vol. VII b, Part 3, 3rd number. H. Salié (Editor).
Published by Akademie-Verlag, Berlin. 160 pp. Price DM 24.00 (1970).

003.035 Astronomy and Astrophysics. D. A. Kemp.
Macdonald, London. 23 + 584 pp. Price £ 10, $25.00 respectively. – Review in Journ. Phys. A,General Phys., Vol. 3, 585 (1970).

003.036 Cosmology. J. Charon.
Translated from the French by P. Moore.
World University Library. Weidenfeld and Nicolson, London. 256 pp. Price 35s., 18s. respectively (1970). – Review in Nature, Vol. 227, 756 (*J. Gribbin*).

003.037 The Theory of Stellar Spectra. C. R. Cowley.
Topics in Astrophysics and Space Physics. Gordon and Breach, London–New York. 11 + 260 pp. Price 195s., $23.40 respectively (1970).

003.038 The Dwarf Novae. J. S. Glasby.
Constable and Co., Ltd., London. 293 pp. Price 60s. (1970).

003.039 The Interstellar Medium. S. A. Kaplan, S. B. Pikelner.
Harvard University Press, Cambridge, Mass.; Oxford University Press, London, 11 + 465 pp. Price $20.00, 190s. respectively (1970).

003.040 The Ionosphere and Its Interaction with Satellites.
M. A. Kasha.
Gordon and Breach, London–New York. 13 + 156 pp. Price 100s., $12.00 respectively (1970).

003.041 Widening Horizons: Man's Quest to Understand the Structure of the Universe. Z. Kopal.
Kahn and Averill, London. 176 pp. Price 30s. (1970).

003.042 Einführung in die spezielle Relativitätstheorie.
C. Kacser.
Translation of English Edition "Introduction to the Special Theory of Relativity". Verlag Berliner Union, Stuttgart. 281 pp. Price DM 28.00 (1970).

003.043 Weltraumfahrt. Physik - Technik - Biologie.
W. Petri.
Hanns Reich Verlag, München 183 pp. Price DM 18.80 (1970).

003.044 Introduction to the Solar Wind. J. C. Brandt.
Freeman and Co., San Francisco. 199 pp. Price $ 10 (1970). – Review in Sky Telescope, Vol. 39, 317 (1970).

003.045 Jordan/Eggert/Kneisel, Handbuch der Vermessungs - kunde. Tenth revised and rearranged edition.
Vol. IIa: **Geodätische Astronomie.** K. Ramsayer.
J. B. Metzlersche Verlagsbuchhandlung, Stuttgart. 19 + 903 pp. Price DM 260.00 (1970).

003.046 Knaurs Lexikon der Naturwissenschaften, Astrono - mie, Chemie, Geologie, Meteorologie, Physik.
J. Ackner.
Verlag Droemer Knaur, München - Zürich. 320 pp. Price DM 16.80 (1969). – Review in Bild Wissenschaft, 7. Jahrgang, No. 4, p. 398 (1970).

003.047 Bogen om Manen. A. Lundbak.
P. Haase & Sons Forlag, Kobenhavn, 128 pp. Price D. Kr. 59.65 (1969). – Review in Astron. Tidssk., Vol. 3, 99 - 100; 1970 (*G. Larsson-Leander*).

003.048 Pictorial Astronomy. D. Alter, C. H. Cleminshaw, J. G. Phillips.

T. Y. Crowell, New York. 8 + 328 pp. Price $10.00 (1969). –
Review in Sky Telescope, Vol. 39, 117 (1970).

003.049 Status of General Relativity.   W. B. Bonnor.
Francis Hodgson, Guernsey. 19 pp. Price 40s.
(1970). – Review in Nature, Vol. 226, 43 - 44:1970 (W. C.
Saslaw).

003.050 Atmospheric Tides. Thermal and Gravitational.
S. Chapman, R. S. Lindzen.
D. Reidel Publishing Company, Dordrecht - Holland. 9 + 200
pp. Price f 38.00 (1970).

**003.051 The Mathematical Theory of Non - Uniform Gases.
An Account of the Kinetic Theory of Viscosity.**
Thermal Conduction and Diffusion in Gases.
S. Chapman, T. G. Cowling, D. Burnett.
At the Cambridge University Press, Cambridge. Third Edition.
24 + 423 pp. Price £5 net. (1970).

003.052 Gravitation and the Universe.   R. H. Dicke.
American Philosophical Society, Philadelphia.
82 pp. Price $2.50 (1969). – Reviews in Nature, Vol. 226,
779 - 780; 1970 (J. Gribbin); Sky Telescope, Vol. 39, 253
(1970); Phys. Today, Vol. 23, No. 4, p. 77.

**003.053 Cosmic Ray Physics: Nuclear and Astrophysical
Aspects.**   S. Hayakawa.
Interscience Monographs and Texts in Physics and Astronomy,
Vol. 22. Interscience Publishers, New York. 14 + 778 pp.
Price $ 39.50, 370 s. respectively (1970). – Review in
Nature, Vol. 227, 311; 1970 (H. Elliot).

**003.054 Astronomy Gnomonics. A catalogue of instruments
of the 15th to the 19th centuries in the collections**
of the National Technical Museum, Prague.
Z. Horský, O. Skopová.
National Technical Museum, Prague, 202 pp. (1968). – Review in Journ. History Astron., Vol. 1, (Part 1), 81 - 82;
1970 (G. L'E. Turner).

**003.055 The Earth: Its Origin, History and Physical Consti-
tution.**   H. Jeffreys.
Cambridge University Press, Cambridge. Fifth Edition.
525 pp. Price $ 22.50 (1970). – Review in Sky Telescope,
Vol. 40, 44 (1970).

003.056 Physics of Stars.
S. A. Kaplan.
"Nauka", Moskva. Revised Edition. 212 pp. Price 40 Kop.
(1970). In Russian.

003.057 Introduction à la Physique des Intérieurs Stellaires.
V. Kourganoff.
Dunod Editeur, Paris. 207 pp. Price F 29 (1970). – Reviews
in Mem. Soc. Astron. Italiana, Nuova Serie, Vol. 41, 143 -
144; 1970 (M. G. Fracastoro); Sky Telescope, Vol. 39,
317 (1970).

003.058 The Solar System and Back.
I. Asimov.
Doubleday and Co., Inc., New York. 246 pp. Price $5.95
(1970). – Review in Sky Telescope, Vol. 39, 184 (1970).

003.059 The Moon as Viewed by Lunar Orbiter.
National Aeronautics and Space Administration,
NASA SP-200. U.S. Government Printing Office, Washington,
D.C. 152 pp. rrice $ 7.75 (1970). – Review in Sky Telescope,
Vol. 40, 15 (1970).

003.060 Nuclear and Relativistic Astrophysics and Nuclidic

Cosmochemistry: 1963 – 1967, Vol. II, III.
B. Kuchowicz.
Nuclear Energy Information Center, Warsaw. Review Rep.
NEIC-RR-37, 38. 304, 264 pp. respectively (1969).

003.061 Problems of Lunar Geology.
"Nauka", Moskva. Geol. in-t AN SSSR, vyp. (No.)
204, 291 pp. Price 2 Rbl. 21 Kop. (1969).   In Russian.

**003.062 Solar and Geophysical Events 1960 – 1965
(Calender Record).**
Compiled by J. V. Lincoln, General Editor: A. C. Stickland.
The M.I.T. Press, Massachusetts Institute of Technology,
Cambridge, Mass. – London. Annals of the IQSY (Interna-
tional Years of the Quiet Sun), Vol. 2. 9 + 297 pp. Price
£7 0s. 0d. (1968). – Review in Solar Physics, Vol. 12,
502; 1970 (Z. Svestka).

003.063 The Times Atlas of the Moon.
H. A. G. Lewis (Editor).
Times Newspapers Limited, London. 37 + 110 pp. Price
95s. (1969). – Reviews in Nature, Vol. 225, 770; 1970
(E. Phillips); Planet. Space Sci., Vol. 18, 950; 1970
(R. J. Fryer); Spaceflight, Vol. 12, 225; 1970 (C. A. Cross).

003.064 Discoverers of Space. A Pictorial Narration.
E. Lessing.
Burns & Oates. London. 4 + 172 + 24 + 1 pp. Price £6 (1969).
Reviews in Journ. Astron. Soc. Victoria, Vol. 23, 26 - 27;
1970 (P. Simon); Journ. History Astron., Vol. 1, (Part 1),
90 (1970); Spaceflight, Vol. 12, 188; 1970 (L. J. Carter).

003.065 Star Performance.   H. P. Lattin.
Whitmore Publishing Co., Philadelphia. 238 pp.
Price $ 4.95 (1969). – Reviews in Sky Telescope, Vol. 39,
184 (1970), Vol. 39, 383 - 384; 1970 (G. Lovi).

**003.066 Statistical Investigation of the Brightness of
Irregular and Semiregular Variables.**
F. I. Lukatskaya.
Izdatel'stvo "Naukova dumka", Kiev. Akademiya Nauk
Ukrainskoj SSR. Glavnaya astronomicheskaya observatoriya.
152 pp. Price 55 Kop. (1969).   In Russian.

**003.067 Geophysical Measurements: Techniques, Observa-
tional Schedules and Treatment of Data.**
C. M. Minnis (Editor).
The M.I.T. Press, Massachusetts Institute of Technology,
Cambridge, Mass. – London. Annals of the IQSY (International Years of the Quiet Sun). Vol. 1. 18 + 398 pp. Price
£9 7s. 0d. (1968). – Review in Solar Physics, Vol. 12, 502;
1970 (Z. Svestka).

003.068 Early Solar Physics.
A. J. Meadows.
Pergamon Press, New York. 8 + 312 pp. Price $ 4.75 (1970).

003.069 The High Firmament.
A. J. Meadows.
Leicester University Press, Leicester; Humanities Press Inc.,
New York. 14 + 207 pp. Price 42s. (1969). – Review in
Journ. British Astron. Ass., Vol. 80, 329; 1970 (C. A. Ronan).

003.070 Thin-film Optical Filters.
H. A. Macleod.
Adam Hilger Ltd., London. 332 pp. Price £7 8s. (1969).
Review in Journ. British Astron. Ass., Vol. 80, 155; 1970
(H. E. Dall).

003.071 Theory of Weak Interactions in Particle Physics.
R. E. Marshak, Riaduzzin, C. P. Ryan.

Wiley-Interscience, New York. 14 + 770 pp. Price $ 29.95
(1969). – Review in Science, Vol. 168, 962 - 963; 1970
(L. Wolfenstein).

003.072 **Cosmology.** W. H. McCrea.
Francis Hodgson, Guernsey. 18 pp. Price 40 s.
(1970). – Review in Nature, Vol. 226, 43 - 44; 1970
(W. C. Saslaw).

003.073 **A History of Japanese Astronomy.**
S. Nakayama.
Harvard-Yenching Institute Monograph Series, Vol. 18.
Harvard University Press, Cambridge, Mass. 329 pp.
Price $ 10.00 (1969). – Reviews in Sterne, 46. Jahrgang,
p. 33 - 34; 1970 *(H. Lambrecht),* SuW, Vol. 9, 134;
1970 *(F. Schmeidler).*

003.074 **Einstein Spaces.** A. Z. Petrov.
Pergamon Press, Oxford. 8 + 411 pp. Price £7,
$ 12.00 respectively (1969). – Review in Journ. Phys. A,
General Phys., Vol. 3, 229; 1970 *(J. S. Dowker).*

003.075 **Cosmic Electrodynamics.**
J. H. Piddington.
Wiley, New York – London. 10 + 305 pp. Price 175 s.
(1970). – Review in Nature, Vol. 226, 471 - 472; 1970
(J. Gribbin).

003.076 **Celestial Mechanics.**
M. de la Place.
Chelsea, Bronx, New York. 9 + 508 pp. Price $ 17.50 (1969).
Review in Nature, Vol. 225, 880 - 881; 1970 *(L. Rosenfeld).*

003.077 **Raum, Zeit, Materie. Vorlesungen über allgemeine
Relativitätstheorie.** H. Weyl.
6th edition. Springer-Verlag, Berlin–Heidelberg–New York.
8 + 338 pp. Price $ 8.20 (1970). – Review in Orion Schaff-
hausen, 28. Jahrgang, p. 96 - 97; 1970 *(H. Müller).*

003.078 **A Biographical Dictionary of Scientists.**
T. I. Williams (Editor).
A. & C. Black, London. 12 + 592 pp. Price £5 (1969). – Review in Journ. History Astron., Vol. 1, (Part 1), 89 - 90
(1970).

003.079 **Der Mond.** R. Proske.
Buch und Zeit Verlagsgesellschaft, Köln. 160 pp.
Price DM 9.80 (1969).

003.080 **Nikolaus Kopernikus.** F. Schmeidler.
Wissenschaftliche Verlagsgesellschaft mbH,
Stuttgart. 246 pp. Price DM 27.50 (1970). – Review in
Sky Telescope, Vol. 39, 317 (1970).

003.081 **Radio Emission of the Sun and Planets.**
V. V. Zheleznyakov.
Pergamon Press, Oxford–London–New York. 14 + 697 pp.
Price 300s, $ 40.00 respectively (1970). – Reviews in
Nature, Vol. 226, 471; 1970 *(R. D. Davies);* Science, Vol.
168, 1506 (1970).

003.082 **Pulsars.** V. L. Ginzburg.
"Znanie", Moskva. 32 pp. Price 6 Kop. (1970). In
Russian.

003.083 **Interstellar Dust.** M. Gershinberg.
Translated from the English edition. "Mir", Moskva.
32 pp. Price 1 Rbl. 34 Kop. (1970). In Russian.

003.084 **Astronomiya 1967. Physics of the Sun.**
M. A. Livshits.

Itogi Nauki. Seriya "Astronomiya". Moskva. 226 pp. Price
1 Rbl. 76 Kop. (1970). In Russian.

003.085 **Erde, unser Planet.** G. Gamow.
Franz Ehrenwirth Verlag KG, München. 264 pp.
Price DM 19.80 (1969). – Review in Phys. Ber., Vol. 49,
1379; 1970 *(E. Bretnütz).*

003.086 **Sir Christopher Wren.**
H. Gould.
Franklin Watts Inc., New York. 216 pp. Price $ 3.95 (1970).
Review in Sky Telescope, Vol. 39, 386 (1970).

003.087 **The Fate of the Solar System. Popular Sketches on
Celestial Mechanics.** V. G. Demin.
"Nauka", Moskva. 255 pp. Price 44 Kop. (1969). In Russian.–
Review in Referativ. Zhurn. 51. Astron., 3.51.73 (1970).

003.088 **The Shadow of the Telescope. A Biography of
John Herschel.** G. Buttmann.
Translated from German by B. E. J. Pagel, D. S. Evans.
Scribners, New York. 219 pp. Price $ 7.95 (1970). – Reviews
in Science, Vol. 168, 731 - 732; 1970 *(W. F. Cannon),* Sky
Telescope, Vol. 39, 184 (1970).

003.089 **Dutton's Navigation and Piloting.**
G. D. Dunlap, H. H. Shufeldt.
United States Naval Institute, Annapolis Md., U.S.A. 715 pp.
Price $ 15.00 (1969). – Review in Journ. Inst. Navigation,
London, Vol. 23, 269 - 270; 1970 *(C. H. Cotter).*

003.090 **The Riddle of Gravitation.** P. Bergmann.
Translation from the English Edition. "Nauka",
Moskva. 215 pp. Price 61 Kop. (1969). In Russian.

003.091 **Man in Inner and Outer Space. Selected Lectures on
the U.S. Manned Moonlanding-Programme, the Sun
and Our Own Planet.** S. T. Butler, H. Messel (Editors).
Pergamon Press, Oxford–New York. 472 pp. Price 45s.,
$ 7.00 respectively (1969). – Reviews in Planet. Space Sci.,
Vol. 18, 445; 1970 *(W. J. Raitt);* Spaceflight, Vol. 12, 142;
1970 *(P. A. Elliott);* Space Sci. Rev., Vol. 10, 737; 1970
(C. de Jager).

003.092 **Wege in den Weltenraum.** A. C. Clarke.
Econ-Verlag, Düsseldorf. 412 pp. Price DM 20.00
(1969).

003.093 **Positions and proper motions of 520 near-pole stars.
Zone +84° to +90° declination.**
V. V. Tel'nyuk-Adamchuk.
Izdatel'stvo Kievskogo Universiteta; Astronomicheskaya
observatoriya, Kiev. 216 pp. Price 53 Kop. (1969).
In Russian.

003.094 **Lunar Surface Photos. Pictures of Apollo 11
Mission.**
Available from Superintendent of Documents, U.S. Govern-
ment Printing Office, Washington, D.C. Set 12 Pictures. Price
$ 1.75. – Review in Spaceflight, Vol. 12, 22 (1970).

003.095 **Nineteenth-Century Spectroscopy.**
W. McGucken.
Johns Hopkins Press, Baltimore, Md. 233 pp. Price $ 11.00
(1969). – Review in Sky Telescope, Vol. 39, 387 (1970).

003.096 **Man on the Moon.**
J. M. Mansfield.
Stein and Day, New York; Constable and Co., Ltd., London.
285 pp. Price $ 6.95 (1969). – Reviews in Sky Telescope,
Vol. 39, 117 (1970); Spaceflight, Vol. 12, 187 - 188; 1970

(J. P. Gilmartin).

003.097 **Mondflug-Atlas.**
    P. Moore.
Hallwag-Verlag, Bern—Stuttgart. 48 pp. Price DM 19.00
(1969).

003.098 **Geology of the Moon.**
    T. A. Mutch.
Princeton University Press, Princeton. 324 pp. Price $ 17.50
(1970). — Review in Sky Telescope, Vol. 40, 44 (1970).

003.099 **Atlas of Cometary Forms.** J. Rahe, B. Donn,
    K. Wurm.
National Aeronautics and Space Administration. NASA Spe-
cial Publication 198. Superintendent of Documents, U.S.
Government Printing Office, Washington, D.C. 128 pp.
Price $ 2.25 (1969). — Review in Sky Telescope, Vol. 39, 117
(1970).

003.100 **Essential Relativity, Special, General, and Cosmolo-
    gical.** W. Rindler.
Van Nostrand Reinhold, New York—London. 14 + 322 pp.
Price $ 11.50, 107s. respectively (1970). — Review in Nature,
Vol. 226, 571; 1970 *(C. W. Kilmister).*

003.101 **Sextant Observations for the Seafarer.**
    G. Richards.
Nautical Publishing Company, Lymington, Hampshire. 172 pp.
Price 32s. (1969). — Review in Journ. Inst. Navigation,
London, Vol. 23, 128 - 129; 1970 *(D. H. Sadler).*

003.102 **The Elements and Structure of the Physical Sciences.**
    J. A. Ripley, Jr., R. C. Whitten.
John Wiley & Sons Inc.,New York. 690 pp. Price $11.50
(1969). — Reviews in Sky Telescope, Vol. 39, 253 (1970);
Vol. 40, 43; 1970 *(W. H. Glenn).*

003.103 **Where the Winds Sleep: Man's Future on the Moon.**
    N. P. Ruzic.
Doubleday and Company, Inc., New York. 236 pp. Price
$5.95 (1970). — Review in Sky Telescope, Vol. 39, 387 (1970)

003.104 **Telescopes for Skygazing.** H. E. Paul.
    Amphoto, New York. 160 pp. Price $5.95 (1970). —
Review in Sky Telescope, Vol. 39, 252 - 253 (1970).

003.105 **Physics of the Earth.** F. D. Stacey.
    John Wiley & Sons, New York. 12 + 324 pp. Price
$11.95 (1969).

003.106 **The Revolution in Astronomy.** P. W. Hodge.
    Holiday House, Inc., New York. 189 pp. Price
$4.95 (1970). — Review in Sky Telescope, Vol. 39, 387 (1970).

003.107 **Advances in Plasma Physics, Vol. 2.**
    A. Simon, W. B. Thompson (Editors).
John Wiley & Sons, New York; John Wiley & Sons, Ltd., Chi-
chester. 7 + 211 pp. Price $ 13.50 (1969). — Reviews in Journ.
Phys. A, General Phys., Vol. 3, 321 - 322; 1970 *(K. Dolder);*
IEEE Spectrum, Vol. 7, No. 5, p. 100; 1970 *(M. B. Bachynski).*

003.108 **Wonder and Glory: The Story of the Universe.**
    C. D. Simak.
St. Martin's Press, New York. 238 pp. Price $5.95 (1969). —
Review in Sky Telescope, Vol. 39, 386 (1970).

003.109 **Statistische Auswertungsmethoden.**
    L. Sachs.
Springer-Verlag, Berlin—Heidelberg—New York. 677 pp.
Price DM 58.00 (1969). — Review in SuW, Vol. 9, 23; 1970

(R. Wielen).

003.110 **Hierarchical Structures.** L. L. Whyte, A. G.
    Wilson, D. Wilson (Editors).
American Elsevier Publishing Company, Inc., New York. 322
pp. Price $12.50 (1969). — Review in Sky Telescope, Vol. 39,
117 (1970).

003.111 **Two-Star Position Finding, 1970.** E. M. Weyer.
    E. M. Weyer, Westbrook, Conn. 286 pp. Price
$10.00 (1969). — Review in Sky Telescope, Vol. 39, 184
(1970).

003.112 **We Reach the Moon.** J. N. Wilford.
    W. W. Norton and Co., Inc., New York. 352 pp.
Price $7.95 (1969). — Reviews in Sky Telescope, Vol. 39, 184
(1970); Vol. 39 , 248 - 249; 1970 *(S. S. Ross).*

003.113 **Principles of Astronomy.**
    S. P. Wyatt.
Allyn & Bacon, Inc., Boston. 561 pp. (1970).

003.114 **The Paradox of Olbers' Paradox: A Case History of
    Scientific Thought.** S. L. Jaki.
Herder and Herder, New York. 269 pp. Price $9.50 (1969). —
Review in Journ. History Astron., Vol. 1, (Part 1), 83 - 84;
1970 *(R. Schlegel).*

003.115 **Mariner—Mars 1969: A Preliminary Report, 1969.**
    National Aeronautics and Space Administration.
NASA SP - 225, [Available from the Clearinghouse for Federal
Scientific and Technical Information, Springfield, Va.], 145 pp
Price $3.00 (1969). — Review in Sky Telescope, Vol. 39, 252
(1970).

003.116 **Course in Theoretical Astrophysics.**
    V. V. Sobolev.
National Aeronautics and Space Administration. NASA TT
F - 531, [Available from Clearinghouse of Federal Scientific
and Technical Information, Springfield, Va.], 493 pp. Price
$3.00 (1969). — Review in Sky Telescope, Vol. 39, 184 (1970)

003.117 **Hallo Erde.** R. Metzler.
    Loewes-Verlag Ferdinand Carl KG, Bayreuth.
336 pp. Price DM 16.80 (1969).

003.118 **Mondlandung.** M. Maegreth.
    Belser-Verlag, Stuttgart. 302 pp. Price 21s. (1969).
Review in Spaceflight, Vol. 12, 224; 1970 *(E. Sellner).*

003.119 **Jupiterbeobachtungen von 1926 bis 1964.**
    W. Löbering.
Verlag Johann Ambrosius Barth, Leipzig. 24 pp. Price
MDN 17.00 (1969). — Reviews in Sky Telescope, Vol. 39,
386 (1970); Sterne, 46. Jahrgang, p. 47 - 48; 1970
(P. Ahnert).

003.120 **Space Research and Technology, Vol. 1 — The
    Space Environment.**
N. H. Langton (Editor).
University of London Press, London; American Elsevier
Publishing Company, New York. 238 pp. Price 35s.,$ 7.00
respectively. (1969). — Review in Journ. British Astron.
Ass., Vol. 80, 243; 1970 *(H. Miles).*

003.121 **Mondflug in Frage und Antwort.**
    E. von Khuon, G. Siefarth.
L. Schwann Verlag, Düsseldorf. 160 pp. Price DM 7.80
(1969). — Review in Umschau, 70. Jahrgang, p. 360; 1970
(H. J. Fahr).

003.122 **Das Weltall.** C. Friedemann.
Urania-Verlag, Leipzig–Jena–Berlin. 224 pp.
Price MDN 6.80 (1969).

003.123 **Project Apollo. The Way to the Moon.**
P. J. Booker, G. C. Frewer, G. K. C. Pardoe.
American Elsevier Publishing Company, Inc., New York. 8 +
216 pp. Price $5.50 (1969). – Review in Spaceflight, Vol. 12,
225; 1970 (*J. M. Tibbs*).

003.124 **Time and Modern Physics (A Compilation of
Articles).** D. A. Frank-Kamenetskij (Editor).
Translated from the French edition. "Mir", Moskva. 152 pp.
Price 35 Kop. (1970). In Russian.

003.125 **Man on the Moon.** P. Fairly.
Arthur Barker Ltd., London. 252 pp. Price 38s.
(1969). – Review in Spaceflight, Vol. 12, 142; 1970
(*S. R. Stebbing*).

003.126 **Das heliozentrische System in der griechischen,
persischen und indischen Astronomie.**
B. L. van der Waerden.
Neujahrsblatt der Naturforschenden Gesellschaft in Zürich,
55 pp. (1970). – Review in Orion Schaffhausen, 28. Jahrgang,
p. 97; 1970 (*N. Hasler-Gloor*).

003.127 **Our Blue Planet: The Story of the Earth's Evolution.**
H. Haber.
Translated from the German Edition (Stuttgart 1965) by E.
Stuhlinger. Charles Scribner's Sons, New York, 8 + 88 pp.
Price $5.95 (1969).

003.128 **Gesetze des Weltalls.** J. Herrmann.
Kosmos-Verlag, Stuttgart. 102 pp. Price DM 9.80
(1969). – Review in Sterne, 46. Jahrgang, p. 94 - 95; 1970
(*S. Marx*).

003.129 **International Physics and Astronomy Directory,
1969 - 1970.**
W. A. Benjamin, New York. 8 + 808 pp. Price $35.00,
$12.50 respectively (1969).

003.130 **Sterrenkunde.** C. de Jager.
Wetenschappelijke Uitgeverij n.v., Amsterdam.
900 pp. Price Vol. 1: hfl 17.50, Vol. 2: hfl 14.50, Vol. 3:
hfl 18.50, Vol. 4: hfl 19.50 – compl. hfl 59.00 (1969).
Review in Hemel en Dampkring, Vol. 68, 8 - 9; 1970
(*T. de Vries*).

003.131 **1970 Celestial Calendar and Handbook.**
C. F. Johnson, Jr.
C. F. Johnson, Jr., Watertown. 38 pp. Price $1.50 (1969). –
Review in Sky Telescope, Vol. 39, 47 (1970).

003.132 **Mathematical Geodesy.** M. Hotine.
Environmental Science Services Administration,
Washington, D. C. 16 + 416 pp. Price $5.50 (1969). – Review
in Science, Vol. 167, 1713; 1970 (*H. Moritz*).

003.133 **Handbook for Telescope Making.**
N. E. Howard.
Faber and Faber, London. 322 pp. Price 50 s. (1969). – Re-
view in Journ. British Astron. Ass., Vol. 80, 332 - 333; 1970
(*H. R. Hatfield*).

003.134 **Les Observatoires Spatiaux.**
J.-C. Pecker.
Presses Universitaires de France, Paris. 180 pp. Price 25 F.
(1969).

003.135 **The Condon Report, Scientific Study of Unidenti-
fied Flying Objects.**
E. P. Dutton & Co., New York. 967 pp. $12.95; also Bantam
Books, New York. $1.95 (1969). – Reviews in Icarus, Vol. 11,
443 - 447; 1970 (*J. E. McDonald*); Vol. 11, 447 - 450; 1970
(*H.-Y. Chiu*).

003.136 **The University of Colorado Report on Unidentified
Flying Objects.**
National Academy of Sciences, Washington, D. C. (1969).
Review in Icarus, Vol. 11, 440 - 443 (1970).

003.137 **Useful Applications of Earth-Oriented Satellites.**
National Academy of Sciences, National Research
Council, Washington, D. C. (1969). – Review in Icarus,
Vol. 11, 436; 1970 (*S. F. Singer*).

003.138 **La Lune des premiers astronomes aux vols Apollo.**
P. Couperie, C. Moliterni.
Edition Serg, Paris. 24 pp. and 99 illustrations (1969).
Reviews in L'Astronomie, 84ᵉ année, p. 137 - 138 (1970);
The Moon, Vol. 1, 426; 1970 (*Z. Kopal*).

003.139 **Elementary Processes for Cosmic Ray Astrophysics.**
V. L. Ginzburg.
Topics in Astrophysics and Space Physics, No. 1. Gordon and
Breach, New York – London. 8 + 131 pp. Price $12.60, 105s.
respectively (1970). – Review in Nature, Vol. 227, 867 - 868;
1970 (*J. Gribbin*).

003.140 **How We Will Reach the Stars.** J. W. Macvey.
Collier, New York; Collier-Macmillan, London. 12 +
244 pp. Price $ 1.25 (1969).

003.141 **Selected Space Goals and Objectives and their
Relation to Natioanl Goals.**
Battelle Memorial Institute, Columbus, Ohio. National Aero -
nautics and Space Administration. NASA CR–105304,
[Available from the Clearinghouse for Federal Scientific and
Technical Information, Springfield, Va.], 232 pp. Price $3.00
(1969). – Review in Sky Telescope, Vol. 39, 184 (1970).

003.142 **An Atlas: Around the World.**
George Philip and Son, London. 128 pp. Price
30s. (1969). – Reviews in Planet. Space Sci., Vol. 18, 126;
1970 (*R. Common*); Spaceflight, Vol. 12, 139; 1970
(*L. J. Carter*). – The atlas contains 124 full-colour illustrations.

003.143 **De Landing op de Maan en de Maanreis van de
Apollo 8.** C. Titulaer.
N. V. Polyvisie, Hilversum. 2 × 12 slides. Price per serie
f 10.50 (1969). – Review in Hemel en Dampkring, Vol. 67,
378; 1969 (*F. P. Israel*).

003.144 **Operatie Maan.** C. Titulaer.
Haagse Courant. 172 pp. (1969). – Review in
Hemel en Dampkring, Vol. 67, 375; 1969 (*M. Minnaert*).

003.145 **Die Natur. (Wissen im Überblick Vol. 1).**
Herder-Verlag, Freiburg–Basel–Wien. 702 pp.
Price DM 80.00 (1968). – Review in Sterne, 46. Jahrgang,
p. 39 - 41; 1970 (*H. Lambrecht*).

003.146 **Astronomía Elemental – Una Introdución al Uni-
verso.** C. M. Varsavsky.
Angel Estrada y Cía. (1969). – Review in Revista Astron.,
Vol. 41, No. 170, p. 87 (1969).

003.147 **The Universe.** D. Bergamini.
Time Life International, Amsterdam. 192 pp.

Price 38s. 6d. (1969). – Review in Spaceflight, Vol. 12, 297; 1970 *(K. Ward)*.

003.148 **The Earth in the Universe.**
V. V. Fedynskii (Editor).
For Israel Program for Scientific Translations H. A. Humphrey Ltd., London. 5 + 402 pp. Price £ 6.00 (1969).

003.149 **Lunar Globe.**
Published by Scan-Globe, Denmark and marketed in England by George Philip & Son, Ltd., London. Price 25s. (1969). – Reviews in Journ. British Astron. Ass., Vol. 80, 330; 1970 *(P. Moore);* Spaceflight, Vol. 12, 225; 1970 *(L. J. Carter).*

003.150 **Colonization of the Moon.** D. S. Halacy, Jr.
D. van Nostrand, Princeton, N. J. 159 pp. Price $3.95 (1969).

003.151 **Modern Astronomy.** D. S. Birney.
Allyn & Bacon, Inc., Boston. 338 pp. (1969).

003.152 **The Young Specialist Looks at Stars and Planets.**
W. Widmann, K. Schütte.
Burke Publishing Company Limited, London. 159 pp. Price 8s. (1969). – Review in Spaceflight, Vol. 12, 301; 1970 *(A. E. Fanning).*

003.153 **Astronomia Geodezyjna.**
W. Opalski, L. Cichowicz.
Second edition. Państwowe Przedsiębiorstwo Wydawnictw Kartograficznych, Warszawa. 518 pp. Price zł 52.00 (1970).

003.154 **Introduction to Ionospheric Physics.**
H. Rishbeth, O. W. Garriott.
Academic Press, New York. 10 + 334 pp. Price $16.00 (1969). Review in Science, Vol. 168, 1336 - 1337; 1970 *(S. J. Bauer).*

003.155 **Radio Astrophysics. Nonthermal Processes in Galactic and Extragalactic Sources.**
A. G. Pacholczyk.
W. H. Freeman & Company, San Francisco. 21 + 269 pp. Price $13.00 (1970).

004 History of Astronomy, Chronology

004.001 **History of astronomy and geodesy.**
G. Zhelnin.
Astron. Geod. Eston. SSR, Tartu, (see 003.004), p. 120 - 126
(1969). In Russian.

004.002 **Sunrise and moonrise at Stonehenge.**
J. H. Robinson.
Nature, Vol. 225, 1236 - 1237 (1970).
In this report the author offers an improvement to
Hawkin's theory of Stonehenge and he proposes an explana-
tion of why the Heel Stone was asymmetrically situated,
south of the centre line of the Avenue.

004.003 **Observing the moon in megalithic times.**
A. Thom.
Journ. British Astron. Ass., Vol. 80, 93 - 99 (1970). – Christ-
mas Lecture, British Astron. Ass. 1969 Dec. 31.

004.004 **The story of Greenwich time.**
H. D. Howse.
Journ. British Astron. Ass., Vol. 80, 208 - 211 (1970).

004.005 **Report on the utilization of an original Sir William
Herschel 7-foot reflector.**
C. Wise.
Journ. British Astron. Ass., Vol. 80, 218 - 219 (1970).

004.006 **Some mathematical curiosities embedded in the
solar system.** M. Gardner.
Sci. American. Vol. 222, No. 4, p. 108 - 112 (1970).

004.007 **Megalithic rings: Their design construction.**
T. M. Cowan.
Science, Vol. 168, 321 - 325 (1970).
It is proposed that prehistoric man used a unique
method to scribe the simple megalithic designs.

004.008 **Before the Principia: The maturing of Newton's
thoughts on dynamical astronomy, 1664 - 1684.**
D. T. Whiteside.
Journ. History Astron., Vol. 1, (Part 1), 5 - 19 (1970).
The author sketches the revised picture of history of the
gravity law which begins to emerge of the growth to maturity
of Newton's ideas on dynamical astronomy during the two
crucial decades, 1664 - 84.

004.009 **The solar eclipse technique of Yaḥyā B. Abī
Manṣūr.** E. S. Kennedy, N. Faris.
Journ. History Astron., Vol. 1 (Part 1), 20 - 38 (1970).
This paper discusses those parts of Yaḥyā's zīj which
have to do with solar eclipses. There are three numerical
tables and a set of cryptic directions for computing eclipses.

004.010 **The cosmology of Thomas Wright of Durham.**
M. Hoskin.
Journ. History Astron., Vol. 1, (Part 1), 44 - 52 (1970).
A reading of Wright's work "An original theory or new
hypothesis of the universe" (1750) shows that he not only
did not but, in a real sense, could not accept the disk-shaped
model of our star system so often ascribed to him.

004.011 **New light on Olbers's dependence on Chéseaux.**
S. L. Jaki.
Journ. History Astron., Vol. 1, (Part 1), 53 - 55 (1970).

004.012 **Die astronomischen Teleskope William Herschels.**
A. Maurer.

Orion Schaffhausen, 28. Jahrgang, p. 5 - 8 (1970).

004.013 **A comet and meteors in the early sixteenth century.**
C. M. Botley.
Journ. British Astron. Ass., Vol. 80, 304 - 305 (1970).

004.014 **The presentation of Caroline Herschel's telescope
to the Society by Sir John Herschel.**
H. R. Calvert.
Quarterly Journ. Roy. Astron. Soc., Vol. 11, 105 - 108
(1970).

004.015 **Sidelights on astronomy during the Society's
history.** H. Dingle.
Quarterly Journ. Roy. Astron. Soc., Vol. 11, 109 - 119
(1970).

004.016 **Die große Konjunktion der Planeten Jupiter und
Saturn im Jahre 7 v. Chr. und ihre Deutung als
Stern der Weisen.** D. Wattenberg.
Vorträge und Schriften Archenhold-Sternw. Berlin-Treptow,
No. 34, 48 pp. (1969).

004.017 **Studies on Zyuzi-Reki or Shou-shih-li (2).**
H. Hirose.
Univ. Tokyo, Tokyo Astron. Obs. Report, Vol. 15, (No. 2),
(No. 57), p. 376 - 383 (1970). In Japanese.

004.018 **Studies on Semmyō-Reki or Hsüan-ming-li (3).**
M. Utida.
Univ. Tokyo, Tokyo Astron. Obs. Report, Vol. 15, (No. 2),
(No. 57), p. 384 - 399 (1970). In Japanese.

004.019 **Die Geschichte der Astronomie als Aufgabe der
Forschung und Popularisierung.**
D. B. Herrmann, D. Wattenberg.
Blick in das Weltall, Archenhold-Sternw. Berlin-Treptow,
No. 5, p. 40 - 50 (1970).

004.020 **150 Jahre Astronomie an der Universität Bonn
1818 – 1968.** F. Becker.
Sonderdruck aus 150 Jahre Rheinische Friedrich-Wilhelms-
Universität zu Bonn 1818 – 1968, (Math. Nat.), p. 72 - 87
(1969).

004.021 **Franz Xaver von Zach und seine „Allgemeinen
Geographischen Ephemeriden".**
D. B. Herrmann.
Sudfhoffs Arch., Vol. 52, 347 - 359 (1969).
This paper described the history of the German geo-
graphical and astronomical journal "Allgemeine Geographische
Ephemeriden" (AGE) founded by F. X. v. Zach and published
by him in the years 1798 - 1800. The suitability of Zach as
an editor of scientific journals is demonstrated. Zach's
principles of edition are illustrated.

004.022 **Das Gothaer Astronomentreffen im Jahre 1798 –
ein Vorläufer heutiger wissenschaftlicher Kongresse.**
D. B. Herrmann.
Sterne, 46. Jahrgang, p. 119 - 123 (1970).

004.023 **The calendar.** F. L. Leroux.
Journ. Astron. Soc. Western Australia, Vol. 21,
June, p. 5 - 8 (1970).

004.024 **Galilei and Huygens.** M. G. J. Minnaert.
Utrechtse Sterrekundige Overdrukken, No. 90,

15 pp. (1969)

004.025 History of astronomy. E. Rybka.
Trans. IAU, Vol. 14A, (see 003.028), 481 - 489 (1970). – Report of Commission 41.

**004.026 La détermination des positions du soleil en astro-
nomie indienne ancienne.** J. Filliozat.
Arch. Internat. Histoire Sci., Vol. 21, 89 - 94 (1968).

**004.027 An early instance of deductive discovery: Tycho
Brahe's lunar theory.** V. E. Thoren.
Isis, Vol. 58, 19 - 36 (1967). – See Zentralblatt Math. Grenz-
gebiete, Vol. 162, 8 (1969).

004.028 A survey of the Toledan tables. G. J. Toomer.
Osiris, 1968, p. 5 - 174 (1969).
The aim of this paper is to give a description of the
collection of mediaeval astronomical tables known as the
"Tabule Toletane" which shall suffice for their identification
and also reveal their affiliation with or derivation from other
known astronomical tables.

004.029 Papyrology and sources in astronomical history.
J.-O. Fleckenstein-Gallo.
Veröff. Forsch.-Inst. Deutsch. Museums Gesch. der Natur-
wiss. Techn., Ser. B, No. 1, p. 151 - 155 (1968).

004.030 Uigur and Tibetan lists of the Indian lunar mansions.
W. Petri.
Veröff. Forsch.-Inst. Deutsch. Museums Gesch. der Natur-
wiss. Techn., Ser. A, No. 38, p. 83 - 90 (1968).

**004.031 The date of invention of Babylonian planetary
theory.** B. L. van der Waerden.
Arch. History Exact Sci., Vol. 5, 70 - 78 (1968).
There are reasons to assume that System A of Babylonian
Planetary Theory was invented under Dareios I, i.e., between
522 and 486 B.C. This conclusion is based upon an analysis
of a set of planetary tables published by Aaboe and Sachs,
Journ. Cuneiform Studies, Vol. 20, 1 - 33 (1966).

**004.032 John Michell and Henry Cavendish: Weighing the
stars.** R. McCormmach.
British Journ. History Sci., Vol. 4, 126 - 155 (1968). – See

Zentralblatt Math. Grenzgebiete, Vol. 177, 3 (1970).

004.033 Did Anaxagoras observe a sunspot in 467 B.C.?
P. J. Bicknell.
Isis, Vol. 59, 87 - 90 (1968). – See Zentralblatt Math. Grenz-
gebiete, Vol. 179, 310 - 311 (1970).

004.034 Galileo's concept of inertia. J. A. Coffa.
Physis, Vol. 10, 261 - 281 (1969).

**004.035 Galileo gleanings. XVII: The question of circular
inertia.** S. Drake.
Physis, Vol. 10, 282 - 289 (1969).

004.036 Kepler's second law of planetary motion.
E. J. Aiton.
Isis, Vol. 60, 75 - 90 (1969).
The aim of the article is to counter some of the false
assertions concerning Kepler's second law of planetary motion
to be found in the literature.

004.037 Das astronomische System der Persischen Tafeln. I.
J. J. Burckhardt, B. L. van der Waerden.
Centaurus, Vol. 13, 1 - 28 (1968). – See Zentralblatt Math.
Grenzgebiete, Vol. 188, 4 (1970).

**004.038 Some medieval reports of Venus and Mercury
transits.** B. R. Goldstein.
Centaurus, Vol. 14, 49 - 59 (1969).

004.039 The error in Kepler's acronychal data for Mars.
C. Wilson.
Centaurus, Vol. 13, 263 - 268 (1969).

Forschungen und Publikationen zur Geschichte der
Astronomie in der Deutschen Demokratischen Republik.
Eine Bibliographie (1949 - 1969). See Abstr. 002.008.

Astronomy Gnomonics. A catalogue of instruments
of the 15th to the 19th centuries in the collections of the
National Technical Museum, Prague.
See Abstr. 003.054.

Bernard Walther: Innovator in astronomical
observation. See Abstr. 005.005.

005 Biography

005.001 **Lockyer as astronomer.** A. J. Meadows.
Nature, Vol. 225, 230 - 232 (1970).
The first editor of "Nature" was a self-taught astronomer whose contributions included the prediction of the existence of helium on the basis of spectroscopic evidence. Although the journal he edited was successful, Lockyer never convinced the scientific world of the ultimate validity of his own scientific work.

005.002 **V. M. Slipher's trailblazing career.** J. S. Hall.
Sky Telescope, Vol. 39, 84 - 86 (1970).

005.003 **Dr. Henry C. King, planetarium director.**
H. C. King.
Spaceflight, Vol. 12, 10 - 13 (1970).

005.004 **How S. W. Burnham became an astronomer.**
J. Ashbrook.
Sky Telescope, Vol. 39, 225 (1970).

005.005 **Bernard Walther: Innovator in astronomical observation.** D. deB. Beaver.
Journ. History Astron., Vol. 1, (Part 1), 39 - 43 (1970).

005.006 **Captain James Cook, R. N., F. R. S. and Australia.**
J. L. Perdrix.
Journ. Astron. Soc. Victoria, Vol. 23, 30 - 34 (1970).

005.007 **S. W. Burnham's Lick and Yerkes years.**
J. Ashbrook.
Sky Telescope, Vol. 39, 363 - 364 (1970).

005.008 **Alexander von Humboldt in seinen Beziehungen zur Astronomie.** D. Wattenberg.
Astron. in der Schule, 7. Jahrgang, p. 65 - 67 (1970).

005.009 **De revolutionibus orbium coelestium. Nikolaus Kopernikus und die Geschichte seines Lebenswerkes.**
D. Wattenberg.
Vorträge und Schriften Archenhold-Sternw. Berlin-Treptow, No. 33, 39 pp. (1969).

005.010 **C. Mayer.** K. Morav.
Říše hvězd, Vol. 51, 89 - 90 (1970). In Czech.

005.011 **Les contributions de Lemaître au développment de la cosmologie actuelle.** O. Godart.
Revue des Questions Scientifiques, Vol. 141, No. 2, 221 - 222 = Publ. Inst. Astron. Géophys. G. Lemaître, Louvain, Vol. 3, No. 4 (1970).

005.012 **The scientific work of Allesandro Piccolomini.**
R. Suter.
Isis, Vol. 60, 210 - 222 (1969).

006 Personal Notes

H. W. Babcock, received the Gold Medal of the Royal Astronomical Society.
Publ. Astron. Soc. Pacific, Vol. 82, 367 (1970).

H. W. Babcock received the Gold Medal of the Royal Astronomical Society.
Phys. Today, Vol. 23, No. 5, p. 91 (1970).

H. W. Babcock received the Gold Medal of the Royal Astronomical Society. B. Lovell.
Quarterly Journ. Roy. Astron. Soc., Vol. 11, 85 - 87 (1970).

A. Ballner, 70th birthday. V. Hýbl.
Říše hvězd, Vol. 51, 72 - 73 (1970) In Czech.

F. Becker, 70. birthday.
SuW, Vol. 9, 151 (1970).

W. P. Bidelman, Director of the Warner and Swasey Observatory.
Sky Telescope, Vol. 40, 16 (1970).

A. Blaauw, is named Director of the European Southern Observatory.
Astron. Nachr., Vol. 292, 48 (1970).

R. P. Cesco is named Director of La Plata Observatory.
Inform. Bull. Southern Hemisphere, No. 15, p. 56 (1969).

C. Hayashi, received the Eddington Medal of the

Royal Astronomical Society.
Publ. Astron. Soc. Pacific, Vol. 82, 367 (1970).

C. Hayashi received the Eddington Medal.
Phys. Today, Vol. 23, No. 5, p. 91 (1970).

C. Hayashi received the Gold Medal of the Royal Astronomical Society. B. Lovell.
Quarterly Journ. Roy. Astron. Soc., Vol. 11, 88 (1970).

J. Klepešta, 70th birthday. O. Hlad.
Říše hvězd, Vol. 51, 115 (1970). In Czech.

V. Petr, 60th birthday.
Říše hvězd, Vol. 51, 58 (1970). In Czech.

R. Rajchl, 60th birthday.
Říše hvězd, Vol. 51, 18 - 19 (1970). In Czech.

R. Rajchl, 60th birthday.
Stud. Geophys. Geod., Vol. 14, 351 - 352 (1970).

N. Richter, 60th birthday.
Blick in das Weltall, Archenhold-Sternw. Berlin-Treptow, No. 2, p. 13 (1970).

I. L. Thomsen is named Astronomer in Charge of the Mount John University Observatory.
Inform. Bull. Southern Hemisphere, No. 15, p. 57 (1969).

A. Unsöld, 65th birthday.

Phys. Blätter, 26. Jahrgang, p. 191 (1970).

H. C. Urey received the Leonard Medal.
Meteoritics, Vol. 4, 301 (1969).

R. L. Waterfield received the Bronce Medal of the

Astronomical Society of the Pacific.
Observatory, Vol. 90, 36 (1970).

R. Waterfield, received the Bronze Medal of the
Astronomical Society of the Pacific.
Publ. Astron. Soc. Pacific, Vol. 82, 370 (1970).

007 Obituaries

N. Ambolt, died 1969 November 23.
G. Larsson-Leander.
Astron. Tidssk., Vol. 3, 52 (1970).

S. Ballarin, died on 1970 March 11.
Mem. Soc. Astron. Italiana, Nuova Serie, Vol. 41, 259 (1970).

A. Ya. Bezrukova, 1905 - 1969 Dec. 3.
Solnechnye Dannye 1969 Byull., No. 11, p. 94 (1970).
In Russian.

J. A. Brown, died 1970 May 2.
Science, Vol. 168, 956 (1970).

L. Cichowicz, 1922 May 9 - 1968 June 2.
W. Dobaczevska.
Byull. Stantsij Optichesk. Nablyud. Iskusstv. Sputnikov Zemli,
No. 53 (1), p. 3 - 4 (1969). In Russian.

E. Davanzo, died on 1970 Jan. 5.
M. Hack.
Mem. Soc. Astron. Italiana, Nuova Serie, Vol. 41, 149 (1970).

M. Davidson, died 1968 June 25.
E. Öpik.
Irish Astron. Journ., Vol. 9, 170 (1969).

A. J. Deutsch, died 1969 Nov. 11.
Sky Telescope, Vol. 39, 33 (1970).

N. S. Domanskaya, 1931 - 1968 July 5.
V. K. Abalakin.
Byull. Inst. Teoret. Astron., *Leningrad,* Vol. 12, 155 - 156 (1970). In Russian.

J.-H. Focas, 1909 – 1969.
A. Dollfus.
L'Astronomie, 84ᵉ année, p. 217 - 219 (1970).

M. Fowler, 1898 April 28 - 1969 May.
Southern Stars, Vol. 23, 116 (1970).

G. Gamow, died 1968 August 19.
E. Öpik.
Irish Astron. Journ., Vol. 9, 170 (1969).

V. N. Golovzin, 1905 - 1968 Oct. 22.
M. Yu. Zeldina.
Vestn. Kiev. Univ., Ser. Astron., No. 11, p. 140 (1969).
In Russian.

L. G. Henyey, 1910 February 3 - 1970 February 18.
Publ. Astron. Soc. Pacific, Vol. 82, 367 - 369 (1970).

L. Henyey, died on Febr. 18, 1970.

Sky Telescope, Vol. 39, 290 (1970).

J. Hickox, died on 1970 April 12.
Observatory, Vol. 90, 123 (1970).

C. Hoffmeister, died 1968 January 2.
E. Öpik.
Irish Astron. Journ., Vol. 9, 169 (1969).

I. I. Iljinsky, 1887 May 5 - 1968 April 21.
N. A. Chernega.
Vestn. Kiev. Univ., Ser. Astron., No. 11, p. 138 - 139 (1969).
In Russian.

L. F. Jenkins, died on 1970 May 9.
Sky Telescope, Vol. 40, 16 (1970).

I. A. Khvostikov, 1910 – 1969 Aug. 7.
Byull. Abastumansk. Astrofiz. Obs. No. 37, p. 185 - 187 (1969). In Russian.

H. Knox-Shaw, died on 1970 April 11.
A. D. Thackeray.
Monthly Notes Astron. Soc. Southern Africa, Vol. 29, 54 - 55 (1970).

H. Knox-Shaw, died on 1970 April 11.
Observatory, Vol. 90, 123 (1970).

A. Kohlschütter, died 1969 May 28.
Astron. Nachr., Vol. 292, 48 (1970).

A. König, died 1969 April 28.
Astron. Nachr., Vol. 292, 48 (1970).

E.-E. Kühne, died 1969 July 30.
Astron. Nachr., Vol. 292, 48 (1970).

W. Löbering, died 1969 Dec. 22.
G. Loibl.
Sterne, 46. Jahrgang, p. 123 - 125 (1970).

D. F. Martyn, 1906 June 17 – 1970 March 6.
Nature, Vol. 226, 780 (1970).

O. Mathias, 1900 March 22 – 1969 Febr. 4.
H. Haupt.
Astron. Nachr., Vol. 292, 47 (1970).

W. Münch, died 1969 July 7.
Astron. Nachr., Vol. 292, 48 (1970).

A. V. Nielsen, died 1970 February 24.
Astron. Tidssk., Vol. 3, 100 (1970).

T. Nuraliev, 1934 - 1968 Dec. 13.

Tsirk. Astron. Inst., *Tashkent,* No. 15 (362), p. 32 (1970). In Russian.

V. M. Slipher, died 1969 Nov. 8.
Phys. Today, Vol. 23, No. 2, p. 101 (1970).

F. Soják, died 1970 Jan. 13.   O. Obůrka.
Říše hvězd, Vol. 51, 115 - 116 (1970).   In Czech.

T. E. Sterne, died 1970 February 7.
Phys. Today, Vol. 23, No. 5, p. 91 - 92 (1970).

T. E. Sterne, died 1970 February 7.
Sky Telescope, Vol. 39, 224 (1970).

I. Thomsen, died on 1969 Dec. 28th.
Sky Telescope, Vol. 39, 83 (1970).

I. L. Thomsen, 1910 Aug. 30 - 1969 Dec. 28.
Southern Stars, Vol. 23, 113 - 116 (1970).

H. Vogt, 1890 Oct. 5 – 1968 Jan 23.
E. Kollnig-Schattschneider.
Astron. Nachr., Vol. 292, 45 - 46 (1970).

H. Vogt, 1890 Oct. 5 - 1968 Jan. 23.
W. Fricke.
Jahrbuch Heidelberger Akad. Wiss. 1969, p. 47 - 49 (1970).

R. Weber, 1903 Febr. 25 - 1969 July 27.
A. Brun.
L'Astronomie, 84ᵉ année, p. 79 - 81 (1970).

M. H. Wrubel, died 1968 October 26.
E. Öpik.
Irish Astron. Journ., Vol. 9, 171 (1969).

008 Observatories, Institutes

Reports, communications and publications of observatories and astronomical institutes are recorded in this section; included are numbered series of reprints. Whenever possible, the numbers of the abstracts refering to the publications are given. Observatories and institutes are listed in alphabetical order of their towns. In some cases observatory publications do not give the name of the town. The following list which gives names and towns of some institutions may serve as an aid in such cases.

Algonquin Radio Observatory	Lake Traverse, Ontario, Canada
Allegheny Observatory	Pittsburgh, Pennsylvania
Arthur J. Dyer Observatory	Nashville, Tennessy
Bosscha Observatory	Lembang, Indonesia
Boyden Observatory	Bloemfontein, South Africa
Bureau International de l'Heure	Paris, France
Cajigal Observatory	Caracas, Venezuela
California Institute of Technology	Pasadena, California
Cape of Good Hope	Cape Town, South Africa
Carter Observatory	Wellington, New Zealand
Cavendish Laboratory	Cambridge, England
Ceskoslovenská Akademie Ved Astronomický Ustav	Praha, Czechoslovakia
Chamberlin Observatory, University of Denver	Denver, Colorado
Commonwealth Observatory	Canberra, Australia
David Dunlap Observatory, University of Toronto	Richmond Hill, Ontario
Dearborn Observatory	Evanston, Illinois
Department of Astronomy and Observatory, Univ. California	Los Angeles, California
Department of Astronomy, University of Texas	Austin, Texas
Division Radiophysics, C.S.I.R.O. University Grounds	Sydney, New South Wales
Dominion Astrophysical Observatory	Victoria, British Columbia
Dominion Observatory	Ottawa, Ontario
Dominion Radio Astrophysical Observatory	Penticton, British Columbia
Dudley Observatory	Albany, New York
Dunsink Observatory	Dublin, Ireland
Enhelhardt Observatory	Kazan, R.S.F.S.R.
European Southern Observatory	Hamburg, West Germany
Florida State University Radio Observatory	Tallahassee, Florida
Flower and Cook Observatories, University of Pennsylvania	Philadelphia, Pennsylvania
Four College Observatories	Amherst, Massachusetts
Fraunhofer Institut	Freiburg, West Germany
Georgetown Observatory	Washington, D.C.
Goddard Space Flight Center	Greenbelt, Maryland
Goethe Link Observatory, University of Indiana	Bloomington, Indiana
Griffith Observatory	Los Angeles, California
Harvard College Observatory	Cambridge, Massachusetts
High Altitude Observatory University of Colorado	Boulder, Colorado
Institut of Theoretical Astrophysics, Blindern	Oslo, Norway
Inter-American Observatory	Cerro Tololo, Chile
International Latitude Observatory	Mizusawa, Japan
Kandilli Observatory	Istanbul, Turkey

Kapteyn Astronomical Laboratory	Groningen, Netherlands
Karl Schwarzschild Observatorium	Tautenburg, German Democratic Republic
Kenneth Mees Observatory	Rochester, New York
Kwasan Observatory	Kyoto, Japan
Leander McCormick Observatory University of Virginia	Charlottesville, Virginia
Lee Observatory	Beirut, Lebanon
Leuschner Observatory	Berkeley, California
Lick Observatory	Mount Hamilton, (Santa Cruz), California
Lindheimer Astronomical Research Center	Evanston, Illinois
Lockheed Solar Observatory	Saugus, California
Lohrmann-Institut für Geodätische Astronomie	Dresden, German Democratic Republic
Louisiana State University Observatory	Baton Rouge, Louisiana
Lowell Observatory	Flagstaff, Arizona
Lunar and Planetary Laboratory	Tucson, Arizona
Max-Planck-Institut für Physik und Astrophysik	München, West Germany
McDonald Observatory	Fort Davis, Texas
McMath Hulbert Observatory	Pontiac, Michigan
Molonglo Radio Observatory, University of Sydney	Sydney, New South Wales
Mount Cuba Observatory	Wilmington, Delaware
Mullard Radio Astronomy Observatory	Cambridge, England
Narrabri Observatory, University of Sydney	Sydney, New South Wales
National Bureau of Standards	Washington, D.C.
National Observatory, USA	Kitt Peak, Arizona
National Radio Astronomy Observatory	Green Bank, West Virginia
Naval Research Laboratory, USA	Washington, D.C.
New Mexico State University Observatory	Las Cruces, Mexico
Nizamiah Observatory	Hyderabad, India
Nuffield Radio Astronomy Laboratories, Jodrell Bank, University of Manchester	Manchester, England
Observatoire Royal de Belgique	Uccle, Belgium
Observatorio de Cartuja	Granada, Spain
Observatorio del Ebro	Tortosa, Spain
Observatorio Fabra	Barcelona, Spain
Observatory, University of Michigan	Ann Arbor, Michigan
Ohio State University Radio Observatory	Columbus, Ohio
Ole Roemer-Observatoriet	Aarhus, Denmark
Perkins Observatory, Ohio State and Wesleyan Universities	Delaware, Ohio
Purple Mountain Observatory	Nanking, China
Radcliffe Observatory	Pretoria, South Africa
Radiophysics Laboratory, C.S.I.R.O.	Sydney, New South Wales
Remeis-Sternwarte	Bamberg, West Germany
Rensselaer Observatory	Troy, New York
Republic Observatory	Johannesburg, South Africa
Royal Radar Establishment, Radio Astronomy Division	Malvern, England
Rutherford Observatory, Columbia University	New York, New York
Sagamore Hill Radio Observatory	Bedford, Massachusetts
Saint-Michel, l'Observatoire	Haute Provence, France

Sternberg Observatory   **Moscow,** R.S.F.S.R.
Smithsonian Astrophysical
    Observatory   **Cambridge,** Massachusetts
Specola Astronomica Vaticana   **Castel Gandolfo,** Italy
Specola di Padova   **Asiago,** Italy
Sproul Observatory   **Swarthmore,** Pennsylvania
Steward Observatory,
    University of Arizona   **Tucson,** Arizona
United States Naval Observatory   **Washington,** D.C.

University of Florida,
    Radio Observatory   **Gainesville,** Florida
University of Illinois Observatory **Urbana,** Illinois
Uttar Pradesh State Observatory **Naini Tal,** India
Van Vleck Observatory   **Middletown,** Connecticut
Warner and Swasey Observatory **Cleveland,** Ohio
Washburn Observatory   **Madison,** Wisconsin
Yale University Observatory   **New Haven,** Connecticut
Yerkes Observatory   **Williams Bay,** Wisconsin

008.001   Abastumani

Abastumanskaya Astrofizicheskaya Observatoriya, **Gora Kanobili. Byulleten'.** Akademiya Nauk Gruzinskoj SSR, Nos. 37, 38.

Chronicle.
Byull. Abastumansk. Astrofiz. Obs. No. 37, p. 189 - 190 (1969). In Russian.

008.002   Albany

Dudley Observatory, State University of New York at Albany, Albany, New York. − Report 1968 - 1969. C. L. Hemenway.
Bull. American Astron. Soc., Vol. 2, 30 - 32 (1970).

Dudley Observatory, *Albany, N. Y.,* **Reprints** Nos. B-34 (H. Patashnick, C. L. Hemenway, 03.034.075), C-14 (A. G. D. Philip, 01.114.104), C-15 (A. G. D. Philip, 02.115.003), C-16 (A. G. D. Philip, 02.112.007).

008.003   Amherst

Four College Observatories: Amherst College, Amherst, Massachusetts; **Mount Holyoke College,** South Hadley, Massachusetts; **Smith College,** Northhampton, Massachusetts; **University of Massachusetts,** Amherst, Mass-achusetts. − Report 1968 - 1969.   W. M. Irvine.
Bull. American Astron. Soc., Vol. 2, 37 - 41 (1970).

Contributions from the Four College Observatories, *Amherst,* Nos. 15 (E. R. Harrison, AJB 68, 145.41), 20 (T. Arny, 01.065.019), 36 (E. R. Harrison, 01.156.005), 49 (D. J. Lovell, 03.082.093).

Department of Physics and Astronomy, University of Massachusetts, Amherst, Mass., Separate prints (W. M. Irvine, T. Simon, D. H. Menzel, J. Charon, G. Lecomte, P. Griboval, A, T. Young, AJB 68, 81.24; W. M. Irvine, T. Simon, D. H. Menzel, C. Pikoos, A. T. Young, AJB 68, 81.25; V. Vanysek, 01.106.002; G. R. Huguenin, J. H. Taylor, 01.141.083; W. T. Plummer, 01.093.012; E. R. Harrison, 02.162.055).

008.004   Ann Arbor

University of Michigan Observatories, Ann Arbor, Michigan. − Report 1968 - 1969.   O. Mohler.
Bull. American Astron. Soc., Vol. 2, 78 - 85 (1970).

008.005   Asiago

Contributi dell'Osservatorio Astrofisico dell Università di Padova in Asiago, Nos. 231 (L. Rosino, L. Pigatto, 02.122.111), 232 (R. Barbon, 02.158.075), 233 (R. Margoni, M. Perinotto, 02.119.017).

008.006   Auckland

Auckland Observatory.
Inform. Bull. Southern Hemisphere, No. 15, p. 23 (1969). Current research report.

008.007   Babelsberg

Sternwarte Babelsberg. Institut für relativistische und extragalaktische Forschung. − Report 1968. H.-J. Treder.
Monatsber. Deutsch. Akad. Wiss. Berlin, Vol. 11, 766 - 768 (1969).

008.008   Bamberg

Dr. Remeis-Sternwarte Bamberg, Astronomisches Institut der Universität Erlangen-Nürnberg. − Report 1969. W. Strohmeier.
Mitt. Astron. Ges., No. 28, p. 9 - 10 (1970).

008.009   Baton Rouge

Louisiana State University Observatory, Baton Rouge, Louisiana. − Report 1968 - 1969. A. U. Landolt.
Bull. American Astron. Soc., Vol. 2, 73 - 75 (1970).

Contributions of the Louisiana State University Observatory, Nos. 14 (A. U. Landolt, 01.122.015), 15 (P. Lee, L. H. Aller, S. J. Czyzak, R. N. Duvall, 01.113.015), 16 (A. U. Landolt, 01.118.007), 17 (C. L. Perry, 01.151.028), 18 (A. U. Landolt, 01.122.027), 19 (C. L. Perry, 01.113.026), 20 (D. J. Mac Connell, C. L. Perry, 01.152.006), 21 (A. U. Landolt, 02.122.040), 22 (A. U. Landolt, 02.113.011), 23 (P. Lee, 02.132.002), 24 (C. L. Perry, G. Hill, 02.153.006), 25 (C. L. Perry, H. E. Bond, 02.153.020), 26 (G. Hill, C. L. Perry, 02.153.011), 27 (H. E. Bond, J. S. Neff, 02.113.037), 28 (H. E. Bond, A. U. Landolt, 02.121.047), 29 (A. U. Landolt, 02.121.048), 30 (D. Sher, 03.151.006), 31 (A. U. Landolt, 03.124.101), 33 (R. J. Dufour, P. Lee, 03.132.018).

008.010 Bedford

Sagamore Hill Radio Observatory, Air Force Cambridge Research Laboratories, Bedford, Massachusetts. Report 1968 - 1969.   J. Aarons. Bull. American Astron. Soc., Vol. 2, 122 - 123 (1970).

008.011 Beirut

Lee Observatory, American University of Beirut, Lebanon, Monthly Bulletin, Astronomical Section, 1969 August - 1970 March (03.075.029).

008.012 Belo Horizonte

Instituto de Ciencias Exatas da Universidade Federal de Minas Gerais. (Institute of Exact Sciences of the Federal University of Minas Gerais). Inform. Bull. Southern Hemisphere, No. 15, p. 12 (1969). Current research report.

008.013 Beograd

Sur l'activité de l'Observatoire Astronomique de Beograd de 1966 à 1969.   P. M. Djurković. Bull. Obs. Astron. Beograd, Vol. 27, No. 2, p. 143 - 156.

Bulletin de l'Observatoire de Beograd, Vol. 27, No. 2 (V. Oskanjan, A. Kubičela, J. Arsenijević, 03.034.050, A. Bacsák, 03.081.037; J. Arsenijević, 03.079.109; B. Popović, 03.042.066; Z. Brkić, D. Djurović, 03.032.022; D. Djurović, 03.034.051; D. Djurović, 03.032.023; G. Teleki, 03.045.012; G. Teleki, R. Grujić, 03.045.013; M. Djokić, 03.032.024; V. Oskanjan, A. Kubičela, J. Arsenijević, 03.122.075; R. Grujić, M. Djokić, 03.045.014; Brkić, Djurović, Momčilović, Jovanović, 03.044.012; 03.096.017; P. M. Djurković, 03.008.013).

Publications de l'Observatoire Astronomique de Beograd, No. 16 (P. M. Djurković, 03.012.021).

008.014 Berkeley

Research Units and Academic Departments, University of California: Berkeley, Los Angeles, San Diego, and Santa Cruz. – Report 1968 – 1969. I. King, G. Abell, E. M. Burbidge, R. P. Kraft. Bull. American Astron. Soc., Vol. 2, 2 - 23 (1970).

Berkeley Reprints, Nos. 457 (L. V. Kuhi, 01.122.110), 458 (F. Schweizer, 02.098.001), 459 (S. A. Kellman, J. E. Gaustad, 02.131.042), 460 (D. W. Goldsmith, H. J. Habing, G. B. Field, 02.131.052), 461 (M. Breger, 02.122.102), 463 (H. Spinrad, B. J. Taylor, 02.114.115), 464 (C. S. Bowyer, G. B. Field, 02.142.002), 465 (R. F. Knacke, D. D. Cudaback, J. E. Gaustad, 02.131.050), 466 (J. E. Gaustad, F. C. Gillett, R. F. Knacke, W. A. Stein, 02.114.064), 467 (G. B. Field, 02.131.022), 468 (L. Anderson, L. V. Kuhi, 02.122.046), 469 (J. G. Phillips, R. S. Freedman, 02.122.092).

008.015 Berlin

Lehrstuhl für Astrophysik der Technischen Universität. – Report 1969.   K. Hunger. Mitt. Astron. Ges., No. 28, p. 11 (1970).

Heinrich-Hertz-Institut für solar-terrestrische Physik der Deutschen Akademie der Wissenschaften zu Berlin. – Report 1968.   F. W. Jäger, H. Daene. Monatsber. Deutsch. Akad. Wiss. Berlin, Vol. 11, 758 - 762 (1969).

Heinrich-Hertz-Institut, Solare Beobachtungsergebnisse.Deutsche Akademie der Wissenschaften zu Berlin, Zentralinstitut für Solar-Terrestrische Physik, Berlin–Adlershof. HHI Solar Data, Vol. 21, 1970 January - April (03.075.020).

Heinrich-Hertz-Institut, Solare Beobachtungsergebnisse. Deutsche Akademie der Wissenschaften zu Berlin, Zentralinstitut für Solar-Terrestrische Physik, Berlin–Adlershof. HHI Supplement Series of Solar Data, Vol. 1, No. 7 (A. Böhme, A. Krüger, H. Künzel, 03.077.042).

008.016 Berlin-Treptow

Archenhold-Sternwarte Berlin-Treptow, Sonderdruck No. 17 (D. Wattenberg, 02.009.002).

Archenhold-Sternwarte Berlin-Treptow,Vorträge und Schriften, No. 33 (D. Wattenberg, 03.005.009), 34 (D. Wattenberg, 03.004.016), 35 (H. Winkler, 03.047.018), 36 (H. Winkler, 03.041.012).

008.017 Bloemfontein

The Lamont-Hussey Observatory, Bloemfontein. F. Holden. Inform. Bull. Southern Hemisphere, No. 15, p. 1 - 3 (1969).

Mazelspoort: Boyden Observatory. A. H. Jarrett. Inform. Bull. Southern Hemisphere, No. 15, p. 25 - 26 (1969). Current research report.

C.S.I.R., National Institute for Telecommunications Research. Inform. Bull. Southern Hemisphere, No. 15, p. 26 (1969).

008.018 Bloomington

Goethe Link Observatory, Indiana University, Bloomington, Indiana. – Report 1968 – 1969. F. K. Edmondson. Bull. American Astron. Soc., Vol. 2, 41 - 43 (1970).

008.019 Bochum

Astronomisches Institut. – Report 1969. T. Schmidt-Kaler. Mitt. Astron. Ges., No. 28, p. 12 - 16 (1970).

Astronomisches Institut: Bereich Extraterrestrische Physik in der Abteilung XII. – Report 1969. R.-H. Giese. Mitt. Astron. Ges., No. 28, p. 17 - 18 (1970).

Tätigkeitsbericht 1968 des Instituts für Weltraumforschung. H. Kaminski. Separate print Sternw. Bochum, 8 pp. (1969).

Jahresbericht 1969 des Instituts für Weltraumforschung. H. Kaminski, Separate print Sternw. Bochum, 17 pp. (1970).

008.020 Bonn

Astronomische Institute Bonn. I. Sternwarte der Universität Bonn mit Observatorium Hoher List; II. Radio-Sternwarte der Universität Bonn; III. Institut für Astrophysik und Extraterrestrische Forschung der Universität Bonn; IV. Max-Planck-Institut für Radioastronomie. – Report 1969. O. Hachenberg, P. G. Mezger, W. Priester, H. Schmidt, R. Wielebinski. Mitt. Astron. Ges., No. 28, p. 19 - 45 (1970).

Beiträge zur Radioastronomie, Max-Planck-Institut für Radioastronomie, Bonn, Band 1, Heft 5 (M. A. F. Thiel, 03.022.090), 6 (D. Ristow, 03.141.178; P. Zimmermann, 03.141.179).

Max-Planck-Institut für Radioastronomie. Bonn. Sonderdrucke, Nos. 1 (K. Rohlfs, J. D. Murray, AJB 68, 133.39), 2 (M. Grewing, K. Rohlfs, 01.141.060), 3 (M. Grewing, U. Mebold, K. Rohlfs, 01.141.065), 4 (G. M. Gruber, 01.157.019), 5 (B.-H. Grahl, M. Grewing, 02.141.040), 6 (K. Rohlfs, U. Mebold, M. Grewing, 02.131.092), 7 (H. J. Wendker, 03.142.031), 8 (B. Jones, R. Wielebinski, 03.141.009).

008.021 Borowiec

Polish Academy of Sciences, Astronomical Latitude Station, Borowiec, Circulars, Nos. 112 - 113 (03.044.026).

Polish Academy of Sciences, Astronomical Latitude Station, Borowiec, Reprints Nos. 32 (K. Schilling, 03.045.019), 33 (W. Jakś, K. Schilling, 03.034.074), 34 (W. Jakś, 03.045.020), 35 (W. Jakś, 03.045.021), 36 (I. Gajderowicz, S. Oszczak, 03.032.040), 37 (J. Moczko, 03.046.017).

008.022 Boulder

Joint Institute for Laboratory Astrophysics of the National Bureau of Standards and the University of Colorado, Boulder, Colorado. – Report 1968 – 1969. D. G. Hummer. Bull. American Astron. Soc., Vol. 2, 59 - 65 (1970).

008.023 Brno

Contributions of the Public Observatory and Planetarium Brno, Nos. 4 (Z. Kvíz, J. Mikušek, 03.104.033; J. Grygar, L. Kohoutek, Z. Kvíz, J. Mikušek, 03.104.034; J. Grygar, L. Kohoutek, 03.104.035; L. Kohoutek, J. Grygar,

03.104.036; Z. Kvíz, 03.106.021), 5 (O. Obůrka, 03.123.014), 6 (O. Obůrka, 03.123.015), 7 (M. Šulc, 03.104.037), 8 (V. Znojil, 03.104.060; V. Znojil, 03.104.061; V. Znojil, 03.104.062; V. Znojil, 03.104.063).

008.024 Bucarest

Observations Solaires, Académie de la République Socialiste de Roumanie, Observatoire de Bucarest, Secteur Solaire. Rotations 1529 - 1542 (03.075.013).

008.025 Byurakan

Byurakan Astrophysical Observatory, Armenia, USSR, Reprints, Nos. 33 (V. V. Leonov, 02.124.005), 34 (V. V. Papoyan, D. M. Sedrakian, E. V. Chubarian, 02.066.021), 35 (D. W. Weedman, E. Ye. Khachikian, 02.158.034), 36 (R. M. Avakian, M. A. Mnatsakanian, 02.066.023), 37 (R. S. Vardanian, N. B. Yengibarian, 02.064.024), 38 (T. D. Antonova, V. V. Vitkevitch, V. G. Panajian, 02.141.078), 39 (V. G. Panajian, 02.141.079), 40 (V. Yu. Terebizh, 02.063.024), 41 (G. A. Gurzadian, 02.122.122), 42 (V. V. Papoyan, D. M. Sedrakian, E. V. Chubarian, 02.126.008), 43 (M. A. Arakelian, 02.141.186), 44 (B. E. Markarian, 02.158.079), 45 (E. S. Parsamian, H. S. Chavushian, 02.122.123), 46 (G. S. Sahakian, M. N. Mnatsakanian, 03.066.014), 47 (B. E. Markarian, 03.158.020), 48 (K. A. Sahakian, 03.158.021), 49 (M. A. Arakelian, 03.141.044), 50 (M. A. Mnatsakanian, 03.066.015), 51 (V. A. Ambartsumian, L. V. Mirzoyan, E. S. Parsamian, H. S. Chavushian, L. K. Erastova, 03.153.010), 52 (V. A. Ambartsumian, 03.152.009), 53 (M. A. Arakelian, E. A. Dibay, V. F. Yesipov, 03.158.059), 54 (R. S. Vardanian, 03.122.046), 55 (M. K. Babadzhanyianz, 03.141.114).

008.026 Cambridge, Engl.

University of Cambridge. Report of the Observatories Syndicate for the year ending 1969 September 30. R. O. Redman. Separate print Cambridge, Engl., 6 pp. (1970).

University of Cambridge, The Observatories, Cambridge, Engl. Separate prints (R. and R. Griffin, 01.114.003; R. and R. Griffin, 01.114.049; R. Griffin, 01.064.037; 02.010.022).

008.027 Cambridge, Mass.

Harvard College Observatory, Cambridge, Massachusetts. – Report 1968 – 1969. L. Goldberg. Bull. American Astron. Soc., Vol. 2, 43 - 49 (1970).

Smithsonian Astrophysical Observatory, Cambridge, Massachusetts. – Report 1968 – 1969. Bull. American Astron. Soc., Vol. 2, 124 - 130 (1970).

Smithsonian Institution. Astrophysical Observatory. Research in Space Science. SAO Special Reports, Nos. 290 (B. Miller, 03.055.009), 291 (N. M. Hall, J. R. Cherniack, 03.021.005), 296 (B. Miller, 03.055.008), 301 (G. H. Rieke, 03.142.020), 305 (R. E. McCrosky, Z. Ceplecha, 03.104.041), 306 (P. W. Hodge, 03.159.016), 307 (J. A. Wood, U. B. Marvin, B. N. Powell, J. S. Dickey, Jr., 03.094.190), 308

(D. E. Brownlee, P. W. Hodge, 03.105.116), 312 (Y. Kozai, 03.081.004), 313 (L. G. Jacchia, 03.082.048), 314 (J. B. Pollack, D. Pitman, B. N. Khare, C. Sagan, 03.097.032), 315 (E. M. Gaposchkin, K. Lambeck, 03.081.045), 316 (M. P. Friedman, 03.082.074).

008.028 Cape Town

Royal Observatory at the Cape of Good Hope. Inform. Bull. Southern Hemisphere, No. 15, p. 28 - 29 (1969). Current research report.

008.029 Carloforte

Circolari della Stazione Astronomica Internazionale di Latitudine, Carloforte-Cagliari, Serie B (1), No. 1 (E. Proverbio, F. Carta, F. Mazzoleni, 03.045.023).

Pubblicazioni della Stazione Astronomica Internazionale di Latitudine, Carloforte-Cagliari, Nuova Serie, Nos. 7 (E. Proverbio, F. Chlistovsky, 02.044.016), 8 (F. Chlistovsky, E. Proverbio, 03.044.029), 9 (E. Proverbio, F. Chlistovsky, 03.044.030).

008.030 Castel Gandolfo

Specola Vaticana. Annual report 1969: Report of the Astronomical Observatory; Report of the Astrophysical Laboratory.   D. J. K. O'Connell, J. Junkes. Printed in Vatican City, 11 pp. (1970).

Ricerche Astronomiche. Specola Vaticana, Città del Vaticano, Vol. 8, Nos. 1 (F. C. Bertiau, J. Grobben, 03.119.015), 2 (J. Grobben, R. P. Michaelis, 03119.016), 3 (W. J. Miller, 03:122.115), 4 (P. J. Treanor, 03.034.069), 5 (G. V. Coyne, 03.121.060), 6 (G. V. Coyne, 03.121.061), 7 (G. V. Coyne, 03.116.020).

008.031 Catania

Relazione annuale per il 1969. G. Godoli, Oss. Astrofis. Catania, Pubbl., Nuova Ser., No. 143, 15 pp. (1970).

Osservatorio Astrofisico di Catania. Pubblicazione, Nuova Serie, Nos. 131 (S. Cristaldi, L. Paternò, 02.034.027), 132 (S. Cristaldi, G. Godoli, M. Narbone, M. Rodonò, 02.122.053), 133 (C. Blanco, F. Catalano, G. Godoli, 02.116.007), 134 (S. Catalano, M. Rodonò, 02.121.020), 135 (G. Godoli, G. L. Tagliaferri, 03.085.010), 138 (S. Catalano, S. Cristaldi, M. Rodonò, 02.124.102), 140 (F. Affronti, C. Blanco, 03.082.094), 141 (R. Campisi Cristaldi, O. Morgante, M. L. Sturiale, G. Celeani, C. D'Arrigo, G. Domina, G. Patti, S. Rifici, G. Sapienza, S. Torrisi, G. Godoli, 03.075.032), 142 (C. Blanco, G. Godoli, L. Paternò, A. Righini, M. Rodonò, 03.082.095).

008.032 Cerro Tololo

Cerro Tololo Inter-American Observatory, Contri-butions, Nos. 77 (J. E. Hesser, B. M. Lasker, D. R. Bochonko, D. E. Mook, 02.141.004), 81 (G. Hill, C. L. Perry, 02.153.011), 89 (N. Sanduleak, 03.159.022).

008.033 Charlottesville

Leander McCormick Observatory, University of Virginia, Charlottesville, Virginia. – Report 1968 - 1969. L. W. Fredrick. Bull. American Astron. Soc., Vol. 2, 65 - 69 (1970).

008.034 Cleveland

Warner and Swasey Observatory, Case Western Reserve University, Reprints, Nos. 169 (C. B. Stephenson, N. Sanduleak, D. Hoffleit, AJB 68, 105.33), 171 (D. J. MacConnell, AJB 68, 143.12), 172 (R. H. Rubin, AJB 68, 133.90), 173 (J. S. Drilling, AJB 68, 144.09), 174 (R. M. Hjellming, AJB 68, 133.74),175 (C. L. Terrill, 01.122.044), 177 (S. B. Parsons, 01.064.031), 178 (W. A. Hiltner, C. B. Stephenson, N. Sanduleak, AJB 68, 104.50), 179 (N. Sanduleak, AJB 68, 145.258), 180 (B. Hidajat, V. M. Blanco, AJB 68, 144.19), 181 (C. B. Stephenson, AJB 68, 23.74),182 (N. Sanduleak, A. G. Davis Philip, AJB 68, 143.15), 183 (M. P. Fitzgerald, 01.115.007), 184 (M. P. Fitzgerald, AJB 68, 133.14), 185 (N. Sanduleak, 01.159.002), 187 (C. B. Stephenson, N. B. Sanwal, 01.115.009), 188 (S. W. McCuskey, 02.115.003), 189 (R. E. Murphy, 02.118.028), 190 (F. M. Stienon, 02.124. 101), 192 (N. Sanduleak, 02.159.001).

008.035 Coimbra

Comunicações do Observatório Astronómico da Universidade de Coimbra, No. 6 (A. Simões da Silva, M. M. Pinheiro, 03.118.008).

008.036 Columbus

The Observatories of The Ohio State and Ohio Wesleyan Universities, Columbus and Delaware, Ohio. A. Slettebak. Bull. American Astron. Soc., Vol. 2, 110 - 115 (1970).

Ohio State University Radio Observatory, Columbus, Ohio. – Report 1968 - 1969.   J. D. Kraus. Bull. American Astron. Soc., Vol. 2, 115 - 116 (1970).

008.037 Cordoba

Observatorio Astronómico. (Astronomical Observatory, National University of Córdoba). J. L. Sérsic. Inform. Bull. Southern Hemisphere, No. 15, p. 4 (1969). Current research report.

Observatorio Astronómico de Córdoba, Separata No. 173 (C. R. Fourcade, J. R. Laborde, 01.154.006).

Observatorio Astronómico (*Universidad Nacional de Córdoba, Argentina***), Tirada Aparte,** Nos. 165 (J. Landi Dessy, A. Niell, A. Puch, 03.031.007), 169 (J. L. Sérsic, 03.066.079),

178 (R. F. Sistero, 02.121.076), 179 (J. L. Sérsic, 02.158.052).

008.038 Delaware

The Observatories of The Ohio State and Ohio Wesleyan Universities, Columbus and Delaware, Ohio. A. Slettebak.
Bull. American Astron. Soc., Vol. 2, 110 - 115 (1970).

Contributions from the Perkins Observatory, The Ohio State University and Ohio Wesleyan University, Series I, Nos. 100 (R. R. Wright, 01.158.034), 101 (A. J. Deutsch, O. C. Wilson, P. C. Keenan, 01.114.046), 102 (A. H. Markowitz, 01.114.105), 103 (W. E. Mitchell, Jr., O. C. Mohler, 02.076.007), 104 (P. C. Keenan, A. J. Deutsch, R. F. Garrison, 02.122.082), 105 (R. F. Wing, W. K. Ford, Jr., 02.114.053), 106 (H. Spinrad, R. F. Wing, 02.114.018), 107 (P. Buerger, 02.117.030), 108 (G. W. Collins II, 03.064.009), 109 (A. Slettebak, 02.114.023).

Contributions from the Perkins Observatory, The Ohio State University and Ohio Wesleyan University, Series II, Nos. 21 (A. Slettebak, P. C. Keenan, R. K. Brundage, 01.114.070), 22 (P. C. Keenan, A. Slettebak, R. L. Bottemiller, 01.114.057).

008.039 Dresden

Lohrmann-Institut für Geodätische Astronomie der Technischen Universität Dresden. – Report 1968. H.-U. Sandig.
Monatsber. Deutsch. Akad. Wiss. Berlin, Vol. 11, 774 - 776 (1969).

Mitteilungen des Lohrmann-Instituts der Technischen Universität Dresden, Nos.15 (H. Potthoff, 03.041.027), 16 (S. Wächter, 03.041.028), 17 (H. Potthoff, K.-G. Steinert, 03.041.029), 18 (H.-U. Sandig, 03.045.016), 19 (H.-U. Sandig, 03.041.030).

008.040 Dublin

Dunsink Observatory. Report for the year ending 1969 March 31. P. A. Wayman.
Quarterly Journ. Roy. Astron. Soc., Vol. 11, 63 - 66 (1970).

008.041 Dushanbe

Byulleten' Instituta Astrofiziki, Akademiya Nauk Tadzhikskoj SSR, Nos. 49 (P. B. Babadzhanov, T. I. Getman, A. F. Zausayev, S. A. Karaselnikova, 03.104.045; Sh. O. Isa - mutdinov, R. P. Chebotarev, 03.104.046; V. M. Kolmakov, 03.104.047; U. Shodiev, 03.104.048; A. M. Bakharev, 03.104.049; O. V. Dobrovolsky, I. Kh. Ibadinov, 03.103.134; A. G. Krylov, 03.055.028), 50 (L. S. Marochnik, 03.151.067; L. S. Marochnik, V. K. Babkov, 03.151.068; O. V. Dobrovols - ky, Kh. Ibadinov, 03.102.017; A. Ja. Filin, 03.155.058; A. Ja. Filin, 03.155.059).

008.042 Edinburgh

Report of the Astronomer Royal for Scotland for the year ending 31st March 1970. H. A. Brück.
The Royal Observatory, Edinburgh, 14 pp. (1970).

Communications from the Royal Observatory, Edinburgh, Nos. 70 (M. J. Smyth, P. W. J. L. Brand, 03.036. 009), 73 (G. C. Sudbury, 02.034.056), 74 (V. C. Reddish, 02.065.038), 75 (J. W. Harris, 02.131.054), 76 (B. N. G. Guthrie, 02.114.087), 77 (W. McD. Napier, M. W. Ovenden, 03.119.008), 78 (J. G. Ireland, 03.131.014), 79 (N. C. Wick - ramasinghe, K. Nandy, 03.131.015), 80 (W. McD. Napier, 03.064.021), 81 (V. C. Reddish, 03.131.028), 82 (M. T. Brück, L. C. Lawrence, K. N. Nandy, A. D. Thackeray, R. Wood, 03.159.003), 83 (K. Nandy, R. D. Wolstencroft, 03.158.109), 84 (K. Nandy, H. Seddon, 03.131.026), 87 (G. C. Sudbury, M. F. Ingham, 03.155.008).

Publications of the Royal Observatory, Edinburgh, Vol. 7, No. 1 (K. Nandy, F. Smriglio, 03.113.058).

008.043 El Leoncito

Programs and experiences at "El Leoncito" 1967 – 1968. J. A. Hughes.
Asoc. Argentina Astron. Bol., No. 14 (see 012.020), p. 83 - 90 (1968). In Spanish.

008.044 El Segundo

The San Fernando Observatory, Aerospace Corporation, El Segundo, California. – Report 1968 – 1969. E. B. Mayfield.
Bull. American Astron. Soc., Vol. 2, 123 - 124 (1970).

008.045 Evanston

Lindheimer Astronomical Research Center and Dearborn Observatory, Evanston, Illinois. – Report 1968 – 1969. J. A. Hynek.
Bull. American Astron. Soc., Vol. 2, 69 - 72 (1970).

008.046 Flagstaff

Lowell Observatory, Flagstaff, Arizona. – Report 1968 – 1969. J. S. Hall.
Bull. American Astron. Soc., Vol. 2, 75 - 78 (1970).

008.047 Frankfurt/M.

Astronomisches Institut der Universität. – Report 1969. W. Gleissberg.
Mitt. Astron. Ges., No. 28, p. 46 - 47 (1970).

008.048 Freiburg

Fraunhofer Institut, mit den Observatorien Schau-insland und Anacapri. – Report 1969.
K. O. Kiepenheuer.
Mitt. Astron. Ges., No. 28, p. 48 - 53 (1970).

Fraunhofer Institut. Map of the Sun. 1970 Jan. 1 - June 30 (03.075.009).

008.049 Gainesville

University of Florida Observatories, University of Florida, Gainesville, Florida. – Report 1968 – 1969.
A. G. Smith, F. B. Wood.
Bull. American Astron. Soc., Vol. 2, 33 - 36 (1970).

008.050 Genève

Activités de l'Observatoire de Genève en 1968.
E. Antonini.
Orion Schaffhausen, 28. Jahrgang, p. 54 - 55 (1970).

Rapport annuel d'activité scientifique de l'Observatoire de Genève pour 1968.
C. R. des Séances, SPHN Genève, NS, Vol. 4, Fasc. 1, p. 18 - 22 = Publ. Obs. Genève, Sér. A, Fasc. 76/I (1969).

Publications de l'Observatoire de Genève, Série A, Fasc. 77 (A. Gaide, 03.113.042; C. Navach, 03.113.043; P. Bouvier, 03.151.052; G. Janin, 03.151.053; G. Janin, 03.151.054; G. Janin, 03.151.055; G. Goy, 03.113.044; B. Hauck, 03.113.045).

008.051 Göttingen

Universitäts-Sternwarte Göttingen, Institut für Sonnenforschung Locarno-Orselina (Tessin). – Report 1969.
W. Deinzer, R. Kippenhahn, E. Schröter, H. H. Voigt.
Mitt. Astron. Ges., No. 28, p. 53 - 60 (1970).

008.052 Grahamstown

Rhodes University Physics Department.
G. M. Gruber.
Inform. Bull. Southern Hemisphere, No. 15, p. 27 (1969).
Current research report.

008.053 Granada

Contribuciones Astronomicas, Observatorio de Cartuja, Granada (Spain), No. 2 (T. Vives Soteras, 01.082.072).

008.054 Green Bank

National Radio Astronomy Observatory, Charlottesville, Virginia, Green Bank, West Virginia, and Tucson, Arizona. – Report 1968 – 1969.

D. S. Heeschen.
Bull. American Astron. Soc., Vol. 2, 100 - 108 (1970).

National Radio Observatory, *Green Bank*, **Reprints,** Series A, Nos. 132 (P. G. Mezger, T. L. Wilson, F. F. Gardner, D. K. Milne, 03.155.015), 133 (Y. Terzian, B. Balick, 02.133.023), 134 (B. G. Clark, G. K. Miley, 02.141.100), 135 (D. E. Hogg, G. H. Macdonald, R. G. Conway, C. M. Wade, 02.141.154), 136 (E. Churchwell, M. Felli, 03.152.010), 137 (B. E. Turner, D. Buhl, E. B. Churchwell, P. G. Mezger, L. E. Snyder, 03.131.065), 138 (T. L. Wilson, W. Altenhoff, 03.157.001), 139 (F. J. Kerr, Aa. Sandqvist, 03.155.001), 140 (D. S. Heeschen, 03.008.054), 141 (E. Churchwell, M. Felli, 03.157.002), 142 (T. L. Wilson, 03.141.117), 143 (D. Buhl, A. Tlamicha, 03.077.029), 144 (R. M. Hjellming, R. D. Davies, 03.131.081), 145 (H. J. Wendker, 03.142.031), 146 (E. C. Reifenstein III, T. L. Wilson, B. F. Burke, P. G. Mezger, W. J. Altenhoff, 03.157.011), 147 (P. G. Mezger, T. L. Wilson, F. F. Gardner, D. K. Milne, 03.159. 002), 148 (T. L. Wilson, P. G. Mezger, F. F. Gardner, D. K. Milne, 03.141.037), 149 (S. Gulkis, 03.099.053), 150 (Aa. Sandqvist, 03.141.033), 151 (K. I. Kellermann, 03.099.054), 152 (E. Churchwell, P. Mezger, E. Reifenstein III, R. Rubin, B. Turner, 03.132.007), 153 (K. J. Gordon, C. P. Gordon, 03.141.060), 154 (T. K. Menon, 03.132.022), 155 (R. L. Brown, 03.022.096), 156 (B. E. Turner, C. E. Heiles, E. Scharlemann, 03.131.035), 157 (E. Churchwell P. G. Mezger, 03.132.011), 158 (W. J. Webster, Jr., W. J. Altenhoff, 03.132.012), 159 (A. C. E. Sinclair, J. P. Basart, D. Buhl, W. A. Gale, M. Liwshitz, 03.093.028).

National Radio Astronomy Observatory, *Green Bank*, **Reprints,** Series B, Nos. 156 (S. J. Goldstein, Jr., J. T. James, 02.141.163), 157 (S. J. Goldstein, Jr., D. D. Meisel, 02.141.115), 158 (G. K. Miley, D. E. Hogg, J. Basart, 03.141.017), 159 (J. M. Sutton, D. H. Staelin, R. M. Price, R. Weimer, 03.141.026), 160 (P. G. Mezger, 03.131.127), 161 (K. J. Gordon, 02.158.064), 162 (K. I. Kellermann, 03.141.205), 163 (B. Zuckerman, P. Palmer, 03.131.037), 164 (K. W. Riegel, S. D. Kilston, 03.157.007), 165 (R. L. Brown, 03.142.016), 166 (D. H. Staelin, M. S. Ewing, R. M. Price, J. M. Sutton, 03.141.123), 167 (W. J. Altenhoff, T. L. Wilson, 03.141.200), 168 (D. Morrison, M. J. Klein, 03.092.010), 169 (D. L. Jauncey, C. C. Bare, B. G. Clark, K. I. Kellermann, M. H. Cohen, 03.141.121), 170 (J. Pfleiderer, 03.141.049), 171 (M. Felli, E. Churchwell, 03.132.017), 172 (E. B. Fomalont, 03.141.127), 173 (C. Heiles, G. K. Miley, 03.141.129), 174 (B. E. Turner, P. Palmer, B. Zuckerman, 03.131.112), 175 (G. K. Miley, B. E. Turner, B. Balick, C. Heiles, 03.131.111), 176 (D. H. Staelin, J. M. Sutton, 03.134.011).

008.055 Greenbelt

Goddard Space Flight Center, Greenbelt. NASA Technical Note, TN D-5709 (P. Musen, 03.042.047).

008.056 Groningen

Nederlandse Vereniging voor Weer- en Sterrenkunde. **Observations of Variable Stars. Report** (Kapteyn Astronomical Laboratory, Groningen – Netherlands), No. 17 (03.123.011).

008.057 Hamburg

Hamburger Sternwarte. – Report 1969.

A. Behr, A. Weigert.
Mitt. Astron. Ges., No. 28, p. 61 - 64 (1970).

De Europese Zuidelijke Sterrenwacht in Chili.
A. Blaauw.
Hemel en Dampkring, Vol. 68, 97 - 105 (1970).

Deutsches Hydrographisches Institut (DHI). – Report 1969.
W. Horn.
Mitt. Astron. Ges., No. 28, p. 60 - 61 (1970).

Jahresbericht No. 23 für das Jahr 1968.
H. U. Roll.
Deutsches Hydrographisches Institut, Hamburg. 2 + 133 pp. (1969).

Deutsches Hydrographisches Institut, Hamburg.
Zeitsignalaufnahmen, Astronomischer Zeit- und Breitenbestimmungen, 1969 July – December (03.044.009).

008.058 Haute Provence

Publications de l'Observatoire de Haute Provence, Vol. 10, Nos. 14 (A. Terzan, 01.124.104), 15 (C. Bertaud, 03.114.065), 16 (K. Serkowski, W. Chojnacki, 01.116.005), 17 (C. Fehrenbach, M. Petit, 01.124.104), 18 (Y. Andrillat, L. Houziaux, 02.124.102), 19 (M. Bloch, J. P. Swings, 01.114.051), 20 (R. Crillon, G. Monnet, 01.158.019), 21 (R. Crillon, G. Monnet, 01.158.029), 22 (G. Carranza, R. Crillon, G. Monnet, 01.158.020), 23 (A. Terzan, 01.124.102), 24 (M. Schneider, 01.122.043), 25 (G. Wlérick, 03.034.059), 26 (M. Duchesne. M. Feissel, B. Guinot, 01.043.006), 27 (Y. Andrillat, M.-O. Baylac, 01.114.075), 28 (R. Louise, 02.133.002), 29 (G. Courtès, R. Louise, G. Monnet, 02.132.004), 30 (R. Louise, G. Monnet, 02.131.091), 31 (C. Bertaud, B. Dumortier, P. Véron, G. Wlérick, G. Adam, J. Bigay, R. Garnier, M. Duruy, 02.122.097), 32 (M. Imbert, 02.119.016), 33 (M. Bloch, N. Jousten, J. P. Swings, 03.114.122), 34 (N. Martin, 02.155.012).

008.059 Heidelberg

Astronomisches Rechen-Institut. – Report 1969.
W. Fricke.
Mitt. Astron. Ges., No. 28, p. 65 - 70 (1970).

Astronomisches Rechen-Institut in Heidelberg.
Mitteilungen, Serie A, Nos. 36 (P. Brosche, 02.158.003), 37 (W. Fricke, 02.02.036), 38 (U. Geyer, 02.115.019), 39 (P. Brosche, J. Sündermann, 02.044.041), 40 (R. Wielen, 02.153.041).

Astronomisches Rechen-Insitut in Heidelberg.
Mitteilungen, Serie B, No. 22 (J. Schubart, 01.098.017).

Astronomy and Astrophysics Abstracts, Vol. 1 (02.002.034), Vol. 2 (03.002.028).

Landessternwarte und Max-Planck-Institut für Astronomie Heidelberg-Königstuhl. – Report 1969.
H. Elsässer.
Mitt. Astron. Ges., No. 28, p. 71 - 77 (1970).

Lehrstuhl für Theoretische Astrophysik. – Report 1969. G. Traving.
Mitt. Astron. Ges., No. 28, p. 70 (1970).

008.060 Honolulu

Institute for Astronomy, University of Hawaii, Honolulu, Hawaii. – Report 1968 – 1969. J. T. Jefferies.
Bull. American Astron. Soc., Vol. 2, 49 - 56 (1970).

Institute for Astronomy, University of Hawaii, Honolulu, Hawaii. Separate prints (F. Q. Orrall, AJB 68, 62.07; G. D. Finn, J. T. Jefferies, 01.063.021; J. T. Jefferies, 01.074.042; J. T. Jefferies, F. Q. Orrall, J. B. Zirker, 01.074.043; R. R. Fisher, G. R. Mann, 03.073.079).

008.061 Jena

Universitäts-Sternwarte der Friedrich-Schiller-Universität Jena. – Report 1968.
H. Lambrecht.
Monatsber. Deutsch. Akad. Wiss. Berlin, Vol. 11, 772 - 774 (1969).

008.062 Kiel

Institut für Theoretische Physik und Sternwarte der Universität. – Report 1969.
A. Unsöld.
Mitt. Astron. Ges., No. 28, p. 77 - 78 (1970).

Sonderdrucke der Sternwarte Kiel, Nos. 146 (V. Weidemann, AJB 68, 105.35), 147 (V. Weidemann, AJB 68, 105.34), 148 (A. Unsöld, AJB 68, 66.157), 149 (A. Unsöld, 03.015.020), 150 (A. Unsöld, 03.065.095), 151 (A. O. J. Unsöld, 01.114.026), 152 (V. Weidemann, 01.066.047), 153 (B. Baschek, D. Reimers, 01.114.069), 154 (T. Garz, H. Holweger, M. Kock, J. Richter, 02.071.003), 155 (D. Reimers, 02.122.008), 156 (T. Garz, M. Kock, J. Richter, B. Baschek, H. Holweger, A. Unsöld, 02.071.023), 157 (V. Weidemann, 01.126.020).

008.063 Kitt Peak

Kitt Peak National Observatory, Contributions Nos. 371 (J. R. Powell, R. Lynds, 03.034.068), 382 (F. A. Franklin, A. F. Cook, II, 02.100.002), 416 (J. D. Rosendhal, 02.064.003), 424 (R. J. W. Henry, N. F. Lane, 03.022.094), 426 (A. A. Hoag, W. C. Miller, 02.036.016), 430 (V. Canuto, H.-Y. Chiu, L. Fassio-Canuto, 02.061.032), 431 (H.-Y. Chiu, L. Fassio-Canuto, 02.061.033), 433 (T. Owen, 02.099.017), 435 (03.012.009), 437 (J. R. Percy, 02.119.012), 438 (D. M. Hunten, R. M. Goody, 02.093.007), 439 (E. J. Weber, 02.080.009), 440 (R. W. Michie, 03.162.054), 441 (J. R. Powell, T. E. Trumbo, C. R. Lynds, 02.031.019), 444 (H. A. Abt, W. P. Bidelman, 02.119.015), 446 (A. Dalgarno, M. B. McElroy, A. I. Stewart, 03.082.091), 447 (C. D. Slaughter, 02.103.101), 448 (J. W. Chamberlain, 02.084.009), 449 (H. Silver, 03.031.031), 450 (W. C. Livingston, 02.080.027), 451 (P. Lee, 02.132.002), 452 (D. L. Crawford, J. V. Barnes, 02.154.005), 453 (N. R. Sheeley, Jr., 02.071.058), 454 (Y.

Öhman, 02.073.044), 455 (R. D. McClure, R. Racine, 02.154.004), 456 (P. Léna, 03.071.028), 457 (H. E. Bond, J. S. Neff, 02.113.037), 458 (T. Owen, H. P. Mason, 02.097.008), 459 (R. S. Fisk, H. A. Abt, 02.119.013), 460 (D. F. Strobel, 03.099.024), 461 (M. J. S. Belton, D. M. Hunten, 02.097.032), 462 (D. M. Hunten, 03.099.016), 463 (R. C. Anderson, J. G. Pipes, A. L. Broadfoot, L. Wallace, 03.093.015) 464 (R. J. W. Henry, 03.091.018), 465 (R. J. W. Henry, M. B. McElroy, 03.099.025), 466 (M. B. McElroy, 03.091.014), 467 (J. D. Rosendhal, 03.065.002), 468 (M. Breger, 02.122.102), 469 (M. Breger, 02.122.103), 470 (D. L. DuPuy, J. B. DeVeny, 02.158.050), 471 (J. L. Schmitt, 02.114.057), 472 (U. Fink, M. J. S. Belton, 03.099.029), 473 (A. U. Landolt, 02.121.048), 474 (L. Binnendijk, 02.121.012), 475 (L. Binnendijk, 02.121.013), 476 (R. E. Murphy, 02.118.028), 477 (M. B. McElroy, D. M. Hunten, 02.097.047), 479 (R. A. Wells, 02.097.043), 480 (J. Bahng, 02.113.061), 481 (Y. Öhman, 02.073.051), 482 (G. A. Dulk, J. A. Eddy, J. P. Emerson, 03.099.012), 483 (D. E. Shemansky, 03.022.095), 484 (O. Engvold, W. Livingston, 02.072.090), 485 (D. S. Hall, L. M. Garrison, Jr., 02.121.093), 486 (H. M. Dyck, R. D. McClure, 02.131.124), 489 (D. N. B. Hall, R. W. Noyes, 02.072.027), 490 (B. B. Bookmyer, 02.121.054).

008.064 København

Publikationer og mindre Meddelelser fra Københavns Observatorium, Nos. 199 (A. Reiz, J. O. Petersen, P. M. Hejlesen, 01.115.010), 200 (H. Kristenson, 01.032.041), 201 (K. Gyldenkerne, Vagn Mejdahl, R. M. West, 03.124.109), 202 (R. M. West, 02.115.004), 206 (K. T. Johansen, 02.121.006), 207 (K. T. Johansen, 03.121.023).

008.065 Kodaikanal

Annual report of the Kodaikanal Observatory for the year 1967.   M. K. V. Bappu.
Printed by the Manager Government of India Press, Delhi. 12 pp. (1969).

Kodaikanal Observatory, Bulletin, Nos. 171 (M. K. V. Bappu, 03.075.007), 173 (M. K. V. Bappu, 03.075.008).

Kodaikanal Observatory, Bulletins, Series A, No. 186 (N. Raghavan, 03.071.050).

008.066 Lake Tekapo

Mount John University Observatory.
(Observatory of the Universities of Pennsylvania and Canterbury).   M. S. Snowden.
Inform. Bull. Southern Hemisphere, No. 15, p. 23 - 24 (1969). Current research report.

008.067 La Plata

Annual report of La Plata Observatory for 1966.
Asoc. Argentina Astron. Bol., No. 12, (see 012.019), p. 40 - 52 (1968).   In Spanish.

Observatorio Astronómico.(Astronomical Observatory, National University of La Plata).
C. Jaschek, M. Jaschek, J. Sahade.

Inform. Bull. Southern Hemisphere, No. 15, p. 5 - 6 (1969). Current research report.

Observatorio Astronómico de la Universidad Nacional de La Plata, Serie Astronómica, Tomo 35 (J. L. Sérsic, 03.042.045).

Separata Astronómica, Observatório Astronómico – La Plata – Argentina, Nos. 90 (J. Sahade, 01.114.089), 91 (M. Jaschek, C. Jaschek, S. Malaroda, 02.114.060), 92 (A. Feinstein, J. C. Muzzio, 02.122.096), 93 (R. C. Cameron, AJB 67, 55.33), 94 (A. E. Ringuelet, 02.114.073), 95 (M. Jaschek, C. Jaschek, M. Arnal, 02.114.056), 96 (R. C. Cameron, AJB 67, 104.61).

008.068 Las Cruces

New Mexico State University Observatory, Las Cruces, New Mexico. – Report 1968 – 1969.
C. W. Tombaugh.
Bull. American Astron. Soc., Vol. 2, 108 - 110 (1970).

008.069 Lausanne

Observatoire Universitaire de Lausanne, Communication, No. 19 (B. Hauck, 03.113.045).

008.070 Lembang

Bosscha Observatory.
Inform. Bull. Southern Hemisphere, No. 15, p. 21 - 22 (1969). Current research report.

Report for the years 1966 – 1969.
B. Hidajat.
Separate print Bosscha Obs., Dep. Sci., Bandung Inst. Technol., 6 pp. (1970).

Contributions from the Bosscha Observatory, Bandung Institute of Technology, Department of Science, No. 42 (T.-K. Tan, B. Hidajat, 02.123.036).

008.071 Leningrad

The 50th anniversary of the Institute of Theoretical Astronomy.   G. A. Chebotarev.
Vestn. AN SSSR, 1969 No. 8, p. 140 - 142.   In Russian.
Abstr. in Referativ. Zhurn. 51. Astron., 3.51.35 (1970).

Byulleten' Instituta Teoreticheskoj Astronomii, Akademiya Nauk SSSR. Izdatel'stvo "Nauka", Leningrad, Vol. 12, No. 1 [134] (G. A. Chebotarev, 03.098.014; N. S. Samoilova-Yakhontova, 03.098.015; V. I. Skripnichenko, 03.042.050; A. M. Finkelstein, 03.066.062; G. A. Chebotarev, N. A. Beljaev, R. P. Yeremenko, 03.098.016; G. A. Chebotarev, M. Ja. Shmakova, 03.098.017; G. A. Chebotarev, 03.103.109; N. S. Chernykh, 03.099.039), No. 2 [135] (V. K. Abalakin, 03.007.000; R. A. Zeinalov, 03.052.023; V. K. Kaysin, 03.042.051; P. P. Loginov, 03.042.052; A. A. Orlov, 03.042.053; R. K. Sadreeyev, 03.042.054; W. M. Chepurova, 03.042.055; L. I. Chernykh, 03.098.018).

Trudy Instituta Teoreticheskoj Astronomii, Akademiya Nauk SSSR. Izdatel'stvo "Nauka", Leningrad. Vyp. (No.) 14, (M. V. Anolik, G. A. Krasinsky, L. J. Pius, 03.042.007; S. G. Sharaf, N. A. Boudnikova, 03.085.001).

008.072 Liège

Université de Liège (Belgique). Institut d'Astrophysique, Cointe-Sclessin, Collection in 4°, Nos. 191 (L. Delbouille, 03.051.033), 192 (L. Delbouille, G. Roland, 02.071.080), 193 (N. Grevesse, J. P. Swings, 01.071.028), 194 (F. Remy, 03.022.099), 195 (R. Simon, 02.065.001), 196 (P. Swings, 02.093.051), 197 (P. Guillaume, 02.042.005), 198 (F. Remy, 03.022.100), 199 (J. M. Vreux, G. Marette, 03.022.101), 200 (P. Renson, 03.116.021), 201 (D. J. Malaise, 03.102.011).

Université de Liège (Belgique). Institut d'Astrophysique, Cointe-Sclessin, Collection in 8°, Nos. 577 (J. C. Gérard, 03.083.091), 578 (A. Delcroix, J. Lemaire, 01.073.041), 579 (R. Malbrouck, R. Duysinx, 03.084.051), 580 (I. Dubois, 03.022.098), 581 (N. Grevesse, G. Blanquet, 01.071.045), 582 (J. M. Vreux, 03.034.077), 583 (J.-C. Gérard, O. E. Harang, 02.084.008), 584 (J.-P. Swings, 02.064.057), 585 (M. Bloch, N. Jousten, J. P. Swings, 03.114.123), 586 (J. Demaret, 03.065.110), 587 (M. Lefranc, J. Mawhin, 02.061.037), 588 (J. Mawhin, 03.021.012), 589 (H. Robe, 03.065.111), 590 (P. Swings, 03.061.028), 591 (J. Mawhin, 03.021.013).

008.073 Lisbonne

Bulletin de l'Observatoire Astronomique de Lisbonne (Tapada), No. 17 (A. P. Botelheiro, 03.041.013).

008.074 London, Canada

The Observatories of the University of Western Ontario, London, Canada. – Report 1968 – 1969. W. H. Wehlau. Bull. American Astron. Soc., Vol. 2, 165 - 166 (1970).

008.075 Los Angeles

Research Units and Academic Departments, University of California: Berkeley, Los Angeles, San Diego, and Santa Cruz. – Report 1968 – 1969. I. King, G. Abell, E. M. Burbidge, R. P. Kraft. Bull. American Astron. Soc., Vol. 2, 2 - 23 (1970).

008.076 Louvain

Publications de l'Institut d'Astronomie et Géophysique Georges Lemaitre, Vol. 3, Nos. 1 (A. Boury, AJB 68, 51.61), 2 (L. G. Bossy, 03.022.089), 3 (O. Godart, 03.081.006), 4 (O. Godart, 03.005.011), 5 (W. H. McCrea, 03.162.053).

008.077 L'vov

Tsirkulyar. Astronomicheskaya Observatoriya, L'vov, No. 44 (I. A. Klimishin, A. F. Novak, 03.066.080; O. S. Bojko, 03.123.037; M. B. Girnyak, 03.123.038; V. V. Golovatyj, 03.123.039; A. T. Dul'tsev, 03.123.040; I. V. Shpychka, 03.122.117; B. T. Babij, Yu. V. Fridel', 03.071.049; P. A. Olijnyk, 03.072.043; A. A. Logvinenko, T. V. Rad'o, I. I. Terebushko, 03.032.036).

008.078 Madison

Washburn Observatory, University of Wisconsin, Madison, Wisconsin. – Report 1968 – 1969.   A. D. Code. Bull. American Astron. Soc., Vol. 2, 157 - 162 (1970).

008.079 Madrid

Universidad de Madrid – Facultad de Ciencias. Seminario de Astronomia y Geodesia, Publicación, Nos. 61 (M. R. Fernández, 03.114.137), 62 (C. Machin, 03.081.049).

008.080 Manchester

University of Manchester Nuffield Radio Astronomy Laboratories Jodrell Bank. Report for the year ending 1969 August 31.   B. Lovell. Quarterly Journ. Roy. Astron. Soc., Vol. 11, 52 - 62 (1970).

008.081 Middletown

Van Vleck Observatory, Wesleyan University, Middletown, Connecticut. – Report 1968 – 1969. S. W. McCuskey. Bull. American Astron. Soc., Vol. 2, 149 - 157 (1970).

008.082 Milano

Osservazioni meteorologiche eseguite nell'Osservatorio Astronomico di Brera in Milano durante l'anno 1969. F. Zagar. Pubbl. Oss. Astron. Milano, 29 pp. (1970).

008.083 Minneapolis

University of Minnesota, Minneapolis, Minnesota, Separate print (W. J. Luyten, 03.112.002).

008.084 Mizusawa

Monthly Notes of the International Polar Motion Service, 1969 Nos. 11 - 12, 1970 Nos. 1 - 4 (03.045.015).

008.085 Mons

Centre Universitaire de l'Etat à Mons. Faculté des Sciences, Département d'Astrophysique. **Communications,** Nos. 9 (E. Blondelot, 03.041.034), 13 (L. Houziaux, J. Blondelot-Lickes, 03.041.035), 14 (L. Houziaux, C. Jamar, A. Monfils, 03.114.138), 15 (A. Delcroix, 03.064.062), 16 (E. Blondelot, 03.114.139).

Communications du Département d'Astrophysique de la Faculté des Sciences de Mons. **Mons Astrophysical Papers,** Nos. 7 (Y. Andrillat, L. Houziaux, 02.124.102), 8 (Y. Andrillat, L. Houziaux, 02.124.102), 10 (Y. Andrillat, L. Houziaux, 03.124.100), 11 (L. Houziaux, 03.114.126).

008.086 Montevideo

Departamento de Astronomia y Fisica, Facultad de Humanidades y Ciencias, Universidad de la República, Montevideo, Uruguay. **Publicación,** Nos. 34 (S. Codina, 03.131.140), 35 (R. Freire, 03.045.022).

008.087 Moskva

Chronicle. (Diss. and theses for habilitation at the Scientific Council of the Sternberg-Institute). G. S. Khromov. Astron. Tsirk., No. 547, p. 7 - 8 (1970). In Russian.

Soobshcheniya Gosudarstvennogo Astronomicheskogo Instituta im.P. K. Shternberga. Izdatel'stvo Moskovskogo Universiteta, No. 159 (V. M. Chepurova, 03.042.004; B. N. Noskov, 03.042.005; B. N. Noskov, 03.042.006).

008.088 Mt. Wilson

Mount Wilson and Palomar Observatories. Annual report of the director, 1968 - 1969. H. W. Babcock. Reprinted from Carnegie Institution, Washington, Year Book, Vol. 68, 67 pp. (1970).

From Mount Wilson and Palomar Observatories. Publ. Astron. Soc. Pacific, Vol. 82, 158 (1970).

008.089 München

Universitäts-Sternwarte. − Report 1969. P. Wellmann. Mitt. Astron. Ges., No. 28, p. 106 - 108 (1970).

Max-Planck-Institut für Physik und Astrophysik, Institut für Astrophysik und Institut für extraterrestrische Physik. − Report 1969. L. Biermann, R. Lüst. Mitt. Astron. Ges., No. 28, p. 79 - 105 (1970).

Max-Planck-Institut für Physik and Astrophysik, München, Separate print (W. F. Huebner, 03.102.012).

008.090 Münster

Astronomisches Institut der Universität. − Report 1969. H. Straßl. Mitt. Astron. Ges., No. 28, p. 109 - 111 (1970).

008.091 Naini Tal

Uttar Pradesh State Observatory, *Naini Tal,***Reprints,** Nos. 27 (P. P. Saxena, 01.082.089), 28 (M. C. Pande, V. P. Gaur, B. M. Tripathi, 01.071.022), 29 (M. C. Pande, V. P. Gaur, B. M. Tripathi, 01.072.035), 30 (S. C. Joshi, S. K. Gurtu, M. C. Joshi, 01.122.047).

008.092 Nashville

Dyer Observatory, Vanderbilt University, Nashville, Tennessee. − Report 1968 – 1969. R. H. Hardie. Bull. American Astron. Soc., Vol. 2, 32 - 33 (1970).

The Arthur J. Dyer Observatory, Vanderbilt Uni-versity, Nashville, Tennessee, Reprints, Nos. 46 (D. S. Hall, 02.121.035), 47 (D. S. Hall, R. H. Hardie, 02.121.092), 48 (D. S. Hall, L. M. Garrison, Jr., 02.121.093), 49 (E. J. Devinney, Jr., D. S. Hall, D. H. Ward, 03.121.003), 52 (A. M. Heiser, 03.082.092).

Clear weather over Nashville. See Abstr. 082.092.

008.093 Neuchâtel

Rapport annuel du Directeur sur l'exercice 1969 et Rapport sur le Concours chronométrique 1969. J. Bonanomi, G. Jornod. Observatoire Cantonal de Neuchâtel. 32 pp. (1970).

Observatoire de Neuchâtel. Bulletin. Série B, 1969 Novembre − 1970 Février (03.044.024), Série D, 1969 Novembre − 1970 Février (03.044.025).

008.094 New Haven

Yale University Observatory, New Haven, Connecticut. − Report 1968 – 1969. P. Demarque. Bull. American Astron. Soc., Vol. 2, 166 - 169 (1970).

Transactions of the Astronomical Observatory of Yale University, Vol. 30 (D. Hoffleit, D. Eckert, P. Lü, K. Paranya, 03.041.005).

008.095 Ondřejov

Solar research at the Ondřejov Observatory. Report from Solar Institute. Z. Švestka. Solar Physics, Vol. 12, 332 - 340 (1970).

008.096 Ottawa

Dominion Observatory, Ottawa, Ontario.
M. W. Grey.
Journ. Royal Astron. Soc. Canada, Vol. 64, 174 - 176 (1970).

Contributions from the Dominion Observatory,
Ottawa, Vol. 8, No. 11 (J. F. Clark, 03.105.142); Nos. 271
(C. S. Beals, 03.105.003), 285 (R. S. Roger, 02.141.068),
287 (J. L. Caswell, 03.125.004), 288 (A. H. Bridle, V. R.
Venugopal, 02.131.047), 291 (J. L. Caswell, 02.141.178),
293 (A. H. Bridle, J. L. Caswell, 03.141.048).

Contributions of the National Research Council,
Ottawa, Ontario, No. 11,215 (J. A. Galt, N. W. Broten,
T. H. Legg, J. L. Locke, J. L. Yen, 03.141.004).

Publications of the Dominion Observatory, Ottawa,
Vol. 17C, Nos. 3 (H. E. Cook, A. B. Cook, 03.084.272),
4 (H. E. Cook, A. B. Cook, R. G. Madill, 03.084.286);
Vol. 38, Nos. 1 (A. E. Evans, 03.084.218), 2 (G. J. van
Beek, 03.084.219), 3 (A. E. Evans, 03.084.220), 4 (A. E.
Evans, 03.084.222), 5 (G. J. van Beek, 03.084.217), 6
(D. R. Auld, D. G. Holmes, 03.084.216); Vol. 39, Nos. 1
(A. B. Cook, S. J. Sprysak, 03.084.270), 4 (C. S. Beals,
A. Hitchen, 03.105.127), 5 (G. V. Haines, W. Hannaford,
P. H. Serson, 03.084.271).

008.097 Padova

Osservatorio Astronomico di Padova. Comunica-
zioni e Rassegne, Nos. 62 (A. Mammano, R. Margoni, L. Rosino, 02.124.100), 63 (F. Bertola, AJB 68, 145.161), 64
(L. Rosino, 01.154.013), 65 (G. Romano, AJB 68, 125.31),
66 (A. Mammano, L. Rosino, 02.114.025), 67 (P. Cazzola,
A. Saggion, 02.061.042).

Pubblicazioni dell'Osservatorio Astronomico di
Padova, Nos. 151 (G. Romano, M. Perissinotto, AJB 68,
125.30), 152 (S. Bonometto, F. Lucchin, 01.151.020),
153 (C. Barbieri, S. Bonometto, A. Saggion, AJB 68,
134.17), 154 (G. Barbaro, N. Dallaporta, G. Fabris,
01.153.001), 155 (G. Barbaro, N. Dallaporta, C. Summa,
02.125.009), 156 (G. Romano, 02.123.018), 157 (G. Romano, M. Perissinotto, 02.123.014).

008.098 Paris

Bureau International de l'Heure. Rapport Annuel
pour 1969. B. Guinot, M. Feissel, F. Laclare.
Printing Office: Observatoire de Paris. 140 pp. (1970).

Bureau International de l'Heure, Circulaires B/C
Nos. 165 - 170 (03.045.018).

Bureau International de l'Heure, Circulaires,
D38 - D44 (03.044.022).

008.099 Philadelphia

Flower and Cook Observatory, University of Penn-
sylvania, Philadelphia, Pennsylvania. – Report 1968 - 1969.
R. H. Koch.
Bull. American Astron. Soc., Vol. 2, 36 - 37 (1970).

Flower and Cook Observatory, *Philadelphia,*
Reprints, Nos. 169 (K.-C. Leung, AJB 67, 104.75),
187 (K.-C. Leung, AJB 68, 123.43), 188 (C. R. Chambliss,
AJB 68, 122.13), 189 (C. P. Olivier, 03.104.058), 190
(C. P. Olivier, 03.104.059).

008.100 Pittsburgh

Allegheny Observatory, University of Pittsburgh,
Pittsburgh, Pennsylvania. – Report 1968 - 1969.
N. E. Wagman.
Bull. American Astron. Soc., Vol. 2, 1 - 2 (1970).

008.101 Porto Alegre

Instituto de Astronomia da Universidade Federal
do Rio Grande do Sul. (Astronomical Institute of the University of Rio Grande do Sul). J. C. Haertel.
Inform. Bull. Southern Hemisphere, No. 15, p. 12 (1969).
Current research report.

008.102 Potsdam

Institut für Sternphysik der Deutschen Akademie
der Wissenschaften zu Berlin; Astrophysikalisches Observato-
rium Potsdam und Sternwarte Sonneberg. – Report 1968.
J. Wempe.
Monatsber. Deutsch. Akad. Wiss. Berlin, Vol. 11, 762 - 766
(1969).

Geodätisches Institut der Deutschen Akademie
der Wissenschaften zu Berlin. – Report 1968.
H. Kautzleben.
Monatsber. Deutsch. Akad. Wiss. Berlin, Vol. 11, 770 - 772
(1969).

008.103 Poznan

Astronomical Observatory of the Adam Mickie-
wicz University. H. Hurnik.
Postępy Astron., Vol. 18, 109 (1970). In Polish.

008.104 Praha

Académie Tchécoslovaque des Sciences, Institut
Astronomique, **Station de l'Heure à Prague,** Série 5, Nos. 4 -
6 (03.044.028).

Contributions and Observations from the People's
Observatory of Prague, Vol. 7, Ser. 2, No. 1 (03.096.021).

008.105 Pretoria

Communications from the Radcliffe Observatory,
Pretoria, Nos. 100 (T. L. Evans, 02.153.019), 102 (A. J.
Wesselink, 02.153.024), 103 (M. L. Clayton, M. W. Feast,
02.122.101).

Radcliffe Observatory, Pretoria, Reprints,
Nos. 73 (A. D. Thackeray, 02.008.097), 74 (A. D. Thackeray,
02.118.033), 75 (R. Wood, 03.159.012), 76 (A. D. Thackeray,
03.159.010), 77 (M. W. Feast, 03.122.024), 78 (A. D.
Thackeray, 03.132.038), 79 (M. W. Feast, 03.122.025).

008.106 Princeton

Princeton University Observatory, Princeton, New
Jersey. — Report 1968 – 1969. L. Spitzer.
Bull. American Astron. Soc., Vol. 2, 116 - 120 (1970).

008.107 Pulkovo

Izvestiya Glavnoj Astronomicheskoj Observatorii v
Pulkove, Akademiya Nauk SSSR. Astrometriya, Astrofizika,
Radioastronomiya, No. 185.

008.108 Richmond Hill

David Dunlap Observatory, University of Toronto,
Richmond Hill, Ontario. — Report 1968 – 1969.
D. A. MacRae.
Bull. American Astron. Soc., Vol. 2, 23 - 30 (1970).

David Dunlap Observatory, Richmond Hill, Ont.
D. A. MacRae.
Journ. Roy. Astron. Soc. Canada, Vol. 64, 60 - 62 (1970).

Communications from the David Dunlap Observa -
tory, University of Toronto, Richmond Hill, Ontario, Canada,
Nos. 212 (P. J. E. Peebles, 01.162.047), 223 (D. P. Hube,
02.119.011), 224 (P. C. Keenan, A. J. Deutsch, R. F. Garrison,
02.122.082), 225 (S. van den Bergh, 02.160.016), 231 (S. P. S.
Anand, 02.062.004), 232 (R. C. Roeder, 02.141.055), 233
(S. van den Bergh, W. W. Dodd, 02.125.001), 234 (J. D. Fernie
02.124.100), 235 (D. L. DuPuy, J. B. de Veny, 02.158.050),
236 (J. R. Percy, 02.119.012), 237 (J. A. Gilbert, R. G.
Conway, P. P. Kronberg, 02.141.046), 238 (J. R. Percy,
03.122.001), 239 (E. R. Seaquist, 02.099.049), 241 (R. F.
Christy, 02.122.131), 242 (R. Racine, 03.122.002), 243
(J. D. Fernie, 02.115.018), 244 (S. van den Bergh,
03.162.001).

008.109 Rio de Janeiro

Observatorio Nacional. (National Observatory).
Inform. Bull. Southern Hemisphere, No. 15, p. 12 - 13 (1969).
Current research report.

Contribuiçoẽs do Observatório do Valongo,
Universidade Federal do Rio de Janeiro, Série I, Nos. 9 - 10
(L. E. da Silva Machado, J. F. Caria Caldeira, 03.075.025).

Contribuiçoẽs do Observatório do Valongo,
Universidade Federal do Rio de Janeiro, Série II, No. 9,
(L. E. da Silva Machado, D. P. Pinto Filho, 03.099.055).

Contribuiçoẽs do Observatório do Valongo,
Universidade Federal do Rio de Janeiro, Série III, Nos. 10 - 16
(H. de Sonza, L. E. da Silva Machado, 03.096.020).

008.110 Rochester

C. E. Kenneth Mees Observatory, University of
Rochester, Rochester, N.Y., Reprints, Nos. 21 (H. L. Helfer,
02.115.012), 22 (H. M. Van Horn, 03.064.035).

008.111 Roma

Monthly Bulletin, Osservatorio Astronomico di
Roma, Nos. 145 - 150 (M. Cimino, 03.075.019).

Photographic Journal of the Sun, Osservatorio
Astronomico di Roma, Nos. 26 - 27 (M. Cimino, 03.075.018).

008.112 Sacramento Peak

Sacramento Peak Observatory, Sunspot, New
Mexico. — Report 1968 – 1969.
Bull. American Astron. Soc., Vol. 2, 120 - 122 (1970).

008.113 San Diego

Research Units and Academic Departments, Uni-
versity of California: Berkeley, Los Angeles, San Diego, and
Santa Cruz. — Report 1968 – 1969.
I. King, G. Abell, E. M. Burbidge, R. P. Kraft.
Bull. American Astron. Soc., Vol. 2, 2 - 23 (1970).

008.114 Santa Cruz

Research Units and Academic Departments, Univer-
sity of California: Berkeley, Los Angeles, San Diego, and
Santa Cruz. — Report 1968 – 1969.
I. King, G. Abell, E. M. Burbidge, R. P. Kraft.
Bull. American Astron. Soc., Vol. 2, 2 - 23 (1970).

008.115 Santiago

Departamento de Astronomĭa, Universidad de Chile.
(Astronomy Department, University of Chile).
Inform. Bull. Southern Hemisphere, No. 15, p. 17 - 20
(1969). — A) National Astronomical Observatory, Cerro
Calán; B) Maipú Radioastronomical Observatory.

Universidad de Chile, Departamento de Astronomia,
Santiago, Separata 5 (S. Tapia, 01.122.040), 6 (C. Anguita,
F. Noël, 02.041.008).

008.116 São José dos Campos

Observatório Astronômico do Instituto Tecnológico
de Aeronáutica. (Astronomical Observatory of the Aeronautical Technical Institute).
Inform. Bull. Southern Hemisphere, No. 15, p. 14 - 15 (1969).
Current research report.

008.117 São Paulo

Centro de Rádioastronomia e Astrofísica da Universidade Mackenzie. (Center of Radioastronomy and Astrophysics, Mackenzie University).
Inform. Bull. Southern Hemisphere, No. 15, p. 15 - 16 (1969).
Current research report.

Instituto Astronômico e Geofísico. (Astronomical and Geophysical Institute).
Inform. Bull. Southern Hemisphere, No. 15, p. 16 (1969).
Current research report.

008.118 Saugus

Lockheed Solar Observatory, Saugus, California.
Report 1968 - 1969. L. W. Acton.
Bull. American Astron. Soc., Vol. 2, 72 - 73 (1970).

008.119 Seattle

Astronomy Department, University of Washington, Seattle, Washington. – Report 1968 - 1969.
G. Wallerstein.
Bull. American Astron. Soc., Vol. 2, 162 - 165 (1970).

008.120 Shemakha

Soobshchenie Shemakhinskoj Astrofizicheskoj Observatorii, Akademiya Nauk Azerbajdzhanskoj SSR, vyp. (No.) 4 (R. Eh. Gusejnov, 03.073.031; R. Eh. Gusejnov, 03.073.032; S. M. Azimov, 03.121.014; Z. A. Ismailov, 03.122.030; M. B. Kerimbekov, 03.071.021; D. M. Kulnizade, 03.071.022; S. G. Mamedov, 03.073.033).

008.121 Sonneberg

Institut für Sternphysik der Deutschen Akademie der Wissenschaften zu Berlin; Astrophysikalisches Observatorium Potsdam und Sternwarte Sonneberg. – Report 1968.
J. Wempe.
Monatsber. Deutsch. Akad. Wiss. Berlin, Vol. 11, 762 - 766 (1969).

Mitteilungen über Veränderliche Sterne, *Sonneberg,* Band 5, Heft 6 (G. A. Richter, 03.123.028; W. Götz, W. Wenzel, 03.123.029; H. Geßner, 03.123.030; H. Geßner, 03.123.031; H. Huth, F. Splittgerber, 03.123.032; L. Meinunger, 03.123.033; L. Meinunger, 03.123.034; P. Ahnert, 03.122.116; K. Kockel, G. Reimann, 03.123.035; H. Huth, 03.123.036).

008.122 Stanford, Calif.

Stanford Center for Radar Astronomy, Stanford University, Stanford, California and Stanford Research Institute, Menlo Park, California. – Report 1968 - 1969.
V. R. Eshleman, R. L. Leadabrand, A. M. Peterson.
Bull. American Astron. Soc., Vol. 2, 132 - 134 (1970).

Stanford Radio Astronomy Institute. – Report from Solar Institute.
R. N. Bracewell.
Solar Physics, Vol. 11, 161 (1970).

Stanford Radio Astronomy Institute, Stanford University, Stanford, California. – Report 1968 - 1969.
R. N. Bracewell.
Bull. American Astron. Soc., Vol. 2, 134 - 135 (1970).

008.123 Stockholm

Stockholms Observatoriums Annaler, Band 23, No. 2 (Y. Öhman, H. Westin, U. Kusoffsky, 03.073.020).

008.124 Strasbourg

Publication de l'Observatoire de Strasbourg, Vol. 1, Fasc. 1 (P. Lacroûte, 03.041.038), 2 (A. Valbousquet, AJB 68, 112.27), 3 (P. Lacroute, 03.041.039), 4 (A. Schmitt, 03.041.040), 5 (A. Schmitt, 03.055.033), 6 (A. Schmitt, 03.055.034).

008.125 Swarthmore

The Sproul Observatory, Swarthmore College, Swarthmore, Pennsylvania. – Report 1968 - 1969.
P. van de Kamp.
Bull. American Astron. Soc., Vol. 2, 131 - 132 (1970).

Sproul Observatory, Swarthmore, Pennsylvania, Reprints, Nos. 177 (P. van de Kamp, 01.008.114), 178 (P. van de Kamp, 01.122.112), 179 (P. van de Kamp, 01.111.004), 180 (S. L. Lippincott, 01.118.009), 181 (J. F. Wanner, 01.118.010), 182 (P. J. Morel, 01.118.011), 183 (P. van de Kamp, S. E. Andersen, 01.119.005), 184 (P. van de Kamp, 01.112.007), 185 (P. van de Kamp, J. L. Warren, 01.118.012), 186 (B. H. Feierman, 01.118.013), 187 (P. J. Morel, 01.118.014), 188 (W. D. Heintz, 01.118.027), 189 (P. van de Kamp, 02.117.002), 190 (W. D. Heintz, 02.118.002), 191 (W. D. Heintz, 02.117.041).

008.126 Sydney

Division of Radiophysics, C.S.I.R.O. (Australian National Radio Astronomy Observatory, Parkes, N.S.W.).
B. J. Robinson.
Inform. Bull. Southern Hemisphere, No. 15, p. 8 - 10 (1969).
Current research report.

Division of Radiophysics, C.S.I.R.O. (Solar Radio Observatory, Culgoora). S. F. Smerd.
Inform. Bull. Southern Hemisphere, No. 15, p. 10 - 11 (1969).
Current research report.

Division of Radiophysics, C. S. I. R. O., Epping, New South Wales, Australia, Separate prints (K. Kai, 02.074.058; J. B. Whiteoak, F. F. Gardner, 02.131.103; J. P. Wild, 02.077.038; D. K. Milne, 02.125.012; W. M. Goss, V. Radhakrishnan, 02.132.020; D. J. Cooke, 02.141.082; R. N. Manchester, J. D. Murray, V. Radhakrishnan, 02.141.102;

V. Radhakrishnan, 02.141.131; D. Wills, J. G. Bolton, 02.141.183; F. F. Gardner, D. Morris, J. B. Whiteoak, 02.141.184; F. F. Gardner, J. B. Whiteoak, D. Morris, 02.141.185; P. A. Shaver, W. M. Goss, 02.157.009; R. A. Batchelor, B. F. C. Cooper, D. J. Cole, A. J. Shimmins, 03.033.029; B. F. C. Cooper, G. A. Wells, 03.033.030; B. McA. Thomas, 03.033.031; B. F. C. Cooper, 03.033.032; K. Kai, 03.077.004; K. Kai, 03.077.006; K. S. Stankevich, 03.097.022; J. B. Whiteoak, F. F. Gardner, 03.131.004; F. F. Gardner, J. B. Whiteoak, 03.131.029; L. F. Smith, R. A. Batchelor, 03.132.029; M. M. Komesaroff, 03.141.005; M. M. Komesaroff, D. Morris, D. J. Cooke, 03.141.013; F. F. Gardner, R. X. McGee, M. W. Sinclair, 03.141.014; D. Morris, U. J. Schwarz, D. J. Cooke, 03.141.062; D. Morris, U. J. Schwarz, O. B. Slee, 03.141.063; R. D. Davies, F. F. Gardner, 03.141.081; J. P. Wild, 03.141.082; R. M. Price, 03.157.024; J. B. Whiteoak, 03.158.001).

Radiophysics Laboratory, C. S. I. R. O. Epping, New South Wales, Australia, Separate prints (J. B. Whiteoak, F. F. Gardner, D. Morris, 01.099.013; F. F. Gardner, J. B. Whiteoak, 02.141.153).

Chatterton Astrophysics Department. (Molonglo Radio Observatory).
Inform. Bull. Southern Hemisphere, No. 15, p. 7 - 8 (1969).
Current research report.

Narrabri Observatory. (School of Physics, University of Sydney).
Inform. Bull. Southern Hemisphere, No. 15, p. 11 (1969).
Current research report.

008.127 Tampa

University of South Florida Observatory, Tampa, Florida. – Report 1968 – 1969.   H. K. Eichhorn-von Wurmb.
Bull. American Astron. Soc., Vol. 2, 130 - 131 (1970).

008.128 Tartu

Tartu Astronoomia Observatorium, Teated, No. 24 (H. Eelsalu, 03.155.063).

008.129 Tashkent

Solar physics at the Tashkent Astronomical Observatory. – Report from Solar Institute.
V. Scheglov, Yu. Slonim.
Solar Physics, Vol. 11, 157 - 160 (1970).

Chronicle.
Tsirk. Astron. Inst., *Tashkent*, No. 8 (355), p. 17 - 18 (1968); No. 15 (362), p. 32 - 35 (1970).

Tsirkulyar Astronomicheskogo Instituta, Akademiya Nauk Uzbekskoj SSR, Nos. 6 (353) - 19 (366).

008.130 Tautenburg

Karl-Schwarzschild-Observatorium Tautenburg der Deutschen Akademie der Wissenschaften zu Berlin. – Report 1968.   N. Richter.

Monatsber. Deutsch. Akad. Wiss. Berlin, Vol. 11, 769 - 770 (1969).

Mitteilungen des Karl-Schwarzschild-Observatoriums Tautenburg der Deutschen Akademie der Wissenschaften zu Berlin, Nos. 45 (N. Richter, 02.158.088), 46 (F. Börngen, 02.158.089), 47 (W. Högner, 02.036.021), 48 (H. Artus, 02.034.084).

008.131 Thessaloniki

Contributions from the Astronomical Department of the University of Thessaloniki, Nos. 25 (J. Hadjidemetriou, AJB 66, 111.10), 28 (G. Contopoulos, 03.151.073), 47 (G. Bozis, AJB 68, 41.12), 48 (B. Barbanis, AJB 68, 43.05), 49 (J. D. Hadjidemetriou, 01.117.001), 50 (J. D. Hadjidemetriou, 01.117.004), 51 (G. Bozis, 01.042.019), 52 (J. D. Hadjideme - triou, 02.042.004), 53 (G. Contopoulos, 03.155.068).

008.132 Tokyo

Annals of the Tokyo Astronomical Observatory, University of Tokyo, Second Series, Vol. 11 No. 4 (S. Iijima, Y. Niimi, 03.041.010; H. Yasuda, 03.101.004); Vol. 12, No. 1 (F. Sato, 03.131.063; F. Sato, 03.131.064).

Bulletin of Solar Phenomena, Tokyo Astronomical Observatory, Vol. 21, Nos. 3 - 4 (03.075.033).

Contributions from the Department of Astronomy, University of Tokyo, Nos. 113 (N. Kaifu, M. Morimoto, 02.131.014), 114 (Y. Fujita, T. Tsuji, H. Maehara, 03.114.124), 115 (F. Kamijo, 01.064.047), 116 (W. Unno, 02.064.038), 117 (M.-a. Kondo, 03.162.024), 118 (H. Maehara, 03.122.036), 119 (H. Hirabayashi, T. Ojima, M. Morimoto, 01.157.016).

Time and Latitude Bulletins, Tokyo Astronomical Observatory, Vol. 43, Nos. 8 - 12; Vol. 44, Nos. 1 - 2 (M. Huruhata, 03.044.027).

Tokyo Astronomical Bulletin, Tokyo Astronomical Observatory, Second Series, Nos. 197 (S. Yajima, K. Mizugaki, K. Yamaguchi, 03.073.002), 198 (K. Ichimura, T. Noguchi, E. Watanabe, 03.122.062), 199 (S. Nagasawa, T. Suzuki, M. Miyashita, 03.072.042), 200 (K. Ishida, M. Ohashi, 03.131.126).

Tokyo Astronomical Observatory, Reprints, Nos. 366 (M. Miyamoto, 02.151.050), 367 (T. Takakura, 02.141.087), 368 (K. Nariai, 03.119.006), 369 (T. Takakura, E. Scalise, Jr., 03.077.005).

University of Tokyo, Tokyo Astronomical Observatory, Report, Vol. 15, No. 1 (No. 56) (H. Morishita, 03.082.018; K. Saito, S. Shinozawa, 03.079.103; S. Okazaki, M. Nasaka, 03.045.003; S. Okazaki, M. Nasaka, T. Yamazaki, 03.045.004; S. Itō, 03.047.005; R. Fukaya, 03.046.005; H. Yasuda, H. Hara, R. Fukaya, T. Ina, 03.032.007; S. Kohno, 03.035.003; H. Kinoshita, R. Nagai, M. Yoneda, 03.042.019; T. Okamoto, 03.034.013; T. Takenouchi, K. Tomita, M. Nukariya, 03.104.014).

008.133 Tonantzintla y Tacubaya

Boletin de los Observatorios Tonantzintla y Tacu -

baya, Vol. 5, No. 32 (G. Haro, E. Chavira, 03.122.028; G. Haro, 03.122.029; B. Iriarte Erro, 03.153.005; B. Iriarte Erro, 03.113.014; E. E. Mendoza V., 03.159.005; E. E. Mendoza V., M. Jaschek.C. Jaschek, 03.114.042), No. 33 (E. E. Mendoza V., T. Gómez, 03.113.015), 34 (G. Haro, E. Chavira, 03.153.021; G. Haro, G. González, 03.153.022; A. D. Andrews, 03.132.032; E. E. Mendoza V., 03.113.049; B. I. Erro, 03.113.050; P. Pişmiş, 03.153.024; R. Costero, M. Peimbert, 03.132.033; L. R. Terrazas, G. González, 03.079.102; M. E. Méndez, 03.103.102; A. Cornejo, J. Castro, D. Malacara, 03.031.026).

008.134 Torino

Contributi dell' Osservatorio Astronomico di Torino, (Pino Torinese), Nos. 50 (M. G. Fracastoro, 02.115.008), 51 (G. Cocito, M. A. Vogliotti, 03.034.076), 52 (T. Tamburini Job, 02.118.031), 53 (F. Job, T. Tamburini, M. A. Zaccone, 02.118.032).

Pubblicazioni Varie Fuori Serie dell' Osservatorio Astronomico di Torino (Pino Torinese), Nos. 36 (03.047.021), 37 (03.014.013), 38 (02.011.026), 39 (03.047.022).

008.135 Tortosa

Publicaciones del Observatorio del Ebro, Miscelánea, Nos. 23 (J. O. Cardus, AJB 67, 75.32), 24 (J. O. Cardus, AJB 68, 78.09).

008.136 Torun

The information on the University Astronomical Center in Torun. W. Iwanowska. Postępy Astron., Vol. 18, 109 - 113 (1970). In Polish.

008.137 Trieste

Publicazione Osservatorio Astronomico di Trieste, Nos. 396 (M. Hack, 02.114.072), 402 (A. Abrami, 03.074.007) 403 (B. Cester, 02.121.059), 404 (B. Cester, 02.122.112), 405 (B. Cester, 02.122.113), 406 (N. Gökkaya, 03.116.005), 409 (P. Zlobec, 03.075.030, 03.075.031), 410 (P. Zlobec, 03.072.044), 412 (A. Abrami, 03.077.053), 414 (A. Abrami, 03.077.054), 416 (A. Abrami, 03.077.055).

008.138 Tübingen

Astronomisches Institut der Universität und Lehrstuhl für Theoretische Astrophysik. — Report 1969. G. Elwert, G. Möllenstedt, K. Walter. Mitt. Astron. Ges., No. 28, p. 111 - 118 (1970).

008.139 Tucson

Steward Observatory and the Department of Astronomy, University of Arizona, Tucson, Arizona. Report 1968 – 1969. B. J. Bok. Bull. American Astron. Soc., Vol. 2, 135 - 144 (1970).

008.140 Uccle

Observatoire Royal de Belgique (Koninklijke Sterrenwacht van Belgie), **Communications** (Mededelingen), Série B, Nos. 40 (J. Dommanget, 03.021.015), 41(R. Dejaiffe, 02.041.003), 42 (P. Melchior, 02.094.005), 44 (Y. Wako, 02.045.014), 45 (J. C. Usandivaras, B. Ducarme, 03.081.050), 46 (J. Dommanget, 02.117.023), 47 (H. Debehogne, 03.041.041), 51 (B. Ducarme, 03.081.009), 52 (P. Melchior, M. Bonatz, J. Blankenburgh, 03.081.007).

008.141 Urbana

University of Illinois Observatory, Urbana, Illinois. Report 1968 – 1969. G. C. McVittie. Bull. American Astron. Soc., Vol. 2, 56 - 59 (1970).

008.142 Utrecht

Utrechtse Sterrekundige Overdrukken, Sterrewacht "Sonnenborgh" — Laboratorium voor Ruimteonderzoek, Utrecht, Nos. 65 (A. B. Underhill, AJB 68, 104.104), 77 (A. D. Fokker, AJB 68, 144.13), 82 (J. R. W. Heintze, AJB 68, 104.140), 83 (E. W. Elst, AJB 68, 51.150), 84 (C. de Jager, L. Neven, AJB 68, 64.53), 85 (R. Snijders, AJB 68, 68.51), 86 (J. Houtgast, M. G. J. Minnaert, AJB 68, 64.50), 87 (J. Houtgast, O. Namba, AJB 68, 68.29), 88 (A. Jacobs, AJB 68, 67.47), 89 (J. Kleczek, M. Kuperus, 01.073.003), 90 (M. G. J. Minnaert, 03.004.024), 91 (R. Snijders, 01.076.010), 92 (C. de Jager, 02.073.071), 93 (J. N. van Gils, H. F. van Beek, L. D. de Feiter, R. V. Hendrickx, 01.084.008), 94 (E. Dekker, 01.022.005), 95 (B. Asselbergh, W. van Rensbergen, 01.133.017), 96 (J. R. W. Heintze, 01.114.042), 97 (O. Namba, W. E. Diemel, 01.071.029), 98 (R. Snijders, 02.076.043), 99 (A. Jacobs, 01.074.008), 100 (J. Degewij, J. van Diggelen, AJB 68, 93.118), 101 (A. B. Underhill, 01.114.006), 102 (E. P. J. van den Heuvel, 03.117.028), 103 (J. R. W. Heintze, 01.071.025), 104 (M. de Groot, 02.114.042), 105 (H. G. van Bueren, 01.034.044), 106 (A. B. Underhill, H. G. Geuverink, 02.114.048), 107 (A. B. Underhill, 01.153.005), 108 (W. van Rensbergen, 01.022.063), 109 (M. de Groot, 02.122.005), 111 (M. A. J. Snijders, 01.114.035), 112 (A. C. Brinkman, P. de Groene, 03.034.060), 115 (R. Faraggiana, 01.114.066), 116 (A. J. F. den Boggende, A. C. Brinkman, W. de Graaff, 03.034.061).

008.143 Victoria

Dominion Astrophysical Observatory, Victoria, B. C. A. H. Batten. Journ. Roy. Astron. Soc. Canada, Vol. 64, 108 - 109 (1970).

008.144 Vilnius

The establishment of the Vilnius Astronomical Observatory and the first five decades of its activity. S. P. Matulajtite. Liet TSR Mokslų Akad darbai, Tr. AN Lit. SSR, 1969, A, No. 2 (30), p. 121 - 130. In Russian. — Abstr. in Referativ. Zhurn. 51. Astron., 4.51.3 (1970).

Astronomijos Observatorijos, Biuletenis (Bulletin of the Vilnius Astronomical Observatory), No. 26 (K. Zdanavicius, J. Sudzius, Z. Sviderskiene, V. Straizys, V. Burnasov, R. Drazdys, A. Bartkevicius, G. Kakaras, G. Kavaliauskaite, V. Jasevicius, 03.113.009; A. Bartkevicius, L. P. Metik, 03.113.010; J. Sudzius, 03.113.011; V. Straizys, 03.113.012; A. Bogdanovic, 03.036.003).

008.145 **Warszawa**

Warsaw University Observatory and Astronomical Institute, Polish Academy of Sciences, Reprints Nos. 281 (S. M. Ruciński, 03.064.022), 282 (K. Stępień, 03.119.004), 283 (J. Ziółkowski, 03.117.022), 284 (J. Smak, 03.119.012), 285 (K. Stępień, 03.121.042), 286 (I. Semeniuk, 03.121.043), 287 (M. Sroczyńska, 03.074.060), 288 (G. Sitarski, 03.103.111), 289 (G. Sitarski, 03.103.112).

Publikacje Działu Geodezji Wyższej i Astronomii, Geodezyjnej Zg. PAN, Nos. 13 (J. Łatka, 02.054.013), 14 (K. Minowski, 03.046.016), 15 (J. B. Zieliński, 02.046.016), 16 (S. Domaradzki, 03.055.032).

008.146 **Washington**

U.S. Naval Observatory, Washington, D.C. Report 1968 - 1969. K. A. Strand. Bull. American Astron. Soc., Vol. 2, 144 - 149 (1970).

Astronomical Papers prepared for the use of the American Ephemeris and Nautical Almanac, Vol. 20, Part II (R. L. Duncombe, 03.098.033).

Publications of the United States Naval Observatory, *Washington,* Second Series, Vol. 18, Part V (K. A. Strand, 03.118.022), Part VII (V. V. Kallarakal, I. W. Lindenblad, F. J. Josties, R. K. Riddle, M. Miranian, B. F. Mintz, A. P. Klugh, 03.118.023), Vol. 19, Part III (A. N. Adams, B. L. Klock, D. K. Scott, 03.041.036).

United States Naval Observatory, *Washington,* Circular, Nos. 127 (B. L. Klock, D. K. Scott, 03.041.037), 128 (S. Elvove, 03.021.014).

U. S. Naval Observatory, *Washington, D. C.,* Reprints. Nos. 97 (P. K. Seidelmann, R. L. Duncombe, W. J. Klepczynski, 02.101.002), 98 (W. J. Klepczynski, 02.099.003), 99 (K. A. Strand, 02.117.003), 100 (C. E. Worley, 02.118.001), 101 (K. A. Strand, R. K. Riddle, 02.111.002), 102 (J. W. Christy, R. L. Walker, Jr., 02.118.027), 103 (R. S. Harrington, 02.117.013), 104 (J. E. Bixby, T. C. van Flandern, 02.101.007), 105 (K. A. Strand, 03.008.000), 106 (T. C. van Flandern, 02.094.159), 107 (B. L. Klock, R. Z. Geller, M. A. Dachs, 03.032.041).

National Aeronautics and Space Administration, Washington, D.C. I. Ames Research Center; II. Goddard Space Flight Center; III. Jet Propulsion Laboratory; IV. Langley Research Center; V. Manned Spacecraft Center; VI. Marshall Space Flight Center; VII. NASA Headquarters: Physics and astronomy programs. – Report 1968 - 1969. N. G. Roman. Bull. American Astron. Soc., Vol. 2, 85 - 98 (1970).

National Bureau of Standards, Washington, D.C. Report 1968 - 1969. C. E. Moore. Bull. American Astron. Soc., Vol. 2, 98 - 100 (1970).

008.147 **Waterloo**

Contributions of the University of Waterloo Observatory, Nos. 3 (M. P. FitzGerald, W. Wilson, J. E. Stegman, 02.113.059), 5 (M. P. FitzGerald, N. Houk, 03.114.034).

008.148 **Wien**

Universitäts-Sternwarte mit Leopold-Figl-Observatorium für Astrophysik. – Report 1969. J. Meurers. Mitt. Astron. Ges., No. 28, p. 118 - 125 (1970).

Das Leopold Figl-Observatorium für Astrophysik der Universität Wien. H. Haupt. Wetter und Leben, Jahrgang 22, p. 89 - 93 (1970). The Leopold Figl-Observatory was officially delivered to the University of Vienna on September 25, 1969. The report reviews the most recent history of the project and the final site selection. Details on the telescope, the building and the dome as well as on the scientific instrumentation are given.

Institut für Theoretische Astronomie. – Report 1969. K. Ferrari d'Occhieppo. Mitt. Astron. Ges., No. 28, p. 125 - 126 (1970).

008.149 **Williams Bay**

Yerkes Observatory, Williams Bay, Wisconsin. Report 1968 - 1969. C. R. O'Dell. Bull. American Astron. Soc., Vol. 2, 170 - 175 (1970).

008.150 **Würzburg**

Astronomisches Institut und Sternwarte. – Report 1969. H. Haffner. Mitt. Astron. Ges., No. 28, p. 127 - 128 (1970).

009 Notes on Observatories, Planetaria, and Exhibitions

009.001 **Das Große Baader-Planetarium.**
W. Jahn.
SuW, Vol. 9, 71 - 72, 74 (1970).

009.002 **L'observatoire «Mira» et ses projets.**
T. Pieraerts.
Ciel et Terre, Vol. 86, 145 - 156 (1970).

009.003 **An ambitious observatory is built by father and son.**
C. L. Stong.
Sci. American, Vol. 222, No. 4, p. 114 - 118, 120 (1970).

009.004 **The Big Bear Solar Observatory.**
H. Zirin.
Sky Telescope, Vol. 39, 215 - 219 (1970).

009.005 **Das Zeiss Planetarium Modell VI.**
E. Übelacker.
SuW, Vol. 9, 92 - 94 (1970).

009.006 **Die neue Beobachtungskuppel der Volkssternwarte München.** H. Oberndorfer.
SuW, Vol. 9, 128 (1970).

009.007 **Sonderprojektoren im Planetarium Recklinghausen.**
J. Herrmann.
SuW, Vol. 9, 130 (1970).

009.008 **Report on work on the organization of the Central Astronomical Observatory.** S. Piotrowski.
Postępy Astron., Vol. 18, 209 - 213 (1970). In Polish.

009.009 **New observatory of the University of Western Ontario.** W. Wehlau.
Journ. Roy. Astron. Soc. Canada, Vol. 64, 1 - 4 (1970).

009.010 **Centennial Planetarium, Calgary, Alta.**
S. Wieser.
Journ. Roy. Astron. Soc. Canada, Vol. 64, 63 (1970).

009.011 **The future of the planetarium.**
L. Gilchrist.
Journ. Roy. Astron. Soc. Canada, Vol. 64, 64 - 65 (1970).

009.012 **H. R. MacMillan Planetarium, Vancouver, B. C.**
D. A. Rodger.
Journ. Roy. Astron. Soc. Canada, Vol. 64, 110 (1970).

009.013 **The McLaughlin Planetarium, Toronto, Ont.**
N. Green.
Journ. Roy. Astron. Soc. Canada, Vol. 64, 111 (1970).

009.014 **Ein Jahrhundert wechselvoller Geschichte der Mannheimer Sternwarte 1783 – 1883.**
G. Klare.
SuW, Vol. 9, 148 - 150 (1970).

009.015 **Die Archenhold-Sternwarte Berlin-Treptow.**
D. Wattenberg.
Separate print Archenhold-Sternw., Berlin-Treptow. 64 pp. (1969).

009.016 **Die neue von Ardenne-Sternwarte in Dresden.**
J. Classen.
Sterne, 46. Jahrgang, p. 85 - 87 (1970).

009.017 **El Observatorio de Juvisy. A. Flammarion in memoriam.** R. Compte Porta.
El Universo, Vol. 24, 24 - 31 (1970).

009.018 **A new amateur observatory in Montreal.**
M. L. E. Coallier.
Journ. Royal Astron. Soc. Canada, Vol. 64, 194 - 196 (1970).

009.019 **A home built planetarium.**
J. B. Gould.
Journ. Royal Astron. Soc. Canada, Vol. 64, 196 - 197 (1970).

009.020 **Prima sesiune ştiinţifică a observatorului astronomic popular din Bucureşti.** I. C. Singeorzan.
Stud. Cerc. Astron., Vol. 15, 113 - 114 (1970).

009.021 **Tätigkeitsbericht 1969 der Privatsternwarte Karls - ruhe.** W. Malsch.
Separate print: Erzbergerstr. 111a, Karlsruhe. 2pp. (1970).

009.022 **Restoration at Greenwich Observatory.** D. Howse.
Sky Telescope, Vol. 40, 4 - 9 (1970).

010 Societies, Associations, Organizations

010.001 American Association of Variable Star Observers (AAVSO)

No publication received.

010.002 American Astronomical Society (AAS)

American astronomers report.
Sky Telescope, Vol. 39, 92 - 94 (1970).
Highlights of the 131st meeting of the American Astronomical Society at New York City, December 8 - 11, 1969: Movie of the milky way's hydrogen clouds; Sharpening optical images; HBV 475: A peculiar emission star; The extragalactic distance scale.

American astronomers report.
Sky Telescope, Vol. 39, 161 - 162 (1970).
Highlights of the 131st meeting of the American Astronomical Society at New York City, December 8 - 11, 1969: Interstellar formaldehyde abundances; Coronal observations from a manned spacecraft; Computer-simulated solar systems; Distribution and motions of supergiant stars.

010.003 Association of Lunar and Planetary Observers (ALPO)

The San Diego convention of the A.L.P.O.
R. E. Wend.
Strolling Astronomer, Vol. 22, 24 - 28 (1970).
San Diego, 1969 Aug. 21 - 23.

A proposed constitution for the A.L.P.O.
R. G. Hodgson.
Strolling Astronomer, Vol. 22, 38 - 43 (1970).

The San Diego business meeting of the A.L.P.O.
Strolling Astronomer, Vol. 22, 43 (1970).

Crisis in the Lunar Section: A program for the 1970's.   H. D. Jamieson, J. E. Westfall, C. R. Chapman, C. L. Ricker, K. J. Delano, H. W. Kelsey.
Strolling Astronomer, Vol. 22, 64 - 65 (1970).

Announcements.
Strolling Astronomer, Vol. 22, 69 - 71, 105 - 106 (1970).

010.004 Astronomical Society of Australia (ASA)

The Fourth Annual General Meeting.– Report of the Council, December 1969.
Proc. Astron. Soc. Australia, Vol. 1, 353 (1970).

010.005 Astronomical Society of Czechoslovakia

No publication received.

010.006 Astronomical Society of the Pacific (ASP)

Activities of the Society.
Publ. Astron. Soc. Pacific, Vol. 82, 160 - 167 (1970).

010.007 Astronomical Society of Southern Africa (ASSA)

Notices.
Monthly Notes Astron. Soc. Southern Africa, Vol. 29, 1 - 2, 17 - 18, 41, 53 (1970).

010.008 Astronomical Society of Victoria (ASV)

Annual report. 1969.
J. B. Trainor, L. G. Foster, T. B. Tregaskis, B. S. Adcock, D. H. Whitehead, W. G. H. Tregear, A. E. Coombs, B. A. J. Clark, L. R. Whitby.
Journ. Astron. Soc. Victoria, Vol. 23, 1 - 16 (1970). – Included are reports on the activities of different sections of the Society.

Society notes.
Journ. Astron. Soc. Victoria, Vol. 23, 52 (1970).

010.009 Astronomical Society of Western Australia (ASWA)

Reports of proceedings – 209th - 214th ordinary meeting.
Journ. Astron. Soc. Western Australia, Vol. 21, January - June (1970).

Annual report of Meteor Section of Junior Section of the Astronomical Society of Western Australia – 1969.
Journ. Astron. Soc. Western Australia, Vol. 21, January p. 2 - 5 (1970).

010.010 Astronomische Gesellschaft (AG)

Mitteilungen der Astronomischen Gesellschaft,
No. 28 (Jahresberichte astronomischer Institute für 1969). Edited by K. Schaifers.

010.011 Astronomisk Selskab Kobenhavn

No publication received.

010.012 British Astronomical Association (BAA)

Notices.
Journ. British Astron. Ass., Vol. 80, 85 - 87, 169 - 172, 253 - 255 (1970).

Meetings of the Association.
Journ. British Astron. Ass., Vol. 80, 88 - 92, 173 - 181, 256 - 262 (1970).

Variable Star Section.
J. S. Glasby, P. Moore.
Journ. British Astron. Ass., Vol. 80, 146 - 147 (1970).

Historical Section.
E. A. Beet, C. M. Botley.
Journ. British Astron. Ass., Vol. 80, 148 (1970).

Lunar Section.
R. C. Maddison.
Journ. British Astron. Ass., Vol. 80, 220 - 222 (1970).

Meteor Section.
K. B. Hindley.
Journ. British Astron. Ass., Vol. 80, 223 - 226 (1970).

Artificial Satellite Section.
H. Miles.
Journ. British Astron. Ass., Vol. 80, 230 - 232 (1970).

010.013 **British Interplanetary Society**

Society news.
Spaceflight, Vol. 12, 47, 136 - 138, 184 - 186, 259 - 260 (1970).

Society meetings.
Spaceflight, Vol. 12, 126 - 127 (1970).

25th annual general meeting.
L. J. Carter.
Spaceflight, Vol. 12, No. 4, p. I - IV (1970).

010.014 **Committee on Space Research (COSPAR)**

COSPAR 1969. W. Zonn.
Postępy Astron., Vol. 18, 117 - 122 (1970). In Polish.
Prag, 1969 May 15 - 24.

XII réunion plénière de COSPAR et la symposium "Dynamique des Satellites". J. B. Zieliński.
Geodezja i Kartografia, Vol. 19, 75 - 77 (1970). In Polish.
Prag, 1969 May 15 - 24.

010.015 **European Space Research Organization (ESRO)**

No publication received.

010.016 **International Astronautical Federation (IAF)**

No publication received.

010.017 **International Astronomical Union (IAU)**

International Astronomical Union, Information Bulletin, No. 23, [Printed by D. Reidel, Dordrecht–Holland], 66 pp. (1970). – Contents: Introduction; Part I. On the celebration of the 50th anniversary of the IAU; Part II. Draft statutes and by-laws of the Union; Part III. Information.

Information Bulletin, No. 24, 24 pp. (1970).

IAU to convene in England.
Sky Telescope, Vol. 39, 90 - 91 (1970).

50th anniversary of the International Astronomical Union. P. G. Kulikovskij.
Zemlya i Vselennaya, No. 1, p. 59 - 64 (1970). In Russian.

010.018 **Meteoritical Society**

Houston meeting – October. Abstracts of papers.
Meteoritics, Vol. 4, 257 - 300 (1969).

010.019 **Nederlandse Vereniging voor Weer- en Sterrenkunde**

Verenigingsnieuws.
Hemel en Dampkring, Vol. 68, 41, 63, 140 - 142 (1970).

Jongerenwerkgroep.
Hemel en Dampkring, Vol. 68, 51 - 52, 73, 96, 124 - 125, 153 - 156 (1970).

010.020 **Polskie Towarzystwo Astronomiczne (PTA)**

Resolutions of the plenary meeting of the Polish Astronomical Society on 27th September 1969.
Postępy Astron., Vol. 18, 247 - 248 (1970). In Polish.

Report of the plenary meeting of the Polish Astronomical Society on 27th September 1969. T. Kwast.
Postępy Astron., Vol. 18, 248 - 252 (1970). In Polish.

Report on the activities of the president of the Polish Astronomical Society for the period 1967 - 1969.
K. Rudnicki, W. Zonn.
Postępy Astron., Vol. 18, 252 - 254 (1970). In Polish.

XIV meeting of the Polish Astronomical Association. K. Ziołkowski, M. Pańków.
Urania Kraków, Vol. 41, 87 - 90 (1970). In Polish.

010.021 **Polskie Towarzystwo Miłośników Astronomii**

PTMA Chronicle.
Urania Kraków, Vol. 41, 22 - 23, 180 - 184 (1970).
In Polish.

010.022 **Royal Astronomical Society (RAS)**

Meetings of the Society.
Observatory, Vol. 90, 1 - 6, 37 - 52, 77 - 85 (1970).

150th Anniversary Meeting of the Royal Astronomical Society. B. Lovell.
Observatory, Vol. 90, 85 - 99 (1970).

Some prominent personalities and events in the early history of the Royal Astronomical Society.
G. J. Whitrow.
Quarterly Journ. Roy. Astron. Soc., Vol. 11, 89 - 104 (1970).

Sesquicentenary exhibition in the library of the Royal Astronomical Society, 1970 February – September.
G. J. Whitrow.
Quarterly Journ. Roy. Astron. Soc., Vol. 11, 132 - 136 (1970).

The Society and the year of revolutions.
P. S. Laurie.
Quarterly Journ. Roy. Astron. Soc., Vol. 11, 120 - 125 (1970).

Sidelights on astronomy during the Society's history. See Abstr. 004.015.

010.023 **Royal Astronomical Society of Canada (RAS Canada)**

Minutes of the annual meeting, May 16, 1970.
M. Fidler.
Journ. Royal Astron. Soc. Canada, Vol. 64, 203 - 204 (1970).

General assembly at Edmonton May 15 - 18, 1970.
M. Fidler.
Journ. Royal Astron. Soc. Canada, Vol. 64, L9 - L10 (1970).

010.024   Royal Astronomical Society of New Zealand (RAS New Zealand)

Annual report of council for the year ended 1969 September 30.   G. A. Eiby, D. J. Cameron.
Southern Stars, Vol. 23, 109 - 113 (1970).

Royal Astronomical Society of New Zealand.
Inform. Bull. Southern Hemisphere, No. 15, p. 24 (1969).
Current research report.

010.025   Schweizerische Astronomische Gesellschaft (SAG)

Aus der SAG und den angeschlossenen Gesellschaften.
Orion Schaffhausen, 28. Jahrgang, p. 25 - 26, 57 - 60, 97 (1970).

010.026   Sociedad Astronómica de México

Actividades de la Sociedad.
El Universo, Vol. 23, 87 - 88, 119 - 120 (1970).

010.027   Società Astronomica Italiana (SAI)

No publication received.

010.028   Société Astronomique de France (SAF)

Les séances de la Société.
L'Astronomie, 84e année, p. 86 - 92, 131 - 134, 176 - 178, 225 - 226, 295 - 297 (1970).

Commission des Instruments.
A. Hamon.
L'Astronomie, 84e année, p. 37 - 43 (1970).

Commission des Surfaces Planétaires.
A. Dollfus, J. Dragesco.
L'Astronomie, 84e année, p. 122 - 123, 227 - 229, 281 - 292 (1970).

Commission de Photographie Astronomique.
A. Hamon.
L'Astronomie, 84e année, p. 124 - 127, 180 - 183 (1970).

010.029   Société Astronomique "R. Bošković"

No publication received.

010.030   Société Chronométrique de France

No publication received.

010.031   Société Belge d'Astronomie, de Météorologie, et de Physique du Globe

Séances mensuelles.
M. Bauduin, M. Ducuroir.
Ciel et Terre, Vol. 86, 81 - 82, 174, 263 - 267 (1970).

010.032   Svenska Astronomiska Sällskapet

Svenska Astronomiska Sällskapet; Astronomiska Sällskapet Tycho Brahe; Göteborgs Astronomiska Klubb.
Styrelsens berättelse för år 1969.
E. Holmberg, B.-A. Lindblad, A. Winnberg.
Separate print, Svenska Astronomiska Sällskapet, Stockholm.
8 pp. (1970).

010.033   VAGO (Astronomical-Geodetical Society of the UdSSR)

No publication received.

010.034   Vereniging voor Sterrenkunde, Belgie

No publication received.

010.035   Argentine Astronomical Association

XV meeting of the Argentine Astronomical Association.
Inform. Bull. Southern Hemisphere, No. 15, p. 52 - 54 (1969).
4 - 7 October 1969.

010.036   Société d'Astronomie d'Anvers. Cinquantième rapport 1969.   J. Storms, R. de Terwangne.
Imprimerie «La Prévoyance», Antwerpen. 47 pp. (1970).
In French and Flemish.

010.037   Création de l'Union International des Astronomes Amateurs (I.U.A.A.).
Gaz. astron., No. 2, p. 16 - 17 (1969). – Bologna, 1969
April 18 - 22.

010.038   The Nantucket Maria Mitchell Association. Sixty-eighth annual report for the year ending December 31, 1969.
The Nantucket Maria Mitchell Association, Nantucket, Mass., 60 pp. (1970). Included is the annual report of Maria Mitchell Observatories by *D. Hoffleit*.

010.039   Asociación Argentina "Amigos de la Astronomia"

Nuestra Asociación.
H. Ottonello.
Revista Astron., Vol. 41, 5 - 14 (1970).

Noticias de la asociacion.
Revista Astron., Vol. 41, No. 170, p. 86 (1969).

011 Reports on Colloquia, Congresses, Meetings, Symposia, and Expeditions

011.001 **The moon at Houston.**
Nature, Vol. 225, 321 - 327 (1970).
An account of the Apollo 11 lunar science conference, based on reports and discussions with investigators who were present.

011.002 **Double star specialists meet.** S. L. Lippincott.
Sky Telescope, Vol. 39, 31 - 33 (1970). – Nice, 1969 Sept. 8 - 10.

011.003 **Report on the XXth International Astronautical Congress. Mar del Plata, Argentina, 6 - 10 October 1969.**
Astronaut. Acta, Vol. 15, 183 - 186 (1970). In French.

011.004 **Vierte astrometrische Konferenz in den USA – Sternparallaxen.** W. Gliese.
SuW, Vol. 9, 42 - 43 (1970). – Charlottesville, 1969 October 7 - 10.

011.005 **Symposium on physical geodesy.**
Bull. Géod., Nouvelle Sér., No. 95, p. 9 - 14 (1970). – Prague, 1969 Sept. 22 - 28.

011.006 **IAU-URSI Symposium on Lunar and Planetary Surfaces and Atmospheres, Woods Hole, August 11 - 15, 1969.** J. V. Evans.
The Moon, Vol. 1, 276 - 277 (1970).

011.007 **Il secondo colloquio dell'UAI sulla polvere interstellare.** M. Perinotto.
Mem. Soc. Astron. Italiana, Nuova Serie, Vol. 41, 131 - 134 (1970). – Jena, 1969 Aug. 22 - 26.

011.008 **Il colloquio internazionale sulla «Rotazione stellare».** P. L. Bernacca.
Mem. Soc. Astron. Italiana, Nuova Serie, Vol. 41, 135 - 137 (1970) – Columbus, Ohio, 1969 Sept. 8 - 11.

011.009 **Bericht über das IAU-Symposium Nr. 38 „The spiral structure of our Galaxy".**
G. Richter.
Sterne, 46. Jahrgang, p. 11 - 15 (1970). – Basel, 1969 Aug. 29 - Sept. 4.

011.010 **Zweites IAU-Kolloquium über interstellaren Staub.**
J. Gürtler.
Sterne, 46. Jahrgang, p. 15 - 17 (1970). – Jena, 1969 Aug. 22 - 25.

011.011 **The international colloquium on classical and relativistic magnetohydrodynamics.**
Yu. M. Mikhajlov.
Geomagn. Aeronom., Vol. 10, 187 (1970). In Russian.
Lille, 1969 June 14 - 20.

011.012 **The improvement of the precision of astrometrical investigations (Conference at Pulkovo).**
G. S. Kosin.
Vestn. AN SSSR, 1969 No. 10, p. 98 - 99. In Russian.
Abstr. in Referativ. Zhurn. 51. Astron., 3.51.30 (1970).

011.013 **Symposium No. 6 of the International Astronomical Union and the International Union for Theoretical and Applied Mechanics on cosmic gas dynamics at the Crimea.**
S. B. Pikel'ner.
Astron. Zhurn. Akad. Nauk SSSR, Vol. 47, 228 - 230 (1970).
In Russian. English translation in Soviet Astron. AJ, Vol. 14, No. 1. – 1969 Sept. 6 - 19.

011.014 **The Sixteenth International Astrophysical Colloquium in Liège.** J. Smak.
Postępy Astron., Vol. 18, 123 - 129 (1970). In Polish.
Liège, 1969, June 30 - July 2.

011.015 **General meeting on the physics of comets. (Kiev, 1969 Oct. 6 - 9).**
G. A. Rubo, K. I. Churyumov.
Astron. Zhurn. Akad. Nauk SSSR, Vol. 47, 454 - 457 (1970).
In Russian. – English translation in Soviet Astron., AJ, Vol. 14, No. 2.

011.016 **Report and discussion of the Third International Biophysics Congress of the International Union of Pure and Applied Biophysics in Cambridge, Massachusetts, 29th August to 3rd September, 1969.**
P. Shapshak.
Icarus, Vol. 11, 432 - 435 (1970).

011.017 **Teaching of astronomy in the secondary schools.**
J. Stodółkiewicz.
Postępy Astron., Vol. 18, 239 - 246 (1970). In Polish.
Chorzów, 1969 Sept. 22.

011.018 **International conference on laboratory astrophysics and opening of the Petrie Science Building, York University, Toronto, November 7 - 8, 1969.**
R. W. Nicholls.
Journ. Roy. Astron. Soc. Canada, Vol. 64, 66 - 68 (1970).

011.019 **Report of the Second Lunar International Laboratory (LIL). Discussion Panel: XXth International Astronautical Congress 6 October 1969, Mar del Plata, Argentina.** F. J. Malina, G. E. Mueller, F. Vinsonneau.
Astronaut. Acta, Vol. 15, 235 - 240 (1970).

011.020 **Vijf en twintigste bijeenkomst van werkende amateurs.** G. A. W. C. van Hemert tot Dingshof.
Hemel en Dampkring, Vol. 68, 58 - 63 (1970). – Utrecht, 1969 Oct. 25 - 26.

011.021 **Brazilian colloquium on site testing.**
Inform. Bull. Southern Hemisphere, No. 15, p. 49 (1969). – Belo Horizonte, Brazil, 28 April – 1 May 1969.

011.022 **International Symposium of Astronomy: Periodic orbits, stability and resonances.**
Inform. Bull. Southern Hemisphere, No. 15, p. 49 - 52 (1969). – São Paulo, Brazil, 4 - 12 September 1969.

011.023 **Le séminaire international de Nancy en 1967.**
Gaz. astron., No. 2, p. 11 - 15 (1969). – 1967 March 31 – April 2.

011.024 **Mond-Symposium in Houston.**
J. Classen.
Sterne, 46. Jahrgang, p. 83 - 84 (1970). – Houston, 1970, Jan.

011.025 **Full Assembly of the Astronomical Soviet of the Academy of Sciences of the USSR concerning the**

investigation of the earth's rotation.
Ya. S. Yatskiv.
Astron. Zhurn. Akad. Nauk SSSR, Vol. 47, 683 - 684 (1970).
In Russian. English translation in Soviet Astron. AJ, Vol. 14,
No. 3. – Kiev, 1969 Sept. 30 - Oct. 3.

011.026 **The symposium on stellar composition and nucleo-
synthesis.** J. L. Greenstein.
Comments Astrophys. Space Phys., Vol. 2, 85 - 91 (1970).
Pasadena, 1970 January 12 - 13.

011.027 **On the expedition of the Pulkovo astronomers to
Chile.** M. S. Zverev.
Izv. Glav. Astron. Obs. v Pulkove, No. 185, p. 3 - 31 (1970).
In Russian.
    A brief account of the history of the Pulkovo astronomers
expedition to Chile is given. The progress of its work during
five years (1962 - 1968) is described.

011.028 **New details about lunar specimens.**
Priroda, No. 5.70, p. 98 - 101 (1970). In Russian.
Conference in Houston, 1970 Jan. 5 - 8.

011.029 **The meeting of the National Committee for Canada
of the IAU at Kingston, March 13 - 14, 1970.**
V. Gaizauskas.
Journ. Royal Astron. Soc. Canada, Vol. 64, 177 - 186 (1970).

011.030 **Report of the Second Lunar International Laboratory (LIL) Discussion Panel, XXth International
Astronautical Congress, Mar del Plata, Argentina, 6 October
1969.** F. J. Malina.
The Moon, Vol. 1, 476 - 480 (1970).

011.031 **Conference on the Origin and Evolution of the
Moon and of the Planets, California Institute of
Technology, Pasadena, Calif., U.S.A., March 24 - 27, 1970.**
Z. Kopal.
The Moon, Vol. 1, 481 - 496 (1970).

011.032 **NATO Advanced Study Institute on the Moon and
Planets,held at the University of Newcastle upon
Tyne, England between April 9th and 16th, 1970.**
M. Moutsoulas, S. K. Runcorn.
The Moon, Vol. 1, 497 - 511 (1970).

011.033 **Scientific session of the Division of General Physics
and Astronomy of the USSR Academy of Sciences.**
Uspekhi fiz. nauk, Vol. 99, 503 - 514 (1969). In Russian.
Abstr. in Referativ. Zhurn. 51. Astron., 5.51.19 (1970).
Baku, 1969 May 19 - 21.

011.034 **Scientific session of the USSR Academy of
Sciences. Division of General Physics and Astronomy.**
Uspekhi fiz. nauk, Vol. 99, 514 - 531 (1969). In Russian.
Abstr. in Referativ. Zhurn. 51. Astron., 5.51.20 (1970).
Moscow, 1969 June 11 - 12.

011.035 **Symposium on optical characteristics of high layers
of the earth's atmosphere and of the circum-terres-
trial space.** T. I. Toroshelidze.
Vestn. AN SSSR, No. 2, p. 86 - 88 (1970). – Abstr. in Referativ. Zhurn. 62. Issled. kosm. prostranstv., 6.62.13 (1970).

011.036 **International conference on cosmic rays.**
N. A. Dobrotin.
Vestn. AN SSSR, No. 1, p. 65 - 68 (1970). In Russian.

Abstr. in Referativ. Zhurn. 51. Astron., 6.51.14 (1970).
Budapest, 1969 Aug. 25 - Sept. 4.

011.037 **Fifth USSR Winter School on cosmophysics.**
L. I. Miroshnichenko.
Geofiz. byull. Mezhduved. geofiz. kom-t pri Prezidiume AN
SSSR, 1969 No. 21, p. 56 - 62. In Russian. – Abstr. in
Referativ. Zhurn. 51. Astron., 6.51.22 (1970). Apatity, 1968
March 21 - April 5.

011.038 **Il simposio internazionale sulle relazioni sole-
terra.**
B. C. Fossi, G. Poletto, G. L. Tagliaferri.
Mem. Soc. Astron. Italiana, Nuova Serie, Vol. 41, 255 (1970).

011.039 **Ein Symposium über die Gasdynamik des interstellaren Mediums.** H. Zimmermann.
Sterne, 46. Jahrgang, p. 97 - 102 (1970). – Miskhor (Krim),
1969 Sept. 8 - 19.

011.040 **Chronicle.** V. I. Varshavskij.
Nauchn. Informatsii, vyp. (No.) 13, p. 145 - 147
(1969). In Russian. – Note on a conference at Riga, 1969
April 10 - 12.

011.041 **Riverside telescope meeting.**
Sky Telescope, Vol. 40, 28 - 30 (1970).

011.042 **First meeting of young friends of astronomy.**
V. A. Bronshtehn.
Zemlya i Vselennaya, No. 2, p. 82 - 88 (1970). In Russian.

011.043 **News on the planets and the moon.**
M. Ya. Marov.
Zemlya i Vselennaya, No. 3, p. 61 - 69 (1970). In Russian.
Symposium held at Woods Hole, Mass., 1969 August.

011.044 **Planetary geophysical investigations.**
N. T. Morozovskij.
Zemlya i Vselennaya, No. 3, p. 70 - 72 (1970). In Russian.
Symposium held in Tbilisi , 1969 October.

011.045 **The sixth international symposium on cosmic gas
dynamics. Crimea, 8 - 19 Sept. 1969.**
S. B. Pikel'ner.
Vestn. AN SSSR, 1969, No. 12, p. 102 - 103. In Russian. –
Abstr. in Referativ. Zhurn. 51. Astron., 7.51.14 (1970).

011.046 **Problems of cosmic-ray physics (Conference in
Leningrad, 1969 Oct. 9 - 20).** G. B. Zhdanov.
Vestn. AN SSSR, No. 2, p. 98 - 100 (1970). In Russian. –
Abstr. in Referativ. Zhurn. 51. Astron., 7.51.17 (1970).

011.047 **Chemical abundance in stars and nebulae (Confe -
rence in Pskov, 1969 Oct. 15 - 17).**
A. K. Kolesov, V. L. Khokhlova.
Vestn. AN SSSR, No. 2, p. 100 - 102 (1970). In Russian. –
Abstr. in Referativ. Zhurn. 51. Astron., 7.51.18 (1970).

011.048 **Protokoll der 115. Sitzung der Schweiz. Geodätischen Kommission vom 14. Juni 1969 im
Bernerhof in Bern mit Auszügen aus den Berichten über
die Tätigkeit im Jahre 1968.**
Société Helvétique des Sciences Naturelles, (Schweiz. Natur-
forschende Gesellschaft), Commission géodésique Suisse,
Ecole polytechnique fédérale, Zürich. 80 pp. (1970).

012 Proceedings of Colloquia, Congresses, Meetings, and Symposia

012.001 **Contributions from the Working Group on Electromagnetic Probing of the Upper Atmosphere set up by the International Union of Radio Science (URSI).**
E. K. Smith (Editor).
Journ. Atmosph. Terr. Phys., Vol. 32, 455 - 736 (1970).
This report contains 15 papers on electromagnetic probing of the upper atmosphere (ionosphere) by members of a new URSI Working Group, presented during the XIVth general assembly, Ottawa, 1969 August.

012.002 **Stars, Nebulae, Galaxies.**
Publications of the symposium held on the occasion of the 60th birthday of the Member of Academy V. A. Ambartsumyan, Byurakan, 16 - 18. September 1968.
A. A. Boyarchuk, V. V. Ivanov, L. V. Mirzoyan, V. V. Sobolev, G. M. Tovmasyan (Editors).
Izdatel'stvo AN Armyanskoj SSR, Erevan. 296 pp. Price 1 Rbl. 21 Kop. (1969).   In Russian.

012.003 **Summary of Apollo 11 Lunar Science Conference.**
Science, Vol. 167, (No. 3918), 449 - 784 (1970).
Houston, Texas, 1970 January 5 - 8. – The summary was prepared by the Lunar Sample Analysis Planning Team.

012.004 **Phase Transformations and the Earth's Interior.**
Proceedings of a symposium held in Canberra, Australia, 6 - 10 January 1969 by the International Upper Mantle Committee and the Australian Academy of Sciences.
A. E. Ringwood, D. H. Green.
Phys. Earth Planet. Interiors, Vol. 3, 11 + 519 pp. = Upper Mantle Project Scientific Report No. 26.

012.005 **Manned Laboratories in Space.** Second International Orbital Laboratory Symposium, organized by the International Academy of Astronautics at the XIXth International Astronautical Congress New York, 18 October 1968.
S. F. Singer (Editor).
Astrophysics and Space Science Library, Vol. 16. D. Reidel Publishing Company, Dordrecht–Holland. 13 + 133 pp. Price f 30.00 (1969).

012.006 **Low-Frequency Waves and Irregularities in the Ionosphere.** Proceedings of the 2nd ESRIN-ESLAB Symposium, held in Frascati, Italy, 23 - 27 September, 1968.
N. D'Angelo (Editor).
Astrophysics and Space Science Library, Vol. 14. D. Reidel Publishing Company, Dordrecht–Holland. 7 + 218 pp. Price f 43.00 (1969).

012.007 **Space Engineering.** Proceedings of the Second International Conference on Space Engineering, held at the Fondazione Giorgio Cini, Isola di San Giorgio, Venice, Italy, May 7 - 10, 1969.   G. A. Partel (Editor).
Astrophysics and Space Science Library, Vol. 15. D. Reidel Publishing Company, Dordrecht–Holland. 11 + 728 pp. Price f 140.00 (1970).

012.008 **Particles and Fields in the Magnetosphere.**
Proceedings of a symposium organized by the Summer Advanced Study Institute, held at the University of California, Santa Barbara, Calif., August 4 - 15, 1969.
B. M. McCormac (Editor).
Astrophysics and Space Science Library, Vol. 17. D. Reidel Publishing Company, Dordrecht–Holland. 11 + 453 pp. Price f 85.00 (1970).

012.009 **The Atmosphere of the Jovian Planets.**
Papers from the Third Arizona Conference on Planetary Atmospheres.
Journ. Atmosph. Sci., Vol. 26, No. 5, Part 1, p. 795 - 1001 = Contr. Kitt Peak National Obs. No. 435 (1969).

012.010 **Problemi evolutivi nel sistema solare.**
Convegni del 12 - 15 novembre 1968. Pubblicazione dell'Istituto Nazionale di Alta Matematica. Academic Press, London – New York. Symposia Mathematica, Vol. 3, 1 - 214 (1970).

012.011 **The Lunar Science Conference, Houston, Texas, U.S.A., January 5 - 8,1970.**
The Moon, Vol. 1, 352 - 402 (1970). – Abstracts of papers.

012.012 **La structure interne des étoiles.** XIe cours de perfectionnement de l'Association Vaudoise des Chercheurs en Physique. Observatoire de Genève, Sauverny. 12 + 420 pp (1970). – Saas-Fee, 1969 March 24 - 29.

012.013 **The Spiral Structure of Our Galaxy.**
International Astronomical Union, Symposium No. 38, held in Basel, Switzerland, August 29 – September 4, 1969.   W. Becker, G. Contopoulos (Editors), with a summary and outlook by B. J. Bok.
D. Reidel Publishing Company, Dordrecht – Holland. 13 + 478 pp. Price Dfl. 75.00 (1970).

012.014 **Ultraviolet Stellar Spectra and Related Ground-Based Observations.**
L. Houziaux, H. E. Butler (Editors).
International Astronomical Union. Symposium No. 36, held in Lunteren, The Netherlands, 24 - 27 June, 1969. D. Reidel Publishing Company, Dordrecht – Holland. 15 + 361 pp. Price Dfl. 58.00 (1970).

012.015 **Papers presented at the Fourth Annual General Meeting held at the Australian National University, Canberra, on 17, 18 and 19 December 1969.**
Proc. Astron. Soc. Australia, Vol. 1, (No. 7), 296 - 353 (1970).

012.016 **Solar-terrestrial Physics: Solar Aspects(Proceedings of Joint IQSY/COSPAR Symposium, London 1967, Part I).**   A. C. Stickland (Editor).
Annals of the IQSY, (International Years of the Quiet Sun), Vol. 4. The M. I. T. Press, Massachusetts Institute of Technology, Cambridge, Mass. – London. 10 + 414 pp. Price £9 2s. 0d. (1969). – Rev. in Solar Physics, Vol. 12, 502, 1970 *(Z. Švestka).*

012.017 **Solar-terrestrial Physics: Terrestrial Aspects (Proceedings of Joint IQSY/COSPAR Symposium, London 1967, Part II).**   A. C. Stickland (Editor).
Annals of the IQSY, (International Years of the Quiet Sun), Vol. 5. The M.I.T. Press, Massachusetts Institute of Technology, Cambridge, Mass., London. 10 + 460 pp. Price £ 10 10s.0d. (1969). – Rev. in Journ. British. Interplanet. Soc., Vol. 23, 532, 1970 *(J. C. Gilbert)*; Solar Physics, Vol. 12, 502, 1970 *(Z. Švestka).*

012.018 **Crab Nebula Symposium**, June 18 - 21, 1969, Flagstaff, Arizona.
Publ. Astron. Soc. Pacific, Vol. 82, (No. 486), 375 - 564, with a panel discussion by L. Woltjer, H.-Y. Chiu, P. Morrison, J. E. Gunn, T. Gold, W. J. Cocke, p. 534 - 562 (1970).

012.019   **12a. reunión de la Asociación Argentina de Astro-**

nomia.   L. A. Milone (Editor).
Asoc. Argentina Astron. Bol., Cordoba, No. 12, 54 pp.
(1968). − 1966 November 24 - 26.

012.020   **14a. reunión de la Asociación Argentina de Astronomia.**   L. A. Milone (Editor).
Asoc. Argentina Astron. Bol., Cordoba, No. 14, 111 pp.
(1968). − 1968 October 10 - 12.

012.021   **Radovi prikazani na IV Kongresu Matematičara, Fizičara i Astronoma Oktobra 1965 u Sarajevu.**
Edited by P. M. Djurković.
Publ. Obs. Astron. Beograd, No. 16, 121 pp. (1969).

012.022   **International Union of Amateur Astronomers. Proceedings of the first general assembly,** (Bologna, 1969, April 19 - 22).   A. Leani (Editor).
I.U.A.A. Publ.−Padus, Cremona (Italy), 109 pp. with an appendix of the "3· Congresso Nazionale" dell'Unione degli Astrofili Italiani, 20 pp. (1969).

012.023   **Theory and Observation of Normal Stellar Atmospheres.** Third Harvard-Smithsonian Conference on Stellar Atmospheres (April 1968).
O. Gingerich (Editor).
The M.I.T. Press, Cambridge, Mass. 472 pp. Price $ 12.50 (1969). − Review in Sky Telescope, Vol. 39, 253 (1970).

012.024   **Proceedings of the International Conference on Investigations of the Interplanetary Medium with the Aid of Cosmic Rays.**
Leningrad. 290 pp. Price 50 Kop. (1969).   In Russian.
Leningrad, 1969 June 3 - 7.

012.025   **Symposium on Unidentified Flying Objects.**
Clearinghouse for Federal Scientific and Technical Information, Springfield, Virginia. 4 + 247 pp. Price $ 3.00 (1968). − Review in Icarus, Vol. 11, 439 - 440; 1970 *(P. M. Millman).*

012.026   **Exploitation of Space for Experimental Research.** Vol. 24: Conference Proceedings, AAS 14th Annual Meeting, Dedham, Mass., 1968 May 13 - 14.
H. Zuckerberg (Editor).
American Astronautical Society, Tarzana, Calif. 363 pp. Price $ 14.25 (1969).

013 Reports on Astronomy in Various Countries and Particular Fields, International Cooperation

013.001 Development of astronomy in the Estonian SSR.
Ya. Ehjnasto.
Astron. Geod. Eston. SSR, Tartu, (see 003.004), p. 5 - 29
(1969).   In Russian.

013.002 A planetary astronomer visits the Soviet Union.
D. P. Cruikshank.
Sky Telescope, Vol. 39, 76 - 79 (1970).

013.003 The prospects for British astronomy.   B. Lovell.
Quarterly Journ. Roy. Astron. Soc., Vol. 11, 71 -
84 (1970). – The Presidential Address, delivered at the Anniversary Meeting of the Royal Astronomical Society on 1970
February 13.

**013.004 Inter-union commission on frequency allocations
for radio astronomy and space science (URSI -
IAU - COSPAR).**
Radio astronomy and space research review of requirements
for allocations of frequencies. (Document IUCAF/142 dated
24 March 1969). U.R.S.I. Inform. Bull. No. 171, p. 21 - 27
(1969).
   An URSI review of frequency allocations for radio
astronomy prepared at the IUCAF meeting. Brussels, Feb.
18 - 19, 1969. – *DNC*

013.005 Revolution in modern astronomy.
V. A. Ambartsumian, V. V. Kaziutinski.
Priroda, No. 4.70, p. 16 - 26 (1970).   In Russian.

013.006 Early New Zealand astronomy.
R. A. McIntosh.
Southern Stars, Vol. 23, 101 - 108 (1970).

**013.007 Contemporary state of cognition of space and the
role of astronomy in modern science.**
A. Tursunov.
Izv. AN Tadzh. SSR. Otd. fiz.-matem. i geol.-khim. n., 1969
No. 2, (32), p. 7 - 13. In Russian. – Abstr. in Referativ.
Zhurn. 51. Astron., 6.51.1 (1970).

013.008 The universe and its development.
V. L. Ginzburg.
Fiz.-matem. spisanie, Vol. 12, No. 3, p. 206 - 216 (1969).

013.009 Advances of astronomy in the year 1969.
J. Grygar.

Říše hvězd, Vol. 51, 41 - 54 (1970).   In Czech.

013.010 Astronomical observations from outside the terrestrial atmosphere.   R. Wilson.
Trans. IAU, Vol. 14A, (see 003.028), 525 - 546 (1970).
Report of Commission 44. – Solar astronomy, (*R. Tousey*);
The interplanetary medium, (*M. Neugebauer*); Ultraviolet
stellar astronomy, (*H. E. Butler*); X-ray sources, (*L. Gratton*);
The X- and gamma-ray background radiation, (*W. Kraushaar*).

013.011 About the work of young astronomers at Uglitsch.
Yu. A. Grishin.
Zemlya i Vselennaya, No. 1, p. 89 - 93 (1970). In Russian.

**013.012 V. I. Lenin and the development of the Soviet
astronomy.**   V. G. Fesenkov.
Zemlya i Vselennaya, No. 2, p. 6 - 10 (1970). In Russian.

013.013 The radio astronomer's universe.
E. G. Bowen.
Navigation. Sydney, Vol. 3, (No. 2), 138 - 147 (1969).
Popular article.

013.014 Prospettive dell' astronomia ottica.
M. G. Fracastoro.
Atti Fondaz G. Ronchi, Anno 23, p. 467 - 474 = Publ. Ist.
Nazionale Ottica, Arcetri–Firenze, Ser. IV, No. 513 = Pubbl.
Varie Fuori Ser. Oss. Astron. Torino, No. 37 (1968).
   The historical developments and the present situation are
discussed, concerning the optical instruments for astronomy.
Some information is given on the most recent technical im -
provements and on the most important achievements that will
probably be accomplished in the near future.

013.015 Ballongastronomi.   D. Dravins.
Astron. Tidssk., Vol. 3, 53 - 68 (1970).

013.016 Die Tiefen der Galaxien.
V. Ambarzumjan.
Ideen exakt. Wissens, (Wiss. Technik Sowjetunion), No. 10.69,
p. 615 - 625 (1969).

**013.017 Position of astronomical science in Yugoslavia
and its perspectives.**   G. Teleki.
Publ. Obs. Astron. Beograd, No. 16, (see 012.021), p. 65 -
81 (1969).   In Serbo-Croatian.

014 Teaching in Astronomy

014.001 Die Fernsehkamera in der Volkssternwarte.
B. Wedel.
SuW, Vol. 9, 74 - 75 (1970).

014.002 **Astronomische Modelle.** M. Schürer.
Orion Schaffhausen, 28. Jahrgang, p. 33 - 37
(1970). In German and French.

014.003 **Artificial stars in a teaching laboratory.**
D. Clarke.
Sky Telescope, Vol. 39, 295 - 297 (1970).

014.004 **Zur Aktualisierung des Lehrstoffes im Astronomie-unterricht.** K. Lindner.
Astron. in der Schule, 7. Jahrgang, p. 17 - 20 (1970).

014.005 **Astronomische Schülerbeobachtungen in gleicher Front.** H. Tauscher.
Astron. in der Schule, 7. Jahrgang, p. 20 - 23 (1970).

014.006 **Anwendungsmöglichkeiten des Modells der helio-zentrischen Planetenbahnen im Astronomieunter-richt.** W. König.
Astron. in der Schule, 7. Jahrgang, p. 42 - 43 (1970).

014.007 **Der Astronomieunterricht in der polnischen Ober-schule.** J. Salabun.
Astron. in der Schule, 7. Jahrgang, p. 58 - 60 (1970).

014.008 **Studieneinführung: Astronomie.** Astronomie, die Extraterrestrische Physik (Weltraumforschung).
T. Schmidt-Kaler.
Aspekte, 3. Jahrgang, No. 1/2, p. 31 - 34 (1970).

014.009 **Teaching of astronomy.** E. A. Müller.
Trans. IAU, Vol. 14A, (see 003.028), 559 - 566 (1970). – Report of Commission 46.

014.010 **The astronomical school of Kazan university.**
A. A. Nefed'ev.
Zemlya i Vselennaya, No. 2, p. 73 - 78 (1970). In Russian.

014.011 **Astronomin i skolan.** Astronomins ställning i skolundervisningen och i lärarutbildningen i Sverige. O. Eklöf.
Astron. Tidssk., Vol. 3, 88 - 91 (1970).

014.012 **Significance of astronomy in upbringing-educational process of pupils at secondary schools.**
S. Sadžakov.
Publ. Obs. Astron. Beograd, No. 16, (see 012.021), p. 59 - 64 (1969). In Serbo-Croatian.

Ein neues Gerät für astronomische Volksbildung.
Umschau, 70. Jahrgang, p. 23 - 24 (1970).

015 Miscellanea

015.001 **The field of Arcturus.** J. Ashbrook.
Sky Telescope, Vol. 39, 87 - 88 (1970).

015.002 **The role of the artist in astronautics.**
D. A. Hardy.
Spaceflight, Vol. 12, 14 - 16 (1970).

015.003 **The philosophical aspects of astronautics (II).**
S. Lubertowicz.
Urania Kraków, Vol. 41, 41 - 50 (1970). In Polish.

015.004 **Einstein und das Weltgerüst. Was bedeutet Relativität? – Die Korrektur von Grundbegriffen ist auch philosophisch von Belang.** E. Verhülsdonk.
SuW, Vol. 9, 121 - 123 (1970).

015.005 **Conditions of life in the universe.**
V. G. Fesenkov.
Priroda, No. 1.70, p. 20 - 27 (1970). In Russian.

015.006 **The Shapley-Curtis debate.** N. S. Hetherington.
Astron. Soc. Pacific, Leaflet No. 490, 8 pp. (1970).

015.007 **The wayward heavens literature.** L. Berman.
Astron. Soc. Pacific, Leaflet No. 488, 8 pp. (1970).

015.008 **Carbon monoxide as a basis for primitive life on other planets: A comment.** J. Postgate.
Nature, Vol. 226, 978 (1970).

015.009 **A guide to stellar objects.**
J. S. Griffith.
Spaceflight, Vol. 12, 213 - 219 (1970).

015.010 **Der Mensch in Raum und Zeit und Ewigkeit.**
H. Kienle.
Phys. Blätter, 26. Jahrgang, p. 193 - 199 (1970).

015.011 **Astronomical notebook.** J. S. Griffith.
Spaceflight, Vol. 12, 289 - 290 (1970).

015.012 **Per una collaborazione nel campo a confine fra scienze astronomiche e geofisiche e scienze della vita.** G. Piccardi.
Coelum, Vol. 38, 1 - 9 (1970).

015.013 **Gibt es einen Zusammenhang zwischen dem Alter kosmischer Körper und Systeme und ihrem spezifischen Volumen?** T. Landscheidt.
Abh. naturwiss. Verein Bremen, Vol. 37, 203 - 225 (1970).
The age of the earth, the planets, the sun, the globular clusters, the galaxies, and the universe can be represented as a logarithmic function of the specific volume of these objects. This connexion seems to indicate that all cosmic bodies expands out of a compact state of development whose point of departure is the specific volume of the proton. The proportional constant of the function can be described as a combination of the three fundamental pure numbers of theoretical physics if calculations are based on natural microphysical reference quantities.

015.014 **Some questions of modern radioastronomy.**
S. Ya. Braude.
Visnik AN URSR, 1969, No. 3, p. 47 - 55. In Ukrainian.
Abstr. Referativ. Zhurn. 51. Astron., 4.51.52 (1970).

015.015 **Outlook for the application of television microscopy for discovering extraterrestrial forms of life.**
B. L. Kozlov.
In-t kosmich. issled. AN SSSR, Moskva. 11 pp. (1970).
In Russian. – Abstr. in Referativ. Zhurn. 62. Issled. kosm. prostranstv., 6.62.388 (1970).

015.016 **On the determination of Easter.** P. Andrle.
Říše hvězd, Vol. 51, 54 - 58 (1970). In Czech.

015.017 **Changes in the astronomical telegraphic code.**
J. Bouška.
Říše hvězd, Vol. 51, 31 - 32 (1970). In Czech.

015.018 **Astronomical telegrams.** B. G. Marsden.
Trans. IAU, Vol. 14A, (see 003.028), 15 - 17 (1970). – Report of Commission 6.

015.019 **Modern astronomy and dialectic.**
V. V. Kazyutinskij.
Zemlya i Vselennaya, No. 2, p. 51 - 57 (1970). In Russian.

015.020 **Astronomie und Astrophysik - ohne Fernrohr.**
A. Unsöld.
"Christiana Albertina" Kieler Universitäts-Zeitschrift, 1968, Heft 6, p. 55 - 58 = Separate print Sternw. Kiel No. 149 (1968).

Applied Mathematics, Physics

021 Mathematics, Computing, Machine Programs

021.001 Mechanized Algebraic Operations (MAO).
A. Rom.
Celestial Mechanics, Vol. 1, 301 - 319 (1970).
A software package for Mechanized Algebraic Operations (MAO) is described. With MAO one is able to manipulate on the computer Poisson series in literal form. The system is operational; it has application in the fields of celestial mechanics, astrodynamics, and nonlinear mechanics. Besides describing the system, the present paper suggests various techniques to prepare problems such that they lend themselves for an automated treatment with MAO. Optimized implementation of the general subroutines is discussed.

021.002 Power series solutions.
D. G. Saari.
Celestial Mechanics, Vol. 1, 331 - 342 (1970).
A means of extending the radius of convergence of a power series solution of a system of differential equations is presented. It is essentially a change of the independent variable by means of a conformal mapping. Conditions on this change of variables which should yield a computational advantage are discussed.

021.003 Trigonometric interpolation.
G. E. O. Giacaglia.
Celestial Mechanics, Vol. 1, 360 - 367 (1970).
We present a new way of representation of trigonometric polynomials by the use of Lagrangian type orthonormal functions. We develop special formulae for trigonometric interpolation of periodic even or odd functions defined in $[0, \pi]$. The method is generalized to obtain explicit form of trigonometric interpolation in case of unequally spaced points.

021.004 Lagrangian systems on manifolds, I.
W. M. Oliva.
Celestial Mechanics, Vol. 1, 491 - 511 (1970).

021.005 Smithsonian Package for Algebra and Symbolic Mathematics. N. M. Hall, J. R. Cherniack.
SAO, *Cambridge, Mass.,* Special Rep., No. 291, 5 + 49 pp. (1969).

021.006 Die Grundbegriffe der Fehlerrechnung.
F. Schmeidler.
SuW, Vol. 9, 88 - 91 (1970).

021.007 Application of computers to the spectrophotometric reduction of stellar spectra.
Z. Turło, J. Hanasz.
Postępy Astron., Vol. 18, 181 - 183 (1970). In Polish. Abstr. Polish Astron. Soc.

021.008 On the determination of the wavelength of spectral lines with the help of an electronic digital computer.
G. I. Abbasov.
Astron. Zhurn. Akad. Nauk SSSR, Vol. 47, 673 - 675 (1970). In Russian. English translation in Soviet Astron. AJ, Vol. 14, No. 3.
A method of spectrogram reductions with the help of an electronic digital computer has been worked out. The computer makes a dispersion curve on the basis of the initial lines and determines the wavelengths of the studied spectral lines by their own positions on the spectrogram.

021.009 The programming of some astronomical calculi for the electronic desk computer "Olivetti-Programma 101". H. Alexandrescu.
Stud. Cerc. Astron., Vol. 15, 73 - 97 (1970). In Roumanian.
The paper presents the way to program some astronomical problems for an electronic desk computer, namely the "Olivetti-Programma 101" computer. Among these programs, are those referring to the calculation for the Astronomical Ephemeris published by the Bucharest Observatory.

021.010 Spherical triangles with integral-degree parts.
D. H. Sadler.
Journ. Inst. Navigation, Vol. 23, 379 - 381 (1970).

021.011 Computer-aided filter design manual. Prepared under contract for NASA by Sperry Rand Corporation, Huntsville, Alabama. S. Gussow, G. Weathers.
U.S. National Aeronautics Space Administration, Office Technol. Utilization Sci. Techn. Inform. Div., NASA SP-3049, 101 pp. (1969).
Includes passive and active filter synthesis, design charts and computer programs. – *BMT*

021.012 Degré topologique et solutions périodiques des systèmes différentiels non linéaires.
J. Mawhin.
Bull. Soc. Roy. Sci. Liège, Vol. 38, 308 - 398 = Univ. Liège, Inst. d'Astrophys., Coll. 8°, No. 588 (1969).

021.013 Equations intégrales et solutions périodiques des systèmes différentiels non linéaires. J. Mawhin.
Bull. Cl. Sci. Acad. Roy. Belgique, *Bruxelles*, 5e Sér., Vol. 55, 934 - 947 = Univ. Liège, Inst. d'Astrophys., Coll. 8°, No. 591 (1969).

021.014 Astronomical data in machine readable form.
S. Elvove.
United States Naval Obs. Circ., No. 128, 11 pp. (1970).
This *Circular* supersedes *Circular* 114 (1967). It contains a revised list, with occasional changes in item number, of the astronomical data available in machine readable form from the U. S. Naval Observatory.

021.015 Considérations sur la solution par la méthode des moindres carrés, d'un système surabondant d'équations linéaires dans le cas d'instabilité.
J. Dommanget.
Obs. Royal Belgique, *Uccle*, Commun., Sér. B, No. 40, 16 pp. (1969).

021.016 Series expansion of the perturbing function in terms of Bessel functions of pure imaginary arguments. C. A. Altavista.
Asoc. Argentina Astron. Bol., No. 12, (see 012.019), p. 9 (1968). In Spanish. – Abstract.

022 Physical Papers Related to Astronomy and Astrophysics

022.001 Hönl-London factors for doublet transitions in diatomic molecules. R. J. M. Bennett.
Monthly Notices, Roy. Astron. Soc., Vol. 147, 35 - 46 (1970).

Hönl-London factors are given for the general doublet transition in a diatomic molecule between two states both of which are intermediate between Hund's coupling case (a) and case (b). As a necessary preliminary the energy levels for the intermediate case are derived.

022.002 Theoretical energy levels in Fe IV.
B. Warner, R. C. Kirkpatrick.
Monthly Notices, Roy. Astron. Soc., Vol. 147, 115 - 121 (1970).

Predicted energy levels are given for the $3d^5$, $3d^4 4s$ and $3d^4 4p$ configurations of Fe IV, based on least squares adjustment of Slater parameters to fit known energy levels.

022.003 A revised radio-frequency spectrum for $H_2{}^+$.
W. B. Somerville.
Monthly Notices, Roy. Astron. Soc., Vol. 147, 201 - 205 (1970).

New calculations of the second order interaction d'' in the hyperfine structure of $H_2{}^+$ were made. The calculations are related to the results of laboratory experiments and revised frequencies and transition probabilities are presented.

022.004 Total transition probabilities for the Bowen levels of O III. H. G. Berry, W. S. Bickel, I. Martinson, R. J. Weymann, R. E. Williams.
Astrophys. Letters, Vol. 5, 81 - 83 (1970).

The methods of beam-foil spectroscopy have been used to measure the radiative lifetimes of certain Bowen levels of O III. In particular, we have found the total transition probability of the $3p\ ^3S_1$ level to be $3.4 \times 10^8\ sec^{-1}$. This value indicates that the Bowen fluorescence mechanism is the only important process which populates the Bowen levels.

022.005 Experimental condensation of silicates.
C. Meyer, Jr.
Meteoritics, Vol. 4, 284 (1969). − Abstract.

022.006 The $1s\ 2s$ states of negative hydrogen.
C. Ingemann-Hilberg, M. Rudkjøbing.
Astrophys. Space Sci., Vol. 6, 101 - 106 (1970).

Term values for the $1s\ 2s$ states of H^- are calculated with the use of simple eigenfunction approximations constructed on the basis of available Hartree-type functions for the $1s^2$ and $2s^2$ states of the ion. The results seem to support a tentative identification of the interstellar diffuse absorption bands at $\lambda 4890$ and $\lambda 6180$ as due to negative hydrogen.

022.007 On the isotropization of electrons in synchrotron sources. D. B. Melrose.
Astrophys. Space Sci., Vol. 6, 321 - 337 (1970).

Electrons radiating synchrotron radiation develop a pitch angle anisotropy, and so become unstable to the coherent emission of hydromagnetic waves. The evolution of the coupled system of anisotropic electrons and waves is studied in the absence of any dissipation of the waves in the ambient medium. The anisotropy of the electrons approaches a steady state in which the anisotropy is energy independent and of order v_A/c (v_A = Alfvén speed). The conditions for this small degree of anisotropy to be maintained are examined.

022.008 Die Physik im 20. Jahrhundert. V. F. Weisskopf.
Phys. Blätter, 26. Jahrgang, p. 64 - 72, 101 - 108 (1970).

022.009 Superdense water ice.
A. H. Delsemme, A. Wenger.
Science, Vol. 167, 44 - 45 (1970).

A new allotropic form of water ice with a density of 2.32 ± 0.15 grams per cubic centimeter has been observed at very low pressures and for temperatures lower than $100°K$. It is most likely amorphous.

022.010 Observation of a plasma polarization shift for the resonance line of ionized helium.
J. R. Greig, H. R. Griem, L. A. Jones, T. Oda.
Phys. Rev. Letters, Vol. 24, 3 - 5 (1970).

The He II 304-Å line emitted from a $N_e \approx 3 \times 10^{17}$-$cm^{-3}$, $T_e \approx 4$-eV plasma in a T-type electric shock tube shows a blue shift of ~0.05 Å. This shift is interpreted as due to the negative space charge from the polarization of plasma regions occupied by positive ions.

022.011 Synchrotron emission at strong radiative damping.
C. S. Shen.
Phys. Rev. Letters, Vol. 24, 410 - 415 (1970).

The motion and radiation of a relativistic classical charged particle in an intense magnetic field is analyzed when the radiative reaction force is comparable with the Lorentz force. Their significance for cosmic problems, especially the acceleration and radiation of cosmic-ray electrons in pulsars, is discussed.

022.012 Microwave detection of $H_2{}^{18}O$.
F. X. Powell, D. R. Johnson.
Phys. Rev. Letters, Vol. 24, 637 (1970).

Laboratory detection of a microwave absorption in $H_2{}^{18}O$ near 5.33 cm is reported. This observed signal has been assigned to a pure rotational transition between the 6_{16} and 5_{23} levels in the ground vibrational state of $H_2{}^{18}O$. Signals from these same two rotational levels for $H_2{}^{16}O$ have been detected in emission from several sources in the galaxy.

022.013 Lifetime of the $2^1 S$ state of He.
A. S. Pearl.
Phys. Rev. Letters, Vol. 24, 703 - 705 (1970).

A travelling Auger detector is used to measure the attenuation of a cooled thermal beam of He $2^1 S$ atoms in a vacuum of 1×10^{-8} Torr. Its velocity is determined by a time-of-flight technique. The lifetime obtained is $(38 \pm 8) \times 10^{-3}$ sec.

022.014 Measurement of the lifetime of the $A^1 \Pi$ state of CO by level-crossing spectroscopy.
W. C. Wells, R. C. Isler.
Phys. Rev. Letters, Vol. 24, 705 - 708 (1970).

The lifetime of the $v = 2$ level of the $A^1 \Pi$ state of CO has been measured by molecular level-crossing spectroscopy.

022.015 Quantization of Wheeler-Feynman electrodynamics.
F. Hoyle, J. V. Narlikar.
Nature, Vol. 225, 1233 - 1234 (1970).

022.016 Continuous absorption coefficients for non-hydrogenic atoms. G. Peach.
Mem. Roy. Astron. Soc., Vol. 73, (Part 1), 1 - 123 (1970).

In two earlier papers free-free and bound-free continuous absorption coefficients have been calculated for the atoms and ions C, N, O, Mg, Mg^+, Si, Cl and Ca^+ in the temperature range $4000 - 13000°K$. These tables have now been revised and extended to include He, Li, C^+, C^{+2}, N^+, N^{+2}, O^+, Na, Al, Al^+, Si^+, and K, and the range of temperature considered has been increased to 48000° K.

022.017 Difficulties with fusion catalysis by quarks.
E. E. Salpeter.
Nature, Vol. 225, 165 - 166 (1970).

The suggestion that quarks can catalyze nuclear fusion reactions in stellar interiors is re-examined. It is shown that medium heavy nuclei act as catalytic poisons by binding the quarks fairly quickly.

022.018 Second bound state for the hydrogen negative ion.
G. W. F. Drake.
Phys. Rev. Letters, Vol. 24, 126 - 127 (1970).

It is frequently stated that the $1s^2\ {}^1S$ state is the only bound state of H⁻, although Holøien estimated that the $2p^2\ {}^3P$ state may also be bound. We show by a variational calculation that the $2p^2\ {}^3P$ state is definitely bound below the $n = 2$ threshold of hydrogen. The best eigenvalue obtained is -0.125350 a.u.

022.019 Energy dependence of the 4610- and 4765-Å lines of Ar⁺ excited in low-energy He⁺ + Ar collisions.
M. Lipeles, R. D. Swift, M. S. Longmire, M. P. Weinreb.
Phys. Rev. Letters, Vol. 24, 799 - 801 (1970).

We have measured the energy dependence of the 4610- and 4765-Å lines of Ar⁺ excited in low-energy He⁺ + Ar collisions from threshold to 700 eV He⁺ energy. The 4610-Å line rises sharply with energy to a peak near threshold, while the rise of the 4765-Å line is more gradual than previously reported and than most other lines excited in this process.

022.020 Absorption spectrum of Cs I in the vacuum ultra-violet. J. P. Connerade.
Astrophys. Journ., Vol. 159, 685 - 694 (1970).

A new investigation of the absorption spectrum of Cs vapor in the 600 - 900 Å region has resulted in revisions and extension of the early work of Beutler and Guggenheimer. In particular, a prominent series has been observed. Transitions are listed; assignments are based on a new analysis of terms from the Cs II spectrum. A comparison between the Rb I and Cs I spectra is made, and two types of two-electron excitation in inner-shell absorption spectra are distinguished.

022.021 Absorption spectrum of Rb I in the vacuum ultra-violet. J. P. Connerade.
Astrophys. Journ., Vol. 159, 695 - 702 (1970).

A new investigation of the absorption spectrum of Rb vapor in the region 600 - 900 Å has yielded considerable extension of Beutler's early results, and the analysis has been revised.

022.022 Probabilities for radiation and predissociation. II. The excited states of CH, CD, and CH⁺, and some astrophysical implications. J. E. Hesser, B. L. Lutz.
Astrophys. Journ., Vol. 159, 703 - 718 = Contr. Cerro Tololo Inter-American Obs., No. 90 (1970).

Decay rates determined from data on absolute phase shift versus frequency are presented for CH, as well as for CH⁺ The spectra were produced by electron-beam excitation of low-pressure CH_4 and CD_4 gases. The formation rates of CH in interstellar space must be significantly increased to account for the observed intensities. It is also pointed out that the lack of agreement between calculations of the solar G-band intensity and equivalent-width measurements is not due to uncertainties in the molecular constants.

022.023 Hunting for quarks. E. L. Feinberg.
Priroda, No. 1.70, p. 28 - 33 (1970). In Russian.

022.024 Gravithermodynamics – III. Phenomenological non-equilibrium theory and finite-time fluctuations.

W. C. Saslaw.
Monthly Notices, Roy. Astron. Soc., Vol. 147, 253 - 278 (1970).

We inquire into the physics of a self-gravitating medium which may be far from equilibrium. A unified theory, based on propagators for macroscopic variables, describes both spatial and temporal fluctuations, and the coupling between them. The implication of the theory for the formation of galaxies is discussed.

022.025 A new description of the gadolinium spectra in the photographic infrared. N. Spector, S. Held.
Astrophys. Journ., Vol. 159, 1079 - 1090 (1970).

The gadolinium spectra excited by electrodeless discharge and a d.c. arc in air were observed in the wavelength region 7300 – 12300 Å. About 950 lines were measured, with an uncertainty of ±0.015 Å. Assignments of the lines to Gd I or Gd II are given for about 900 lines.

022.026 Revised absolute f-values for λ3247 of neutral copper and λ3720 of neutral iron.
G. D. Bell, E. F. Tubbs.
Astrophys. Journ., Vol. 159, 1093 - 1100 (1970).

Recent atomic-beam measurements of absolute f-values for the resonance lines of neutral copper and iron yield values which are about 30 percent higher than atomic-beam values reported in the past. In the case of copper, the new value is in excellent agreement with recent measurements of lifetimes. In the case of iron, some discrepancies between results obtained by different investigations still exist.

022.027 Transition probabilities for the cyanogen $B^2\Sigma^+ - X^2\Sigma^+$ transition. H. S. Liszt, J. E. Hesser.
Astrophys. Journ., Vol. 159, 1101 - 1105 (1970).

Radiative lifetimes for the first three vibrational levels of the $B^2\Sigma^+$ state in CN have been determined by using the Lawrence phase-shift experiment, with the result that $\langle\tau(v = 0, 1, 2)\rangle = 59.3 \pm 6.0$ nsec.

022.028 De-excitation of singly ionized barium ions by electron impact. J. Davis, S. Morin.
Astrophys. Journ., Vol. 159, 1125 - 1128 (1970).

We present calculations for de-excitation cross-sections of singly ionized barium ions by electron impact over the energy range 0.5 – 30 eV.

022.029 Ionization equilibria for high ions of Fe and Ni.
C. Jordan.
Monthly Notices, Roy. Astron. Soc., Vol. 148, 17 - 23 (1970).

Calculations of ionization equilibria, ionization and recombination rates for high ions of Fe and Ni are presented.

022.030 Electrodynamic broadening of spectral lines with linear Stark–effect. Yu. I. Galushkin.
Astron. Zhurn. Akad. Nauk SSSR, Vol. 47, 375 - 383 (1970). In Russian. – English translation in Soviet Astron., AJ, Vol. 14, No. 2.

The broadening of hydrogen spectral lines is considered for the case, when Doppler, Zeeman and electrodynamical effects of broadening are present simultaneously.

022.031 Quadrupole transitions in the H_2, HD and D_2 molecules. A. Birnbaum, J. D. Poll.
Journ. Atmosph. Sci., Vol. 26, 943 - 945 (1969). – Conference paper (see 012.009).

Matrix elements of the quadrupole moment, calculated in the adiabatic approximation, are given for various transitions in the 0-0, 1-0, 2-0, 3-0, 4-0 and 5-0 bands of the H_2, HD, and D_2 molecules in the ground electronic state.

022.032 Rotation-vibration quadrupole matrix elements

and quadrupole absorption coefficients of the ground electronic states of H_2, HD and D_2.
A. Dalgarno, A. C. Allison, J. C. Browne.
Journ. Atmosph. Sci., Vol. 26, 946 - 951 (1969). – Conference paper (see 012.009).

022.033 Some gf values for X-ray lines of Fe XVIII.
R. D. Chapman, Y. Shadmi, J. Oreg.
Bull. American Astron. Soc., Vol. 2, 186 - 187 (1970).
Abstr. AAS.

022.034 Relative f values from astrophysical sources.
S. J. Little, L. H. Aller.
Bull. American Astron. Soc., Vol. 2, 206 (1970). – Abstr. AAS.

022.035 Electron impact excitation and ionization rates for hydrogen. D. H. Sampson, L. B. Golden.
Bull. American Astron. Soc., Vol. 2, 216 (1970). – Abstr. AAS.

022.036 Broadening and shift of calcium lines by microfields.
H. J. Kusch, H. P. Pritschow.
Astron. Astrophys., Vol. 4, 31 - 35 (1970). In German.
Line broadening and shift by microfields of 4 neutral lines and 4 ionic lines of calcium were measured using side-on-spectroscopy.

022.037 Ionization energies and oscillator strengths for Fe XVI, Co XVII, and Ni XVIII.
R. P. McEachran, C. E. Tull, M. Cohen.
Astron. Astrophys., Vol. 4, 152 - 155 (1970).
Theoretical ionization energies for a number of doublet S, P, D, and F states of three highly ionized members of the sodium isoelectronic sequence have been calculated by means of the frozen-core approximation. The frozen-core orbital wave-functions have been employed to calculate oscillator strengths for all permitted electric dipole transitions between the various states, using both length and velocity formulations of the transition matrix element. The resulting values are generally in good agreement with each other, and may be helpful in the analysis of some Fe XVI lines observed in the solar corona.

022.038 Experimental oscillator strengths of Fe II lines and the solar iron abundance.
B. Baschek, T. Garz, H. Holweger, J. Richter.
Astron. Astrophys., Vol. 4, 229 - 233 (1970). In German.
The wall-stabilized arc burning in argon with a small admixture of iron chloride used by Garz and Kock (1969) to measure oscillator strengths of Fe I lines also allows the determination of oscillator strengths of Fe II lines. We present oscillator strengths of 14 Fe II lines based on the absolute scale of Garz and Kock for Fe I lines. A solar abundance log ϵ (Fe) = 7.63 ± 0.20, normalized to log ϵ (H) = 12.00, is derived out of ionized iron lines. A comparison with previous experimental and stellar oscillator strengths of Fe II lines is carried out and some implications on analyses of stellar spectra are discussed.

022.039 Free-free absorption coefficient of the negative hydrogen ion. J. L. Stilley, J. Callaway.
Astrophys. Journ., Vol. 160, 245 - 260 (1970).
The dipole adiabatic-exchange wave functions of polarized-orbital theory have been employed to obtain the $S \leftrightarrow P$ and $P \leftrightarrow D$ contributions to the free-free absorption coefficient of H^- in the dipole-length and dipole-velocity formulations. These results were found to be in good agreement with other recent calculations and with solar observations.

022.040 A line list for an iron-spark spectrum (10–18 Å).

L. Cohen, U. Feldman.
Astrophys. Journ. *(Letters)*, Vol. 160, L105 - L106 (1970).
A low-inductance 17–19 kV vacuum spark was used to generate spark spectra of iron electrodes. Wavelengths and visually estimated intensities are given for the region 10–18 Å.

022.041 Discrete absorption and photodissociation of molecular hydrogen. A. Dalgarno, T. L. Stephens.
Astrophys. Journ. *(Letters)*, Vol. 160, L107 - L109 (1970).
Accurate calculations are reported of the radiative lifetimes of the different vibrational levels of excited states of molecular hydrogen and of the fractions of the absorptions that terminate in the vibrational continuum of the ground electronic state leading to dissociation.

022.042 On the accuracy of machine programs for calculating oscillator strengths by Coulomb approximation.
H. Friedrich, K. Katterbach, E. Trefftz.
Journ. Quant. Spectrosc. Radiat. Transfer, Vol. 10, 11 - 16 (1970).
A comparison of the results of three different programs for the calculation of Coulomb approximation oscillator strengths is given.

022.043 Inverse bremsstrahlung in the field of the lithium atom. B. Ya'akobi.
Journ. Quant. Spectrosc. Radiat. Transfer, Vol. 10, 61 - 63 (1970).
The absorption coefficient due to inverse bremsstrahlung in the field of the lithium atom is calculated for different temperatures and wavelengths employing s-waves phase shifts, including exchange and correlation.

022.044 Relative intensity calculations for carbon dioxide. V. Relative intensities of the more abundant isotopes. L. D. G. Young.
Journ. Quant. Spectrosc. Radiat. Transfer, Vol. 10, 99 - 105 (1970).
The ratio of the rotational line intensity to the total band intensity is given for temperatures of 200°, 250° and 300° K. Only sigma-sigma transitions are considered for the isotopes $^{13}C^{16}O_2$, $^{12}C^{16}O^{18}O$, $^{12}C^{16}O^{17}O$ and $^{13}C^{16}O^{18}O$.

022.045 Measurement of the absorption strength of the methane $3\nu_3$, J-manifolds at 9050 cm^{-1}
J. S. Margolis.
Journ. Quant. Spectrosc. Radiat. Transfer, Vol. 10, 165 - 174 (1970).
The absorption strengths of the J-manifolds in the R branch of the $3\nu_3$ band of methane at 9050 cm^{-1} have been measured. A technique has been developed which satisfactorily treats the problem of the complex blending of the fine structure in the J-manifolds. The rotational temperature of the methane sample is calculated and is in good agreement with the laboratory temperature. The uncertainties in the results introduced by blending of extraneous lines and unaccounted J dependence of the half-width is discussed.

022.046 Total radiative intensity calculations for 100% CO_2 and 90% CO_2 – 10% N_2. G. H. Stickford, Jr.
Journ. Quant. Spectrosc. Radiat. Transfer, Vol. 10, 249 - 270 (1970).

022.047 The shift of the CN violet bands in the presence of argon. P. K. Henry, S. Y. Ch'en.
Journ. Quant. Spectrosc. Radiat. Transfer, Vol. 10, 293 - 295 (1970).

022.048 On the oscillator strength of the resonance transitions of Sr$^+$.
M. Hashmi, A. J. van der Houven van Oordt, J.-G. Wegrowe.

Journ. Quant. Spectrosc. Radiat. Transfer, Vol. 10, 297 - 298 (1970).

Density measurements by the resonance fluorescence scattering method and other diagnostic techniques in an Sr plasma, produced by contact ionization, yield for the ratio of the oscillator strengths of both resonance transitions of Sr^+ a value of about unity.

022.049 New molecular constants for the Phillips system of
C_2. I. R. Marenin, H. R. Johnson.
Journ. Quant. Spectrosc. Radiat. Transfer, Vol. 10, 305 - 309 = Publ. Goethe Link Obs., Indiana Univ., *Bloomington*, No. 101 (1970).

New rotational, vibrational and equilibrium constants for the Phillips system of C_2 were obtained by re-analyzing the data of Ballik and Ramsay. The agreement, even to high J values, between observed and computed wavenumbers is about 0.02 cm^{-1} when using the new rotational constants and about 0.03 cm^{-1} when using the new equilibrium constants.

022.050 Measurement of intensities of the α and β band
systems of TiO. C. Linton, R. W. Nicholls.
Journ. Quant. Spectrosc. Radiat. Transfer, Vol. 10, 311 - 314 (1970).

The intensities of the emission bands of the systems of TiO emitted in a shock tube are compared with computed synthetic band profiles in which self absorption effects have been included. It is shown that, apart from a small increase for the 3,0 sequence of the α system, the variation of the electronic transition moment over each system is negligible.

022.051 Étude des contours des raies d'absorption de quel-
ques multiplets de l'atome neutre de nickel dans le
proche ultra-violet. J. Laurent, S. Weniger.
Journ. Quant. Spectrosc. Radiat. Transfer, Vol. 10, 315 - 334 (1970).

022.052 Radiative transition probabilities between the
$3p^5 4p$ and $3p^5 4s$ configurations of neutral argon.
R. A. Nodwell, J. Meyer, T. Jacobson.
Journ. Quant. Spectrosc. Radiat. Transfer, Vol. 10, 335 - 340 (1970).

022.053 Cross sections, oscillator strengths and inelastic
widths for some NV lines.
J. Davis, S. Morin.
Journ. Quant. Spectrosc. Radiat. Transfer, Vol. 10, 357 - 364 (1970).

We present calculations of the cross sections, oscillator strengths and the inelastic Stark widths for several NV lines, using Burgess' semi-classical theory of electron-ion collisions. This method of calculating ion-broadening parameters is compared with other existing methods.

022.054 A relativistically corrected total absorption coef-
ficient and a corrected upper bound for Rosseland
opacities. H. O. Dogliani.
Journ. Quant. Spectrosc. Radiat. Transfer, Vol. 10, 511 - 513 (1970).

First order relativistic effects in the Schrödinger Hamiltonian modify the Thomas–Kuhn sum rule such that $\Sigma_c f_{ac} = Z - \delta$, where Z is the nuclear charge, f_{ac} is the oscillator strength for a dipole transition between states a and c, and δ results from relativistic effects. With this modification in the sum rule, the Bernstein–Dyson limit for the Rosseland opacity is lowered as is also the total integrated mass absorption coefficient.

022.055 Measurement of silicon II relative f-values.
J. P. Barach.
Journ. Quant. Spectrosc. Radiat. Transfer, Vol. 10, 519 -

521 (1970).

$32gf$ values for silicon II (and 4 for silicon III) have been determined to 20 per cent precision by Boltzmann plots at temperatures from 2.40 to 1.75 eV. A pulsed arc within a pyrex tube is the source. A Kerr cell shuttered spectrograph detects the radiation.

022.056 A comparison of electron impact ionization rates
for N and O ions. J. L. Kulander.
Journ. Quant. Spectrosc. Radiat. Transfer, Vol. 10, 299 - 303 (1970).

Rates of ionization for N and O ions due to electron collisions are calculated using several widely accepted approximate and semi-empirical cross sections for a wide range of electron temperatures.

022.057 Identification and classification of the
$3p^6 3d$–$3p^5 3d4s$ transitions in Co IX, Ni X, and
Cu XI.
S. Hoory, S. Goldsmith, B. S. Fraenkel, U. Feldman.
Astrophys. Journ., Vol. 160, 781 - 784 (1970).

In this work the identification and classification of the transitions above mentioned are extended from Fe VIII to Cu XI.

022.058 Theoretical energy levels in Fe V.
B. Warner.
Astron. Astrophys., Vol. 5, 1 - 3 (1970).

Predicted energy levels are given for the configurations $3d^4$ and $3d^3 4p$ in Fe V.

022.059 Partition functions and equilibrium constants for
ScO, YO and LaO. M. S. Vardya.
Astron. Astrophys., Vol. 5, 162 - 164 (1970).

Partition functions and equilibrium constants for ScO, YO and LaO have been computed for temperatures between $1000 - 8000$ °K in steps of 500 °K. The excited electronic states $A^2 \Pi$ and $B^2 \Sigma$ have also been considered and their contribution to partition function given.

022.060 Thermodynamics of strong interactions at high
energy and its consequences for astrophysics.
R. Hagedorn.
Astron. Astrophys., Vol. 5, 184 - 205 (1970).

Statistical thermodynamics in a particular form derived from high energy physics is used to describe the thermodynamical properties of what might have been our universe before its energy density became much lower than nuclear density.

022.061 On the determination of oscillator strengths from
free burning arcs. R. J. Takens.
Astron. Astrophys., Vol. 5, 244 - 263 (1970).

A number of discrepancies between the oscillator strengths given by Corliss and Bozman (1962) and by other authors has been found in recent years. To solve these problems the line emission of free burning arcs is studied in detail. The following effects are discussed: persistence of the atoms in the arc, departures from local thermodynamic equilibrium, arc structure and self-absorption. The most striking result is the large correction for the absolute scale of FeI and many other elements. These corrections remove the notorious discrepancy between coronal and photospheric abundance of Fe and Ni in the sun.

022.062 Physics in the twentieth century.
V. F. Weisskopf.
Science, Vol. 168, 923 - 930 (1970). – Review article.

022.063 Collision strengths for transitions in ions with con-
figurations $_3p^3$. S. J. Czyzak, T. K. Krueger,

P. de A. P. Martins, H. E. Saraph, M. J. Seaton.
Monthly Notices, Roy. Astron. Soc., Vol. 148, 361 - 365 (1970).

Collision strengths are given for electron-impact transitions between the levels $^4S_{3/2}$, $^2D_{3/2}$, $^2D_{5/2}$, $^2P_{1/2}$ and $^2P_{3/2}$ in ions with configuration $_3p^3$. At energies below the threshold for excitation of ^2P the collision strengths are averaged over resonances in series converging onto ^2P.

022.064 Semi-classical calculations of the collisional broadening of ion lines. D. E. Roberts.
Astron. Astrophys., Vol. 6, 1 - 17 (1970).

We report further semi-classical calculations of the collisional ("Stark") broadening of isolated lines emitted by singly charged ions perturbed by electrons and ions. The calculations are relevant to low temperature (~1 eV to 3 eV) astrophysical and laboratory plasmas.

022.065 Absolute measurement of mean lifetimes of excited states using the method of correlated photons in a cascade. C. Camhy-Val, A. M. Dumont.
Astron. Astrophys., Vol. 6, 27 - 50 (1970). In French.

The method of correlated photons in a cascade consists in measuring the time interval t_i between the spontaneous emission of two correlated photons of an excited atom in cascading deexcitation, using a time-to-pulse-height converter and a pulse-height analyser. The time intervals t_i follow the statistical law $\exp(-t_i/\tau)$ where τ is the mean lifetime of the intermediate state. We have used the present method rather than the pulsed electron beam method because of the cascading effects which may, in this case, involve unravelling several superimposed exponential decays to obtain a lifetime.

022.066 Fine structure proton excitation rates for positive ions in the $2p$, $2p^5$, $3p$, $3p^5$ series.
O. Bely, P. Faucher.
Astron. Astrophys., Vol. 6, 88 - 92 (1970).

Semi-classical collision theory is used to compute cross-sections and excitation rates for proton impact excitation of the fine structure transition in positive ions with the following configurations: $2p$, $2p^5$, $3p$, and $3p^5$. Fifteen ions are considered in each iso-electronic sequence. Excitation rates are presented in four tables for a wide range of temperatures.

022.067 The generalized Ohm's law with radiation pressure.
Y. Nakagawa, S. Kato.
Astron. Astrophys., Vol. 6, 254 - 260 (1970).

A generalized Ohm's law has been derived on the basis of the kinetic theory of an ionized gas mixture. The effect of radiation pressure is incorporated in the definition of the current density. It is shown that the effect of radiation pressure enters in the Ohm's law through the term proportional to the temperature gradient. The application of the results in relation to the thermal generation of magnetic fields by the "battery" mechanism proposed firstly by Biermann (1950) in rotating stars, is briefly discussed.

022.068 Excitation of lithium like ions by electron impacts. III. Transitions $2p \rightarrow ns$, $2p \rightarrow np$ and $2p \rightarrow nd$.
O. Bely, D. Petrini.
Astron. Astrophys., Vol. 6, 318- 321 (1970).

Coulomb-Born calculations are performed for the transitions $2p \rightarrow 3l$, $2p \rightarrow 4l$ and $2p \rightarrow 7l$ ($l = 0, 1, 2$) in Be II, N V, Ne VIII and the limiting hydrogenic case $Z = \infty$. Graphs are given which enable to get a cross-section $2p \rightarrow nl$ ($l = 0, 1, 2$) for any n and any lithium like ion. The results are discussed.

022.069 Absolute intensity calibration methods in the vacuum UV region. G. Boldt.
IAU Symposium No. 36, (see 012.014), p. 5 - 11 (1970).

As a summary of the principal results presented at the ESRO symposium on Calibration Methods in the Vacuum Ultra Violet (Munich, 1968) a description is given of three different absolute intensity calibration methods.

022.070 Comments on recent Stark-broadening calculations for the Hα and Hγ lines of hydrogen.
M. E. Bacon, D. F. Edwards.
Journ. Quant. Spectrosc. Radiat. Transfer, Vol. 10, 563 - 569 (1970).

Recent Stark-broadening calculations of the lineshapes for the Hα and Hγ hydrogen lines have been re-evaluated. Consideration is given to the apparent disagreement with respect to the role played by lower-state broadening and "strong collision" effects in accounting for the improved agreement between measured and calculated halfwidth values. The results indicate that further work is necessary on the strong collision and quadrupole contributions to the linewidth.

022.071 Franck-Condon factors and r-centroids for NO$^+$ CP, SiF, BF, BCl and BBr.
T. Wentink, Jr., R. J. Spindler, Jr.
Journ. Quant. Spectrosc. Radiat. Transfer, Vol. 10, 609 - 619 (1970).

Franck-Condon factors, based on the Morse potential, are given for various electronic transitions in the molecules NO$^+$, CP, SiF, BF, BCl and BBr. The corresponding r-centroids are reported or discussed.

022.072 Franck-Condon factors and r-centroids for the gamma system of NO.
R. J. Spindler, Jr., L. Isaacson, T. Wentink, Jr.
Journ. Quant. Spectrosc. Radiat. Transfer, Vol. 10, 621 - 628 (1970).

Franck-Condon factors and r-centroids for the gamma bands of NO have been recalculated using a more recent RKR-potential curve for the $X^2\Pi$ ground state than was employed in an earlier study. The resulting Franck-Condon factors are appreciably different from the earlier values and should lead to good correlation of absolute-intensity measurements with theoretical predictions.

022.073 Computed profiles of He I 5016 Å at high electron densities. A. J. Barnard, J. Cooper.
Journ. Quant. Spectrosc. Radiat. Transfer, Vol. 10, 695 - 702 (1970).

Profiles of He I 5016 Å and its forbidden component have been calculated and are tabulated for electron densities from 3×10^{16} to 6×10^{17} cm^{-3} and temperatures from 10^4 to 8×10^4 °K. The ion broadening is treated in the quasi-static approximation and the electron broadening of the overlapping Stark components in the impact approximation.

022.074 Experimental measurements of some ArII transition probabilities and a comparison of published values.
N. M. Nerheim, H. N. Olsen.
Journ. Quant. Spectrosc. Radiat. Transfer, Vol. 10, 755 - 773 (1970).

Transition probabilities of a number of ArII lines have been determined experimentally from observing a 1.1 atm free-burning arc. Techniques used in previous work were modified to allow observation of lines in the UV spectral region.

022.075 Absorption coefficient of H$^-$ and interparticle interaction – I. Debye-Hückel approximation.
S. P. Tarafdar, M. S. Vardya.
Journ. Quant. Spectrosc. Radiat. Transfer, Vol. 10, 789 - 797 (1970).

We have attempted here to estimate the effect of the perturbation due to charged particles under the Debye-Hückel approximation on the bound-free absorption coefficient of

H⁻, because of its importance in late-type dwarf stars.

022.076 **Investigation of pressure broadening of a neon line using a Zeeman scanning technique.**
J. Meyer, J. C. Burnett, B. Stansfield.
Journ. Quant. Spectrosc. Radiat. Transfer, Vol. 10, 799 - 804 (1970).
Pressure broadening of the neon line λ6074.3 Å by neon atoms is investigated using a highly resolving Zeeman scanning technique. The results indicate that the broadening can be explained by perturbing van der Waals interactions with neighbouring ground state atoms.

022.077 **Production of seemingly Doppler–shifted radiation by time- and space-dependent radiant-energy sources.**
S. S. Penner.
Journ. Quant. Spectrosc. Radiat. Transfer, Vol. 10, 831 - 833 (1970).
An optical arrangement is sketched which will produce monochromatic beams with frequencies that may be interpreted as corresponding to Doppler–shifted radiation.

022.078 **Measurement of nitrogen II and carbon II resonance multiplet line shapes from a plasma.**
J. D. E. Fortna.
Naval Res. Lab., Washington, D. C., NRL Report 6950, 82 pp. (1969).
The spectral line shapes of the nitrogen II and carbon II resonance multiplets, which occur in the vacuum ultraviolet region of the spectrum, have been measured. The Stark widths were deduced from the optically thin Lorentz wings.

022.079 **Experimental and theoretical electron impact broadening of some Mg II and Ca II lines of astrophysical interest.** J. Chapelle, S. Sahal-Bréchot.
Astron. Astrophys., Vol. 6, 415 - 422 (1970).
The electron impact profiles of the Mg II ($3s - 3p$, 2802.7 Å and $3d - 4f$, 4481.3 Å) and Ca II ($4s - 4p$, 3933.7 Å and $3d - 4p$, 8542 Å) lines have been measured in an argon plasma jet and have been calculated within the impact and semi-classical approximations. The agreement between experiments and calculations is good.

022.080 **Bremsstrahlung, synchrotron radiation, and Compton scattering of high-energy electron traversing dilute gases.** G. R. Blumenthal, R. J. Gould.
Rev. Modern Phys., Vol. 42, 237 - 270 (1970).
Expressions are derived for the total energy loss and photon-production spectrum by the processes of Compton scattering, bremsstrahlung, and synchrotron radiation from highly relativistic electrons. For Compton scattering, the general case, the Thomson limit, and the extreme Klein-Nishina limit are considered. Bremsstrahlung is treated for the cases where the electron is scattered by a pure Coulomb field and by an atom. For the latter case the effects of shielding are discussed extensively.

022.081 **Expansion formulas for harmonic oscillators in three-dimensional space.** H. Kinoshita.
Univ. Tokyo, Tokyo Astron. Obs. Report, Vol. 15, (No. 2), (No. 57), p. 361 - 372 (1970). In Japanese.

022.082 **Une approximation pour le traitement des collisions proches dans l'élargissement Stark des raies d'hydrogène.** Nguyen-Hoe, M. Caby.
Comptes Rendus Acad. Sci. Paris, Sér. B, Vol. 270, 1657 - 1660 (1970).

022.083 **Inverse Compton-effect of radiation on relativistic electrons.** V. M. Charugin.
Astron. Zhurn. Akad. Nauk SSSR, Vol. 47, 664 - 666 (1970).

In Russian. English translation in Soviet Astron. AJ, Vol. 14, No. 3.
The expression for the Compton cross section is generalized in the case, when the energy of a scattered quantum is smaller than the energy of a falling quantum. An expression is obtained for the emission flux produced by scattering of its synchrotron radiation on relativistic electrons.

022.084 **A global theory of relativity under the 5-dimensionally doubly extended Lorentz transformation group with oriented II-geodesic spheres as particle models and its Hamilton's canonical formalism.**
T. Takasu.
Sci. Rep. Tôhoku Univ. Sendai, Japan, First Ser., Vol. 52, 154 - 253 (1969).

022.085 **Astrospectra in laboratory.**
L. A. Mitrofanova.
Priroda, No. 5.70, p. 31 - 38 (1970). In Russian.

022.086 **A theoretical model for the interaction between excited and ground state atoms. Application to pressure broadening.** E. Roueff.
Astron. Astrophys., Vol. 7, 4 - 9 (1970).
A theoretical model for the long range interaction between excited and ground state atoms is derived after a paper from Smirnov (1967) in a more reliable form. Applications have been made to pressure broadening of sodium lines by helium.

022.087 **Quarks and astronomy.**
F. Nichitiu.
Stud. Cerc. Astron., Vol. 15, 105 - 111 (1970).
The properties of quarks and the actual methods for finding quarks especially the astronomical methods under the assumption of quark-atoms and quark-ions are discussed in a qualitative way. The second part deals with the problem of quarks participating in quasar's dynamics.

022.088 **Radio frequencies for $N^{14}O^{16}$.**
R. Neumann.
IAU Circ. No. 2203 (1970).

022.089 **Le tracé des rayons dans un milieu anisotrope, non permanent et légèrement absorbant dans le cadre d'une théorie hamiltonienne à quatre dimensions.**
L. G. Bossy.
U. F. S. I. Symposium on Electromagnetic Waves, 1968, 4 pp. = Publ. Inst. Astron. Géophys. G. Lemaître, Louvain, Vol. 3, No. 2 (1969).

022.090 **Darstellungs- und Transformationstheorie quasi-monochromatischer Strahlungsfelder. – Zusammenfassender Bericht.** M. A. F. Thiel.
Beiträge Radioastronomie, Max-Planck-Inst. Radioastronomie, Bonn, Vol. 1, (No. 5), 111 - 146 (1970).
The purpose of this report is to give a coherent derivation of several representations of the polarization-state of quasi-monochromatic radiation fields, the relations between these representations, and the transformation of the polarization-state when the radiation is transmitted through an instrument.

022.091 **Pressure and internal energy of relativistic partially degenerate electron gas.** U. Uus.
Nauch. Informatsii, vyp. (No.) 13, p. 110 - 115 (1969). In Russian.
Density, pressure and internal energy per cm³ have been obtained numerically from the Fermi-Dirac integrals as a function of temperature and the degeneracy parameter Ψ. Effects of relativity and pair-creation are included.

022.092 **Fundamental spectroscopic data.**
M. V. Migeotte.
Trans. IAU, Vol. 14A, (see 003.028), 125 - 140 (1970).
'Report of Commission 14.

022.093 **Synchrotron radiation from mildly relativistic electrons.** J. P. Wild.
Proc. Astron. Soc. Australia, Vol. 1, 348 - 350 (1970).

022.094 **Polarization and exchange effects in low-energy electron-H_2 scattering.** R. J. W. Henry, N. F. Lane
Phys. Rev., Second Series, Vol. 183, 221 - 231 = Contr. Kitt Peak National Obs., No. 424 (1969).

In this paper, we investigate the effects of polarization and exchange terms on elastic and rotational excitation cross sections for scattering by molecular hydrogen of electrons with energies less than 10 eV.

022.095 **Transition probabilities and collision broadening cross section of the N_2 Lyman−Birge−Hopfield system.** D. E. Shemansky.
Journ. Chemical Phys., Vol. 51, 5487 - 5494 = Contr. Kitt Peak National Obs., No. 483 (1969).

022.096 **Double photoionization of helium.** R. L. Brown.
Phys. Rev., Ser. A, Vol. 1, 586 - 590 = National Radio Astron. Obs., *Green Bank*, Repr. Ser. A, No. 155 (1970).

The helium double-photoionization cross section has been evuluated using a Hylleraas wave function for the ground state in order to include correlation between the atomic elec - trons more directly than in previous calculations.

022.097 **The hydrogen molecule − comparison between theory and experiment.**
P. G. Wilkinson, H. O. Pritchard.
Canadian Journ. Phys., Vol. 47, 2493 - 2508 (1969).

022.098 **Structure géométrique de l'état électronique $\tilde{C}$ de la molécule SO_2.**
I. Dubois.
Journ. Molecular Structure, Vol. 3, 269 - 281 = Univ. Liège, Inst. d'Astrophys., Coll. 8°, No. 580 (1969).

022.099 **Quelques remarques relatives à un fonctionnement particulier d'une cathode creuse en graphite.**
F. Remy.
Comptes Rendus, Acad. Sci. Paris, Sér. B, Vol. 268, 1229 - 1232 = Univ. Liège, Inst. d'Astrophys., Coll. 4°, No. 194 (1969).

022.100 **Développement d'une cathode creuse en graphite a température contrôlable destinée à l'étude spectroscopique de molécules carbonées et silicées d'intérêt astrophysique.** F. Remy.
Rev. Phys. Appliquée, Vol. 4, 375 - 377 = Univ. Liège, Inst. d'Astrophys., Coll. 4°, No. 198 (1969).

022.101 **Sur la possibilité d'utilisation de sources phosphores-centes excitées par un élément radioactif pour la calibration en campagne.** J. M. Vreux, G. Marette.
Rev. Phys. Appliquée, Vol. 4, 421 - 422 = Univ. Liège, Inst. d'Astrophys., Coll. 4°, No. 199 (1969).

Instruments and Astronomical Techniques

031 Optics, Methods of Observation and Reduction

031.001 **A test for figuring Cassegrain secondary mirrors.**
J. L. Richter.
Sky Telescope, Vol. 39, 49 - 53 (1970).

031.002 **Die Informationsleistung astronomischer Teleskope.**
H.-U. Keller.
SuW, Vol. 9, 9 - 11 (1970).

031.003 **Die visuelle Beobachtung schwach leuchtender Objekte.**   R. Eckert.
SuW, Vol. 9, 18 - 19 (1970).

031.004 **Theorie und Anwendung des Ringmikrometers.**
L. Brandt.
SuW, Vol. 9, 47 - 48, 50 (1970).

031.005 **Ein wellenoptisches Kriterium für Fokussierungstoleranzen.**   W. Weiss.
Sterne, 45. Jahrgang, p. 226 - 230 (1969).

031.006 **Die Selbstherstellung eines leistungsfähigen Teleskopspiegels für Amateurastronomen.**
R. Jetschke, G. Jetschke.
Sterne, 45. Jahrgang, p. 230 - 238 (1969).

031.007 **New parametric tables for reflecting telescopes for aplanats, anastigmats, flat-fielded, distortion free and third order errors free systems.**
J. Landi Dessy, A. Niell, A. Puch.
Bol. Inst. Mat. Astron. Fis. Univ. Nacional Córdoba, Vol. 3, No. 1, p. 3 - 43 = Tirada aparte Obs. Astron. Córdoba, No. 165 (1969).

031.008 **The telescopic resolution of disk light sources.**
G. J. Kirby.
Journ. British Astron. Ass., Vol. 80, 130 - 132 (1970).

031.009 **Moon illusion explained on the basis of relative size.**   F. Restle.
Science, Vol. 167, 1092 - 1096 (1970).
The moon looks small overhead not because it seems close but because of the broad extent to the horizon.

031.010 **Optimal conditions for registration of star transit moments.**   R. Kalnina.
Uch. Zap., *Riga,* Vol. 121, 11 - 27 (1969).   In Russian.
A new method of lag determination for a photoelectric device for registration of star transits is set up. Light intensity distribution in star image is taken into account.

031.011 **Objective with increased spherical aberration for the photoelectric transit instrument APM-10.**
M. Ābele.
Uch. Zap., *Riga,* Vol. 121, 28 - 34 (1969).   In Russian.
An objective with increased spherical aberration but corrected chromatic aberration and coma is calculated by the method of ray-tracking analysis.

031.012 **On methods of lag determination for photoelectric registration of star transits.**
K. Šteins.
Uch. Zap., *Riga,* Vol. 121, 35 - 48 (1969).   In Russian.
A comparison of a method in which a photocurrent is approximated by means of harmonic analysis, and another in which the star image is dispersed into uniformly lighted circles, yields equal results.

031.013 **L'astronomie à travers les siècles. IX. – Les yeux de verre des astronomes. B. – La plaque photographique.**   E.-H. Geneslay.
L'Astronomie, 84e année, p. 153 - 175 (1970).

031.014 **On the limiting magnitude of telescopes.**
G. Chincarini.
Mem. Soc. Astron. Italiana, Nuova Serie, Vol. 41, 121 - 124 (1970). – Letter.

031.015 **Notas para el aficionado el telescopio - su funcionamiento.**   J. C. Forte.
Revista Astron., Vol. 41, No. 171, p. 36 - 40 (1969).

031.016 **On the application of the photographic method of equidensites to spectrophotometry of solar formations.**   A. A. Rybina.
Solnechnye Dannye 1969 Byull., No. 10, p. 111 - 115 (1970). In Russian.
The photographic method of equidensites as applied to photometry of spectral lines is described. The profiles of three lines were constructed using the equidensity maps and the records with the Mϕ–4. They have been found in good agreement.

031.017 **Some methodic problems of astrophotography.**
A. B. Paley.
Astron. Zhurn. Akad. Nauk SSSR, Vol. 47, 206 - 210 (1970). In Russian. English translation in Soviet Astron. AJ, Vol. 14, No. 1.
Since the lunar light scattered by the atmosphere is polarized to a certain extent it is proposed to photograph the sky (at moon shine) with a polaroid, orientated perpendicularly to a plane of the scattered light polarization. A method is proposed for the determination of stellar magnitudes by a microphotometer with linear scale.

031.018 **Kern Objektive auf dem Mond.**
H. Labhart.
Orion Schaffhausen, 28. Jahrgang, p. 20 - 21 (1970).

031.019 **Teleskopspiegel in Skelett-Bauweise.**
A. Hoffmann.
Orion Schaffhausen, 28. Jahrgang, p. 43 - 44 (1970).

031.020 **How to construct an amateur telescope.**
L. Newelski.
Urania Kraków, Vol. 41, 168 - 172 (1970).   In Polish.

031.021 **The imaging properties of the rotation collimator.**
A. P. Willmore.
Monthly Notices, Roy. Astron. Soc., Vol. 147, 387 - 403 (1970).

Properties of the rotation collimator as a method of detecting point sources at infinity, such as cosmic X-ray sources, are discussed. Gratings of various types are considered, and it is shown that the image formed has a central and satellite peaks, each surrounded by circular fringes. Varying the mask-space ratio affects the amplitude of the satellite peaks; by super-imposing gratings with ratios of 1 : 1 and 1 : 2 a maximum number of them are suppressed. The angular size of the central image is of the order of a/l where a is the ruling period and l the separation of the masks. Next, possible two-dimensional patterns are analysed, and it is shown that both a chequerboard pattern and a grid of circular holes have suitable propterties. To obtain the necessary flexibility image reconstruction in a computer is to be preferred to an analogue method.

031.022 Attainment of diffraction limited resolution in large telescopes by Fourier analysing speckle patterns in star images. A. Labeyrie.
Astron. Astrophys., Vol. 6, 85 - 87 (1970).
Resolution in excess of the limitation set by seeing, and reaching the diffraction value, can be obtained on star features by laser processing the speckle pattern observed in short exposures made with a large telescope. The technique may be considered as an extension of the Michelson stellar interferometry; it is applicable to star diameter measurements and stellar system studies.

031.023 An amateur-built precision mirror tester. J. J. Woerner.
Sky Telescope, Vol. 39, 389 - 393 (1970).

031.024 The limiting photographic magnitudes of stars for modern telescopes.(The penetrating power of a telescope is without reduction for aberrations).
O. A. Melnikov, A. N. Gerashchenko.
Izv. Glav. Astron. Obs. v Pulkove, No. 185, p. 248 - 267 (1970). In Russian.
A review of literature is given and the formulation of the problem is discussed. The data on the limiting stellar magnitude (treshold) obtained from observations are given and numerical evaluations of the limiting magnitude in dependence on the parameters of the telescopes and the sizes of the tremor disc are made.

031.025 On the resolving power of telescopes. E. V. Morozova.
Izv. Glav. Astron. Obs. v Pulkove, No. 185, p. 268 - 278 (1970). In Russian.
A brief account is made of all literature dealing with the problem of resolving power of telescopes. The practical resolving power of telescopes was evaluated from binaries with different magnitude of the components and different tremor discs caused by turbulence of the earth's atmosphere. For the sake of simplicity it is supposed that the intensities in the star images are simply summed up. The results of the calculation are given in tables and figures. It follows from these that turbulence of the earth's atmosphere limits the theoretical resolving power of large telescopes.

031.026 Note on the design of two Ross type photographic objectives.
A. Cornejo, J. Castro, D. Malacara.
Bol. Obs. Tonantzintla y Tacubaya, Vol. 5 (No. 34), 241 - 245 (1970).

031.027 Investigations on the division diameters of the meridian circle. I. Rusu.
Stud. Cerc. Astron., Vol. 15, 17 - 27 (1970).
In this work the author wants to find, theoretically and practically at a divided circle of a meridian instrument, the diameter correction η. This represents for two diametrical opposite divisions the difference of a division to the diameter passing through another division and its dividing center.

031.028 The influence of diameter correction η in determining eccentricity elements of a divided circle.
I. Rusu.
Stud. Cerc. Astron., Vol. 15, 29 - 36 (1970). In Roumanian.
In this work the author estimates the precision with which are determined the eccentricity elements φ, 2μ and γ of a divided circle of a meridian telescope if the diameter correction η is considered.

031.029 Vision in astronomical instruments and the observation of planetary surfaces. J. Dragesco.
Journ. Astron. Soc. Western Australia, Vol. 21, May,p. 5 - 6; June, p. 3 - 5 (1970).

031.030 Making a 6-inch air-spaced visual objective.
O. R. Knab.
Sky Telescope, Vol. 40, 46 - 53 (1970).

031.031 Optics: Yesterday's science — today's engineering.
H. Silver.
Design News, Vol. 24, No. 14, 6 pp. = Contr. Kitt Peak National Obs., No. 449 (1969).

031.032 Objective with increased spherical aberration for the photoelectric transit instrument APM-10.
M. K. Abele.
Zinatn. rakstu karjums. Latv. univ., Uch. zap. Latv. un-t, Vol. 121, 28 - 34 (1969). In Russian. — Abstr. in Referativ. Zhurn. 51. Astron., 7.51.80 (1970).

031.033 On the determination of the optical center of astronomical photos and of the focal length of a camera. D. B. Uvarov.
Trudy In-t prikl. geofiz. Gl. upr. gidrometeorol. sluzhby pri Sov. Min. SSSR, 1969, vyp. (No.) 15, p. 202 - 207. In Russian. Abstr. in Referativ. Zhurn. 51. Astron., 7.51.188 (1970).

031.034 Determination of the diameters of star images by means of an oscillograph. A. I. Yazev.
Trudy metrol. in-tov SSSR, 1969, vyp. (No.) 106 (166), p. 168 - 171. In Russian. — Abstr. in Referativ. Zhurn. 51. Astron., 7.51.189 (1970).

031.035 X-ray imaging with multiple-pinhole cameras using a posteriori holographic image synthesis.
G. W. Stroke, G. S. Hayat, R. B. Hoover, J. H. Underwood.
Optics Commun., Vol. 1, (No. 3), 138 - 140 (1969).
A new form of 'extended-source' Fourier-transform holography may be used to synthesize into a single image the multiplicity of N identical images obtained by a multiple-pinhole camera in X-ray astronomy.

031.036 The role of optics in the Apollo program.
L. R. Lankes.
Optical Spectra, Vol. 3, No. 5, p. 35 - 57 (1969).
Some of the more significant and unusual aspects of the application of optics and optical devices are described, including radio telescopes, lunar photographic cartography, and cameras and other optical instruments for investigating lunar structure and phenomena in situ.

031.037 Optical polishing of metal surfaces.
A. Grigórieff, A. De Palo.
Asoc. Argentina Astron. Bol., No. 12, (see 012.019), p. 35 - 37 (1968). In Spanish. — Abstract.

031.038 The effect of artificial dark clouds on mean star

counts.   H. Wilkens.
Asoc. Argentina Astron. Bol., No. 14 (see 012.020), p. 33 - 36 (1968).   In Spanish.

Zur Objektivierung der Beobachtung von Sterndurchgängen durch den Meridian.
See Abstr. 041.027.

Some results of clock correction determinations from observations of pairs of stars situated symmetrically around the zenith.   See Abstr. 044.007.

A control method of the determination of standard time.   See Abstr. 044.008.

On the reduction of astronomical determinations of latitude, longitude and azimuth to a common epoch.
See Abstr. 045.008.

A method of observation and computation for approximate determination of astronomic latitudes and longitudes.   See Abstr. 046.006.

A method for higher precision of the coordinates of the observational place.   See Abstr. 046.013.

I. Holographic image-deblurring and aperture-synthesis methods: Applications in X-ray astronomy. II. High-resolution synthesized-aperture imaging using superposition of partial-resolution images in single photograph.
See Abstr. 076.016.

A method of measuring the slope of the solar radio emission spectrum and its application to observations during the eclipse of September 22, 1968 in Gorki.
See Abstr. 077.010.

Schilt's method: Application to stellar photometry with the electronic camera.   See Abstr. 113.030.

032 Astronomical Instruments

032.001 **The 236-inch Soviet reflector.**   V. Lutsky.
Sky Telescope, Vol. 39, 99 (1970).

032.002 **A folded Herschelian off-axis reflector.**
J. R. Pawlick.
Sky Telescope, Vol. 39, 191 - 192 (1970).

032.003 **Die astronomische Ballonsonde THISBE.**
D. Lemke.
SuW, Vol. 9, 29 - 32 (1970).

032.004 **The construction of astronomical instruments.**
U. Vejsmann.
Astron. Geod. Eston. SSR, Tartu, (see 003.004), p. 99 - 109
(1969).   In Russian.

032.005 **Le télescope de 137 cm à miroir métallique de
l'Observatoire de Milan-Merate.**
G. de Mottoni.
L'Astronomie, 84ᵉ année, p. 69 - 74 (1970).

032.006 **A Phoenix amateur's 12 1/2-inch Schmidt-Casse-
grain.**   M. Kaufman.
Sky Telescope, Vol. 39, 254 - 260 (1970).

032.007 **The division errors of the meridian circle.**
H. Yasuda, H. Hara, R. Fukaya, T. Ina.
Tokyo Astron. Obs. Rep., Vol. 15, No. 1 (No. 56), p. 135 -
157 (1970).   In Japanese.

032.008 **Photoelectric zenith tube for observations of stars
of equal zenith distances.**   M. Ābele.
Uch. Zap., *Riga*, Vol. 121, 49 - 105 (1969).   In Russian.
The technical and optical parameters of the new
zenith tube are given. First observational results show the
suitability of the instrument for the purpose of time service.

032.009 **Grandes reflectores modernos.**   J. Landi Dessy.
Revista Astron., Vol. 41, No. 170, p. 15 - 26
(1969).

032.010 **Scattered light in a solar telescope and the solar
halo.**
V. I. Bulavina, N. I. Kozhevnikov, E. A. Makarova.
Solnechnye Dannye 1969 Byull., No. 12, p. 79 - 88 (1970).
In Russian.
The results of measurements of scattered light in the
focal plane of a solar telescope are given. It is found that the
brightness of the scattered light is mainly due to scattering by
the telescopic mirrors. For λ 6300 Å the brightness of the
scattered light at a distance of 5' from the solar limb is less
than 5×10^{-4} of the brightness of the solar disk center.

032.011 **Universal three-axial instrument.**
Z. Kordylewski.
Artificial Satellites Polish Acad. Sci., Vol. 5, No. 1, p. 27 -
31 (1969).
The instrument has three axes, each provided with a
divided circle: a vertical axis, a horizontal one and a third one,
perpendicular to the latter, which holds a small AT-1 telescope, rotating around it.

032.012 **Joint determination of the distortion and the centre
of distortion in the satellite camera.**
I. Gajderowicz, S. Oszozak.
Artificial Satellites Polish Acad. Sci., Vol. 5, No. 1, p. 37 -
42 (1969).

The method consists in comparing the measured coordinates with the standard coordinates of stars. After transforming the standard coordinates the differences O–C are computed.

032.013 **The meridian circle of the Kiev University Observa-
tory.**   N. A. Chernega.
Vestn. Kiev. Univ., Ser. Astron., No. 11, p. 120 - 128 (1969).
In Russian.

032.014 **Optimum size of infrared photometric telescopes.**
H. L. Johnson, W. L. Richards.
Astrophys. Journ. *(Letters)*, Vol. 160, L111 - L116 (1970).
We have estimated the costs of construction for alu-
minum-mirror photometric telescopes ranging up to 120
inches in diameter, based upon our combined experience
in constructing such telescopes.

032.015 **The latest flight of Stratoscope II.**
J. Ashbrook.
Sky Telescope, Vol. 39, 365 - 367 (1970).

032.016 **A portable 6-inch Dall-Kirkham and celestial
camera.**   K. Moll.
Sky Telescope, Vol. 39, 394 - 396 (1970).

032.017 **The spectroheliograph of the Observatorio de
Física Cósmica, San Miguel, Argentina.**
T. Paneth, J. Seibold.
Inform. Bull. Southern Hemisphere, No. 15, p. 30 - 34 (1969).

032.018 **La grande lunette de l'Observatoire de Nice.**
P. Couteau.
L'Astronomie, 84ᵉ année, p. 213 - 216 (1970).

032.019 **On the progress of balloon heliophysics in the
Soviet Union.**
V. A. Krat, V. N. Karpinsky, V. M. Sobolev, L. Z. Dulkin,
B. N. Motenko, P. A. Khalezov.
Izv. Glav. Astron. Obs. v Pulkove, No. 185, p. 124 - 131
(1970).   In Russian.
The advantages of astronomical observations by means
of stratospheric balloons as compared with ground observations are discussed. The importance of obtaining uniform
series of photographs of the solar photosphere is stressed,
which is necessary for investigations of dynamics of the physical processes on the sun and for eliminating the telluric lines
of water vapour from the spectrum of celestial bodies. The
first Soviet balloon station is described; results of observations
are given.

032.020 **The six-canal solar magnetograph of the Pulkovo
Observatory.**   R. N. Ikhsanov, Y. P. Platonov.
Izv. Glav. Astron. Obs. v Pulkove, No. 185, p. 132 - 158
(1970).   In Russian.
A description of a new magnetograph of the Pulkovo
Observatory is given. The magnetograph permits to get information, directly and simultaneously, on polarization in the
line under investigation in the Stokes parameters (J, V, Q) and
azimuth χ. Besides, it is used for recording the radial velocities
and brightness in the continuous spectrum.

032.021 **Some problems of the theory of the alt-azimuthal
mounting of a telescope. II. The influence of errors
in the initial data and alt-azimuthal mounting on the precision
of operating a telescope.**   N. N. Mikhelson.
Izv. Glav. Astron. Obs. v Pulkove, No. 185, p. 279 - 304

(1970).   In Russian.

The influence of instrumental and methodical errors on the precision of setting and guiding the telescope, installed on the alt-azimuthal mounting are considered.

032.022 Étude des variations de la collimation de l'instru-ment des passages "Bamberg" No. 63131 d'après des observations astronomiques.   Z. Brkić, D. Djurović. Bull. Obs. Astron. Beograd, Vol. 27, No. 2, p. 33 - 42 (1969).

032.023 Examen du niveau à bulle de l'instrument des passages "Bamberg" et du niveau No. 65672 de l'instrument des passages "Hildebrand". D. Djurović. Bull. Obs. Astron. Beograd, Vol. 27, No. 2, p. 52 - 57 (1969).

032.024 Quelques problèmes sur la détermination de la correction du tour de la vis micrométrique de la lunette zénithale à Belgrade.   M. Djokić. Bull. Obs. Astron. Beograd, Vol. 27, No. 2, p. 82 - 86 (1969).

032.025 Report on the 2.7-meter reflector. H. J. Smith. IAU Circ. No. 2209 (1970).

032.026 Die kleinen Fernrohre.   R. Brandt. Orion Schaffhausen, 28. Jahrgang, p. 84 - 90 (1970).

032.027 New Soviet cameras for photographic observations of artificial celestial bodies. A. G. Masevich, A. M. Lozinskij. Vestn. AN SSSR, No. 2, p. 38 - 44 (1970).   In Russian. Abstr. in Referativ. Zhurn. 62. Issled. kosm. prostranstv., 6.62.250 (1970).

032.028 The horizontal solar telescope ATSU-5 with a double diffraction spectrograph. E. A. Gurtovenko. Astrometriya i Astrofiz., *Kiev*, No. 8, (see 003.024), p. 77 - 84 (1969).   In Russian.

The optical system and the work on the rearrangement of the telescope and spectrograph are described briefly.

032.029 Der Selbstbau einfacher astronomischer Meßinstru-mente.   E. Lehmann. Sterne, 46. Jahrgang, p. 109 - 119 (1970).

032.030 The spar telescope of Lockheed Solar Observatory. G. A. Carroll. Sky Telescope, Vol. 40, 10 - 13 (1970).

032.031 The great Melbourne telescope.   L. A. Jones. Journ. Astron. Soc. Western Australia, Vol. 21, March,p. 3 - 7 (1970).

032.032 Instruments and techniques. J. Rösch, G. Wlerick. Trans. IAU, Vol. 14A, (see 003.028), 53 - 70 (1970). Report of Commission 9.

032.033 Photoelectric double-beam telescope with changing base.   S. B. Novikov. Astron. Tsirk., No. 554, p. 6 - 8 (1970).   In Russian.

032.034 How to build a Cassegrain telescope. S. D. Chuvakhin. Zemlya i Vselennaya, No. 3, p. 86 - 90 (1970). In Russian.

032.035 Some results of an examination of the satellite came-ra AFU-75 at the Uzhgorod tracking station. M. V. Bratijchuk,, A. G. Kirichenko.

Byull. Stantsij Optichesk. Nablyud. Iskusstv. Sputnikov Zemli, No. 53 (1), p. 29 - 33 (1969).   In Russian.

032.036 Investigation of the stability of the new azimuthal mounting of the NAFA-3c/25-c camera. A. A. Logvinenko, T. V. Rad'o, I. I. Terebushko. Tsirk. L'vov. Astron. Obs., No. 44, p. 43 - 46 (1970). In Russian.

032.037 New Soviet photographic cameras for tracking artificial celestial bodies.   A. G. Masevich, A. M. Lozinskij. Vestn. AN SSSR, No. 2, p. 38 - 44 (1970). In Russian. – Ab - str. in Referativ. Zhurn 51. Astron., 7.51.186 (1970).

032.038 Attempt to determine the distortion of the objec - tive of the AFR-1-type wide-angle astrograph at the Sternberg Astronomical Institute.   L. I. Khrushchev. Trudy metrol. in-tov SSSR, 1969, vyp. (No.) 106 (166), p. 160 - 162. In Russian. – Abstr. in Referativ. Zhurn. 51. Astron., 7.51.187 (1970).

032.039 Quantitative star-gazing or measuring the size of stars.   R. Hanbury Brown. Australian Journ. Sci., Vol. 32, (No. 4), 117 - 125 (1969).

Description of Narrabri intensity interferometer, and a discussion of results obtained with it. – *FFG*

032.040 Determination of the distortion of the satellite camera in Borowiec.   I. Gajderowicz, S. Oszczak. Materiały i Prace, Vol. 32 (Publ. Inst. Geophys. Polish Acad. Sci.), 35 - 49 = Polish Acad. Sci. Astron. Latitude Station Borowiec, Repr. No. 36 (1969).

032.041 Inductosyn angular readout system of the U. S. Naval Observatory six-inch transit circle. B. L. Klock, R. Z. Geller, M. A. Dachs. Proc. of Electro-Optical Systems Design Conference, 1969, p. 633 - 641 = U. S. Naval Obs. Repr., No. 107 (1970).

032.042 Modern ideas in the construction of large reflecting telescopes.   J. Landi Dessy. Asoc. Argentina Astron. Bol., No. 12, (see 012.019), p. 9 (1968).   In Spanish. – Abstract.

032.043 An astrometrical aplanatic telescope with a figurated flat secondary mirror.   J. Landi Dessy. Asoc. Argentina Astron. Bol., No. 14 (see 012.020), p. 18 - 28 (1968).

032.044 Temperature effects on the large transit instrument. Lj. Mitić, I. Pakvor. Publ. Obs. Astron. Beograd, No. 16, (see 012.021), p. 27 - 33 (1969).   In Serbo-Croatian.

Variations of instrumental parameters are dictated by temperature variations of corresponding periods.

032.045 About the telescope flexure as the one of most important problems of vertical circles. M. Mijatov. Publ. Obs. Astron. Beograd, No. 16, (see 012.021), p. 35 - 39 (1969).   In Serbo-Croatian.

The report points out to the importance of a study of vertical circles flexure giving a short outlook about the observational methods and results obtained in research on flexure of existing vertical circles.

Bahnverfolgungs-Radargeräte von Thomson-CSF. Weltraumfahrt, 21. Jahrgang, p. 54 - 55 (1970).

Telescopes for Skygazing.   See Abstr. 003.104.

033 Radio Telescopes and Equipment

033.001 **Radioteleskope zur Erforschung des Weltraums.**
O. Hachenberg, T. Schmidt-Kaler.
Umschau, 70. Jahrgang, p. 65 - 69 (1970). − Popular article.

033.002 **Refractive corrections in high-accuracy radio interferometry.** D. F. Dickinson, M. D. Grossi,
M. R. Pearlman.
Journ. Geophys. Res., Vol. 75, 1619 - 1621 (1970).
We have calculated from models using a ray-tracing program differential phase delays (geometric length minus electromagnetic length) for paths through the atmosphere and ionosphere. Typical values are given.

033.003 **Optimal value for the antenna spacing of a three antenna interferometer.** V. V. Lebedev.
Astrophys. Space Sci. Library, Vol. 15, 640 - 642 (1970).
Conference paper (see 012.007).

033.004 **Solar radiospectrograph for the dynamical spectra in the 10 - 70 MHz band.** J. Hanasz.
Postępy Astron., Vol. 18, 185 - 187 (1970). In Polish.
Abstr. Polish Astron. Soc.

033.005 **Radiotelescope for observations of the sun at 50 cm wavelength.**
A. T. Nesmyanovich, Yu. A. Chesnok, V. V. Chmil, A. M. Sviridov.
Vestn. Kiev. Univ., Ser. Astron., No. 11, p. 46 - 52 (1969).
In Russian.

033.006 **Complex coherent radar station with two frequencies.** G. I. Kolomiets, Yu. V. Chumak.
Vestn. Kiev. Univ., Ser. Astron., No. 11, p. 110 - 113 (1969).
In Russian.

033.007 **Dicke Panel scores effects of fund freeze on U.S. radio astronomy.** S. Tilson.
IEEE Spectrum, Vol. 7, No. 1, p. 20, 22 (1970).

033.008 **New radio telescopes in Europe.**
P. P. Kronberg.
Journ. Roy. Astron. Soc. Canada, Vol. 64, 105 - 107 (1970).
Concerning the radio telescopes at Bonn (West Germany) and Westerbork (Netherlands).

033.009 **On the sensitivity and NEP in coherent and incoherent detection.** A. Baudry.
Astron. Astrophys., Vol. 6, 325 - 326 (1970).
This letter is a short comparison of coherent and incoherent detection based chiefly upon the approximate derivation of a radio Noise Equivalent Power.

033.010 **A new approach to aperture synthesis processing.**
W. R. Burns, S. S. Yao.
Astron. Astrophys., Vol. 6, 481 - 485 (1970).
In practice a complete measurement of the visibility function is not always available. One is thus faced with estimating the source brightness distribution based on the poorly sampled visibility function. A technique has been investigated in which the available measurements are used to interpolate over the missing regions. The brightness distribution is then obtained from the interpolated visibility function. This technique is described and example results are shown. The results are compared to those obtained without the use of the interpolation.

033.011 **Das 100-m-Teleskop des Max-Planck-Instituts**
für Radioastronomie in Bonn. K. Rohlfs.
SuW, Vol. 9, 140 - 145 (1970).

033.012 **The delay line system for the Molonglo radio telescope.**
T. W. Clarke, H. S. Murdoch, M. I. Large.
Proc. Instn. Radio Electronics Engineers Australia, Vol. 30, 236 - 240 (1969).
The circuits of the delay line system of the Molonglo radio telescope are described in detail. − BMT

033.013 **Large 22-MHz array for radio astronomy.**
C. H. Costain, J. D. Lacey, R. S. Roger.
IEEE Trans. Antennas Propagation, Vol. AP-17, No. 2, p. 162 - 169 (1969).
General description of the layout, dipole elements, receiving system and performance of the T-array of the Dominion Radio Astrophysical Observatory, Penticton. −BMT

033.014 **The beam forming system for the Molonglo radio telescope.** M. I. Large, R. H. Frater.
Proc. Instn. Radio Electronics Engineers Australia, Vol. 30, 227 - 235 (1969).
The paper describes in detail the beam forming system of the north–south arm of the Molonglo radio telescope. The tolerance on the phase and delay required to form a number of beams is discussed. − BMT

033.015 **A guided tour of the fast Fourier transform.**
G. D. Bergland.
IEEE Spectrum, Vol. 6, No. 7, p. 41 - 52 (1969).
This article is intended as a primer on the fast Fourier transform, which has revolutionized the digital processing of waveforms. − JWB

033.016 **A frequency–stabilized microwave band-rejection filter using high dielectric constant resonators.**
M. A. Gerdine.
IEEE Trans. Microwave Theory and Techn., Vol.MTT-17, 354 - 359 (1969).
The high permittivity dielectric resonator is a low-loss microwave filter element whose size is substantially less than metal waveguide cavities. Temperature stabilising their resonant frequency increases the applicability of such elements.
JWB

033.017 **Another realization of an all-pass transfer function using an operational amplifier.** R. Khera.
Proc. IEEE, Vol. 57, 1337 - 1338 (1969).
A phase shift network is described which adjusts the phase from 0 to 360 deg of an incoming signal without change in amplitude with respect to phase. − JWB

033.018 **A near-constant-phase variable attenuator.**
G. B. Shelton.
Proc. IEEE, Vol. 57, 1345 - 1346 (1969).
A circuit was developed which permits variable attenuation of an RF signal while holding the phase shift to a minimum (less than 2 deg phase shift over a 50 db dynamic range).
JWB

033.019 **Minimisation of R.M.S. phase error in large reflector antennas.** T. B. Vu.
Proc. Instn. Electrical Engineers, Vol. 116, 1165 - 1167 (1969).
A method of minimising the R.M.S. phase error in a large paraboloidal reflector antenna is presented. The problem

involves the search for the axis, the focus and the focal length of the paraboloid which best fit the actual reflector, in the least square sense. Although standard techniques may be successfully employed. The method presented in this paper is simple, and results can be quickly obtained with the aid of a digital computer. As an illustration of the technique, the required parameters of the best fit paraboloid are computed for the 210-foot Parkes radio telescope, when the dish is tilted at an angle of 40 deg from the zenith. – *MWS*

033.020   On possibilities of increasing the resolving power of the large Pulkovo radio telescope.
O. N. Shivris, Y. N. Parijsky.
Izv. Glav. Astron. Obs. v Pulkove, No. 185, p. 191 - 201 (1970).   In Russian.
It is found possible to decrease essentially a systematic error of the reflector of variable profile antennas at the expense of a corresponding selection of the form of reflecting elements and their specific mounting. It allows for using the Pulkovo large radio telescope at mm-wavelengths, which would increase considerably its resolving power and enlarge the sphere of astronomical problems to be solved by means of this instrument.

033.021   An investigation of electric characteristics of the large Pulkovo radio telescope at 8 mm.
G. B. Gelfreich, V. Y. Golnev, Y. K. Zverev, Y. N. Parijsky, A. A. Stotsky, O. N. Shivris.
Izv. Glav. Astron. Obs. v Pulkove, No. 185, p. 202 - 218 (1970).   In Russian.
In the autumn 1966 the Pulkovo large radio telescope was investigated and put in operation after the reconstruction. The reconstruction was undertaken in order to increase the precision of the antenna surface and to extend the range of the radio telescope up to the wavelength of 0.8 cm. A high precision of the antenna area was achieved by means of a complex use of geodetical and radiotechnical methods of adjustment.

033.022   The reconstruction of the large Pulkovo radio telescope.
Y. K. Zverev, A. I. Kopylov, O. N. Shivris.
Izv. Glav. Astron. Obs. v Pulkove, No. 185, p. 219 - 235 (1970).
Several technical problems, solved in connection with a reconstruction of the large Pulkovo radio telescope for a spreading of its frequency range towards short waves, are discussed. The main changes and additions in the construction of the radio telescope are described as well as the special geodetic works made during the reconstruction.

033.023   Adjustment and mounting of the variable profile antenna.   A. A. Stotsky, O. N. Shivris.
Izv. Glav. Astron. Obs. v Pulkove, No. 185, p. 236 - 241 (1970).   In Russian.
A method of mounting and adjusting for operating the Pulkovo large radio telescope (PLRT) is described, which is based on using a special phase-comparison X-band radar system (phase comparator). This method was used for adjusting the main reflector of the PLRT during the observations of radio brightness distribution on the Venus, the sun and the moon at 0.8 cm wavelength with a resolution of 15".

033.024   A phase comparator of electric lengths.
A. A. Stotsky, V. M. Solovjev.
Izv. Glav. Astron. Obs. v Pulkove, No. 185, p. 242 - 247 (1970).   In Russian.
A phase X-band radar system is described. This permits to compare the electric lengths of two paths in the air of ~100 m length with a precision more than 10^{-6}. This device is designed for geodetic measurements and investigations of phase fluctuations in radio waves, which propagate through the atmosphere near to the earth's surface.

033.025   Some possible modifications of large radio antennas.   V. I. Prishlin.
Radiotekhn. i ehlektronika, Vol. 14, 2097 - 2107 (1969). In Russian. – Abstr. in Referativ. Zhurn. 51. Astron., 4.51.90 (1970).

033.026   The multi-mirror antenna design of the RT-22 radiotelescope of the Physical Institute of the Academy of Sciences.
L. D. Bakhrakh, M. I. Grigor'eva, A. D. Kuz'min, L. I. Matveenko.
Izv. vyssh. uchebn. zavedenij. Radiofizika, Vol. 12, 1109 - 1114 (1969).   In Russian. – Abstr. in Referativ. Zhurn. 51. Astron., 4.51.91 (1970).

033.027   Improvement of the efficiency of the radiotelescope RT-22 at 8 mm wavelength.
L. D. Bakhrakh, M. I. Grigor'eva, V. I. Zagatin, P. D. Kalachev, A. D. Kuz'min, L. I. Matveenko, G. S. Misezhnikov, V. A. Nikitin, V. A. Puzanov, A. E. Salomonovich, R. L. Sorochenko, V. B. Shtejnshlejger.
Izv. vyssh. uchebn. zavedenij. Radiofizika, Vol. 12, 1115 - 1120 (1969).   In Russian. – Abstr. in Referativ. Zhurn. 51. Astron., 4.51.92 (1970).

033.028   On the problem of the diffraction correction to measures with the "artificial moon".
T. V. Tikhonova.
Izv. vyssh. uchebn. zavedenij. Radiofizika, Vol. 12, 1121 - 1131 (1969).   In Russian. – Abstr. in Referativ. Zhurn. 51. Astron., 4.51.95 (1970).

033.029   The Parkes interferometer.   R. A. Batchelor, B. F. C. Cooper, D. J. Cole, A. J. Shimmins.
Proc. Instn. Radio Electronics Engineers, Australia, Voi. 30, (No. 10), 305 - 313 = Separate print Division Radiophys. C. S. I. R. O., Sydney (1969).
Description of variable baseline interferometer installed at Parkes. The instrument employs a number of novel features including observations with one telescope in motion and simultaneous operation at two frequencies 1402.8 and 467.6 MHz. The maximum resolution is better than one minute of arc at its present frequency of 1402.8 MHz. – *JWB*

033.030   Six-times multiplier with two watts output at 2700 MHz.   B. F. C. Cooper, G. A. Wells.
Proc. Instn. Radio Electronics Engineers, Australia, Vol. 30, (No. 10), 340 - 341 = Separate print Division Radiophys. C. S. I. R. O., Sydney (1969).
A multiplier is described which uses a step recovery diode to deliver an output of two watts at 2.7 GHz from an input of five watts at 450 MHz. – *JWB*

033.031   Bandwidth properties of corrugated conical horns.   B. McA. Thomas.
Electronics Letters, Vol. 5, (No. 22), 561 - 563 = Separate print Division Radiophys. C. S. I. R. O., Sydney (1969).
It is shown that the aperture fields of circumferentially corrugated conical horns remain virtually unchanged over a very wide frequency range, and the bandwidth improves as the length of the horn is increased. The theory is applicable over a large range of flare angles. – *MWS*

033.032   Post-detector filtering in radiometry.   B. F. C. Cooper.
Proc. Instn. Radio Electronics Engineers, Australia, Vol. 31, 41 - 48 = Separate print Division Radiophys. C. S. I. R. O. Sydney (1970).

033.033 Electro-optic spectrograph for radio astronomy.
T. W. Cole.
Optics Technol.,Vol. 1, (No. 1), 31 - 35 (1968).
Describes an experimental investigation of a type of spectrograph that employs electro-optic processing and its application to radio astronomy. – *MWS*

033.034 Dual adjacent directional coupler.
P. C. Tan.
Electronics Letters, Vol. 5, (No. 13), 283 - 285 (1969).
Design of a compact broadband directional coupler is presented. Experimental results of a 0.3 – 2.0 GHz 23db coupler are presented. – *JWB*

033.035 An improvement of the phase stabilization of a simple microwave interferometer.
B. van der Sijde, L. P. M. van Run.
Journ. Sci. Instruments, Ser. 2, Vol. 2, 584 - 586 (1969).
It is shown in this paper that by deriving the output signal from the point where the first derivative of the phase to frequency is equal to zero one can achieve a considerable improvement in phase stability of an interferometer. – *JWB*

033.036 Design of a multimode cavity for a parametric amplifier. N. K. Kapoor.
J. Instn. Telecommun. Engineers, Vol. 15, 605 - 608 (1969).
A brief description of the design of a triple mode cavity for use in a negative resistance parametric amplifier. – *DNC*

033.037 Use of atomic frequency standards for phase calibration of large aerial arrays.
K. V. Sheridan.
Electronics Letters, Vol. 5, (No. 16), 363 - 365 (1969).
Describes the use of a rubidium standard in the phase calibration of the Culgoora 80 MHz radioheliograph. – *BMT*

033.038 Radiometric evaluation of antenna-feed component losses. C. T. Stelzried, T. Y. Otoshi.
IEEE Trans. Instrum. Meas., Vol. IM-18, (No. 3), 172 - 183 (1969).
A radiometric method for calibrating the loss of multimode antenna feed components, in which the field is linearly or circularly polarized,is presented. – *DNC*

033.039 Phase-locked degenerate parametric amplifier.
P. A. Watson, M. E. Butcher.
Electronics Letters, Vol. 5, (No. 17), 392 - 393 (1969).
The phase-locked degenerate parametric amplifier realized in relatively simple form by using the phase-locking properties of Gunn oscillators. – *DNC*

033.040 Experimental investigations of the input-impedance characteristics of an antenna in a rectangular waveguide. M. J. Al-Hakkak.
Electronic Letters, Vol. 5, (No. 21), 513 - 514 (1969).
The input-impedance of a monopole antenna in a rectangular waveguide is investigated experimentally. It is found that theoretical results based on the assumption of a sinusoidal current distribution are in good agreement for monopole lengths up to about 0.6 of the guide height and about one quarter lambda. For larger monopoles are a more accurate theory should be sought. – *JWB*

033.041 Microwave locating reflectometer.
P. I. Somlo, D. L. Hollway.
Electronic Letters, Vol. 5, (No. 20), 468 - 469 (1969).
A microwave locating reflectometer has been described which is capable of producing a plot of the complex reflection coefficient gamma against distance along a waveguide. No sharp pulses or step functions are required and a plot is produced in real time without digital computation. – *JWB*

033.042 Miniature microstrip circulators using high-dielectric-constant substrates. B. Hershenov, R. L. Ernst.
RCA Rev., Vol. 30, (No. 3), 541 - 543 (1969).
Describes a broad-band miniature circulator using a high dielectric constant substrate, (relative dielectric constant = 83), for quarter-wavelength impedance transformers. Good performance over the frequency band of approximately 2.8 to 3.7 GHz has been measured. – *MWS*

033.043 A phase-measuring scheme for a large radiotelescope.
A. G. Little.
IEEE Trans. Antennas Propagation, Vol. AP-17, 547 - 550 (1969).
Describes a rapid method for measuring the phase of the Molonglo array. – *BMT*

033.044 *KU*-band interferometer.
G. D. Papadopoulos, B. F. Burke.
Mass. Inst. Technol. Res. Lab. Electronics Quarterly Progr. Rep., No. 95, p. 10 - 14 (1969).
Brief description of the back end of a *KU*-band interferometer installed at MIT. – *BFC*

033.045 Large low-frequency orbiting radio telescope.
H. U. Schuerch, J. M. Hedgepeth.
Report NASA-CR-1201, Astro Res. Corp., Santa Barbara, Calif., 56 pp. (1968). – See Phys. Abstr., Vol. 73, No. 19023 (1970).

033.046 Application of pressurized liquid nitrogen inside parametric-amplifier structures for input-noise-temperature improvement. H. J. C. Blume.
U.S. National Aeronautics and Space Administration, NASA Techn. Note, NASA TN D-5509, 30 pp. (1969).
This paper describes a technique whereby liquid nitrogen under helium gas pressure is used as the dielectric material surrounding the lossy diode within the parametric amplifier structure. With this technique the heat conduction from the lossy diode is sufficient to maintain the diode at ambient temperature. – *DNC*

033.047 Computer aided design of microwave circuits.
F. E. Emery, G. J. Policky.
Electronic Engineering, Vol. 41, (No. 502), 41 - 45 (1969).
Describes the development of a general purpose computor program for the analysis of ladder type transmission networks. – *MWS*

033.048 Modular construction allows precise amplifier alignment. R. H. Frater, C. T. Murray.
Electronic Engineering, Vol. 41, (No. 502), 46 - 49 (1969).
Description of a modular construction technique which allows multistage amplifiers to be aligned stage by stage to give precise overall bandpass characteristics. As an example of the application of the modular approach, the I. F. pre-amplifier designed for the Mills cross radio telescope at Molonglo is described. – *MWS*

033.049 Proposed high-efficiency spherical-reflector antenna.
P. J. B. Clarricoats, S. H. Lim.
Electronics Letters, Vol. 5, (No. 26), 709 - 711 (1969).
Proposes a technique for raising the efficiency of spherical reflector antennas embodying the Gregorian method of phase correction. The technique requires an additional subreflector, introduced between the Gregorian connector and the spherical reflector, in order to cause the power from the feed horn to be uniformly distributed over the main reflector aperture. Two examples are given and a theoretical efficiency of 93% is predicted in one case. – *MWS*

033.050 **Computer-predicted performance of corrugated conical feeds using experimental primary-radiation patterns.** D. D. Booker, P. A. Mcinnes.
Electronics Letters, Vol. 6, (No. 1), 18 - 20 (1970).

Five X-band corrugated conical feeds of varying half flare angle and constant flare length have been built and their radiation patterns measured. These data have been fed into a computer program to produce curves of gain factor and spillover for various F/D ratio parabolas. – *JWB*

033.051 **Low-loss varactor diode switches for radio astronomy receivers.** T. L. Landecker, R. Wielebinski.
Proc. Instn. Radio Electronics Engineers Australia, Vol. 31, (No. 3), 73 - 76 (1970).

Details of varactor diode switches for 400, 700 and 1400 MHz are given. – *DNC*

033.052 **Electronically controllable primary feed for profile-error compensation of large parabolic reflectors.**
A. W. Rudge, D. E. N. Davies.
Proc. Instn. Electr. Engineers, Vol. 117, 351 - 358 (1970).

The authors propose a multielement feed array in the focal plane of large parabolic reflectors. They present a method of compensating profile errors by phase control of the array. The technique is restricted to long period profile errors. – *DNC*

033.053 **Simple active filters: Design procedure.**
M. Bronzite.
Wireless world, Vol. 77, (No. 1413), 117 - 119 (1970).

A simple design procedure for low frequency high pass and low pass filters using integrated circuits. The paper also contains tables of element values for use with the design procedures. – *JWB*

033.054 **A high-performance line source feed for the AIO spherical reflector.**
L. M. Lalonde, D. E. Harris.
IEEE Trans. Antennas Propagation, Vol. AP-18, (No. 1), 41 - 48 (1970).

An aberration-correcting line source feed has been designed, modelled, constructed and tested in the Arecibo reflector.

033.055 **Radio frequency performance of a 210-ft ground antenna: X-band.** D. A. Bathker.
California Inst. Technol. Jet Propulsion Lab. Techn. Rep. 32-1417, 25 pp. (1969).

On the performance at X-band of the 210-ft paraboloidal ground antenna at Goldstone. – *MWS*

033.056 **Beam switching Cassegrain feed system and its applications to microwave and millimeterwave radioastronomical observations.**
S. D. Slobin, W. V. T. Rusch, C. T. Stelzried.
Rev. Sci. Instruments, Vol. 41, 439 - 443 (1970).

A beam switching Cassegrain antenna configuration was developed for the purpose of making radioastronomical measurements at a wavelength of 3.33 mm. The subreflector periodically switches the antenna beam between the astronomical source and the adjacent sky reference position. The tilting subdish system is used in conjunction with a synchronous detection radiometer, and a complete description of both the rf and electronic system is given.

033.057 **Only a countable number of brightness temperature distributions could produce an observed intensity interferogram.** R. H. T. Bates.
New Zealand Journ. Sci., Vol. 12, 467 - 469 (1969).

It is known that there are only a finite number of bright - ness temperature distributions of finite extent which can give rise to an observed intensity interferogram, measured with directive antennas. It is proved here that these distributions of finite extent are the only physically meaningful distribu - tions which can give rise to the intensity interferogram.

033.058 **The multiple beam phasing of the Molonglo radio telescope.** R. E. B. Munro, H. S. Murdoch, M. I. Large.
Proc. Instn. Radio Electronics Engineers, Australia, Vol. 31, (No. 1), 10 - 23 (1970).

Description of the technique used and the performance of the beam phasing networks of the Molonglo radio telescope. – *BMT*

033.059 **Optimisation of efficiency of reflector antennas: Approximate method.** T. B. Vu.
Proc. Instn. Electr. Engineers, Vol. 117, (No. 1), 30 - 34 (1970).

A brief discussion of the theoretical problem of obtaining maximum efficiency for both paraboloid and spherical reflector antennas. – *JWB*

033.060 **Measurement of microwave parameters by the ratio method.** B. O. Weinschel.
Microwave Journ., Vol. 12, (No. 8), 69 - 73 (1969).

The author discusses ratio methods for determining reflection coefficient, transmission coefficient, SWR and insertion loss. Accurate fixed frequency measurements of small SWR using a slotted line and the display of swept frequency slotted line measurements are also discussed. – *DNC*

Radio astronomy.   See Abstr. 157.027.

034 Astronomical Accessories

034.001 Interféromètre embarquable pour spectroscopie stellaire. A. Girard.
Comptes Rendus Acad. Sci. Paris, Sér. B, Vol. 270, 463 - 466 (1970).
Il s'agit d'un interféromètre de Michelson dans lequel un des deux miroirs est, par construction et une fois pour toutes, légèrement incliné par rapport à l'axe, selon le réglage bien connu en «franges de coin d'air».

034.002 A fast and accurate guiding system. G. Pålsgård, J. O. Stenflo.
Solar Physics, Vol. 11, 155 - 156 (1970). — Research note.

034.003 A solar flare videometer. P. E. Tallant.
Solar Physics, Vol. 11, 263 - 275 (1970).
A new instrument, called a videometer, has been developed to measure solar flare area, peak intensity and integrated intensity in real time. The videometer uses a closed circuit television system to convert an optical Hα image into electrical signals for measurement. Observations of two class I flares with the videometer are discussed.

034.004 Some developments of the magnetic beam absorption filter. M. Cimino, A. Cacciani, M. Fofi.
Solar Physics, Vol. 11, 319 - 333 (1970).
In the present paper some experimental arrangements are shown which utilize the magnetic filter described by Cimino et al. (1968). For a single cell we have elaborated an elementary theory in the following cases: (1) absorption by an atomic beam in a uniform magnetic field (i.e. pure damping profile); (2) atomic beam in a non-uniform magnetic field; (3) vapours in a uniform magnetic field (i.e. gaussian distribution); (4) vapours in a non-uniform magnetic field.

034.005 Der Laser in der Astronomie. W. Schlosser.
Umschau, 70. Jahrgang, p. 81 (1970).

034.006 An easily built solar viewer. D. C. Lemmon.
Sky Telescope, Vol. 39, 89 (1970).

034.007 A spectroscope attachment for viewing solar prominences. J. B. Newton.
Sky Telescope, Vol. 39, 120 - 123 (1970).

034.008 Combining camera lenses for solar photography. A. Boyko.
Sky Telescope, Vol. 39, 134 - 135 (1970).

034.009 On making a channeled pitch lap. J. T. Carle.
Sky Telescope, Vol. 39, 186 - 190 (1970).

034.010 Isotrop-homogenes Quarzglas in Laser-Reflektoren auf dem Mond. P. Bäumler.
SuW, Vol. 9, 4 - 7 (1970).

034.011 Ein gutes lichtelektrisches Photometer für Privat- und Schulsternwarten. W. Alt.
SuW, Vol. 9, 46 (1970).

034.012 Celestial photography with fiber-optics image tubes. P. W. Hodge.
Sky Telescope, Vol. 39, 234 - 235 (1970).

034.013 A photographic measurement of filter with a densitometer. T. Okamoto.
Tokyo Astron. Obs. Rep., Vol. 15, No. 1 (No. 56), p. 188 - 194 (1970). In Japanese.

034.014 Sur une caméra électronique à grand champ destinée à la photométrie astronomique.
A. Lallemand, L. Renard, B. Servan.
Comptes Rendus Acad. Sci. Paris, Sér. B, Vol. 270, 385 - 388 (1970).

034.015 Erratum: On polarimetry in solar active regions. I. The new Locarno polarimeter; observing procedures. [Solar Physics, Vol. 9, 225 - 234 (1969)].
E. Wiehr.
Solar Physics, Vol. 11, 172 (1970).

034.016 About the choice of parameters in photoelectric registration of star transits.
K. Šteins.
Uch. Zap., Riga, Vol. 121, 3 - 10 (1969). In Russian.
The advantage of a grating is shown, based on a linear relation between time required for settling of level gauge and length of bubble. A model for a movable slit is described.

034.017 Artificial light source with periodically changing light flux. M. Ābele, A. Rubāns.
Uch. Zap., Riga, Vol. 121, 106 - 110 (1969). In Russian.
The light source is designed for lag determination of photoelectric devices. Its flux may be changed from 10^{-11} to 10^{-8} lm.

034.018 Quantitative one-emulsion or one-TV-tube colorimetry of the earth. G. Courtès.
Journ. British Interplanet. Soc., Vol. 23, 357 - 362 (1970).
The experiments described for colorimetric studies of the earth's surface open up new fields of planetary research.

034.019 Use of highly reflecting crystals for spectroscopy and polarimetry in X-ray astronomy.
J. R. P. Angel, M. C. Weisskopf.
Astron. Journ., Vol. 75, 231 - 236 = Columbia Astrophys. Lab. Contr. No. 12 (1970).
The way in which Bragg reflection can most efficiently be used for spectroscopy and polarimetry in X-ray astronomy is discussed, and it is shown that both types of measurement are best carried out with crystals of high integrated reflectivity. Graphite, lithium hydride, and tungsten disulphide crystals exhibit this property, provided they have the correct domain structure. Two spectrometers making use of these crystals are described.

034.020 Spettroscopia nell'estremo ultravioletto all'Osservatorio di Arcetri. A. M. Cantù.
Mem. Soc. Astron. Italiana, Nuova Serie, Vol. 41, 77 - 88 (1970).
The extreme ultraviolet spectroscopy laboratory equipments at the Arcetri Observatory are described. First results of line identifications in the range 130-215 Å are also given.

034.021 Theorie und Anwendung des Ringmikrometers. L. Brandt.
SuW, Vol. 9, 100, 102 (1970).

034.022 Ein Blink-Komparator für Sternfreunde. A. Schnitzer.
SuW, Vol. 9, 102, 104 (1970).

034.023 Como construimos una cupula de 3 metros de diametro. A. J. Camponovo, V. S. Brena.
Revista Astron., Vol. 41, No. 170, p. 71 - 77 (1969).

034.024   A method of determining the optical parameters
for "AφC" *(antifading system)* of meteor recorders.
Yu. I. Voloshchuk, V. A. Nechitajlenko.
Vestn. Khar'kovsk. politekhn. in-ta, 1969 No. 36 (84), p. 32 -
38.   In Russian. — Abstr. in Referativ. Zhurn. 51. Astron.,
3.51.351 (1970).

034.025   Fadenkreuzokulare und ihre Beleuchtungsein-
richtungen. 2. Teil.
H. G. Ziegler.
Orion Schaffhausen, 28. Jahrgang, p. 44 - 49 (1970).

034.026   Optical search techniques for pulsars.
M. J. Disney, W. J. Cocke.
Bull. American Astron. Soc., Vol. 2, 190 (1970). — Abstr.
AAS.

034.027   A quantitative interferometer for double-star
measures.   J. L. Elliot, I. S. Glass.
Bull. American Astron. Soc., Vol. 2, 191 (1970). — Abstr.
AAS.

034.028   OSO-VI: The Harvard experiment.
M. C. E. Huber, A. K. Dupree, L. Goldberg,
R. W. Noyes, W. H. Parkinson, E. M. Reeves, G. L. Withbroe.
Bull. American Astron. Soc., Vol. 2, 200 (1970). — Abstr.
AAS.

034.029   An echelle spectrograph for astronomical use.
D. J. Schroeder, C. M. Anderson.
Bull. American Astron. Soc., Vol. 2, 217 (1970). — Abstr.
AAS.

034.030   Millimeter-wavelength observations of solar activity,
the moon, and Venus with a Josephson junction
detector.   B. T. Ulrich.
Bull. American Astron. Soc., Vol. 2, 222 - 223 (1970).
Abstr. AAS.

034.031   A slitless spectrograph for the flash spectrum.
W. C. Atkinson.
Sky Telescope, Vol. 39, 318 - 323 (1970).

034.032   An instrument for determining the personal dif-
ference in observations of artificial earth satellites.
V. L. Afanasyev.
Vestn. Kiev. Univ., Ser. Astron., No. 11, p. 131 - 135 (1969).
In Russian.
The dependence of the personal error on the speed and
brightness of the object and on the background brightness is
investigated.

034.033   A theory of a photoelectric multislit micrometer.
E. Høg.
Astron. Astrophys., Vol. 4, 89 - 95 (1970).
A formula is presented for the mean error of the difference of two discrete observations of a stochastic variable with
any given power spectrum. The influence of image motion,
scintillation, and shot noise on the accuracy of star positions
obtained by a photoelectric multislit micrometer with different slit systems is derived and checked by observations. The
limiting mean error for meridian observations set by the mean
image motion $0.''15$ can be obtained by a slit system for simultaneous observation of α and δ.

034.034   A proposal for a solar magnetograph.
M. Semel.
Astron. Astrophys., Vol. 5, 330 - 332 (1970).
A proposal for a magnetograph which should solve the
whole problem of calibration is advanced. In fact calibration
becomes almost banal, and is almost independent of line formation. A standard magnetograph could be modified for use
of this method with very little difficulty.

034.035   An investigation of the properties of vacuum ultra-
violet radiation detectors.   V. Tiyt.
IAU Symposium No. 36, (see 012.014), p. 12 (1970).
Abstract.

034.036   Densitomètre à transistor.
Y. Grandjean.
L'Astronomie, 84e année, p. 206 - 212 (1970).

034.037   Tripel-Prismen aus Quarzglas auf dem Mond.
P. Bäumler.
Umschau, 70. Jahrgang, p. 383 - 384 (1970).

034.038   Two economical transistor inverter circuits for
speed control of synchronous motor clock drives.
T. C. Platt.
Journ. British Astron. Ass., Vol. 80, 270 - 281 (1970).

034.039   On the shifting error of the moveable slide.
R. Fukaya.
Univ. Tokyo, Tokyo Astron. Obs. Report, Vol. 15, (No. 2),
(No. 57), p. 347 - 360 (1970).   In Japanese.

034.040   A new tunable filter with a very narrow pass-band.
J. V. Ramsay, H. Kobler, E. G. V. Mugridge.
Solar Physics, Vol. 12, 492 - 501 (1970).
A new tunable optical filter incorporated in a solar
magnetograph consists of an interference filter followed by
three automatically controlled Fabry-Pérot interferometers.
It has a spectral resolving power of 10^5, an angular resolution
of 5" arc and a field angle of 30" arc for this spectral resolving power. By using auxiliary optics the angular resolution
of the object may be improved at the expense of the angular
field. It is operational over the range 400 – 650 nm and may
be tuned through ± 0.2 nm about the selected transmission
band.

034.041   The B. A. A. blink device. — A description.
H. W. Kelsey.
Strolling Astronomer, Vol. 22, 68 - 69 (1970).

034.042   Passive radiation coolers for infrared sensors.
D. K. Anand, S. A. Jeter.
Journ. British Interplanet. Soc., Vol. 23, 495 - 508 (1970).

034.043   A far ultraviolet photometer for space research.
C. S. Bowyer, F. Paresce, M. Lampton, J. Mack.
Planet. Space Sci., Vol. 18, 835 - 845 (1970).
A photometer for use in the wavelength region from
100 to 1300 Å has been developed. Specific bands within
this region can be isolated by combining the spectral response
of individual solar blind photocathodes with the transmission
qualities of various thin metallic films.

034.044   Visuele magnitudemeter.
J. Klinkspoor.
Hemel en Dampkring, Vol. 68, 147 - 150 (1970).

034.045   Reduction for the errors of mounting of a photoelectric grid with inclined slits.   G. I. Pinigin.
Izv. Glav. Astron. Obs. v Pulkove, No. 185, p. 83 - 92 (1970).
In Russian.
A brief account is made of the photoelectric ocular
micrometers suggested for registrations of star transits. Methods for taking into account the influence of inaccuracies
in the manufacture and mounting of the grid with inclined
slits, used for observations with the Sukharev horizontal
meridian circle, on the observational results are investigated.

034.046 On the TV spectrophotometer.
  N. F. Kuprevitch.
Izv. Glav. Astron. Obs. v Pulkove, No. 185, p. 305 - 315
(1970). In Russian.
  The photographical and photoelectrical methods of recording the spectrum are discussed. It is shown that the TV
system can combine sometimes the advantages of both these
methods. The possible fields of application of such systems are
analysed. A description is given of the experimental TV
spectrophotometer with an image orthicon LI 201, mounted
at the Pulkovo Observatory for investigations of the spectral
response of colar divider of two canal television telescope.
The results of laboratory recordings of spectra and measurements of the light characteristic linearity, made by means of
this developed device, are given.

**034.047 On the analysis of the thermal stability of the
  vacuum high - resolution ultraviolet spectrograph.**
L. A. Kamionko, G. V. Kirjan, Yu. P. Platonov, Y. S. Stre-
letsky, Y. L. Shakhbazjan.
Izv. Glav. Astron. Obs. v Pulkove, No. 185, p. 316 - 322
(1970). In Russian.
  The value of the thermal stability of different construc-
tive schemes of mounting the vacuum ultraviolet spectrograph
is given for finding and calculating those instrumental elements
which are most susceptible to thermal influences. The reasons
for the selected scheme and necessary steps for satisfactory
operation of the instrument with respect to thermostability
are given.

034.048 Direct photography with a Carnegie image tube.
  S. Jeffers.
Journ. Royal Astron. Soc. Canada, Vol. 64, 121 - 128 (1970).
  A Carnegie image tube has been tested and installed at
the Nasmyth focus of the 24-inch telescope and used for
direct photography. Results are presented which contrast
the performance of the 74-inch telescope used in direct
photography to that of the 24-inch telescope with the
Carnegie image tube.

**034.049 The construction of an Hα solar prominence
  spectroscope.** J. B. Newton.
Journ. Royal Astron. Soc. Canada, Vol. 64, 197 - 200 (1970).

**034.050 Polarimètre photoélectrique de l'Observatoire
  Astronomique de Belgrade.**
V. Oskanjan, A. Kubičela, J. Arsenijević.
Bull. Obs. Astron. Beograd, Vol. 27, No. 2, p. 1 - 11 (1969).
  Dans cet article nous présenterons tout d'abord un
exposé sommaire des premières phases du développement
de ce photomètre, puis nous donnerons une description plus
détaillée de sa construction actuelle en tant que polarimètre.

**034.051 Erreurs systématiques de la détermination de
  l'épasseur du contact d'un micromètre impersonnel.** D. Djurović.
Bull. Obs. Astron. Beograd, Vol. 27, No. 2, p. 43 - 51 (1969).

**034.052 Photometric calibration of direct photographs of
  star fields using a calcite-polaroid filter.**
M. T. Brück.
Observatory, Vol. 90, 104 - 107 (1970).

034.053 An improved Hα filter-prism combination.
  M. F. McCarthy, P. J. Treanor.
Observatory, Vol. 90, 108 - 110 (1970).

**034.054 A device for solving problems involving culmination
  of celestial bodies.** M. G. Grigor'ev.
Uch. zap. Vladimirsk. gos. ped. in-t. Ser. "Fizika". Vladimir,
1969, p. 224 - 228. In Russian.

**034.055 Computation of the principal parameters of the
  photoelectric photometer for measuring the atmospheric transparency.** V. I. Goryshin.
Trudy Gl. geofiz. observ., 1969, vyp. (No.) 237, p. 55 - 61.
In Russian. – Abstr. in Referativ. Zhurn. 51. Astron.,
6.51.241 (1970).

034.056 Bildverstärker "vergrößert" Teleskop.
  G. V. Schultz, D. Ellinger.
Umschau, 70. Jahrgang, p. 478 - 479 (1970).

034.057 Der Tautenburger Himmelshintergrund-Monitor.
  R. Ziener.
Sterne, 46. Jahrgang, p. 106 - 108 (1970).

**034.058 Investigation of the screw of the eye-piece micro-
  meter of the Tashkent meridian circle from right
ascension.** M. F. Bykov, I. G. Zaugol'nikova.
Tsirk. Astron. Inst., *Tashkent,* No. 14 (361), p. 19 - 23
(1969). In Russian.

**034.059 Etudes d'astres faibles en lumière totale avec la
  caméra électronique.** G. Wlérick.
Advances Electronics and Electron Physics, Vol. 28, 787 -
800 = Publ. Obs. Haute Provence, Vol. 10, No. 25 (1969). –
Conference paper.

034.060 Proportional counter with automatic gain control.
  A. C. Brinkman, P. de Groene.
Nuclear Instruments and Methods, Vol. 66, 316 - 320 =
Utrechtse Sterrekundige Overdrukken No. 112 (1968).
  In aid of a solar X-ray experiment aboard the earth satellite ESRO-II a proportional counter was developed, at the
Utrecht Space Research Laboratory, of which the gas gain can
be monitored. There are different factors that can influence
the gas gain of a proportional counter, of these the gas pressure and supply voltage are the most important. This paper
describes the electronic circuitry that makes use of this monitor voltage to control the high voltage supply of the proportional counter so that a constant output pulse height distribution of the X-ray detector can be achieved.

**034.061 Comments on the ageing effect of gas-filled propor-
  tional counters.**
A. J. F. den Boggende, A. C. Brinkman, W. de Graaff.
Journ. Sci. Instruments (Journ. Phys. E), Series 2, Vol. 2,
5 pp. = Utrechtse Sterrekundige Overdrukken, No. 116
(1969).

034.062 The electronic pulsarium.
  S. P. Maran, F. C. Hallberg, E. W. Nyberg.
Sky Telescope, Vol. 40, 17 - 19 (1970).

**034.063 On the air cooling of astronomical photosensitive
  devices.** A. N. Abramenko.
Astron. Tsirk., No. 555, p. 4 - 5 (1970). In Russian.

**034.064 On the effectiveness of cooling for multialkali
  photocathods of a cascade image converter.**
A. N. Abramenko, V. V. Prokofjeva.
Astron. Tsirk., No. 555, p. 5 - 7 (1970). In Russian.

**034.065 Investigation of optical characteristics of the polari-
  zation interference filter with variable pass-band.**
T. A. Vinogradova-Smirnova, M. V. Kushnir, A. A. Spitalnaja.
Solnechnye Dannye 1970 Byull., No. 2, p. 95 - 101 (1970).
In Russian.

034.066 A small-size additional device for transistors.
  I. D. Reiljan.

Byull. Stantsij Optichesk. Nablyud. Iskusstv. Sputnikov Zemli, No. 53 (1), p. 36 - 37 (1969). In Russian.

034.067 A plate-holder for the camera NAFA–3c/25.
K. Havliček.
Byull. Stantsij Optichesk. Nablyud. Iskusstv. Sputnikov Zemli, No. 53 (1), p. 38 - 40 (1969). In Russian.

034.068 Methods of increasing the storage capacity of high - gain image intensifier systems.
J. R. Powell, R. Lynds.
Advances Electronics Electron. Phys., Vol. 28, 745 - 752 = Contr. Kitt Peak National Obs., No. 371 (1970).

034.069 The image-tube spectrograph of the Vatican Obser - vatory.   P. J. Treanor.
Ric. Astron. Specola Vaticana, Vol. 8, (No. 4), 61 - 83 (1970).
The performance of conventional slit-spectrographs and objective prisms used with telescopes of small aperture is com - pared with the potential performance of an image-tube slit-spectrograph adapted to a small telescope. The characteristics and design problems of such image-tube spectrographs are discussed with special reference to the problems of weight, space, and optical efficieny. The mechanical and optical fea - tures of an image-tube slit-spectrograph designed and con - structed at the Vatican Observatory are described.

034.070 On the study of the grating spectrographs ASP-20 and DFS-3 of the horizontal solar telescope ATSU-5.
F. G. Rozhavskij, A. M. Kumantsev, Z. N. Shukstova.
Uch. zap. Ural'skogo un-ta, 1969, No. 70, p. 206 - 223. In Russian. – Abstr. in Referativ. Zhurn. 51. Astron. 7.51.100 (1970).

034.071 Experimental investigation of the delay of a photo - electric device.   A. I. Yazev.
Trudy metrol. in-tov SSSR, 1969, vyp. (No.) 106 (166), p. 150 - 159. In Russian. – Abstr. in Referativ. Zhurn. 51. Astron., 7.51.194 (1970).

034.072 Optics and the Mariner imaging instrument.
D. R. Montgomery, L. A. Adams.
Applied Optics, Vol. 9, (No. 2), 277 - 287 (1970).
A television instrument was developed to observe the features of Mars at resolutions and accuracies greater than those obtainable from earth-based telescopes and from photographs obtained by Mariner IV. The television instru-ments developed for Mariners VI and VII are of a two-camera configuration, exposed alternately during the planetary encounter and providing both analog and digital data to two tape recorders on each spacecraft.

034.073 Recording of star spectra with a fiber optic electro-static image intensifier.
J. R. Hansen, J. DeJonge, W. R. Beardsley.
Image Technol., Vol. 11, No. 10, p. 25 - 34 (1969).
Electrostatic image intensifiers with fiber-optic faceplates offer potential advantages over two-stage magnetically-focused devices currentlv used in astronomical spectroscopy. The image intensifier is used at Allegheny Observatory to record star spectra. The spectrum of the star HD 213014 of photo-graphic magnitude 8.5 is possible to record in 6 minutes. Without the intensifier this magnitude would require an expo-sure of about three hours, a reduction in exposure time of 30 times.

034.074 The micrometergraph - a device for the automatic recording of readings from the micrometer of visual zenital telescopes.   W. Jakś, K. Schilling.
Materiały i Prace, Vol. 32 (Publ. Inst. Geophys. Polish Acad. Sci.), 15 - 23 = Polish Acad. Sci. Astron. Latitude Station Borowiec, Repr. No. 33 (1969).

034.075 Oscillating fiber microbalance.   H. Patashnick, C. L. Hemenway.
Rev. Sci. Instruments, Vol. 40, 1008 - 1011 = Dudley Obs. Repr., No. B-34 (1969).
In conjunction with the micrometeorite research at Dudley Observatory, a microbalance has been developed to provide mass measurements of particles in the range $10^{-5} - 10^{-11}$ g.

034.076 Costanti strumentali del misuratore di lastre Gaert-ner dell'Osservatorio Astronomico di Torino.
C. Cocito, M. A. Vogliotti.
Contr. Oss. Astron. Torino, No. 51, 18 pp. (1969).
Periodic and progressive corrections of Gaertner screw have been studied again. The angle between the screw and the guide on which the holder plate is moving is determined. Some formulas are given for transforming instrumental coordinates in orthogonal ones.

034.077 Tracés de rayons dans le montage Wadsworth inversé.   J. M. Vreux.
Bull. Cl. Sci., Acad. Roy. Belgique, *Bruxelles,* 5e Sér., Vol. 55, 181 - 204 = Univ. Liège, Inst. d'Astrophys., Coll. 8°, No. 582 (1969).

034.078 Study of spectrograph I at Bosque Alegre.
L. A. Milone.
Asoc. Argentina Astron. Bol., No. 12, (see 012.019), p. 10 (1968).   In Spanish. – Abstract.

034.079 Calculation of a Cassegrain type inverted collimator for a grating spectrograph.
A. Grigóreff, R. Banilis.
Asoc. Argentina Astron. Bol., No. 12, (see 012.019), p. 33 (1968). – Abstract.

034.080 Fabry-Pérot interferential technique.
G. Carranza.
Asoc. Argentina Astron. Bol., No. 14 (see 012.020), p. 8 - 12 (1968).   In Spanish.

034.081 New views at the examination of one-second levels.
G. Teleki, S. Sadžakov, M. Mijatov.
Publ. Obs. Astron. Beograd, No. 16, (see 012.021), p. 22 - 26 (1969).   In Serbo-Croatian.

Josephson detectors make astronomical observations.
Phys. Today, Vol. 23, No. 4, p. 55 - 56 (1970).

The horizontal solar telescope ATSU–5 with a double diffraction spectrograph.   See Abstr. 032.028.

Instruments and techniques.
See Abstr. 032.032.

Isophotometry of astronomical photographs by a photometric analyser for TV cameras.
See Abstr. 036.006.

035 Clocks and Frequency Standards

035.001 Note sur le cadran solaire de Brou. L. Janin.
L'Astronomie, 84e année, p. 83 - 86 (1970).

035.002 Explanation of apparent diurnal variation in caesium clock rates.
A. H. Cribbens, L. M. Stephenson.
Nature, Vol. 226, 139 (1970).

035.003 On the bearing of radio time signals on short waves as arrive. S. Kohno.
Tokyo Astron. Obs. Rep., Vol. 15, No. 1 (No. 56), p. 158 - 167 (1970). In Japanese.

035.004 Frequency comparison of five commercial cesium standards with a NASA experimental hydrogen maser. A. R. Chi, F. G. Major, J. E. Lavery.
Proc. IEEE, Vol. 58, 142 - 143 (1970).

Recent long-term phase comparisons between five commercial cesium beam frequency standards and a NASA experimental hydrogen maser, NX-1, have yielded a value for the frequency of the hydrogen maser of 1 420 405 751.7767 ± 0.0024 Hertz which is 6.77 × 10^{-12} ± 1.66 lower than that given by Vessot et al. in 1966.

035.005 Continuous clock comparison between Koganei and Mitaka with telephone lines.
N. Matsunami, M. Torao, K. Fujiwara, T. Hara, T. Sakai, T. Kato.
Univ. Tokyo, Tokyo Astron. Obs. Report, Vol. 15, (No. 2), (No. 57), p. 232 - 244 (1970). In Japanese.

035.006 On a high stability phase shifter using sine condenser. K. Fujiwara, T. Hara, T. Sakai, T. Kato.
Univ. Tokyo, Tokyo Astron. Obs. Report, Vol. 15, (No. 2), (No. 57), p. 245 - 252 (1970). In Japanese.

035.007 Analysis of perturbed rates of astronomical pendulum clocks. K. Bielicka.
Acta Astron., Vol. 20, 163 - 193 (1970).

Theoretical and numerical investigations on the problem of the mutual influence of pendulum clocks were performed. The mutual perturbations of the clock rates were observed in the Laboratory of Time and Frequency Measurement at the Central Office of Quality and Measures in Warsaw. The results obtained this way are not only of a theoretical importance, but they are also applicable in the Time Service based on the rate of astronomical pendulum clocks.

035.008 Corrections to Czechoslovak time signals. V. Ptáček.
Říše hvězd, Vol. 51, 23, 36, 60, 77, 99, 118, 138 (1970). 1969 Oct. - 1970 April.

035.009 The quartz clock QC–I No. 0110 at the tracking station No. 1055 of the Uzhgorod university.
A. G. Kirichenko, Yu. S. Nakonechny.
Byull. Stantsij Optichesk. Nablyud. Iskusstv. Sputnikov Zemli, No. 53 (1), p. 34 - 35 (1969). In Russian.

035.010 An electronic clock for a coded time channel on a magnetic tape recorder. J. W. Curtis.
Journ. Sci. Instruments, (Journ. Phys. E), Vol. 3, 173 - 176 (1970).

An electronic clock that can be used to provide a coded time signal for a timing channel on a magnetic tape recorder is described. The clock provides a direct means of determining the elapsed time, whereas more commonly employed methods of timing do not. The construction of the clock is made practical by the use of integrated circuitry which ensures that it is both compact and low on power consumption.

036   Photographic Auxiliaries

036.001   **Anscochrome 500 – ein Farbfilm hoher Empfindlichkeit für den Sternfreund.**
W. Hänig.
VdS Nachrichtenblatt, 19. Jahrgang, p. 7 (1970).

036.002   **Photographie astronomique. La latensification.**
R. Bucaille.
L'Astronomie, 84e année, p. 115 - 121 (1970).

036.003   **Spectral sensitivity of some sorts of "ORWO" photographic plates.**   A. Bogdanovič.
Bull. Vilnius Astron. Obs., No. 26, 36 - 46 (1969).
In Russian.
   The spectral sensitivity curves for some sorts of "ORWO" photographic plates are obtained and analyzed.

036.004   **Commentaar aangaande het fotograferen in kleuren van sterrenbeelden.**   L. van Mellaert.
Gaz. astron., No. 2, p. 9 - 10 (1969).
   Colour photography of star trails give very good results with some presently available emulsions. The trails are beautiful and well reproduced, they follow nearly the spectral class of the different objects. By means of a projector, the dias will enable the amateur to identify each star by comparing the drawing with a star atlas.

036.005   **Dunkelkammerarbeit an einer Kometenaufnahme.**
F. Seiler.
Orion Schaffhausen, 28. Jahrgang, p. 80 - 81 (1970).

036.006   **Isophotometry of astronomical photographs by a photometric analyser for TV cameras.**
H. B. Bredohl.
Astron. Astrophys., Vol. 7, 167 - 168 (1970).

Isophotes of comets or nebulae are obtained very easily from astronomical plates by injecting onto a TV-screen marking points corresponding to a fixed value of photographic density.

036.007   **On the necessity of control tests of astronomical films of increased speed.**   O. D. Dokuchaeva.
Astron. Tsirk., No. 543, p. 6 - 8 (1970).   In Russian.

036.008   **Hints on planetary photography for amateurs – I.**
R. B. Minton.
Sky Telescope, Vol. 40, 56 - 59 (1970).

036.009   **Linearity of electronographic emulsions.**
M. J. Smith, P. W. J. L. Brand.
Advances Electronics Electron Phys., Vol. 28B, 737 - 743 = Commun. Royal Obs., Edinburgh, No. 70 (1969).

036.010   **New kinds of photographic films for astronomical and spectroscopic use.**   P. V. Mejklyar,
O. D. Dokuchaeva.
Vestn. AN SSSR, No. 2, p. 45 - 48 (1970). In Russian. – Abstr. in Referativ. Zhurn. 51. Astron., 7.51.253 (1970).

036.011   **Experiences with the Sabattier effect.**
H. A. Dottori.
Asoc. Argentina Astron. Bol., No. 14 (see 012.020), p. 97 - 100 (1968).   In Spanish.

   La photographie planétaire à haute résolution au Pic du Midi.   See Abstr. 091.037.

   Quantitative analytic composite photography.
See Abstr. 113.008.

Positional Astronomy. Celestial Mechanics

041 Positional Astronomy, Star Catalogues and Atlases

041.001 Probleme der langbrennweitigen Astrometrie.
P. van de Kamp.
Sterne, 45. Jahrgang, p. 209 - 217 (1969). — Review article.

041.002 Planetary occultations 1970.
G. E. Taylor.
Journ. British Astron. Ass., Vol. 80, 216 (1970).

041.003 A catalog of positions for 502 stars in the region of the Pleiades.
H. K. Eichhorn, W. D. Googe, C. F. Lukac, J. K. Murphy.
Department of the Army, U.S. Army Topographic Command, Washington, D.C., TOPOCOM Techn. Rep. No. 70, 9 + 141 pp. (1969).

This report contains the right ascensions and declinations of 502 stars in the region of the Pleiades cluster and describes the unconventional procedures used to obtain the extremely high relative accuracy (on the order of 0''01) of these positions. Relative proper motions for the reference stars were computed by comparison of the positions given in this paper with those previously determined by König. Other results of astrometric significance are discussed.

041.004 Das Inertialsystem und seine Bedeutung für die Astronomie und die Geodäsie. H. U. Sandig.
Sterne, 46. Jahrgang, p. 1 - 11 (1970).

041.005 Catalogue of the positions and proper motions of stars between declinations -40° and -50°, reduced to the equinox of 1950 without applying proper motions.
D. Hoffleit, with collaboration of D. Eckert, P. Lü, K. Paranya.
Trans. Astron. Obs. Yale Univ., *New Haven*, Vol. 30, 68 + 269 pp. (1970).

041.006 Resultados de las observaciones para el catálogo fundamental de estrellas debiles del hemisferio sur (CFED). Posiciones preliminares de 378 estrellas, de declinaciones comprendidas entre +9° y −85° para equinoccio y época 1950.0. S. S. Slaucitajs.
Separate print Obs. Astron. Univ. Nacional La Plata, Dep. Astrometria Meridiana, La Plata. 13 pp. (1969).

041.007 Definitive plate constants for the Astrographic Catalogue north of +40° declination.
A. Günther, H. Kox.
Astron. Astrophys., Vol. 4, 156 - 158 (1970).
The data of the AGK 3 are used for a study of systematic errors inherent in the published Catalogues of the Carte du Ciel in the zones Greenwich, Rome-Vatican, Catania, Helsingfors. The iterative procedure involving least-squares-solutions for individual plates and for a representation of the residuals by power series of the coordinates of the reference stars taken from AGK 3 converge after three to four steps. The members of the power series are used up to the third order in the coordinates and the diameter (or magnitude) of stars. Tables present mean values of residuals sorted according to position in the plate area, spectral type, and magnitude.

041.008 Identification précise des étoiles d'un cliché photographique. P. Paquet.
Ciel et Terre, Vol. 86, 187 - 194 (1970).
A simple non-homographic method is given for the identification of stars with an accuracy of a few hundredths of millimeter when the orientation and approximate coordinates of a photographic plate are known.

041.009 Mutual occultations of planets.
J. Meeus.
Journ. British Astron. Ass., Vol. 80, 282 - 287 (1970).

041.010 On the star system and the observation program for the Tokyo PZT.
S. Iijima, Y. Niimi.
Ann. Tokyo Astron. Obs., Second Series, Vol. 11, 157 - 201 (1969).
At the beginning, distribution of PZT stars in zenith distance was rather uniform for the whole range of right ascension. However, the uniformity has become worse and stars in the marginal zone have escaped from the field of view from year to year according to the precessional motion of equator. Here we have tried to supplement some new stars to the PZT star list in order to recover the total number of stars and the uniformity of star distribution in zenith distance. Apparent places of the PZT stars for daily observation were computed. In the present report, source of proper motions for the PZT stars are reviewed, as well as examination of the star system as compared with the Smithsonian Astrophysical Observatory Catalogue and with the results of meridian circle observations.

041.011 The SPF 1 catalogue of right ascensions.
C. Anguita, G. Carrasco, P. Loyola, D. D. Polojentsev, K. N. Tavastsherna, M. S. Zverev.
Inform. Bull. Southern Hemisphere, No. 15, p. 36 - 38 (1969).

041.012 Finsternisse und Bedeckungen.
H. Winkler.
Vorträge und Schriften Archenhold-Sternw. Berlin-Treptow, No. 36, 48 pp. (1969).

041.013 Positions d'étoiles en déclinaison. A. P. Botelheiro.
Bull. Obs. Astron. Lisbonne (Tapada), No. 17, 10 pp. (1970).

041.014 Photographic positions of Mars obtained with the short-focus double astrograph of the Pulkovo Observatory during 1962 - 1965. T. P. Kiseleva.
Izv. Glav. Astron. Obs. v Pulkove, No. 185, p. 98 - 102 (1970). In Russian.
The results of photographic observations of Mars made during 1962 - 1965 with the short-focus double astrograph (AKD) of the Pulkovo Observatory are given. An accidental error of one position of Mars (mean square error) is equal to ± 0''34 in right ascension and ± 0''26 in declination.

041.015 Study of causes of dispersions in the observations with the meridian circle. A. Acker.
Astron. Astrophys., Vol. 7, 65 - 67 (1970). In French.
Using the results of 445 series of observations with the

meridian circle, in which each star had been observed several times, it has been possible to determine the qualities of the different evenings, by the value of the mean dispersion ϵ^2 of the observations. We try in this paper to explain the great disparity of the values of ϵ^2, by studying the correlations between ϵ^2 and different climatic factors.

041.016 **AGK2–AGK3.**
P. Lacroute, A. Valbousquet.
IAU Circ. No. 2221 (1970).

041.017 **Criterio di riduzione del catalogo S.I.L. e indice di disomogeneità nelle** $\Delta \delta_\delta$ **(FK4–GC).**
E. Fichera, A. Pugliano.
Mem. Soc. Astron. Italiana, Nuova Serie, Vol. 41, 183 - 188 (1970).
    The general principles for reducing the I.L.S. catalogue to the star positions established by the FK4 are discussed. As all the stars of the I.L.S. catalogue are present in the GC, we used this catalogue as an intermediary. We have so obtained an index of discontinuity in the $\Delta \delta_\delta$ (FK4 – GC) between the star positions in the northern hemisphere and in the southern one.

041.018 **A catalogue of early-type stars whose spectra have shown emission lines.** L. R. Wackerling.
Mem. Roy. Astron. Soc., Vol. 73, (Part 3), 153 - 319 (1970).
    A catalogue of 5326 early-type emission-line stars is presented and its content is briefly analysed. The catalogue provides collated designations, positions, and an indication of the availability of spectral and photometric data for stars of classes W, O, B, A and F. The surface distribution of the objects along the Milky Way and the details of concentrations near some OB associations are briefly presented.

041.019 **Differential catalogue of the right ascensions of 292 bright stars between declinations +50° and +60°.** G. G. Khodak.
Tsirk. Astron. Inst., *Tashkent*, No. 7 (354), 24 pp. (1968). In Russian.

041.020 **Differential catalogue of the right ascensions of 167 bright stars between declinations +70° and +90°.** I. G. Zaugol'nikova.
Tsirk. Astron. Inst., *Tashkent*, No. 9 (356), 14 pp. (1968). In Russian.

041.021 **Differential catalogue of the right ascensions of 354 bright stars between declinations +30° and +40°.** I. M. Boroditskij.
Tsirk. Astron. Inst., *Tashkent*, No. 11 (358), 23 pp. (1969). In Russian.

041.022 **Differential catalogue of the right ascensions of 450 bright stars between declinations +10° and +20°.** K. Kh. Kasymov, I. M. Boroditskij.
Tsirk. Astron. Inst., *Tashkent*, No. 13 (360), 28 pp. (1969). In Russian.

041.023 **Differential catalogue of the right ascensions of 306 bright stars between declinations +40° and +50°.** K. Kh. Kasymov. I. M. Boroditskij.
Tsirk. Astron. Inst., *Tashkent*, No. 17 (364), 23 pp. (1970). In Russian.

041.024 **The right ascensions of the sun and Venus obtained from observations with the Tashkent meridian circle.** I. G. Zaugol'nikova.
Tsirk. Astron. Inst., *Tashkent*, No. 12 (359), p. 16 - 19 (1969). In Russian.

041.025 **Absolute catalogue of right ascensions of 284 stars obtained in Tashkent.** Eh. A. Sanakulov.
Tsirk. Astron., Inst., *Tashkent*, No. 15 (362), p. 14 - 31 (1970). In Russian.

041.026 **Determination of the systematic errors** Δa_a **for zenith, equatorial and northern stars from materials of the Time Service.** Eh. A. Sanakulov.
Tsirk. Astron. Inst., *Tashkent*, No. 16 (363), p. 13 - 29 (1970). In Russian.

041.027 **Zur Objektivierung der Beobachtung von Sterndurchgängen durch den Meridian.**
H. Potthoff.
Wiss. Zeitschr. Techn. Univ. Dresden, Vol. 17, 1477 - 1484 = Mitt. Lohrmann-Inst. Techn. Univ. Dresden, No. 15 (1968).
    In der Arbeit wird eine systematische Übersicht über die Verfahren der Beobachtung von Sterndurchgängen durch den Meridian unter besonderer Berücksichtigung der photoelektrischen Verfahren gegeben. Die bei den Beobachtungen auftretenden wesentlichen Störeinflüsse und Fehlerquellen werden behandelt, um daraus und aus der Analyse der bisher verwirklichten Beobachtungsverfahren ein neues Verfahren abzuleiten, das deren Nachteile nicht besitzt.

041.028 **Beitrag zur Objektivierung von Meridiandurchgangsbeobachtungen.** S. Wächter.
Arbeiten Vermessungs– und Kartenwesen Deutsche Demokratische Republik [Editor: Geodätischer Dienst, Leipzig], Vol. 17, 87 pp. = Mitt. Lohrmann-Inst. Techn. Univ. Dresden No. 16 (1969). – Thesis Techn. Univ. Dresden.

041.029 **Beobachtung genauer Almukantaratdurchgangszeiten von Sternen ohne Mikrometer.**
H. Potthoff, K.-G. Steinert.
Vermessungstechnik, 17. Jahrgang, No. 2, 4 pp. = Mitt. Lohrmann-Inst. Techn. Univ. Dresden, No. 17 (1969).

041.030 **Anschluß von Einzelobjekten an Nachbarsterne auf Himmelsaufnahmen.** H.-U. Sandig.
Vermessungstechnik, 17. Jahrgang, No. 11, p. 408 - 411 = Mitt. Lohrmann-Inst. Techn. Univ. Dresden, No. 19 (1969).

041.031 **Positional astronomy.** A. A. Nemiro.
Trans. IAU, Vol. 14A, (see 003.028), 39 - 51 (1970). – Report of Commission 8.

041.032 **Carte du Ciel.** W. Dieckvoss.
Trans. IAU, Vol. 14A, (see 003.028), 225-226 (1970). – Report of Commission 23.

041.033 **A Messier album.** J. H. Mallas, E. Kreimer.
Sky Telescope, Vol. 39, 26 - 27, 104, 172 - 173, 236, 297 - 298, 371 - 372; Vol. 40, 31, (1970).

041.034 **Positional uncertainties in the Smithsonian "Star Catalog" in ecliptic coordinates.**
E. Blondelot.
Centre Univ. Mons,Fac. Sci., Dép. Astrophys. Commun., No. 9, 7 pp. (1970).

041.035 **Erreurs d'impression et anomalies dans quelques catalogues d'étoiles fréquemment utilisés.**
L. Houziaux, J. Blondelot-Lickes.
Centre Univ. Mons, Fac. Sci., Dép. Astrophys., Commun., No. 13, 6 pp. (1970).

041.036 **Washington meridian observations of the moon. Six-inch transit circle results, 1925 – 1968.**
A. N. Adams, B. L. Klock, D. K. Scott.

Publ. United States Naval Obs., Second Ser., Vol. 19, (Part III), 439 - 498 (1969).

The astronomical results contained in this publication represent meridian observations of the moon made with the six-inch transit circle from 1925 to 1968, and reduced to the system of the FK4 catalog.

041.037　Observations of the sun, moon, and planets. Six-inch transit circle results.
B. L. Klock, D. K. Scott.
United States Naval Obs. Circ., No. 127, 13 pp. (1970).

This *Circular* contains positions of the sun, moon, and planets observed with the six-inch transit circle between 1968 June 30 and 1969 July 17. These results are provisional.

041.038　Etudes sur l'emploi de recouvrements de plaques pour l'établissement de catalogues photographiques.
P. Lacroute.
Publ. Obs. Strasbourg, Vol. 1, Fasc. 1, [Reprinted from "Highlights in Astronomy", (D. Reidel Publishing Company, Dordrecht—Holland), 1968, p. 319 - 337], 19 pp. (1968).

041.039　Etude des solutions optima pour des résolutions de clichés séparés.　P. Lacroute.
Publ. Obs. Strasbourg, Vol. 1, Fasc. 3, [Extrait du 87e Congrès de l'Association française pour l'Avancement des Sciences, Nancy, 1968, p. 11 - 23], 13 pp. (1969).

041.040　Méthode de réduction de la position d'un astre errant par rattachement à *n* étoiles de référence dont le centre de gravité est choisi voisin de l'astre.
A. Schmitt.
Publ. Obs. Strasbourg, Vol. 1, Fasc. 4, [Extrait du 87e Congrès de l'Association française pour l'Avancement des Sciences, Nancy, 1968, p. 33 - 38], 6 pp. (1969).

041.041　Réduction des positions photographiques d'objets célestes. Discussion de diverses méthodes.
H. Debehogne.

Obs. Roy. Belgique, *Uccle,* Commun., Sér. B, No. 47 [Extrait du 87e Congrès de l'Association française pour l'Avancement des Sciences, Nancy, 1968, p. 1 - 7], 7 pp. (1969).

041.042　Positionsastronomi, III. Anvendelse af positioner og egenbevaegelser.　J. Andersen, H. J. Fogh Olsen.
Astron. Tidssk., Vol. 3, 18 - 36 (1970).

041.043　Systematic errors in the catalogues N 30 and FK 4.
G. Carrasco, P. Loyola.
Asoc. Argentina Astron. Bol., No. 14 (see 012.020), p. 32 - 33 (1968).　In Spanish. — Abstract.

041.044　Execution of instruments system out of Küstner's series.　Lj. Mitić, I. Pakvor.
Publ. Obs. Astron. Beograd, No. 16, (see 012.021), p. 45 - 47 (1969).　In Serbo-Croatian.

041.045　Determination of stars absolute right ascension out of Time Service observations.　D. Djurović.
Publ. Obs. Astron. Beograd, No. 16, (see 012.021), p, 97 - 112 (1969).　In Serbo-Croatian.

041.046　Atlas Stellarum 1950.0.
H. Vehrenberg.
Treugesell-Verlag KG, Düsseldorf. Deliveries 9 and 10. 35, respectively 28 charts (1970).

Rigorous computation of proper motions and their effects on star positions.　See Abstr. 112.006.

Some results of the determination of absolute proper motions of stars referred to galaxies.
See Abstr. 112.014.

Accurate positions of 502 stars in the region of the Pleiades.　See Abstr. 153.006.

042 Celestial Mechanics

042.001 **Poor man's guide to celestial mechanics.**
H. Alfvén.
Nature, Vol. 225, 229 (1970).
A simple approximate method can be used for some problems in celestial mechanics.

042.002 **Some simple results regarding gravitational disturbance by exterior planets – with historical applications.** D. Rawlins.
Monthly Notices, Roy. Astron. Soc., Vol. 147, 177 - 186 (1970).
Because of the present apparent difficulties with Neptune's motion, Brown's transformation of the longitudinal perturbation problem is revived and past papers on the subject are criticized. The awkward behaviour of the traditional perturbation function is remarked, for purposes of contrast, but then is examined in relation to the disturbed elements and thereby shown to represent no physical difficulty, thus incidentally invalidating Peirce's case against the Adams-Leverrier discovery of the planet Neptune.

042.003 **Sur une généralisation de la solution analytique du système d'équations canoniques de Hori résultant de l'élimination des termes à courte période d'une théorie planétaire générale du premier ordre.** J. Meffroy.
Comptes Rendus Acad. Sci. Paris, Sér. A, Vol. 270, 166 - 168 (1970).
On a obtenu, dans une note précédente, une solution analytique des équations canoniques de Hori résultant de l'élimination des termes à courte période de la partie principale F_{1p} de la fonction perturbatrice d'une théorie planétaire générale du premier ordre lorsque F_{1p} est réduit à ses termes de degrés 0, 1, 2 en X, Y, P, Q. On généralise cette solution et on calcule complètement les constantes qui la définissent.

042.004 **Determination of the elements of an intermediary orbit from initial dates in the hyperbolic motion.**
V. M. Chepurova.
Soobshch. Gos. Astron. Inst. Shternberga, No. 159, p. 3 - 13 (1969). In Russian.

042.005 **Orbits of hyperbolic type in the problem of two fixed centers.** B. N. Noskov.
Soobshch. Gos. Astron. Inst. Shternberga, No. 159, p. 14 - 42 (1969). In Russian.

042.006 **Improvement of the elements of an intermediary orbit in the hyperbolic case.**
B. N. Noskov.
Soobshch. Gos. Astron. Inst. Shternberga, No. 159, p. 43 - 54 (1969). In Russian.

042.007 **The trigonometrical theory of secular perturbations of major planets.**
M. V. Anolik, G. A. Krasinsky, L. J. Pius.
Trudy Inst. Teoret. Astron., *Leningrad,* Vyp. (No.) 14, p. 3 - 47 (1969). In Russian.
The Lagrange–Laplace method for obtaining secular perturbations has been extended to the terms of the fourth order in the eccentricities and inclinations in the development of the disturbing function. The secular perturbations of eight major planets, Pluto excluded, have been represented in purely trigonometric form, using for this aim the method of canonical transformations. The obtained numerical results permit to study the evolution of the planetary orbits during a very long time interval.

042.008 **Mean motions and characteristics exponents at the libration points.**
V. Szebehely, D. S. Ingram, J. E. Cochran.
Astron. Journ., Vol. 75, 92 - 95 (1970).
The roots of the characteristic equation at the equilibrium points of the restricted problem of three bodies are presented as functions of the mass parameter, μ. The explicit solution of this problem is given in an approximate form by a few members of the appropriate Taylor series. The series expansions are in terms of $\mu^{1/3}$, $\mu^{1/2}$, and μ depending on the location of the equilibrium point and on the nature of the root.

042.009 **An algorithm for integrating stepwise the restricted problem in Thiele's coordinates.**
R. H. Estes, E. R. Lancaster.
Celestial Mechanics, Vol. 1, 297 - 300 (1970).
Recurrence formulas are developed for the coefficients in the Taylor Series expansions of the solution of the planar restricted three body problem in the regularized form of Thiele.

042.010 **Investigation of the hydrodynamic analogy in the restricted problem of three bodies.**
J. D. Mulholland.
Celestial Mechanics, Vol. 1, 320 - 330 (1970).
The hydrodynamic analogy concept is examined from the standpoint of the possibility of physical realization of an analog device. The conditions that must be satisfied, conservation of mass and momentum and uniqueness of physical properties, are discussed in detail and applied to examples, including the Birkhoff formulation.

042.011 **On oscillatory motion in the problem of three bodies.** D. G. Saari.
Celestial Mechanics, Vol. 1, 343 - 346 (1970).
In the case of oscillatory motion in the problem of three bodies it is shown that as $t \to \infty$ the mutual distances between particles cannot separate faster than $Ct^{2/3}$ where C is some positive constant. As bounding functions of time exist for the other classifications of motion in the three body problem, it follows in general that the mutual distances between particles is $0\,(t)$ as $t \to \infty$.

042.012 **Escape from a gravitational system of positive energy.** H. Pollard, D. G. Saari.
Celestial Mechanics, Vol. 1, 347 - 350 (1970).
It is a still unproved conjecture that a Newtonian gravitational system of positive energy must decrease in the sense that at least one particle must escape from the center of mass of the system as $t \to \infty$. Some partial results were obtained in Pollard (1967). An improvement is offered in this one.

042.013 **The stability of the triangular Lagrangian points for commensurability of order two.**
K. T. Alfriend.
Celestial Mechanics, Vol. 1, 351 - 359 (1970).
The two variable expansion method is used to show that the triangular equilibrium points in the planar restricted problem of three bodies are unstable at $\mu = \mu_2 = 0.024293 \dots$. It is also shown that even though the equilibrium points are stable for $\mu \approx \mu_2$ finite motions in the neighborhood of equilibrium points may be unstable.

042.014 **Periodic orbits emanating from a resonant equilibrium.** J. Henrard.

Celestial Mechanics, Vol. 1, 437 - 466 (1970).

For a conservative Hamiltonian system with two degrees of freedom, in the case where the two frequencies at an equilibrium of the elliptic type are commensurable or close to being so, completely canonical transformations can be formally constructed in explicit terms under the form of Lie transforms to the effect that it renders one angle coordinate ignorable and gives to the transformed Hamiltonian the form of what Garfinkel calls an ideal problem of resonance. For the problem so reduced, the unnormalized residual being omitted, natural families of periodic orbits are analyzed, their emergence from the equilibrium is discussed as well as their characteristic exponents. Special attention is given to the evolution of the system of natural families under a continuous transition through the resonance band.

042.015　**Orbit determination on the basis of the integrals of the two-body problem in spherical coordinates.**
M. P. Imnadze.
Byull. Abastumansk. Astrofiz. Obs. No. 37, p. 165 - 176 (1969). In Russian.

042.016　**The pseudo-anomaly and its application to ephemeris calculation.**　M. P. Imnadze.
Byull. Abastumansk. Astrofiz. Obs. No. 37, p. 177 - 184 (1969).　In Russian.

042.017　**Prüfung der Theorie der Erdbahn durch numerische Integration.**　J. Schubart.
SuW, Vol. 9, 13 - 14 (1970).

042.018　**Sur un terme séculaire mixte de la fonction déterminante d'une théorie planétaire générale du premier ordre construite à l'aide des variables canoniques de Hori.**
J. Meffroy.
Comptes Rendus Acad. Sci. Paris, Sér. A, Vol. 270, 911 - 913 (1970).

On met en évidence un terme séculaire mixte en λ dans l'expression de la fonction déterminante S_{1p} qui élimine les termes à courte période provenant de la partie principale F_{1p} de la fonction perturbatrice d'une théorie planétaire générale du premier ordre construite à l'aide des variables canoniques de Hori H, X, P, λ, Y, Q et réduite à ses termes de degrés 0, 1, 2 en X, Y, P, Q.

042.019　**Geometry of Keplerian motion.**
H. Kinoshita, R. Nagai, M. Yoneda.
Tokyo Astron. Obs. Rep., Vol. 15, No. 1 (No. 56), p. 168 - 187 (1970).　In Japanese.

042.020　**The motion of a satellite of a spheroidal planet in case of small eccentricity and inclination.**
G. T. Arazov.
Dokl. AN Azerb. SSR, Vol. 25, No. 6, p. 10 - 14 (1969).
In Russian. – Abstr. in Referativ. Zhurn. 51. Astron., 3.51.136 (1970).

042.021　**The integral of bipolar momenta in some problems of celestial mechanics.**
V. V. Radzievsky, E. F. Brazhnikova.
Astron. Zhurn. Akad. Nauk SSSR, Vol. 47, 211 - 216 (1970). In Russian. English translation in Soviet Astron.,AJ, Vol. 14, No. 1.

The idea of the integral of bipolar momenta, equal to a product of kinetic momenta of a particle relative to two poles, is introduced. The case, when these poles are bodies with final masses in the restricted problem of three bodies and in the problem of two fixed centres, is investigated. In the last case the integral of bipolar momenta is easily obtained in spherical coordinates. It is proposed to use surfaces of zero momenta for the qualitative analysis of the motion.

042.022　**The photo-gravitational restricted problem of three bodies.**　Y. A. Chernikov.
Astron. Zhurn. Akad. Nauk SSSR, Vol. 47, 217 - 223 (1970). In Russian. English translation in Soviet Astron.,AJ, Vol. 14, No. 1.

The restricted problem of three bodies: the sun – planet – particle is considered taking into account the solar radiation pressure. The differential equations of the motion of the particle are obtained and it is shown that in spite of the lack of Jacobi's integral they permit particular solutions, corresponding to six points of libration. A. M. Ljapunov's first method proves the instability of the solutions.

042.023　**A method of constructing an analytical trigonometric theory of the motion of resonance asteroids.**
E. A. Grebenikov.
Astron. Zhurn. Akad. Nauk SSSR, Vol. 47, 431 - 440 (1970). In Russian. – English translation in Soviet Astron., AJ, Vol. 14, No. 2.

The basis of a method, allowing to construct the analytical theory of the motion of asteroids, the mean motions of which are commensurable or almost commensurable with the mean motion of Jupiter, is stated. This theory does not contain secular inequalities. The point of this method consists in using a solution of the restricted problem as an intermediate orbit.

042.024　**Numerical investigations of the stability of periodic solutions in the restricted three-body problem.**
K. Ziołkowski.
Postępy Astron., Vol. 18, 193 - 196 (1970).　In Polish.
Abstr. Polish Astron. Soc.

042.025　**Stability of the triangular points in the elliptic restricted problem of three bodies.**
A. H. Nayfeh, A. A. Kamel.
AIAA Journ., Vol. 8, 221 - 223 (1970).

A perturbation scheme is used to study the stability of infinitesimal motions about the triangular points in the elliptic restricted problem of three bodies. Fourth-order analytical expressions for the transition curves that separate stable from unstable orbits in the μ–e plane are given. These power series are recast into rational fractions to extend their validity to larger values of eccentricity.

042.026　**Necessary condition for applicability of a unified perturbation theory.**　P. D. Usher.
Bull. American Astron. Soc., Vol. 2, 223 (1970). – Abstr. AAS.

042.027　**How to assemble a Keplerian processor.**
R. Broucke.
Celestial Mechanics, Vol. 2, 9 - 20 (1970).

This article describes how a set of computer programs has been constructed to perform the literal series expansions of the two-body problem. The different steps of the approach are outlined, from the basic generation of fundamental Kepler functions with the aid of Bessel series to the construction of derived Kepler functions by elementary Poisson series operations. The different tests and checks which have been made are also described. The most extensive test application of the package of programs, the expansion of the lunar disturbing function, is included at the end of the article.

042.028　**A discussion of Hill's method of secular perturbations and its application to the determination of the zero-rank effects in non-singular vectorial elements of a planetary motion.**　P. Musen.
Celestial Mechanics, Vol. 2, 41 - 59 (1970).

We have developed a set of symmetric formulas for

computing the zero-rank perturbations in the framework of Hill's theory. This symmetry facilitates the optimization of programming. The vectorial elements of motion are chosen in such a manner that the system can also be used in those cases when the eccentricity or the inclination of the orbit becomes small but oscillates in a wide interval.

042.029 Solution of Lambert's problem for short arcs.
        E. R. Lancaster.
Celestial Mechanics, Vol. 2, 60 - 63 (1970).
    Approximation formulas are found for $\dot{x}(0)$ and $\dot{x}(1)$, where $x(t)$ satisfies $\ddot{x} = f(x, t)$, $x(0) = x_0$, $x(1) = x_1$. The results are applied to an example of two-body motion.

042.030 Recurrence formulae for the Hansen's developments.
        N. X. Vinh.
Celestial Mechanics, Vol. 2, 64 - 76 (1970).
    In this paper we derive some recurrence formulae which can be used to calculate the Fourier expansions of the functions $(r/a)^n \cos mv$ and $(r/a)^n \sin mv$ in terms of the eccentric anomaly E or the mean anomaly M. We also establish a recurrence process for computing the series expansions for all n and m when the expansions of two basic series are known.

042.031 An Encke-type special perturbation method.
        G. H. Born.
Celestial Mechanics, Vol. 2, 103 - 113 (1970).
    In this paper, a combination analytical-numerical integration method for solving the differential equations of a modified set of Lagrange's planetary equations is described. The integration method is an Encke-type method because it involves integrating the deviations between the actual trajectory and a reference trajectory. The reference trajectory is obtained from an analytical solution containing the dominant secular and periodic effects of the gravitational field of the primary body. A set of nonsingular elements is used so that the method will be valid for all circular and elliptical motions. It is shown that the method is an accurate and efficient means of satellite ephemeris generation.

042.032 The equivalence of von Zeipel mappings and Lie transforms.   H. Shniad.
Celestial Mechanics, Vol. 2, 114 - 120 (1970).
    An order independent method is used to identify the transformation obtained from von Zeipel's method with that obtained from Lie transforms. The correspondence between the generators is given in explicit form.

042.033 On the ideal resonance problem – II.   A. H. Jupp.
        Monthly Notices, Roy. Astron. Soc., Vol. 148, 197 - 210 (1970).
    For problems with an ideal resonance structure, the author has previously published a paper which presented a new method of solution for the region of libration. In the present publication the general nature of the higher order solution is discussed. It is found that, to any power of $\mu^{1/2}$, an asymptotic solution can be constructed, free from singularities and mixed secular terms.

042.034 Resonance in the restricted problem.
        W. H. Jefferys.
Symposia Mathematica, Vol. 3, 13 - 20 (1970). – Conference paper (see 012.010).

042.035 Double resonance in the motion of a satellite.
        G. E. O. Giacaglia.
Symposia Mathematica, Vol. 3, 45 - 63 (1970). – Conference paper (see 012.010).

042.036 Numerical study of dynamical systems with three degrees of freedom. I. Graphical displays of four-dimensional sections.   C. Froeschlé.
Astron. Astrophys., Vol. 4, 115 - 128 (1970).
    Numerical and graphical methods have been developed in order to study dynamical systems with three degrees of freedom. Using the method of the "Surface of Section" the problem is reduced to the study of a set points in a four-dimensional space. Two graphical methods are described and used. These methods have been tested in a particular case of the three-dimensional restricted three-body problem with perturbations. For orbits close to the one of the bodies, the points appear to lie on a two-dimensional manifold, for more distant orbits they appear to fill a manifold with three dimensions at least.

042.037 On one first vector integral of planetary motion.
        B. Vujanovic.
Astronaut. Acta, Vol. 15, 231 - 233 (1970).
    In this note an independent first vector integral of planetary motion is derived. The connection with the well known Runge-Lenz vector and the vector of moment of momentum is also discussed. One two parameter infinitesimal transformation group, which leaves the corresponding Hamiltonian invariant, is found.

042.038 Concerning the genealogy of long period families at L_4.   J. Henrard.
Astron. Astrophys., Vol. 5, 45 - 52 (1970).
    The family of long period orbits at L_4 does not evolve in a continuous manner with the mass ratio. Extended numerical continuations carried with very great accuracy enable us to follow the evolution of the long period family from the inner resonance 1/4 to the inner resonance 1/1 through collapses at the equilibrium, mergers with bridges and redistribution of branches. This study stresses the extreme complexity of the phase space around an equilibrium for a one-parameter family of conservative Hamiltonian systems with two degrees of freedom.

042.039 Semi-numerical method for solving Hill's problem. – Application to Phoebe.   B. Elmabsout.
Astron. Astrophys., Vol. 5, 68 - 83 (1970).   In French.
    We intend to find an orbit in the neighbourhood of the variational orbit which will include all the perturbations depending on eccentricity and inclination. As far as the effects of the first order in eccentricity are concerned we have to solve Hill's equation. A method using successive approximations gives the general solution of this equation and the value of constant c which characterizes the motion of the periastron. The convergence of this method is demonstrated rigorously for small positive and negative values of parameter m.

042.040 Numerical study of dynamical systems with three degrees of freedom. II. Numerical displays of four-dimensional sections.   C. Froeschlé.
Astron. Astrophys., Vol. 5, 177 - 183 (1970).
    Further numerical methods of a non-graphical nature have been used to study dynamical systems with three degrees of freedom: 1) a local fitting of the points in the four-dimensional "surface of section" by a quadratic surface; 2) a computation of the divergence of two orbits with nearly identical initial conditions. These methods have been tested in a particular case of the three-dimensional restricted three-body problem. The results obtained previously by graphical methods are confirmed.

042.041 Characteristic exponents at L_4 in the elliptic restricted problem.   A. Deprit, A. Rom.
Astron. Astrophys., Vol. 5, 416 - 425 (1970).
    The systems considered here are Hamiltonian, linear with periodic coefficients and depending on a small parameter. A perturbation method based on Lie transforms is proposed to develop the characteristic exponents of the monodromy

matrix in power series of the small parameter. The algorithm is applied to the triangular Lagrangian solutions in the elliptic restricted problem. The principal parts of their characteristic exponents have been obtained in literal form as irrational functions of the mass ratio. The stability of the Lagrangian solutions is discussed in relation to the eccentricity of the primaries.

042.042 **Periodic solutions close to commensurabilities in the three-body problem.** A. T. Sinclair.
Monthly Notices, Roy. Astron. Soc., Vol. 148, 325 - 351 (1970).

Various families of periodic solutions are shown to exist in the three-body problem in which the two secondary bodies are close to a commensurability in mean motions. Both the restricted problem and the planar non-restricted problem are considered. In the restricted problem the disturbing body has either an eccentric orbit in the plane of the disturbed body's orbit, or a circular orbit inclined to this plane. The existence of some of these solutions is verified numerically.

042.043 **On the matrizant of the two-body problem.**
R. A. Broucke.
Astron. Astrophys., Vol. 6, 173 - 182 (1970).

The present article develops a general theory for the construction of the matrizant of the two-body problem, or any other completely integrable dynamical system. This theory essentially constructs the fundamental matrix of the variational equations and then shows how its inverse is obtained. The matrizant is in this way decomposed in a product of two matrices. The matrices which are thus developed have essentially two main applications: computation of perturbations and differential correction of orbits with observations. Our results include the partial derivatives of the velocity components and thus make possible the applications to radar Doppler observations.

042.044 **Some periodic orbits in the elliptic restricted problem of three bodies.** P. J. Shelus, S. S. Kumar.
Astron. Journ., Vol. 75, 315 - 318 (1970).

This paper gives a preliminary report of the numerical exploration into the effects of the ellipticity of the orbits of the primaries on the linear stability characteristics of certain periodic orbits in the circular case. Our results show that period orbits of the infinitesimal body which are unstable in the circular case remain unstable in the elliptic case. Further we find that for high enough eccentricity of the orbits of the primaries, periodic orbits which are stable in the circular case become unstable in the elliptic case.

042.045 **Aplicaciones de un cierto tipo de transformaciones canónicas a la mecánica celeste.**
J. L. Sérsic.
Obs. Astron. Univ. Nacional La Plata, Ser. Astron., Vol. 35, 30 pp. (1969).

A formalism based on a particular kind of canonical transformations is developed and used to discuss the general solutions of the differential equations of the theory of perturbations in celestial mechanics.

042.046 **The geometry of the Roche coordinates.**
M. Kitamura.
Astrophys. Space Sci., Vol. 7, 272 - 358 (1970).

The exact geometry of the Roche curvilinear coordinates (ξ, η, ζ) in which ξ corresponds to the zero-velocity surfaces is investigated numerically in the plane, as well as in the spatial, case for various values of the mass-ratio between the two point-masses (m_1, m_2) constituting a binary system.

042.047 **A discussion of Hill's method of secular perturbations and its application to the determination of**

the zero-rank effects in non-singular vectorial elements of a planetary motion. P. Musen.
Goddard Space Flight Center, Greenbelt, Maryland. NASA Technical Note, NASA TN D-5709, 28 pp. (1970).

We have developed a set of symmetric formulas for computing the zero-rank perturbations in the framework of Hill's theory. This symmetry facilitates the optimization of programming. The vectorial elements of motion are chosen such that the system can also be used when the eccentricity or the inclination of the orbit becomes small but oscillates in a wide interval.

042.048 **Stability of the float-type solution for axial symmetric bodies.** H. Kinoshita.
Univ. Tokyo, Tokyo Astron. Obs. Report, Vol. 15, (No. 2), (No. 57), p. 330 - 340 (1970). In Japanese.

042.049 **On the non-linear equations of secular perturbations of minor planets.** H. Kinoshita.
Univ. Tokyo, Tokyo Astron. Obs. Report, Vol. 15, (No. 2), (No. 57), p. 341 - 346 (1970). In Japanese.

042.050 **Hansen's method of partial anomalies. I. Theory.**
V. I. Skripnichenko.
Byull. Inst. Teoret. Astron., *Leningrad*, Vol. 12, 16 - 54 (1970). In Russian.

The method of partial anomalies devised by Hansen in 1856 is described. The formulas suitable for application of the method using computers are presented.

042.051 **On a case of generalization of the problem of two fixed centres.** V. K. Kaysin.
Byull.Inst. Teoret. Astron., *Leningrad*, Vol. 12, 163 - 171 (1970). In Russian.

This article deals with the generalization of the problem of two fixed centres in the case, where a material point is attracted to the origin of coordinates with a force proportional to the radius. The solution of the equations of motion of the material point was obtained by means of Hamilton-Jacobi's method. The possible application of the solution to some problems of stellar dynamics is shown.

042.052 **About free motion of mechanical systems under the supposition of a retardment of inner forces action.** P. P. Loginov.
Byull. Inst. Teoret. Astron., *Leningrad*, Vol. 12, 172 - 194 (1970).

Expansion and disintegration of mechanical systems without outer forces are explained by means of the retardment of the inner forces action. An interpretation of the retardment and the estimates of the values of the retardment are given too. The values of the retardment proved to be negligible; at the distance of 10 a.u. they are within one second.

042.053 **On the determination of intermediate orbits of the planetary satellites disturbed by the solar attraction.** A. A. Orlov.
Byull. Inst. Teoret. Astron., *Leningrad*, Vol. 12, 195 - 205 (1970). In Russian.

The system of formulas for approximative computation of spatial motion in Hill's problem has been given. Two numerical examples are considered. The results of computations by means of the system of formulas are compared with those obtained by numerical integration.

042.054 **A comparison of perturbations from Neptune in the motion of Ceres as determined by the methods of Laplace – Newcomb and Hill.** R. K. Sadreeyev.
Byull. Inst. Teoret. Astron., *Leningrad*, Vol. 12, 206 - 215 (1970).

The perturbations in the motion of Ceres from Neptune were determined by the analytical method of Laplace – Newcomb. Calculations performed by V. F. Proskurin by Hill's method were used for comparison and control.

042.055 The solution of the problem of two fixed centres in the hyperbolic case. W. M. Chepurova.
Byull. Inst. Teoret. Astron., *Leningrad,* Vol. 12, 216 - 233 (1970). In Russian.
The motion in the hyperbolic case of the problem of two fixed centres is considered. The final formulae for the elliptical coordinates of the moving point are deduced. The coordinates are functions of the time and some constants, which are the elements of the constructed intermediate orbit.

042.056 A surface of zero moments in the problem of two fixed centres.
E. F. Brazhnikova, V. V. Radzievsky.
Astron. Zhurn. Akad. Nauk SSSR, Vol. 47, 650 - 659 (1970). In Russian. English translation in Soviet Astron.,AJ, Vol. 14, No. 3.
A new case of the applicability of the model of two fixed centres, corresponding to the fast revolution of massive components along an elliptic orbit and to the slow movement of a very remote dust particle, is discussed.

042.057 The existence of an integral of equations of the translational satellite motion in the circular problem of three bodies. G. F. Osipov.
Astron. Zhurn. Akad. Nauk SSSR, Vol. 47, 676 - 678 (1970). In Russian. English translation in Soviet Astron.,AJ, Vol. 14, No. 3.
The integral analogous to Jacobi's is obtained in the restricted problem of three bodies. One of the bodies has three planes of dynamical symmetry.

042.058 On the canonical system of solutions in celestial mechanics. S. Kikuchi.
Sci. Rep. Tôhoku Univ. Sendai, Japan, First Ser., Vol. 52, 142 - 148 (1969).
S. Lie's theorem concerning the canonical feature of the infinitesimal transformation is generalized to the case when the infinitesimal generator depends on the time. A formal solution is obtained by the inverse operator from the non-conservative integral obtained in the preceding paper. An approximate solution of the equation of perturbation is obtained up to the third order of small σ.

042.059 On the mixed-secular terms in perturbed intermediaries. B. Garfinkel, G.-I. Hori, K. Aksnes.
Astron. Journ., Vol. 75, 651 - 656 (1970).
Three results are established in this paper: (i) mixed-secular terms arise in consequence of the use of the Delaunay elements in a non-degenerate intermediary, (ii) no such terms appear in the calculated position coordinates, and (iii) such terms can be entirely avoided in the analysis by the use of the so-called "natural" elements, derived by a linear combination of the action and angle variables.

042.060 The moon's influence on the location of the sun-earth exterior libration point.
R. W. Farquhar.
Celestial Mechanics, Vol. 2, 131 - 133 (1970).
The moon's influence of the location of the sun–earth exterior libration point affects a shift of about 20 km relative to the dynamic equilibrium point of the restricted four-body problem (sun, earth, moon, satellite) in the direction of x-axis.

042.061 An iterative method of general perturbations programmed for a computer.
P. K. Seidelmann.
Celestial Mechanics, Vol. 2, 134 - 146 (1970).
The article describes the iterative Hansen's method together with some of the necessary subroutines which have been programmed to handle the series on a digital computer. The method has been tested by generating the theory of a major planet and the theory of a minor planet having a near commensurability with Jupiter and some of the results have been described in the last section of the paper.

042.062 A separable potential in triaxially ellipsoidal coordinates satisfying the Laplace equation.
S. J. Madden, Jr.
Celestial Mechanics, Vol. 2, 217 - 227 (1970).
This paper derives the most general potential function which allows separation of the Hamilton-Jacobi equation in orthogonal coordinates and which satisfies the Laplace equation. The resulting potential is then specialized to the case of interest for near-earth satellites, where the proper behavior of the potential at infinity is obtained and singularities in the region of interest are eliminated. The Vinti potential is found as a special case.

042.063 The stability of motion in a periodic cubic force field. J. H. Bartlett, C. A. Wagner.
Celestial Mechanics, Vol. 2, 228 - 236 (1970).
The non-linear differential equation $\ddot{x} + p(t) x^3 = 0$, where $p(t)$ is a periodic square wave function of time with period τ, has been integrated by using a table of Jacobian elliptic functions. In the neighborhood of a typical elliptic fixed point, namely that for 11τ, 12-decimal accuracy has been used to determine a region which is stable.

042.064 Variable constant of gravitation and celestial mechanics. A. M. Finkelstein.
Celestial Mechanics, Vol. 2, 237 - 252 (1970).
In the present paper the celestial mechanics consequences (light deflection, radar ranging of the planet, geodetic precession and secular effects in the orbital elements in the two-body problem) for the class of the theories based on the vacuum Jordan's Lagrangian has been considered.

042.065 On the motion of test masses in the gravitational field of a rotating body.
N. V. Mitskievič, I. Pulido Garcia.
Dokl. Akad. Nauk SSSR, Ser. Mat. Fiz., Vol. 192, 1263 - 1265 (1970). In Russian.

042.066 Méthode de Laplace modifiée pour la détermination des orbites à l'aide des O – C.
B. Popović.
Bull. Obs. Astron. Beograd, Vol. 27, No. 2, p. 29 - 32 (1969). In Esperanto.

042.067 Methods for computing the perturbations in the rectangular coordinates of planets. II.
L. K. Babadzhanyanz.
Vestn. Leningr. un-ta, 1969, No. 19, p. 134 - 145. In Russian. Abstr. in Referativ. Zhurn. 51. Astron., 5.51.141 (1970).

042.068 "Comment on canonical transformation applications to optimal trajectory analysis". A. Deprit.
AIAA Journ., Vol. 8, 1182 - 1183, with a reply by W. F. Powers, B. D. Tapley.

042.069 A note concerning an application of the generalized Huang's model of the restricted four-body problem.
V. Matas.
Bull. Astron. Inst. Czechoslovakia, Vol. 21, 139 - 152 (1970).
Motion of an infinitesimal body is examined in the vicinity of the smallest primary of the generalized Huang's

model. The disturbing function of the problem given is shown to consist of four simple terms if $\mu^{1/4}$, μ– mass ratio of the moon and earth, taken as a small parameter. Their first-order disturbing effect is studied separately by an analytical method. A possible application to nearlunar satellites is mentioned, too.

042.070 Errata: Perturbation of libration points of the restricted three-body problem due to gravitational and radiative influence of a fourth body. Existence of a periodic solution in the vicinity of the libration points.
[Bull. Astron. Inst. Czechoslovakia, Vol. 20, 322 - 326 (1969), see 02.042.028]. V. Matas.
Bull. Astron. Inst. Czechoslovakia, Vol. 21, 193 (1970).

042.071 Celestial mechanics. W. J. Eckert.
Trans. IAU, Vol. 14A, (see 003.028), 19 - 37 (1970). – Report of Commission 7.

042.072 Collisions in the reduced 3-body problem.
L. Markus, C. Weaver.
American Journ. Math., Vol. 91, 385 - 394 (1969). – See Zentralblatt Math. Grenzgebiete, Vol. 184, 498 (1969).

042.073 Die Gröbnersche Methode und ihre Anwendung auf die numerische Bahnberechnung.
H. Knapp.
Bul. Inst. politehn. Iaşi, Nouvelle Sér., Vol. 14 (18), No. 3/4, p. 79 - 82 (1969).

042.074 The indirect effect of solar attraction on the motion near the triangular libration points of the earth-moon-system. L. G. Luk'yanov.
Vestn. Mosk. un-ta. Fiz., astron., 1969, No. 6, p. 9 - 15. In Russian.– Abstr. in Referativ. Zhurn. 51. Astron. 7.51.114 (1970).

042.075 Construction of an optimum elliptic orbit with per - manent revolution in the equatorial plane of an axisymmetric planet. V. S. Novoselov.
Vestn. Leningr. un-ta, 1969, No. 19, p. 146 - 152. In Russian. - Abstr. in Referativ. Zhurn. 51. Astron., 7.51.118 (1970).

042.076 Period of low orbits. R. H. Good.
American Journ. Phys., Vol. 38, 540 - 541 (1970).
See Phys. Abstr., Vol. 73, No. 37929 (1970).

042.077 Regularization theory for the elliptic restricted three body problem. R. F. Arenstorf.
Differential Equations, Vol. 6, 420 - 451 (1969).
This paper transfers to the elliptic case the known results from the restricted three body problem about the dynamical meaning (collisions) and the character (algebraic branch points) of real singularities and the 'existence' of the solutions (after analytic continuation through singularities) on the real time axis.

042.078 Close-in orbits in the restricted problem of three bodies. R. B. Barrar.
Quarterly Applied Math., Vol. 27, (No. 3), 396 - 398 (1969).
Utilizing a transformation due to Birkhoff (1915) the author establishes for the restricted problem of three bodies the existence of conditionally periodic orbits that move in a small neighborhood of one of the primaries. These orbits result from perturbation of elliptic orbits and are valid for all mass ratios of the two primaries.

042.079 Validity of Hansen's method of partial anomalies. P. E. Nacozy.
Thesis, Yale Univ., New Haven, Conn. [Available from Univ. Microfilms, Ann Arbor, Mi.], 173 pp. (1969).
In 1847, Hansen described and partially applied a method of general perturbations, called the method of partial anomalies. This study presents an application of the method to comet Encke as perturbed by the earth so as to complete Hansen's numerical example and to present a definitive numerical proof of the method.

042.080 On non–linear prediction in dynamics. C. A. Altavista.
Asoc. Argentina Astron. Bol., No. 12, (see 012.019), p. 16 - 18 (1968). In Spanish. – Abstract.

042.081 A numerical solution of Kepler's problem in universal variables. P. E. Zadunaisky, R. C. Blanchard.
Asoc. Argentina Astron. Bol., No. 14 (see 012.020), p. 47 - 48 (1968). In Spanish.

042.082 An application of Hansen's method of general perturbations to a minor planet. P. K. Seidelmann.
Thesis, Univ. Cincinnati, Ohio. [Available from Univ. Micro - films, Ann Arbor, Mi.], 79 pp. (1969).
Hansen's classical method of general perturbations wave revised from its basis on a Taylor series expansion to an iterative method, which can start with the best theory known, or with only the mean, or osculating, elements of the planet. The method and the necessary sub-routines to handle the Fourier series were programmed for a digital computer.

Principles of Celestial Mechanics.
See Abstr. 003.002.

Anwendung der Hammersteinschen Methode der unendlich vielen Variablen auf Probleme der Satellitengeodäsie und Himmelsmechanik. See Abstr. 046.018.

The thesis by I. N. Uljanov "Olbers' method for the determination of parabolic orbits and its application to the comet 1853 Klinkerfues". See Abstr. 103.109.

043 Astronomical Constants

043.001   Secular changes in the lunar elements.
C. F. Martin, T. C. Van Flandern.
Science, Vol. 168, 246 - 247 (1970).
   Corrections to the adopted values for centennial rates
of change of four elements of the lunar orbit, the location of
the FK4 equinox, and the obliquity of the ecliptic are presented. They are derived from analyses of lunar occultations
distributed over several centuries. Generally, these corrections
help to resolve existing discrepancies between theory and observations.

043.002   On the secular change of the obliquity of the
          ecliptic.   J. H. Lieske.
Astron. Astrophys., Vol. 5, 90 - 101 (1970).
   The motion of the earth-moon barycenter is examined
as a possible source of the discrepancy between the observed
and theoretical rates of change of the obliquity of the
ecliptic. Analysis of the barycenter's components of orbital
angular momentum indicates that Newcomb's rate of change
of obliquity is essentially correct. Observations of Eros are
re-examined for the determination of a secular error in the
obliquity and it is concluded that the data do not indicate
the existence of a secular error in obliquity which is
significantly different from zero.

043.003   La précession—nutation et la structure de la
          Terre.   R. O. Vicente.
Ciel et Terre, Vol. 86, 177 - 186 (1970).

043.004   Using one pendulum and a rotating mass to measure
          the universal gravitational constant.
D. M. Sheppard.
American Journ. Phys., Vol. 38, 380 (1970).

043.005   A determination of the astronomical unit from
          hydrogen-line radial-velocity measurements.
S. H. Knowles.
Thesis, Yale Univ., New Haven, Conn. [Available from Univ.
Microfilms, Ann Arbor, Mich., USA Order No. 69 - 8375],
155pp. (1969)
   The scale of the earth's orbit has been determined by
measuring the radial velocity of a radio wavelength spectral
line. The use of 21-centimeter neutral hydrogen absorption
features has allowed a determination considerable more accu -
rate than previous experiments using optical-wavelength lines.
From a year's series of measurements, the astronomical unit
is determined to be 149.588.000 km ± 10.000 km., in satisfac -
tory agreement with the currently accepted value.

   **Variable constant of gravitation and celestial
mechanics.**   See Abstr. 042.064.

   **On a possible connection between solar activity and
quasi-periodical variability of the gravitational constant.**
See Abstr. 072.015.

   **Martian mass and earth-moon mass ratio from
coherent S-band tracking of Mariners 6 and 7.**
See Abstr. 097.012.

   **Some results of the determination of absolute
proper motions of stars referred to galaxies.**
See Abstr. 112.014.

   **Preliminary results of determinations of the proper
motions of stars with reference to galaxies.**
See Abstr. 112.016.

044 Time, Rotation of the Earth

044.001 **De tijdvereffening.** J. Meeus.
Hemel en Dampkring, Vol. 68, 21 - 27 (1970).

044.002 **Tijdschalen.** W. de Rop.
Hemel en Dampkring, Vol. 68, 44 - 45 (1970).

044.003 **Aspectos del progreso en el Servicio Nacional de la hora.** J. E. Marpegán.
Revista Astron., Vol. 41, No. 170, p. 59 - 63; No. 171, p. 21 - 33 (1969).

044.004 **The science of time.** D. A. Frank-Kamenetski.
Priroda, No. 3.70, p. 11 - 19 (1970). In Russian.

044.005 **Résultats des observations faites à Paris avec l'astrolabe impersonnel A. Danjon. Temps et latitude 1967.** G. Billaud.
Astron. Astrophys., Suppl. Series, Vol. 1, 263 - 279 (1970).
On donne les résultats des observations à l'astrolabe OPL n° 35 effectuées pendant 1967. Les résultats sont exprimés dans le système FK4.

044.006 **Time determination and secular variation of longitude.** M. Torao.
Univ. Tokyo, Tokyo Astron. Obs. Report, Vol. 15, (No. 2), (No. 57), p. 203 - 231 (1970). In Japanese.

044.007 **Some results of clock correction determinations from observations of pairs of stars situated symmetrically around the zenith.** N. V. Pyshnenko.
Astron. Zhurn. Akad. Nauk SSSR, Vol. 47, 633 - 640 (1970). In Russian. English translation in Soviet Astron., AJ, Vol. 14, No. 3.
It is shown that in the determinations of clock corrections by means of pairs of stars situated symmetrically around the zenith (A. A. Nemiro's method), the system of corrections is obtained better, and the mean error of clock corrections is decreased more considerably in comparison with usual methods.

044.008 **A control method of the determination of standard time.** G. P. Pilnik, G. S. Sitnik.
Astron. Zhurn. Akad. Nauk SSSR, Vol. 47, 641 - 649 (1970). In Russian. English translation in Soviet Astron., AJ, Vol. 14, No. 3.
Results of observations with transit instruments and photographic zenith tubes are compared. A precision estimate of definite moments of the International Time Service and Standard Time for 1960 - 1963 is given. Shortcomings of definite moments of the International Time Service and Standard Time for 1960 - 1963 is given. Shortcomings of methods of deriving definite systems are noted.

044.009 **Astronomische Zeit- und Breitenbestimmung. Empfangszeiten von Zeitsignalen.**
Edited by Deutsches Hydrographisches Institut, Hamburg. 1969 July - December. (1970).

044.010 **Réception des ondes de basse fréquence (VLF) à l'Observatoire de Bucarest.** V. Stavinschi.
Stud. Cerc. Astron., Vol. 15, 37 - 41 (1970). In Roumanian.

044.011 **Sur l'amélioration de la précision dans les enregistrements des observations astronomiques de l'heure.**
V. Stavinschi, R. Dorobanțu.
Stud. Cerc. Astron., Vol. 15, 43 - 58 (1970). In Roumanian.

044.012 **Détermination astronomique de l'heure en 1965 - 1968. Heure demi-definitive (systeme TU₂) en** $0!0001$**, 1965 - 1968.**
Z. Brkić, D. Djurović, M. Jovanović, Simić.
Bull. Obs. Astron. Beograd, Vol. 27, No. 2, p. 99 - 132 (1969).

044.013 **Some results of time determination with the Moscow photographic zenith tube.**
A. A. Tochilina.
Vestn. Mosk. un-ta. Fiz., astron., 1969, No. 5, p. 125 - 127. In Russian. – Abstr. in Referativ. Zhurn. 51. Astron., 4.51.149 (1970).

044.014 **Standard time and time zones in Canada.**
M. M. Thomson.
Journ. Royal Astron. Soc. Canada, Vol. 64, 129 - 162 (1970).

044.015 **Flutreibung und Akzeleration des Mondes.**
K. Ferrari d'Occhieppo.
Sternenbote, 13. Jahrgang, p. 54 - 59 (1970).

044.016 **Determination of time (TU 1).**
N. A. Omelina.
Tsirk. Astron. Inst., *Tashkent,* 1968 Nos. 6 (353), 8 (355), 10 (357); 1969 Nos. 12 (359), 14 (361); 1970 Nos. 15 (362), 16 (363), 18 (365), 19 (366). In Russian. – 1967 July – 1969 September.

044.017 **The rate of the earth's rotation in 1969 and the offset values of the standard frequencies from the nominal SI values to be adopted in 1970.**
D. Yu. Belotserkovsky.
Astron. Tsirk., No. 548, p. 7 - 8 (1970). In Russian.

044.018 **Rotation of the earth.** P. Melchior, S. Yumi.
Trans. IAU, Vol. 14ʌ, (see 003.028), 177 - 185 (1970). – Report of Commission 19.

044.019 **Time.** F. Zagar.
Trans. IAU, Vol. 14A, (see 003.028), 343 - 355 (1970). – Report of Commission 31.

044.020 **Synchronization of distant clocks by television pulse comparisons.** M. J. Miller.
Proc. Astron. Soc. Australia, Vol. 1, 352 (1970).

044.021 **The use of television signals for time and frequency dissemination.**
D. D. Davis, J. L. Jespersen, G. Kamas.
Proc. IEEE, Vol. 58, 931 - 933 (1970).

044.022 **Universal time and coordinates of the pole; Emission time of time signals; Coordinated time; Informations.**
Bureau International de l'Heure, Paris, Circ. D38 – D44 (1970).
Circular D of the BIH provides users with the current results relative to the universal time, the coordinates of the pole, the time of emission of time-signals, the coordination of time maintained by the laboratories.

044.023 **Determination of short-period tidal waves from observations with photoelectric transit instruments.**
T. K. Nikol'skaya, L. A. Solov'eva.
Trudy metrol. in-tov SSSR, 1969, vyp. (No.) 106 (166), p. 144 - 149. In Russian. – Abstr. in Referativ. Zhurn. 51. Astron., 7.51.145 (1970).

044.024 **Détermination astronomique de l'heure et de la latitude.**
Obs. Neuchâtel, Bull. (B), 1969 Novembre – 1970 Février (1970).

044.025 **L'heure astronomique définitive de l'Observatoire de Neuchâtel.**
Obs. Neuchâtel, Bull. (D), 1969 Novembre – 1970 Février (1970).

044.026 **Time and Latitude Service.**
Polish Acad. Sci., Astron. Latitude Station, Boro - wiec, Circ. Nos. 112 - 113 (1970). – 1969 October – 1970 March.

044.027 **International Time and Latitude Service at the Tokyo Astronomical Observatory.**
M. Huruhata.
Time and Latitude Bull., Tokyo Astron. Obs., Vol. 43, Nos. 8 - 12; Vol. 44, Nos. 1 - 2 (1969/70). – 1969 August – 1970 February. – Astronomical observations; System of universal times; System of UTC; Time keeping; International comparison of time; Radio time signals emitted from Japan.

044.028 **Détermination astronomique de l'heure et heures demi-définitives de réception des signaux horaires.**
Acad. Tchécoslov. Sci., Inst. Astron., Station de l'Heure, Prague, Sér. 5, Nos. 4 - 6 (1969). – 1969 July – December.

044.029 **Synchronisation des échelles de temps locales au moyen de signaux horaires dans la détermination des positions géographiques.**
F. Chlistovsky, E. Proverbio.
Proc. Internationale Navigationstagung, Hamburg, 1969, 19 pp. = Pubbl. Stazione Astron. Internazionale Latitudine, Carloforte-Cagliari, Nuova Ser., No. 8 (1970).

044.030 **Variations des velocités apparentes de phase et de groupe dans la régione des émissions LF.**
E. Proverbio, F. Chlistovsky.
Proc. Internationale Navigationstagung, Hamburg, 1969, 12 pp. = Pubbl. Stazione Astron. Internazionale Latitudine, Carloforte-Cagliari, Nuova Ser., No. 9 (1970).

044.031 **La note de précision du Service de l'Heure de l'Observatoire Astronomique de Beograd après l'installation des horloges à quartz.** Z. Brkić.
Publ. Obs. Astron. Beograd, No. 16, (see 012.021), p. 6 - 8 (1969). In Serbo-Croatian.
Une augmentation remarquable résultait d'installation des horloges à quartz dans le Service de l'Heure à l'Observatoire de Beograd. A l'aide de la précision des données des autres service de l'heure des observatoires participants au BIH et les données de notre Observatoire on a deduit la note de précision pour 1964.

044.032 **Contemporary Time Service in the world and at us.**
D. Šaletić.
Publ. Obs. Astron. Beograd, No. 16, (see 012.021), p. 83 - 89 (1969). In Serbo-Croatian.

Time signal service VNG.
Journ. Astron. Soc. Victoria, Vol. 23, 22 - 24 (1970).

Results of measuring the positions of the moon by the method of equal altitudes. See Abstr. 041.001.

Beitrag zur Objektivierung von Meridiandurchgangs-beobachtungen. See Abstr. 041.028.

Magnetohydrodynamic twisting oscillations in the core of the earth and variations of the length of the day. See Abstr. 081.020.

045 Latitude Determination, Polar Motion

045.001 **Comments on paper be E. Irving and W. A. Robertson, 'Test for polar wandering and some possible implications'.** J. Francheteau, J. G. Sclater.
Journ. Geophys. Res., Vol. 75, 1023 - 1027, with a reply by E. Irving and W. A. Robertson, p. 1027 (1970). – Letter.

045.002 **The Chandlerian nutation.**
R. G. Hipkin.
Quarterly Journ. Roy. Astron. Soc., Vol. 11, 43 - 48 (1970).
Report of a geophysical discussion held at Burlington House on 1969 January 29.

045.003 **Determination of coordinates of the instantaneous pole with both of the data of time and latitude observations.** S. Okazaki, M. Nasaka.
Tokyo Astron. Obs. Rep., Vol. 15, No. 1 (No. 56), p. 55 - 76 (1970). In Japanese.

045.004 **Variation of difference between coordinates of the instantaneous pole derived from the data of time observations and those of latitude observations.**
S. Okazaki, M. Nasaka, T. Yamazaki.
Tokyo Astron. Obs. Rep., Vol. 15, No. 1 (No. 56), p. 77 - 98 (1970). In Japanese.

045.005 **Effects of major seismic events on the rotation of the earth.** A. Ben-Menahem, M. Israel.
Geophys. Journ. Roy. Astron. Soc., Vol. 19, 367 - 393 (1970).
A spherical theory is advanced for perturbations of the earth's rotation by major earthquakes and explosions. Numerical results show that a single shallow earthquake of magnitude 8.5, occurring at a suitable latitude and with a favourable strike-azimuth, may suffice to maintain the Chandler wobble for about one year.

045.006 **On the Chandlerian motion of the earth's pole.**
V. V. Nesterov, L. V. Rykhlova.
Astron. Zhurn. Akad. Nauk SSSR, Vol. 47, 426 - 430 (1970).
In Russian. – English translation in Soviet Astron., AJ, Vol. 14, No. 2.
Melchior's laws of the Chandlerian motion were checked on the basis of calculated polar coordinates (1846.0 – 1891.5) and already known coordinates (1891.5 – 1965.0). Dependences between the period of Chandlerian motion, its amplitude and amplitude of annual motion were analysed by statistical methods. The authors did not find any real dependence between these parameters.

045.007 **Doppler satellite observations of polar motion.**
R. J. Anderle, L. K. Beuglass.
Bull. Géod., Nouvelle Série, No. 96, p. 125 - 141 (1970).
Observations on Navy navigation satellites made by thirteen Doppler receiving stations have been used to determine the position of the earth's pole daily for a six month period of time. A precision of one meter has been obtained on the basis of forty-eight hours of observations on one satellite. No bias is apparent between computations based on different satellites, but differences of about a meter exist with respect to values published by the Bureau International de l'Heure on the basis of astronomical observations.

045.008 **On the reduction of astronomical determinations of latitude, longitude and azimuth to a common epoch.** A. A. Mikhailov.
Astron. Zhurn. Akad. Nauk SSSR, Vol. 47, 613 - 618 (1970).
In Russian. English translation in Soviet Astron. AJ, Vol. 14, No. 3.
The annual reports of the International Polar Motion Service show that besides periodic variations of latitude a secular or nonperiodic motion of the mean pole on the surface of the earth, probably caused by a drift of the earth's crust, exists. This motion of the pole cannot be ignored in precise astronomical determinations of latitude, longitude and azimuth made at different times.

045.009 **The results of latitude observations made with the ZTF-135 at the Pulkovo Observatory during September 1948 – December 1967.** V. I. Sakharov.
Izv. Glav. Astron. Obs. v Pulkove, No. 185 p. 32 - 48 (1970). In Russian.
Observations according to the latitude program, intended for a nutation cycle of 18.6 years, have been completed. The purpose of these observations is the prolongation of the study of short-period variations of latitude, the so-called "diurnal term". In tables the variations of the Pulkovo latitude for each tenth of a year, read from the smoothed curve, and also the Z-term are given correspondingly.

045.010 **A new program of latitude observations with the ZTF-135 at Pulkovo for 1968 - 1987.**
L. D. Kostina.
Izv. Glav. Astron. Obs. v Pulkove, No. 185, p. 49 - 69 (1970). In Russian.
A new program of latitude observations with the ZTF-135 at Pulkovo, worked out for one nutation cycle from 1968 to 1987, is described.

045.011 **A refinement of the latitude series obtained in Mizusawa and Paris in connection with investigations of low-frequency noise.** N. R. Persijaninova.
Izv. Glav. Astron. Obs. v Pulkove, No. 185, p. 70 - 82 (1970). In Russian.
The latitudes from observations in Mizusawa with the Bamberg zenith telescope, the Cookson zenith telescope and the PZT during 1957.0 - 1962.0 and also in Paris with the astrolabe during 1956.5 - 1962.0 were reduced in connection with investigations of their low-frequency noises. Being reduced, the latitudes were corrected by the chain method. The curves of latitude variations were plotted by a graphical method and also by analytical smoothing. A comparison is made of latitude variation curves, plotted by different methods.

045.012 **Comparison of latitude variations obtained from observations in Belgrade and Jozefoslaw (place near Warsaw).** G. Teleki.
Bull. Obs. Astron. Beograd, Vol. 27, No. 2, p. 58 - 67 (1969).
Observing material of two stations for period 1959 - 1964 was used to execute the change of mean latitude of Belgrade due to the programme alteration in year 1960. The obtained size of change came to +0.″069, being near to the previously calculated data +0.″092.

045.013 **Caractéristiques de variation de la latitude de Belgrade de 1960.0 à 1965.5.**
G. Teleki, R. Grujić.
Bull. Obs. Astron. Beograd, Vol. 27, No. 2, p. 68 - 81 (1969).

045.014 **Observations à la lunette zénithale (de 110 mm) du Service de Latitude de l'observatoire en 1965 – 1968.** R. Grujić, M. Djokić.
Bull. Obs. Astron. Beograd, Vol. 27, No. 2, p. 91 - 98 (1969).

045.015 **Monthly Notes of the International Polar Motion**

Service.
IPMS Monthly Notes, International Latitude Obs. Mizusawa (Japan). 1969 Nos. 11 - 12, p. 99 - 114; 1970 Nos. 1 - 4, p. 1 - 32 (1970). – Announces the values of latitudes observed at the collaborating stations during 1969 November until 1970 April.

045.016 Die Bestimmung der Lotabweichung η für Potsdam aus Sternbedeckungen. H.-U. Sandig.
Acta Techn. Hung. Vol. 52, 405 - 411 (1965) = Mitt. Lohrmann-Inst. Techn. Univ. Dresden, No. 18 (1969).

045.017 Variations in the secular motions of the pole.
H. J. M. Abraham.
Proc. Astron. Soc. Australia, Vol. 1, 350 - 351 (1970).

045.018 Coordonnées du pôle instantané rapportées à l'origine conventionnelle internationale et corrections de longitude TU 1 – TU 0, à 0h TU.
Bureau International de l'Heure, Paris, Circ B/C, Nos. 165 - 170 (1970). – Valeurs interpolées et extrapolées.

045.019 Variations in the latitude of Borowiec for the period 1963.2 - 1968.0. K. Schilling.
Materiały i Prace, Vol. 32 (Publ. Inst. Geophys. Polish Acad. Sci.), 3 - 13 = Polish Acad. Sci. Astron. Latitude Station Borowiec, Repr. No. 32 (1969).

045.020 Coordinates of the pole from PZT observations during the years 1962.1 - 1967.9. W. Jaks.
Materiały i Prace, Vol. 32 (Publ. Inst. Geophys. Polish Acad. Sci.), 25 - 30 = Polish Acad. Sci. Astron. Latitude Station Borowiec, Repr. No. 34 (1969).

045.021 Motion of the pole during the years 1966.0 - 1976.5 from observations made on Danjon astrolabia.
W. Jaks.
Materialy i Prace, Vol. 32 (Publ. Inst. Geophys. Polish Acad. Sci.), 31 - 33 = Polish Acad. Sci. Astron. Latitude Station Borowiec, Repr. No. 35 (1969).

045.022 Un metodo general para la determinación de latitud y meridiano de un lugar. R. Freire.
Dep. Astron. Fis., Fac. Humanidades Ciencias, Univ. Montevideo, Publ. No. 35, 6pp. (1969).
The method is based on the measurement of the altitude of several stars that cross a given vertical of known relative azimuth or some reference point.

045.023 Filtered latitude system for analysis of the polar motion. E. Proverbio, F. Carta, F. Mazzoleni.
Circ. Stazione Astron. Internazionale Latitudine, Carloforte-Cagliari, Ser. B (1), No. 1, 19 pp. (1970).

045.024 Problems of Z-member in latitude observations.
G. Teleki, R. Grujić.
Publ. Obs. Astron. Beograd, No. 16, (see 012.021), p. 11 - 16 (1969). In Serbo-Croatian.

045.025 Catalogue of latitude stars of Beograd Observatory. V. Radogostić.
Publ. Obs. Astron. Beograd, No. 16, (see 012.021), p. 40 - 44 (1969). In Serbo-Croatian.

Résultats des observations faites à Paris avec l'astrolabe impersonnel A. Danjon. Temps et latitude 1967.
See Abstr. 044.005.

Universal time and coordinates of the pole; Emission time of time signals; Coordinated time; Informations.
See Abstr. 044.022.

Time and Latitude Service. See Abstr. 044.026.

International Time and Latitude Service at the Tokyo Astronomical Observatory. See Abstr. 044.027.

Astronomisch-geodätische Arbeiten in der Schweiz, Vol. 28: Längen-, Azimut- und Breitenbestimmungen.
See Abstr. 046.014.

046 Geodetic Astronomy, Navigation

046.001 Geodesy. G. Zhelnin.
Astron. Geod. Eston. SSR, Tartu, (see 003.004), p.
110 - 119 (1969). In Russian.

046.002 Scaling a satellite triangulation net.
F. Halmos.
Bull. Géod., Nouvelle Sér., No. 95, p. 87 - 90 (1970).

046.003 Distance off by vertical sextant angle.
J. W. Crosbie.
Journ. Inst. Navigation, *London,* Vol. 23, 253 - 255 (1970).

046.004 Captain Mário Gama's direct method for star-sight reduction. C. H. Cotter.
Journ. Inst. Navigation, *London,* Vol. 23, 255 - 258 (1970).

046.005 On the observations with the precision theodolite with photographic reading after Gigas.
R. Fukaya.
Tokyo Astron. Obs. Rep., Vol. 15, No. 1 (No. 56), p. 113 - 134 (1970). In Japanese.

046.006 A method of observation and computation for approximate determination of astronomic latitudes and longitudes. P. Bencini.
Boll. Geod. Sci. Affini, Vol. 28, 369 - 415 (1969).
In Italian.

046.007 Bestimmung der Wechsellage zweier geodätischer Systeme aus Synchronbeobachtungen von künstlichen Erdsatelliten. W. Dobaczewska.
Artificial Satellites Polish Acad. Sci., Vol. 4, No. 1, p. 10 -15 (1968).

Die Verbindung von geodätischen Netzen beruht auf ihrem Anschluß an ein gemeinsames Koordinatensystem. Die Lösung dieser Aufgabe erfolgt durch Bestimmung der Neigung der Achsen beider Referenzellipsoide, der gegenseitigen Lage ihrer Zentren sowie des Skalenunterschiedes beider Systeme.

046.008 Techniques and errors of balloon location by low-orbit meteorological satellites. P. N. Denbigh.
Journ. British Interplanet. Soc., Vol. 23, 409 - 421 (1970).

This paper discusses the use of terrestrial navigation schemes for balloon location and also active location techniques by the satellite itself, namely range-range, range-Doppler and Doppler-Doppler.

046.009 On the effect of the lateral refraction upon the astronomical determinations of the longitude.
F. Andersson.
Bull. Géod., Nouvelle Série, No. 96, p. 163 - 167 (1970).

046.010 On the possibility of the determination of longitude variations from radioastronomical observations.
E. N. Fedoseev, N. S. Blinov.
Astron. Zhurn. Akad. Nauk SSSR, Vol. 47, 671 - 673 (1970).
In Russian. English translation in Soviet Astron. AJ, Vol. 14, No. 3.

The problem of the determination of longitude variations between two points on the earth's surface is discussed. A very long base radio interferometer is proposed for this purpose. For the comparison of received frequencies a hydrogen maser should be used.

046.011 Air-navigation astronomy. N. Ya. Kondrat'ev.
Voenizdat, Moskva. 311 pp. Price 79 Kop. (1969).
In Russian.

046.012 Simultaneous observations of a celestial body and control of the angle measurements.
V. A. Kovalenko.
Geod., kartogr. i aehrofotos"emka. Mezhved. resp. nauchno-tekhn. sb., 1969, vyp. (No.) 8, p. 14 - 21. In Russian. – Abstr. in Referativ. Zhurn. 52. Geod. Aehros"emka, 6.52.153 (1970).

046.013 A method for higher precision of the coordinates of the observational place. A. B. Palej.
Uch. zap. Ivanovsk. gos. ped. in-t, Vol. 78, 104 - 105 (1969).
In Russian. – Abstr. in Referativ. Zhurn. 52. Geod. Aehros"-emka, 6.52.152 (1970).

046.014 Astronomisch-geodätische Arbeiten in der Schweiz, Vol. 28: Längen-, Azimut- und Breitenbestimmungen. Prepared by N. Wunderlin.
Edited by the Schweizerischen Geodätischen Kommission, Zürich. 151 pp. (1970).

046.015 A preliminary determination of the direction Nikolajev–Helvan. A. Dinescu.
Byull. Stantsij Optichesk. Nablyud. Iskusstv. Sputnikov Zemli, No. 53 (1), p. 27 - 28 (1969). In Russian.

046.016 Numerical analysis of the adjustment of satellite triangulation in the rectangle "Organ Pass–Arequipa–Curacao–Jupiter". K. Minowski.
Geodezja i Kartografia, Vol. 18, 145 - 163 = Publ. Działu Geod. Wyższej i Astron. Geod. Zg. PAN, No. 14 (1969).
In Polish.

046.017 Bestimmung der Längendifferenzen Borowiec–Dresden–Potsdam im Jahre 1966. J. Moczko.
Materiały i Prace, Vol. 32 (Publ. Inst. Geophys. Polish Acad. Sci.), 51 - 117 = Polish Acad. Sci. Astron. Latitude Station Borowiec, Repr. No. 37 (1969).

046.018 Anwendung der Hammersteinschen Methode der unendlich vielen Variablen auf Probleme der Satellitengeodäsie und Himmelsmechanik.
R. Sigl, M. Schneider, C. Reigber, H. Ludwig.
Bundesministerium für Bildung und Wissenschaft, Forschungsber. W 70-33, (Weltraumforschung), 39 pp. (1970).

Several problems of satellite geodesy and celestial mechanics can be formulated as boundary value problems. To solve the integral equations equivalent to them, one can use a method given by Hammerstein. This paper is concerned with the determination of satellite orbits, field parameters and station coordinates.

046.019 Izvodjenje polazne vrednosti longitude Astronomsko–Geofizicke Opservatorije u Ljubljani.
P. Ranzinger.
Publ. Obs. Astron. Beograd, No. 16, (see 012.021), p. 9 - 10 (1969). In Serbo-Croatian.

046.020 Micrometer value in latitude observations and problems of its determination. M. Djokić.
Publ. Obs. Astron. Beograd, No. 16, (see 012.021), p. 17 - 21 (1969). In Serbo-Croatian.
The report emphasised the importance of the micrometer screw revolution or a zenith-telescope, showing methods by which the micrometer value had been determined.

Das Inertialsystem und seine Bedeutung für die Astronomie und die Geodäsie. See Abstr. 041.004.

Beitrag zur Objektivierung von Meridiandurchgangs- beobachtungen.   See Abstr. 041.028.

earth.   See Abstr. 081.003.

The contribution of satellite geodesy to the deter- mination of the figure and the internal constitution of the

Problems of astronomical navigation on the lunar surface.   See Abstr. 094.207.

047 Ephemerides, Almanacs, Calendars

047.001 **Astronomical Yearbook of the USSR for the year 1972.**   V. K. Abalakin (Editor).
Institut Teoreticheskoj Astronomii Akademii Nauk SSSR. Izdatel'stvo "Nauka", Leningradskoe Otdelenie, Leningrad. 719 pp. Price 7 Rbl. 6 Kop. (1970).   In Russian.

047.002 **Events of 1970 in the graphic time table.**
Sky Telescope, Vol. 39, 35 - 37 (1970).

047.003 **The Handbook of the British Astronomical Association, 1970.**
C. Dinwoodie (Editor).
British Astronomical Association, Hounslow West, Middlesex. 82 pp. Price $ 1.50 (1969).

047.004 **Ephémérides Astronomiques pour 1970.**
Publiées par la Société Astronomique de France. L'Astronomie, 84e année, Suppl. pour Janvier 1970. 65 pp. price F 8.00.

047.005 **Check on numerical values appearing in official calendar after Meiji's calendar amendment.**
S. Itō.
Tokyo Astron. Obs. Rep., Vol. 15, No. 1 (No. 56), p. 99 - 112 (1970).   In Japanese.

047.006 **Connaissance des Temps** ou des mouvements célestes pour l'an 1971, à l'usage des astronomes et des navigateurs. Publiée par le Bureau des Longitudes. Gauthier-Villars Editeur, Paris. 42 + 643 pp. (1970).

047.007 **Hvězdářská ročenka 1970.**
Compiled by J. Bouška, B. Onderlička, J. Ruprecht.
Ročnik 46. Academia, nakladatelstvi Československé akademie věd, Praha. 219 pp. Price kčs 13.- (1970).

047.008 **Philippine Astronomical Handbook 1970.**
Prepared under the supervision of E. V. Calpo.
Republic of the Philippines – Department of Commerce and Industry – Weather Bureau, Manila. 9 + 51 pp. (1969).

047.009 **Tables of Sunrise, Sunset, Twilight, Moonrise and Moonset 1970.** Prepared under the supervision of E. V. Calpo.
Republic of the Philippines – Department of Commerce and Industry – Weather Bureau, Manila. 6 + 56 pp. (1969).

047.010 **Astronomical Ephemeris for the Year 1971.**
Issued by Her Majesty's Nautical Almanac Office, London; Nautical Almanac Office, United States Naval Observatory, Washington. Her Majesty's Stationery Office, London. 17 + 520 pp. Price £2 0s. 0d. net (1970).

047.011 **The Air Almanac 1970, September – December.**
Her Majesty's Stationery Office, London; United States Naval Observatory, Washington. 246 + A82 + F4 pp. Price £ 1 10s. 0d. (1970).

047.012 **The Heavens in 1970.**
Separate print Astron. Soc. Pacific, 8 pp. (1970).

047.013 **Almanac for geodetic engineers 1970.** Prepared under the supervision of E. V. Calpo.
Republic of the Philippines – Department of Commerce and Industry – Weather Bureau, Manila. 8 + 22 pp. (1969).

047.014 **Anuário do Observatório de S. Paulo para 1970.**
Published by Instituto Astronômico e Geofisico, Universidade de São Paulo, Brasil. 10 + 228 + 29 pp. (1969).

047.015 **Almanaque Náutico ano 1971, publicado de orden de la Superioridad por el Instituto y Observatorio de Marina de San Fernando (Cádiz).**
Imprenta del Observatorio de Marina, San Fernando. 418 pp. Price 200 pesetas (1970).

047.016 **Astronomische Grundlagen für den Kalender 1972.**
Edited by Astronomisches Rechen-Institut, Heidelberg. Verlag G. Braun GmbH., Karlsruhe. 88 pp. Price DM 22.50 (1970).

047.017 **Kalendarium des 20. Jahrhunderts.**
H. Haupt.
Druck und Verlag: A. Grünsfeld & Co., Wien. 63 pp. (1969).

047.018 **Blick in die Sternenwelt 1970.**
H. Winkler.
Vorträge und Schriften Archenhold-Sternw. Berlin-Treptow, No. 35, 48 pp. (1969).

047.019 **The Star Almanac for Land Surveyors for the Year 1971.**
Prepared by *H. M. Nautical Almanac Office*, published by Order of *The Science Research Council.* Her Majesty's Stationery Office, London. 70 pp. Price 7s.0d. (1970).

047.020 **Ephemerides.**
G. A. Wilkins, V. K. Abalakin, J. Kovalevsky, D. H. Sadler, W. Fricke, A. M. Sinzi, N. C. Lahiri, M. Rodriguez, R. L. Duncombe.
Trans. IAU, Vol. 14A, (see 003.028), 1 - 9 (1970). – Report of Commission 4.

047.021 **Annuario 1969.**
Edited by Osservatorio Astronomico di Torino, [Printed by Scuola Salesiana del Libro, Catania]. Pubbl. Varie Fuori Ser. Oss. Astron. Torino, No. 36, 43 pp. (1968).

047.022 **Annuario 1970.**
Edited by Osservatorio Astronomico di Torino, [Printed by Scuola Salesiana del Libro, Catania]. Pubbl.Varie Fuori Ser. Oss. Astron. Torino, No. 39, 56pp. (1969).

The programming of some astronomical calculi for the electronic desk computer "Olivetti-Programma 101".
See Abstr. 021.009.

Space Research

051 Extraterrestric Research, Spaceflight Related to Astronomy

051.001 **A wide field astronomical project.**
F. di Benedetto, L. Gratton.
Journ. British Interplanet. Soc., Vol. 23, 227 - 236 (1970).
The authors give a broad description of the aims of a wide field astronomical project, which is at present, under consideration by ESRO (European Space Research Organization) and give a short historical account of the action taken up to now in Europe on this project.

051.002 **Roster of space activity.** R. N. Watts, Jr.
Sky Telescope, Vol. 39, 81 - 82 (1970).

051.003 **Space report.**
Spaceflight, Vol. 12, 29 - 33, 67 - 74, 124 - 126, 158 - 162 (1970).
(1) Lunar laser reflector; Eagle's moon radar; Apollo's volcanic probe; Moon house; Research on stars; Observations of Pluto. (2) Titanium on the moon; Expedition to Mars; Wasteland on Mars; First Mercury probe study; Student satellite tracker. (3) Lunar reflectivity; Glazed rocks on the moon; Moon buggy. (4) Astronomy board report; Space-to-air navigation; Boreas reenters (*1969-83A*).

051.004 **Status report on space programmes and policies of the Federal Republic of Germany.**
A. Spaeth.
Spaceflight, Vol. 12, 56 - 59 (1970).

051.005 **Which highways to the stars?**
J. Strong.
Spaceflight, Vol. 12, 174 - 177 (1970).
Concerning the nearest double star systems within 20 light years having highest probability to be planetary systems.

051.006 **Flüge mit Stratosphärenballonen zu wissenschaftlichen Untersuchungen.** H. Röhrs.
SuW, Vol. 9, 33 - 36 (1970).

051.007 **Weltraumprogramm der Bundesrepublik Deutschland 1969 - 1973.** C. Leinert.
SuW, Vol. 9, 43 - 44 (1970).

051.008 **After the moon, the earth!**
W. O. Roberts.
Science, Vol. 167, 11 - 16 (1970).
AAAS presidential address, delivered 28 December 1969 at the Boston meeting.

051.009 **Mission to an asteroid.**
H. Alfvén, G. Arrhenius.
Science, Vol. 167, 139 - 141 (1970).
We shall confine ourselves to discussing missions − manned or unmanned − to asteroids in our close environment. There are a number of asteroids which at regular intervals come close to the earth. A landing on such an asteroid would be of special significance to the investigation of the early history of the solar system.

051.010 **Les «îles cosmiques».** I. S. Chklovski.
L'Astronomie, 84ᵉ année, p. 67 - 68 (1970).

051.011 **Mission to Martian satellites.**
D. J. Milton.
Science, Vol. 167, 1758 (1970). − Remarks to a paper by H. Alfvén, G. Arrhenius, Science, Vol. 167, 139 - 141 (1970).

051.012 **Die erfolgreich gestarteten künstlichen Erdsatelliten und Raumsonden.** U. Güntzel-Lingner.
SuW, Vol. 9, 95 - 97, 126 - 127 (1970).

051.013 **Outer space in laboratory.** I. M. Podgorni.
Priroda, No. 2.70, p. 61 - 65 (1970). In Russian.

051.014 **Manned space stations − gateway to our future in space.** R. R. Gilruth.
Astrophys. Space Sci. Library, Vol. 16, 1 - 10 (1969).
Conference paper (see 012.005).

051.015 **Astronomical research with a large orbiting telescope.** L. Spitzer, Jr.
Astrophys. Space Sci. Library, Vol. 16, 88 - 98 (1969).
Conference paper (see 012.005).

051.016 **Space programme in India.** H. G. S. Murthy.
Astrophys. Space Sci. Library, Vol. 15, 645 - 649 (1970). − Conference paper (see 012.007).

051.017 **Space report.**
Spaceflight, Vol. 12, 248 - 252 (1970).
Viking 75 (*1975 mission to Mars*); Atmosphere Explorer; Comet in a hydrogen cloud; Lunar meteoroids; Lunar volcanoes.

051.018 **Sixth Orbiting Solar Observatory.**
S. P. Maran, J. M. Thole, J. L. Donley.
Bull. American Astron. Soc., Vol. 2, 207 - 208 (1970).
Abstr. AAS.

051.019 **Space astronomy.** M. L. Kratage, D. B. Wood.
Astron. Soc. Pacific, Leaflet No. 489, 8 pp. (1970).

051.020 **The American space program for the 1970's.**
R. N. Watts, Jr.
Sky Telescope, Vol. 39, 294, 298 (1970).

051.021 **Kurzberichte.**
Weltraumfahrt, 21. Jahrgang, p. 23 - 29 (1970).
Neue Radaruntersuchungen der Marsoberfläche.

051.022 **Space report.**
Spaceflight, Vol. 12, 200 - 205, 219 (1970).
(5) Australis in orbit; Four-man Apollo? Rings of Saturn; Anglo-Australian telescope.

051.023 **Kurzberichte.**
Weltraumfahrt, 21. Jahrgang, p. 60 - 62 (1970).
Mondmaterial für deutsche Wissenschaftler; Apollo 13: Missionsziel nicht erreicht; Satellitenbahn durch Laser markiert; Osumi: Japans Eintritt in den Weltraum; Mit Pioneer F und G zum Jupiter.

051.024 **Zu einigen astronomischen Ergebnissen der Raumfahrt.** O. Günther.
Astron. in der Schule, 7. Jahrgang, p. 13 - 16 (1970).

051.025 **Ein Programm zur Erforschung des äußeren Sonnensystems.** W. Kokott.
SuW, Vol. 9, 151 (1970).

051.026 **Space report.**
Spaceflight, Vol. 12, 278 - 281 (1970).
LM effects on Surveyor III; New OSO 'eye'.

051.027 **Space in Japan, 1969 - 1970.**
Editorial Office: Space Development Division, Research Coordination Bureau, Science and Technology Agency, Tokyo, Japan. 163 pp. – Contents: Outline of space activities; Space-related organizations; Industry's role in space research.

051.028 **Les satellites artificiels de l'année 1969.** J. Thurnheer.
Orion Schaffhausen, 28. Jahrgang, p. 91- 94 (1970).

051.029 **Astronautics in the year 1969.** J. Bouška.
Říše hvězd, Vol. 51, 105 - 111, 131 - 136 (1970).
In Czech.

051.030 **Advances of space astronomy.** M. Grün, P. Koubský.
Pokroky, Vol. 15, 62 - 76 (1970). In Czech.

051.031 **American and Soviet projects of manned orbital laboratories.** M. Grün.
Vesmir, Vol. 49, 16 - 17 (1970). In Czech.

051.032 **Development of the Soviet guided space ships.** K. P. Feoktistov.
Zemlya i Vselennaya, No. 2, p. 12 - 20 (1970). In Russian.

051.033 **Project to map the sky in the infrared from an E.S.R.O. satellite.** L. Delbouille.
Phil. Trans. Roy. Soc. London, Ser. A, Vol. 264, 319 - 320 = Univ. Liège, Inst. d'Astrophys., Coll. 4°, No. 191 (1969).

051.034 **Interplanetarische Raumforschung. Das heutige Wissen über Venus und Erde.** W. Moros.
Universitas, Vol. 24, (No. 10), 1051 - 1054 (1969).

051.035 **Künstliche Erdsatelliten und Raumsonden: Situa – tionsbericht.**
Weltraumfahrt, 21. Jahrgang, p. 22 (1970).

052 Astrodynamics and Navigation of Space Vehicles

052.001 **Spherical coordinate intermediaries for an artificial satellite.** B. Garfinkel, K. Aksnes.
Astron. Journ., Vol. 75, 85 - 91 (1970).

The publication of a new intermediary by Aksnes (1965) has opened the way for a relatively simple calculation of a complete second-order theory for an artificial satellite. Previous attempts at such a theory have been only partially successful. The paper discusses some interesting properties of the Aksnes intermediary, and places it in a historical relation to the intermediaries of Garfinkel (1958, 1964) and of Sterne (1958). A general algorithm that includes both the Aksnes and the Garfinkel orbits as special cases is constructed.

052.002 **The librational dynamics of deformable bodies.** T. P. Mitchell, J. Lingerfelt.
Celestial Mechanics, Vol. 1, 289 - 296 (1970).

This paper aims to study the effects of material elasticity on the librational motion of an arbitrary shaped satellite. In particular, the influence of the elastic behavior on the librational frequency is determined. The approach adopted is quite general in that the specific shape of the satellite is not prescribed other than to assume that the orbit plane of its center of mass coincides with a principal plane of the satellite and that the body is symmetrical with respect to the principal planes.

052.003 **The motion of a lunar satellite.** C. Oesterwinter.
Celestial Mechanics, Vol. 1, 368 - 436 (1970).

Presented in this theory is a semianalytical solution for the problem of the motion of a satellite in orbit around the moon. The major part of the problem is solved by means of the von Zeipel method. After eliminating from the Hamiltonian all terms with the period of the satellite and those with the period of the moon, it is suggested to solve the re-maining problem with the aid of numerical integration of the modified equations of motion. The final results, however, are incomplete in the lunar gravitational perturbations. Nevertheless, the theory does give the largest such variations and it does present the methods by which perturbations may be derived for any gravity terms not actually developed.

052.004 **Satellite prediction formulae for Vinti's model.** D. O'Mathuna.
Celestial Mechanics, Vol. 1, 467 - 478 (1970).

In this paper we give a new form of solution for the dynamical problem associated with Vinti's potential. The feature of the solution proposed here is that the three coordinates are expressed in terms of a single independent variable: the prediction scheme is then completed by the formula relating this independent variable to the time. The formulae are presented in a form that is ready for computation.

052.005 **Optimal rocket trajectories in a general force-field.** C. J. Brookes, J. Smith.
Astronaut. Acta, Vol. 15, 129 - 132 (1970).

The problem of determining the necessary flight characteristics of an optimal rocket trajectory was investigated for general case $g = g(r)$ and in particular $g = -\gamma/r^n$ (Newtons law).

052.006 **Satellite orbit prediction in the presence of atmospheric drag.** A. G. Lubowe.
Astronaut. Acta, Vol. 15, 143 - 148 (1970).

A method of orbit prediction for satellites with perigee altitudes above 100 miles and eccentricities up to 0.1 is described. A computer program based upon the addition of the first order secular node-to-node drag perturbations to the high accuracy orbit prediction method developed for the Telstar satellites is used to obtain numerical results.

052.007 Effects of the variation of drag coefficient on the ephemeris of earth satellites.   R. R. Hunziker.
Astronaut. Acta, Vol. 15, 161 - 167 (1970).

The variations after one revolution, in semimajor axis, period and eccentricity for an orbit using a variable drag coefficient with respect to those for an orbit using a constant drag coefficient have been computed by a first order integration of the variational equations which follow from those in the form due to Gauss. Numerical results are given.

052.008 Orbit determination using analytic partial derivatives of perturbed motion.
J. L. Arsenault, K. C. Ford, P. E. Koskela.
AIAA Journ., Vol. 8, 4 - 12 (1970).

It is shown in this paper that the state transition matrix of a satellite orbit may be calculated analytically including the effects of perturbations. This is accomplished by generating each perturbation in the Gaussian form (using three mutually perpendicular components of the perturbing acceleration) to find the transition matrix from epoch state to the state at any other time. This analytic method is considerably faster than the numerical integration of the variational equations and it is shown that the methods are in good agreement.

052.009 Investigation of the orbits of a close revolution round the moon with return to the earth's atmosphere. I.   V. A. Il'in, V. V. Demeshkina, N. A. Istomin.
Kosmich. Issled., Vol. 8, 48 - 58 (1970).   In Russian.

052.010 Navigation for Apollo lunar landings.
J. P. Mayer, R. K. Osburn.
Journ. Inst. Navigation, *London*, Vol. 23, 131 - 148 (1970).

The paper gives a general outline of the operational use of the space navigation systems which contributed to the success of the first manned lunar exploration mission. The ground and onboard navigation systems, their interfaces with the Apollo guidance system, and their uses in various phases of lunar landing are reviewed.

052.011 GEOS-II and 13th order terms of the geopotential.   B. C. Douglas, J. G. Marsh.
Celestial Mechanics, Vol. 1, 479 - 490 (1970).

The resonance of GEOS-II (1968-002A) with 13th-order terms of the geopotential is analyzed. The odd-degree geopotential coefficients (13, 13), (15, 13), and (17, 13) given by Yionoulis most accurately model the resonance effects on GEOS-II of any of the published sets of the 13th-order coefficients. Values of $C_{14, 13}$ ($= 0.57 \times 10^{-21}$) and $S_{14, 13}$ ($= 6.5 \times 10^{-21}$) to be used with the odd-degree set of Yionoulis were obtained from an analysis of the observed along-track position variation of GEOS-II. These coefficients, when used with those of Yionoulis, yield greatly improved 'fits' to the data and orbital prediction capability. However, further refinement is possible after the small effects of the remaining even-degree resonant terms have been modeled.

052.012 Limitations of the standard method for the midcourse correction of lunar trajectories.
W. K. Hrushow.
AIAA Journ., Vol. 8, 424 - 432 (1970).

A method for computation of the midcourse correction velocity for lunar impact trajectories is analyzed. Only the magnitude of the impact miss, but not the impact time, is considered for correction. It is shown that this method has certain limitations. When applied for correction of trajectories intended to impact a material point in space, or for correction of lunar fly-by trajectories, this method performs flawlessly. It also performs reasonably well when applied on standard trajectories.

052.013 Explicit form of the disturbing function in the

problem of motion of a synchronous satellite in the gravitational field of the centrally symmetric earth, of the moon and the sun.   S. G. Zhuravlev.
Sb. nauchn. rabot aspirantov. Un-t druzhby naradov im. Patrisa Lumumby. Fak. fiz.-matem. i estestv. n., 1969 vyp. (No.) 6, p. 118 - 129.   In Russian. – Abstr. in Referativ. Zhurn. 51. Astron., 3.51.140 (1970).

052.014 About some second order inequalities in the motion of distant earth satellites.
V. P. Dolgachev.
Vestn. Mosk. un-ta. Fiz., astron., 1969 No. 5, p. 74 - 80.
In Russian. – Abstr. in Referativ. Zhurn. 51. Astron., 3.51.142 (1970).

052.015 Construction of Krotov function for some problems of optimalizing orbital transitions.
V. V. Ivashkin.
Kosm. Issled., Vol. 8, 189 - 200 (1970).   In Russian.

052.016 Optimal deceleration of momentum at reentry into the atmosphere.   Yu. G. Sikharulidze.
Kosm. Issled., Vol. 8, 201 - 205 (1970).   In Russian.

052.017 Gravitational short-period perturbations of the artificial satellite motion.   J. B. Zieliński.
Artificial Satellites Polish Acad. Sci., Vol. 4, No. 1, p. 21 - 32 (1968).

In this paper the influence of tesseral and sectorial harmonics is investigated by the method of numerical integration. Special attention is given to the changes of the length of radius-vector.

052.018 Errors in orbit determination.   G. Bressanin.
Celestial Mechanics, Vol. 2, 77 - 87 (1970).

The gaussian noise which affects tracking measurements causes an error in the computed value of the orbit parameters. This study provides a method for evaluating: (a) the length of the arc over which the satellite must be tracked; (b) the number of measurements to be made along this arc; (c) the position of the arc with respect to the orbit, necessary to reach the desired accuracy of the calculated orbit parameters for a given pointing error of the tracking antenna.

052.019 Calculation of oblateness perturbations using Hansen's method.   A. G. Lubowe.
Celestial Mechanics, Vol. 2, 88 - 102 (1970).

The basic principles and equations of Hansen's method are briefly stated. These equations are integrated to first order accuracy by analytic quadratures for the case of perturbations due to the earth's oblateness and the procedure for obtaining the position of a satellite acted upon by such a perturbation is described in detail. Comparison of accuracies of one day predictions made with this Hansen's method and a first order variation of parameters method indicates comparable accuracies can be achieved with approximately $^1/_{10}$ of the computation time when using Hansen's method.

052.020 The critical inclination problem: A simple treatment.   R. R. Allan.
Celestial Mechanics, Vol. 2, 121 - 122 (1970).

A short analysis is presented in the hope of clarifying the situation.

052.021 A terminal guidance theory using dynamic programing formulation.   C. S. Chang.
AIAA Journ., Vol. 8, 912 - 916 (1970).

An optimal guidance law in the neighborhood of a nominal (optimal or not) trajectory is developed. This guidance law is optimal in the sense of guiding a vehicle from an arbitrary starting point to an arbitrary fixed-end point along a neighborhood trajectory with minimum control energy consumption.

052.022 **Numerical verification of analytic expressions for the perturbations due to an arbitrary zonal harmonic of the geopotential.** A. G. Lubowe, R. E. Jenkins.
Celestial Mechanics, Vol. 2, 21 - 40 (1970).

Expressions are given for the first order node-to-node perturbations in the orbital elements of a satellite due to an arbitrary zonal harmonic of the geopotential. Comparison with a double precision numerical integration is made for an intermediate altitude satellite, Telstar I. Discrepancies in semi-major axis after 1 period are of the order of 0.1 mm. Discrepancies in timing are of the order of 0.03 msec. A detailed discussion of computational efficiency is included.

052.023 **On the determination of the circular orbit of an AES** (*artificial earth satellite*) **using non-complete observations.** R. A. Zeinalov.
Byull. Inst. Teoret. Astron., *Leningrad*, Vol. 12, 157 - 162 (1970). In Russian.

The present paper deals with a method of computation of a preliminary circular orbit using three observations, two observational times being unknown. The classical method for the circular orbit determination has been modified for the case where the non-complete observations are used, provided that the observations cover a small arc of one revolution. The basic equations for three unknowns (the orbital radius, and two observational times) have been derived. The system of equations has been solved using Newton's method.

052.024 **Elements of the theory of the bearings of a celestial body from a plane with movable platform.**
L. M. Vorob'ev.
Kosmich. Issled., Vol. 8, 360 - 364 (1970). In Russian.

052.025 **Investigation of the orbits of a close revolution round the moon with return to the earth's atmosphere. II.** V. A. Il'in, V. V. Demeshkina, N. A. Istomin.
Kosmich. Issled., Vol. 8, 365 - 376 (1970). In Russian.

052.026 **Construction of an intermediary orbit with consideration of large perturbations in the motion of a space ship near the boundary of the sphere of activity of a planet.** S. V. Al'shevskij.
Kosmich. Issled., Vol. 8, 377 - 382 (1970). In Russian.

052.027 **Investigation of the geometry of the paths of artificial earth satellites.**
V. L. Kalachev.
Kosmich. Issled., Vol. 8, 383 - 396 (1970). In Russian.

Abgeleitet wird eine Gleichung zur Berechnung der Trasse (= scheinbaren Bahn) eines künstlichen Erdsatelliten der sich auf einer Kepler-Ellipse bewegt. Die durch die Erdabplattung bedingte säkulare Bewegung des aufsteigenden Knotens der Bahn wird berücksichtigt. Eine Klassifikation der Trassen für verschiedene Bahnen wird gegeben.

052.028 **Attitude stability of a gravity-stabilized gyrostat satellite.** M. R. M. Crespo da Silva.
Celestial Mechanics, Vol. 2, 147 - 165 (1970).

This paper treats analytically the problem of the stability of the attitude motions of a gravity-stabilized gyrostat satellite that is in a circular orbit around a spherical planet. The vehicle considered consists of a body with no special symmetries that has any number of rotors attached to it. Stability (both infinitesimal and in the sense of Liapunov) of the attitude motions of the vehicle can be quickly predicted by using the results derived here, which are summarized in the form of a continuous, three-dimensional, stability diagram.

052.029 **The main problem of artificial satellite theory for small and moderate eccentricities.**
A. Deprit, A. Rom.
Celestial Mechanics, Vol. 2, 166 - 206 (1970).

Perturbation techniques based on Lie transforms as suggested by Deprit were used as the theoretical foundation for programming the analytical solution of the main problem in satellite theory (all gravitational harmonics being zero except J_2). The collection of formulas necessary and sufficient to construct an ephemeris is given in the exposition. Short and long period displacements, as well as the secular terms, have been obtained up to the third order in J_2 as power series of the eccentricity. The determination of the constants of motion from the initial conditions has been given an elementary solution that is both complete and explicit without being iterative. Reliability tests have been run in two instances: in-track errors for ANNA 1B are only 20 cm after 210 days in orbit, while for RELAY II, they are 2.4 m, even after 350 days in orbit.

052.030 **On the computation of the spherical harmonic terms needed during the numerical integration of the orbital motion of an artificial satellite.** L. E. Cunningham.
Celestial Mechanics, Vol. 2, 207 - 216 (1970).

A method is presented for the accurate and efficient computation of the forces and their first derivatives arising from any number of zonal and tesseral terms in the earth's gravitational potential. The basic formulae are recurrence relations between some solid spherical harmonics, associated with the standard polynomial ones.

052.031 **On error bounds and initialization in satellite orbit theories.** J. V. Breakwell, J. Vagners.
Celestial Mechanics, Vol. 2, 253 - 264 (1970).

The order of magnitude of the error is investigated for a first-order von Zeipel theory of satellite orbits in an axisymmetric force field, i.e., first-order long period and short-period effects are included along with second order secular rates. The treatment is valid for zero eccentricity and/or inclination. The results are specifically applicable to accuracy comparisons of the Brouwer orbit prediction method with numerical integration. A modified calibration is presented for the general asymmetric force field which includes tesseral harmonics.

052.032 **Construction of an optimum elliptic orbit of long revolution in the equatorial plane of an axisymmetric planet.** V. S. Novoselov.
Vestn. Leningr. un-ta, 1969 No. 19, p. 146 - 152. In Russian.
Abstr. in Referativ. Zhurn. 62. Issled. kosm. prostranstv., 4.62.248 (1970).

052.033 **Some probable characteristics of the motion of an artificial earth satellite around the center of mass.**
N. F. Martynova.
Vestn. Leningr. un-ta, No. 1, p. 146 - 153 (1970). In Russian.
Abstr. in Referativ. Zhurn. 62. Issled. kosm. prostranstv., 6.62.239 (1970).

052.034 **Branched trajectory optimization by the projected gradient technique.** J. Gera.
AIAA Journ., Vol. 8, 1121 - 1126 (1970).

A numerical procedure based on the projected gradient technique is developed for the optimization of branched trajectories. For example, the flight path of a rocket vehicle that breaks into two or more parts, each with a separate mission, is a branched trajectory. It is shown that only minor modifications to the standard projected gradient technique are needed in order to obtain optimal branched solutions.

Principles of Celestial Mechanics.
See Abstr. 003.002.

On the mixed-secular terms in perturbed intermediaries. See Abstr. 042.059.

A separable potential in triaxially ellipsoidal coordinates satisfying the Laplace equation. See Abstr. 042.062.

A note concerning an application of the generalized Huang's model of the restricted four-body problem. See Abstr. 042.069.

Die Gröbnersche Methode und ihre Anwendung auf die numerische Bahnberechnung. See Abstr. 042.073.

053 Lunar and Planetary Probes and Satellites

053.001 **The send-off Apollo 12.** R. Hillenbrand.
Sky Telescope, Vol. 39, 9 - 10 (1970).

053.002 **Apollo 11: A systems analysis.**
D. Baker.
Spaceflight, Vol. 12, 35 - 40 (1970).

053.003 **Manned Mars landing missions using electric propulsion.** J. S. MacKay.
Spaceflight, Vol. 12, 41 - 46 (1970).

053.004 **The triumph of Apollo 12.**
P. J. Parker.
Spaceflight, Vol. 12, 77 - 80, 81, 118 - 120 (1970).

053.005 **Objects in heliocentric orbit.** G. Falworth.
Spaceflight, Vol. 12, 92 - 95 (1970).
Contains technical and astrodynamic data of all planetary probes until 1968 December 21.

053.006 **Objects in selenocentric orbit.**
G. Falworth.
Spaceflight, Vol. 12, 143 - 144 (1970). — Contains technical and astrodynamic data of all moon probes and satellites until 1969 July 16.

053.007 **Orbital laboratory — stepping stone to interplanetary flight.** W. M. Hollister.
Astrophys. Space Sci, Library, Vol. 16, 99 - 109 (1969).
Conference paper (see 012.005).

053.008 **Manned flight to the nearer planets.**

S. F. Singer.
Astrophys. Space Sci. Library, Vol. 16, 110 - 115 (1969).
Conference paper (see 012.005).

053.009 **An analysis and simulation of Mars landing.**
E. K. Casani.
Astrophys. Space Sci. Library, Vol. 15, 584 - 598 (1970).
Conference paper (see 012.007).

053.010 **Mission to Jupiter.** P. J. Burinskas.
Spaceflight, Vol. 12, 237 - 238 (1970).

053.011 **Apollo 13 makes it back to earth.**
R. N. Watts, Jr.
Sky Telescope, Vol. 39, 350 (1970).

053.012 **Exploits de la N.A.S.A. avec Apollo VIII.**
Gaz. astron., No. 2, p. 23 - 24 (1969).

Planning the grand tour.
Spaceflight, Vol. 12, 104 - 105 (1970).
Describes the plan for multiple-planet missions to explore the outer solar system by a space probe. The best outer planet alignment in 179 years occurs in the 1976 to 1980 time period.

Lunar orbit rendezvous.
Spaceflight, Vol. 12, 152 - 154 (1970).

Intermediary orbits of planetary satellites.
See Abstr. 091.040.

054 Artificial Earth Satellites

054.001 Sun study satellite.
P. J. Parker.
Spaceflight, Vol. 12, 21 - 22 (1970). – Describes the Orbiting Solar Observatory (OSO) programme and its six launched vehicles.

054.002 The small astronomy satellite.
M. R. Townsend, C. E. Fichtel.
Spaceflight, Vol. 12, 63 - 66 (1970).

054.003 Eye on aurorae.   P. J. Parker.
Spaceflight, Vol. 12, 131 (1970). – Concerning satellite ESRO 1 B = Boreas = 1969–83 A.

054.004 The ESRO IV satellite.
P. J. Conchie.
Spaceflight, Vol. 12, 132 - 136 (1970).

054.005 AZUR – der erste deutsche Forschungssatellit.
W. Kokott.
SuW, Vol. 9, 12 (1970).

054.006 Le satellite Pageos 1 (1966-56 A).
R. Futaully.
L'Astronomie, 84ᵉ année, p. 75 - 78 (1970).

054.007 Considerations in choosing the orbit for an earth resources survey satellite.
J. Otterman, B. T. Bachofer.
Journ. British Interplanet. Soc., Vol. 23, 369 - 383 (1970).

054.008 Japan's first satellite.
R. N. Watts, Jr.
Sky Telescope, Vol. 39, 227 (1970). – Concerning 1970-11 A.

054.009 Methods of determining the orientation of an artificial earth satellite from telemetric measurements.   Eh. K. Lavrovskij, S. I. Trushin.
Kosm. Issled., Vol. 8, 218 - 228 (1970).   In Russian.

054.010 Determination of the orientation of the artificial earth satellites Elektron 2 and 4.
Eh. K. Lavrovskij, S. I. Trushin.
Kosm. Issled., Vol. 8, 229 - 242 (1970).   In Russian.

054.011 Telemetry from the first Chinese artificial earth satellite.   G. E. Perry.
Nature, Vol. 226, 829 (1970). – Concerning 1970–34A.

054.012 Orbiting Solar Observatory.   A. W. L. Ball.
Spaceflight, Vol. 12, 244 - 247 (1970). – Concerning the solar research satellites OSO 1 to OSO 6 and its programs.

054.013 Induction drag on large Echo-type satellites: Observational evidence.
V. J. Slabinski.
Bull. American Astron. Soc., Vol. 2, 218 (1970). – Abstr. AAS.

054.014 Elementi orbitali del satellite Echo 2 calcolati dagli ultimi passaggi su Milano.
L. Buffoni, A. Manara.
Atti Accad. Nazionale Lincei Rend., Cl. Sci. fis., mat., nat., Ser. Ottava, Vol. 47, 189 - 191 (1969).
   Orbital elements of Echo 2 satellite are derived from optical observations at Brera Observatory covering the last

passages of the satellite. The resulting orbital parameters are briefly discussed.

054.015 Analysis of geodetic satellite (Geos I) observations in North America.
I. I. Mueller, C. R. Schwarz, J. P. Reilly.
Bull. Géod., Nouvelle Série, No. 96, p. 143 - 162 (1970).
   Geos I observations made at thirty optical and four Secor stations were analyzed in the geometric and short-arc modes for the purpose of detecting systematic distortions in the North American Datum.

054.016 Chinese satellite.
R. N. Watts, Jr.
Sky Telescope, Vol. 39, 350 (1970).

054.017 Sur la rotation des satellites ballons.
F. Link.
Ciel et Terre, Vol. 86, 195 - 198 (1970). – Conference paper.

054.018 New generation TIROS.
P. J. Parker.
Spaceflight, Vol. 12, 221 - 222 (1970). – Concerning new weather satellite ITOS.

054.019 Intercosmos 1 in orbit.
I. P. Tindo, I. A. Zhitnik.
Priroda, No. 4.70, p. 78 - 87 (1970).   In Russian.

054.020 The satellite Intercosmos 1 and its scientific program. (Cooperation of scientists of the socialist countries in investigations of the cosmos).
L. A. Vedeshin, M. G. Kroshkin.
Vestn. AN SSSR, 1969 No. 12, p. 19 - 27.   In Russian.
Abstr. in Referativ. Zhurn. 62. Issled kosm. prostranstv., 4.62.119 (1970).

054.021 Satellite digest.   G. Falworth.
Spaceflight, Vol. 12, 8 - 9, 75 - 76, 122 - 123, 163 - 165, 206 - 208, 253, 255, 261 - 267, 282 - 283, 304 - 308 (1970). – Listing of all known artificial earth satellites on a month-by-month basis.

054.022 Kunstmanen.   J. Meeus.
Hemel en Dampkring, Vol. 68, 6 - 8, 88 - 90 (1970).

054.023 Czechoslovak instruments on board Intercosmos I.
B. Valniček.
Říše hvězd, Vol. 51, 81 - 85 (1970).   In Czech.

054.024 Computation of the coordinates of a satellite with small eccentricities.   V. V. Terentyev.
Astron. Tsirk., No. 547, p. 5 - 6 (1970).   In Russian.

054.025 Dispersions of fast fluctuations in the deceleration of satellites at different heights.
N. P. Slovokhotova, V. E. Chertoprud.
Byull. Stantsij Optichesk. Nablyud. Iskusstv. Sputnikov Zemli, No. 53 (1), p. 4 - 6 (1969). In Russian.

054.026 Intensity distribution in the Echo 2 spectrum.
V. A. Smirnov, J. D. Russo.
Byull. Stantsij Optichesk. Nablyud. Iskusstv. Sputnikov Zemli, No. 53 (1), p. 6 - 9 (1969). In Russian.
   Curves were determined which characterize spectral features of the photographic system used for artificial satellite ob-

servations. The comparison of similar curves shows a satisfactory coincidence of the intensity distribution in the satellite spectrum and in the continuous solar spectrum.

054.027 Evaluation of the time when a satellite crosses a certain celestial pole. M. Ill.
Byull. Stantsij Optichesk. Nablyud. Iskusstv. Sputnikov Zemli, No. 53 (1), p. 23 - 26 (1969). In Russian.

OAO probes 'larger' universe.
Spaceflight, Vol. 12, 230 - 231 (1970). – Concerning satellite OAO II (1968–110A) and its astronomical programs.

Japan: Fourth into orbit.
Spaceflight, Vol. 12, 274, 293 (1970). – Concerning first Japanese satellite Ohsumi (*1970-11A*).

DIAL erforscht die hohe Atmosphäre.
Weltraumfahrt, 21. Jahrgang, p. 39 - 42 (1970).

Determination of short-periodic atmospheric density variations from quasi-simultaneous visual observations of artificial satellites. See Abstr. 082.090.

055 Observations of Earth Satellites, Lunar and Planetary Probes

055.001 Zur Sichtbarkeit kleinster Objekte auf dem Mond. V. Happach.
VdS Nachrichtenblatt, 19. Jahrgang, p. 15 (1970).

055.002 Cosmos 71 – 1965 53 F (rocket). Equatorial coordinates (January – December 1966).
Rezul'taty Nablyud. Sovet. Iskusstv. Sputnikov Zemli No. 112, 60 pp. (1968). In Russian.

055.003 Cosmos 71 – 1965 53 F (rocket). Horizontal coordinates (January – December 1966).
Rezul'taty Nablyud. Sovet. Isskustv. Sputnikov Zemli No. 113, 71 pp. (1968). In Russian.

055.004 Poljot 1 – 1963 43 A (sputnik). Horizontal coordinates (January – December 1967).
Rezul'taty Nablyud. Sovet. Iskusstv. Sputnikov Zemli No. 117, 58 pp. (1969). In Russian.

055.005 Optical tracking of lunar probes. G. E. Taylor.
Journ. British Astron. Ass., Vol. 80, 232 - 233 (1970).

055.006 De gaswolken van Apollo-12. A. G. Jansen.
Hemel en Dampkring, Vol. 68, 34 - 40 (1970).

055.007 Results of observations of Soviet artificial earth satellites according to the INTEROBS program for the year 1967. M. Ill, K. Sütö (Editors).
Rezul'taty Nablyud. Sovet. Iskusstv. Sputnikov Zemli No. 115 (8), 107 pp. (1968). In German. – Concerning Poljot 1 satellite = 1963 43 A, Cosmos 54 rocket = 1965 11 D, Cosmos 71 rocket = 1965 53 F.

055.008 Catalog of precisely reduced observations. No. P-18. B. Miller.
SAO, *Cambridge, Mass.,* Special Rep., No. 296, 4 + 196 pp. (1969).

055.009 Satellite orbital data No. E-8. B. Miller.
SAO, *Cambridge, Mass.,* Special Rep., No. 290, 3 + 27 pp. (1968).

055.010 Osservazioni radio-ottiche di satelliti artificiali. A. Manara.
Mem. Soc. Astron. Italiana, Nuova Serie, Vol. 41, 1 - 7 (1970).
    A new apparatus for receiving the signals sent by artificial satellites and time signals utilized for optical tracking is described. Then some Doppler shifts of artificial satellites are

studied; finally future methods for orbit calculations are exposed.

055.011 The influence of refraction anomalies upon the observations of satellites. J. Bieniewski.
Artificial Satellites Polish Acad. Sci., Vol. 5, No. 1, p. 33 - 35 (1969).
    The author deals with the problem of refraction errors and various refraction anomalies, as well as their influence on the observed position of a satellite. Some ways are proposed as regards the diminution of that influence.

055.012 Photographical positions of the artificial earth satellite "Pageos". S. Świerkowska.
Artificial Satellites Polish Acad. Sci., Vol. 5, No. 1, p. 43 - 46 (1969).
    Results obtained from observations of Pageos, made during August 1966 are presented. The 34 photographs processed embrace the period from 3.8.to 10.8.66. The observations were made with the photographic camera of the Astronomical Observatory in Poznań, with the "Tessar" lens (f = 360 mm).

055.013 Analyse directe numerique de la précision de position et de temps dans la série d'observation.
M. Bielicki.
Artificial Satellites Polish Acad. Sci., Vol. 4, No. 1, p. 3 - 10 (1968).
    C'est le calcul de variance qui lie les erreurs moyennes quadratiques de deux données d'observation du type: des observations pures, des observations élaborées, des observations déjà notées et enfin, l'erreur générale quadratique d'observation. On applique cette méthode aux observations des satellites artificiels à la station 1155.

055.014 The deep space network. An instrument for radio navigation for the Mariner mission to Mars, 1969.
J. P. Fearey, N. A. Renzetti.
Astrophys. Space Sci. Library, Vol. 15, 599 - 639 (1970).
Conference paper (see 012.007).

055.015 Observations en Belgique de phénomènes lumineux se rapportant au vol lunaire d'Apollo 12.
M. Ackerman, E. Aerts, J. Vercheval, H. Debehogne.
Ciel et Terre, Vol. 86, 199 - 211 (1970).
    Ground based observations made in Belgium of luminous phenomena related to the Apollo 12 mission are described. Precise sky coordinates are given for some events as well as spectrographic informations on their light emission.

055.016 **Das Bodenbetriebssystem "Azur".**
M. Schurer.
Weltraumfahrt, 21. Jahrgang, p. 56 - 57 (1970).

055.017 **Phénomènes aeronomiques dus à Apollo VIII.**
Gaz. astron., No. 2, p. 24 - 26 (1969).

055.018 **Poljot 1 – 1963 43 A (sputnik). Equatorial coordinates (January – December 1967).**
Rezul'taty Nablyud. Sovet. Iskusstv. Sputnikov Zemli
No. 116, 82 pp. (1969). In Russian.

055.019 **Data on results of observations of Soviet artificial earth satellites for 1967.**
Rezul'taty Nablyud. Sovet. Iskusstv. Sputnikov Zemli
No. 115, Appendix, 5 pp. (1969). In Russian.

055.020 **Cosmos 55 – 1965 11 D (rocket). Equatorial coordinates (January – May 1967).**
Rezul'taty Nablyud. Sovet. Iskusstv. Sputnikov Zemli
No. 118, 72 pp. (1969). In Russian.

055.021 **Cosmos 54 – 1965 11 D (rocket). Equatorial coordinates (January – December 1967).**
Rezul'taty Nablyud. Sovet. Iskusstv. Sputnikov. Zemli
No. 119, 56 pp. (1969). In Russian.

055.022 **Cosmos 54 – 1965 11 D (rocket). Horizontal cooordinates (January – December 1967).**
Rezul'taty Nablyud. Sovet. Iskusstv. Sputnikov Zemli
No. 120, 82 pp. (1969). In Russian.

055.023 **Cosmos 71 – 1965 53 F (rocket). Equatorial coordinates (January – December 1967).**
Rezul'taty Nablyud. Sovet. Iskusstv. Sputnikov Zemli
No. 121, 93 pp. (1969). In Russian.

055.024 **Cosmos 71 – 1965 53 F (rocket). Horizontal coordinates (January – December 1967).**
Rezul'taty Nablyud. Sovet. Iskusstv. Sputnikov Zemli
No. 122, 101 pp. (1969). In Russian.

055.025 **A survey of the southern sky at 55 MHz.**
P. Rohan, L. B. Soden.
Australian Journ. Phys.,Vol. 23, 223 - 225 (1970).
For radar measurements aimed at investigating the physical processes accompanying the re-entry of high speed objects into the earth's atmosphere it is important to maximize the signal-to-noise ratio of the observed radar data.

055.026 **Détermination d'un vecteur spatial à l'aide de deux mesures laser et d'une observation optique.**
A. Dinescu.
Stud. Cerc. Astron., Vol. 15, 3 - 8 (1970).
On propose une méthode pour déterminer les vecteurs spatiaux en fonction de deux côtes mesurés par laser à partir de deux stations vers le satellite et aussi une direction entre une station et le satellite. Ces formules sont appliquées pour la direction San Fernando – Haute Provence avec 12 paires, pour vérifier la méthode.

055.027 **Zur astronomischen Auswertung photographischer Satellitenaufnahmen.** G. Zimmermann.
Nachr. Karten–, Vermessungswesen, Reihe I, Heft No. 44, p. 1 - 34 (1970).
In this report the author describes the working methods of evaluating satellite photographs by the astrometric procedure. Two details, viz. the automatical identification of stars and the approximation of ideal image coordinates are illustrated by means of arithmetical examples.

055.028 **Photographic tracking of artificial satellites with the camera NAFA-3c/25.** A. G. Krylov.
Byull. Inst. Astrofiz., *Dushanbe*, No. 49, p. 32 - 42 (1969).
In Russian.
The method of observations of various artificial earth satellites with the camera NAFA-3c/25 is described.

055.029 **Photographic observations of satellites with the aerial camera K-24 with polaroid filter.**
T. M. Vermeesch, L. J. Lantwaard.
Byull. Stantsij Optichesk. Nablyud. Iskusstv. Sputnikov Zemli, No. 53 (1), p. 41 - 43 (1969). In Russian.

055.030 **Experience of tracking of the satellite Pageos with the camera AFU-75 in Ulan-Bator.**
R. Radnaa, S. Sanžžav, V. T. Groschev.
Byull. Stantsij Optichesk. Nablyud. Iskusstv. Sputnikov Zemli, No. 53 (1), p. 44 - 45 (1969). In Russian.

055.031 **Determination of summary errors for visual tracking.** V. I. Kuryshev.
Byull. Stantsij Optichesk. Nablyud. Iskusstv. Sputnikov Zemli, No. 53 (1), p. 46 - 52 (1969). In Russian.

055.032 **Détermination des coordonnées topocentriques sphériques des satellites artificiels de la Terre à la** base d'observations photographiques. S. Domaradzki.
Geodezja i Kartografia, Vol. 18, 283 - 298 = Publ. Działu
Geod. Wyższej i Astron. Geod. Zg. PAN, No. 16 (1969).
In Polish.

055.033 **Discussion provisoire des résultats obtenus à l'Observatoire de Strasbourg dans l'observation** des satellites géodésiques Echo I, Echo II et Pageos du Programme Géodésique Européen et Mondial.
A. Schmitt.
Publ. Obs. Strasbourg, Vol. 1, Fasc. 5 [Extrait du 87e Congrès de l'Association française pour l'Avancement des Sciences, Nancy, 1968, p. 28 - 32], 5 pp. (1969).

055.034 **Précision des positions de satellites réduites à l'Observatoire de Strasbourg dans le Programme** Géodesique Ouest Européen. A. Schmitt.
Publ. Obs. Strasbourg, Vol. 1, Fasc. 6, 7 pp. (1970).

Optical observations of Apollo 12.
Sky Telescope, Vol. 39, 127 - 130 (1970).

Keeping track of earth satellites.
Spaceflight, Vol. 12, 2 - 7 (1970).
The article describes the extensive satellite tracking network of the North American Air Defense Command (NORAD). More than 10000 satellite observations made by radio, radar, and optical sensors are processed daily.

Beobachtung von Apollo 12, 1969 November 14.
SuW, Vol. 9, 8 (1970).

Beobachtungen der Startphase von Apollo 12.
VdS Nachrichtenblatt, 19. Jahrgang, p. 41 - 43 (1970).

Theoretical Astrophysics

061 General Theoretical Problems of Astrophysics, Gravitational Instability, Neutrino Astronomy, X Ray- and Gamma Ray-Astronomy, Frequency and Origin of Elements etc.

061.001 High-energy gamma-ray astronomy.
G. G. Fazio.
Nature, Vol. 225, 905 - 911 (1970).
Gamma-ray astronomy above 10 MeV is difficult from the experimental point of view, but promising data are beginning to appear, as this survey progress report shows.

061.002 Numerical studies of penetrative convective instabilities. A. J. Faller, R. Kaylor.
Journ. Geophys. Res., Vol. 75, 521 - 530 (1970).
The stability of thermal convection for a statically unstable layer surmounted by a statically stable layer has been attacked by direct numerical integration of the linearized equations of motion and heat conduction. Two types of vertical temperature variation are considered, namely a piecewise-linear distribution and a piecewise-parabolic distribution.

061.003 Physics of stars and nebulae.
A. Sapar, G. Kuzmin.
Astron. Geod. Eston. SSR, Tartu, (see 003.004), p. 30 - 51 (1969). In Russian.

061.004 Solare Neutrinos. H. Ruhm.
SuW, Vol. 9, 60 - 62 (1970).

061.005 Astrophysical determination of the coupling constant for the electron-neutrino weak interaction.
R. B. Stothers.
Phys. Rev. Letters, Vol. 24, 538 - 541 (1970).
This letter reports the results of an unambiguous test for an upper limit on the $(\bar{e}\nu_e)(\bar{\nu}_e e)$ coupling constant using the statistics of white-dwarf stars, and concludes with a summary of the results of tests providing a lower limit.

061.006 Evidence for the existence of heavy nuclei prior to the s-process. H. D. Zeh.
Nature, Vol. 225, 361 - 362 (1970).
Abundances and neutron-capture cross sections particularly of the lead isotopes show evidence, that "s-nuclei" have been formed by neutron irradiation of "r-nuclei", thus giving information about the chronological order of nuclear formation processes. The data furthermore give a rather rigorous upper limit of about 8×10^9 yr for the age of the majority of the "r-nuclei".

061.007 Primordial synthesis of superheavy nuclei in the galaxy. B. Kuchowicz.
Nature, Vol. 225, 440 - 441 (1970).
If massive bulks of superdense prestellar matter are exploding, synthesis of heavy and even superheavy nuclei (from the hypothetical new island of stability in the vicinity of Z = 110 or 114) may proceed in the external layers. As the valley of beta stability in the (Z, N) plane is shifted toward the N-axis at higher densities, the alpha-instability region may be bypassed. This may give an essential contribution to the abundance of superheavy elements, in addition to their possible production in supernovae. Evidence for extinct superheavy elements comes both from isotopic anomalies of Kr and Xe in meteorites, and from the relative overabundance of odd isotopes of even elements in the region from Te to rare earths.

061.008 On the frequency dependence of the line source function. R. C. Canfield.
Solar Physics, Vol. 12, 63 - 65 (1970).
The author gives a comment on a paper by G. Worrall (Solar Physics, Vol. 8, 18 - 19, 1969).

061.009 Properties of low frequency oscillatory convection in a strong magnetic field. J. D. Zugzda.
Astron. Zhurn. Akad. Nauk SSSR, Vol. 47, 340 - 350 (1970). In Russian. – English translation in Soviet Astron., AJ, Vol. 14, No. 2.
Low frequency oscillatory convection in a polytropic atmosphere in the presence of a strong vertical magnetic field is investigated in linear approximation. Nonadiabatic oscillations are considered and it is shown that the maximum increment has oscillations with a determinate size of the cells. An instability criterion for quasi-adiabatic and quasi-isothermal oscillations, its dependence from the temperature gradient and the ratio of boundary temperatures are obtained. Flow of heat (convective) and wave energy transported by oscillatory convection is investigated. Properties of oscillatory convection and sunspot theory are discussed.

061.010 Smoothing mechanism of the odd-even structure of elemental abundance curve.
T. Ohnishi, T. Nishi.
Progress Theor. Phys. Japan, Vol. 43, 558 - 559 (1970).
The r-process which takes place in the carbon burning shell of supernovae or later-stage stars, where carbon flash may occur, produces nuclides having a lesser odd-even effect with regard to mass number and these nuclides contribute to the smoothing of the odd-even structure of overall r-process nuclides in the universe.

061.011 Gravitational contraction of a spheroidal cloud.
Ya. B. Zeldovich, Ya. M. Kazhdan.
Astrofizika, Vol. 6, 109 - 122 (1970). In Russian. – English translation in Astrophysics, Vol. 6, No. 1.
Two problems of dynamical motion of gas under the action of gravitation are discussed. 1. The compression of a gas with finite pressure uniformly distributed inside the sphere at the initial moment. 2. The compression of cold gas or dust (p_o = 0) with nonuniform initial density distribution.

061.012 The origin of magnetic fields. E. N. Parker.
Astrophys. Journ., Vol. 160, 383 - 404 (1970).
The Henry Norris Russell lecture 1969 August 12, the 130th meeting of the American Astronomical Society, State University of New York, Albany.

061.013 Thermal-convective instability. R. J. Defouw.
Astrophys. Journ., Vol. 160, 659 - 669 (1970).

The Schwarzschild criterion for convection is generalized to include departures from adiabatic motion. It is demonstrated that a thermally unstable atmosphere is also convectively unstable, regardless of the atmospheric temperature gradient. The effects of conduction, viscosity, opacity, and rotation are evaluated. In this paper the assumption is made that the radiative cooling (or the source function) depends only on the local values of density and temperature.

061.014 The effect of an arbitrary law of slow rotation on the oscillations and the stability of gaseous masses.
N. R. Lebovitz.
Astrophys. Journ., Vol. 160, 701 - 723 (1970).

The operator formulation of stability theory is adapted to the case in which the rotation is slow (of order λ, say) and both the rotating configuration and the time-dependent disturbances are axisymmetric. It is shown that, to find the effect of rotation on the square of an oscillation frequency to first order in λ, one need only know the eigenfunction to order zero, i.e., one need only know the corresponding eigenfunction in the spherical case. The formulae obtained are applied to the case of radial pulsations and dynamical instability, where it is found that any law of rotation has a stabilizing influence, certainly if γ is constant, and probably in many cases when γ is variable.

061.015 The oscillations and the stability of rotating masses with toroidal magnetic fields.
Y. Nakagawa, S. K. Trehan.
Astrophys. Journ., Vol. 160, 725 - 733 (1970).

The oscillations and the stability of rotating masses for axisymmetric equilibrium configurations with toroidal magnetic fields are examined on the basis of the second-order virial equations. It is shown that the equilibrium state is, in general, not spherically symmetric except for the equipartition state with no surface rotation. It is also shown that the transverse shear modes and the toroidal modes of oscillation could become unstable for large values of surface rotation whereas these modes remain stable for the equipartition state. The pulsation mode which is coupled with the nonradial modes of oscillation is shown to become unstable either for large values of surface rotation or for strong magnetic fields.

061.016 Gravitational instability: An approximate theory for large density perturbations.
Ya. B. Zeldovich.
Astron. Astrophys., Vol. 5, 84 - 89 (1970).

An approximate solution is given for the problem of the growth of perturbations during the expansion of matter without pressure. The solution is qualitatively correct even when the perturbations are not small. Infinite density is first obtained on disc-like surfaces by unilateral compression. The following layers are compressed first adiabatically and then by a shock wave. Physical conditions in the compressed matter are analysed.

061.017 On a cosmic background of low-energy neutrinos.
T. de Graaf.
Astron. Astrophys., Vol. 5, 335 - 340 (1970).

A review is given of the theoretical concepts which have been proposed concerning a cosmic background of low-energy neutrinos.We find a new value for this upper limit from recent results for the beta-decay-spectrum and consider the scattering process between neutrinos and nucleons, for which new experimental data and theoretical ideas have been given recently.

061.018 Calculation of stellar opacity taking into account the light absorption in spectral lines.
A. F. Nikiforov, V. B. Uvarov.

Dokl. Akad. Nauk SSSR, Ser. Mat. Fiz., Vol. 191, 47 - 49 (1970). In Russian.

Methods applied in this paper to opacity calculations are described. As a starting approximation for the determination of wave functions of electrons Schrödinger's equation with the Thomas-Fermi potential was used. It has been shown that electrons of the so-called intermediate group can be effectively accounted for as electrons of the continuous spectrum. For the determination of discrete energy levels at high temperatures a trial-potential method was applied. A quasi-classical approximation proved to be sufficiently accurate for finding wave functions of the continuous spectrum. Effective cross sections for the absorption of photons have been calculated in a non-relativistic dipole approximation. Critical remarks on similar calculations of opacity by Cox and Carson have been expressed. The results of this paper are closer to those of Cox, and differ sometimes from those of Carson more then twice. *B. Onderlička*

061.019 Cosmic investigations of the radiation of stars, nebulae and galaxies. A. B. Severnyj.
Vestn. AN SSSR, 1969 No. 11, p. 91 - 95. In Russian.
Abstr. in Referativ. Zhurn. 62 Issled. kosm. prostranstv., 4.62.17 (1970).

061.020 Possibilities of observing neutrinos emitted from collapsing stars and neutrino oscillations.
G. V. Domogatskij, G. T. Zatsepin.
Izv. AN SSSR. Ser. fiz., Vol. 33, 1796 - 1799 (1969). In Russian. – Abstr. in Referativ. Zhurn. 51. Astron., 5.51.463 (1970).

061.021 Lepton charge and neutrino astrophysics.
B. M. Pontekorvo.
Izv. AN SSSR. Ser. fiz., Vol. 33, 1787 - 1791 (1969). In Russian. – Abstr. in Referativ. Zhurn. 51. Astron., 5.51.464 (1970).

061.022 Extra-atmospheric submillimeter astronomy.
A. E. Salomonovich.
Uspekhi fiz. nauk, Vol. 99, 417 - 438 (1969). In Russian. Abstr. in Referativ. Zhurn. 51. Astron., 4.51.224 (1970).

061.023 The instability of the congruent Darwin ellipsoids. II.
S. Chandrasekhar.
Astrophys. Journ., Vol. 160, 1043 - 1048 (1970).

The resonant oscillations of one of two congruent Darwin ellipsoids, forced by the natural oscillations of the other, are considered; and the instability of the ellipsoids to synchronous coupled oscillations is traced to this resonance.

061.024 Electromagnetodynamic stability of viscous fluid cylinders. G. Bhowmik, S. P. Talwar.
Progr. Theor. Phys. Japan, Vol. 43, 1413 - 1422 (1970).

The motivation of the present paper is to investigate the joint influence of viscosity and magnetic field on the stability of cylindrical configurations, and to also study the influence of a uniform axial electric field on the stability of viscous liquid columns.

061.025 Radiative opacities and Compton scattering in strong magnetic fields. V. Canuto.
Astrophys. Journ. (*Letters*), Vol. 160, L153 - L155 (1970).

The wave function of a nonrelativistic electron in a magnetic field has been used to compute the Compton cross-section by the standard method. We report here the result of such a computation.

061.026 *r*-process production ratios of chronologic importance. P. A. Seeger, D. N. Schramm.
Astrophys. Journ. (*Letters*), Vol. 160, L157 - L160 (1970).

We have found values of the r-process production ratios of galactic chronologic importance, taking into account the uncertainties in the mass law as well as other calculation parameters. The values prior to any reactions during freezing out and ejection are: $^{232}Th/^{238}U = 1.96 \pm 0.25$, $^{244}Pu/^{238}U = 0.96 \pm 0.21$, $^{237}Np/^{238}U = 1.89 \pm 0.26$, and $^{235}U/^{238}U = 1.89 \pm 0.36$.

061.027 **Superfluidity and superconductivity in the universe.**
V. L. Ginzburg.
Journ. Statist. Phys., Vol. 1, (No. 1), 3 - 24 (1969).

The paper aims to elucidate the current status of the problem concerning the existence and observation of superfluid and superconducting states in the universe, that is, under cosmic conditions. Following an introduction the paper discusses Bose-Einstein condensation, superfluidity, and superconductivity; possibilities for the occurrence of superfluidity and superconductivity under cosmic conditions; superconductivity of dense, degenerate electron plasma (large planets, white dwarfs); superfluidity and superconductivity in neutron stars; and finally superfluidity in a cosmological neutrino 'sea'.

061.028 **Free radicals in astrophysics.**
P. Swings.
Mém. Soc. Roy. Sci. Liège, 5e Sér., Vol. 18, Fasc. 3, 26 pp. = Univ. Liège, Inst. d'Astrophys., Coll. 8°, No. 590 (1969).

On the muon neutrino emission in gravitational collapse of massive stars. See Abstr. 065.089.

Neutrino astronomy of the sun. Part I. Neutrino production. See Abstr. 080.006.

Solar-neutrino fluxes with recent corrections to opacity. See Abstr. 080.013.

On the spectrum of solar B^8-neutrinos. See Abstr. 080.018.

Influence of a horizontal magnetic field on the criterion of convective instability. See Abstr. 080.020.

Cosmic electrons and related astrophysics. See Abstr. 143.010.

062 Magneto-Hydrodynamics, Plasma

062.001 **Radiation from plasma systems moving with relativistic velocities.** J. J. Monaghan.
Monthly Notices, Roy. Astron. Soc., Vol. 147, 207 - 213 (1970).
The radiation produced when a plasma system, with a centre of mass velocity close to the velocity of light, passes through a spatially varying, but static magnetic field, is studied. The analysis is performed in the rest frame of the plasma system where the electromagnetic field approximates closely that of a plane wave. In this reference frame the radiation may be viewed as a reflection process. The spectrum is determined for both large and small systems. Applications to astrophysical phenomena are discussed briefly.

062.002 **Über ein Verfahren zur gleichzeitigen Bestimmung von Elektronenkonzentration und Elektronenstoßzahl in einem homogenen isotropen oder in einem langsam veränderlichen Plasma.** W. Muschler.
Zeitschr. Naturforschung, Vol. 25a, 106 - 114 (1970).
A method is described, which allows determination of the complex refractive index of a plasma by separate measurement of the E- and H-component of an electromagnetic wave. By means of the complex refractive index simultaneously electron concentration and electron collision frequency of the medium can be stated.

062.003 **Errata to the paper: "Tables for computing dielectronic recombination rates from Burgess' general formula".** [Bull. Astron. Inst. Czechoslovakia, Vol. 20, 159 - 163 (1969)]. V. Letfus.
Bull. Astron. Inst. Czechoslovakia, Vol. 21, 65 (1970). – See Astron. Astrophys. Abstr., 02.062.002.

062.004 **A finite-amplitude solution of the self-consistent Vlasov equations.** W. Lünow.
Zeitschr. Naturforschung, Vol. 25a, 164 - 169 (1970).
The paper deals with the propagation of a finite-amplitude circularly polarized electromagnetic plane wave through a fully ionized hot collisionless plasma, composed of Q ($\geqq 1$) types of ions and of electrons, as a strict solution of the nonrelativistic self-consistent Vlasov equations within the scope of some given applicability criterions.

062.005 **On the theory of acceleration of charged particles in a cosmic plasma.**
M. E. Katz, A. K. Yukhimuk.
Geomagn. Aeronom., Vol. 10, 328 - 331 (1970).
In Russian. – Brief information.

062.006 **Stability of a heterogeneous conducting fluid with a radial gravitational force.**
N. Rudraiah.
Publ. Astron. Soc. Japan, Vol. 22, 41 - 55 (1970).
The stability of a heterogeneous incompressible nonviscous perfectly conducting fluid between two fixed coaxial cylinders with an applied magnetic field in the azimuthal direction and with a radial gravitational force is examined under the assumption of axial symmetry. With the aid of Sturm's comparison theorem, a necessary condition for instability is established.

062.007 **Transformation of plasma waves in a magnetoactive plasma.** W. K. Yip.
Planet. Space Sci., Vol. 18, 479 - 494 (1970).
Using the system of hydrodynamic equations, we obtain the coefficient for transformation of high frequency weakly damped plasma waves in a magnetoactive plasma. The ex-

pressions obtained are studied numerically with parameters appropriate to the active solar corona.

062.008 **A numerical model of hydromagnetic turbulence.**
D. L. Moss.
Bull. American Astron. Soc., Vol. 2, 210 (1970). – Abstr. AAS.

062.009 **Toroidal magnetic fields and differentially rotating sun.** Y. Nakagawa.
Bull. American Astron. Soc., Vol. 2, 211 (1970). – Abstr. AAS.

062.010 **Elliptical polarization of a synchrotron source in the case of quasitransverse propagation.**
A. G. Pacholczyk, T. L. Swihart.
Bull. American Astron. Soc., Vol. 2, 212 (1970). – Abstr. AAS.

062.011 **Sur l'instabilité magnétohydrodynamique d'un plasma qui possède une pression anisotrope. (I). L'équation de dispersion.** M. Vasiu.
Studia Univ. Babeş–Bolyai, Sci. Phys., Fasc. 1, p. 87 - 93 (1970).
Dans le mémoire présent nous voulons déduire l'équation de dispersion d'un modèle de plasma infini, homogène, compressible, non visqueux, doué d'une conductivité électrique finie et d'une pression anisotrope.

062.012 **Magnetic fields of a moving conductive liquid. II.**
G. A. Rubo.
Vestn. Kiev. Univ., Ser. Astron., No. 11, p. 41 - 45 (1969).
In Russian.
The magnetic field of a moving conductive liquid with spiral symmetry is investigated. It is shown that the density is independent on the spiral coordinate in the case of nonviscous adiabatic motion. Conditions for a magnetic field in the case of viscous motion are obtained.

062.013 **A numerical model of hydromagnetic turbulence.**
D. L. Moss.
Monthly Notices, Roy. Astron.Soc., Vol. 148, 173 - 191 (1970).
A kinematic numerical model of two-dimensional hydromagnetic turbulence is developed to investigate the amplification of an externally maintained seed magnetic field under the action of fluid motions of varying magnetic Reynolds number. The model is applied to the problem of the expulsion of flux from a turbulent region in a conducting fluid, and it is suggested that some flux may always remain in such a region. The turbulent dynamo problem is briefly discussed.

062.014 **Erratum: Impact theory of the broadening and shift of spectral lines due to electrons and ions in a plasma (continued).** [Astron. Astrophys., Vol. 2, 322 - 354 (1969)]. S. Sahal-Bréchot.
Astron. Astrophys., Vol. 4, 162 (1970).

062.015 **Relaxation times for establishing steady-state populations in optically thin and thick plasmas.**
H. W. Drawin.
Journ. Quant. Spectrosc. Radiat. Transfer, Vol. 10, 33 - 48 (1970).
Relaxation times have been calculated for the ground and the excited states of hydrogen atoms and hydrogen-like ions on the basis of the collisional-radiative model. At low and medium electron densities (non-collisional-dominated plasmas),

these times become a sensitive function of the number of reabsorbed photons in the resonance lines and in the resonance continuum.

062.016 **Astrophysical implications of the continuity equation plasma oscillation.** J. R. Roth.
Nature, Vol. 226, 626 - 628 (1970).

The continuity equations for neutral and charged particles in a partly ionized gas can possess periodic solutions. These oscillations give rise to a pulsed efflux of charged particles from an oscillating region, and may provide a clock mechanism for such periodic astrophysical phenomena as pulsars and geomagnetic micropulsations. The electron number density waveform displays the pulsar-like characteristics of sharp, narrow, periodic peaks between broad, flat minima. If a turbulent heating process acts on the pulses of electrons, pulsed r-f radiation might result. Some results are presented from an experimental study of a laboratory plasma subject to the continuity-equation oscillation.

062.017 **On the spectrum of relativistic particles accelerated by plasma turbulence.**
V. N. Tsytovich, A. S. Chikhachev.
Astron. Zhurn. Akad. Nauk SSSR, Vol. 47, 479 - 482 (1970).
In Russian. English translation in Soviet Astron. AJ, Vol. 14, No. 3.

Spectra of particles accelerated by high-frequency turbulence of plasma are analysed; it is shown that the spectra are of power type.

062.018 **Investigations of cosmic plasma using the Yakutsk cosmophysical and aeronomical devices.**
Yu. G. Shafer.
Izv. AN SSSR. Ser. fiz., Vol. 33, 1877 - 1889 (1969).
In Russian. – Abstr. in Referativ. Zhurn. 51. Astron., 5.51.250 (1970).

062.019 **Hydrogen Stark broadening calculations with the unified classical path theory.**
C. R. Vidal, J. Cooper, E. W. Smith.
NBS Monograph 116, [Superintendent of Documents, U.S. Government Printing Office, Washington, D.C.], 143 pp. (1970).

This publication presents a new unified theory of hydrogen line profiles which are broadened by the electric micro-fields produced by electrons and ions in a plasma. The theoretical approach presented is the first one providing for calculations of normalized profiles covering the entire line profile from the impact broadening domain in the line center out to the distant quasistatic line wings. Both limits had until now, been treated independently by different theories. A computer program (Fortran IV), that allows calculation of Lyman lines for electron densities and temperatures of practical interest is included.

062.020 **Plasmas and magnetohydrodynamics in astrophysics.**
L. Davis, Jr.
Trans. IAU, Vol. 14A, (see 003.028), 515 - 524 (1970).
Report of Commission 43. – Plasma in the sun and stars, (*L. Mestel*); Interplanetary plasma,(*R. Lüst*); Major problems in plasma astrophysics and magnetohydrodynamics,(*S. A. Kaplan*); Cosmic rays, (*L. Davis*).

062.021 **Effect of radiation upon the post-shock state.**
P. Koch, R. A. Gross.
Phys. Fluids, Vol. 12, 1182 - 1192 (1969).

062.022 **Expansion of a plasma shell into a vacuum magnetic field.** J. W. Poukey.
Phys. Fluids, Vol. 12, 1452 - 1458 (1969).

062.023 **Interaction of magnetoacoustic and entropy waves with normal magnetohydrodynamic shock waves.**
K. O. Westphal, J. F. McKenzie.
Phys. Fluids, Vol. 12, 1228 - 1236 (1969).

062.024 **Atomic spectroscopy with the shock tube.**
T. D. Wilkerson, D. W. Koopman, M. Miller, R. Bengtson, G. Charatis.
Phys. Fluids, Vol. 12, No. 5, part II, p. 22 - 29 (1969).

Semi-classical calculations of the collisional broadening of ion lines. See Abstr. 022.064.

Determination of electron density in plasma by the number of the extreme resolved line.
See Abstr. 073.005.

Le problème des N corps en astronomie et en physique des plasmas. See Abstr. 151.015.

063 Radiative Transfer

063.001 Diffuse reflection from a semi-infinite atmosphere.
G. C. Pomraning.
Astrophys. Journ., Vol. 159, 119 - 126 (1970).

The classical problem of computing the albedo from a half-space is considered. It is shown that one can derive an appropriate variational principle for this problem, and that the variational estimates of the albedo based upon asymptotic trial functions are remarkably accurate. This study suggests that in certain work on radiative transfer the complexities introduced by accounting for polarization effects and the anisotropy of the Rayleigh phase function can be avoided. It may be sufficient, depending upon the accuracy required, to assume an isotropic phase function averaged over polarization. A byproduct of the variational analysis is an analytic approximation to Chandrasekhar's H-function which may be useful in other problems of radiative transfer.

063.002 Nonlinear integral equations in the theory of radiative transfer. V. V. Sobolev.
Stars, Nebulae, Galaxies, Symposium Byurakan 1968, (see 012.002), p. 9 - 19 (1969). In Russian.

063.003 Decomposition of Ambartsumyan's function.
Eh. G. Yanovitskij.
Stars, Nebulae, Galaxies, Symposium Byurakan 1968, (see 012.002), p. 21 - 26 (1969). In Russian.

063.004 The mean scattering number of a photon in a homogeneous optically thick sphere.
V. V. Ivanov.
Stars, Nebulae, Galaxies, Symposium Byurakan 1968, (see 012.002), p. 27 - 39 (1969). In Russian.

063.005 Transfer of resonance radiation in an optically thick layer. D. I. Nagirner.
Stars, Nebulae, Galaxies, Symposium Byurakan 1968, (see 012.002), p. 41 - 44 (1969). In Russian.

063.006 Polychromatic scattering of light in a half-space.
V. Yu. Terebizh.
Stars, Nebulae, Galaxies, Symposium Byurakan 1968, (see 012.002), p. 45 - 50 (1969). In Russian.

063.007 On the theory of an instationary radiation field.
I. N. Minin.
Stars, Nebulae, Galaxies, Symposium Byurakan 1968, (see 012.002), p. 51 - 55 (1969). In Russian.

063.008 On the theory of generation and transfer of radiation in an instationary cosmic plasma.
S. A. Kaplan, V. N. Tsytovich.
Stars, Nebulae, Galaxies, Symposium Byurakan 1968, (see 012.002), p. 57 - 63 (1969). In Russian.

063.009 Diffuse reflexion of quanta on a turbulent medium.
N. B. Engibaryan, A. G. Nikogosyan.
Stars, Nebulae, Galaxies, Symposium Byurakan 1968, (see 012.002), p. 65 - 72 (1969). In Russian.

063.010 A generalized Milne–Eddington–model in the theory of spectral line formation. Sh. A. Sabashvili.
Stars, Nebulae, Galaxies, Symposium Byurakan 1968, (see 012.002), p. 73 - 79 (1969). In Russian.

063.011 Transfer of resonance radiation in an optically thick layer. D. I. Nagirner.
Astrofizika, Vol. 5, 507 - 524 (1969). In Russian. – English

translation in Astrophysics, Vol. 5, No. 4.

Transfer of resonance radiation in a plane-parallel layer of optical thickness τ_0 is considered. The scattering is assumed to be conservative, with complete frequency redistribution. The rigorous asymptotic ($\tau_0 \gg 1$) formula expressing the resolvent function $\Phi(\tau, \tau_0)$ for arbitrary τ ($0 \leqslant \tau \leqslant \tau_0$) in terms of $\Phi(\tau) \equiv \Phi(\tau, \infty)$ is found. Hence the solution of a general problem of conservative light scattering in an optically thick layer is found.

063.012 Line formation in a magnetic field. I. Scattering matrix. H. Domke.
Astrofizika, Vol. 5, 525 - 537 (1969). In Russian. – English translation in Astrophysics, Vol. 5, No. 4.

A method is proposed which enables to obtain the scattering matrix of a resonance line with arbitrary type of splitting of levels in a magnetic field.

063.013 The direct calculation of the H-function for completely noncoherent scattering. R. F. Warming.
Astrophys. Journ., Vol. 159, 593 - 604 (1970).

The generalized H-function for completely noncoherent isotropic scattering is expressed in terms of a real integral which is admirably suited for computational purposes. The resulting integral is then modified so that it may be evaluated numerically by using Legendre-Gauss quadrature. Six-place tables of the H-function are given for a Lorentz scattering profile and a wide range of the albedo parameter ω.

063.014 Moments of Chandrasekhar's functions H_l and H_r.
G. R. Bond, C. E. Siewert.
Astrophys. Journ., Vol. 159, 731 - 732 (1970).

The first twenty-two moments of Chandrasekhar's functions H_l and H_r are given.

063.015 On Zheleznyakov's equation of radiative transfer in a magnetoactive plasma. D. Walsh.
Astrophys. Journ., Vol. 159, 733 - 734 (1970). – Note.

063.016 Theory of radiative transfer in inhomogeneous atmospheres. III. Extension of the perturbation method to azimuth-independent terms.
K. D. Abhyankar, A. L. Fymat.
Astrophys. Journ., Vol. 159, 1009 - 1018 (1970).

In the case where the phase matrix, corresponding to the azimuth-independent term of the radiation field scattered by an innomogeneous plane-parallel atmosphere, is separable, the simplified matrix equations of the problem are treated by the perturbation method of Fymat and Abhyankar. The nonlinear singular integral equations for the matrix functions K, L, K^*, and L^*, given by Sekera (1963), are solved by linearizing the equation for K. This linearized equation is written in an operator form, and its N-solution is given.

063.017 Theory of radiative transfer in inhomogeneous atmospheres. IV. Application of the matrix perturbation method to a semi-infinite atmosphere.
K. D. Abhyankar, A. L. Fymat.
Astrophys. Journ., Vol. 159, 1019 - 1028 (1970).

In order to describe the transfer of partially polarized radiation through an inhomogeneous semi-infinite atmosphere, a matrix N-function has been introduced by considering the limit of the K-function as the optical thickness of a finite atmosphere becomes infinite. The nonlinear singular integral equation satisfied by N is given. This equation is solved by our perturbation method in the form of a Neumann series. The region of convergence of this series solution is delimited for the Rayleigh law of scattering. An iteration scheme for computing the solution is described.

063.018 A solution of a time-dependent equation of radiative transfer. A. D. Code.
Astrophys. Journ., Vol. 159, 1029 - 1039 (1970).

The time-dependent equation of radiative transfer for a plane-parallel isotropic scattering medium is solved in the first Gaussian approximation. The time dependence of the transmitted and reflected radiation is characterized by a time constant equal to $1/\kappa\rho c$. In any physical situation in which the intensity varies significantly during the time required for a photon to travel a unit optical depth, the time-dependent characteristics will be important. This is the case for novalike variables, novae, variable nebulae, and presumably quasi-stellar sources, among others.

063.019 Time-dependent radiation transfer in a semi-infinite atmosphere. A. D. Code, G. Eason.
Astrophys. Journ., Vol. 159, 1041 - 1046 (1970).

The time-dependent equation of radiative transfer is treated in the first Gaussian approximation for the case of isotropic scattering. The equations are solved for a semi-infinite atmosphere by the method of Laplace transforms. The reflected radiation initially decays exponentially with time, with a time constant proportional to $1/\kappa\rho c$. The results provide an accurate solution for a specific time-dependent transfer problem that may be used to test other approximate treatments.

063.020 Probabilistic model for radiative transfer problems for inhomogeneous infinite cylindrical shell medium. T. K. Leong, K. K. Sen.
Publ. Astron. Soc. Japan, Vol. 22, 57 - 74 (1970).

A probabilistic model for deriving a family of fundamental equations governing the diffusion of photon in an inhomogeneous infinite cylindrical shell medium has been proposed. A set of four integro-differential equations for the scattering, transmission, back scattering and back transmission functions are derived.

063.021 Some mathematical aspects of multiple scattering in a finite inhomogeneous slab with anisotropic scattering. S. Ueno, R. E. Kalaba, H. H. Kagiwada, R. E. Bellman.
Publ. Astron. Soc. Japan, Vol. 22, 75 - 83 (1970).

The aim of this paper is to discuss the problem of radiative transfer in a finite inhomogeneous plane-parallel atmosphere with anisotropic scattering, whose albedo for single scattering varies in the vertical direction. Some integro-differential equations with initial conditions are derived for the internal intensity, the source function, and the probability of photon emergence, with the aid of the invariant imbedding technique.

063.022 Milne's problem for anisotropic light scattering. V. V. Sobolev.
Astron. Zhurn. Akad. Nauk SSSR, Vol. 47, 246 - 253 (1970). In Russian. – English translation in Soviet Astron., AJ, Vol. 14, No. 2.

The function $u(\eta)$ which gives the relative intensity of radiation transmitted by a semi-infinite atmosphere at the angle arccos η with the normal is considered. Formulae for $u(\eta)$ are given.

063.023 Transfer of partially polarized radiation through finite inhomogeneous atmospheres.
A. L. Fymat, K. D. Abhyankar.
Bull. American Astron. Soc., Vol. 2, 193 (1970). – Abstr. AAS.

063.024 Light scattering in a medium with a moving boundary. V. V. Leonov.
Astrofizika, Vol. 6, 89 - 100 (1970). In Russian. – English translation in Astrophysics, Vol. 6, No. 1.

The problem of light scattering in a one-dimensional semi-infinite homogeneous medium with a moving boundary is considered. The differential equation for the probability of escape of the quantum from the medium is obtained. Explicit expressions of the reflection probability and of the emergence probability from the medium are given.

063.025 Monte-Carlo calculation of line profiles in expanding and rotating atmospheres. C. Magnan.
Journ. Quant. Spectrosc. Radiat. Transfer, Vol. 10, 1 - 9 (1970).

Line formation in expanding and rotating atmospheres is investigated by use of Monte-Carlo techniques. Typical profiles are shown in some simple cases: spherical shells, expanding and rotating disks. Non-coherent scattering is taken into account; a two-level atom and an isothermal atmosphere are assumed.

063.026 An alternate form of Ueno's X- and Y-equations for inhomogeneous atmospheres.
A. L. Fymat, K. D. Abhyankar.
Journ. Quant. Spectrosc. Radiat. Transfer, Vol. 10, 49 - 54 (1970).

Starting from the global form of the equations of transfer, the nonlinear integral equations in the X-, Y-, X^*-, Y^*-functions for inhomogeneous atmospheres, similar to the Chandrasekhar equations for homogeneous atmospheres, are derived. Their relations to Ueno's and Chandrasekhar's equations are discussed and two approaches to the treatment of the problem of uniqueness of the solutions of these equations are suggested.

063.027 Effect of band or line shape on the radiative transfer in a nongray planar medium.
A. L. Crosbie, R. Viskanta.
Journ. Quant. Spectrosc. Radiat. Transfer, Vol. 10, 487 - 509 (1970).

Radiative transfer through an absorbing, emitting and heat generating nongray medium confined between two parallel walls is studied. Specific attention is directed toward the evaluation of the effect of band or line shape on the temperature distribution and radiative flux.

063.028 Erratum: Imperfect Rayleigh scattering in a semi-infinite atmosphere. [Vol. 4, 101 - 110 (1970)].
K. D. Abhyankar, A. L. Fymat.
Astron. Astrophys., Vol. 5, 491 (1970).

063.029 A perturbation solution of the Milne problem. J. J. Monaghan.
Monthly Notices, Roy. Astron. Soc., Vol. 148, 353 - 360 (1970).

The Milne problem in radiative transfer is solved approximately by a perturbation method. The integral equation formulation is used and the kernel is replaced by a single exponential term containing an arbitrary parameter b. The perturbation term involves the difference between the actual and approximate kernels.

063.030 Transfer of resonance radiation in purely scattering media – I. Semi-infinite medium.
V. V. Ivanov.
Journ. Quant. Spectrosc. Radiat. Transfer, Vol. 10, 665 - 680 (1970). In Russian.

The transfer of resonance radiation in a purely scattering halfspace is studied under the assumption of complete frequency redistribution.

063.031 **Transfer of resonance radiation in purely scattering media – II. Optically thick layer.**
V. V. Ivanov.
Journ. Quant. Spectrosc. Radiat. Transfer, Vol. 10, 681 - 694 (1970).   In Russian.

The transfer of resonance radiation in a purely scattering plane layer, of large optical thickness, is studied under the assumption of complete frequency redistribution. The absorption coefficient is assumed to have a Doppler profile.

063.032 **Photon polarization and frequency change in multiple scattering.**
A. Z. Dolginov, Yu. N. Gnedin, N. A. Silant'ev.
Journ. Quant. Spectrosc. Radiat. Transfer, Vol. 10, 707 - 754 (1970).

Maxwell's equations were used to derive a system of equations for the Stokes parameters and for the intensity of multiply-scattered radiation in a medium, taking into consideration the change of the photon frequency.

063.033 **Radiative transfer and spectra of celestial bodies.**
V. V. Ivanov.
"Nauka", Moskva. 472 pp. Price 2 Rbl. 24 Kop. (1969). In Russian. – Review in Referativ. Zhurn. 51. Astron., 4.51.195 (1970).

063.034 **The radiation regime in deep layers of a turbid medium with large scattering.**   V. M. Loskutov.
Vestn. Leningrad. un-ta, 1969, No. 13, p. 143 - 149. In Russian. – Abstr. in Referativ. Zhurn. 51. Astron., 4.51.205 (1970).

063.035 **Some nonlinear problems of radiative transfer.**
R. S. Vardanyan, N. B. Engibaryan.
Dokl. AN Arm. SSR, Vol. 49, 135 - 140 (1969).   In Russian. Abstr. in Referativ. Zhurn. 51. Astron., 5.51.224 (1970).

064 Stellar Atmospheres , Stellar Envelopes

064.001 **Spectral line formation by noncoherent scattering with a dipole phase function.**
D. G. Hummer.
Astrophys. Letters, Vol. 5, 1 - 4 (1970).

Results are presented for lines formed by atoms that scatter photons with complete redistribution in frequency and a dipole (Rayleigh) phase function.

064.002 **Nonlinear limb darkening for early-type stars.**
D. A. Klinglesmith, S. Sobieski.
Astron. Journ., Vol. 75, 175 - 182 (1970).

A set of coefficients has been obtained by fitting an empirical nonlinear limb-darkening law to values of $I(\mu)/I(1)$ computed from a grid of hydrogen line-blanketed model atmospheres. In addition, a set of coefficients for the conventional linear law has been calculated. The comparison with previous theoretical values for the linearized law shows good agreement with Grygar's (1965) results but systematically smaller values than those found by Hosokawa (1967).

064.003 **Stellar wind for non-negligible gravitational potential of the atmosphere.** R. D. Weidelt.
Astrophys. Space Sci., Vol. 6, 205 - 216 (1970).

We consider a spherically symmetric, isothermal and stationary stellar atmosphere whose gravitational potential cannot be neglected. If κ, the ratio of the density on the base of the corona to the mean density of the star is not zero, the density vanishes at infinity for any solution of the hydrodynamic equations. We deduce the maximum mass loss depending on two dimensionless parameters, the ratio of the gravitational to the thermal energy and κ. This mass loss has itself a maximum for $\kappa = 1/3$.

064.004 **Orbits of charged bodies.**
D. K. Sarvajna.
Astrophys. Space Sci., Vol. 6, 258 - 262 (1970).

An investigation into the possibility that material drawn out of a star goes into orbit around that star if electromagnetic effects are included, has been made. It is found that if the body has an initial charge of some 10^{37} e.s.u., and decreasing with time then sufficient angular momentum can be transferred to make orbits not intersecting the stellar surface possible.

064.005 **On the structure and effective temperature of shock waves moving in stellar atmospheres.**
I. A. Klimishin.
Stars, Nebulae, Galaxies, Symposium Byurakan 1968, (see 012.002), p. 145 - 148 (1969). In Russian.

064.006 **The investigation of the influence of radiation on the convection in a polytropic atmosphere.**
N. S. Petruchin.
Astrofizika, Vol. 5, 615 - 622 (1969). In Russian. – English translation in Astrophysics, Vol. 5, No. 4.

The effect of radiation on convection of an unstable polytropic atmosphere is analytically investigated by the method of perturbations. Disturbances with optical thickness $\tau \geqslant 1$ are considered. Some matrix elements which allow to determine the growth-rate and the functions characterizing convective motions for any mode at the first approximation are found. The expression for the growth-rate of the main mode is obtained and investigated.

064.007 **The atmosphere of the hydrogen-deficient star Sigma Orionis E.**
D. A. Klinglesmith, K. Hunger, R. C. Bless, R. L. Millis.

Astrophys. Journ., Vol. 159, 513 - 523 (1970).

A fine analysis of the hydrogen-deficient star σ Ori E is performed using a grid of constant-flux model atmospheres. The following quantities are computed: the profiles and equivalent widths of Hγ, Hδ, and five He I lines; the Balmer and helium discontinuities; the slope of the Paschen continuum; UBV colors; and bolometric corrections. The mass and radius are determined to be $M = 9.8\,M_\odot$ and $R = 4.6\,R_\odot$. The fairly rapid rotation of σ Ori E makes it impossible to determine the microturbulent velocity. The abundances of five elements are determined for $v_t = 0$ and 10 km sec^{-1}. Except for the deficiency of hydrogen, there appear to be no conspicuous abundance anomalies.

064.008 **Models for the envelopes of Be stars. II.**
J. M. Marlborough.
Astrophys. Journ. Vol. 159, 575 - 581 (1970).

Hα line profiles are presented for reasonable values of all parameters characterizing the model envelopes described in an earlier paper. Two computational difficulties which arose in that paper are resolved.

064.009 **Intrinsic polarization in nongray atmospheres.**
G. W. Collins II.
Astrophys. Journ., Vol. 159, 583 - 591 (1970).

This paper presents a method for solving the problem of polarized radiation arising in a plane-parallel nongray atmosphere from Thomson scattering by free electrons. The results are explicitly applied to the atmospheres of early-type stars.

064.010 **Mass loss by hot stars.** L. B. Lucy, P. M. Solomon.
Astrophys. Journ., Vol. 159, 879 - 893 (1970).

A mechanism is proposed to explain the mass loss observed for luminous, hot stars. We show that the ultraviolet resonance lines of ions such as Si IV, C IV, N V, and S VI can give strongly negative effective gravities in the outer parts of the reversing layers of hot stars. We argue that a static reversing layer is then no longer possible and that a continuous outflow of mass occurs.

064.011 **Shock wave propagation in stellar atmospheres.**
N. Virgopia.
Mem. Soc. Astron. Italiana, Nuova Serie, Vol. 41, 21 - 32 (1970).

Propagation of shock waves through the gravitational atmosphere of a 3 M$_\odot$ star in the giant stage has been considered by an approximate analytical method first elaborated by Bird. As initial values, disturbances in the form of a shock Mach number are assumed at some point in the stellar atmosphere and allowed to propagate outward. It is shown that, owing to the behaviour of sound velocity in the static atmospheric model, the shock front speed and particle velocity of the perturbed gas decrease outward. Other approximate results concerning radiative shock flux and the possibility of mass ejection are also briefly discussed.

064.012 **Determinación de la presencia de compuestos químicos en atmosferas estelares y planetarias, cometas y espacio interestelar.** F. P. Huberman.
Revista Astron., Vol. 41, No. 170, p. 41 - 54; No. 171, p. 5 - 16 (1969).

064.013 **A fine analysis of two yellow supergiants. II. The continuous energy distributions.** S. B. Parsons.
Astrophys. Journ., Vol. 159, 951 - 962 (1970).

Continuum energy distributions from model atmospheres,

with appropriate corrections for line absorption, are compared with stellar continuum observations. Deviations from LTE are included in the calculation of continuous opacity, but the effects on the emergent fluxes are small. Scanner observations of α Per show good agreement with the models, although there is ambiguity in the implied effective temperature and color excess. The angular diameter is well determined, and the distance modulus suggests a radius of about 60 $R_\odot$. Broad-band color indices are computed in the UBV and six-color systems. The color-color plot in the UBV system shows definite separation of the models, but not of temperature versus reddening effects.

064.014 Emission in stellar atmospheres from behind a shock wave. V. I. Golinko.
Astron. Zhurn. Akad. Nauk SSSR, Vol. 47, 145 - 148 (1970). In Russian. English translation in Soviet Astron. AJ, Vol. 14, No. 1.
The gradient of temperatures was determined behind the front of a shock wave. This method can be used for a quantitative interpretation of the emission spectra of stars.

064.015 The atmospheres of Tau Scorpii (B0 V) and Lambda Leporis (B0.5 IV).
J. Hardorp, M. Scholz.
Astrophys. Journ., Suppl. Series, Vol. 19, 193 - 233 (1970).
New observations of the continuous and line spectra of τ Sco and λ Lep are reported. These spectra are studied by using model-atmosphere techniques in the LTE approach. The present investigation intends: (i) to demonstrate that it is indeed not possible to account for all relevant observations at one time; (ii) to show the problems in deriving the surface gravity from profiles of Balmer lines; (iii) to give a modern interpretation of the helium spectrum; (iv) to investigate more closely the problem of occurrence of microturbulence in early-type stars; and (v) to apply the most recent data of f-values and damping constants in an analysis of both old and new observations of the "metallic" line spectra.

064.016 Neutral-helium line strengths. I. Line profiles for a grid of approximate line-blanketed model atmospheres. J. Norris, B. Baschek.
Astrophys. Journ., Suppl. Series, Vol. 19, 305 - 326 (1970). Abstr. in Astrophys. Journ., Vol. 159, 1129 - 1130 (1970).
Line profiles of neutral helium are presented for a grid of approximate line-blanketed model atmospheres. The isolated lines are computed with the line-broadening theory of Griem, Baranger, Kolb, and Oertel. The wings (long-wavelength halves) of the strong lines of the $2P-nD$ series are obtained from the quasi-static theory of Pfennig and Trefftz; the cores are computed from an adiabatic-impact treatment following Traving.

064.017 Neutral-helium line strengths. II. The B-type subdwarf HD 205805.
B. Baschek, J. Norris.
Astrophys. Journ., Suppl. Series, Vol. 19, 327 - 336 (1970). Abstr. in Astrophys. Journ., Vol. 159, 1130 (1970).
Spectra and continuum colors have been obtained for the B-type subdwarf HD 205805. On the assumption of LTE, an analysis of the helium line strengths leads to the conclusion that the atmospheric-helium abundance of HD 205805 is smaller by a factor of 10 than that of the Population I star γ Peg. The helium singlet-triplet anomaly is explained by both the higher gravity and the lower helium abundance of the subdwarf. There is no evidence for departures from LTE. An analysis, using the data of Sargent and Searle, of three other sdB stars, shows that the helium deficiency is a common property of these stars.

064.018 Neutral-helium line strengths. III. The singlet-triplet

anomaly of Population I stars.
J. Norris.
Astrophys. Journ., Suppl. Series, Vol. 19, 337 - 341 (1970). Abstr. in Astrophys. Journ., Vol. 159, 1130 - 1131 (1970).
The singlet-triplet anomaly of main-sequence Population I stars is explained within framework of LTE.

064.019 Galactic cosmic ray origin of Li, Be and B in stars. H. Reeves, W. A. Fowler, F. Hoyle.
Nature, Vol. 226, 727 - 729 (1970).
A study of the implications of the hypothesis that high energy processes involving cosmic rays acting on the interstellar medium are the sources of the elements Li, Be and B present in stellar atmospheres and in the solar system.

064.020 Relativistic stellar winds and the Crab pulsar. R. N. Henriksen, D. R. Rayburn.
Bull. American Astron. Soc., Vol. 2, 198 (1970). – Abstr. AAS.

064.021 A method for the construction of model stellar atmospheres. W. McD. Napier.
Monthly Notices, Roy. Astron. Soc., Vol. 147, 287 - 293 (1970).
A non-iterative variational technique for constructing model stellar atmospheres is described. There are no restrictions on the physical state of the atmosphere other than hydrostatic equilibrium and energy conservation. The temperature within the atmosphere is expressed as a series in pressure the coefficients of which are determined by a search technique.

064.022 An upper limit to the Chandrasekhar polarization in early type stars.
S. M. Ruciński.
Acta Astron., Vol. 20, 1 - 12 (1970).
Arguments are presented that the degree of polarization for the greater part of the stellar disc is much smaller than it was predicted by Chandrasekhar. This reduction reflects the contribution of the pure thermal absorption to the total opacity in stellar atmospheres; it varies with the wavelength and the effective temperature. For the atmospheres of B5V and O8V stars the reduction in yellow is at least by factor of 16 and 4, respectively.

064.023 Model atmospheres for the central stars of planetary nebulae. D. G. Hummer, D. Mihalas.
Monthly Notices, Roy. Astron. Soc., Vol. 147, 339 - 354 (1970).
Approximately 70 model atmospheres for the central stars of planetary nebulae have been computed under the assumption of hydrostatic, radiative and local thermodynamic equilibrium and of plane-parallel stratification. These models have effective temperatures and surface gravities in the range $30\,000\,°K \leqslant T_{eff} \leqslant 200\,000\,°K$ and $3.4 \leqslant \log g \leqslant 7.5$. The atmospheres have been taken to consist of hydrogen, helium, oxygen, nitrogen, carbon and neon, and the opacity included contributions from both ground and excited states of each ion. The transfer equation is solved and the temperature corrections are calculated.

064.024 Expanding atmospheres in OB supergiants – V. The ultra-violet resonance lines and radiative accleration in the Orion supergiants. J. B. Hutchings.
Monthly Notices, Roy. Astron. Soc., Vol. 147, 367 - 376 (1970).
Line profiles are computed for the observed ultra-violet resonance lines in the Orion supergiants. Initial velocity and ion density distributions are deduced which lead to approximate numerical estimates of radiative acceleration processes affecting the entire atmospheric mass. Recomputation of the line profiles yields a very close fit with observation. A

mass loss rate of 10^{-6} $M_\odot$ yr^{-1} is deduced for these stars. Some general remarks are made on mass loss from early type supergiants.

064.025 Line-blanketed model atmosphere for ϵ Vir (HD 113226).   T. E. Morgan.
Bull. American Astron. Soc., Vol. 2, 210 (1970). – Abstr. AAS.

064.026 Outer boundary conditions in low-mass stars.   W. C. Straka.
Bull. American Astron. Soc., Vol. 2, 220 (1970). – Abstr. AAS.

064.027 Convection in envelopes of white dwarfs.   H. M. van Horn.
Bull. American Astron. Soc., Vol. 2, 223 - 224 (1970). Abstr. AAS.

064.028 Convective energy transport in stellar atmospheres. I: A convective thermal model.
R. K. Ulrich.
Astrophys. Space Sci., Vol. 7, 71 - 86 (1970).
This paper discusses and modifies the meteorological model known as the convective thermal for the purpose of obtaining the strength of coupling between adjacent layers with different properties. As in the standard mixing length theory the principle uncertainty remains the average initial radius of the cells. This initial radius is determined to be between 700 and 950 km by a comparison to the solar granulation.

064.029 An invariant imbedding approach to spectral-line formation by noncoherent scattering.
F. B. Fuller, R. F. Warming.
Astrophys. Space Sci., Vol. 7, 93 - 118 (1970).
Spectral-line formation under the assumption of complete frequency redistribution is studied for a plane-parallel atmosphere containing a nonuniform distribution of internal emission sources. A description of the numerical calculations is given and several graphical solutions are included to illustrate emergent intensity profiles of Doppler broadened lines.

064.030 Thermal conductivity in stellar atmospheres. I. Without magnetic field.   P. Ulmschneider.
Astron. Astrophys., Vol. 4, 144 - 151 (1970).
The coefficient of thermal conductivity for pure hydrogen, pure helium and a gas mixture appropriate for a stellar atmosphere has been computed in the temperature range 3000 to 100000°K and the range of gas pressure from 10^{-3} to 10^6 dyn/cm^2. The translational as well as reactive contributions to the coefficient are given. The error is less than 10% (except for pure He) in the neutral region but may be more than 35% in the ionizing region. In the fully ionized region there is complete agreement with Spitzer and Härm (1953).

064.031 Supersonic stellar winds in early-type stars.   J. M. Marlborough, J.-R. Roy.
Astrophys. Journ., Vol. 160, 221 - 224 (1970).
The effect on stellar winds of the mechanical force due to radiation is considered in general. If the flow velocity is initially subsonic in or near the star, it is shown that a supersonic stellar wind cannot arise as a result of the mechanical force due to radiation balancing or exceeding the gravitational force.

064.032 Non-LTE model atmospheres. IV. Results for multi-line computations.
L. H. Auer, D. Mihalas.
Astrophys. Journ., Vol. 160, 233 - 243 (1970).

Results are presented for model atmospheres in hydrostatic, radiative, and steady-state statistical equilibrium, including bound-bound transitions. These calculations allow for the effects of Hα, Hβ, Hγ, Pα, Pβ, and Bα. Stark profiles of Edmonds, Schlüter, and Wells are used to compute detailed line profiles. With the full non-LTE calculation we have been able to predict correctly the entire line profile, including the line core.

064.033 Comments on stellar mass loss.   G. S. Kutter.
Astrophys. Journ., Vol. 160, 369 - 371 (1970).
The purpose of this note is threefold: (1) to point out a limitation of the hydrostatic equilibrium equation in stellar atmospheres, (2) to suggest momentum transfer from the radiation field to the gas as a mechanism for mass loss in early-type supergiants, and (3) to estimate the physical conditions determining whether this mechanism or coronal heating is responsible for mass loss.

064.034 On the instability of a stellar envelope due to radiation pressure.   D. G. Wentzel.
Astrophys. Journ., Vol. 160, 373 - 377 (1970).
The gradient of radiation pressure in a stellar envelope may cause a density inversion locally. The main instability of such an inverted layer is normal convection, not a Rayleigh-Taylor instability. Radiation also may damp the convection in some supergiant atmospheres that are normally considered to be convective.

064.035 Convection in envelopes of white dwarfs.   H. M. Van Horn.
Astrophys. Journ. *(Letters)*, Vol. 160, L53 - L56 (1970).
Incomplete ionization in the envelopes of white dwarfs leads to extensive surface convection zones which may modify the spectra and drastically reduce the lifetimes of these stars.

064.036 Neutral helium lines and the helium anomaly in hot stars.   A. I. Poland.
Astrophys. Journ., Vol. 160, 609 - 617 = Publ. Geothe Link Obs., Indiana Univ., *Bloomington*, No. 98 (1970).
In this paper we use the departures from LTE given in our previous paper to calculate non-LTE line profiles and equivalent widths for the $2S$-$4P$, $2P$-$4S$, $2P$-$5S$, and $2P$-$4D$ transitions in the singlet and triplet systems of neutral helium. We use these results to explain the helium anomaly as observed in hot main-sequence stars.

064.037 Diffusion processes in peculiar A stars.   G. Michaud.
Astrophys. Journ., Vol. 160, 641 - 658 (1970).
We discuss the possibility that the atmospheres of Ap stars are stable enough for diffusion to become important. We discuss some general aspects of diffusion in A and B stars. The radiation force transferred through bound-free and bound-bound transitions is found to be frequently larger than the gravitational force. Autoionization levels may play an important role in transferring momentum from the radiation field to certain preferred elements.

064.038 Meridional circulation in rotating stellar atmospheres.   R. C. Smith.
Monthly Notices, Roy. Astron. Soc., Vol. 148, 275 - 312 (1970).
Expressions are found for the meridional circulation velocities in the Roche envelope of a uniformly rotating star, using a non-local equation for the radiative transfer. An order-of-magnitude model of the turbulent region is developed, in which turbulent viscosity prevents the flow speeds from becoming too large. The effect of the turbulence on the emergent flux of radiation is discussed.

064.039 **The energy transfer in the convective envelope of a rotating star and the solar differential rotation.**
Y. Osaki.
Monthly Notices, Roy. Astron. Soc., Vol. 148, 391 - 406 (1970).

In this paper, we simplify the problem by formulating and solving the equation governing convective energy transport under the assumption of solid body rotation with no meridional circulation. We then use the resulting thermal structure to study the rotation law and the meridional circulation. Finally, the results are applied to the sun.

064.040 **A model atmosphere analysis of the Ap star κ Cancri.**
M. F. Aller.
Astron. Astrophys., Vol. 6, 67 - 84 (1970).

A detailed model atmosphere analysis is presented for the peculiar A star κ Cancri based on high dispersion spectrograms secured at the coudé focus of the Lick 120-inch telescope and a published energy distribution. The observed energy distribution and hydrogen-line profiles indicate that the atmosphere can be represented by a model with $\theta = .36$ at $\log \tau_{5000} = 0.0$ and with $\log g = 3.6$. The metal-line equivalent widths, half widths, and central line depths served to determine the stellar abundances and to further verify the model. A detailed description of element abundances is given.

064.041 **Review of ultraviolet and visual continuum observations and comparisons with models.** R. C. Bless.
IAU Symposium No. 36, (see 012.014), p. 73 - 82 (1970).

This paper first briefly describes model atmosphere grids now available for comparison with observations. The recent recalibration of the absolute energy distribution of α Lyr substantially improves the agreement of models and observations in the visual. Temperature scales determined by various methods agree reasonably well except for the hottest stars. Recent ultraviolet results suggest that earlier observations of O- and B-type stars indicating large flux deficiencies were probably in error. However, late B- and A-type stars may emit less energy in the UV than that predicted by models which do not include the opacities caused by silicon, magnesium, and carbon.

064.042 **Stellar-wind theory for O and B stars.**
P. M. Solomon.
IAU Symposium No. 36, (see 012.014), p. 236 - 237 (1970). Abstract.

064.043 **Convective energy transport in stellar atmospheres. II: Model atmosphere calculation.**
R. K. Ulrich.
Astrophys. Space Sci., Vol. 7, 183 - 200 (1970).

The motion of convective cells in an environment which changes rapidly with depth is examined. In such an environment a cell may move through regions with different levels of ionization and with associated differences in heat capacity. The energy equation is cast in a manner which is independent of the history of the cells. The convective flux at a given level of the atmosphere is written as an average over an ensemble of cells originating at a range of other levels. A procedure for correcting the temperature gradient for these non-local effects is described and results for a model solar atmosphere are given.

064.044 **Atmospheric structure and chemical composition of main-sequence K-stars.** P. Strohbach.
Astron. Astrophys., Vol. 6, 385 - 405 (1970). In German.

Differential fine analyses relative to the sun are performed for HD 192310, HD 156026, 61 Cyg A, HD 190404, and HD 219134. Model atmospheres are computed using empirical relative temperature stratifications which are determined from the profiles of broad metal lines, taking into account the in-

fluence of convection. By fitting computed continua, colors, Balmer lines, and equivalent widths to the observations we obtain the temperature and the microturbulence.

064.045 **On the influence of stellar wind on accretion.**
V. F. Schwartzman.
Astron. Zhurn. Akad. Nauk SSSR, Vol. 47, 660 - 662 (1970). In Russian. English translation in Soviet Astron. AJ, Vol. 14, No. 3.

There exists a critical intensity of the stellar wind, below which the ejection of particles from a star must be replaced by accretion of the surrounding gas onto it.

064.046 **On the use of variable Eddington factors in non-LTE stellar atmospheres computations.**
L. H. Auer, D. Mihalas.
Monthly Notices, Roy. Astron. Soc., Vol. 149, 65 - 74 (1970).

It is shown that by use of variable Eddington factors, the accuracy of difference-equation solutions of transfer problems may be greatly improved with only small additional computational effort. It is found that a direct iterative calculation of the Eddington factors leads to a strongly convergent procedure. The resulting set of equations is of wide applicability to problems involving non-coherent radiative transfer. The method is illustrated by application to the classical grey problem, and to a non-LTE stellar atmospheres computation.

064.047 **Magnetic braking by a stellar wind – III. The oblique rotator with a quasi-radial field.**
L. Mestel, C. S. Selley.
Monthly Notices, Roy. Astron. Soc., Vol. 149, 197 - 220 (1970).

In this paper, the magnitude and sign of the precessional torque component are computed for the special case with the magnetic field outside the star the sum of a basic, split monopole field, with radial field-lines and an equatorial current-sheet, a perturbation field, of order ϵ, due to a small angle-dependent flux distribution over the stellar surface, and further perturbations due to the stellar rotation a, of order a and ϵa respectively. The field of order a yields a braking torque, but has too much symmetry to yield a precessional component, which results only from the part of order $a\epsilon$.

064.048 **Stellar winds.** P. Goldreich, W. H. Julian.
Astrophys. Journ., Vol. 160, 971 - 977 (1970).

We present relativistic analytic solutions for the model stellar-wind problem posed by Weber and Davis. These solutions are consistent with the hypothesis of minimum torque proposed by Michel, but do not require it as an additional assumption. The equation of radial momentum has three critical points which occur where the radial flow velocity is equal to the propagation speed of infinitesimal disturbances. A unique continuous solution is determined by the requirement that it pass through all these critical points.

064.049 **Effective temperatures, gravities, and the mass determination of A and F stars.**
M. Breger, L. V. Kuhi.
Astrophys. Journ., Vol. 160, 1129 - 1139 (1970).

The method of obtaining and the sources of error in obtaining effective temperatures and gravities by fitting observed energy distributions to theoretical model fluxes are discussed. Measured blanketing values are shown to be dependent on rotational velocity, $v \sin i$. Energy distributions have been measured with a spectrum scanner for eight variable and twelve nonvariable A and F stars in and near the lower instability strip. Values of effective temperature and gravity were obtained. There is excellent agreement between effective temperatures derived from the infrared/red slope and the energy distribution in the blue. An attempt is also made to reconcile previously published results.

064.050 A curve-of-growth analysis of the super-metal-rich G dwarf HR 72. H. Spinrad, W. R. Luebke, Jr.
Astrophys. Journ., Vol. 160, 1141 - 1148 (1970).

A conventional curve-of-growth comparison was made between HR 72, a strong-line G dwarf, and the sun. HR 483 and β Com were also observed as controls. HR 72 is only slightly hotter than the sun ($\Delta\theta = -0.015$); it is very metal-rich, with [Fe/H] = +0.43. Apparently, all metals are over-abundant from Na to possibly La. Recently noted supermetal-licity in giants is then very likely due to their original composition rather than to some form of self-enrichment of heavy elements.

064.051 Non-LTE model atmospheres. V. Multi-line hydrogen-helium models for O and early B stars.
D. Mihalas, L. H. Auer.
Astrophys. Journ., Vol. 160, 1161 - 1176 (1970).

Results of calculations of non-LTE model atmospheres including the effects of $L\alpha$, $L\beta$, $L\gamma$, $H\alpha$, $H\beta$, and $P\alpha$ are presented for models on the range $25000°K \leq T_{eff} \leq 50000°K$, with $\log g = 4$ and $N(He)/N(H) = 0.10$. These results should prove of value in the analysis of rocket and satellite observations of early-type stars and of emission nebulae.

064.052 Model envelope of FU Orionis.
B. E. Zhilyaev.
Astrometriya i Astrofiz., *Kiev,* No. 8, (see 003.024), p. 27 - 30 (1969). In Russian.

A convective model envelope of FU Ori has been computed. Dispersion and distribution functions of star brightness are calculated on the basis of a simple model. The results are in satisfactory agreement with the observations.

064.053 Model atmospheres and envelopes of A0–G5 stars.
V. I. Golinko, N. S. Komarov, G. S. Krasnova.
Astrometriya i Astrofiz., *Kiev,* No. 8, (see 003.024), p. 35 - 51 (1969). In Russian.

The models of stellar atmospheres and envelopes were calculated for different chemical compositions.

064.054 Propagation of a strong shock wave into the extended envelope of supergiant stars and the light curves of supernovae. E. K. Grasberg, D. K. Nadezhin.
Nauchn. Informatsii, vyp. (No.) 13, p. 96 - 109 (1969). In Russian.

On the base of hydrodynamical equations (radiative conductivity included) the propagation of a strong shock wave into the atmosphere of a supergiant star (with radius 5000 – 10000 $R_\odot$) has been computed. The results obtained suggest that the light curves of the type II supernovae near maximum can be explained by a thermal mechanism. The presence of a large envelope before explosion in this case is necessary. It may be supposed also that the same mechanism in extended envelopes can be applied for type I supernovae light curves preceding the exponential luminosity decrease region.

064.055 Theory of stellar atmospheres. A. B. Underhill.
Trans. IAU, Vol. 14A, (see 003.028), 425 - 436 (1970). – Report of Commission 36.

064.056 Non-L.T.E. analysis of spectral lines.
D. Mugglestone.
Proc. Astron. Soc. Australia, Vol. 1, 296 - 301 (1970).

064.057 On the propagation of a shock wave through a stellar atmosphere. V. E. Panchuk.
Astron. Tsirk., No. 553, p. 6 (1970). In Russian.

064.058 On the ejection of a neutron star shell.
P. R. Amnuel, O. H. Guseinov. F. K. Kasumov.
Astron. Tsirk., No. 560, p. 3 - 5 (1970). In Russian.

064.059 A study of the atmospheres of the hot subdwarfs HD 127493, HD 128220 and HD 113001.
L. J. Tomley.
Thesis, Univ. Washington, St. Louis, Mo. [Available from Univ. Microfilms, Ann Arbor, Mi.], 243 pp. (1969).

An analysis of three hot subdwarfs, HD 127493 and the subdwarf components of the binary systems HD 128220 and HD 113001, was performed to gain information about these previously unstudied objects. Moderately high dispersion spectrograms have been used to determine the atmospheric parameters.

064.060 Spectrum formation in steady-state extended atmospheres. A. B. Underhill.
Earth Extraterrestr. Sci., Vol. 1, No. 2, p. 69 - 71 (1969).

This colloquium dealt with the following topics: problems encountered in the interpretation of spectra from stars with extended atmospheres; mathematical methods for handling the problem of spectral line formation in a stellar atmosphere; an assessment of what is known about the chromospheres and coronas of stars; an evaluation of suggestions for improving our knowledge.

064.061 Convective energy transport in stellar atmospheres. R. K. Ulrich.
Thesis, Univ. California, Berkeley. [Available from Univ. Microfilms, Ann Arbor, Mi.], 121 pp. (1969).

The problem of convective energy transport in realistic model stellar atmospheres is considered. An iterative series of models has been computed to ensure that the non-locally determined convective flux together with the radiative flux are constant with depth. Two atmospheric cases are studied in detail: a solar case and a case matching Arcturus.

064.062 Diffusion in stars with convective envelope.
A. Delcroix.
Centre Univ. Mons, Fac. Sci., Dep. Astrophys., Commun., No. 15, 8 pp. (1970).

Diffusion phenomena are studied in stars with important convective envelope, using multicomponent and fourth order approximation formulae. The results are compared to those obtained by two-component and first order approximation relations.

064.063 The carbon, nitrogen and oxygen abundances of the K giants Alpha Bootis, Alpha Serpentis, Beta Geminorum and Epsilon Pegasi. T. F. Greene.
Thesis, Univ. Washington, Seattle. [Available from Univ. Microfilms, Ann Arbor, Mi.], 160pp. (1968). – See Phys. Abstr., Vol. 73, No. 45262 (1970).

064.064 Theoretical profiles of emission lines I in V/R variables. A. Ringuelet-Kaswalder.
Asoc. Argentina Astron. Bol., No. 12, (see 012.019), p. 38 (1968). – Abstract.

Partition functions and equilibrium constants for ScO, YO and LaO. See Abstr. 022.059.

Thermal-convective instability.
See Abstr. 061.013.

Monte-Carlo calculation of line profiles in expanding and rotating atmospheres. See Abstr. 063.025.

Nuclear reactions in stellar surfaces and their relations with stellar evolution. See Abstr. 065.053.

Convection regions and coronas of sun and stars.
See Abstr. 074.001.

The solar (stellar) wind as a one-dimensional flow.
See Abstr. 074.023.

The calibration of narrow band photometry – I.
Cambridge observations of G and K giants.
See Abstr. 113.016.

Expanding atmospheres in OB supergiants – IV.
A mass-loss survey.   See Abstr. 114.004.

Ultraviolet fluxes and bolometric corrections for
late B to F main-sequence stars.   See Abstr. 114.028.

Spectroscopic investigation of B 3 V stars: ι Herculis, η Hydrae, HD 58343.   See Abstr. 114.080.

The stellar temperature scale from O5 to A0.
See Abstr. 114.086.

Mass loss from early-type stars.
See Abstr. 114.096.

A discussion of the theory for interpreting ultraviolet stellar spectra.   See Abstr. 114.097.

Spectral line formation in Wolf-Rayet envelopes.
See Abstr. 114.113.

Problems in the formation of neutral helium lines
in early type stars.   See Abstr. 114.134.

Harmonic analysis of rigidly rotating Ap stars.
See Abstr. 116.007.

Stellar rotation: Uniformly rotating stars with
hydrogen-line-blanketed model atmospheres. Comparison
with observations of A-type stars.   See Abstr. 116.012.

On the obscuration hypothesis of R Coronae
Borealis.   See Abstr. 122.119.

Intrinsic polarization of early-type stars with
extended atmospheres.   See Abstr. 131.110.

Fast motions in nebulae, and the stellar wind.
See Abstr. 132.003.

065 Stellar Structure, Stellar Evolution, Stellar Nucleosynthesis

065.001 **The influence of a poloidal magnetic field on convection in stellar cores.**   D. L. Moss, R. J. Tayler.
Monthly Notices, Roy. Astron. Soc., Vol. 147, 133 - 138 (1970).

It is shown that a strong poloidal magnetic field can suppress convection in a star with a small convective core but that a strong enough field to suppress convection must be largely confined within the star. This in turn means that no presently observed surface magnetic field extended through the interior of a star is by itself strong enough to suppress core convection. If convection is suppressed by a magnetic field which does not seriously distort the spherical symmetry of a star, the surface properties of the star are essentially unaffected.

065.002 **Evolutionary effects in the rotation of supergiants.**   J. D. Rosendhal.
Astrophys. Journ., Vol. 159, 107 - 118 (1970).

A statistical investigation has been made of rotation and macroturbulence in early-type Ia and Iab supergiants. The principal results are: (1) At all spectral types in the range B0–A5 both rotation and macroturbulence contribute to the observed line broadening. In the early B stars, rotation is as important a broadening agent as large-scale mass motion. In the middle-B and A stars turbulence dominates, but there still is an appreciable contribution from rotation. (2) In spite of the fact that the stars observed seem to be losing mass, there is no strong evidence for significant loss of angular momentum.

065.003 **Hyperonic equation of state.**   W. D. Langer, L. C. Rosen.
Astrophys. Space Sci., Vol. 6, 217 - 227 (1970).

An equation of state for cold matter at neutron star densities, $\rho > 10^{14}$ gm/cm^3, is evaluated. The gas is considered to be a degenerate mixture of neutrons, protons, leptons, hyperons and massive baryons. We derive the equilibrium equations including the effects of nuclear interactions among all the hadrons.

065.004 **Neutron star models based on an improved equation of state.**   J. M. Cohen, W. D. Langer, L. C. Rosen, A. G. W. Cameron.
Astrophys. Space Sci., Vol. 6, 228 - 239 (1970).

Using an equation of state for cold degenerate matter which takes nuclear forces and nuclear clustering into account, neutron star models are constructed. Stable models were obtained in the mass range above 0.065 $M_\odot$ and density range $10^{14.08}$ to $10^{15.4}$ gm/cm^3. All of these models were found to be bound. The outer crystalline layer of the star was found to have a thickness of 200 m or more depending on the mass of the model.

065.005 **The stability of the equilibrium of supermassive and superdense stars.**   G. Calamai.
Astrophys. Space Sci., Vol. 6, 240 - 257 (1970).

The second variation of the mass-energy (by constant baryon number) is analysed to inquire into the stability of the equilibrium for a spherical mass under the action of its own gravitation. The equations of general relativity are used.

065.006 **Maximal masses of Fe^{56} and Ni^{56} stars.**   A. Kovetz, G. Shaviv.
Astrophys. Space Sci., Vol. 6, 396 - 399 (1970).

Upper bounds are derived for the masses of pure Ni^{56} and Fe^{56} stars, which are dynamically stable with respect to photodisintegration. Possible effects on stellar evolution are discussed.

065.007 **A polytropic model for the helium shell flash.**   H.-C. Thomas.
Astrophys. Space Sci., Vol. 6, 400 - 414 (1970).

A model consisting of two polytropes is constructed, to represent a helium core of a star during the helium shell flash occurring at the onset of helium burning in a degenerate core. The maximum temperature reached during the flash can be predicted as a function of core mass and mass inside the helium burning shell. This temperature will generally be too low for the production of neutrons out of ^{14}N. Some additional results on the helium shell flash in a star of 1.3 $M_\odot$ are also presented.

065.008 **Perturbation theory for stellar interior problems.**   T. W. Edwards.
Astrophys. Space Sci., Vol. 6, 436 - 449 (1970).

A rather general first order linearized perturbation theory is derived with which one can calculate a quasi-static equilibrium stellar interior model if given an initial approximation thereto. The process developed is applicable to both static and evolutionary models, wherein slightly different physical assumptions or physical conditions are imposed.

065.009 **On the evolution speed of young stars in associations.**
L. V. Mirzoyan, Eh. S. Kazaryan, O. S. Chavushyan.
Stars, Nebulae, Galaxies, Symposium Byurakan 1968, (see 012.002), p. 153 - 157 (1969).   In Russian.

065.010 **New source of intense magnetic fields in neutron stars.**
V. Canuto, H. Y. Chiu, C. Chiuderi, H. J. Lee.
Nature, Vol. 225, 47 - 48 (1970).

The magnetization equation M = M (H) which gives rise to the DeHaas-Van Alphen effect is shown to possess self-consistent quasi-stable solution when H → B = H + 4πM|$_{H = 0}$, i.e. M = M (4πM). For a degenerate electron gas at a given density this equation possesses many solutions due to the oscillatory quantum behavior of M (4πM). The highest of these solutions when plotted vs. ρ goes like $\rho^{2/3}$ as the flux conservation law would predict. It is found that for white dwarfs ($\rho_6 \simeq$ 1g/cc) the magnetic field is about $10^7 - 10^8$ gauss, while for neutron stars $\rho = 10^{14}$ g/cc the corresponding field is about $10^{12} - 10^{13}$ gauss.

065.011 **Anisotropic superfluidity in neutron star matter.**   M. Hoffberg, A. E. Glassgold, R. W. Richardson, M. Ruderman.
Phys. Rev. Letters, Vol. 24, 775 - 777 (1970).

The magnitude of the expected superfluid transition temperature and the form of the gap are obtained for regimes of neutron density expected in neutron-star cores. For neutron densities exceeding 1.5×10^{14} g cm^{-3} the gap is anisotropic but nodeless, leading to thermodynamic properties of a conventional BCS-type superfluid.

065.012 **Metal-poor stars. I. Evolution from the main sequence to the giant branch.**
I. Iben, Jr., R. T. Rood.
Astrophys. Journ., Vol. 159, 605 - 617 (1970).

The evolution of metal-poor stars during early hydrogen-burning phases is described as a function of all composition variables. It is shown than, on the basis of a comparison with observed turnoff colors alone, the relative ages of stars in M13, M3, M15, and M92 cannot be determined to better than several billion years.

065.013 On the numerical stability of computations of stellar evolution.   D. Sugimoto.
Astrophys. Journ., Vol. 159, 619 - 627 (1970).

We discuss the numerical instability arising from the coupling between hydrostatic equilibrium and thermal processes in a star. Two alternative physical pictures are possible: the heat wave does or does not propagate through the adjacent shells in the star in a given time step (slow or rapid evolution). Correspondingly, we have two alternative approaches to the mathematical formulation. If the physical picture is wrong, we encounter a numerical instability. After analyzing the nature of the instability, we show that a single mathematical scheme is possible which always meets the necessary physical picture required by rapid or slow evolution in the stellar core or envelope.

065.014 Differential rotation in stellar convection zones.   F. H. Busse.
Astrophys. Journ., Vol. 159, 629 - 639 (1970).

The dynamical effects of rotation on thermal convection in a fluid layer of spherical shape induce an equatorial acceleration. Special properties of the convection zone assumed in earlier theories on the differential rotation of the sun are not required. To demonstrate the mechanism, the mathematical problem of convection in a rotating Boussinesq fluid subject to temperature and gravity fields of spherical symmetry is considered.

065.015 The influence of ion correlations on the free-free opacity of stellar interiors.   W. D. Watson.
Astrophys. Journ., Vol. 159, 653 - 658 (1970).

The influence of ion correlations on the free-free contribution to the opacity of stellar interiors is calculated. This effect is found to be of most importance under conditions in which the flow of energy by radiation competes with conduction by degenerate electrons. For example, at the edges of degenerate cores in red giants this is the case. Here the radiative opacity is reduced by roughly 10 percent.

065.016 The effect of mass loss on a contracting star.   L. V. Kuhi, J. E. Forbes.
Astrophys. Journ., Vol. 159, 871 - 877 (1970).

An attempt has been made to simulate the T Tauri phase of pre-main-sequence evolution by including a simple scheme for mass loss. Four stellar models have been calculated for intercomparison: two with initial mass 1.5 $M_\odot$ and final mass 1.2146 $M_\odot$, with rates of mass loss calculated by different prescriptions; and two with constant masses of these same two values. The final zero-age main-sequence position of the cases with mass loss corresponds exactly to the final mass; more surprisingly, the time scale for contraction to the main sequence also nearly corresponds to the final mass, the higher mass loss during the early Hayashi phase having little effect.

065.017 Studies in stellar evolution. IX. Theoretical isochrones for early-type clusters.
E. Simpson, R. E. Hills, W. Hoffman, S. A. Kellman, E. Morton, Jr., F. Parese, C. Peterson.
Astrophys. Journ., Vol. 159, 895 - 901 (1970).

Theoretical isochrones are constructed for stars of intermediate mass with a solar-type chemical composition of $X = 0.730$, $Y = 0.245$, and $Z = 0.025$. The effects of small changes in helium abundance are investigated by means of other isochrones. The effect on the isochrones is negligible. The isochrones are then fitted to some diagrams of galactic clusters in an effort to determine their ages and distances.

065.018 Final evolution of a low-mass star. I.
W. K. Rose, R. L. Smith.

Astrophys. Journ., Vol. 159, 903 - 912 (1970).

In the present paper, it is shown that the density of the helium-burning shell increases as the star ascends the red-giant branch. The shell flashes that occur during the early and less luminous stages of evolution are not sufficiently violent to lead to mass ejection. However, near the tip of the red-giant branch ($L \sim 10^4 L_\odot$ for the stellar models in question) the shell flashes may be sufficiently violent to lead directly to mass ejection. This point will be studied more carefully in a future paper.

065.019 Stellar reaction rates for ^{28}Si.
P. B. Lyons, J. W. Toevs, C. A. Barnes, W. A. Fowler, D. G. Sargood.
Astrophys. Journ., Vol. 159, 913 - 917 (1970).

Using the data given in two earlier papers, we present calculations of the photodisintegration rates, and empirical fits to these rates, for ^{28}Si in the environment postulated for the silicon-burning phase.

065.020 The development of a cocoon star.
K. Davidson.
Astrophys. Space Sci., Vol. 6, 422 - 435 (1970).

A newly-formed massive star is likely to be surrounded by dense gas and dust as it approaches the main sequence. Radiation pressure must push some of the inner material outward before the star begins to produce ionizing radiation; this affects the formation of the H II region. A remarkably dense 'dust front' may precede the ionization front. The observable radio and infrared spectra are discussed. If the dust cloud is composed of small graphite grains, extraordinarily large far-infrared fluxes are possible.

065.021 Supermassive stars formed by the non-elastic evolution of stellar systems.   L. E. Gurevich.
Astron. Zhurn. Akad. Nauk SSSR, Vol. 47, 32 - 42 (1970). In Russian. English translation in Soviet Astron. AJ, Vol. 14, No. 1.

A concentrated stellar system, evaporating and contracting leads to a state in which non-elastic "contact" collisions of stars are more essential than their gravitational interactions in the case of encounters. The emitted gas probably condenses into one central star. The structure of such a supermassive star is investigated and it is shown that in its main mass convective intermixing must take place. Its influence on the structure and luminosity of a supermassive star is estimated.

065.022 Improved opacity tables for Population I stellar interiors.   W. D. Watson.
Astrophys. Journ., Suppl. Series, Vol. 19, 235 - 242 (1970). Abstr. in Astrophys. Journ., Vol. 159, 377 (1970).

Recent improvements in the calculation of the line contribution to stellar interior opacities are employed to compute Rosseland opacities for ten Population I compositions. Opacities at thirty-eight temperature-density points are presented for each element mixture. The compositions are chosen to provide a basis for interpolation and extrapolation in stellar interior calculations.

065.023 Rosseland opacity tables for Population I compositions.   A. N. Cox, J. N. Stewart.
Astrophys. Journ., Suppl. Series, Vol. 19, 243 - 259 (1970). Abstr. in Astrophys. Journ., Vol. 159, 377 (1970).

Detailed tables of Rosseland mean opacities, without allowance for electron conduction, are presented for ten Population I compositions. These tables provide a basis for interpolation and extrapolation in stellar structure calculations. Changes from the methods employed in previous publications are described. A discussion and calculations concerning the accuracy of the free-free absorption are given.

065.024 Rosseland opacity tables for Population II compositions. A. N. Cox, J. N. Stewart.
Astrophys. Journ., Suppl. Series, Vol. 19, 261 - 279 (1970).
Abstr. in Astrophys. Journ., Vol. 159, 377 (1970).

Compact opacity tables, not including the effects of electron conduction, are given for nineteen Population II mixtures, including four representing advanced stages of evolution.

065.025 Some general relativistic inequalities for a star in hydrostatic equilibrium – II.
J. N. Islam.
Monthly Notices, Roy. Astron. Soc., Vol. 147, 377 - 386 (1970).

Two theorems are proved involving the variables that describe the interior of a star in hydrostatic equilibrium in general relativity. These theorems are analogues of two classical theorems.

065.026 Non-radial oscillations: Lagrange's formalism.
T. Ishizuka.
Publ. Astron. Soc. Japan, Vol. 22, 125 - 127 (1970).

The equations governing linear non-radial oscillations are formulated in Lagrange's form.

065.027 Evolution of stars of large masses and Strömgren zones. G. S. Bisnovaty-Kogan, R. A. Sunyaev.
Astron. Zhurn. Akad. Nauk SSSR, Vol. 47, 441 - 442 (1970).
In Russian. – English translation in Soviet Astron., AJ, Vol. 14, No. 2.

The time of recombination and cooling of Strömgren zones of massive stars, which have left the main sequence, turns out larger than the time of decreasing of the surface temperature of a star.

065.028 Thermal instability and star formation.
K. Kossacki.
Postępy Astron., Vol. 18, 235 - 237 (1970). In Polish.
Short report.

065.029 On density inversion in the convective zones of stars. S. Chapman, M. S. Vardya.
Observatory, Vol. 90, 69 - 70 (1970). – Letter.

065.030 Nonlinear periodic pulsations of stars.
N. H. Baker, K. von Sengbusch.
Bull. American Astron. Soc., Vol. 2, 181 (1970). – Abstr. AAS.

065.031 Pre-main-sequence evolution of low-mass stars.
A. S. Grossman.
Bull. American Astron. Soc., Vol. 2, 195 (1970). – Abstr. AAS.

065.032 Nonradial oscillations of white dwarfs, hot white dwarfs, and 10 $M_\odot$ models.
R. Harper, W. K. Rose.
Bull. American Astron. Soc., Vol. 2, 196 - 197 (1970).
Abstr. AAS.

065.033 Vibration of a Coulomb crystal at high pressure.
W. B. Hubbard.
Bull. American Astron. Soc., Vol. 2, 200 (1970). – Abstr. AAS.

065.034 On the termination of He shell burning in solar-mass stars. G. S. Kutter.
Bull. American Astron. Soc., Vol. 2, 204 (1970). – Abstr. AAS.

065.035 Stellar evolution at 50 solar masses.
S. C. Morris.
Bull. American Astron. Soc., Vol. 2, 210 (1970). – Abstr. AAS.

065.036 Grey and nongrey boundary conditions for pre-main-sequence low-mass stars. T. A. Pauls, J. P. Mutschlecner.
Bull. American Astron. Soc., Vol. 2, 213 - 214 (1970).
Abstr. AAS.

065.037 Models for partially mixed stars.
R. T. Rood.
Bull. American Astron. Soc., Vol. 2, 215 - 216 (1970).
Abstr. AAS.

065.038 Nonlinear pulsations and stability of slowly rotating stars. J.-L. Tassoul.
Bull. American Astron. Soc., Vol. 2, 221 (1970). – Abstr. AAS.

065.039 Calculation on the neutron-matter star.
C.-G. Wang, W. K. Rose.
Bull. American Astron. Soc., Vol. 2, 224 (1970). – Abstr. AAS.

065.040 Stellar stability and the upper end of the main sequence. K. Ziebarth.
Bull. American Astron. Soc., Vol. 2, 227 (1970). – Abstr. AAS.

065.041 Evolution of iron stars. Gravitational contraction and the decomposition of iron.
K. Nakazawa, T. Murai, R. Hōshi, C. Hayashi.
Progress Theor. Phys. Japan, Vol. 43, 319 - 333 (1970).

The evolution of iron stars in the phase of gravitational contraction and collapse is studied with an automatic computation method for the stellar masses 2.6 $M_\odot$ and 10 $M_\odot$. The computation is made for the two alternative cases with and without neutrino production by electron-neutrino interaction. For the stages before the stars become unstable because of the iron-helium phase transition, the features of evolution are essentially the same as those of carbon stars.

065.042 A perturbed red-giant model. The influence of relativistic gas characteristics.
T. W. Edwards.
Astrophys. Space Sci., Vol. 7, 151 - 159 (1970).

A population II, 1.3 $M_\odot$, pre-helium flash, red-giant model is investigated with respect to the influence of inclusion of relativistic gas characteristics, i.e., the equations of state, entropy. specific heats, and the adiabatic gradient. Little change is found in the observable properties of the model, but slightly larger changes are found in the interior properties, the most important of which is the narrowing of the already thin radiative zone between the hydrogen burning shell and the extensive outer convective envelope from 9.6 to 8.0 density scale heights.

065.043 Neutron stars of small masses.
Y. L. Vartanian.
Astrofizika, Vol. 6, 167 - 170 (1970). In Russian. – English translation in Astrophysics, Vol. 6, No. 1.

The cooling time and interior characteristics of hot neutron stars of small masses are considered.

065.044 Microturbulence in main sequence stars.
F. H. Chaffee, Jr.
Astron. Astrophys., Vol. 4, 291 - 301 (1970).

High dispersion spectra have been obtained for 23 main sequence stars in the spectral range from A 2 to G 0. Microturbulent velocities (ξ) have been determined for each of these stars from a curve of growth analysis of approximately 90 Fe I lines in each stellar spectrum. The variation of ξ with effective temperature (T_{eff}) is such that its mean value is 2 km/s at A 2, rises to 4 km/s in the late A stars, back to 2 km/s at F 5 and 4 km/s at G 0. It is also found that ξ correlates weakly with the Strömgren index. The shape of the ξ versus T_{eff} curve can be explained by assuming a general behavior for the velocity field in and above the stellar convection zone.

065.045 Secular stability of uniformly rotating polytropes.
J.-L. Tassoul, J. P. Ostriker.
Astron. Astrophys., Vol. 4, 423 - 427 (1970).   In French.

We show that uniformly rotating polytropes of index n less than 0.808 are secularly overstable, beyond the first bifurcation point on a given sequence, when viscosity is taken into account. Secular stability for $n \gtrsim 0.8$ is due to the fact that for such polytropes, when constrained to rotate uniformly, $T/|W| \lesssim 0.14$ (T = kinetic energy, W = potential energy); whereas secular overstability appears only when $T/|W| > 0.14$, the latter critical value being almost independent of n.

065.046 Nucleosynthesis in explosive oxygen burning.
J. W. Truran, W. D. Arnett.
Astrophys. Journ., Vol. 160, 181 - 192 (1970).

A preliminary investigation of the conditions appropriate to the synthesis of nuclei of intermediate mass ($28 \lesssim A \lesssim 42$) by explosive oxygen burning is presented. These calculations make use of the recent measurements of the $^{16}O + {}^{16}O$ reaction rate by Patterson, Winkler, and Spinka. The resulting abundance distributions are found to be sensitive to the temperature and density of the medium, to the time scale of expansion, and to the degree of neutron enrichment. Encouragingly good fits to both the elemental and isotopic abundance features in the silicon-to-calcium region are obtained for explosive oxygen burning in a rather restricted temperature range ($3.5 \times 10^9 \lesssim T \lesssim 3.7 \times 10^{9\,\circ}K$) for densities in the range $10^5 \lesssim \rho \lesssim 10^6 g\,cm^{-3}$.

065.047 Neutron star crusts and alignment of magnetic axes in pulsars.   P. Goldreich.
Astrophys. Journ. *(Letters)*, Vol. 160, L11 - L15 (1970).

The electromagnetic torque which brakes the rotation of a magnetic neutron star also tends to align the magnetic axis with the rotation axis. For fluid stars the time scale for alignment is comparable to the time scale for rotational braking. However, the presence of a crystalline mantle impedes the alignment of the magnetic axis with the rotation axis. Actual rates of alignment are probably determined by creeping or cracking of the solid mantle.

065.048 Models for neutron-core stars based on realistic nuclear-matter calculations.
C. G. Wang, W. K. Rose, S. L. Schlenker.
Astrophys. Journ. *(Letters)*, Vol. 160, L17 - L20 (1970).

A realistic nuclear-matter calculation has shown that, at densities below 10^{15} g cm^{-3}, the stable mass range for neutron-core stars is between 0.13 and 0.26 $M_\odot$. At densities greater than 10^{15} g cm^{-3}, hyperons become important; and an accurate equation of state for hyperons is beyond the knowledge of present-day physics.

065.049 Rapidly rotating supermassive stars.
G. G. Fahlman.
Astrophys. Journ. *(Letters)*, Vol. 160, L87 - L90 (1970).

The recent suggestion that the high energy needed for the activity of quasi-stellar objects is derived from the kinetic energy of rotating supermassive stars is examined. On the basis of this study, it is concluded that objects like 3C 345 and 3C 454.3 cannot be uniformly rotating supermassive stars.

065.050 Introduction générale à l'étude de la structure interne des étoiles.   P. Bouvier.
Structure interne des étoiles, Saas-Fee, (see Abstr. 012.012), p. 1 - 43 (1970).

065.051 Oscillations et stabilité stellaires.
P. Ledoux.
Structure interne des étoiles, Saas-Fee, (see Abstr. 012.012), p. 44 - 211 (1970).

065.052 Stellar evolution according to numerical models.
A. Weigert.
Structure interne des étoiles, Saas-Fee, (see Abstr. 012.012), p. 212 - 282 (1970).

065.053 Nuclear reactions in stellar surfaces and their relations with stellar evolution.
H. Reeves.
Structure interne des étoiles, Saas-Fee, (see Abstr. 012.012), p. 283 - 419 (1970).

065.054 Star formation from shock waves.
I. Halliday.
Journ. Roy. Astron. Soc. Canada, Vol. 64, 51 - 52 (1970).

065.055 Microturbulence in A-type supergiants.
J. D. Rosendhal.
Astrophys. Journ., Vol. 160, 627 - 639 = Contr. Kitt Peak National Obs., No. 508 (1970).

A survey has been undertaken of microturbulence in high-luminosity stars. In this paper, we present in detail the results for the A-type Ia supergiants. We also present the results of the combination of our observations with published data in order to give a preliminary picture of the overall behavior of the microturbulence along a broad baseline of spectral types.

065.056 The evolution of a red giant after rapid mass loss, and the Harman-Seaton sequence.
W. Deinzer, K. von Sengbusch.
Astrophys. Journ., Vol. 160, 671 - 683 (1970).

The evolution of a stellar model is investigated numerically in order to account for the consequences of mass loss preceding the observed evolution of the central stars of planetary nebulae. A red giant of 1.3 $M_\odot$ is stripped of almost all of its envelope, and hydrostatic equilibrium is restored by quasi-static, adiabatic expansion. The evolution of this model depends on the mass that remains in the envelope. If $\Delta M_{env}/M > 10^{-3}$, thermal equilibrium is restored, i.e., after the thermal-relaxation phase, the luminosity of the stellar model is produced entirely by the hydrogen-burning shell source. If $\Delta M_{env}/M \leq 10^{-3}$, thermal equilibrium cannot be restored, and after evolving through a luminosity maximum the model becomes a white dwarf. The evolutionary tracks obtained in the latter case are compared with the Harman-Seaton sequence.

065.057 Homogeneous models for population I and population II compositions.   H. Copeland, J. O. Jensen, H. E. Jørgensen.
Astron. Astrophys., Vol. 5, 12 - 34 (1970).

Zero-age models of intermediate and low masses with chemical composition parameters $0.60 \leq X \leq 0.90$ and $0.001 \leq Z \leq 0.030$ have been studied. Coulomb interaction between the particles is taken into account and also the H_2 molecule is included in the equation of state, the adiabatic gradient and the specific heats. Especially the effects on the zero-age lines and on the mass-luminosity relation caused by the H_2 molecules are studied.

065.058 Rotation in evolving stars.
R. Kippenhahn, E. Meyer-Hofmeister, H. C. Thomas.

Astron. Astrophys., Vol. 5, 155 - 161 (1970).

With a simplified treatment of stellar rotation the evolutionary track of a star of 9 solar masses has been computed from the main sequence to the exhaustion of helium. Two different assumptions on the distribution of angular momentum during evolution were made. In both cases the star ends up with a fast spinning-core surrounded by a slowly rotating envelope. Some possible consequences of the rapid rotation of the central regions are discussed.

065.059 **Non-linear limiting of the Goldreich–Schubert instability.** R. A. James, F. D. Kahn.
Astron. Astrophys., Vol. 5, 232 - 239 (1970).

The development of one of the unstable modes of the Goldreich–Schubert instability leads to changes in the angular momentum distribution in a star. After a certain stage, the mode is limited by a rapid instability involving motions along isentropic surfaces. All unstable modes are limited by this effect, and it seems unlikely that the angular momentum distribution set up can be destroyed. The time scale for redistribution of angular momentum in stars is probably much longer than the Kelvin–Helmholtz time.

065.060 **The evolution of a vibrationally unstable main-sequence star of 130 $M_\odot$.** I. Appenzeller.
Astron. Astrophys., Vol. 5, 355 - 371 (1970).

Nonlinear pulsation calculations have been carried out for a 130 $M_\odot$ main-sequence star. The scope of the new calculations was to find out if such a star will be destroyed by rapid mass loss during its main-sequence evolution, or if the growth of the pulsation amplitude is halted by nonlinear effects before the star starts to eject large amounts of mass. In addition, it has been tried to determine how the observable properties of such a very massive main-sequence star are changed by the vibrational instability in order to make possible the identification of massive main-sequence stars from observations without determining directly their mass.

065.061 **Evolution of rapidly rotating stars with mass loss.** P. A. Strittmatter, J. W. Robertson, D. J. Faulkner.
Astron. Astrophys., Vol. 5, 426 - 430 (1970).

A method is described for calculating evolutionary sequences of uniformly rotating stars including the effects of equatorial mass loss. Results are presented for a 9 $M_\odot$ model with initial rotational velocity near maximum. The rate of mass loss during the hydrogen burning phase is consistent with observational evidence from Be stars.

065.062 **Consequences of weak interactions for stellar evolution.** H. Heintzmann.
Astron. Astrophys., Vol. 5, 488 - 489 (1970).

On the basis of a classical model field theory, upper limits for possible weak interactions between baryons are derived. The consequences of such forces are investigated for stellar evolution. It turns out that such weak forces are only important for white dwarf- and neutron-stars. Gravitational collapse of stars can be stopped and the pulsation rates of pulsars can be reproduced easily and a simple explanation for supernovae can be given.

065.063 **Mass loss from rapidly rotating evolving B-stars.** E. Meyer-Hofmeister, H.-C. Thomas.
Astron. Astrophys., Vol. 5, 490 (1970).

Results on mass loss caused by rotational instability are presented for a nine solar mass star evolving off the main-sequence.

065.064 **Non-similarity linear wave cloud collapse solutions.** A. E. Wright.
Monthly Notices, Roy. Astron. Soc., Vol. 148, 455 - 461 (1970).

A set of analytical non-similarity solutions are presented for the equations of gravito-gas dynamics with thermal gas pressure and radiative cooling. It is suggested that, as well as being of interest in themselves, they afford a particularly stringent method of testing numerical gas cloud collapse computer programs for computational inaccuracies and instabilities.

065.065 **On the mechanism of formation of semi-convective zone in stars.** M. Gabriel.
Astron. Astrophys., Vol. 6, 124 - 129 (1970).

After a criticism of the ability of overstable convection to give rise to semi-convective zones in massive stars, a new mechanism is presented. In the frame of this theory, the models for massive stars are identical to those proposed by Schwarzschild and Härm.

065.066 **Evolution of very massive stars with pulsational mass loss.** N. R. Simon, R. Stothers.
Astron. Astrophys., Vol. 6, 183 - 192 (1970).

Evolution with moderate mass loss has been investigated for hydrogen-burning stars whose mass exceeds the critical mass for pulsational stability at the top of the main sequence. A prescription based on minimizing the degree of pulsational instability is used to specify the rate of mass loss. In contrast with the possible cases of no significant mass loss or very rapid mass loss, the assumption of moderate mass loss seems to yield results which are consistent with the relevant observational data, particularly for the most luminous group of Wolf-Rayet stars.

065.067 **Shell source burning stars with highly condensed cores.** S. Refsdal, A. Weigert.
Astron. Astrophys., Vol. 6, 426 - 440 (1970).

Simple analytical relations are derived – similar to the homology relations for homogeneous stars – which describe the behaviour of shell source burning models with highly condensed cores. Many features which are known from evolutionary calculations (and partly from observations) of such stars can be understood as coming essentially from the increase of the mass of the core. Changes of the radius of the core, of the radiation pressure, and of the chemical composition are also considered. For a comparison, sequences of numerically calculated equilibrium models are presented. They show that the analytical relations give, even quantitatively, good approximations to the numerical results.

065.068 **On homogeneous stellar models.** J. O. Jensen, H. E. Jørgensen.
Astron. Astrophys., Vol. 6, 488 - 490 (1970).

New opacities have been published by Cox and Stewart. They are here compared with those previously used in our model computations and the influence on zero-age models is discussed.

065.069 **Rapid contraction of protostars to the stage of quasi-hydrostatic equilibrium. II. 10, 10^2 , 10^3 and 10^4 solar masses without radiation flow.**
T. Nakano, N. Ohyama, C. Hayashi.
Progr. Theor. Phys. Japan, Vol. 43, 672 - 683 (1970).

The collapse of opaque protostars is computed up to the onset stage of quasi-hydrostatic equilibrium in the same way as for 1 $M_\odot$ in a previous paper. It is found that a mass ejected from the surface by the shock wave is negligibly small for the protostars of 1 and 10 $M_\odot$, while it amounts to 10 or 15 per cent of the total mass for the protostars of greater masses.

065.070 **Neutronen-sterren.** J. Heise.
Hemel en Dampkring, Vol. 68, 75 - 84 (1970).

065.071 **URCA shells in dense stellar interiors.**

S. Tsuruta, A. G. W. Cameron.
Astrophys. Space Sci., Vol. 7, 374 - 406 (1970).

Thermal and vibrational energy losses due to URCA shells in stellar interiors are calculated. Analytic expressions are derived for semidegenerate, relativistic electrons. Results are given for more general cases calculated with a computer. The calculations are carried out for a large number of nuclei that may contribute to URCA energy losses in various stages of stellar evolution. An illustration is given of the cooling and vibrational damping of a white dwarf.

065.072 **Plasmon neutrinos emission in a strong magnetic field. I: Transverse plasmons.**
V. Canuto, C. Chiuderi, C. K. Chou.
Astrophys. Space Sci., Vol. 7, 407 - 415 (1970).

In this paper we generalize the Adams, Ruderman, Woo and Zaidi plasmon decay process to include the presence of a strong magnetic field. Two cases are studied; propagation parallel and perpendicular to the magnetic field. In either case we found that relevant changes only show for $H \simeq 10^{12} - 10^{13}$ G.

065.073 **Relativistic isentropic stellar models.**
G. Shaviv, A. Kovetz.
Astrophys. Space Sci., Vol. 7, 416 - 423 (1970).

Relativistic, isentropic, homogeneous models are constructed by a method that automatically detects instabilities, and evolutionary tracks of central conditions are shown on a (T, ρ) diagram. It is pointed out that general relativistic instability will prevent the formation of neutron stars through hydrostatic evolution and may be relevant in setting off low-mass supernovae.

065.074 **Formation of stellar associations.**
S. B. Pikel'ner.
Astrophys. Space Sci., Vol. 7, 489 - 493 (1970).

Because of the difficulties encountered by considering formation of stellar clusters as compression and fragmentation of clouds, we shall discuss some other mechanism of stellar formation which may work in the spiral arms.

065.075 **Electrical and thermal conductivities in neutron stars.** A. M. Gentile.
Astrophys. Letters, Vol. 5, 245 - 246 (1970).

Expressions for the electrical and thermal conductivity in the deep interior of a neutron star are given. They were obtained by solving the relativistic Boltzmann equation by a variational method. The electrical conductivity is only slightly higher than that obtained in the nonrelativistic calculation of Baym, Pethick, and Pines.

065.076 **Investigation of low frequency instability of a rotating magnetic neutron star.**
Yu. V. Vandakurov.
Astrophys. Letters, Vol. 5, 267 - 269 (1970).

Numerical integration of equations of non-radial oscillation for a rotating neutron star with internal toroidal magnetic field shows that self-excitation of oscillation is possible.

065.077 **L'evoluzione delle stelle. II – Le stelle doppie.**
P. Giannone, M. A. Giannuzzi.
Coelum, Vol. 38, 10 - 21 (1970).

065.078 **Red giant mass loss and cepheid variables.**
R. J. Tayler.
Monthly Notices, Roy. Astron. Soc., Vol. 149, 17 - 24 (1970).

It is suggested in this paper that, if substantial mass loss occurs in the red giant phase of stellar evolution, this might be accompanied by an increase in the helium content of the outer layers of the star, so that the predicted low masses and high helium contents of cepheids might be explained simultaneously. In addition, mass loss could modify subsequent evolutionary tracks so that, in contrast with the result obtained by Hofmeister, stars with normal metal and helium content might cross the cepheid strip several times.

065.079 **Evolution of single stars. I. Stellar evolution from main sequence to white dwarf or carbon ignition.**
B. Paczyński.
Acta Astron., Vol. 20, 47 - 58 (1970).

Model evolutionary calculations have been made for population I stars ($X = 0.7, Z = 0.03$) with masses of 0.8, 1.5, 3, 5, 7, 10, and 15 $M_\odot$. Neutrino losses were taken into account. First the computations were made under the assumption that the stars do not lose mass. These computations were started on the main sequence and carried out through the phases of hydrogen and helium exhaustion in the center.

065.080 **Chemical compositions and nucleosynthesis on the galactic scale.** E. M. Burbidge, G. R. Burbidge.
Comments Astrophys. Space Phys., Vol. 2, 92 - 98 (1970).

The authors show that there are composition differences in different galaxies, which are expected from the theory of stellar and galactic nucleosynthesis. They discuss here what data are available, and show that recent statements that the relative abundances of all the common elements are the same in all galaxies are not correct.

065.081 **Influence of opacity on the pulsational stability of massive stars with uniform chemical composition.**
R. Stothers, N. R. Simon.
Astrophys. Journ., Vol. 160, 1019 - 1029 (1970).

The maximum mass of homogeneous stars which are stable against nuclear-energized pulsations has been redetermined by using a full opacity formula and an accurate treatment of the equilibrium structure of the outer layers where nonscattering sources of opacity are important damping agents. Sixteen composition mixtures were used, covering the range $0 \le Y \le (1 - Z)$ and $0 \le Z \le 0.05$. The nuclear-energy sources were taken to be the CN cycle and the triple-α process.

065.082 **Nonlinear pulsations and stability of slowly rotating stars.** J.-L. Tassoul.
Astrophys. Journ., Vol. 160, 1031 - 1042 (1970).

This paper deals with finite-amplitude oscillations of slowly rotating, slightly distorted stars. Lagrangian variables and the scalar virial theorem are employed to describe, in a first approximation, nonlinear adiabatic motions which, in the limit of zero rotation, are purely radial.

065.083 **Rapid contraction of protostars to the stage of quasi-hydrostatic equilibrium. III. – Stars of 0.05, 1.0 and 20 $M_\odot$ with energy flow by radiation and convection.**
S. Narita, T. Nakano, C. Hayashi.
Progr. Theor. Phys. Japan, Vol. 43, 942 - 964 (1970).

The evolution of collapsing protostars of 0.05, 1 and 20 $M_\odot$ is studied with computations which include radiative as well as convective energy flow, which has been entirely neglected in previous papers. It is found that, when a shock wave which has been generated at the center reaches the outermost layers, the protostars flare up suddenly.

065.084 **Su alcune proprietà termodinamiche della materia in condizioni di degenerazione elettronica.**
V. Castellani, I. Mazzitelli, A. Renzini.
Mem. Soc. Astron. Italiana, Nuova Serie, Vol. 41, 215 - 219 (1970).

Correct analytical expressions are derived for the fundamental thermodynamic parameters of stellar matter in electronic degeneration conditions. In comparison with the usual approximations they show an appreciable agreement with the values of the adiabatic gradient, but some discrepancies in the evaluation of the energy balances can be noticed.

065.085 r-process nucleosynthesis of C^{12} seed nucleus.
T. Ohnishi.
Progr. Theor. Phys. Japan, Vol. 43, 1480 - 1490 (1970).

The produced amount of nuclides by r-process nucleosynthesis is calculated under the condition of high temperature and high density. In this case the seed nucleus of this process is mainly C^{12}, which is pumping-out from the equilibrium system, and absorbs enough free neutrons to succeed the rapid-neutron-capture process. The existence of these free neutrons is due to neutronization by the weak interaction at the time of quasi-stational, hydrodynamic contraction of later stage stars, i.e. presupernovae, and therefore one of the possible astronomical sites of this process is the inner shell of type-I supernovae.

065.086 Computation of the structure of thin nuclear burning shells in stellar models. U. Uus.
Nauchn. Informatsii, vyp. (No.) 13, p. 126 - 144 (1969).
In Russian.

The method for the calculation of the structure of a thin burning shell, allowing to determine the chemical composition in the shell independently of the previous evolution of the star, is improved. The method is generalized for the case of two burning shells moving one after another in stellar models with carbon cores and hydrogen-rich envelopes. Results of the computations of a set of double-shell models are presented.

065.087 On thermonuclear reactions in the interior of stars similar to the sun.
V. A. Dergachev, G. E. Kocharov.
Cosmic Rays No. 11, Moscow, p. 146 - 148 (1969).
In Russian.

The paper presents calculations of the equilibrium values of concentrations of different particles participating in the proton-proton cycle in a wide range of temperatures ($10^6 \leqslant T \leqslant 5 \times 10^7$), density and chemical composition of the stellar substance. The possibilities of different variants of the p-p-cycle are determined. The obtained results are applied for the real model of the sun.

065.088 Stellar constitution.   A. G. Massevitch.
Trans. IAU, Vol. 14A, (see 003.028), 405 - 423 (1970). – Report of Commission 35.

065.089 On the muon neutrino emission in gravitational collapse of massive stars.   G. V. Domogatsky.
Nauchn. Informatisii, vyp. (No.) 13, p. 94 - 95 (1969).
In Russian.

065.090 Adiabatic oscillations of stars.   R. Van der Borght.
Proc. Astron. Soc. Australia, Vol. 1, 325 - 326 (1970).

065.091 Radial oscillations of stellar models: Density distributions which lead to three-term recurrence relations.   J. O. Murphy, A. C. Smith.
Proc. Astron. Soc. Australia, Vol.1, 328 - 329 (1970).

065.092 OH emission from protostars.   I. D. Johnston.
Proc. Astron. Soc. Australia, Vol. 1, 336 - 337 (1970).

065.093 Possible masses of neutron stars.
O. H. Guseinov, F. K. Kasumov.
Astron. Tsirk., No. 562, p. 7 - 8 (1970).   In Russian.

065.094 Long period oscillations in rotating neutron stars.

M. Ruderman.
Nature, Vol. 225, 619 - 620 (1970).

If pulsars contain a superfluid neutron core, long-period torsional oscillations of this core may exist. A star quake in the solid crust may then excite a periodic wobble with a period of a few months as is observed in the Crab nebula.

065.095 Evolution kosmischer Materie.
A. Unsöld.
Verhandl. Schweiz. Naturforsch. Gesellschaft, 1968, p. 35 - 50 = Separate print Sternw. Kiel, No. 150 (1968). – Review article.

065.096 Hyperon star matter.
K. Koebke, E. Hilf, R. Ebert.
Nature, Vol. 226, 625 - 626 (1970).

The approximation of the smooth abundance distribution (the number of different hyperons per mass interval): H(m) = exp (mc^2/kT$_0$ + 52), $T_0 = 2 \times 10^{12}$ K fits the experimental data well up to 2 GeV. Using these equations and the relativistic Fermi-integrals and integrating over the whole hyperon spectrum, we obtain the equation of state of the multi-component gas to be found in the interior of hyperon stars. The internal structure of hyperon and (for comparison) of neutron stars is implicitly given by two figures.

065.097 Radial oscillations of composite polytropes. I.
M. Singh.
Proc. National Inst. Sci. India, Ser. A, Vol. 35, (No. 4), 586 - 589 (1969).

Composite polytropic configurations have been obtained by taking polytropic indices as 1.5 and 3 respectively in the core and envelope. Some numerical solutions of the Emden equation have also been developed.

065.098 Microturbulence in main sequence stars.
H. T. Chaffee, Jr.
Thesis, Univ. of Arizona, Tucson. [Available from Univ. Microfilms, Ann Arbor, Mi.]. 148 pp. (1968).

065.099 An exact analytic solution for explosion waves in stellar interiors.   G. D. Ray.
Proc. National Inst. Sci. India, Ser. A, Vol. 35, No. 2, Suppl., p. 182 - 194 (1969).

An exact solution for propagation of explosion waves in stellar models has been obtained. The disturbance is headed by a shock surface of variable strength. The total energy of the wave also varies with time. Constancy of total energy, however, follows as a particular case.

065.100 A theoretical model of pulsating neutron stars (pulsars).   N. Kumar.
Phys. Letters, Vol. 30A, No. 3, p. 199 - 201 (1969).

Angular momentum and moment-of-inertia considerations peculiar to heavy nuclei when applied to neutron stars predict a highly elongated ellipsoidal object, spinning about a minor axis. Preferential emission of the major-axis ends may account for the observed pulsations.

065.101 The fifth state of matter: On super-dense matter and neutron stars.   B. Kuchowicz.
Sci. Progress, Vol. 56, (No. 224), 531 - 539 (1969).

065.102 Stable masses and radii of super dense baryon stars.
L. M. Libby, F. J. Thomas.
Phys. Letters, Vol. 30B, (No. 6), 400 - 401 (1969).

Stable masses and radii of super dense baryon stars are computed from the isostatic relativistic equations of star equilibrium, using an equation of state for 45 baryon species and using the proton-proton interaction energy derived from high energy elastic scattering up to 30 GeV. Masses for an interval of radius 5×10^4 cm $< r < 10^5$ cm, are unstable, so that a degenerate star losing mass and approaching r of 5×10^4 cm suffers a partial collapse to half radius.

065.103 **A simplified method for calculating the internal structure of neutron stars.** D. Kisdi.
Acta Phys. Acad. Sci. Hungaricae, Vol.27, 289 - 298 (1969).
A very economical method is proposed for solution of the hydrostatic equilibrium equations of neutron stars. As a simple example the model of homogeneous internal distribution is constructed.

065.104 **Effects of three-body forces in a neutron star.** G. Chanmugam.
Phys. Rev.,Second Series, Vol. 186, 1384 - 1386 (1969).
It is shown that three-body forces give significant contri - butions to the equation of state of a neutron-star at high densities.

065.105 **Supermassive stars and neutrino emission.** P. Bandyopadhyay, P. R. Chaudhuri.
Nuovo Cimento Lettere, Vol. 3, 269 - 271 (1970).
The role of neutrino emission from supermassive stars taking into account the photon-neutrino weak-coupling theory is studied.

065.106 **Massive main-sequence stars in rapid differential rotation.** J. W. K. Mark.
Thesis, Princeton Univ., Princeton, N. J. [Available from Univ. Microfilms, Ann Arbor, Mi.], 51 pp. (1969).
The effects of rapid differential rotation on the structure and observable properties of main-sequence stars have been studied. In general, the effects of differential rotation are an order of magnitude larger than those of uniform rotation for the same equatorial velocity. Furthermore, differential rotation shifts the position of the star in the H-R diagram almost downwards along the main sequence, so that a star with rapid differential rotation may as much as twice as massive as a non-rotating star of the same colour and absolute magnitude.

065.107 **A third family of stable equilibria?** U. H. Gerlach.
Thesis, Princeton Univ., Princeton, N. J. [Available from Univ. Microfilms, Ann Arbor, Mi.], 106 pp. (1969).
Basic aspects of degenerate superdense stars are con - sidered within the framework of general relativity, current evidence pointing towards the existence of these stars is reviewed.

065.108 **Evolution of a 30 $M_\odot$ population I star.** A. M. Schindler.
Thesis, Brandeis Univ., Waltham, Mass. [Available from Univ. Microfilms, Ann Arbor, Mi.], 174 pp. (1969).
The evolution of a population I star of 30 $M_\odot$ is followed from the wholly convective gravitationally con - tracting stage.

065.109 **Plasma physics in an astrophysical environment.** I. W. Roxburgh.
Report AFCRL-69-0134 [Available from CFSTI, Springfield, Va.]. 14 pp. (1969).
Problems involving the stability and equilibrium of astrophysical systems have been investigated, particularly the structure of rotating turbulent convective zones and the effect of rotation of the shape of a star.

065.110 **Les équations relativistes régissant la structure de configurations radiatives à symétrie sphérique et en mouvement lent.** J. Demaret.
Bull. Soc. Roy. Sci. Liège, Vol. 38, 219 - 239 = Univ. Liège, Inst d'Astrophys., Coll. 8°, No. 586 (1969).
We derive the equations of stellar structure in general relativity for spherically symmetric configurations in slow motion, from Einstein's gravitational field equations and conservation equations, written for Schwarzschild's coordinates.

065.111 **Étude des oscillations et de la stabilité du sphéroide de Jeans et de l'ellipsoide de Roche.** H. Robe.
Mém. Soc. Roy. Sci. Liège, 5e Sér., Vol. 18, Fasc. 1, 69 pp. = Univ. Liège, Inst. d'Astrophys., Coll. 8°, No. 589 (1969).

065.112 **Massetab fra stjerner.** E. H. Olsen.
Astron. Tidssk., Vol. 3, 72 - 81 (1970).

The symposium on stellar composition and nucleo-synthesis. See Abstr. 011.026.

Difficulties with fusion catalysis by quarks. See Abstr. 022.017.

The effect of an arbitrary law of slow rotation on the oscillations and the stability of gaseous masses. See Abstr. 061.014.

The oscillations and the stability of rotating masses with toroidal magnetic fields. See Abstr. 061.015.

The instability of the congruent Darwin ellip-soids. II. See Abstr. 061.023.

Outer boundary conditions in low-mass stars. See Abstr. 064.026.

Pulsation damping by the nonleptonic weak interaction in hyperon stars. See Abstr. 066.001.

General relativistic rotational properties of stellar models. See Abstr. 066.004.

The optical appearance of a collapsing star in terms of the scalar-tensor theory of gravitation. See Abstr. 066.043.

On the stability of ultrarelativistic stars. See Abstr. 066.061.

The helium-rich stars of early spectral type. See Abstr. 114.013.

Stellar rotation: Uniformly rotating stars with hydrogen-line-blanketed model atmospheres. Comparison with observations of A-type stars. See Abstr. 116.012.

Échange de masse dans les étoiles doubles serrées. See Abstr. 117.008.

On the evolution of close binary systems with a total mass 2.5 $M_\odot$. 2. See Abstr. 117.013.

Forced oscillations in close binaries. The adiabatic approximation. See Abstr. 117.014.

On the structure of close binary members. See Abstr. 117.016.

Evolution of close binary systems with initial com-ponents of 2 and 1.5 $M_\odot$. See Abstr. 117.019.

Evolution of close binaries. VI. A program for computation of evolution with mass exchange. See Abstr. 117.022.

Evolution of close binaries. VI. Case B of mass exchange in systems $4 + 3.2$ cm$_\odot$ and $4 + 1.6$ m$_\odot$. See Abstr. 117.030.

Pulsation analysis for stars in thermal imbalance. See Abstr. 122.019.

Nonequilibrium beta-processes as a source of the thermal energy of white dwarfs.   See Abstr. 126.007.

On the gamma and radio radiation from neutron stars in the state of accretion.   See Abstr. 142.029.

Large-scale galactic shock phenomena and the implications on star formation.   See Abstr. 151.048.

Stellar evolution in globular clusters. See Abstr. 154.005.

066 Relativistic Astrophysics (without Cosmology), Background Radiation, Gravitation Theory

066.001 **Pulsation damping by the nonleptonic weak interaction in hyperon stars.** P. B. Jones.
Astrophys. Letters, Vol. 5, 33 - 36 (1970).

In a hyperon star, departures from equilibrium neutron and hyperon number densities are restored by the nonleptonic weak interaction, with a relaxation time of the order of the expected period of pulsation. This relaxation time and the large neutron-hyperon mass difference cause pulsation damping in times comparable with those for gravitational radiation.

066.002 **Anisotropie du champ gravifique et ondes ellipsoïdales de la gravité.** J. Carstoiu.
Comptes Rendus Acad. Sci. Paris, Sér. A, Vol. 270, 689 - 692 (1970).

066.003 **An orbiting clock experiment to determine the gravitational red shift.**
D. Kleppner, R. F. C. Vessot, N. F. Ramsey.
Astrophys. Space Sci., Vol. 6, 13 - 32 (1970).

Plans are presented for an experiment to measure the gravitational red shift of the earth by comparing a ground-based and satellite-borne hydrogen maser clock. The limiting accuracy is estimated to yield a determination of the red shift to 1 part in 10^5, corresponding to a clock stability of 3 parts in 10^{15}.

066.004 **General relativistic rotational properties of stellar models.** J. M. Cohen.
Astrophys. Space Sci., Vol. 6, 263 - 274 (1970).

In this paper we give general relativistic expressions for the angular momentum and rotational kinetic energy of slowly rotating stars. These expressions contain contributions from the pressure, gravitational red shift and Doppler shift, and the motion of inertial frames. These contributions are not negligible, e.g., there are stable neutron star models for which the angular velocity of inertial frames at the center is about 70% the angular velocity of the star. These expressions are useful in the study of pulsars if pulsars are rotating neutron stars.

066.005 **Maximal red shifts of neutron stars.**
A. Kovetz.
Astrophys. Space Sci., Vol. 6, 293 - 296 (1970).

It has been recently established that there exists a maximal red shift z_{max} for a homogeneous star of given mass M. The relationship $z_{max}(M)$ is obtained for neutron stars in the mass range $0.71 \leqslant M/M_\odot \leqslant 12.06$.

066.006 **Oscillating gyroscopes in a satellite.**
A. Blokland.
Astrophys. Space Sci., Vol. 6, 352 - 357 (1970).

It is shown that a set of three gyroscopes in a satellite can test vital aspects of general relativity in a period of a few days.

066.007 **Gravitationssignale, eine galaktische Botschaft?**
W. Kundt.
Naturwissenschaften, Vol. 57, 6 - 10 (1970).

066.008 **The phenomenon of time dilation.**
D. F. Lawden.
Spaceflight, Vol. 12, 178 - 179, 183 (1970).

066.009 **General cosmological solution of the gravitational**

equations with a singularity in time.
I. M. Khalatnikov, E. M. Lifshitz.
Phys. Rev. Letters, Vol. 24, 76 - 79 (1970).

A way is indicated to construct a general solution of the Einstein equations with a singularity, starting from a previously known solution of a lesser degree of generality. A qualitative description is given of the evolution of the metric in this general solution towards the singularity, which is of a complex, oscillatory nature.

066.010 **Gravitational radiation experiments.**
J. Weber.
Phys. Rev. Letters, Vol. 24, 276 - 279 (1970).

A summary is given of the statistics and coincidences of the Argonne-Maryland gravitational-radiation-detector array. New experiments have been carried out. These include a parallel coincidence experiment in which one coincidence detector had a time delay in one channel and a second coincidence detector operated with no time delays. Other experiments involve observations to rule out the possibility that the detectors are being excited electromagnetically. These results are evidence supporting an earlier claim that gravitational radiation is being observed.

066.011 **Gravitational collapse with asymmetries.**
V. de la Cruz, J. E. Chase, W. Israel.
Phys. Rev. Letters, Vol. 24, 423 - 426 (1970).

Two idealized collapse models, involving a magnetic dipole and a gravitational quadrupole, are analyzed, treating departures from sphericity as small perturbations. Radiative leakage causes externally observable asymmetries to decay to zero in an oscillatory fashion, with a period of the order of the Schwarzschild characteristic time $2Gm/c^3$.

066.012 **Solutions of two problems in the theory of gravitational radiation.** S. Chandrasekhar.
Phys. Rev. Letters, Vol. 24, 611 - 615 (1970).

The evolution of an elongated rotating configuration by gravitational radiation and the possibility of a secular instability being induced by it are considered in the context of the classical homogeneous figures of Maclaurin and Jacobi. The triaxial Jacobian ellipsoid evolves in the direction of increasing angular velocity and approaches (exponentially) the point of bifurcation where it ceases to radiate. Further, radiation reaction does not make the Maclaurin spheroid secularly unstable past the point of bifurcation.

066.013 **The slow rotation of relativistic polytropes.**
V. V. Papoyan, D. M. Sedrakyan, Eh. V. Chubaryan.
Stars, Nebulae, Galaxies, Symposium Byurakan 1968, (see 012.002), p. 273 - 278 (1969). In Russian.

066.014 **On the relativistic generalized theory of gravitation. II. Baryon configurations.**
G. S. Sahakian, M. A. Mnatsakanian.
Astrofizika, Vol. 5, 555 - 579 (1969). In Russian. – English translation in Astrophysics, Vol. 5, No. 4.

Further transformations of the relativistic generalized theory of equations for a static spherically symmetric mass distribution are presented. The conditions on the surface and in the center are discussed, under which configurations can have finite central pressures. Some important properties of gravitars are established. The results for incompressible liquid and real baryon gas are compared with those of other theories.

066.015 Polytropic models in the relativistic generalized theory of gravitation.
M. A. Mnatsakanian.
Astrofizika, Vol. 5, 645 - 649 (1969). In Russian. – English translation in Astrophysics, Vol. 5, No. 4.
The polytropic models (n = 0, 1.5, 3) in the relativistic generalized theory of gravitation are investigated.

066.016 Interaction of vortex and potential motions in relativistic hydrodynamics. I.
A. D. Chernin, E. D. Eidelman.
Astrofizika, Vol. 5, 654 - 656 (1969). In Russian. – English translation in Astrophysics, Vol. 5, No. 4.
Approximative formulas for the generation of a potential motion by a vortex (and vice versa) are found on the basis of relativistic hydrodynamics.

066.017 Interaction of vortex and potential motions in relativistic hydrodynamics. II.
A. D. Chernin.
Astrofizika, Vol. 5, 656 - 658 (1969). In Russian. – English translation in Astrophysics, Vol. 5, No. 4.
The hydrodynamical interaction of motions due to nonlinearity of gravitation equations in general relativity is analysed in post-Newtonian approximation.

066.018 Relativity theory and astronomy (III).
A. Zięba.
Urania Kraków, Vol. 41, 8 - 17 (1970). In Polish.

066.019 On the motion of the perihelion of Mercury. I. Classical theory. M. Abramowicz.
Urania Kraków, Vol. 41, 34 - 41 (1970). In Polish.

066.020 Gravitation dans un milieu cristallin.
J. Carstoiu.
Comptes Rendus Acad. Sci. Paris, Sér. A, Vol. 270, 921 - 924 (1970).

066.021 Gravitational waves: The evidence mounts.
G. L. Wick.
Science, Vol. 167, 1237 - 1239 (1970).

066.022 An oscillating state as an alternative to gravitational collapse. C. Leibovitz.
Nature, Vol. 225, 711 - 712 (1970).
As an over-critical mass is undergoing the process of collapse the ratio of matter with over-nuclear density to matter with under-nuclear density increases. It is shown that in these conditions we cannot neglect the possibility of restoring forces and therefore of an oscillatory state as an alternative to gravitational collapse.

066.023 A generalized Hoyle-Narlikar particle theory.
C. B. G. McIntosh.
Nature, Vol. 226, 339 - 340 (1970).

066.024 Measurement of the isotropic background radiation in the far infrared.
D. Muehlner, R. Weiss.
Phys. Rev. Letters, Vol. 24, 742 - 746 (1970).
Measurements made at an altitude of 40 km with a balloon-borne radiometer that has three separate spectral responses in the spectral region below 20 cm^{-1} indicate that the spectrum of the isotropic background radiation may not be thermal.

066.025 High-precision tests of general relativity.
K. S. Thorne, C. M. Will.
Comments Astrophys. Space Phys., Vol. 2, 35 - 41 (1970).
The purpose of this comment is to review what the possibilities are to test general relativity, and to call attention to the efforts that are necessary for a full exploitation of them. We consider all theories of gravity which (i) are compatible with special relativity locally; (ii) are compatible with the equivalence principle for laboratory-sized bodies; and (iii) agree with Newtonian gravity to the experimentally verified accuracy of about one part in 10^7 in the solar system.

066.026 Electrical conductivity and conductive opacity of a relativistic electron gas. V. Canuto.
Astrophys. Journ., Vol. 159, 641 - 652 (1970).
The electrical conductivity and the conductive opacity are computed for a system of relativistic degenerate electrons in the presence of a system of ions. The numerical values of the electron conduction opacities are given for different values of the parameter Γ, which characterizes the strength of the ion-ion interaction, and for different values of $2 \leq Z \leq 26$.

066.027 Nonradial pulsation of general-relativistic stellar models V. Analytic analysis for l = 1.
A. Campolattaro, K. S. Thorne.
Astrophys. Journ., Vol. 159, 847 - 858 (1970).
The theory of small, adiabatic, dipole perturbations of a star away from hydrostatic equilibrium is developed within the framework of general relativity. The analysis is linearized in the perturbation amplitudes. The "odd-parity" perturbations describe differential rotation, while the "even-parity" perturbations describe pulsation. In both the pulsational and rotational cases, the Einstein field equations are put into simple forms suitable for numerical integration.

066.028 The equation of state of matter at ultra-high densities. G. Chanmugam.
Journ. Phys. A, General Phys., Vol. 3, L19 - L21 (1970).
In this work it is shown that Zeldovich's results for the behavior of matter at ultra-high densities are not valid when many-body forces are included.

066.029 Charged-dust distributions in general relativity.
A. K. Raychaudhuri, U. K. De.
Journ. Phys. A, General Phys., Vol. 3, 263 - 268 (1970).
The paper presents some simple theorems and relations for charged-dust distributions in general relativity.

066.030 Gravitationswellen. J. Rahe.
Umschau, 70. Jahrgang, p. 251 - 252 (1970).

066.031 Lunar mascons as detector of gravitational waves from pulsars. V. de Sabbata.
Mem. Soc. Astron. Italiana, Nuova Serie, Vol. 41, 65 - 68 (1970).
It is proposed to exploit the lunar mascons for the detection of gravitational radiation in the 1-Hz band, as that presumably emitted by pulsars. If the mascons give rise to a resonance effect in the seismic response, the possibility of the detection may be not too far from the existing instruments.

066.032 Die isotrope kosmische 3°-Kelvin-Strahlung.
T. Schmidt.
SuW, Vol. 9, 84 - 88 (1970).

066.033 The first observable relativistic objects.
K. S. Thorne.
Priroda, No. 1.70, p. 62 - 65 (1970). In Russian.

066.034 Insensitivity to cosmic rays of the gravity radiation detector. D. H. Ezrow, N. S. Wall, J. Weber, G. B. Yodh.
Phys. Rev. Letters, Vol. 24, 945 - 947 (1970).
A search for coincidences between cosmic-ray shower signals in a large scintillation counter and gravitational-

radiation detector signals indicates that the gravitational radiation detector does not produce a signal when hit by showers of particle density of the order of 100 particles/m²

066.035 A paradox in the interaction of the gravitational and electromagnetic fields?
J. F. Woodward, W. Yourgrau.
Nature, Vol. 226, 619 - 621 (1970).
    The assumption that the extent of the interaction of the electromagnetic and gravitational fields is fundamentally dependent on the frequency of the electromagnetic field is consistent with data on the deflexion of radiation by massive bodies.

066.036 Gravitation.   W. Thirring.
        Plenarvorträge, 34. Physikertagung 1969 Salzburg (B. G. Teubner, Stuttgart), p. 17 - 25 (1969). – Conference paper.

066.037 Die experimentelle Prüfung der allgemeinen Relativitätstheorie.   R. U. Sexl.
Plenarvorträge, 34. Physikertagung 1969 Salzburg (B. G. Teubner, Stuttgart), p. 471 - 489 (1969). – Conference paper.

066.038 Half-closed spaces in relativistic astrophysics.
A. Zięba.
Postępy Astron., Vol. 18, 101 - 107 (1970).   In Polish.
    The article contains a short survey of basic properties of equations of general relativity theory solutions: the solutions of Schwarzschild, Friedmann, as well as of the solution produced by amalgamation of these two solutions – describing the so called half-closed spaces.

066.039 The Michelson-Morley-Miller experiments before and after 1905.   L. S. Swenson, Jr.
Journ. History Astron., Vol. 1, (Part 1), 56 - 78 (1970).
    The purpose of this paper is to recount the record left by the experimental work of Albert A. Michelson, Edward W. Morley, and Dayton C. Miller with instruments called aetherdrift interferometers between 1880 and 1930. Specifically the author attempts to narrate precisely how Michelson's experiment was first performed, partially repeated, and finally completed by being challenged nearly half a century after its conception.

066.040 Neutronensterne, Gravitationswellen und kosmische Hintergrundstrahlung.   W. Kundt.
Umschau, 70. Jahrgang, p. 344 - 345 (1970).

066.041 Relativistic spheres and their application to cosmic body models.   B. Kuchowicz.
Postępy Astron., Vol. 18, 197 - 199 (1970).   In Polish.
Abstr. Polish Astron. Soc.

066.042 Gravitational radiation in relativistic collisions.
S. Zaromb.
Bull. American Astron. Soc., Vol. 2, 226 - 227 (1970).
Abstr. AAS.

066.043 The optical appearance of a collapsing star in terms of the scalar-tensor theory of gravitation.
H. Nariai.
Progress Theor. Phys. Japan, Vol. 43, 334 - 346 (1970).
    The optical appearance of a collapsing star in terms of the scalar-tensor theory of gravitation is studied in comparison with Ames and Thorne's general relativistic treatment. It is shown that the expressions for the spectral shift, the intensity and the net flux in the rim region of its optical disk which is brightest and bluest are qualitatively of the same form as their counterparts in Ames and Thorne's analysis. This means that the optical appearance is not different from that pre-

dicted by them, in spite of the fact that the hypersurface $R = 2M$ is observable to a distant observer in our space-time.

066.044 On some astronomical effects of variable gravitation.   A. F. Bogorodsky.
Vestn. Kiev. Univ., Ser. Astron., No. 11, p. 3 - 9 (1969).
In Russian.
    Some astronomical effects caused by a decreasing gravitational constant are considered. Some of them may be of great interest for the theory of stellar evolution.

066.045 On the dynamics of self-gravitating spheres.
A. Barnes, G. J. Whitrow.
Monthly Notices, Roy. Astron. Soc., Vol. 148, 193 - 195 (1970).
    It is shown that the analytical results obtained by M. V. Penston for the Newtonian pressure-free collapse of a spherically symmetrical sphere of gas can be extended to general relativity.

066.046 The $2\frac{1}{2}$-post-Newtonian equations of hydrodynamics and radiation reaction in general relativity.
S. Chandrasekhar, F. P. Esposito.
Astrophys. Journ., Vol. 160, 153 - 179 (1970).
    In this paper the equations of hydrodynamics in the $2\frac{1}{2}$-post-Newtonian approximation to general relativity are derived. In this approximation all terms of $O(c^{-5})$ are retained consistently with Einstein's field equations; it is also the approximation in which terms representing the reaction of the fluid to the emission of gravitational radiation by the system first make their appearance.

066.047 On the motion of the perihelion of Mercury. II. Relativity theory.   M. Abramowicz.
Urania Kraków, Vol. 41, 136 - 146 (1970).   In Polish.

066.048 Streuung homogen-isotroper Strahlungsfelder an Fluktuationen des Gravitationspotentials.
G. Dautcourt.
Monatsber. Deutsch. Akad. Wiss. Berlin, Vol. 11, 483 - 495 (1969).
    The gravitational fields are treated in terms of a homogeneous and isotropic random process, the intensity variations are calculated using the general-relativistic equation of radiative transfer. A provisional application to the cosmic blackbody radiation suggests, that intensity variations in the 3°K radiation produced by gravitational fields of clusters of galaxies are of the order $\approx 10^{-7}$.

066.049 Geodätische Kurven-Parameter in Riemannschen Normalkoordinaten.   H.-J. Treder.
Monatsber. Deutsch. Akad. Wiss. Berlin, Vol. 11, 826 - 827 (1969).

066.050 Die Havassche Photoneninvariante im Gravitationsfeld.   G. Dautcourt.
Monatsber. Deutsch. Akad. Wiss. Berlin, Vol. 11, 828 (1969).

066.051 Elementare Betrachtungen zum Gravitationskollaps.
H.-J. Treder.
Monatsber. Deutsch. Akad. Wiss. Berlin, Vol. 11, 856 - 872 (1969).

066.052 Generalization of the Lorentz transformation.
J. Strnad.
Nature, Vol. 226, 137 - 138 (1970).
    A generalization of the Lorentz transformation was proposed recently by Alway in which the homogeneity of space-time is abandoned. Comparing the results of this transformation with experimental data it is shown that a constant appearing in factors like $(1 - v^2/c^2)^{-1/2(1 + \text{const.})}$ is zero

within the experimental error. Thus, measurements strongly favour the usual Lorentz transformation. This transformation has to be generalized in another way if the effect of the gravitational field should be taken into account.

066.053 "Gravars". P. Kafka.
Nature, Vol. 226, 436 - 439 (1970).

If J. Weber observes gravitational radiation, a universal background with the corresponding energy density would conflict with the observed extent of space-time, even with cosmological constant $\neq 0$. Hence, possible sources have to lie in our galaxy. Essentially "pure" gravitational effectivity hints to collisions of already collapsed objects. Former quasar-like activity in the galactic center may have left clusters of "black holes", which in their final evolution convert a considerable part of their rest-mass into pulses of gravitational radiation. Some of their properties are considered.

066.054 **Determination of the radial velocity of a relativistic star.** A. C. le Floch.
Nature, Vol. 226, 737 - 738 (1970).

The radial velocity of an atom is calculated from the relativistic Doppler equation. A graphical construction indicates the connexion between the radial velocity, the transverse velocity and the ratio ν_0/ν of the frequencies. It is shown that if $\nu_0 \leqslant \nu$, we may conclude that the atom moves nearer to the observer. If $\nu_0 > \nu$, it is impossible to state the atom moves nearer to or away from the observer. Therefore, if we consider the redshift of the spectral lines of extragalactic nebulae, this redshift, alone, cannot be a proof of the expansion of the universe.

066.055 **Interplanetary radar time delays in general relativity.**
G. C. McVittie.
Astron. Journ., Vol. 75, 287 - 296 (1970).

The time taken for a radar signal to travel from the earth to another planet and back is calculated in the Schwarzschild space-time for the gravitational field of the sun. The calculation is based entirely on general relativity, the equations of motion of the signal being given by the null geodesics of the space-time and those of the earth and the planet by the ordinary geodesics. The time delay may be measured either in coordinate time or in terrestrial proper time. After a number of approximations, the time delay is expressed in terms of the relativistic spherical polar coordinates of the earth and the planet.

066.056 **Sample-modelling of the gravitational field of the sun.** A. Z. Petrov.
Dokl. Akad. Nauk SSSR, Ser. Mat. Fiz., Vol. 190, 305 - 308 (1970). In Russian.

Modelling of the gravitational field is defined as a mapping of the Riemann space V_4, describing the field, into a certain space where the V_4 geodesics turn into curves determined by given equations. In that sense Schwarzschild's V_4 space of the gravitational field of the sun is mapped into a flat space where planetary orbits and paths of light rays are represented by curves determined by equations analogous to Lorentz's electro-dynamic equations. Modelling of only a certain set of geodesics is called sample-modelling. Einstein's theory of the gravitational field of the sun admits approximate sample-modelling into a flat space. The parameters of the model are determined by the observed deflection of rays at the solar limb. Modelling of the gravitational field is useful in certain problems. *B. Onderlička*

066.057 **On the gravitational collapse of a slightly non-spherical mass.**
A. Z. Patashinskii, A. A. Kharkov.
Dokl. Akad. Nauk SSSR, Ser. Mat. Fiz., Vol. 190, 1074 -

1077 (1970). In Russian.

A more detailed analysis of the problem studied in 1965 by Doroshkevich, Zeldovich and Novikov (Zhurn. eksp. teor. fiz., Vol. 49, 170) is presented. Axial symmetry is supposed. The deviation from spherical symmetry being supposed to be small, linear approximation is used. Perturbations in Schwarzschild's metric are developed in series of generalized spherical functions. The transformation of Lemaître is then applied to the solution of field equations. The neighbourhood of the gravitational radius appears to be an analogy of a wave zone. For low frequencies of gravitational radiation a zone of quasistatic radiation will exist at large distances. For high frequencies the "wave zone" near the gravitational radius turns into a wave zone of gravitational radiation extending to infinity. The effects of non-sphericity decay exponentially with time in the outer space. *B. Onderlička*

066.058 **Solar system Eotvos experiments.**
K. Nordtvedt, Jr.
Icarus, Vol. 12, 91 - 100 (1970).

Several types of experiments to measure celestial body m_g/m_i ratios are proposed and analyzed. It is pointed out that measurements of the m_g/m_i ratio of celestial bodies have the significance of being new and more extensive than past tests of Einstein's gravitational theory even if the experiments result in $m_g/m_i = 1$.

066.059 **On the cosmological red shift.**
S. Bellert.
Astrophys. Space Sci., Vol. 7, 211 - 230 (1970).

The paper is an extension of this author's hypothesis, presented in Astrophys. Space Sci., Vol. 3, 268 (1969), which explains the red shift in terms of a geometry of static space and stationary observers. The author introduces here the notions of 'metric with an observer', 'observed distance' and 'space of observations'; he considers the problem of the equivalence of stationary observers and discusses the relation between the hypothesis, the special theory of relativity and Maxwell equations. Attention is drawn to the agreement between the hypothesis and the experimental results discussed by Shamir and Fox, and those discussed by Kennedy and Thorndike.

066.060 **Stellar ages and an extended theory of gravitation.**
J. O'Hanlon, K.-K. Tam.
Progr. Theor. Phys. Japan, Vol. 43, 684 - 688 (1970).

Some consequences on stellar ages of the time-varying physical constants as deduced from a modified form of Dirac's hypothesis are discussed. The variation of the gravitational "constant" is shown to be consistent with a Jordan theory of gravitation.

066.061 **On the stability of ultrarelativistic stars.**
J. R. Ipser.
Astrophys. Space Sci., Vol. 7, 361 - 373 (1970).

This paper presents two new theorems on the theory of the stability of highly relativistic stars. The first theorem states that a highly relativistic, spherical star is stable if and only if its adiabatic index is greater than a certain critical value. The second theorem shows that, at high central densities, the curves of $-$ (binding energy) vs. radius, for certain hot, isentropic sequences of stellar models must exhibit damped clockwise spirals.

066.062 **A gravitational theory with variable gravitational constant.** A. M. Finkelstein.
Byull. Inst. Teoret. Astron., *Leningrad*, Vol. 12, 55 - 81 (1970). In Russian.

The class of gravitational theories based on Jordan's Lagrangian of the field has been considered. For this class the principal effects related to the motion of a zero as well as non-zero restmass have been examined. In the post-Newtonian approximation this gravitational theory is completely equivalent to the general theory of relativity.

**066.063 Prüfung der Einsteinschen Theorie mittels Raum-
sonden.   W. Kundt.**
Umschau, 70. Jahrgang, p. 448 (1970).

**066.064 The story told by the anisotropy of the relict
radiation.   A. A. Ruzmaikin.**
Priroda, No. 5.70, p. 68 - 70 (1970).   In Russian.

**066.065 Time measurement using a realizable atomic clock
in a moving frame of reference.**
L. M. Stephenson.
Journ. Phys. A, General Phys., Ser. 2, Vol. 3, 368 - 377 (1970).

Firstly, it will be shown that the usual analysis which is made to deduce the time elapse in a moving frame of reference is a legitimate analysis that may be related to a meaningful experiment; but it will be proved that the result of this analysis does not relate to the reading that would be obtained on a physically realizable atomic clock situated in a moving frame. Secondly, it will be shown that a complete, physically realizable, atomic clock does not show any discrepancy when subjected to unaccelerated motion, as compared with an identical stationary clock.

**066.066 On a certain approximate analytical solution
of the equations of the non-relativistic theory of
gravitation.   M. A. Mnatsakanyan.**
Dokl. AN Arm. SSR, Vol. 49, No. 2, p. 78 - 81 (1969). In Russian. – Abstr. in Referativ. Zhurn. 51. Astron., 4.51.219 (1970).

**066.067 On the collapse–anticollapse problem in the
multi-time theory of relativity.**
G. V. Shishkin.
Dokl. AN BSSR, Vol. 13, 980 - 983 (1969).   In Russian. Abstr. in Referativ. Zhurn. 51. Astron., 4.51.220 (1970).

066.068 The theory of the gravitational field.
N. M. Polievktov-Nikoladze.
Zhurn. ehksperim i teor. fiz., Vol. 57, 2010 - 2020 (1969). In Russian. – Abstr. in Referativ. Zhurn. 51. Astron., 4.51.221 (1970).

**066.069 Three-dimensional chronometric-invariant two-
metric formalism in general relativity.**
V. A. Barynin.
Vestn. Mosk. un-ta Fiz., astron., 1969,No. 6, p. 92 - 97 (1969).   In Russian. – Abstr. in Referativ. Zhurn. 51. Astron., 4.51.829 (1970).

**066.070 General solution of the gravitational equations
with a physical singularity.**
V. A. Belinskij, I. M. Khalatnikov.
Zhurn. ehksperim. i teor. fiz., Vol. 57, 2163 - 2175 (1969). In Russian. – Abstr. in Referativ. Zhurn. 51. Astron., 4.51.830 (1970).

**066.071 On the reception of gravitational radiation of
extraterrestrial origin.**
Ya. B. Zel'dovich, V. N. Rudenko, V. B. Braginskij.
Pis'ma v ZhEhTF, Vol. 10, 437 - 441 (1969).   In Russian. Abstr. in Referativ. Zhurn. 51. Astron., 4.51.843 (1970).

066.072 On the anomal rotation of spherical celestial bodies.
K. Stiegler.
Mem. Soc. Astron. Italiana, Nuova Serie, Vol. 41, 221 - 225 (1970).

066.073 Relation between mass and frequency of radiation.
V. Vanýsek.
Říše hvězd, Vol. 51, 65 - 66 (1970).   In Czech.

066.074 Relativistic behaviour of moving terrestrial clocks.
J. C. Hafele.
Nature, Vol. 227, 270 - 271 (1970).
The purpose of this note is to discuss the question: What would be the rate of a standard clock that is moving relative to stationary clocks on the geoid?

**066.075 Does Römer's method yield a unidirectional speed
of light?   L. Karlov.**
Australian Journ. Phys., Vol. 23, 243 - 253 (1970).
It is shown that any finite number of clocks at rest in an inertial reference system can be synchronized in an infinity of epistemologically permissible ways. Römer's method of finding the speed of light is examined from the viewpoint of a reference system in which the coordinate clocks are in a general (non-Einsteinian) state of synchronization. It is shown that the quantity c which the method yields is expressly not the one-way speed of light, but the average speed over a closed path.

**066.076 Static central-symmetrical scalar-tensor gravita-
tional field effects.**
N. A. Zaitsev, S. M. Kolesnikov, A. G. Radinov, K. P. Stanyukovich.
Astron. Tsirk., No. 562, p. 5 - 7 (1970).   In Russian.

066.077 Submillimeter astronomy.
V. I. Lapshin.
Zemlya i Vselennaya, No. 1, p. 47 - 53 (1970). In Russian.

**066.078 Redshift fluctuations arising from gravitational
waves.   W. J. Kaufmann.**
Nature, Vol. 227, 157 - 158 (1970).
If the space between a source of photons and an observer is filled with gravitational radiation, we then expect fluctuations in the appearance of the source analogous to the twinkling of starlight as seen through the Earth's atmosphere. Here I shall concern myself with fluctuations in the redshift.

066.079 Masa gravitacional y estabilidad relativista.
J. L. Sérsic.
An. Soc. Científica Argentina, Vol. 187, Ser. 1, Ciencias, No. 20, p. 111 - 117 = Obs. Astron. Cordoba, Tirada Aparte, No. 169 (1969).

066.080 Relativistic non-stationary shock waves.
I. A. Klimishin, A. F. Novak.
Tsirk. L'vov. Astron. Obs., No. 44, p. 3 - 5 (1970). In Russian.

**066.081 On the problem of clock synchronization in general
relativity.   L. Ya. Arifov, N. S. Bespalova.**
Zhurn. ehksperim. i teor. fiz., Vol. 58, 568 - 572 (1970). In Russian. – Abstr. in Referativ. Zhurn. 51. Astron., 7.51.748 (1970).

**066.082 Inertial forces in the three-dimensional invariant
two-metric formalism.   V. A. Barynin.**
Vestn. Mosk. un-ta. Fiz., astron., No. 1, p. 43 - 46 (1970). In Russian. – Abstr. in Referativ. Zhurn. 51. Astron., 7.51.749 (1970).

**066.083 A gravitional resonance detector with two degrees
of freedom.   G. Ya. Lavrent'ev.**
Pis'ma v ZhEhTF, Vol. 10, 495 - 499 (1969). In Russian. – Abstr. in Referativ. Zhurn. 51. Astron., 7.51.755 (1970).

066.084 **Pulsar yields new test of general relativity.**
New Scientist, Vol. 44, (No. 671), 113 (1969).
A report on tests of general relativity based on the ability to accurately time a pulsar's period of oscillation. — *JBW*

066.085 **Kerr metric black holes.**
J. M. Bardeen.
Nature, Vol. 226, 64 - 65 (1970).
The spherically symmetric Schwarzschild metric used to describe the space−time in the vicinity of collapsed objects is only valid if this object has a zero angular momentum. As this is not the case the metric describing a collapsed object with angular momentum is the Kerr metric, a stationary, axially symmetric solution to the vacuum Einstein equations with two free parameters.

066.086 **New exact static solutions of Einsteins field equations.** H. Heintzmann.
Zeitschr. Physik, Vol. 228, 489 - 493 (1969).
A method is described for deriving new solutions for an ideal fluid from old and give some new solutions which may be of interest in astrophysics.

066.087 **Gravitional radiation and the 2-body problem.**
M. Signore.
Ann. Inst. Poincaré, Ser. A, Vol. 11, 81 - 130 (1969). In French.
In this paper the author investigates the energy change of a 2-body systems gravitional radiation using the fast approxi - mation method and a Lorentz-invariant theory of gravitation.

066.088 **Planetary perihelion precession with velocity-dependent gravitational mass.**
R. Engelke, C. Chandler.
American Journ. Phys., Vol. 38, (No. 1), 90 - 93 (1970).
Planetary perihelion precession is theoretically examined within the framework of special relativity. The Eötvös experiment is used to motivate an assumption that the inertial and gravitational masses have the same velocity dependence. The resultant orbit equation is then approximated by one that is soluble in terms of trigonometric functions, and the accuracy of the approximation is examined.

066.089 **Relativistic stability.** J. L. Sérsic.
Asoc. Argentina Astron. Bol., No. 14 (see 012.020), p. 76 (1968). In Spanish. — Abstract.

Essential Relativity, Special, General, and Cosmolo - gical. See Abstr. 003.100.

Hyperonic equation of state. See Abstr. 065.003.

Neutron star models based on an improved equation of state. See Abstr. 065.004.

The stability of the equilibrium of supermassive and superdense stars. See Abstr. 065.005

New source of intense magnetic fields in neutron stars. See Abstr. 065.010.

Some general relativistic inequalities for a star in hydrostatic equilibrium − II. See Abstr. 065.025.

Models for neutron-core stars based on realistic nuclear-matter calculations. See Abstr. 065.048.

Relativistic isentropic stellar models. See Abstr. 065.073.

Atomic-beam study of the solar 7699Å potassium line and the solar gravitational red-shift. See Abstr. 071.041.

Expansion of the Milky Way and the interpretation of Weber's gravitation signals. See Abstr. 155.066.

Cosmology and relativistic astrophysics. See Abstr. 162.007.

Can discrete sources produce the cosmic microwave radiation? See Abstr. 162.018.

Possible nonequilibrium nature of cosmic-background radiation. See Abstr. 162.019.

On some solutions of Einstein's equations from the point of view of Petrow's classification. See Abstr. 162.026.

Machian interpretation of general relativity. See Abstr. 162.034.

Sun

071 Solar Photosphere, Spectrum

071.001 Abundance of iron in the solar photosphere.
J. E. Ross.
Nature, Vol. 225, 610 - 611 (1970).

Recent measurements of iron f-values indicate that the iron abundance derived from neutral iron lines formed in the solar photosphere based on older f-values should be increased by a factor of ~5. This increase is partially offset if the damping constant is taken into account.

071.002 Damping constants for infrared Fraunhofer lines.
C. de Jager, L. Neven.
Solar Physics, Vol. 11, 3 - 10 (1970).

Empirical values of solar damping constants and their variations with optical depth were derived according to a method developed earlier by the authors. The damping constants refer to six infrared multiplets. The average optical depths range from $\tau_0 = 0.5$ to 2.2. Corresponding theoretical damping constants were computed, mainly on the basis of Van der Waals damping, and with the help of detailed computations of the mean square radii of the atomic levels by Van Rensbergen. The empirical values are systematically larger than the theoretical ones, with factors ranging between 1.8 and 4.9. Some speculations about the source of this discrepancy are given.

071.003 Some oscillator strengths in the spectra of CI, OI, SiI, CaII and SrII. W. van Rensbergen.
Solar Physics, Vol. 11, 11 - 16 (1970).

A number of wave functions for bound electrons have been calculated in scaled Thomas-Fermi ion potentials in order to evaluate some oscillator strengths and mean square radii. The results differ only slightly from those obtained by the well-known method of Bates and Damgaard. After that, the values of the oscillator strengths for the transition $4s \rightarrow 4p$ in SiI have been evaluated in intermediate coupling, showing large differences with the LS-coupling results.

071.004 The isotope ratio of europium in the solar atmosphere. Ö. Hauge.
Solar Physics, Vol. 11, 17 - 21 (1970).

Europium has two stable isotopes, Eu 151 and Eu 153. The high isotope shift and the different hyperfine splitting of the energy levels of the two isotopes make it possible to study the isotope ratio by an analysis of the spectral line profiles. From five spectral lines the solar isotope ratio is found to be equal to the terrestrial ratio within an error limit of about 10%.

071.005 Depth-dependent line blanketing by neutral and ionized metals in a homogeneous model solar photosphere. T. E. Margrave, Jr.
Solar Physics, Vol. 11, 22 - 25 (1970).

Line blanketing due to both the neutral and ionized metals is treated here by the modified picket-fence method and applied to the case of a homogeneous model solar photosphere in strict radiative equilibrium. Improvement in fitting the observed photospheric radiation field is noted over a model in which the line absorption coefficient is constant with depth.

071.006 Short period oscillations and Doppler velocity gradients. J. W. Harvey.
Solar Physics, Vol. 11, 26 - 28 = Contr. Kitt Peak National Obs. No. 522 (1970). – Research note.

071.007 Recherches sur le rayonnement continu du soleil entre 1.5 et 500 microns. P. Léna.
Thesis, Doct. Sci. Phys., Paris. [Centre Document. C. N. R. S., No. 3211], 208 pp. (1969).

071.008 Can the ion H_3^+ account for missing opacity in the solar ultraviolet? J. L. Linsky.
Solar Physics, Vol. 11, 198 - 207 (1970).

Limb darkening and specific intensity data imply more continuous opacity in the solar photosphere between 2000 Å and 3500 Å than has been predicted theoretically. The temperature dependence and wavelength dependence of this missing opacity are in qualitative agreement with those deduced for the ion H_3^+, but it is unlikely that H_3^+ is sufficiently abundant to account for this opacity.

071.009 Observations of the infrared triplet of singly ionized calcium.
J. L. Linsky, R. G. Teske, C. W. Wilkinson.
Solar Physics, Vol. 11, 374 - 383 (1970).

Observations are presented of the CaII infrared triplet at three positions on the solar disk to make possible direct analyses of the lines and comparisons with theoretical computations. The source functions for the two strongest lines (8542 Å and 8662 Å) are equal at those heights corresponding to the wings of the lines ($|\Delta\lambda| > 0.4$ Å) but not to those of the cores.

071.010 New $C^{13}N^{14}$ search regions in the solar spectrum.
T. D. Fay, A. A. Wyller.
Solar Physics, Vol. 11, 384 - 387 (1970).

Ten new rotational line positions, due to the (0,0) red $C^{13}N^{14}$ band, are calculated to fall squarely within continuum regions 1 - 2 Å wide in the near infrared solar spectrum, $\lambda\lambda 10990 - 11630$. Precision observations of the isotopic line strengths in this spectral region, albeit difficult, should resolve the present ambiguity in the blue-violet observations of whether or not the solar C^{12}/C^{13} ratio is equal to or larger than the terrestrial ratio.

071.011 Empirical damping, and the solar abundance of iron. C. Cowley.
Astrophys. Letters, Vol. 5, 149 - 150 (1970).

A solar iron-to-hydrogen ratio, as well as a new system of oscillator strengths for Fe I lines, has been determined at Kiel. The derived values are not consistent with laboratory measurements of line broadening by neutral atoms. If the gf-values and the solar iron-to-hydrogen ratio determined at Kiel are correct, the simple impact theory of van der Waals broadening is also a fair approximation. For Fe I lines we show that this agreement must be accidental.

071.012 On the center-to-limb variation of some Fraunhofer lines in the solar photosphere.

A. Donati Falchi.
Mem. Soc. Astron. Italiana, Nuova Serie, Vol. 41, 69 - 76 (1970).

The equivalent widths of 9 lines, measured at different positions on the solar disc, are compared with those given by Elste, Holweger and B.C.A. models. Holweger model seems to fit better the experimental data but the microturbulence velocity proposed by him seems to have to be diminished of 0.7 km/sec.

071.013   Fraunhofer lines without Zeeman splitting.
G. Sistla, J. W. Harvey.
Solar Physics, Vol. 12, 66 - 68 = Contr. Kitt Peak National Obs. No. 548 (1970).

This note presents the results of a fairly thorough search for magnetically unaffected lines in the solar spectrum.

071.014   Line formation in a magnetic field and the interpretation of magnetographic measurements.
II. The influence of different atmosphere models and of a magnetic field gradient.   J. Staude.
Solar Physics, Vol. 12, 84 - 94 (1970).

The solution of the equations of transfer for the Stokes parameters in a magnetic field given by Unno (1956) has been generalized for a non-linear source function. Model calculations have been carried out for empiric atmosphere models of the photosphere and spots both for homogeneous magnetic fields and for a field strength gradient. Their influence on line contours and magnetographic calibration curves had been studied for the line FeI $\lambda 5250.2$ Å.

071.015   The Herschel effect and solar photography.
R. Jayanthan.
Solar Physics, Vol. 12, 163 - 166 (1970).

A method of photographing the solar image in the UV part of the solar spectrum is described. The resulting images show the bright photospheric network of Sheeley and Chapman; these have been recorded at the small central distance of 20°. During a flare of importance 1b near a spot group, no detectable changes in the spots, their relative positions or the bright photospheric network could be observed.

071.016   Some results of solar cinematography in integral light.   Yu. I. Il'yasov.
Uzv. AN Turkm. SSR. Ser. fiz. tekhn., khim. i geol. n., 1969 No. 5, p. 73 - 78.   In Russian. – Abstr. in Referativ. Zhurn. 51. Astron., 3.51.412 (1970).

071.017   Polarization in dashes as determined from the D_3 helium line.
G. F. Vjalshin, A. A. Shpitalnaja.
Solnechnye Dannye 1969 Byull., No. 12, p. 74 - 79 (1970). In Russian.

The results of observing the dashes in the D_3 helium line using a Wollastone prism are represented. The value of linear polarization P_s and the difference of radial velocities for mutually orthogonal directions of polarization ΔV_0 are given. The variations of ΔV_0 and P_s point to the existence of a fine structure in dashes.

071.018   Observations of solar granulation at the Shemakha Astrophysical Observatory. III.
M. B. Kerimbekov.
Solnechnye Dannye 1969 Byull., No. 12, p. 88 - 90 (1970). In Russian.

The results of observations of solar granulation by the method of high-speed filming are given. The problem of the quality of images obtained is discussed.

071.019   On the mechanism of the formation of active regions.   Yu. B. Ponomarenko.
Astron. Zhurn. Akad. Nauk SSSR, Vol. 47, 98 - 102 (1970). In Russian. English translation in Soviet Astron. AJ, Vol. 14, No. 1.

In the mechanism under discussion the magnetic field of the active region is formed when the general toroidal solar field is carried out to the surface. It is supposed that the carrying out of the toroidal field is realized by the motion of giant granules—convective cells with dimension of ≈ 180000 km and life-time of the order of one month. The velocity of motion near the surface is ≈ 80 m/sec, gas ascends at the centre of giant granules and descends near the edges. The asymmetry of spots and the inclination of the axis of the active region to the line East—West are connected with the influence of the Coriolis force on the motion in giant granules.

071.020   A mechanism of the formation of lines of the CN molecule in the solar photosphere.
A. I. Khlystov.
Astron. Zhurn. Akad. Nauk SSSR, Vol. 47, 103 - 110 (1970). In Russian. English translation in Soviet Astron. AJ, Vol. 14, No. 1.

It is established from the consideration of elementary processes that the mechanism of formation of lines of a violet system of CN molecule bands in the solar photosphere is a completely incoherent scattering in presence of the effect of intercommunication of electron and rotational levels by means of the emission and intercommunication effect of oscillating levels through the emission and oscillating transitions in excited electron state owing to collisions with hydrogen atoms.

071.021   On the consideration of the influence of atmospheric disturbances on the investigation of the granulation.
M. B. Kerimbekov.
Soobshch. Shemakhinsk. Astrofiz. Obs., vyp. (No.) 4, p. 66 - 70 (1969).   In Russian.

071.022   On the contours of weak Fraunhofer lines.
D. M. Kuli-Zade.
Soobshch. Shemakhinsk. Astrofiz. Obs., vyp. (No.) 4, p. 71 - 83 (1969).   In Russian.

071.023   New classification for Gd II solar lines.
N. Spector.
Astrophys. Journ., Vol. 159, 1091 - 1092 (1970).

The recent discovery of a new system of energy levels of singly ionized gadolinium obtained by adding an outer electron to the $4f^8$ core permitted completion of the classification of all Gd II lines observed in the solar spectrum.

071.024   10830 Å line profile on the solar disc.
Nguen Ngan.
Astron. Zhurn. Akad. Nauk SSSR, Vol. 47, 351 - 356 (1970). In Russian. – English translation in Soviet Astron., AJ, Vol. 14, No. 2.

Results of photoelectric observations of the helium line 10830 Å in absorption on the solar disc are given. The line profile has a depression in the blue wing. Apparently, this depression can be explained by the presence of a weak component of the triplet 10829 Å. Various interpretations are considered.

071.025   The flow pattern within solar granules.
J. M. Beckers, R. A. Morrison.
Bull. American Astron. Soc., Vol. 2, 182 (1970). – Abstr. AAS.

071.026   Limb darkening and solar rotation – II. Observed darkening in 1966.   H. H. Plaskett.
Monthly Notices, Roy. Astron. Soc., Vol. 148, 149 - 171 (1970).

Spectrophotometric measures of surface brightness at 6263 Å for diametrically opposite limbs of the sun confirm that the poles are at least five per cent brighter than the equatorial limb. Solutions by Radau integration for the source function and computation of photospheric models give temperature as a function of linear height for a number of latitudes. The currents due to the resulting meridional temperature-gradient occur in a photospheric Ekman-layer. These thermal currents are stable and are maintained against viscous loss by heat absorbed in the photosphere.

071.027 Depression of the solar continuous spectrum by line absorption and the Balmer continuum.
H. Holweger.
Astron. Astrophys., Vol. 4, 11 - 17 (1970).

In the following sections we try to determine the line absorption in the solar spectrum at a number of wavelengths in order to get agreement between the observed radiation flux and that calculated from a solar model. It is shown that line absorption can account for the difference between the observed flux of solar radiation and model predictions in the wavelength range 3300 to 5000 Å. Shortward of 4000 Å several faint lines may lay within 0.1 Å, thus producing a quasi-continuous absorption of 10% or more.

071.028 Continuum infrared radiation of the solar photosphere. P. Léna.
Astron. Astrophys., Vol. 4, 202 - 219 = Contr. Kitt Peak National Obs. No. 456 (1970). In French.

The center to limb variation of the solar brightness has been measured on the sun's disc at wavelengths of 5, 10 and 20 microns from the center of the disc to 2" from the limb ($\mu > 0.06$). Data collection and reduction have been considerably improved over the previous results and serious gains in resolving power and in signal to noise ratio have been achieved. Effects of atmospheric transmission fluctuations are minimized. Source functions at 5, 10 and 18 - 22 microns are computed from the limb darkening curves, using Delache's method: approximation up to $n = 3$ is computed. A homogeneous, LTE model in hydrostatic equilibrium is built which completes and somewhat improves the Bilderberg Continuum Atmosphere in the transition zone. Pressure, electronic pressure, densities, altitudes and $\tau_{0.5}$ are computed in the two extreme cases of a plateau at $4500°$ K and a temperature drop to $4200°$ K.

071.029 Velocity effects on the profiles of Hα and two FeI lines. R. G. Athay.
Solar Physics, Vol. 12, 175 - 185 (1970).

Profiles are computed for Hα and two FeI lines for a differentially moving atmosphere. The results show that the profiles are asymmetric and that velocity measurements made in the Doppler cores will often lead to erroneous results when the velocity gradient is significant in the region of the atmosphere where the core forms.

071.030 Magnetic fields in quiescent solar prominences. II. Photospheric sources. D. M. Rust.
Astrophys. Journ., Vol. 160, 315 - 324 (1970).

Magnetograph observations of the photospheric magnetic fields 100000 km on either side of quiescent filaments have been made by using the 5250 Å line of iron. A comparison of observations with the Kippenhahn and Schlüter magnetohydrostatic model of prominences tends to confirm the correctness of the model. Sections of observed filaments are unstable and appear to be subject to disappearance when a nascent bipolar feature appears in the neutral band.

071.031 Erratum: Solar C II resonance-line profiles.
[Astrophys. Journ. *(Letters)*, Vol. 155, L115 - L116 (1969)]. R. A. Berger, E. C. Bruner, Jr.

Astrophys. Journ. *(Letters)*, Vol. 160, L61 (1970).

071.032 The solar H and K lines. J. L. Linsky, E. H. Avrett.
Publ. Astron. Soc. Pacific, Vol. 82, 169 - 248 (1970). – Review article.

The main features of the solar and stellar observations and their interpretation in terms of the development of line formation theory are outlined. We construct a one-component model of the chromosphere most nearly consistent with the calcium lines and other observed data. The emphasis in this paper is on the sun. An ultimate aim is to understand line formation in stellar chromospheres, for which an understanding of the solar case is a necessary prerequisite.

071.033 An attempt to measure the solar limb darkening using the lunar profile during the eclipse of May 20, 1966. F. Magnant.
Astron. Astrophys., Vol. 5, 382 - 390 (1970). In French.

An experiment was performed at the solar eclipse of May 20, 1966 in order to measure the integrated brightness distribution of the solar limb in the continuous spectrum, on the violet side of the Balmer discontinuity (3650 Å) and in the red at 6045 Å. We describe and discuss the observing methods which are new in several respects, and we analyse the available data.

071.034 Study of the solar continuum in the intermediate infra-red spectral range 3.5 - 24.4 μ.
S. Koutchmy, R. Peyturaux.
Astron. Astrophys., Vol. 5, 470 - 487 (1970). In French.

Several series of measurements of the solar continuum intensity of the centre of the disc have been made in the Pyrénées Mountains at an altitude of 1600 m in excellent meteorological conditions. The chosen spectral resolution was sufficient to make measurements in the water vapor atmospherical windows of the studied wavelength, the signal to noise ratio being kept > 100. A new method for checking the variations of the vertical water vapor equivalent during the observations is described and its exploitation is discussed.

071.035 Variations in line profiles from photosphere to chromosphere. J. Houtgast, O. Namba, R. J. Rutten, T. de Graauw.
Nature, Vol. 226, 1144 - 1145 (1970).

071.036 Photospheric magnetic fields on March 7, 1970.
W. Livingston, J. Harvey, C. Slaughter.
Nature, Vol. 226, 1146 - 1148 (1970).

071.037 The SiH $A^2\Delta - X^2\Pi$ (o, o) band in the Fraunhofer spectrum. D. L. Lambert, E. A. Mallia.
Monthly Notices, Roy. Astron. Soc., Vol. 148, 313 - 324 (1970).

Low noise photoelectric scans of the solar disc and sunspot spectra are examined for lines attributable to the (o, o) band of the SiH system $A^2\Delta - X^2\Pi$. Satisfactory identifications are established. A model atmosphere calculation provides a result for the band oscillator strength: $f_{oo} = 0.0045$. Isotope shifts are calculated for the less abundant species, $Si^{29}H$ and $Si^{30}H$. It is shown that isotopic abundances are probably normal.

071.038 Fraunhofer line statistics.
C. W. Allen.
Monthly Notices, Roy. Astron. Soc., Vol. 148, 435 - 453 (1970).

In two earlier papers statistical methods have been presented for deriving the densities of spectrum lines per frequency and intensity range. In the present paper the statistics are applied for the first time to a well measured complete spectrum, the sun's Fraunhofer spectrum.

071.039 **Balloon observations in the submillimeter region: An absolute measurement of the brightness of the sun and the transparency of the stratosphere above 25 km.** J. Gay.
Astron. Astrophys., Vol. 6, 327 - 348 (1970).

Several experiments have been successfully launched by a group involved in infrared space studies at the Meudon Observatory. The present description deals with solar flights, aimed at absolute spectrophotometry of the sun's radiation in the wavelength range from 50 to 200μ. The purpose of the measurements is to obtain a better determination of the solar brightness temperature in the transition zone between the photosphere and the chromosphere.

071.040 **Line formation in magnetic fields. Comments on the role of atomic level polarization.** F. K. Lamb.
Solar Physics, Vol. 12, 186 - 201 (1970).

The question of magnetograph errors caused by atomic level polarization effects is re-examined using the results of a density matrix treatment of radiative processes. The basic conclusions are as follows: (1) Longitudinal magnetograph measurements using optically thin lines are not influenced by level polarization. (2) Present theories of line formation in magnetic fields do not include the Hanle effect, but this omission is generally unimportant for lines formed in the photosphere and lower chromosphere due to rapid collisional depolarization of atomic levels. (3) For the same reason, other level polarization effects are probably too small to cause significant erros in magnetograph measurements of all but the very weakest magnetic fields, when photospheric and lower chromospheric lines are used. (4) By contrast, level polarization is important in most prominences.

071.041 **Atomic-beam study of the solar 7699 Å potassium line and the solar gravitational red-shift.**
J. L. Snider.
Solar Physics, Vol. 12, 352 - 369 (1970).

The shape and red-shift of the solar potassium resonance line at 7699 Å have been studied by an atomic-beam resonant scattering technique. The line profile was obtained by measuring the scattered light intensity as a function of magnetic field. The mean red-shift of the 40 profiles which showed small or no asymmetry was $(\Delta\lambda)_{mean} = 10 \pm 1 m\text{Å} = (0.61 \pm 0.06) (\Delta\lambda)_{grav}$, where $(\Delta\lambda)_{grav}$ is the gravitational red-shift predicted on the basis of the principle of equivalence. This result, together with those of other recent experiments, is consistent with the previously observed correlation between the red-shift of a solar line and its strength. Various checks of the experimental method are discussed, including preliminary measurements on the solar sodium D1 line.

071.042 **Solar CII resonance line profiles.**
R. A. Berger, E. C. Bruner, Jr., R. J. Stevens.
Solar Physics, Vol. 12, 370 - 378 (1970).

The profiles of the solar CII resonance doublet $\lambda 1334.6$ Å are presented. Possible broadening mechanisms are discussed.

071.043 **On the problem of the temperature constancy in the molecular CO layer of the solar atmosphere.**
G. F. Sitnik, V. V. Polonsky.
Astron. Zhurn. Akad. Nauk SSSR, Vol. 47, 516 - 519 (1970). In Russian. English translation in Soviet Astron. AJ, Vol. 14, No. 3.

On the basis of registrograms of the 1st overtone spectrum of the CO molecule in the region λ 23000 – 24000 Å, equivalent widths of spectral lines are determined. It is shown that for a thin molecular layer of CO in the solar photosphere the temperature is a constant within the limits of the precision of the determination of the ratio of the equivalent widths.

071.044 **Interpretation of infrared spectrophotometric solar measurements.** J. Gay.
Astron. Astrophys., Vol. 7, 24 - 34 (1970).   In French.

We try to explain the rather conspicuous discrepancies between our measurements of the solar brightness temperature in the far infrared spectral region and the theoretical values which can be computed from the most recent models.

071.045 **Study of the turbulence in the solar photosphere from the Ti I-line profiles.**
E. A. Gurtovenko, V. I. Troyan.
Astrometriya i Astrofiz., Kiev, No. 8, (see 003.024), p. 85 - 91 (1969). In Russian.

The study of the turbulence in the solar photosphere was carried out based on the observations of the four Ti I line profiles at different centre-limb distances.

071.046 **Photographic investigation of solar photosphere.** M. Neubauer.
Říše hvězd, Vol. 51, 125 - 129 (1970).   In Czech.

071.047 **Radiation and structure of the solar atmosphere.** R. G. Athay, A. K. Pierce, M. Rigutti.
Trans. IAU, Vol. 14A, (see 003.028), 111 - 124 (1970). Report of Commission 12.

071.048 **A study of the influence of photographical factors on the physical parameters of the solar atmosphere determined by the curve of growth method.** G. D. Poljakova.
Solnechnye Dannye 1970 Byull., No. 1, p. 106 - 111 (1970). In Russian.

The influence of photographical factors on the values of equivalent widths of Fraunhofer lines was investigated by the method of least squares. The method of the curves of growth was used to determine some physical parameters of the solar photosphere.

071.049 **Convection and asymmetry of weak lines.**
B. T. Babij, Yu. V. Fridel'.
Tsirk. L'vov. Astron. Obs., No. 44, p. 33 - 35 (1970). In Russian.

071.050 **Line profile analysis of carbon molecules in the sun.** N. Raghavan.
Kodaikanal Obs. Bull., Ser. A, No. 186, p. A121 - A148 (1968).

Thirteen molecular lines of CN, C_2 and CH have been photoelectrically observed at six disc positions each. Detailed profile calculations have been made for six selected lines.

Experimental oscillator strengths of Fe II lines and the solar iron abundance.   See Abstr. 022.038.

On the progress of balloon heliophysics in the Soviet Union.   See Abstr. 032.019.

On the frequency dependence of the line source function.   See Abstr. 061.008.

Determinación de la presencia de compuestos quimicos en atmosferas estelares y planetarias, cometas y espacio interestelar.   See Abstr. 064.012.

Isotopes of magnesium in the sun.
See Abstr. 072.002.

On polarimetry in solar active regions. II. Selection of lines; interpretation of polarimetric data.
See Abstr. 072.028.

072 Sunspots, Faculae, Solar Activity

072.001 **Why are sunspots dark and faculae bright?**
  R. H. Dicke.
Astrophys. Journ., Vol. 159, 25 - 38 (1970).

It is shown that the darkening or brightening of a disturbed spot on the sun's surface is uniquely determined by the abnormal stress distribution in the "seen layers" of the spot. The magnetic and velocity fields are the sources of the anomalous stress. A crude sunspot model based on the general theory is shown to give results in satisfactory agreement with the observations.

072.002 **Isotopes of magnesium in the sun.**
  D. Branch.
Astrophys. Journ.,Vol. 159, 39 - 49 (1970).

Coarse analysis indicates that in the sunspot region the isotopic ratios of magnesium, ^{24}Mg:^{25}Mg:^{26}Mg, are nearer to 60:20:20 than to the terrestrial and meteoritic values of 79:10:11. Because of various uncertainties involving sunspot models, etc., the conclusion is limited to the statement that the heavy isotopes of magnesium appear to be enhanced in the sun relative to the earth.

072.003 **The occurrence of high-energy releases during flares in localized areas of an active center and time-associated changes in spot umbrae.**
S. M. P. McKenna-Lawlor.
Astrophys. Journ., Vol. 159, 51 - 59 (1970).

Two localized areas in active center HAO-59Q were associated with the commencement of significant energy releases. The positions of persistent flare-sensitive regions, calcium plages, filaments, and spot umbrae in HAO-59Q have been compared with the daily magnetograms of this center of activity. The parts of the chromosphere associated with high-energy flares were directly over areas of the photosphere which contained transient umbrae. Flares with high energy occurred in intervals within which such umbrae disappeared or diminished in area.

072.004 **Dissymétrie Est-Ouest dans la naissance des centres d'activité solaire.** M. Trellis.
Comptes Rendus Acad. Sci. Paris, Sér. B, Vol. 270, 122 - 124 (1970).

La dissymétrie Est–Ouest dans la naissance des centres d'activité solaire signalée déjà par plusieurs auteurs est indépendante de la latitude à laquelle se forment ces centres d'activité et est restée invariable au cours des huit derniers cycles solaires.

072.005 **On the magnetic configuration of sunspot groups which produce solar proton flares.**
K. Sakurai.
Planet. Space Sci., Vol. 18, 33 - 40 (1970).

The magnetic configuration of sunspot groups where solar proton flares take place is examined by using the data on the center-to-limb change of characteristics of type IV radio bursts and Hα-brightening areas on the solar disk.

072.006 **Penumbral magnetic field strengths.**
  E. A. Mallia.
Solar Physics, Vol. 11, 31 - 32 (1970). – Research note.

072.007 **The diatomic molecules BH, BN, and BO in sunspots and the solar abundance of boron.**
O. Engvold.
Solar Physics, Vol. 11, 183 - 197 = Contr. Kitt Peak National Obs. No. 488 (1970).

Absorption band spectra of BH and BO have been sear-

ched for and not found in spectra of sunspots. Electronic oscillator strengths are available only for the $A^1\Pi - X^1\Sigma^+$ system of the BH molecule. The absence of the (0,0) band of BH at 4332 Å reflects a solar abundance of boron log $A_B < 2.5$. The band spectra of BN are several orders of magnitude weaker in sunspots than those of BH and BO.

072.008 **Quelques effets de l'interaction des centres actifs solaires.** M.-J. Martres.
Solar Physics, Vol. 11, 258 - 262 (1970).

Following the study of anomalies in spot groups correlated to the existence of an older active center (A.C.) in the same region, we compared the evolution of 50 A.C. of simple magnetic configuration β or α. We find that a group situated to the west of an older one evolves to the type β_f and shows an eastward motion of the following spot greater than in the case of an isolated group. Symmetrically, the evolution to β_p of a group situated to the east of the original formation is associated with a westward motion of the leading spot greater than in the case of an undisturbed group.

072.009 **Messung von Magnetfeldern auf der Sonne.**
  E. Wiehr.
SuW, Vol. 9, 65 - 68 (1970).

072.010 **The assessment of 'white light' observations of the sun.** E. Budding.
Journ. British Astron. Ass., Vol. 80, 125 - 129 (1970).

072.011 **Magnetic fields in solar active regions.**
  J. H. Reid.
Journ. British Astron. Ass., Vol. 80, 200 - 207 (1970).

Solar flares do always occur in regions where there are measurable magnetic fields, even though there may be no sunspots visible. This paper briefly surveys the types of magnetic measures made and the results that have been deduced from them.

072.012 **On the dependence of sunspot minimum intensity on area.** M. Rossbach, E. H. Schröter.
Solar Physics, Vol. 12, 95 - 100 (1970).

Highly resolved photographs of 25 sunspots and pores of different areas in 2 continuous wavelengths, obtained in summer 1966 at the Sacramento Peak Observatory, have been used to reinvestigate the dependence of sunspot minimum intensity on area. We find a very small, if any, dependence of I_{min} on sunspot area in contradiction to older measurements but in agreement with Sitnik's and Zwaan's statements.

072.013 **The possible existence of HOH lines in the sunspot spectrum.** E. A. Mallia, D. E. Blackwell.
Solar Physics, Vol. 12, 101 - 103 (1970).

Some sunspot umbrae spectra have been studied in the region of 5881 Å to prove the existence of HOH lines. The suggestion of the presence of these lines has not been demonstrated conclusively and their detection is a difficult problem.

072.014 **Electric current in a sunspot.**
  R. Jayanthan.
Solar Physics, Vol. 12, 104 - 105 (1970).

An observation of a total electric current of magnitude 3.8×10^8 amperes per kilometer height in a δ-type sunspot is described.

072.015 **On a possible connection between solar activity**

and quasi-periodical variability of the gravitational constant. N. Kalitzin.
Izv. fiz. in-t s ANEB, Vol. 18, 115 - 118 (1969). In Russian.
Abstr. in Referativ. Zhurn. 51. Astron., 3.51.454 (1970).

072.016 **On the relation betweeen the length of the rising branch and the height of the maximum of the 11-year solar cycle.** Y. I. Vitinsky.
Solnechnye Dannye 1969 Byull., No. 10, p. 107 - 110 (1970). In Russian.

On the basis of the Zürich data and the Schove series it is shown that a long-range variation exists in the correlation between the height of the maximum and the length of the rising branch of the 11 year cycle with a period of the order of 600 - 700 years. Some considerations are given on the application of the results obtained to forecasts of solar cycles.

072.017 **Long-range forecasts of the total area of sunspots.** Y. I. Vitinsky, V. S. Shkljarnik.
Solnechnye Dannye 1969 Byull., No. 12, p. 90 - 94 (1970). In Russian.

The regression method has been used for forecasts of the mean annual values of the total area of sunspots for the next year within the current 11-year cycle. The connection of the length of the rising branch of the cycle with its maximum mean annual total area of sunspots is considered.

072.018 **The structure of the magnetic field of a spot with a photospheric bridge at two levels of the solar atmosphere.** H. I. Abdussamatov.
Astron. Zhurn. Akad. Nauk SSSR, Vol. 47, 82 - 90 (1970). In Russian. English translation in Soviet Astron. AJ, Vol. 14, No. 1.

From simultaneous photographic observations in two lines (Hα and Fe λ 6302.499 Å) the space structure of the magnetic field of a follower spot with photospheric bridge on the solar disk is investigated. It is concluded that the intensity of the magnetic field in the observed light bridge of type II is just the same as that of faculae.

072.019 **Water vapour in sunspots.**
E. A. Mallia, D. E. Blackwell, A. D. Petford.
Nature, Vol. 226, 735 - 737 (1970).

The purpose of this note is to present evidence for the existence of water vapour in sunspots.

072.020 **A contribution to the form of butterfly diagrams in connection with the differential rotation of the sun.** M. Kopecký.
Bull. Astron. Inst. Czechoslovakia, Vol. 21, 73 - 76 (1970).

There is shown in this paper that the overall shape of "butterfly diagrams" of sunspots can be closely connected with the differential rotation of the sun. This fact could evidence in favour of those theories of the periodicity of solar activity that proceed from the differential rotation of the sun as one of the chief factors of the spot-formation process.

072.021 **Physics of the sunspots. Part II.**
J. Jakimiec.
Postępy Astron., Vol. 18, 3 - 18 (1970). In Polish.

In this article basic results of the observations of sunspot magnetic fields are confronted. Moreover main results of the theory of convection in the magnetic fields are quoted and their application to sunspots is discussed.

072.022 **Comparison of smoothing of observed monthly Wolf numbers using sliding means and Whittaker's operator.** L. V. Zhukov, Y. S. Muzalevsky.
Astron. Zhurn. Akad. Nauk SSSR, Vol. 47, 357 - 374 (1970).

In Russian. — English translation in Soviet Astron., AJ, Vol. 14, No. 2.

For the correct extrapolation of a series of relative sunspot numbers it is proposed to estimate the mathematical expectation of the Zürich series of Wolf numbers by Whittaker's operator. Some statistical results are derived. A comparison of the frequency characteristics obtained from sliding means and Whittaker's operator is made.

072.023 **The ionization of atoms in the umbra by the radiation field of the penumbra.**
W. Fullerton, G. Elste.
Bull. American Astron. Soc., Vol. 2, 193 (1970). — Abstr. AAS.

072.024 **Rapid mapping of solar magnetic fields.**
R. Michard, J. Rayrole.
Astron. Astrophys., Vol. 4, 36 - 39 (1970). In French.

We describe a spectroheliographic technique giving images of Active Regions, where upon the longitudinal field appears as a system of interference fringes. The lower level of detection is about 20 Gauss, the resolution 3″, the angular field registered in 5 minutes is 10′ × 12′. This method is useful when one needs immediately a qualitative information on the magnetic structure of A.R., for instance in short term forecasting of flare-activity.

072.025 **Magnetic field configurations in active regions as derived from perspective effects.** E. Ribes.
Astron. Astrophys., Vol. 4, 70 - 74 (1970).

We use the perspective effect to determine the vector field of an active region from a sequence of longitudinal component observations. For this purpose, we study the perspective effect on a magnetic configuration calculated by the current-free approximation. We compare the longitudinal component observed every day and the corresponding calculated one. The reasonably good agreement between them and the stability of the spots in white light lead to the conclusion that the magnetic changes are chiefly due to the center-to-limb effect.

072.026 **Forecast or presage?** W. Szymański.
Urania Kraków, Vol. 41, 184 - 187 (1970).
In Polish.

072.027 **The latitude distribution of magnetically classified sunspot groups and their associated solar flares.**
G. R. Greatrix.
Astron. Astrophys., Vol. 5, 171 - 176 (1970).

A statistical analysis is carried out of data relating to more than 6500 solar flares and their associated spot groups for the maxima of the 18th and 19th solar cycles. There is strong evidence to support the hypothesis that the latitude distributions of solar flares are skewed relative to their associated spot group latitude distributions and that the direction of skew is dependent on the magnetic classification of the spot groups. For a given magnetic classification the direction of skew appears to remain the same for successive cycles. The degree of relative skew is independent of flare intensity.

072.028 **On polarimetry in solar active regions. II. Selection of lines; interpretation of polarimetric data.**
E. Wiehr.
Solar Physics, Vol. 11, 399 - 408 (1970).

It is shown that the line Fe λ 5250.2, generally used for solar Zeeman polarimetry, should not be used for this purpose because of its strong temperature sensitivity (multipl. No. 1). Solar magnetic fields are correlated with areas which have temperatures different from that of the undisturbed photosphere (faculae, spots). The photosphere, however, is used for the calibration procedure. This leads to highly erroneous results. Therefore, the line Fe λ 6302.5 is suggested for solar polarimetry because it is independent on temperature variations. Thus, a Milne-Eddington model represents a sufficient approximation and avoids complicated theories of "line formation in magnetic fields" as well as accurate models for the different parts of active regions (faculae, penumbral filaments, umbrae with dots). Outside the disc center the ambiguity of the azimuth prevents a determination of the zenith angle of solar magnetic fields. As a consequence, the magnetic flux and (in some cases) the polarity remains uncertain.

072.029 **Evaluation of the electric current in a sunspot by the study of the observed transverse component of the magnetic field.** J. Rayrole, M. Semel.
Astron. Astrophys., Vol. 6, 288 - 293 (1970).
The method employed for the evaluation of electric current seems to give significant values. The analysis of these results leads to arguments unfavourable for a force free model. At a certain point, a direct estimation of the Lorentz force is made. Finally we discuss the difficulties and errors in the reduction of the data.

072.030 **Overlapping sunspots.** R. A. Miller.
Journ. British Astron. Ass., Vol. 80, 296 - 298 (1970).

072.031 **Nebenmaxima in den Sonnenfleckenzyklen 0 bis 20.** W. Schulze.
Sterne, 46. Jahrgang, p. 76 - 80 (1970).

072.032 **The determination of the azimuthal electric currents in the surface layers of sunspots.**
J. Jakimiec.
Astron. Zhurn. Akad. Nauk SSSR, Vol. 47, 520 - 532 (1970). In Russian. English translation in Soviet Astron. AJ, Vol. 14, No. 3.
The problem of the determination of the magnetic force in sunspots from measurements of the magnetic field is examined. It is expedient to consider the problem as that of the determination of the electric currents in a sunspot; then the difficulty of determining j φ the azimuthal component of current, is encountered in the first place. A method of determining j φ by solving a boundary value problem for differential equations has been worked out. Detailed calculations have been carried out for the observational data obtained by Stepanov and Gopasyuk.

072.033 **Two new effects in the observation of magnetic fields of sunspots.** V. F. Chistjakov.
Astron. Zhurn. Akad. Nauk SSSR, Vol. 47, 533 - 540 (1970). In Russian. English translation in Soviet Astron. AJ, Vol. 14, No. 3.
Results of measurements of the magnetic splitting of the line Fe λ 6302.508 Å on 306 spectrograms of 59 sunspots, observed in Ussurijsk 1968 - 1969, are given. The description of two new magneto-optical effects is given: 1) an effect of a shift of circular components of Zeeman triplet along the scale of wavelengths and 2) an effect of inequality of the magnetic splitting value in spectra with different types of the circular polarization.

072.034 **On fluctuations of the monthly Wolf numbers.**
Yu. S. Muzalevsky, L. V. Zhukov.
Astron. Zhurn. Akad. Nauk SSSR, Vol. 47, 541 - 550 (1970). In Russian. English translation in Soviet Astron. AJ, Vol. 14, No. 3.
Problems connected with the determination and analysis of fluctuations are discussed. The histogram of the distribution of the fluctuation amplitudes is given.

072.035 **The history and morphology of solar activity, 1964 - 1965.** H. W. Dodson, E. R. Hedeman.
Ann. IQSY, (see 012.016), Vol. 4, 3 - 35 (1969).
The level of activity and the time of minimum between solar cycles 19 and 20 are discussed. The principal intervals of solar calm and solar activity in 1964 - 1965 are identified and flare data are evaluated. The dates of birth of major centers of activity on the visible hemisphere are tabulated and a roughly five-month cadence in the occurrence of critical changes in solar phenomena in these years is pointed out.

072.036 **On some results of observations of sunspots made with the Pulkovo magnetograph.**
R. N. Ikhsanov.
Izv. Glav. Astron. Obs. v Pulkove, No. 185, p. 159 - 166 (1970). In Russian.
The method of the reduction of observational data, obtained with the Pulkovo magnetograph, is discussed. The results are given of the preliminary reduction of observations of two sunspot groups made during 1967. The peculiarities of a distribution of the magnetic field and radial velocities in sunspots are noted.

072.037 **A remark on two representations of the butterfly diagram.**
G. Belvedere, G. Godoli, M. L. Sturiale.
Mem. Soc. Astron. Italiana, Nuova Serie, Vol. 41, 235 - 237 (1970).

072.038 **On the origin of sunspots.** M. Kopecký.
Vesmír, Vol. 49, 146 - 147 (1970). In Czech.

072.039 **Observation of the Evershed effect in polarized light.** S. G. Lisitza, V. F. Tshystjakov.
Astron. Tsirk., No. 548, p. 4 - 6 (1970). In Russian.

072.040 **Frequency–time spectrum of the Zürich series of Wolf numbers (1701–1964).**
O. B. Vasilyev.
Solnechnye Dannye 1970 Byull., No. 1, p. 92 - 99 (1970). In Russian.
The frequency–time spectra of the Zürich series of Wolf numbers (1701–1964) were calculated using 96, 48, and 16 years moving series for ordinary Wolf numbers and using 96 and 32 years moving series for Wolf numbers with variable sign.

072.041 **Statistical long-range forecast of the Zürich series of Wolf numbers.** O. B. Vasilyev, K. A. Kandaurova.
Solnechnye Dannye 1970 Byull., No. 2, p. 106 - 112 (1970). In Russian.
A statistical twelve-year (1970–1981) forecast of the Zürich series of Wolf numbers is made by the reconstitution of the process on basis of its frequency spectrum.

072.042 **On sunspot relative numbers.**
S. Nagasawa, T. Suzuki, M. Miyashita.
Tokyo Astron. Bull., Second Series, No. 199, p. 2307 - 2315 (1970).
In order to test the accuracy of the observations of the sunspot relative number, the values of k and its accuracy in the formula of sunspot relative number have been calculated by the method of least square for several observatories.

072.043 **Investigation of the curves of growth of sunspots.**
P. A. Olijnyk.
Tsirk. L'vov. Astron. Obs., No. 44, p. 36 - 42 (1970). In Russian.

072.044 **Misure di aree delle macchie solari nell' anno 1969.**

P. Zlobec.
Pubbl. Oss. Astron. Trieste, No. 410 (Osservazioni Solari, No. 16 Suppl.), 48 pp. (1970).

072.045 The abundance of lithium in sunspots.
W. A. Traub.
Thesis, Univ. Wisconsin, Madison. [Available from Univ. Microfilms, Ann Arbor, Mi.], 137 pp. (1969).

Describes high-resolution, interferometric observations of the Zeeman effect of the 6708 Å lithium line as seen in sunspots. A theory of line formation is developed and applied to the observed profiles, yielding a lithium abundance which of course depends upon the sunspot model chosen.

072.046 Flare occurrence tomorrow as a function of area and flariness of sunspot today.
A. E. Reilly.
Report AFCRL-69-0148, Air Force Cambridge Res. Labs., Bedford, Mass. [Available from CFSTI, Springfield, Va.], 15 pp. (1969).

An objective technique has been developed for estimating the probability of occurrence of solar flares in a particular sunspot group using both size of the sunspot group and its 'flariness'. Tables are given which indicate the probability that at least one flare of importance two or greater will occur tomorrow, given the area and flariness of the sunspot group today.

072.047 Noen problemer i forbindelse med solflekkenes oppbygning. E. Jensen.
Astron. Tidssk., Vol. 3, 7 - 17 (1970).

A naked-eye sunspot group seen in October and November.
Sky Telescope, Vol. 39, 57 - 59 (1970).

Properties of low frequency oscillatory convection in a strong magnetic field. See Abstr. 061.009.

On the mechanism of the formation of active regions. See Abstr. 071.019.

The SiH $A^2 \Delta - X^2 \Pi$ (o, o) band in the Fraunhofer spectrum. See Abstr. 071.037.

The defect of the number of solar flares and sunspots near the central meridian. See Abstr. 073.030.

The Hα plage associated with the proton flare of July 1966. See Abstr. 073.060.

Birth and development of the sunspot group associated with the proton flare of July 1966. See Abstr. 073.062.

Evolution of the sunspot group after the proton flare of 7 July 1966. See Abstr. 073.080.

Magnetic field decay in the group 21034 during the proton flare period of July 1966. See Abstr. 073.092.

Errata: The defect of the number of solar flares and sunspots near the central meridian. See Abstr. 073.126.

Solar activity. See Abstr. 073.128.

Sur les marées exercées par les planètes sur le soleil et la prévision de l'activité solaire. See Abstr. 091.009.

073 Solar Chromosphere, Flares, Prominences

073.001 On the relative shifts of the D_3 helium emission line in the spectrum of the solar chromosphere, I.
E. D. Khilov.
Solnechnye Dannye 1970 Byull., No. 1, p. 100 - 105 (1970). In Russian.

The relative shifts of the helium emission line D_3 in the solar chromosphere were measured in four points at the solar limb. The shifts were found to increase with height achieving the maximum value at the height 5.''64, and then decrease.

073.002 Large flare of October 30, 1968 and active dark filaments associated with it.
S. Yajima, K. Mizugaki, K. Yamaguchi.
Tokyo Astron. Bull., Second Ser., No. 197, p. 2283 - 2297 (1969).

Observation of the large flare (14S 37W) of October 30, 1968 is reported. The development of the flare and the associated arch shape prominences surrounding the whole sunspot group are described.

073.003 The heliocentric longitude intensity profile of 15-Mev protons from the February 5, 1965, solar flare. J. J. O'Gallagher.
Journ. Geophys. Res., Vol. 75, 1163 - 1171 (1970).

Simultaneous observations of 15-Mev protons from the solar flare of February 5, 1965, on Mariner 4 and IMP 2 show that the particle intensity decays faster near earth than at Mariner. It is shown that, if the longitudinal intensity profile for energetic flare particles is represented by a Gaussian in heliocentric longitude, then corotation of such a distribution past the two separated points of observation will account for this behavior. Furthermore, these observations provide a measure of the longitudinal scale of the distribution.

073.004 On the proton flare observation of September 2, 1966 at Abastumani Observatory.
Ts. S. Khetsuriani, A. S. Tskhovrebadze.
Byull. Abastumansk. Astrofiz. Obs. No. 37, p. 147 - 150 (1969).   In Russian.

The results of proton flare observations of September 2, 1966 carried out by means of Abastumani Observatory solar telescopes are given. The curves of flare development are drawn. The variations of active regions are described.

073.005 Determination of electron density in plasma by the number of the extreme resolved line.
L. N. Kurochka, L. B. Maslennikova.
Solar Physics, Vol. 11, 33 - 41 (1970).

The paper shows that the dependence of the number of the extreme resolved line on the electron density obtained for the Balmer series considering the broadening action of the electrons, remains true for the Lyman and Paschen series as well. For these series a formula which gives the dependence of the number of the extreme resolved line on the velocity of hydrogen atoms when the lines are broadened by the Doppler effect, has been derived. The electron density in certain chromospheric flares, prominences, and in the chromosphere has also been determined.

073.006 Lifetime of the dark and bright mottles of the solar chromosphere.
C. J. Macris, C. E. Alissandrakis.
Solar Physics, Vol. 11, 59 - 60 (1970). – Research note.

073.007 A model for quiescent prominences with helical structure. U. Anzer, E. Tandberg-Hanssen.
Solar Physics, Vol. 11, 61 - 67 (1970).

We present a model for quiescent prominences with helical structure. The model is described by two magnetic fields, one produced by photospheric or subphotospheric currents, the other due to currents along the cylindrical model prominence.

073.008 Long term observations of the Hα chromospheric network. T. J. Janssens.
Solar Physics, Vol. 11, 222 - 242 (1970).

Frequent filtergrams of the quiet sun at Hα + 0.65 Å were taken from above the arctic circle during a period of 62 hr. Features observed in individual filtergrams or movies are described with the dynamic changes they undergo. Thirty filtergrams taken at 2-hr intervals are presented and the development of a typical supergranule is shown in some detail. A study of supergranules shows that they lose their identity in about 21 hr on the average.

073.009 On the asymmetry of moustaches.
A. N. Koval, A. B. Severny.
Solar Physics, Vol. 11, 276 - 284 (1970).

A careful photometry of moustaches is carried out to reveal asymmetry of their emission in the far wings ($\Delta\lambda \geqslant 1$ Å), if it exists. The presence of background continuous emission in spectra of moustaches demands special care with the photometry, and makes the current method of comparison of rest-intensities inadequate. The other sources of errors are also discussed. The blue asymmetry, as a systematic difference of intensities between the blue and the red wing shows itself in some cases, being 2 - 3 times larger than the probable errors.

073.010 Identification of the hard X-ray pulse in the flare of September 11 - 12, 1968.
J. Vorpahl, H. Zirin.
Solar Physics, Vol. 11, 285 - 290 (1970).

A hard X-ray pulse in the 11 - 12 September 1968 flare is identified with the formation of a brilliant kernel. Each stage in the X-ray event corresponds to a definite phase in flare development.

073.011 An example of radio and optical homologous flares. K. P. White, III, T. J. Janssens.
Solar Physics, Vol. 11, 291 - 298 (1970).

Fokker (1967) has raised the question of whether the optical and radio homologies of flares are correlated. Two 2b flares occurring nearly 54 h apart (July 9 and 12, 1968) were observed at 29 wavelengths from Hα + 4.00 Å to Hα – 4.26 Å and at 10-cm radio. Adjacent pictures were spaced 0.295 Å and 2 sec apart. The time resolution of the radio traces was about 10 sec. Detailed comparison of the pictures showed near-perfect similarities in the two events.

073.012 Description of mass motions and brightenings in a class 2b flare, August 8, 1968.
T. J. Janssens, K. P. White, III.
Solar Physics, Vol. 11, 299 - 309 (1970).

Filtergrams spanning Hα ± 4.1 Å, supplemented with observations at 2.8 and 3 GHz, are used variously to describe the onset, dormancy, and flash phase of this 2b flare. Among the phenomena observed are 17- and 23-sec periodic pulsations in the microwave data early in the flare and formation of perpendicular, overlapping threads of red-shifted and blue-shifted material late in the flare.

073.013 Emission cores in H and K lines. V: Asymmetries in K_2 and K_3. R. G. Athay.
Solar Physics, Vol. 11, 347 - 354 (1970).

Enhancement of the violet K_2 emission peak results when the atmospheric layers at heights where K_3 forms are moving downward with velocities of 10 - 20 km/sec or when the K_2 layers and those immediately below are moving upward with velocities of 3 -7 km/sec. Evidence favoring the former alternative is cited.

073.014 On the relative residual intensities of the calcium H and K lines. J. L. Linsky.
Solar Physics, Vol. 11, 355 - 373 = Contr. Kitt Peak National Obs. No. 530 (1970).

We have observed the solar Ca II H and K lines to obtain well-calibrated ratios of their core residual intensities. We conclude that the residual intensity ratio $r(K_3)/r(H_3)$ is 1.048 ± 0.03 in the quiet chromosphere and 1.20 ± 0.03 in a plage region. These ratios correspond closely to those observed in stars with quiet and active chromospheres, respectively. For a chromospheric model suggested by the calcium lines and a four-level Ca II ion, we compute H and K line profiles varying the direct collisional coupling and indirect radiative and collisional coupling via the $3\,^2D$ level.

073.015 Motion of ascending prominences.
H. Westin, L. Liszka.
Solar Physics, Vol. 11, 409 - 424 (1970).

Apparent motions of a number of ascending prominences of the limb-SD type (sudden disappearances) are investigated. The direction, the velocity of ascent and the correlation with flares are studied. The maximum velocity, which seems to deviate systematically from the radial direction, increases with height and shows a clear dependence on the distance to an initiating flare and its importance.

073.016 The flares of July 6 and 8, 1968.
T. Fortini, M. Torelli.
Solar Physics, Vol. 11, 425 - 433 (1970).

A detailed analysis of observations of two flares in different regions of the electromagnetic spectrum (K_{232}, $H\alpha$ and radio) and of the corpuscular radiation (190 MeV, Pioneer 6 and 7) is given.

073.017 The chromosphere-corona transition region.
G. W. Pneuman.
Sky Telescope, Vol. 39, 148 - 151 (1970).

073.018 The effect of asymmetry of the emission of solar eruptions. V. G. Buslavskij, A. B. Severnyj.
Stars, Nebulae, Galaxies, Symposium Byurakan 1968, (see 012.002), p. 129 - 134 (1969). In Russian.

073.019 Couplage du transfert radiatif et de la conduction thermique dans la chromosphère solaire.
H. Frisch.
Comptes Rendus Acad. Sci. Paris, Sér. B, Vol. 270, 918 - 921 (1970).

On étudie l'influence du transport d'énergie radiative sur la distribution de température d'une couche plane chauffée par conduction. La température intérieure peut atteindre, dans certains cas, des valeurs très inférieures à celles des surfaces limites. Des applications à la remontée de température dans la chromosphère sont considérées.

073.020 Solar flares and surges observed at the Swedish Astrophysical Station in Anacapri in the years 1965 - 1968. Y. Öhman, H. Westin, U. Kusoffsky.
Stockholms Obs. Ann., Vol. 23, No. 2, 67 pp. (1970).

In the years 1965 - 1968 solar flare observations were carried out at the Swedish Astrophysical Station in Anacapri for, on the average, 333 days per year. In the present publication a catalogue is presented of all flares and surges observed as well as the number of observing hours for the different

months. The publication also contains some statistical data based on said observations and earlier data collected at the Swedish Astrophysical Station in Anacapri.

073.021 The oblique shock of the proton flare of 7 July 1966. E. W. Greenstadt, I. M. Green,
G. T. Inouye, C. P. Sonett.
Planet. Space Sci., Vol. 18, 333 - 347 (1970).

The proton flare of 7 July 1966 initiated a shock in the solar wind which arrived at the earth on the 8th, and was observed by three satellite magnetometers on Explorer 33, Imp 3, and Vela 3A, all outside the magnetosphere.

073.022 Comments on the discovery of the dark band in the $H\alpha$ solar chromosphere.
O. R. White, R. Bhavilai.
Astrophys. Letters, Vol. 5, 137 - 139 (1970).

The dark band in the low chromosphere described by Loughhead (1969) is shown to arise from unwanted photospheric light.

073.023 Simultaneous measurements of magnetic fields and brightness fields using a 4-image spectroheliograph.
N. R. Sheeley, Jr., O. Engvold.
Solar Physics, Vol. 12, 69 - 83 = Contr. Kitt Peak National Obs. No. 538 (1970).

This paper presents the preliminary results of a study to see whether a quantitative relation between brightness and field strength exists. A new technique is described whereby the strength of the magnetic field and the brightness of the photospheric network can be compared accurately over an entire two-dimensional field of view by means of a suitable photographic cancellation between a Zeeman spectroheliogram and a brightness spectroheliogram.

073.024 New observational results for the solar chromosphere. G. J. Banos, C. J. Macris.
Solar Physics, Vol. 12, 106 - 114 (1970).

A study of $H\alpha$ observations leads to the following results: (a) the bright filaments of the disturbed chromosphere as well as the penumbral ones appear to consist of knots. (b) inside the cells of the chromospheric network of the quiet chromosphere, bright roundish granule-like formations are present, their mean size being of the order of 2500 km. (c) the bright fine mottles seem to lie at the root of the elongated dark ones, each pair of them giving rise to a spicule.

073.025 Sudden disappearance of a large quiescent prominence on the solar disk, April 28, 1967.
M. McCabe.
Solar Physics, Vol. 12, 115 - 124 (1970).

Observations are described of the sudden disappearance of a long quiescent filament whose development could be traced through four rotations from its first appearance on the disk between two well-separated active centers. Following the disappearance, plage-like chromospheric brightenings were observed on either side of the filament axis, with the separation distance increasing with time.

073.026 Regularity in the distribution of spicule groups and their time characteristics.
Y. V. Platov, N. S. Shilova.
Solnechnye Dannye 1969 Byull., No. 12, p. 102 - 106 (1970). In Russian.

The problem of regularity in the distribution of spicule groups over the limb and their time variations is investigated. It is found that the groups have a size of $\sim 10''$ and are apart; an increase of their brightness takes place periodically at distances of $\sim 20''$ and $\sim 40''$. The time variations of these groups nave the character of slow oscillations, which are superimposed by beats with a characteristic time interval (5 - 7, and 2.5 minutes). The rate of the changes of the visible border of the chromosphere is on the average 15 km/sec.

073.027 Energy balance in spicules.
Nguyen-Ngan.
Astron. Zhurn. Akad. Nauk SSSR, Vol. 47, 91 - 97 (1970).
In Russian. English translation in Soviet Astron. AJ, Vol. 14, No. 1.

The problem of the energy balance in spicules is considered. The loss of the gas energy for the excitation of transitions of H, He, Ca$^+$ atoms, for hydrogen ionization and free-free emission is calculated when $n_e = 3 \times 10^{11}$ cm^{-3} and $7500 < T < 20000$ °K. Various possible energy sources are considered for the compensation of losses.

073.028 The electron concentration and structure of chromospheric flares.   L. N. Kurochka.
Astron. Zhurn. Akad. Nauk SSSR, Vol. 47, 111 - 121 (1970).
In Russian. English translation in Soviet Astron. AJ, Vol. 14, No. 1.

The electron concentration in chromospheric flares is determined from half-widths as well as from numbers of extremely resolved lines of Balmer series ($10^{12} \leqslant n_e \leqslant 5 \times 10^{13}$ cm^{-3}).

073.029 Development and spatial structure of proton flares near the limb and coronal phenomena. III. Cosmic rays flare on Nov. 18, 1968.   L. Křivský.
Bull. Astron. Inst. Czechoslovakia, Vol. 21, 67 - 70, 112a - 112b (1970).

On the flare with the ejection of the cosmic rays from 18 November 1968 one shows some typical features of the flare with the ascent of the magnetic tube in the system of the opposite magnetic fields of the spot group. By means of the application of known mechanism of interaction one can explain the origin and space development of these flares with the characteristic emissions. The energy of ejected particles and photons draws from the energy of the magnetic fields in the process of their annihilation.

073.030 The defect of the number of solar flares and sunspots near the central meridian.
L'. Pajdušáková.
Bull. Astron. Inst. Czechoslovakia, Vol. 21, 71 - 72 (1970).

In the paper, there is repeatedly stated that the number of observed flares as well as their measured area near the central meridian show a decrease. The defect is manifested in spot areas to which especially large groups contribute. Besides, it is shown in the paper that the number of microflares as well as individual nuclei of spots and of groups do not exhibit a defect toward the central meridian.

073.031 Spectrophotometric investigation of an intense solar eruption with over-excited upper energy levels of the hydrogen atoms.   R. Eh. Gusejnov.
Soobshch. Shemakhinsk. Astrofiz. Obs., vyp. (No.) 4, p. 3 - 14 (1969).   In Russian.

073.032 On a cause of the over-excitement of the upper energy levels of hydrogen atoms in an intense solar eruption.   R. Eh. Gusejnov.
Soobshch. Shemakhinsk. Astrofiz. Obs., vyp. (No.) 4, p. 15 - 29 (1969).   In Russian.

073.033 Investigation of the weak chromospheric eruption of June 14, 1964.   S. G. Mamedov.
Soobshch. Shemakhinsk. Astrofiz. Obs., vyp. (No.) 4, p. 84 - 97 (1969).   In Russian.

073.034 On the existence of a purely radiative temperature rise in the solar chromosphere.   A. Skumanich.
Astrophys. Journ., Vol. 159, 1077 - 1078 (1970).

The approximations implicit in Cayrel's statistical-equilibrium or non-LTE analysis of the H$^-$ bound-free continuum in the sun are examined and are shown to be consistent and independent of any assumption regarding the nature of the energy equilibrium. It is also shown that Jordan's contention that radiative equilibrium imposes a constraint on the non-LTE state of H$^-$ and thus removes any non-LTE temperature rise in the solar chromosphere is in error.

073.035 Supergranulation at the center of the solar disk.   E. N. Frazier.
Bull. American Astron. Soc., Vol. 2, 192 (1970). – Abstr. AAS.

073.036 Some comments on shock-wave heating of the low solar chromosphere.   S. D. Jordan.
Bull. American Astron. Soc., Vol. 2, 202 (1970). – Abstr. AAS.

073.037 OSO-VI: Surges, flares, and the development of active regions.   E. M. Reeves, A. K. Dupree, L. Goldberg, M. C. E. Huber, R. W. Noyes, W. H. Parkinson, G. L. Withbroe.
Bull. American Astron. Soc., Vol. 2, 215 (1970). – Abstr. AAS.

073.038 Curves of growth taking into account the diffusion of the solar radiation in prominences.
N. A. Yakovkin, M. Yu. Zeldina.
Vestn. Kiev. Univ., Ser. Astron., No. 11, p. 18 - 26 (1969).
In Russian.

The curve of growth of the Hα emission line has been constructed by calculating the diffusion of the solar radiation in prominences. The curves of growth of the other Balmer lines coincide with that for the Hα line if the observed equivalent widths are multiplied by tabulated coefficients.

073.039 Spectrophotometry of a limb flocculus.
P. N. Polupan.
Vestn. Kiev. Univ., Ser. Astron., No. 11, p. 27 - 40 (1969).
In Russian.

The altitudes of emission in hydrogen and Ca$^+$ lines, the parameters of their profiles and other characteristics of a flocculus spectrum were determined. As a result of an analysis of the profiles the occupation of atomic levels, the turbulent and tangential velocities of motion of gas masses were obtained. It is shown that there is a circulatory motion of matter in a flocculus. During increasing brightness the flocculus becomes an intermediate object between an ordinary flocculus and a flare.

073.040 On radiative relaxation of chromospheric oscillations.   M. Stix.
Astron. Astrophys., Vol. 4, 189 - 201 (1970).

The problem of radiative relaxation of oscillations generated by a granule is studied. The variation of the characteristic radiative relaxation time with height is treated in a three layer model atmosphere. The radiative decay time τ_R is assumed to be constant within each layer. Four types of eigenmodes occur in such a model: 1. modified sound waves; 2. pure horizontal sound waves; 3. surface gravity waves; 4. internal gravity waves. The results may be summarized as follows: a single granule can excite the atmosphere to oscillations lasting 3 to 4 pulsation periods in an area of ca. 3″ to 6″ in diameter. Concerning such oscillations one finds agreement with observations in variation of amplitude, frequency spectrum and phase relations of the velocity as well as in amplitude and phase of the temperature fluctuation.

073.041 Fine structure in Ca II on the solar disc.
J. M. Pasachoff.
Solar Physics, Vol. 12, 202 - 215 (1970).

High-dispersion spectra of the core of the K line of Ca II as seen at the center of the solar disc have been reduced. Resolution on the spectra approach 1 arc sec. Line profiles of individual elements are very asymmetric and often are peaked on only one side of the line center. Variations of the line profiles and the emission peaks are discussed. The doubly reversed mean profile of the K line is explained as a spatial average of individual profiles, and it is suggested that single peaks may be caused by Doppler–shifted discrete elements in the chromosphere.

073.042 The emission and propagation of ~40 keV solar flare electrons. I. The relationship of ~40 keV electron to energetic proton and relativistic electron emission by the sun. R. P. Lin.
Solar Physics, Vol. 12, 266 - 303 (1970).

In this paper we examine the emission of $\gtrsim 40$ keV energy electrons by the sun during solar flares and the relationship of these low energy electrons to other energetic solar particle emissions and energetic flare phenomena such as radio and X-ray emission.

073.043 The association of solar optical flares with type III solar bursts from 4 to 2 MHz observed by OGO-III.
T. E. Graedel.
Astrophys. Journ., Vol. 160, 301 - 307 (1970).

Burst radiation in the (4–2)-MHz band, originating from a radial distance range of approximately $5.3 - 6.3\ R_{\odot}$, offers the opportunity of studying the development of coronal streamers at radial distances well beyond those accessible to optical study outside of eclipse. The method adopted here is to associate a radio burst with an observed flare and, hence, with a specific plage region. The plage regions which are most frequently associated with radio bursts are then examined in an attempt to deduce the relationships between plages and coronal streamers.

073.044 Hα filtergrams of the sun on March 7, 1970.
H. Caulk, R. W. Hobbs.
Nature, Vol. 226, 1148 - 1149 (1970).

073.045 Radio observations of the distribution of chromospheric brightness temperature.
M. Simon, D. Buhl, J. R. Cogdell, F. I. Shimabukuro, C. Zapata.
Nature, Vol. 226, 1154 - 1155 (1970).

073.046 A study of monochromatic images of solar prominences in the light of He I D3 and He II 4686 lines.
J. Kubota, J. L. Leroy.
Astron. Astrophys., Vol. 6, 275 - 287 (1970).

Monochromatic observations of prominences in the light of He II 4686 Å and He I D3 were performed with a coronagraph and filter systems at the Pic-du-Midi Observatory. Five sets of monochromatic images (D3 and 4686 Å) of quiescent prominences were studied in detail. It is the purpose of the present investigation to detect differences that may be present between the He II and He I regions by comparing their monochromatic images.

073.047 Resonance lines in the solar chromosphere.
P. Lemaire.
IAU Symposium No. 36, (see 012.014), p. 250 - 255 (1970).

Recent observations of solar Mg II lines by means of a balloon borne instrument are presented in this paper and their correlation with observations of other resonance lines of chromospheric origin is studied.

073.048 Fabry-Pérot interferograms of the solar Mg II resonance lines.
B. Bates, D. J. Bradley, C. D. McKeith, N. E. McKeith, W. M. Burton, H. J. B. Paxton, D. B. Shenton, R. Wilson.
IAU Symposium No. 36, (see 012.014), p. 274 - 276 (1970).

073.049 Photoelectrically scanning device of Hα line profile on solar flares.
M. Miyazawa, H. Miyazaki, K. Higashi.
Univ. Tokyo, Tokyo Astron. Obs. Report, Vol. 15, (No. 2), (No. 57), p. 291 - 316 (1970).   In Japanese.

073.050 The observation of the chromospheric fine structure by the 53-cm Lyot coronagraph.
G. M. Nikolsky.
Solar Physics, Vol. 12, 379 - 390 (1970).

Method for the observation of chromospheric spicules with the large coronagraph mounted at the High Altitude Astronomical Station near Kislovodsk are described. The Hα-spectrograms and the filtergrams obtained by an Hα-Halle birefringent filter are shown. Some results of the observations and some conclusions on the structure of the chromosphere are reported.

073.051 The chromospheric magnetograph.
G. J. Veeder, H. Zirin.
Solar Physics, Vol. 12, 391 - 402 (1970).

By comparison of photoelectric magnetograms with high resolution Hα pictures it is possible to formulate a set of rules by which the magnetic field may be derived directly from the filtergrams. This is possible because of the regularities of magnetic field configurations on the sun and because chromospheric morphology is determined by the magnetic field.

073.052 On frequency and strength of shock waves in the solar atmosphere. P. Ulmschneider.
Solar Physics, Vol. 12, 403 - 415 (1970).

Comparison of computed radiative energy losses of several new empirical chromospheric models with heating by shock wave dissipation gives information on the frequency and strength of shock waves in the solar chromosphere. A mechanical flux of around 2.5×10^6 erg/cm^2 sec is found for the base of the chromosphere. The shocks are weak and the wave period is around 10 sec.

073.053 On the behaviour of the Fe II 3938-line in plages near the limb. O. Bergqvist.
Solar Physics, Vol. 12, 416 - 418 (1970). – Research note.

073.054 Thermal and dynamical stability of prominences.
Y. Nakagawa.
Solar Physics, Vol. 12, 419 - 437 (1970).

A comprehensive examination of the stability of prominences is presented, and the gross behavior of prominences is considered in terms of the stability of an optically thin plasma supported by a magnetic field against gravity, including thermal effects on the energy balance.

073.055 The 3.3-mm brightness distribution of the quiet sun.
F. I. Shimabukuro.
Solar Physics, Vol. 12, 438 - 446 (1970).

This paper reports hour-angle and declination scans of the quiet sun at a wavelength of 3.3 mm taken with an antenna having a half-power beamwidth of 2′.8. On the basis of these measurements it is possible to make a few general remarks about the solar chromosphere.

073.056 Protuberanzen 1969.
G. Klaus, E. Moser, J. Schaedler.
Orion Schaffhausen, 28. Jahrgang, p. 69 - 74 (1970).

073.057 **The proton event of 7 July 1966: Introductory.**
P. Simon.
Ann. IQSY, (see 003.001), Vol. 3, 3 - 7 (1969).
The purpose of the proton flare project was to obtain detailed observations of at least one proton flare event from all possible aspects and to publish the results and conclusions from this co-operative study. Over 100 workers from 54 institutions and 18 countries collaborated in the project.

073.058 **The magnetic fields and proton flare of 7 July 1966.**
A. Severny.
Ann. IQSY, (see 003.001), Vol. 3, 11 - 23 (1969).
Longitudinal $H\|$ and transverse $H\perp$ components of magnetic fields inside the active region which produced the proton flare of 7 July 1966, recorded with the aid of the double magnetograph of the Crimean Astrophysical Observatory, are considered. The event is described in detail.

073.059 **Distribution of magnetic fields in photospheric and chromospheric layers in the active region of 7 July 1966 and their correlation with the flare event of 8 July, 1253 - 1340 UT.** G. Brückner, M. Waldmeier.
Ann. IQSY, (see 003.001), Vol. 3, 24 - 30 (1969).
Longitudinal and transverse field components inside the very active region are determined with a spatial resolution of $2''$ of arc by $2''$ of arc measuring circular and linear polarization of the Fe I 5250 Å line. Also chromospheric fields are measured using the centre of the Hα line.

073.060 **The Hα plage associated with the proton flare of July 1966.** C. Popovici, A. Dimitriu.
Ann. IQSY, (see 003.001), Vol. 3, 31 - 34 (1969).
The Hα plage was a new-born plage between two old Hα plages. Its development was slow and irregular at the beginning and accelerated rapidly on 3 July, when within some eight hours the plage took on its compact, oval form along the axis of the sunspot group.

073.061 **Birth and development of the calcium plage associated with the proton flare of July 1966.**
T. Fortini, M. Torelli.
Ann. IQSY, (see 003.001), Vol. 3, 35 - 39 (1969).

073.062 **Birth and development of the sunspot group associated with the proton flare of July 1966.**
P. S. McIntosh.
Ann. IQSY, (see 003.001), Vol. 3, 40 - 43 (1969).
The birth and early development of the sunspot group consisted of the successive formation of three pairs of spot clusters on the borders of three adjoining network cells. The event is described in detail.

073.063 **Observation des éruptions à protons: L'évènement du 7 juillet 1966: Photométrie des raies coronales 5303 Å et 6374 Å.** J.-L. Leroy.
Ann. IQSY, (see 003.001), Vol. 3, 44 - 46 (1969).
From observations collected simultaneously at the stations of Lomnický štit, Kislovodsk, Pic-du-Midi and Wendelstein, we have plotted synoptic charts for the 5303 Å and 6374 Å radiations. The isophotes display the distribution and time evolution of these coronal emissions in the solar active region where a proton flare appeared on 7 July 1966.

073.064 **Coronal data for the period of the proton flare event of July 1966.** M. N. Gnevyshev.
Ann. IQSY, (see 003.001), Vol. 3, 47 - 48 (1969).
Distribution curves of the coronal line intensity at 5303 Å and 6374 Å are given for the period of the July 1966 proton flare.

073.065 **Coronal electron densities during the period of the July 1966 proton flare.**
G. Newkirk, Jr., R. T. Hansen, S. Hansen.
Ann. IQSY, (see 003.001), Vol. 3, 49 - 62 (1969).
K-coronameter observations during the interval June– September 1966 are used to construct five electron density models of the corona above the proton flare region. These models are compared with others for the corona above active regions. The observations following the proton flares of 10 July and 5 September show the presence of a unique low elevation condensation in the corona.

073.066 **The slowly varying component of the radio emission during the period of the July 1966 proton flare.**
H. Tanaka, T. Kakinuma, S. Enome.
Ann. IQSY, (see 003.001), Vol. 3, 63 - 69 (1969).
The paper presents the results of observations made over the period of the proton flare of July 1966 with radio interferometers. A radio source was observed on 1 July and increased in intensity until the time of the proton event, after which it decayed gradually.

073.067 **Remarks on the slowly varying component of solar radio emission during the proton flare period of July 1966.** A. Krüger.
Ann. IQSY, (see 003.001), Vol. 3, 70 - 77 (1969).
The centre of activity which produced the proton event of 7 July 1966 was characterized by an unusual increase of the slowly varying (S) component of solar radio emission at short centimetre waves.

073.068 **The slowly varying component of solar X-ray emission in the period 1 - 15 July 1966.**
H. Friedman, R. W. Kreplin.
Ann. IQSY, (see 003.001), Vol. 3, 78 - 81 (1969).
Solar X-ray monitoring by the Naval Research Laboratory's SOLRAD 8 satellite (1965–93A) has provided a history of the sun's X-ray emission for the particularly interesting period of early July 1966. By 6 July, X-ray flux levels had increased by factors of 15, 5, and 1.6 in the 0–8, 8–20, and 44–60 Å bands respectively.

073.069 **Flares in the active region during the proton flare period of July 1966.** A. Bruzek.
Ann. IQSY, (see 003.001), Vol. 3, 82 - 91 (1969).
Flare activity began on 3 July and had maximum intensity from 6 July to 9 July. A total of 78 subflares, 35 importance 1 flares and three importance 2B flares were observed in this active region from 2 July to 11 July.

073.070 **Photographie en Hα, Hβ et D₃ de la protubérance active du 9 juillet 1966.** J.-L. Leroy.
Ann. IQSY, (see 003.001), Vol. 3, 92 - 95 (1969).
We have obtained photographs of the large active solar prominence of 9 July 1966, through different monochromatic filters, which enable us to study separately characteristic emissions from hydrogen (Hα and Hβ) and helium (D₃). One cannot find any important difference between prominence images obtained from these various radiations.

073.071 **Active prominences of 9 and 11 July 1966.**
E. A. Gurtovenko, N. N. Morozhenko, A. S. Rakhubovsky.
Ann. IQSY, (see 003.001), Vol. 3, 96 - 109 (1969).
The spectra of active prominences on 9 and 11 July 1966 have been analysed. The physical conditions in the prominences are found to be the same as in quiescent prominences, but there is a sharp difference between them for turbulent velocities in the prominences of 9 and 11 July 1966.

073.072 **Polarization measurements of the proton flare on 11 July 1966.** G. Stiber.

Ann. IQSY, (see 003.001), Vol. 3, 110 - 112 (1969).

The spectacular event on 11 July 1966 was found to have a polarization lower than that predicted by the theory of electron scattering. By special instrumental arrangements, the probable electron density was measured to be about 3×10^{11} electrons cm^{-3}.

073.073 The west-limb activity on 9, 10, 11 July 1966 as observed in the $H\alpha$ line.
B. Valniček, G. Godoli, F. Mazzucconi.
Ann. IQSY, (see 003.001), Vol. 3, 113 - 123 (1969).

An analysis of the limb activity is given. The most important phenomena are the loop system, observed on 9 July, and the eruptive prominence of 11 July.

073.074 Radio bursts associated with active region McMath No. 8362 of July 1966.
O. Yudin, K. Kai.
Ann. IQSY, (see 003.001), Vol. 3, 124 - 134 (1969).

Observational data of radio bursts associated with an active region McMath No. 8362 in the range 1.76 cm – 13.5 m are given for the period 2 - 11 July 1966. After the proton flare on 7 July in the active region 8362 some comparatively greater bursts were observed.

073.075 Remarks on the development and activity of the active region during the proton flare event of July 1966. L. Křivský.
Ann. IQSY, (see 003.001), Vol. 3, 135 - 136 (1969).

By using the method of summation curves, some parameters of the development of the active region which produced the proton flare of July 1966 are derived.

073.076 Ionospheric effects of X-ray emission from an active region with a proton flare (30 June – 11 July 1966). L. Křivský, G. Nestorov.
Ann. IQSY, (see 003.001), Vol. 3, 137 - 143 (1969).

To illustrate the time distribution trend of the occurrence of sudden ionospheric disturbances (SID) effects of all kinds caused by flares in this active region, a summation curve was constructed.

073.077 X-ray emission events preceding the proton flare of 7 July 1966. H. Friedman, R. W. Kreplin.
Ann. IQSY, (see 003.001), Vol. 3, 144 - 153 (1969).

Solar X-ray emission on 6 July was characterized by a "storminess" in the 0-8 and 8-20 Å flux present above background levels on almost all telemetry records. This variable emission is found to originate from subflare activity in the plage region which produced the polar-cap absorption flare of 7 July.

073.078 The behavior of the active region prior to the proton flare 7 July 1966, based on λ-sweep records. H. W. Dodson.
Ann. IQSY, (see 003.001), Vol. 3, 154 - 162 (1969).

The eighteen λ-sweeps secured between 1330 and 2232 UT on 6 July 1966 provided no evidence of $H\alpha$ flares with unusually wide emission or great intensity.

073.079 Variations in the active region, McMath #8362 prior to the proton flare of 7 July 1966.
R. R. Fisher, G. R. Mann.
Ann. IQSY, (see 003.001), Vol. 3, 163 - 168 (1969).

The active region around McMath plage #8362 was extensively observed by several observatories prior to the proton flare of 7 July 1966. This plage was associated with a Mount Wilson γ-type spot region.

073.080 Evolution of the sunspot group after the proton flare of 7 July 1966. P. S. McIntosh, C. Sawyer.

Ann. IQSY, (see 003.001), Vol. 3, 169 - 174 (1969).

The total area of the sunspot group continues to increase for at least two days following the proton flare. However, the large umbrae near the center begin to decay, even as rapid growth continues at both the leading and following ends of the group.

073.081 Optical observations of the proton flare of 7 July 1966. M. K. McCabe, P. A. Caldwell.
Ann. IQSY, (see 003.001), Vol. 3, 175 - 180 (1969).

The flare occurred in a magnetically complex sunspot group – Mount Wilson type γ. It showed a flash phase lasting about two minutes during the rise of intensity to a maximum.

073.082 The dynamic spectrum of the proton event of 7 July 1966. R. T. Stewart.
Ann. IQSY, (see 003.001), Vol. 3, 181 - 183 (1969).

073.083 Decametric radio spectra and positions during the flare of 7 July 1966, 0041 UT. J. W. Warwick.
Ann. IQSY, (see 003.001), Vol. 3, 184 - 185 (1969).

073.084 Single-frequency bursts observed in Japan during the proton event of 7 July 1966. S. Enome.
Ann. IQSY, (see 003.001), Vol. 3, 186 - 189 (1969).

A table and figures give the characteristics of the type IV solar radio burst.

073.085 The complete type IV burst associated with the 7 July 1966 proton event. K. Kai.
Ann. IQSY, (see 003.001), Vol. 3, 190 - 192 (1969).

It is aimed to provide all available data of the solar radio burst associated with the 7 July 1966 proton flare and to describe the characteristics of the burst as completely as possible.

073.086 Very high energy solar X-rays observed during the proton event of 7 July 1966.
T. L. Cline, S. S. Holt, E. W. Hones, Jr.
Ann. IQSY, (see 003.001), Vol. 3, 193 - 197 (1969).

The time history and spectral intensity of solar X-rays of energies from 80 to more than 500 keV were observed during the flare event of 7 July 1966. Three intensity peaks coincide with the times of microwave intensity maxima.

073.087 The solar X-ray flare of 7 July 1966.
J. A. Van Allen.
Ann. IQSY, (see 003.001), Vol. 3, 198 - 201 (1969).

The maximum intensity of the X-ray flare occurred at 0042 UT and reached a value 28 times that of the pre-flare ambient value. The X-ray flare is attributed to the McMath plage region 8362.

073.088 X-ray emission from the proton flare of 7 July 1966. H. Friedman, R. W. Kreplin.
Ann. IQSY, (see 003.001), Vol. 3, 202 - 203 (1969).

073.089 The X-ray and extreme ultraviolet radiation of the 7 July 1966 proton flare as deduced from sudden ionospheric disturbance data. R. F. Donnelly.
Ann. IQSY, (see 003.001), Vol. 3, 204 - 208 (1969).

073.090 The geomagnetic crochet of 7 July 1966.
S. Pintér.
Ann. IQSY, (see 003.001), Vol. 3, 209 - 214 (1969).

073.091 Later development of the center of activity of the proton flare, 7 July 1966: Optical observations.
H. W. Dodson, E. R. Hedeman.
Ann. IQSY, (see 003.001), Vol. 3, 215 - 221 (1969).

073.092 **Magnetic field decay in the group 21034 during the proton flare period of July 1966.**
E. I. Mogilevsky, L. B. Demkina, Yu. N. Dolginova, B. A. Ioshpa, V. N. Obridko, B. D. Shelting, I. A. Zhulin.
Ann. IQSY, (see 003.001), Vol. 3, 222 - 228 (1969).

It has been noted that while considering magnetic field decay after the flare, one can make only detailed photographic measurements of the spot-group field because the region of Hα radiation in the proton flare is closely connected with the distribution of the magnetic field of the spot group.

073.093 **On the development and activity of the active region associated with the proton flare event of July 1966: Summary of observations and conclusions.**
V. Banin, L. D. de Feiter, A. D. Fokker, M. J. Martres, M. Pick.
Ann. IQSY, (see 003.001), Vol. 3, 229 - 241 (1969).

073.094 **The ground-level solar proton event of 7 July 1966 recorded by neutron monitors.**   H. Carmichael.
Ann. IQSY, (see 003.001), Vol. 3, 245 - 253 (1969).

073.095 **The increase in low-energy cosmic ray intensity on 7 July 1966.**
H. S. Ahluwalia, L. V. Sud, M. Schreier.
Ann. IQSY, (see 003.001), Vol. 3, 254 - 266 (1969).

073.096 **Balloon measurements of solar protons in northern Scandinavia on 7 July 1966.**
D. Heristchi, J. Kangas, G. Kremser. J. P. Legrand, P. Masse, M. Palous, G. Pfotzer, W. Riedler, K. Wilhelm.
Ann. IQSY, (see 003.001), Vol. 3, 267 - 281 (1969).

Following the flare of importance 2B of 7 July 1966, 0020 UT, proton measurements were obtained from two balloon flights in northern Scandinavia.

073.097 **The proton flares on 7 July, 2 September 1966, and 28 January 1967.**
P. N. Ageshin, V. V. Boyarevich, Yu. I. Stozhkov, A. N. Charakhchyan, T. N. Charakhchyan.
Ann. IQSY, (see 003.001), Vol. 3, 282 - 287 (1969).

The energy spectrum and time dependences of the solar proton intensity were measured.

073.098 **Sea-level observations of the 7 July 1966 solar proton event.**   A. Fréon, J. Berry, J. Folques.
Ann. IQSY, (see 003.001), Vol. 3, 288 - 294 (1969).

073.099 **Relativistic solar electrons detected during the 7 July 1966 proton flare event.**
T. L. Cline, F. B. McDonald.
Ann. IQSY, (see 003.001), Vol. 3, 295 - 298 (1969).

The 7 July 1966 flare event provided the first opportunity to detect relativistic solar electrons in interplanetary space. The solar electrons have energies between 3 and 12 MeV, nearly two orders of magnitudes higher than any previously studied in space.

073.100 **Spatial gradients of energetic protons and electrons observed after the 7 July 1966 solar flare.**
S. W. Kahler, R. P. Lin.
Ann. IQSY, (see 003.001), Vol. 3, 299 - 312 (1969).

Simultaneous observations of the 7 - 9 July 1966 solar particle event by energetic particle detectors on three satellites are utilized to show that large spatial gradients are present in the fluxes of 0.5 - 20 MeV protons and ≳45 keV electrons.

073.101 **Observations of the solar particle event of 7 July 1966 with University of Iowa detectors.**
T. P. Armstrong, S. M. Krimigis, J. A. Van Allen.
Ann. IQSY, (see 003.001), Vol. 3, 313 - 328 (1969).

Nearly complete time histories from 7 July to 17 July of the intensities of 0.31 - 10 MeV protons and of 2.1 - 17 MeV alpha particles emitted from the solar flare of 0023 UT on 7 July 1966 have been obtained with University of Iowa particle detectors on Explorer 33.

073.102 **Pioneer 6 observations of the solar flare particle event of 7 July 1966.**
U. R. Rao, K. G. McCracken, R. P. Bukata.
Ann. IQSY, (see 003.001), Vol. 3, 329 - 336 (1969).

In this paper we present detailed information on the time dependence, the degree of anisotropy and the spectral composition of cosmic radiation generated during a series of flares which occurred during the period 5 - 20 July 1966, with the major flare occurring on 7 July 1966.

073.103 **The polar-cap absorption on 7 - 10 July 1966.**
Y. Hakura.
Ann. IQSY, (see 003.001), Vol. 3, 337 - 352 (1969).

073.104 **The 7 July 1966 solar cosmic-ray event.**
A. J. Masley, A. D. Goedeke.
Ann. IQSY, (see 003.001), Vol. 3, 353 - 355 (1969).

073.105 **Very low radio frequency observations of the solar flare and polar-cap absorption event of 7 July 1966.**
J. H. Crary, D. D. Crombie.
Ann. IQSY, (see 003.001), Vol. 3, 356 - 365 (1969).

073.106 **Observations of the interplanetary magnetic field, 4 - 12 July 1966.**   N. F. Ness, H. E. Taylor.
Ann. IQSY, (see 003.001), Vol. 3, 366 - 374 (1969).

This note discusses simultaneous observations of the interplanetary magnetic field by three widely separated satellites. These data establish the general macrostructure of the field in cislunar space and include the microstructural feature of the shock wave associated with the geomagnetic sudden commencement at 2102 on 8 July 1966.

073.107 **Observed particle effects of an interplanetary shock wave on 8 July 1966.**   J. A. Van Allen, N. F. Ness.
Ann. IQSY, (see 003.001), Vol. 3, 375 (1969). – Abstract.

073.108 **The Forbush decrease associated with the proton event of 7 July 1966 as recorded by neutron monitors.**   H. Carmichael.
Ann. IQSY, (see 003.001), Vol. 3, 376 - 377 (1969).

Observations of cosmic-ray intensity variations in different directions in space were made by neutron monitors at a number of stations, and are presented graphically.

073.109 **Observations of the interplanetary plasma subsequent to the 7 July 1966 proton flare.**
A. J. Lazarus, J. H. Binsack.
Ann. IQSY, (see 003.001), Vol. 3, 378 - 385 (1969).

In this note we report preliminary results from observations of the interplanetary plasma during a time period that includes the class 2 solar flare of 7 July 1966. The plasma conditions were measured by two spacecraft that were widely separated in solar longitude.

073.110 **Perturbations décelées dans la magnétosphère, au moyen des sifflements radioélectriques, à la suite de l'événement à proton du 7 juillet 1966.**   Y. Corcuff.
Ann. IQSY, (see 003.001), Vol. 3, 386 - 388 (1969).

073.111 **Evidence of contraction of the earth's thermal plasmasphere subsequent to the solar flare events of 7 and 9 July 1966.**
H. A. Taylor, Jr., H. C. Brinton, M. W. Pharo III.
Ann. IQSY, (see 003.001), Vol. 3, 389 - 394 (1969).

073.112 **Solar particle observations inside the magneto-sphere during the 7 July 1966 proton flare event.**
S. M. Krimigis, J. A. Van Allen, T. P. Armstrong.
Ann. IQSY, (see 003.001), Vol. 3, 395 - 407 (1969).

073.113 **Preliminary analysis of micropulsations of the earth's electromagnetic field in connection with the proton flare of 7 July 1966.**
E. T. Matveeva, V. A. Troitskaya.
Ann. IQSY, (see 003.001), Vol. 3, 408 - 412 (1969).

073.114 **The variations of the geomagnetic field in middle and high latitudes during the proton flare event of July 1966.**
A. Best, G. Fanselau, A. Grafe, H.-R. Lehmann, C.-U. Wagner.
Ann. IQSY, (see 003.001), Vol. 3, 413 - 425 (1969).

073.115 **Ionospheric conditions following the proton flare of 7 July 1966 as deduced from topside soundings.**
L. Herzberg, G. L. Nelms.
Ann. IQSY, (see 003.001), Vol. 3, 426 - 436 (1969).

073.116 **Ionospheric conditions following the proton flare event of 7 July 1966 as measured at ground-based stations. I. Low-energy particle effects in the lower ionosphere at medium latitudes.** R. Knuth, E. A. Lauter.
Ann. IQSY, (see 003.001), Vol. 3, 437 - 442 (1969).

073.117 **Ionospheric conditions following the proton flare event of 7 July 1966 as measured at ground-based stations: II. F-region effects.** H. Lange, J. Taubenheim.
Ann. IQSY, (see 003.001), Vol. 3, 443 - 447 (1969).

073.118 **Ionospheric disturbances after the proton flare of 7 July 1966.** N. P. Benkova, R. A. Zevakina.
Ann. IQSY, (see 003.001), Vol. 3, 448 - 456 (1969).

073.119 **Summary on energetic particles observed during the July 1966 proton flare event.**
S. M. Krimigis.
Ann. IQSY, (see 003.001), Vol. 3, 457 - 461 (1969).

073.120 **Summary on low-energy particle events in the ionosphere associated with the July 1966 proton flare event.** W. Dieminger.
Ann. IQSY, (see 003.001), Vol. 3, 462 - 465 (1969).

073.121 **General summary on the results of the first Proton Flare Project period, July 1966.**
P. Simon, Ž. Svestka.
Ann. IQSY, (see 003.001), Vol. 3, 469 - 499 (1969).

073.122 **Temporal correlation between flares and surges.**
F. Mazzucconi, A. Righini.
Mem. Soc. Astron. Italiana, Nuova Serie, Vol. 41, 243 - 246 (1970).
In the present paper the distributions of the time intervals that occur between analogous phases of the evolution curves of surges and of associated flares are examined.

073.123 **The Zeeman effect in metallic lines of solar flares.**
K. V. Alikayeva.
Astrometriya i Astrofiz., *Kiev*, No. 8 , (see 003.024), p. 92 - 95 (1969). In Russian.
Line splitting in the metals Na, Al, Mg in the flare spectra may occur due to Zeeman effect. The magnetic field strength capable of causing the observed splitting is equal to 4000 - 10000 gauss.

073.124 **Generation spectrum for the flare of 23 February 1956.** L. I. Miroshnichenko.

Cosmic Rays No. 11, Moscow, p. 57 - 59 (1969). In Russian.
The paper analyses the data on the nucleon component for 6 stations which have recorded the flare of 23 February 1956 with consideration of the dependence of the diffusion coefficient on the rigidity of particles and the distance to the sun.

073.125 **Observations of chromospheric flares at Ondřejov Observatory during the years 1964 - 1968.**
F. Hřebik, J. Kvičala, L. Křivský, J. Olmr.
Bull. Astron. Inst. Czechoslovakia, Vol. 21, 170 - 192 (1970).
The data about 447 flares, associated 9400, 808, 536 and 260 MHz events and 27 kHz S. E. A. observed and recorded during the years 1964 - 1968 are summarized. The corresponding curves of the H_α line-width changes are plotted.

073.126 **Errata: The defect of the number of solar flares and sunspots near the central meridian. [Bull.** Astron. Inst. Czechoslovakia, Vol. 21, 71 - 72 (1970), see 03.073.030]. L'. Pajdušáková.
Bull. Astron. Inst. Czechoslovakia, Vol. 21, 194 - 197 (1970).

073.127 **Observation of coronal prominences by means of a special mounting.** A. A. Karaev.
Astron. Tsirk., No. 548, p. 1 - 2 (1970). In Russian.

073.128 **Solar activity.** Z. Švestka.
Trans. IAU, Vol. 14A, (see 003.028), 71 - 110 (1970). – Report of Commission 10.

073.129 **Sonneneruptionen und Magnetfelder.**
A. Sewerny.
Ideen exakt. Wissens (Wiss. Technik Sowjetunion), No. 11.69, p. 745 - 752 (1969).

Magnetic fields in quiescent solar prominences. II. Photospheric sources. See Abstr. 071.030.

The solar H and K lines.
See Abstr. 071.032.

Balloon observations in the submillimeter region: An absolute measurement of the brightness of the sun and the transparency of the stratosphere above 25 km.
See Abstr. 071.039.

Line formation in magnetic fields. Comments on the role of atomic level polarization. See Abstr. 071.040.

Radiation and structure of the solar atmosphere.
See Abstr. 071.047.

On the magnetic configuration of sunspot groups which produce solar proton flares. See Abstr. 072.005.

Magnetic fields in solar active regions.
See Abstr. 072.011.

Rapid mapping of solar magnetic fields.
See Abstr. 072.024.

The latitude distribution of magnetically classified sunspot groups and their associated solar flares.
See Abstr. 072.027.

Solar corona, interplanetary plasma and prominences. See Abstr. 074.036.

Flare-associated coronal expansion phenomena.
See Abstr. 074.054.

A small solar X-ray flare on August 17, 1966.
See Abstr. 076.002.

Solar XUV limb brightening observations. II: Lines formed in the chromospheric-coronal transition region. See Abstr. 076.005.

Extreme ultraviolet observations of active regions in the chromosphere and the corona. See Abstr. 076.007.

Intensity distribution in the Lyman-α line at the solar limb.   See Abstr. 076.023.

The equilibrium anisotropy in the flux of 10-Mev solar flare particles and their convection in the solar wind. See Abstr. 078.008.

Rafaga solar durante el eclipse del 7 de Marzo de 1970.   See Abstr. 079.102.

Thermally driven motions in a gravitational atmosphere.   See Abstr. 080.011.

074   Solar Corona, Solar Wind

074.001   Convection regions and coronas of sun and stars.
C. de Loore.
Astrophys. Space Sci., Vol. 6, 60 - 100 (1970).

Photospheric models were calculated for 90 stars with effective temperatures between 2500 K and 41600 K for five log *g*-values ranging from 1 to 5. Molecule formation was taken into account. It turned out that all the investigated stars contain unstable layers, including the hottest. For all stars the convective velocities were calculated and also the generated mechanical fluxes in the convection zones were tabulated. Under the hypothesis that this mechanical energy flux is responsible for the heating of the corona, coronal models were constructed for the sun and for some stars with effective temperatures between 5000 K and 8320 K for log *g*-values of 4 or 5.

074.002   Instabilities associated with heat conduction in the solar wind and their consequences.
D. W. Forslund.
Journ. Geophys. Res., Vol. 75, 17 - 28 (1970).

Associated with the large heat conduction in the solar wind is a skewing of the ion and electron distribution functions. It is shown that this collisional skewing of the electron distribution function can linearly excite collisionless ion-acoustic, electrostatic ion cyclotron, magnetoacoustic, and ion cyclotron waves in the steady-state solar wind even though the net equilibrium current parallel to B is zero. The initial growth rates for these unstable waves are derived, and the effectiveness of the wave-particle interactions in heating the ions and in altering the thermal and electrical conductivities is discussed.

074.003   Discussion of the paper by K. W. Ogilvie and N. F. Ness, 'Dependence of the lunar wake on solar wind plasma characteristics'.   F. C. Michel.
Journ. Geophys. Res., Vol. 75, 233, with a reply by K. W. Ogilvie, N. F. Ness, p. 234 (1970). — Letter.

074.004   Helium in the solar wind.
D. E. Robbins, A. J. Hundhausen, S. J. Bame.
Journ. Geophys. Res., Vol. 75, 1178 - 1187 (1970).

Data obtained from electrostatic analyzers on the Vela 3A and 3B satellites from July 1965 to July 1967 have been analyzed to obtain relative helium abundances and plasma properties.

074.005   The excitation equilibrium of coronal ions.
J. B. Zirker.
Solar Physics, Vol. 11, 68 - 81 (1970).

The relative populations of levels of highly ionized Fe, Ni and Ca ions have been calculated for physical conditions appropriate to the solar corona. The results are presented in the form of tables. Line intensity ratios in the EUV and visible that are sensitive to electron density are discussed and compared with observations.

074.006   Coronal polarization and intensity at the November 12, 1966 solar eclipse.
W. N. Arnquist, D. H. Menzel.
Solar Physics, Vol. 11, 82 - 91 (1970).

The 1966 Douglas Solar Eclipse Expedition obtained photographic records of the intensity and polarization of the solar corona on November 12, from a site at Chiguata, Peru. Here we shall give a more complete description of the equipment, its calibration, methods of reduction, and the results obtained. We compare the observed intensities and polarizations with those predicted by van de Hulst and point out

structure indicated by the polarization data. Further, we suggest that the existence of structure, such as streamers, rays, and so on, in the outer corona indicates that the current models tend to underestimate the importance of the K corona in this region.

074.007   Pulsating radio emissions from the solar corona.
A. Abrami.
Solar Physics, Vol. 11, 104 - 116 (1970).

Three solar outbursts which show pulsating radio emissions at metric waves (239 MHz) are examined. The behaviour of the single frequency, high-time resolution records and the spectral diagrams seem to indicate that such phenomena are peculiar phases of type IV radiation, perhaps connected with absorptions in the solar corona. The spectral analysis of the low-frequency modulation of the emissions show a very definite spectral line with a period ranging from $1\overset{s}{.}7$ to $3\overset{s}{.}1$.

074.008   The excitation of the forbidden coronal lines. II: [CaXV] λλ 5694 and 5446.
R. A. Chevalier, D. L. Lambert.
Solar Physics, Vol. 11, 243 - 257 (1970).

The excitation of the $2s^2\,2p^2$ ground configuration of Ca XV is calculated for coronal densities and temperatures. The calculations include electron and proton excitation of the forbidden transitions and electron excitation via the first excited $(2s2p^3)$ configuration. It is shown that measurements of the line intensity radio $I(\lambda 5694)/I(\lambda 5446)$ are in good agreement with the predictions. The line to continuum observations for limb flares and coronal condensations are discussed. It is suggested that the calcium abundance in condensations is enhanced owing to diffusion processes.

074.009   Contribution à l'étude de la température et de la densité coronales.   Y. Leblanc.
Thesis, Doct. Sci. Phys., Paris. [Centre Document. C. N. R. S., No. 3686], 91 pp. (1969).

074.010   Deux nouvelles méthodes de calcul des températures et des densités des condensations coronales observées sur ondes métriques et en lumière blanche.
Y. Leblanc.
Comptes Rendus Acad. Sci. Paris, Sér. B, Vol. 270, 922 - 925 (1970).

Nous développons deux méthodes permettant de calculer la température et la densité électronique des condensations coronales. Les résultats obtenus par ces deux méthodes sont en bon accord. On trouve que la température des condensations est en moyenne de 1.8×10^6 °K et que la densité, selon les cas, est 2 à 4 fois plus élevée que dans la couronne calme.

074.011   A three-fluid model of solar winds.
T. Yeh.
Planet. Space Sci., Vol. 18, 199 - 215 (1970).

A three-fluid model of solar winds is studied, in which the three constituents, electrons, protons and alpha particles, are coupled by an electric field. Each constituent expands isothermally under the action of a pressure gradient and solar gravitation. It is found that the three-fluid hydrodynamic equations possess a critical solution, in which all three constituents expand supersonically to infinity. In this hydrodynamic equilibrium the abundance of alpha particles compared to protons decreases as the solar plasma expands.

074.012   Ion-temperature anisotropies and the structure of the solar wind.   A. Eviatar, M. Schulz.
Planet. Space Sci., Vol. 18, 321 - 332 (1970).

The introduction of a kinetic equation for the interplanetary medium makes possible a quantitative test of several recently proposed mechanisms for limiting ion-temperature anisotropies in the solar wind. The results suggest that Coulomb collisions and an ion-cyclotron instability contribute comparably to the establishment of this limit under solar-wind conditions usually observed. Conditions favorable to the firehose instability seem to occur only on rare occasions.

074.013 Erratum: Hydromagnetic shocks in the solar wind.
[Solar Physics, Vol. 8, 422 - 434 (1969)].
K. W. Ogilvie, L. F. Burlaga.
Solar Physics, Vol. 11, 180 (1970).

074.014 Two new plasma instabilities in the solar wind.
J. V. Hollweg, H. J. Völk.
Nature, Vol. 225, 441 - 443 (1970).

The left-hand mode in a hot plasma, propagating in the direction of the average magnetic field is considered. For frequencies near the proton gyro-frequency two new instabilities are found. The first is driven by electron anisotropy in the same manner as is the low-frequency firehose instability; it is suggested that it represents a new branch of the firehose. This mode results in a non-collisional coupling between electrons and protons, and can contribute to proton heating. The second new instability is driven by proton anisotropy when the temperature perpendicular to the magnetic field exceeds the parallel termperature.

074.015 Hydromagnetic aspects of solar wind flow past the moon. J. R. Spreiter, M. C. Marsh, A. L. Summers.
Cosmic Electrodynamics, Vol. 1, 5 - 50 (1970).

Data from Explorer 35 satellite showing the presence of a plasma-free cavity behind the moon, often with a slightly enhanced magnetic field, and the lack of a bow wave upstream of the moon are used to establish an idealized mathematical model for the interaction of the solar wind with the moon. Exact numerical results for the density, velocity, temperature, and magnetic fields throughout the region influenced by the moon are presented for representative conditions in the incident solar wind. These results are shown to be in accordance with observations at times when the interplanetary magnetic field is approximately aligned with the direction of flow. A semiquantitative analysis of the properties of solutions to be anticipated for other alignments of the interplanetary magnetic field is also included. The results are used to account for several features of the data obtained by the magnetometers and plasma probe on Explorer 35.

074.016 The solar wind problem with fluctuations.
G. L. Siscoe.
Cosmic Electrodynamics, Vol. 1, 51 - 66 (1970).

The conservation equations appropriate to the solar wind problem are integrated over spherical volumes centered on the sun. There results a set of five equations for spherical surface averages of the various solar wind energies and certain covariances. The equations are used to indicate qualitatively the effect of fluctuations on the averaged quantities whose behavior we know in the case of no fluctuations from previous calculations. An equation is also given which governs the coupling between energies of fluctuations in regions of hypersonic solar wind flow.

074.017 Solar wind interactions – hypersonic analogue.
M. Dryer.
Cosmic Electrodynamics, Vol. 1, 115 - 142 (1970).

This paper extends the 'Knudsen number' concept, borrowed from ordinary fluid mechanics, to the various obstacles in the solar wind to give predictions of solar wind interaction for all planetary bodies. A simple exercise allows

us to organize some of the known interaction characteristics for earth, moon, and to some extend, Mars and Venus. Then it will be shown that Mars, Venus, Jupiter, comets, Saturn and Mercury are likely to be similarly characterized, if earth's case is referred to as a 'continuum' characterization. Moon, Neptune, Uranus, and Pluto will fall within the transition zone between continuum and free particle regions, an interaction referred to as one wherein the shocks become merged with the magnetosheath (or transition) flow.

074.018 Heating of the solar wind.
L. F. Burlaga, K. W. Ogilvie.
Astrophys. Journ., Vol. 159, 659 - 670 (1970).

This paper describes a study in which satellite observations, made in 1967 by the Goddard Space Flight Center on Explorer 34, are used to determine the relative importance of various energy sources which have been proposed to heat the solar wind. Both the instrument and the methods of data reduction used have been described in an earlier paper. The proposed energy sources can be classified according to the region where the heating is assumed to occur: (a) at the base of the solar corona, (b) in an extended region up to 50 $R_\odot$ from the sun, and (c) in the interplanetary medium.

074.019 Solar wind structure determined by corotating coronal inhomogeneities. 1. Velocity-driven perturbations. G. L. Siscoe, L. T. Finley.
Journ. Geophys. Res., Vol. 75, 1817 - 1825 (1970).

We extend the previous work on non-spherically symmetric solar wind models, which involved only two independent variables, by deriving the perturbation equations for the three independent variables, r, θ, and η, of a spherical polar coordinate system corotating with the sun. The results can be applied to an arbitrary distribution of coronal inhomogeneities with lifetimes long as compared with a flow time of several days and, hence, are restricted to large scale solar wind variations.

074.020 About the regime of plasma outflow from the sun.
M. V. Konyukov.
Geomagn. Aeronom., Vol. 10, 13 - 22 (1970).   In Russian.

074.021 The ionospheric effect of low-energy solar plasma radiated during proton flares.
E. E. Goncharova, R. A. Zevakina, E. V. Lavrova, L. A. Yudovitch.
Geomagn. Aeronom., Vol. 10, 67 - 72 (1970).   In Russian.

074.022 Heating of the solar corona.
J. Krełowski.
Postępy Astron., Vol. 18, 19 - 42 (1970).   In Polish.

The article presents a concise relation of the most important theories concerning the heating of the solar corona, developed in the last twenty years. It contains a discussion of as well the already historical studies of Schwarzschild, Schatzman and Alfvén, as of the modern, recently published works of Osterbrock, Uchida, Whitaker and Kuperus.

074.023 The solar (stellar) wind as a one-dimensional flow. Part II. S. Grzędzielski.
Postępy Astron., Vol. 18, 43 - 57 (1970).   In Polish.

The article discusses some simple consequences resulting from the introduction into the description of the solar wind phenomenon of the effects induced 1. by the magnetic field and 2. by the thermal conductivity of the gas.

074.024 Observations of the brightness and polarization of the outer corona during the 1966 November 12 total eclipse of the sun. T. J. Pepin.
Astrophys. Journ., Vol. 159, 1067 - 1075 (1970).

Photographs of the solar corona from a camera flown

on the NASA CV-990 at an altitude of 40000 feet were obtained on 1966 November 12. The absolute brightness, polarization, and direction of polarization of the solar corona between 3.5 and 13 $R_\odot$ were determined from these photographs. A significant polarized component was found in this region, the structure of which can be closely correlated with features in the inner corona. Comparisons are made in the 1963 July 20 and 1965 May 30 eclipse, and evidence is given for changes in brightness of the coronal light between eclipses.

074.025 Solar wind stimulation of the magnetosphere.
S. J. Bame.
Astrophys. Space Sci. Library, Vol. 17, 75 - 78 (1970).
Conference paper (see 012.008).

074.026 Shock waves in the solar wind.   A. J. Hundhausen.
Astrophys. Space Sci. Library, Vol. 17, 79 - 81 (1970). – Conference paper (see 012.008).

074.027 Hydromagnetic observations in the solar wind.
K. W. Ogilvie, L. F. Burlaga.
Astrophys. Space Sci. Library, Vol. 17, 82 - 94 (1970).
Conference paper (see 012.008).

**074.028 Collisionless solar wind. 1. Constant electron
        temperature.**   J. V. Hollweg.
Journ. Geophys. Res., Vol. 75, 2403 - 2418 (1970).
    A 2-fluid model for the solar wind is discussed, in which the electrons, at constant temperature, are treated hydrodynamically and the protons are assumed to become collisionless, at a distance from the sun at which they are already supersonic. A radial magnetic field is included, with radial gravitational and electric fields; the electric field arises from the condition of quasi-neutrality and is an important feature of the model. Quantities such as bulk velocity, concentration, and proton temperatures parallel and perpendicular to the magnetic field are discussed directly in terms of the proton distribution function; viscosity and heat conduction are thus automatically included.

074.029 The effect of viscosity and anisotropy in the pressure on the azimuthal motion of the solar wind.
E. J. Weber, L. Davis, Jr.
Journ. Geophys. Res., Vol. 75, 2419 - 2428 (1970).
    A steady state axially symmetrical model for the flow of the solar wind near the equatorial plane is developed. The radial motion is regarded as known, and the azimuthal motion is investigated, allowing for magnetic forces, viscosity (as modified by the magnetic field), and anisotropic pressure. We obtain an azimuthal velocity of approximately 6 km/sec at 1 AU, which is approximately five times larger than the value obtained from the nonviscous model.

074.030 The proton mean free path in the solar wind.
J. C. Brandt, N. Nichols.
Bull. American Astron. Soc., Vol. 2, 184 (1970). – Abstr. AAS.

**074.031 Solar corona electron density profile from pulsar
        occultation.**   C. C. Counselman, J. M. Rankin, D. W. Richards.
Bull. American Astron. Soc., Vol. 2, 189 (1970). – Abstr. AAS.

**074.032 Investigation of solar coronal electron distribution
        using the pulsar CP 0950.**   D. A. Guidice.
Bull. American Astron. Soc., Vol. 2, 195 (1970). – Abstr. AAS.

074.033 Analysis of the Apollo 11 corona photographs.
R. P. Kovar, N. S. Kovar.

Bull. American Astron. Soc., Vol. 2, 203 (1970). – Abstr. AAS.

**074.034 Solar-wind streams deduced from amplitude
        scintillation in Jovian decameter emission.**
L. Pataki.
Bull. American Astron. Soc., Vol. 2, 213 (1970). – Abstr. AAS.

**074.035 Erratum: "The effect of galactic cosmic rays upon
        the solar wind"** [Astrophys. Journ., Vol. 158, 781
        795 (1969)].   S. F. Sousk, A. M. Lenchek.
Astrophys. Journ., Vol. 159, 1133 (1970).

**074.036 Solar corona, interplanetary plasma and prominen-
        ces.**   A. Unsöld.
Astron. Astrophys., Vol. 4, 220 - 228 (1970).
    The structure and dynamics of the solar corona and the interplanetary plasma are reviewed. We discuss separately, insofar as possible, the Bernoulli equation and the equation of continuity, as well as the equation of state and the equation connecting the temperature T with either the distance from the sun r or the density ρ. Beginning with a brief historical introduction, we then deal with dynamical models. Solar wind and the opposite case of accretion are considered, emphasizing some elementary numerical estimates. Finally, we discuss the dynamics of prominences considered as cool clouds drifting within the much hotter corona. The very high viscosity of the coronal plasma turns out to be essential.

**074.037 Etude comparée des centres métriques et des
        renforcements coronaux observés dans le domaine
        optique. Propriétés, température et densité électroniques.**
Y. Leblanc.
Astron. Astrophys., Vol. 4, 315 - 330 (1970).
    Radio coronal enhancements (S-component centres) have been observed with the east-west interferometer at Nançay at a frequency of 169 MHz, during all the last solar cycle (1957 - 1968). The main properties of these enhancements have been deduced: the lifetime is larger than four solar rotations, the diameter is 9' to 12', and the typical flux density is 0.5×10^{-22} W m^{-2} Hz^{-1}; they are slightly directional; the main part of the emission of these centres comes from a typical altitude of 125.000 km above the photosphere, near the solar minimum. A comparison is made between the characteristics of radio and optical coronal enhancements: they appear to be associated with the same region.

**074.038 Solar coronal streamers. I: Observed locations,
        general evolution, and classification.**
J. D. Bohlin.
Solar Physics, Vol. 12, 240 - 265 (1970).
    The solar disk locations of 13 coronal streamers were determined from a combination of eclipse, K-coronameter and balloon-borne coronagraph observations taken during 1964 and 1965. Of this sample, three were observed twice on photographs taken over intervals of four and 28 days. Most of these streamers could be structurally associated with K-coronameter enhancements to establish their disk locations. Those features having known disk locations all lay above some stage of chromospheric disk activity in the form of active regions and prominences.

074.039 The solar wind.   M. Sroczyńska.
Urania Kraków, Vol. 41, 130 - 135 (1970).
In Polish.

**074.040 The characteristic sizes and the electron density of
        coronal enhancements observed in white light.**
Y. Leblanc, J. L. Leroy, P. Poulain.
Astron. Astrophys., Vol. 5, 391 - 399 (1970).

We are developing here a method which will allow the calculation of the characteristic sizes. the vertical density gradient and the electron density of coronal enhancements. This method is then applied to observations collected with the K-coronameters of Haleakala-Mauna Loa Observatory (1964) and Pic-du-Midi Observatory (1968 - 1969). Twenty enhancements have been studied.

074.041 Preliminary comparison of predicted and observed structure of the solar corona at the eclipse of March 7, 1970. S. M. Smith, K. H. Schatten.
Nature, Vol. 226, 1130 - 1131 (1970).

074.042 Prediction and observation of the coronal structure for the solar eclipse of March 7, 1970.
M. Waldmeier.
Nature, Vol. 226, 1131 - 1132 (1970).

074.043 Rocket observations of the corona on March 7, 1970. M. J. Koomen, J. D. Purcell, R. Tousey.
Nature, Vol. 226, 1135 - 1138 (1970).

074.044 Photometric study of the solar corona.
S. Koutchmy, M. Laffineur.
Nature, Vol. 226, 1141 - 1142 (1970).

074.045 Doppler temperature of the solar corona.
J. G. Hirschberg, A. Wouters, W. I. Fried, F. N. Cooke, Jr., D. Duke, M. Read.
Nature, Vol. 226, 1142 - 1143 (1970).

074.046 The outer corona at the eclipse of March 7, 1970.
D. H. Menzel, J. M. Pasachoff.
Nature, Vol. 226, 1143 - 1144 (1970).

074.047 Association of coronal structures with chromospheric structure. R. G. Teske, A. K. Hutchinson, J. R. Iwanski, T. Soyumer.
Nature, Vol. 226, 1145 - 1146 (1970).

074.048 Anomalous rhenium isotopic ratio in the solar wind: Detection at the nanogram level. J. W. Morgan.
Nature, Vol. 225, 1037 - 1038 (1970).
The possibility of the detection of solar wind Re of anomalous isotopic composition in lunar material is discussed. In the absence of measurements of normal Re in lunar rocks, the abundance is estimated from terrestrial rock values, and from some new achondrite analyses, to be about 10^{-9} g Re g^{-1}. It is concluded that the enrichment by anomalous Re may be sufficient to be detected, and a simple neutron activation technique is suggested.

074.049 Coronagraph observations of the coronal condensation of 4 February 1962. H. Zirin.
Solar Physics, Vol. 11, 497 - 512 (1970).
Coronagraph observations at the time of the total eclipse show that the ionization distribution is substantially unchanged from one region to another. Although the Fe XV line 7059 increases greatly in active regions, this is due to density increase. The iron abundance in a corona is found to be 10^{-4}, that of hydrogen. The peak electron density in the coronal condensation is 8 times 10 to 9 and the peak value of the line to continuum ratio for 5694 is 2.5. The ratio of the Fe XIII line is found to vary with electron density as expected, ranging from 2 : 1 in the condensation to 7 : 1 on the outside.

074.050 Study of ionized iron under coronal conditions. Coronagraph observations of the six coronal ions of the iron group. J. P. Rozelot.
Astron. Astrophys., Vol. 6, 18 - 26 (1970).
In this paper, we present some newly-acquired observa-

tions. The plates were obtained at the Pic-du-Midi Observatory with the 15 cm coronagraph. They cover the end of the most recent solar minimum, as well as the beginning of solar cycle No. 20. The use of a radial slit enables us to make exploratory observations from 1' to 7' from the edge of the photosphere. Spectroscopic analysis of the forbidden lines of Fe X to Fe XV gives results concerning their wavelengths, gradient values in the radial direction or along characteristic lines of fine structure, half-widths, and kinetic temperatures.

074.051 Solar wind models based on exospheric theory.
K. Jockers.
Astron. Astrophys., Vol. 6, 219 - 239 (1970).
In the models the spherically symmetric outflow of an electron-proton plasma is considered in which the protons are collisionless. In one model the electrons are also assumed to be collisionless while in some others an isotropic electron temperature was explicitly prescribed. From comparison of the models with observation one can draw two conclusions about the real solar wind.

074.052 Evaluation of the structure of the solar corona from the rocket experiment made on February 15, 1961. A. A. Dmitriev, R. G. Indzhgia, A. E. Mikirov, S. M. Poloskov.
Dokl. Akad. Nauk SSSR, Ser. Mat. Fiz., Vol. 190, 803 - 804 (1970). In Russian.
During the total solar eclipse on February 15, 1961 several thousands photoelectric brightness measurements of different points of the corona have been obtained from a geophysical rocket at a height of 100 km. Allowing for the angular sensitivity distribution of the photometer the true brightness distribution in the corona has been calculated supposing an error of 9.5% of measured data. The dimensions of inhomogeneities observed in the outer corona range from 24' to 70'. *B. Onderlička*

074.053 Observation of the green corona with a Lyot filter (I).
H. Imai.
Univ. Tokyo, Tokyo Astron. Obs. Report, Vol. 15, (No. 2), (No. 57), p. 280 - 290 (1970). In Japanese.

074.054 Flare-associated coronal expansion phenomena.
A. Bruzek, H. L. Demastus.
Solar Physics, Vol. 12, 447 - 457 = Mitt. Fraunhofer Inst., *Freiburg*, No. 96 (1970).
Two classes of coronal expansion phenomena have been studied in Sacramento Peak coronal movies: Slow, slightly decelerated expansion phenomena ($v \sim 10 - \sim 2$ km/sec) and fast, accelerated, quasi-exploding arches ($v \sim 10 - > 100$ km/sec). The various phenomena were found to be associated with flares in different ways. These expansions may be considered as evidence for corresponding flare associated changes in the coronal magnetic field.

074.055 On acceleration and motion of ions in corona and solar wind. J. Geiss, P. Hirt, H. Leutwyler.
Solar Physics, Vol. 12, 458 - 483 (1970).
Assuming a stationary, radial, spherically symmetric solar wind and a radial magnetic field direction in the vicinity of the sun, an equation of motion for ions heavier than protons in the solar wind is derived. The general properties of this equation are discussed and the results of numerical integrations are given. These results are based on the assumption of maxwellian velocity distribution functions for electrons, protons and ions, but the effects of first order deviations from such distributions are also presented and discussed. Formulae comparing the minimum fluxes for different ions are given. The effects of ion motion and escape on abundances in the corona and in the outer convective zone of the sun are discussed.

074.056 Evidence for a coronal magnetic bottle at 10 solar radii. K. H. Schatten.
Solar Physics, Vol. 12, 484 - 491 (1970).

The Faraday rotation of a radio source (Pioneer 6) occulted by the solar corona has been measured by Levy et al. (1969). During the course of these measurements, three large-scale transient phenomena were observed. These events were preceded by subflares and class 1 flares. These transient events are interpreted as evidence for a coronal magnetic bottle at $10 R_\odot$.

074.057 Determination of the velocity profile of the solar wind in regions of rapid acceleration of the coronal plasma. I. M. Gordon.
Astrophys. Letters, Vol. 5, 247 - 249 (1970).

Temperature estimates of the disturbed coronal plasma above plages, derived from the duration of solar bursts of types I, III, and V, are in complete agreement with those based on the interpretation of radio echoes from the sun. A method is proposed for the determination of the velocity profile of the solar wind in regions of rapid acceleration above the plages, based on the interdependence between the velocity and temperature of the accelerated coronal plasma. The temperature can be determined from the lifetime of solar bursts of types I, III, and V, which are presumed to be equal to the collision damping time.

074.058 Eine einfache Berechnung der Abbremsung schneller H-Atome durch kaltes H-Gas.
J. Schäfer, E. Trefftz.
Zeitschr. Naturforschung, Vol. 25a, 863 - 867 (1970).

In order to make a simple estimate of the thermalization of neutralized solar wind particles, classical elastic energy transfer cross sections are calculated for the H-H scattering process. A very simple approximation is also used for the macroscopic deceleration process. The estimated distance over which the hydrogen atoms lose half of their kinetic energy is 0.32×10^{18} cm and 24×10^{18} cm, for a starting velocity of 50 km/sec and 400 km/sec, resp., and for interstellar hydrogen density of 0.1 cm^{-3}. It appears that generally there is no thermalization shell around a star emitting a stellar wind.

074.059 Origin of the solar wind and its astrophysical aspects. E. R. Mustel.
Ann. IQSY, (see 012.016), Vol. 4, 67 - 87 (1969).

The problem of the outflow of gases from the quiet sun is reviewed. It is concluded that the solar corona is composed of two components: the first component (the streamers) is characterized by relatively low outward velocities and the second one (the solar wind itself) is characterized by relatively high velocities. It is concluded that certain processes of non-thermal origin take place at the base of quasi-stationary corpuscular streams.

074.060 The velocity field in a non-spherical solar wind. M. Sroczyńska.
Acta Astron., Vol. 20, 137 - 147 (1970).

The aim of the present paper is to investigate the effects of the non-sphericity of the solar wind due to solar rotation. The energy equation is left in its simplest, polytropic form, and the flow is assumed to be time independent.

074.061 The 1969 occultation of Taurus A by solar corona. B. Krygier.
Acta Astron., Vol. 20, 149 - 154 (1970).

The paper presents the results of the observations of the occultation of the radio source Taurus A by the solar supercorona, obtained at the Astronomical Institute of the Nicholas Copernicus University in Toruń during June 1969. The results obtained in this year are discussed together with the results from previous years. It is found that in the present year which is close to the maximum of the solar activity the solar supercorona is exceptionally compact.

074.062 Theory of polarized type III and U burst emissions from the solar corona. W. K. Yip.
Planet. Space Sci., Vol. 18, 867 - 885 (1970).

The generation of Cerenkov plasma waves by fast electron streams in the solar corona and their subsequent transformation into electromagnetic waves are considered taking the effect of the coronal magnetic field into account. It is shown that the observed properties of polarized type III and U burst emissions may be accounted for by this process.

074.063 Radiation of plasma waves by an electron stream in the solar corona. W. K. Yip.
Australian Journ. Phys., Vol. 23, 161 - 176 (1970).

The radiation of plasma waves by an electron stream moving in the active solar corona is investigated with the spatial dispersion taken into account. The angular and frequency spectra radiated by a single electron and the rate of growth for the plasma wave in the stream-plasma system are obtained. The expressions obtained are studied numerically with parameters appropriate to the active solar coronal region.

074.064 Deformation velocity of the solar wind, large-scale variations of the interplanetary magnetic field and the K_p-index of magnetic activity.
K. G. Ivanov, N. V. Mikerina.
Geomagn. Aeronom., Vol. 10, 417 - 422 (1970). In Russian.

074.065 Spectrum of a coronal condensation. F. G. Rozhavsky.
Astron. Tsirk., No. 548, p. 2 - 4 (1970). In Russian.

074.066 Photometry of coronal condensations during the solar eclipse of September 22, 1968.
V. P. Vasiljev, V. I. Bistritski.
Astron. Tsirk., No. 549, p. 3 - 5 (1970). In Russian.

074.067 Radio observations of the structure of the solar corona. K. V. Sheridan.
Proc. Astron. Soc. Australia, Vol. 1, 304 - 305 (1970).

In this preliminary study the first two-dimensional pictures showing detailed features of the quiet sun and weak but moderately stable structure at metre wavelengths are presented.

074.068 Radioastronomy from space.
R. G. Stone.
Science Journ., Vol. 6, No. 3, p. 68 - 74 (1970).

Popular article on radio astronomy Explorer I. This satellite contains a 0.2 to 6 MHz swept-frequency receiver. Solar type III bursts observed with the receiver provide temperature and density gradients of the solar corona at distances up to 60 solar radii. Low-frequency galactic background measurements are also possible. – ACR

074.069 The structure, dynamics, and evolution of solar coronal streamers. J. D. Bohlin.
Thesis. Univ. of Colorado, Boulder. [Available from Univ. Microfilms, Ann Arbor, Mi.], 237 pp. (1968).

074.070 The electric field in the solar coronal exosphere and the solar wind. H. K. Sen.
Journ. Franklin Inst., Vol. 287, (No. 6), 451 - 456 (1969).
See Phys. Abstr., Vol. 73, No. 13175 (1970).

Fine structure proton excitation rates for positive ions in the $2p$, $2p^5$, $3p$, $3p^5$ series.   See Abstr. 022.066.

The chromosphere-corona transition region. See Abstr. 073.017.

Observation des éruptions à protons: L'évènement du 7 juillet 1966: Photométrie des raies coronales 5303 Å et 6374 Å.   See Abstr. 073.063.

Coronal data for the period of the proton flare event of July 1966.   See Abstr. 073.064.

Coronal electron densities during the period of the July 1966 proton flare.   See Abstr. 073.065.

Solar activity.   See Abstr. 073.128.

Solar XUV limb brightening observations. I: The lithium-like ions.   See Abstr. 076.003.

Solar XUV limb brightening observations. II: Lines formed in the chromospheric-coronal transition region. See Abstr. 076.005.

Extreme ultraviolet observations of active regions in the chromosphere and the corona. See Abstr. 076.007.

Solar-X-ray observation of forbidden lines in the helium isoelectronic sequence.   See Abstr. 076.019.

Solar radiation from 1 to 100 Å. See Abstr. 076.025.

Evidence of type II and type IV solar radio emission from a common flare-induced shock wave. See Abstr. 077.026.

Interdependence between solar radio bursts of types I and III and non-linear processes in the coronal plasma. See Abstr. 077.036.

The equilibrium anisotropy in the flux of 10-Mev solar flare particles and their convection in the solar wind. See Abstr. 078.008.

Some observed characteristics of solar radar echoes and their implications.   See Abstr. 080.005.

Thermally driven motions in a gravitational atmosphere.   See Abstr. 080.011.

The magnetic reciprocal effect between the magnetosphere and the filamentary inhomogeneity of the solar wind. See Abstr. 084.221.

On the equilibrium of the magnetopause. See Abstr. 084.264.

On the temperature of the exospheres of the earth, Mercury, Venus, Mars, Jupiter, and of the solar corona. See Abstr. 091.035.

Lunar conducting islands and formation of a lunar limb shock wave.   See Abstr. 094.011.

Ionic comet tails and the direction of the solar wind. See Abstr. 102.013.

Observation of a solar flare induced interplanetary shock and helium-enriched driver gas. See Abstr. 106.003.

Analysis of observations of interplanetary scintillations.   See Abstr. 106.011.

Interplanetary scintillations and the structure of solar-wind fluctuations.   See Abstr. 106.017.

Interaction between interstellar hydrogen and the solar wind.   See Abstr. 131.067.

80 MHz observations of the coronal broadening of the Crab nebula.   See Abstr. 134.016.

On the convection, diffusion, and adiabatic deceleration of cosmic rays in the solar wind. See Abstr. 143.031.

On a reverse effect of cosmic rays on the solar wind. See Abstr. 143.046.

075 Solar Patrol

075.001 **Die Sonnentätigkeit im ersten Halbjahr 1969.**
R. Müller.
Sterne, 45. Jahrgang, p. 244 - 247 (1969).

075.002 **Solar activity changes in the years 1968 - 1969.**
W. Szymański.
Urania Kraków, Vol. 41, 57 - 58 (1970).   In Polish.

075.003 **Daily charts of the sun and geophysical graphs.**
Solnechnye Dannye 1969 Byull., No. 10, p. 2 -
106; No. 11, p. 2 - 93; No. 12, p. 2 - 73 (1970).
In Russian.

075.004 **Solar activity in 1969.**
J. Mergentaler.
Urania Kraków, Vol. 41, 151 - 153 (1970).   In Polish.

075.005 **Activité solaire en 1968.**
G. Evrard-Kesteloot, C. Gonze-Delys, M. Hovaere,
A. Koeckelenbergh.
Ciel et Terre, Vol. 86, 212 - 228 (1970).

075.006 **Solar activity during 1969.**
W. M. Baxter.
Journ. British Astron. Ass., Vol. 80, 307 - 312 (1970).

075.007 **Summary of prominence and calcium flocculus
observations, magnetic data, and ionospheric data
for the first half of 1963.**   M. K. V. Bappu.
Kodaikanal Obs. Bull., No. 171, 317 pp. (1969).

075.008 **Summary of prominence and calcium flocculus
observations, magnetic data, and ionospheric data
for the second half of 1963.**   M. K. V. Bappu.
Kodaikanal Obs. Bull., No. 173, 315 pp. (1969).

075.009 **Map of the Sun.**
Edited by Fraunhofer Institut, Freiburg. − 1970
January 1 - June 30.

075.010 **Indices of geomagnetic activity of the observatories
Hartland, Eskdalemuir, Lerwick.**
Journ. Atmosph. Terr. Phys., Vol. 32, 453, 975, 1165, 1343
(1970).

075.011 **Fenomeni solari.**
F. Mazzucconi, S. Delli Santi, M. L. Sturiale,
A. Abrami.
Coelum, Vol. 38, 46 - 53, 93 - 102, 142 - 149, 151 (1970).
1969 September − 1970 February.

075.012 **Osservatorio Magnetico de L'Aquila. Bolletino
magnetico.**   F. Molina.
Coelum, Vol. 38, 54 - 55, 103, 150 (1970). − 1969 September − 1970 February.

075.013 **Observations solaires.**
C. Popovici, E. Tifrea, V. Dinulescu, A. Dimitriu,
S. Nicolescu, G. Mariş.
Obs. Bucarest, Section d'Astrophys., Secteur Solaire, Acad.
Republique Socialiste Roumanie. 55 pp. Price Lei 2.50
(1969). Les rotations 1529 - 1542 (20 décembre 1967 −
4 janvier 1969).

075.014 **Definitieve zonnevlekkengetallen voor het jaar
1969.**
Hemel en Dampkring, Vol. 68, 126 (1970).

075.015 **Zonnevlekkengetallen.**
Hemel en Dampkring, Vol. 68, 18, 73, 96, 126,
156 (1970). − 1969 October − 1970 March.

075.016 **The solar activity in 1968, from optical and radio
observations of the Trieste Astronomical Observatory.**   P. Zlobec.
Mem. Soc. Astron. Italiana, Nuova Serie, Vol. 41, 205 - 214
(1970).
The solar optical and radio data obtained at the Trieste
Astronomical Observatory in the year 1968 are considered.
Comments are made on the behaviour of the present solar
cycle and on the correlation between the optical phenomena
and the radio activity.

075.017 **Solar photosphere charts.**   L. Schmied.
Říše hvězd, Vol. 51, 21, 36, 98, 117 (1970). −
Rotations Nos. 1548 - 1555.

075.018 **Daily Hα chromosphere pictures, daily K_{232} chromosphere pictures, daily white light photosphere
pictures.**   M. Cimino (Editor).
Photographic Journal of the Sun. Oss. Astron. Roma, N.26 -
27 (1969). − 1969 October 31 - December 24.

075.019 **Solar phenomena.**   M. Cimino (Editor).
Oss. Astron. Roma, Monthly Bull. Nos. 145 - 150
(1970). − 1970 January - June: Daily total areas of sunspot-
groups; Heliographic position, classification and area of sunspot-groups; Longitudinal sunspot magnetic fields; Hours of
K-line cinematographic patrol; Hours of H_a cinematographic
patrol; Sudden cosmic noise absorption S.C.N.A. and sudden
enhancement of atmospherics S.E.A.

075.020 **Solare Beobachtungsergebnisse (Solar Data).**
E. A. Lauter, H. Daene, F. W. Jäger, F. Fürstenberg,
H. Künzel, D. Scholz, W. Dittmar.
Zentralinstitut für Solar-Terrestrische Physik, (Heinrich-Hertz-
Inst.), Deutsche Akad. Wiss. Berlin, HHI Solar Data, Vol. 21,
January - April (1970). − Solar radio emission; Sunspot
magnetic data.

075.021 **Solar activity.**   F. G. Mustaeva.
Tsirk. Astron. Inst., *Tashkent*, 1968 Nos. 6 (353),
8 (355), 10 (357); 1969 Nos. 12 (359), 14 (361); 1970 Nos.
15 (362), 16 (363), 18 (365), 19 (366).   In Russian. −
1967 July − 1969 September.

075.022 **Daily solar maps and geophysical diagrams.**
Solnechnye Dannye 1970 Byull., No. 1, p. 1 - 91;
No. 2, p. 1 - 94; No. 3, p. 1 - 106 (1970). In Russian.

075.023 **Magnetic fields of sunspots.**
Prilozheniya k Byulletenyu "Solnechnye Dannye",
1970 Nos. 1 - 3.   In Russian.

075.024 **Sunspot numbers.**
Sky Telescope, Vol. 39, 59, 131, 201, 268, 328,
405; Vol. 40, 62 (1970). − 1969 Nov. − 1970 May.

075.025 **Sunspots and sunspot-groups.**   L. E. da Silva
Machado, J. F. Caria Caldeira.
Contr. Obs. Valongo, Univ. Federal Rio de Janeiro, Sér. I.,
Nos. 9 - 10 (1970). − 1969 October - 1970 March.

075.026 **L'activité solaire.**
M.-J. Martres.

L'Astronomie, 84ᵉ année, p. 49 - 53, 94 - 95, 127 - 130, 178 - 179, 229 - 230, 300 - 301 (1970). – Rotations 1547 - 1554.

075.027 Solar and solar system activity.
R. J. J. Langton, J. R. Smith.
Journ. British Astron. Ass., Vol. 80, 149 - 151, 227 - 229, 323 - 325 (1970).

075.028 Actividad solar.
J. S. Guzmán.
El Universo, Vol. 23, 96, 119 - 120 (1969).

075.029 Solar photospheric observations.
F. Bruin, H. Hourani, T. Assaf, N. G. Bustati.
Lee Obs., American Univ. Beirut, Monthly Bull., Astron. Section, 1969 August – 1970 March (1969/70).

Sunspot relative numbers; Heliographic mean position and classification of the sunspot groups; Number of facular zones.

075.030 Visual observations of the solar photosphere.
P. Zlobec.
Pubbl. Oss. Astron. Trieste, No. 409 (Osservazioni Solari, No. 16), p. 3 - 12 (1970). – 1969 October – December.

075.031 Measures of the solar flux at 235 and 237 MHz.
P. Zlobec.
Pubbl. Oss. Astron. Trieste, No. 409 (Osservazioni Solari, No. 16), p. 13 - 20 (1970). – 1969 October – December.

075.032 Solar observations made at Catania Astrophysical Observatory during 1968.
R. Campisi Cristaldi, O. Morgante, M. L. Sturiale, G. Celeani, C. D'Arrigo, G. Domina, G. Patti, S. Rifici, G. Sapienza, S. Torrisi, G. Godoli.
Oss. Astrofis. Catania, Pubbl., Nuova Ser., No. 141, 173 pp. (1969).

075.033 Solar phenomena.
Tokyo Astron. Obs., Bull. Solar Phenomena, Vol. 21, Nos. 3 - 4, p. 53 - 100 (1970). – Sunspots; Evolution table of sunspot groups; Map of sunspots; Hα flocculi and Hα dark filaments; Solar flares; Hours of Hα patrol; Prominences and filaments; Intensity of coronal emission line 5303 Å; Solar radio emission (1969 July – December).

075.034 Provisional sunspot-numbers for December 1969 – May 1970.
Yamamoto Circ., Nos. 1711, 1716, 1717, 1719, 1721 (1970).

076 Solar UV, X Rays, Gamma Radiation

076.001 **Characteristics of the X-ray and Hα emissions associated with solar proton flares.**
K. Sakurai.
Journ. Geophys. Res., Vol. 75, 225 - 227 (1970).

It is shown that the rise time of both the X-ray and Hα emissions is closely related to the acceleration efficiency of solar cosmic-ray particles; namely, as this time becomes shorter, the accelerated energy of solar cosmic rays tends to become higher. In this paper, we will show a clear-cut relation between this rise time and characteristics of solar cosmic rays.

076.002 **A small solar X-ray flare on August 17, 1966.**
H. M. Horstman, E. H. Moretti.
Journ. Geophys. Res., Vol. 75, 1157 - 1162 (1970).

A detector sensitive to X rays in the 7- to 50-kev energy range was flown on a rocket on August 17, 1966. An increase in counting rate was observed from a region of sky that contained the sun. No particular solar activity was reported for that time, but a solar flare, possibly in the region of the limb, appears to be the most likely explanation.

076.003 **Solar XUV limb brightening observations. I: The lithium-like ions.** G. L. Withbroe.
Solar Physics, Vol. 11, 42 - 58 (1970).

OSO-IV observations of the equatorial limb brightening of XUV resonance lines of N V, O VI, Ne VIII, Mg X and Si XII are interpreted with a modified version of a coronal model developed by Dupree and Goldberg (1967). Good agreement is obtained between the observed limb brightening and that predicted by the model. The sensitivity of the predicted limb-brightening curves to changes in parameters describing the model is discussed. Coronal abundances for N, O, Ne, Mg, and Si are obtained.

076.004 **Solar X-ray bursts at energies less than 10 keV observed with OSO-4.**
J. L. Culhane, K. J. H. Phillips.
Solar Physics, Vol. 11, 117 - 144 (1970).

Using data from a proportional counter spectrometer, sensitive in the wavelength range 1–20 Å, on OSO-4, X-ray bursts in the energy band 3.0 to 4.5 keV have been studied. 150 events have been identified between October 27, 1967 and May 8, 1968, mostly of an impulsive nature. Some gradual rise and fall bursts occur, but there is a selection bias against such long-enduring events. A study of the profiles of these events reveals no basis for identifying different types of impulsive event. The results are discussed in detail.

076.005 **Solar XUV limb brightening observations. II: Lines formed in the chromospheric-coronal transition region.** G. L. Withbroe.
Solar Physics, Vol. 11, 208 - 221 (1970).

Limb brightening of XUV lines of the ions CIII, NIII, NIV, OIII, OIV, OV and SiIV is compared with that predicted by a modified version of a coronal model developed by Dupree and Goldberg. Systematic differences between the predicted and observed limb brightening are found. These differences can be eliminated by introducing into the model the effects of spicules that extend up into the chromospheric-coronal transition region. The spicules are assumed to be opaque to radiation between 500 and 900 Å because of absorption in the hydrogen Lyman continuum.

076.006 **Comments on a paper by D. Meisel entitled 'Identification of a solar X-ray source using D-layer ionization behavior during an eclipse.'**

A. C. Aikin, J. H. Underwood.
Solar Physics, Vol. 11, 334 - 337, with a reply by D. Meisel,
Solar Physics, Vol. 11, 338 - 341 (1970). – Research note.

076.007 **Extreme ultraviolet observations of active regions in the chromosphere and the corona.**
R. W. Noyes, G. L. Withbroe, R. P. Kirshner.
Solar Physics, Vol. 11, 388 - 398 (1970).

New observations of solar active regions have been obtained by the Harvard College Observatory EUV spectroheliometer aboard the OSO-IV spacecraft. From the observations we have determined the enhancement in active regions of the emission from ions formed at various temperatures in the chromosphere and corona. The results are in accord with a simple model of active regions, for which the active region pressure is about 5 times the quiet sun pressure; the temperature gradient in the transition zone is about 5 times the quiet sun value; and the coronal temperature above active regions is slightly increased.

076.008 **High resolution search for solar gamma-ray lines.**
E. A. Womack, J. W. Overbeck.
Journ. Geophys. Res., Vol. 75, 1811 - 1816 (1970).

A balloon-borne directional spectrometer for gamma rays from 30 kev to 6.3 Mev is described, based on a high-resolution Ge (Li) detector. The sun was observed from 4.7 g cm^{-2} on May 5, 1968. Seven lines present in the background are attributed to neutron capture reactions in the detector. No evidence of monochromatic solar gamma-ray emission was detected.

076.009 **The UV continuum 1450 – 2100 Å and the problem of the solar temperature minimum.**
K. G. Widing, J. D. Purcell, G. D. Sandlin.
Solar Physics, Vol. 12, 52 - 62 (1970).

The photometry of the far UV continuum based on a set of 10 rocket spectra illustrates the transition of the solar temperature minimum at 1700 Å in the solar spectrum – (a) the continuum intensity decreases by 30 – 50% between 1700 Å and the ^1D limit of silicon at 1682 Å, and (b) the equivalent brightness temperature shows minimum values throughout the spectral range 1540 – 1682 Å, which average just under 4700 ± 100 K.

076.010 **X-ray and microwave emission of local sources on the sun.**
G. B. Gelfreich, I. A. Zhitnik, M. A. Livshitz.
Astron. Zhurn. Akad. Nauk SSSR, Vol. 47, 329 - 339 (1970).
In Russian. – English translation in Soviet Astron., AJ, Vol. 14, No. 2.

A comparison is made of the emission of sources above 3 sunspot groups in X-ray and microwave regions of the spectrum during the period from June 16 to October 3, 1967. X-ray heliograms with 3' resolution were obtained in 8 - 14 and 6 - 10 Å regions using Cosmos 166. The spectra of the same sources in microwave regions were constructed using observations made with the large Pulkovo radiotelescope with a resolution 1' – 3'. The connection between development of sunspot groups, X-ray and microwave emissions proved to be very complicated.

076.011 **Theoretical intensities of Fe XIV in the solar EUV spectrum.** M. Blaha.
Bull. American Astron. Soc., Vol. 2, 182 - 183 (1970).
Abstr. AAS.

076.012 **OSO-VI: The EUV spectrum of solar-active**

regions.   A. K. Dupree, L. Goldberg, M. C. E. Huber, R. W. Noyes, W. H. Parkinson, E. M. Reeves, G. L. Withbroe.
Bull. American Astron. Soc., Vol. 2, 191 (1970). – Abstr. AAS.

076.013 **Observations of hard solar X rays by the fifth Orbiting Solar Observatory.**
K. J. Frost.
Bull. American Astron. Soc., Vol. 2, 193 (1970). – Abstr. AAS.

076.014 **The bound-free continua of silicon in the ultraviolet solar spectrum.**
W. Henze, Jr.
Bull. American Astron. Soc., Vol. 2, 198 (1970). – Abstr. AAS.

076.015 **Observations of the solar X-ray spectrum from OSO-V.**   W. M. Neupert, M. Swartz.
Bull. American Astron. Soc., Vol. 2, 211 (1970). – Abstr. AAS.

076.016 **I. Holographic image-deblurring and aperture-synthesis methods: Applications in X-ray astronomy. II. High-resolution synthesized-aperture imaging using superposition of partial-resolution images in single photographs.**   G. W. Stroke.
Bull. American Astron. Soc., Vol. 2, 220 (1970). – Abstr. AAS.

076.017 **Measurements of the solar X-ray flux in selected emission lines.**
H. V. Argo, J. A. Bergey, W. D. Evans.
Astrophys. Journ., Vol. 160, 283 - 292 (1970).
Curved-crystal Bragg spectrometers have been flown in low-altitude sounding rockets with two-axis attitude control to measure absolute fluxes from the whole solar disk in selected prominent emission lines in the region 13 - 34 Å. These measurements were made under a variety of solar conditions (but non-flare) and in general indicate a significantly higher flux, integrated over the disk, than has been reported.

076.018 **Iron line emission at 1.9 Å during solar flares.**
J. L. Culhane, K. J. H. Phillips.
Astrophys. Journ., Vol. 160, 309 - 313 (1970).
Solar X-ray spectra have been obtained with a proportional-counter spectrometer on board the Orbiting Solar Observatory 4. Observations of an iron line at around 1.9 Å are presented which suggest that the line is nonthermally excited.

076.019 **Solar-X-ray observation of forbidden lines in the helium isoelectronic sequence.**
A. B. C. Walker, Jr., H. R. Rugge.
Astron. Astrophys., Vol. 5, 4 - 11 (1970).
We have obtained spectral scans of coronal emissions in the wavelength region between 1.5 and 25 Å with 3 crystal spectrometers on board the satellite OVI–17 (1969–025 A) which was launched in March of 1969. We have observed the forbidden $1s^2 S_0 - 1s2s^3 S_1$ transition in O VII, Ne IX, Al XII, Si XIII, and SXV. In this paper we shall discuss the identification of these lines, present the observed values of wavelength, and the relative fluxes for the resonance, intercombination, and forbidden lines.

076.020 **Solar UV flash spectrum 1400 Å – 1960 Å.**
G. E. Brueckner, J. F. Bartoe, K. R. Nicolas, R. Tousey.
Nature, Vol. 226, 1132 - 1135 (1970).

076.021 **Review of astrophysical conclusions from the UV solar spectra.**   S. R. Pottasch.
IAU Symposium No. 36, (see 012.014), p. 241 - 249 (1970).

076.022 **On the contribution of solar activity to the ultraviolet spectrum of the sun.**
A. V. Bruns, V. K. Prokofiev, A. B. Severny.
IAU Symposium No. 36, (see 012.014), p. 256 - 259 (1970).
As measured from space, the contribution of one moderate flare to the emission spectrum of the sun in the far-ultraviolet (304 Å, Lyman continuum, etc.) is comparable with the emission of the whole undisturbed solar disc.

076.023 **Intensity distribution in the Lyman-α line at the solar limb.**   J. C. Vial.
IAU Symposium No. 36, (see 012.014), p. 260 - 270 (1970).
The distribution of the solar intensity in the Lyman α line has been measured close to the visible limb. It is compared to a computation including LTE departures (as evaluated by Y. Cuny). As far as the profile of the line and the intensity (integrated over the line) are concerned, the interspicular model of Coates is the only which seems to agree with the observations.

076.024 **A high-resolution solar spectrum 2000 Å – 2200 Å.**
B. B. Jones, B. C. Boland, R. Wilson, S. T. F. Engstrom.
IAU Symposium No. 36, (see 012.014), p. 271 - 273 (1970).

076.025 **Solar radiation from 1 to 100 Å.**
M. Landini, B. C. Fossi.
Astron. Astrophys., Vol. 6, 468 - 475 (1970).
The emission from the solar corona in the spectral region 1 - 100 Å is computed for temperatures ranging from 10^6 to $10^{7}°$K and emission measure $10^{49}\,cm^{-3}$. The ionization equilibrium calculations of Jordan are used; they include processes of autoionization and dielectronic recombination. Both continuum and line emission are considered.

076.026 **Basic experiments for measuring VUV spectral region of solar radiation.**
K. Nishi, K. Higashi, A. Yamaguchi.
Univ. Tokyo, Tokyo Astron. Obs. Report, Vol. 15, (No. 2), (No. 57), p. 253 - 279 (1970).   In Japanese.

076.027 **Energetic solar radiations.**
H. Friedman.
Ann. IQSY, (see 012.016), Vol. 4, 36 - 56 (1969).
Solar X-ray spectra exhibit evidence of nonthermal processes which indicate that acceleration processes are continuously operative to produce electrons with energies from the keV to the MeV range, even under "quiet" sun conditions.

076.028 **Solar X-ray and diffuse cosmic X-ray spectra measured with a satellite-borne instrument.**
C. Cunningham, D. Groves, R. Price, R. Rodrigues, C. Swift.
Astrophys. Journ., Vol. 160, 1177 - 1183 (1970).
On 1967 August 7 - 8, the spectrum and intensity of the solar X-ray flux and the diffuse cosmic X-ray flux were measured. An instrument using an NaI (Tl) scintillation spectrometer as the detector was carried aboard an Air Force polar-orbiting satellite. The instrument recorded the flux of X-rays with energies between 2.6 and 19 keV. Comparisons are made with similar data published by Hudson et al. We shall show that our data and Hudson's are consistent with a decrease of emission measure with increasing temperature during a flare.

076.029 **Resonance and satellite lines of highly ionized iron in the solar spectrum near 1.9 Å.**
W. M. Neupert, M. Swartz.
Astrophys. Journ. (*Letters*), Vol. 160, L189 - L192 (1970).

A prominent emission feature frequently observed in the solar X-ray spectrum near 1.9 Å at times of chromospheric flares is found, from data obtained by the Fifth Orbiting Solar Observatory, to consist of at least six spectral lines.

076.030   Localization of solar X-ray emission at energies above 3 keV.
R. C. Catura, L. W. Acton, P. C. Fisher.
Nature, Vol. 227, 55 - 56 (1970).
The purposes of the experiment described here were to provide a first step in the investigation of higher energy X-ray emission, to aid the interpretation of other X-ray data which were acquired simultaneously, and to prove the feasibility of this type of measurement for a satellite experiment.

OSO-VI: The Harvard experiment.
See Abstr. 034.028.

Can the ion H_3^+ account for missing opacity in the solar ultraviolet?   See Abstr. 071.008.

Identification of the hard X-ray pulse in the flare of September 11 - 12, 1968.   See Abstr. 073.010.

The emission and propagation of ~40 keV solar flare electrons. I. The relationship of ~40 keV electron to energetic proton and relativistic electron emission by the sun. See Abstr. 073.042.

The slowly varying component of solar X-ray emission in the period 1 - 15 July 1966. See Abstr. 073.068.

X-ray emission events preceding the proton flare of 7 July 1966.   See Abstr. 073.077.

The behavior of the active region prior to the proton flare 7 July 1966, based on λ-sweep records. See Abstr. 073.078.

Very high energy solar X-rays observed during the proton event of 7 July 1966.   See Abstr. 073.086.

The solar X-ray flare of 7 July 1966. See Abstr. 073.087.

X-ray emission from the proton flare of 7 July 1966. See Abstr. 073.088.

Characteristic γ- and X-radiation in the planetary system.   See Abstr. 091.013.

077 Solar Radio Radiation

077.001 Reversals of circular polarization during microwave solar bursts. S. Marques dos Santos, P. Kaufmann, O. T. Matsuura, B. Chu.
Astrophys. Letters, Vol. 5, 85 - 88 (1970).

Rare cases of solar noise bursts observed at 7 GHz show circular polarization that reverses with time. The effect is caused by bursts that are generated simultaneously at two different plages in the sun, each with a different sense of polarization. Multi-burst sources can also show the effect. Observations of sources located in different hemispheres suggest the possibility of a physical connection between the coincident flares.

077.002 Solar bursts at millimeter wavelengths. G. Feix.
Journ. Geophys. Res., Vol. 75, 211 - 218 (1970).

Three limb bursts of exceptional intensity in the millimeter region are presented; the enhanced intensity of their flash phase showed a departure from those received from the solar disk. This behavior of a center of activity suggests a brightening at the limb. On the solar disk, average temperatures 2.5×10^5 °K were obtained, whereas at the limb burst temperatures from the same center of activity on the order of 2×10^8 °K were obtained.

077.003 Trajectories followed by U-like solar radio bursts. A. D. Fokker.
Solar Physics, Vol. 11, 92 - 103 (1970).

Radiospectrographic observations of some U-like bursts have been employed in combination with a model coronal condensation due to Waldmeier to derive trajectories along which the disturbing agency, which excites the radio emission, may have travelled. Such trajectories as connect regions of opposite magnetic polarity within one centre of activity should have a parachute-like shape in order to account for the observations. Travelling velocities are of the order of 35000 to 55000 km/sec. The distribution of U-like bursts in heliographic longitude is investigated.

077.004 Expanding arch structure of a solar radio outburst. K. Kai.
Solar Physics, Vol. 11, 310 - 318 (1970).

A flare-associated complex outburst was observed on 1968, October 23 - 24 with the 80 MHz Culgoora radioheliograph. The height-time plots derived from both the radioheliograph and spectrum observations suggest that two shock waves of different propagation velocities were initiated at the flash phase of the flare: the faster one was responsible for the first type II burst and the first radio-emitting arch; the slower one for the second type II burst and the second arch whose expansion advanced with the shock front.

077.005 Gyro-synchrotron emission in a magnetic dipole field for the application to the center-to-limb variation of microwave impulsive bursts. T. Takakura, E. Scalise, Jr.
Solar Physics, Vol. 11, 434 - 455 (1970).

In order to interpret the observed center to limb variations of spectrum and polarization of microwave impulsive bursts, gyro-synchrotron emission from nonthermal electrons trapped in a magnetic dipole field is computed. The theoretical spectrum and polarization are consistent with observed ones if we put an outer boundary of the radio source at a layer of 100-60 G or $(7 - 9) \times 10^4$ km in height.

077.006 The structure, polarization, and spatial relationship of solar radio sources of spectral types I and III. K. Kai.
Solar Physics, Vol. 11, 456 - 466 (1970).

A description is given of recent high-resolution observations of the sun made at 80 MHz with the Culgoora radioheliograph. In many cases the sources of type I storms show bi-polar structure, the two components being spatially separated by about 3' arc and strongly circularly polarized in opposite senses. Similar behaviour has been found for stationary type IV bursts which appear indistinguishable from type I storms on the heliographic records. On the other hand the sources of the weakly polarized type III bursts show uni-polar structure. The latter avoid the precise locations of type I sources even when they are associated with the same active centre.

077.007 The polarization of solar radio emission at 74 MHz: May 18 – 26, 1967. G. A. Harvey, L. R. McNarry.
Solar Physics, Vol. 11, 467 - 496 (1970).

The passage of McMath plage region 8818 over the visible solar disk resulted in extensive meter-wavelength activity of spectral types I, II, III and IV. The activity at 74 MHz and its polarization have been observed with a narrow-band (10 kHz) timesharing radio polarimeter. The polarization patterns for all the major activity in the May 18 – 26 period are described. Methods used to determine the polarization characteristics of the solar emission in the presence of galactic emission and to resolve the complexities of the solar emission itself are illustrated.

077.008 High resolution observations of solar microwave bursts. H. Tanaka, S. Enomé.
Nature, Vol. 225, 435 - 437 (1970).

Two microwave impulsive bursts with double structure in brightness distribution were observed on Dec. 24, 1968 and Jan. 18, 1969 by the Toyokawa 8-cm quick-scan interferometer. The time delay of the start of enhancement by about 10 s between one source and another of the pair was found in both cases, and the parallel activity after the main phase was remarkable on Jan. 18. A hydromagnetic shock wave model is suggested to explain both phenomena. The sense reversal with frequency of circular polarization was observed on Jan. 18 for both sources.

077.009 Methods of defining the value of radio radiation of the "quiet" sun by the method of statistical processing. I. F. Belov, A. I. Korshunov, V. M. Fridman.
Geomagn. Aeronom., Vol. 10, 136 (1970). In Russian.
Brief information.

077.010 A method of measuring the slope of the solar radio emission spectrum and its application to observations during the eclipse of September 22, 1968 in Gorki. I. F. Belov, M. M. Kobrin, A. I. Korshunov, B. V. Timofeev.
Solnechnye Dannye 1969 Byull., No. 11, p. 95 - 101 (1970). In Russian.

A method is proposed for constructing a design for accurate measurements of the slope of the frequency spectrum of solar radio emission. The parameters of the apparatus designed are presented, as well as preliminary results of measurements of the slope of the radio emission spectrum made at $\lambda = 3.3$ cm during the solar eclipse on 22.IX. 1968 in Gorki.

077.011 Brightness distribution in the polar zone of the solar limb at 4.0 cm as determined from observations of the solar eclipse on May 9, 1967.

G. B. Gelfreich, A. N. Korzhavin.
Solnechnye Dannye 1969 Byull., No. 11, p. 102 - 112 (1970).
In Russian.

From observations of the partial solar eclipse at Pulkovo the brightness distribution in the polar zone of the solar limb was found at 4.0 cm. Besides the bright ring within the optical disc, a dip of brightness (narrow dark ring) superimposing a total brightening near the limb has been found. The possibility of interpretation of such a complicated distribution of radio brightness on the basis of an inhomogeneous (spicular) model of the solar atmosphere is discussed.

077.012 Variations of the intensity of circularly polarized radio emission during solar flares.
N. N. Erjushev, L. I. Tsvetkov.
Solnechnye Dannye 1969 Byull., No. 12, p. 94 - 98 (1970).
In Russian.

The behaviour of circularly polarized radio emission observed at 3.15 cm during three solar flares on July 10 and 12, 1968 has been investigated. Variations of the intensity of polarized emission from local sources were detected several hours before the appearance of a radio burst as well as the intensity variations (an oscillation pattern) during the bursts.

077.013 Longitudinal distribution of the sources of the S-component.
A. T. Nesmjanovich, Y. A. Khomenko.
Solnechnye Dannye 1969 Byull., No. 12, p. 98 - 101 (1970).
In Russian.

Longitudinal distributions of the important sources of the S-component at the frequencies of 9400, 3750, 2000 and 1000 MHz during the period from 1957 to July 1966 were determined from data of the Japanese station Toyokawa. It is shown that the flux maxima of the S-component in the centimeter-decimeter range are connected with the active longitudes according to Vitinsky and correspond (in time) to their central meridian passage.

077.014 On the theory of the type III solar radio bursts. 1.
V. V. Zheleznjakov, V. V. Zaitsev.
Astron. Zhurn. Akad. Nauk SSSR, Vol. 47, 60 - 75 (1970).
In Russian. English translation in Soviet Astron. AJ, Vol. 14, No. 1.

The influence of non-linear effects on the dynamics of the development of beam instability on conditions typical for radio radiation sources of type III is investigated. The necessity of taking into account a quasi-linear relaxation of the distribution function of fast electrons is shown.

077.015 The determination of the two-dimensional structure of the noise storm on May 20, 1966 from data of eclipse observations at 4 stations.   A. A. Gnezdilov.
Astron. Zhurn. Akad. Nauk SSSR, Vol. 47, 76 - 81 (1970).
In Russian. English translation in Soviet Astron. AJ, Vol. 14, No. 1.

Results of the analysis of observations at 4 stations of the solar eclipse on May 20, 1966 at frequencies close to 200 MHz are given. In the analysis a refraction of radio emission in the solar corona is taken into account; the assumption on the radial position of the source of the noise storm above the spot group No. 60, observed at that day on the North-East edge of the solar disk, is grounded.

077.016 On the distribution of radio bursts of type III on heliographic longitude.
V. V. Fomichev, I. M. Chertok.
Astron. Zhurn. Akad. Nauk SSSR, Vol. 47, 226 - 227 (1970).
In Russian. English translation in Soviet Astron. AJ, Vol. 14, No. 1.

It is shown that the distributions of the harmonic type III radio bursts on heliographic longitude observed ex-

perimentally are satisfactorily explained taking into account a combinative mechanism for the formation of the second harmonic.

077.017 Long-wave cosmic radio radiation in the space close to moon.
V. P. Grigor'eva, V. I. Slysh.
Kosm. Issled., Vol. 8, 284 - 289 (1970).   In Russian.

077.018 Flare identification associated with coronal disturbances.
K. H. Schatten.
Science, Vol. 168, 395 - 396 (1970).

077.019 On the theory of bursts of the solar radio radiation of type III. II.
V. V. Zheleznjakov, V. V. Zaitsev.
Astron. Zhurn. Akad. Nauk SSSR, Vol. 47, 308 - 321 (1970).
In Russian. – English translation in Soviet Astron., AJ, Vol. 14, No. 2.

Effects of the energy conversion of plasma waves, generated by a stream of fast electrons, on the radio emission of the dominant mode and the second harmonic of the type III bursts are analyzed.

077.020 On the polarization of solar radio emission at meter waves observed in reflected rays.
V. V. Fomichev, I. M. Chertok.
Astron. Zhurn. Akad. Nauk SSSR, Vol. 47, 322 - 328 (1970).
In Russian. – English translation in Soviet Astron., AJ, Vol. 14, No. 2.

Polarization characteristics of solar radio emission at meter waves observed in reflected rays are considered taking into account the reflection of ordinary and extraordinary waves from appropriate layers of the corona. It is shown that the reflected radio emission can be polarized.

077.021 Type-III solar radio storm: Determination of exciter speed, direction, and plasma distance scale from 11–40 solar radii.   J. Fainberg, R. G. Stone.
Bull. American Astron. Soc., Vol. 2, 191 - 192 (1970).
Abstr. AAS.

077.022 Fast-drift solar burst observed from 3.5 MHz to 50 kHz from the OGO-V satellite.
F. T. Haddock, H. Alvarez.
Bull. American Astron. Soc., Vol. 2, 196 (1970). – Abstr. AAS.

077.023 The slowly varying component of the sun's radiation at millimeter wavelengths.
M. R. Kundu.
Bull. American Astron. Soc., Vol. 2, 203 - 204 (1970).
Abstr. AAS.

077.024 Noise storms in the IYQS period.
A. T. Nesmyanovich, O. S. Popov.
Vestn. Kiev. Univ., Ser. Astron., No. 11, p. 53 - 58 (1969).
In Russian.

The records of solar radio emission at a wavelength of 1.5 m during the noise storms from January 1965 to June 1966 are treated. The connection between noise storms and the optical features of the sun's activity is studied.

077.025 Relation between metric and decametric noise storm activity.
A. Boischot, J. de la Noë, B. Møller-Pedersen.
Astron. Astrophys., Vol. 4, 159 - 160 (1970).

The metre type I storms are generally accompanied by decametre continuums interpreted as type III storms. It is suggested that the two kinds of storms are generated by the same high energetic electrons streaming outwards from the photosphere or low chromosphere.

077.026 Evidence of type II and type IV solar radio emission from a common flare-induced shock wave.
R. T. Stewart, K. V. Sheridan.
Solar Physics, Vol. 12, 229 - 239 (1970).

A solar radio outburst is described in which a moving type IV burst is observed to break up into several components. A close association is found to exist between this source and a type II burst which occurred during the same period and detailed analysis indicates that both bursts were excited by a common shock wave ejected from the flare region.

077.027 Dynamic spectra of type III solar bursts from 4 to 2 MHz observed by OGO-III.
F. T. Haddock, T. E. Graedel.
Astrophys. Journ., Vol. 160, 293 - 300 (1970).

In this paper we present the results of a detailed analysis of solar-burst data for the 16-month period 1966 June 7–1967 September 30, from the University of Michigan Radio Astronomy Observatory experiment aboard the spacecraft Orbiting Geophysical Observatory III. The information resulting from the analysis enables us to specify the characteristics of type III bursts over the frequency range 4–2 MHz, which we conclude originates from a distance range of 5.3–6.3 solar radii from the center of the sun.

077.028 Relative positions of radio sources on the solar disk at 2800 MHz and 408 MHz.   D. Basu.
Journ. Roy. Astron. Soc. Canada, Vol. 64, 90 - 97 (1970).

Inspection of the drift curves of the sun made with high resolution fan beams operating on a frequency of 408 MHz and 2800 MHz for a period 1966–68 provide 10 isolated radio sources for a comparative study. The analysis consists in determining the day-to-day observed positions of a source separately for the two frequencies and comparing with each other. These are also compared with trajectories computed from an equation which gives the position of a source as a function of various solar parameters.

077.029 The mapping of the sun at 3.5 mm.
D. Buhl, A. Tlamicha.
Astron. Astrophys., Vol. 5, 102 - 112 (1970).

Maps of the sun at a wavelength of 3.5 mm having a resolution of 1.'2 were made from May 15 to June 1, 1968, using the NRAO 36-foot telescope. Comparisons are made with Hα photographs and magnetograms which indicate that the intense peaks in the radio emission coincide with bright plage regions and depressions with dark filaments.

077.030 Characteristic properties of small chains of type I solar radio bursts.   Ø. Elgarøy, O. Ugland.
Astron. Astrophys., Vol. 5, 372 - 381 (1970).

During solar radio noise storms, bursts of spectral type I are emitted in large numbers. Sometimes several bursts occur in close succession and form a chain in the frequency-time plane. Using high-resolution dynamic records obtained at the Oslo Solar Observatory, properties of short-lasting chains such as lifetime, bandwidth, frequency drift and fine structure were determined. A possible interpretation of the observational results is discussed.

077.031 Observations at the Sagamore Hill Solar Radio Observatory.   R. M. Straka, J. P. Castelli.
Nature, Vol. 226, 1149 - 1152 (1970).

077.032 16 GHz observations of the eclipse.
K. C. O'Brien, T. E. Graedel, G. J. Owens,
W. H. Ierley, J. H. Doles III, E. R. Nagelberg.
Nature, Vol. 226, 1152 - 1153 (1970).

077.033 Observations with a 7 GHz polarimeter.
O. T. Matsuura, E. Scalise, Jr., P. Marques dos Santos, P. Kaufmann.
Nature, Vol. 226, 1153 - 1154 (1970).

077.034 Interpretation of type I- and IV mB-bursts and noise storms by mode coupling in the warm plasma.
L. Mollwo.
Solar Physics, Vol. 12, 125 - 142 (1970).

It is shown from the tensor conductivity that in the warm plasma the conditions of a coupling point are attained very nearly for plasma parameters $X \approx 1$, $Y \approx 1$ and a very small ϑ even if the very small collision frequency is neglected. The ordinary polarization of the bursts and the other observational facts are explained by mode coupling from the originally emitted extraordinary mode excited by Cerenkov-radiation (type I) or gyroradiation (type IV mB and noise storms).

077.035 Sources of type III solar bursts observed at 169 MHz with the Nançay radioheliograph.
J.-L. Bougeret, C. Caroubalos, C. Mercier, M. Pick.
Astron. Astrophys., Vol. 6, 406 - 414 (1970).

Results of observations of type III solar bursts with the Nançay radioheliograph – space resolution 3.'4, time resolution used 10^{-1} s – are presented. 45 bursts have been studied and classified in different categories: "stationary", "slipping", "double". Measurements of the diameter, lifetime and position are given. An asymmetrical halo is observed in most cases. The interpretation is briefly discussed.

077.036 Interdependence between solar radio bursts of types I and III and non-linear processes in the coronal plasma.   I. M. Gordon.
Astrophys. Letters, Vol. 5, 251 - 255 (1970).

There is ample evidence, not interpreted as yet, that solar radio bursts of types III and V are observed in association with bursts of type I but at the lower frequencies. To explain this phenomenon it is suggested that Langmuir waves with low phase velocities, excited by the suprathermal electrons which are responsible for the bursts of type I, are accelerated by non-linear scattering on thermal electrons to phase velocities of approximately 0.1 to 0.6 C. The resulting K-spectrum of the Langmuir waves $W_k \propto K^{-5/2}$ leads to the rapid acceleration of the electrons to velocities V_0 of approximately 0.1 to 0.6 C, and it is suggested that these electrons are exciters of the bursts of types III and V.

077.037 A study of local sources of the slowly-varying component of solar radio emission in the cm-wavelength range.
G. B. Gelfreikh, S. B. Akhmedov, V. H. Borovik, V. Y. Golnev, A. N. Korzhavin, V. G. Nagnibeda, N. G. Peterova.
Izv. Glav. Astron. Obs. v Pulkove, No. 185, p. 167 - 182 (1970).   In Russian.

A connection of local sources (l. s.) of slowly varying components of solar radio emission with the sunspot groups is observed reliably in the cm-wavelength range. The large series of solar observations were made with the Pulkovo large radio telescope during 1964 - 1966 at 2.0, 3.2, 4.4, 6.6, 7.0 and 8.9 cm. The spectra of several l. s. and regularities of their development were investigated.

077.038 A study of the «constant» and slowly-varying components of solar radio emission by statistical methods.   N. S. Soboleva.
Izv. Glav. Astron. Obs. v Pulkove, No. 185, p. 183 - 190 (1970).   In Russian.

Results are given of a statistical analysis of the integrated solar radio emission at four frequencies from the data obtained during 1957 - 1964, and also of the polarized integrated emission at 3.2 cm (Pulkovo) obtained during the maximum of solar activity (1957 - 1958).

077.039 On two very peculiar phases of the solar type IV event on March 2, 1970. A. Abrami.
Mem. Soc. Astron. Italiana, Nuova Serie, Vol. 41, 231 - 234 (1970).

077.040 Observation of a solar radio burst before the solar eclipse of March 7, 1970.
M. Felli, G. Tofani.
Mem. Soc. Astron. Italiana, Nuova Serie, Vol. 41, 251 - 253 (1970).

077.041 Solar investigations on the wavelength 3.5 mm and 2 cm. A. Tlamicha.
Říše hvězd, Vol. 51, 66 - 69 (1970). In Czech.

077.042 Report on the solar radio events between May 24 and May 28, 1967 and photospheric conditions.
A. Böhme, A. Krüger, H. Künzel.
Heinrich-Hertz-Inst. für Solar-Terrestrische Physik, Deutsche Akad. Wiss. Berlin, HHI Suppl. Ser. Solar Data, Vol. 1, (No. 7), p. 125 - 166 (1969).
The large proton event of 1967 May 23 was investigated by many authors and a compilation was given in a special report of the World Data Center A. The present report contains further informations on other events of this active region and some discussions concerning the development of solar activity with main emphasis to the phenomenon of solar type IV radio bursts. The report is based on optical and radio observations of the HHI and some supplementary informations taken from other data bulletins.

077.043 Radio evidence for the propagation of magneto - hydrodynamic waves along curved paths in the solar corona. S. F. Smerd.
Proc. Astron. Soc. Australia, Vol. 1, 305 - 308 (1970).
In this paper 80 MHz heliograph observations are des - cribed of a remarkable solar outburst on 1969 March 30 initiated by a flare on the invisible hemisphere of the sun.

077.044 Positions of the fundamental and harmonic sources of a type II solar burst. G. A. Dulk.
Proc. Astron. Soc. Australia, Vol. 1, 308 - 310 (1970).

077.045 Position measurements of a harmonic type II solar burst. A. C. Riddle.
Proc. Astron. Soc. Australia, Vol. 1, 310 - 312 (1970).

077.046 Evidence of type II and moving type IV solar bursts excited by a common shock wave.
R. T. Stewart, K. V. Sheridan, K. Kai.
Proc. Astron. Soc. Australia, Vol. 1, 313 - 315 (1970).

077.047 Scatter in the positions of sources of associated type III solar bursts. D. J. McLean.
Proc. Astron. Soc. Australia, Vol. 1, 315 - 316 (1970).

077.048 The solar U-burst at metre wavelengths.
N. R. Labrum, R. T. Stewart.
Proc. Astron. Soc. Australia, Vol. 1, 316 - 318 (1970).

077.049 Determination of the radio brightness over the solar limb during eclipse by means of a radio telescope with narrow directional diagram.
G. P. Apushkinsky, T. A. Vitkovskaja.
Solnechnye Dannye 1970 Byull., No. 2, p. 102 - 105 (1970). In Russian.

077.050 A radio burst on the sun at 0.8 cm in the sunspot group No. 282 on August 23, 1967.

A. N. Tsyganov.
Solnechnye Dannye 1970 Byull., No. 3, p. 111 - 114 (1970). In Russian.

077.051 Polarization reversal of solar microwave bursts.
R. Ramaty, S. S. Holt.
Nature, Vol. 226, 68 - 69 (1970).
Tanaka *et al.* have recently reported that some microwave bursts can be resolved into two components with opposite circular polarizations. This article suggests that this bipolar structure is best explained in terms of gyrosynchrotron re-absorption by the radiating electrons themselves.

077.052 The measurement of complete polarization characteristics of solar radio emission.
L. R. McNarry, G. A. Harvey.
Bull. Radio Electr. Engineering Div., National Res. Council, Canada, Vol. 19 (No. 1), 35 - 47 (1969).
Describes the procedure by which the Stokes parameters are used to determine complete polarization characteristics of solar radio emission at 74 MHz. – *MWS*

077.053 Some important radio bursts observed at metric waves with high temporal resolution. A. Abrami.
Pubbl. Oss. Astron. Trieste, No. 412 (Osservazioni Solari, No. 17 Suppl.), 30 pp. (1970).

077.054 Some radio bursts in the period October 29 – November 3, 1968 observed at metric waves with high temporal resolution. A. Abrami.
Pubbl. Oss. Astron. Trieste, No. 414 [Reprinted from Report UAG - 8 - World Data Center A, p. 163 - 168], 6 pp. (1970).

077.055 The radio outburst of November 18, 1968 at 1026 UT observed at metric waves with high temporal resolution. A. Abrami.
Pubbl. Oss. Astron. Trieste, No. 416 [Reprinted from Report UAG - 8 - World Data Center A, p. 20 - 22], 3 pp. (1970).

An example of radio and optical homologous flares.
See Abstr. 073.011.

Description of mass motions and brightenings in a class 2b flare, August 8, 1968. See Abstr. 073.012.

The association of solar optical flares with type III solar bursts from 4 to 2 MHz observed by OGO-III.
See Abstr. 073.043.

The slowly varying component of the radio emission during the period of the July 1966 proton flare.
See Abstr. 073.066.

Remarks on the slowly varying component of solar radio emission during the proton flare period of July 1966.
See Abstr. 073.067.

Radio bursts associated with active region McMath No. 8362 of July 1966.
See Abstr. 073.074.

The dynamic spectrum of the proton event of 7 July 1966. See Abstr. 073.082.

Decametric radio spectra and positions during the flare of 7 July 1966, 0041 UT. See Abstr. 073.083.

Single-frequency bursts observed in Japan during the proton event of 7 July 1966. See Abstr. 073.084.

The complete type IV burst associated with the

7 July 1966 proton event.   See Abstr. 073.085.

Solar activity.   See Abstr. 073.128.

Pulsating radio emissions from the solar corona.
See Abstr. 074.007.

Theory of polarized type III and U burst emissions
from the solar corona.   See Abstr. 074.062.

X-ray and microwave emission of local sources on
the sun.   See Abstr. 076.010.

Radio astronomy.   See Abstr. 157.027.

078 Solar Cosmic Radiation

078.001 **Effects of sudden commencements on solar protons
at the synchronous orbit.**
G. A. Paulikas, J. B. Blake.
Journ. Geophys. Res., Vol. 75, 734 - 742 (1970).

Increases in solar proton flux coincident with sudden
commencements have been observed by detectors aboard the
geostationary ATS-1 satellite during 1967 and 1968. The flux
increases are of short duration (<30 min) and are of magnetospheric origin. Decay of the proton flux after the sudden
commencement enhancement (sce) is consistent with pitch
angle scattering by hydromagnetic waves.

078.002 **Discussion of paper, 'A comparison of energetic
storm protons to halo protons'.**
L. J. Lanzerotti.
Solar Physics, Vol. 11, 145 - 147 (1970), with a reply by
S. W. Kahler, Solar Physics, Vol. 11, 148 - 150 (1970).
Research note.

078.003 **Solar activity, 27-day variation and long term
modulation of cosmic ray intensity.**
V. K. Balasubrahmanyan, D. Venkatesan.
Solar Physics, Vol. 11, 151 - 154 (1970). – Research note.

078.004 **La propagation des rayons cosmiques solaires.**
E. Barouch.
Comptes Rendus Acad. Sci. Paris, Sér. B, Vol. 270, 973 -
976 (1970).

L'interaction des rayons cosmiques solaires avec les
ondes de choc interplanétaires est étudiée. Les prévisions du
modèle résultant sont en bon accord avec les données expéri-
mentales.

078.005 **About the angular sizes of solar corpuscular
streams.**   A. A. Danilov, I. Ya. Plotnikov.
Geomagn. Aeronom., Vol. 10, 134 - 136 (1970).   In Russian.
Brief information.

078.006 **Solar particle observations over the polar caps.**
G. A. Paulikas, J. B. Blake, A. L. Vampola.
Astrophys. Space Sci. Library, Vol. 17, 141 - 147 (1970).
Conference paper (see 012.008).

078.007 **Anisotropic solar cosmic ray propagation in an
inhomogeneous medium, II.**   L. F. Burlaga.
Solar Physics, Vol. 12, 317 - 327 (1970).

In the present paper a mathematical model for anisotropic solar cosmic ray propagation is discussed further with
emphasis on the limiting results for a continuous medium, and
it is applied to examine qualitatively some basic effects of the
diffusing region (solar shell) which is believed to extend from
somewhere near the earth to a few or several AU. Whereas
most of the earlier work emphasized the characteristics of the
total cosmic ray flux measured at 1 AU, the present work
examines some of the qualitative results that might be seen by

detectors which measure the anisotropy and intensity of solar
particle fluxes over a wide range of distances from the sun.

078.008 **The equilibrium anisotropy in the flux of 10-Mev
solar flare particles and their convection in the
solar wind.**   M. A. Forman.
Journ. Geophys. Res., Vol. 75, 3147 - 3153 (1970).

We conclude that, when the equilibrium anisotropy of
solar cosmic ray particles in the interplanetary medium is
observed, low-energy solar flare particles are being transported
mainly by convection in the solar wind. The ~ 14-hour decay
time for the particle density, calculated by assuming only
convective transport and adiabatic deceleration in a
spherically expanding solar wind, is in reasonable agreement
with observed values.

078.009 **Energetic solar particles.**
E. N. Parker.
Ann. IQSY, (see 012.016), Vol. 4, 57 - 61 (1969).

The several reasons for interest in energetic solar particles are reviewed, including general interest in acceleration
mechanisms, solar abundances, interplanetary magnetic field
configurations, and cosmic-ray propagation through interplanetary fields.

078.010 **Underground cosmic-ray measurements.**
H. Elliot.
Ann. IQSY, (see 012.016), Vol. 4, 175 - 180 (1969).

The present situation with regard to cosmic-ray solar
and sideral time variations underground is reviewed. The 12-
and 24-hour solar waves observed at depths of 60 mwe or less
can be broadly understood in relation to the known properties of the interplanetary magnetic field.

078.011 **Electrons and protons in long-lived streams of
energetic solar particles.**   K. A. Anderson.
Report NASA-CR-97611, California Univ., Berkeley.
[Available from CFSTI, Springfield, Va.], 45 pp. (1968).

In 1966 and 1967 many long-lived streams of low-energy solar electrons and protons were observed near earth.
These streams were sometimes associated with bright flares
which occurred many hours earlier and sometimes no individual flare could be found. The long-lived solar events discussed
here include energetic storm particles, delayed events and
fluxes associated with solar active regions.

078.012 **The energy spectrum of the cosmic ray emission on
28 January 1967.**   P. Chaloupka, J. Ilencik.
Fyz. Casopis (Czechoslovakia), Vol. 19, (No. 4), 236 - 243
(1969).

On the 28th of January 1967 there occurred an emission
of cosmic rays from the Sun. The time dependence of the
particles emitted from the Sun of the exponent of the energy
spectrum is dealt with. Data obtained at 16 stations were used.

The emission and propagation of ~40 keV solar flare electrons. I. The relationship of ~40 keV electron to energetic proton and relativistic electron emission by the sun. See Abstr. 073.042.

The increase in low-energy cosmic ray intensity on 7 July 1966.   See Abstr. 073.095.

Pioneer 6 observations of the solar flare particle event of 7 July 1966.   See Abstr. 073.102.

Generation spectrum for the flare of 23 February 1956.   See Abstr. 073.124.

Energetic solar radiations. See Abstr. 076.027.

A theoretical investigation of corpuscular radiation effects on the F-region of the ionosphere. See Abstr. 083.011.

The altitude dependence of the quiettime cosmic ray ionization over the polar regions at solar minimum.

See Abstr. 083.061.

Observations of the cosmic ray knee with a polar orbiting ionization chamber.   See Abstr. 083.062.

Large-scale structure of the interplanetary magnetic field according to the data of the annual variations of cosmic-ray intensity during the cycle of solar activity. See Abstr. 106.027.

Ground-based synoptic observations of cosmic rays.   See Abstr. 143.036.

IQSY observations of low-energy galactic and solar cosmic rays.   See Abstr. 143.038.

Cosmic ray variations: Theory. See Abstr. 143.039.

Calculations of the rate of formation of electrons by cosmic rays in the earth's atmosphere with allowance for the changes of energy of the falling primary particles. See Abstr. 143.062.

079   Solar Eclipses

079.001   **Atmospheric gravity waves induced by a solar eclipse.**   G. Chimonas, C. O. Hines.
Journ. Geophys. Res., Vol. 75, 875 (1970).
   We wish to point out that a solar eclipse, by interfering with the heat balance in the shadowed portion of the atmosphere, can be expected to generate a signal comprising internal gravity waves of a magnitude that may be detected both at ground level and at ionospheric heights, even at considerable distance from the eclipse path.

079.002   **Eclipses.**   P. Couderc.
El Universo, Vol. 24, 7 - 14 (1970). – Popular article.

079.100   **Solar eclipse, 1965 November 23**

   Ionospheric and magnetic observations during the annular solar eclipse of November 23, 1965.
J. E. van der Laan.
Journ. Geophys. Res., Vol. 75, 1312 - 1318 (1970).
   Vertical-incidence (VI) ionograms, magnetograms, and relative VI signal-amplitude (7-MHz) data were obtained in Bangkok, Thailand, during the annular solar eclipse on 23 November 1965. Marked changes in F_2-layer virtual height and considerable stratification of the F layer, including formation of an $F_{1.5}$ layer, were observed. Magnetograms indicate a large (approximately 25%) diminution in ionospheric current relative to normal solar daily (S_q) variations.

079.101   **Solar eclipse, 1966 November 12**

   Determination of moon–sun conjunction time, based on photographic observations of the November 12, 1966 eclipse with the 43 cm Gautier telescope.
F. Muñoz, C. Mondinalli.
Asoc. Argentina Astron. Bol., No. 12, (see 012.019), p. 38 - 39 (1968).   In Spanish. – Abstract.

   Coronal polarization and intensity at the November 12, 1966 solar eclipse.   See Abstr. 074.006.

   Observations of the brightness and polarization of the outer corona during the 1966 November 12 total eclipse of the sun.   See Abstr. 074.024.

079.102   **Solar eclipse, 1970 March 7**

   Rafaga solar durante el eclipse del 7 de Marzo de 1970.   L. R. Terrazas, G. González C.
Bol. Obs. Tonantzintla y Tacubaya, Vol. 5 (No. 34), 237 - 238 (1970).

   Observation de la lumière cendrée au cours d'une éclipse totale.   S. Koutchmy.
Comptes Rendus Acad. Sci. Paris, Sér. B, Vol. 271, 259 - 261 (1970).
   On estime, à partir d'hypothèses simplificatrices, la brillance des parties claires de la lumière cendree (phase totale) en fonction de la brillance moyenne du soleil.

   Eclipse 70, Mexico.   F. Diego Q.

El Universo, Vol. 24, 15 - 21 (1970).

   International solar flash-spectrum experiment.
Journ. Royal Astron. Soc. Canada, Vol. 64, 201 - 202 (1970).

   Rocket UV flash spectra from the solar eclipse of March 7, 1970.   R. J. Speer, W. R. S. Garton, J. F. Morgan, R. W. Nicholls, L. Goldberg, W. H. Parkinson, E. M. Reeves, T. J. L. Jones, H. J. B. Paxton, D. B. Shenton, R. Wilson.
Nature, Vol. 226, 249 - 250 (1970).
   Flash spectra of the total solar eclipse throughout all its phases have been obtained in the extreme ultraviolet for the first time.

   Prediction of the coronal structure for the solar eclipse of March 7, 1970.   K. H. Schatten.
Nature, Vol. 226, 251 (1970).

   Eclipse not spoilt by shortage of funds.
G. Newkirk, Jr., L. Lacey.
Nature, Vol. 226, 1097 - 1098 (1970).

   Eclipse effects in the ionosphere.
H. Rishbeth.
Nature, Vol. 226, 1099 - 1100 (1970).
   This article surveys some recent trends in the study of the ionospheric effects of solar eclipses.

   Changes in the lower ionosphere during the eclipse– A preliminary report of the Canadian programme.
J. S. Belrose, A. G. McNamara, J. E. Hall, L. R. Bodé, R. Bunker, D. B. Ross, P. H. G. Dickinson, A. J. Hall.
Nature, Vol. 226, 1100 - 1107 (1970).
   Ground-based radio investigations and four rocket launches were carried out in Canada to study the effect of the eclipse on the solar radiation and electron densities in the lower ionosphere (below about 150 km). Four articles describe the experiments.

   Preliminary results from a meteorological rocket experiment.   R. M. Henry, R. S. Quiroz.
Nature, Vol. 226, 1108 - 1110 (1970).

   Rocket density measurements at Eglin, Florida.
A. C. Faire.
Nature, Vol. 226, 1110 - 1111 (1970).

   Changes in the electron content of the ionosphere.
B. J. Bienstock, R. T. Marriott, D. E. St. John, R. M. Thorne, S. V. Venkateswaran.
Nature, Vol. 226, 1111 - 1112 (1970).

   Comparison of changes in total electron content along three paths.   J. A. Klobuchar, C. Malik.
Nature, Vol. 226, 1113 - 1114 (1970).

   Synoptic preview of ionospheric data taken at Fort Monmouth, New Jersey, during the eclipse.
P. R. Arendt, F. Gorman, Jr., H. Soicher.
Nature, Vol. 226, 1114 (1970).

   Observations of ionospheric electron content during the March 7, 1970, solar eclipse.
O. G. Almeida, A. V. da Rosa.
Nature, Vol. 226, 1115 - 1116 (1970).

Variation of electron density in the D-region.
J. N. Rowe, A. J. Ferraro, H. S. Lee.
Nature, Vol. 226, 1116 - 1117 (1970).

Steep-incidence D-region reflexions of low frequency pulses during the March 7 eclipse.
B. Wieder, R. H. Espeland.
Nature, Vol. 226, 1117 - 1118 (1970).

Preliminary results of a radio absorption study at 3.5 MHz.   J. R. Kennedy, J. J. Schauble.
Nature, Vol. 226, 1118 - 1119 (1970).

Electron energy loss factor in the D-region during the eclipse of March 7, 1970.   W. A. Kissick.
Nature, Vol. 226, 1119 - 1121 (1970).

Response of the F-region ionosphere to a solar eclipse.   B. J. Flaherty, H. R. Cho, K. C. Yeh.
Nature, Vol. 226, 1121 - 1123 (1970).

Possible detection of atmospheric gravity waves generated by the solar eclipse.
M. J. Davis, A. V. da Rosa.
Nature, Vol. 226, 1123 (1970).

Eclipse observations at Arecibo, Puerto Rico, on March 7, 1970.   H. C. Carlson, R. Harper, V. Wickwar, R. L. Showen, R. Behnke, T. F. Trost, L. R. Cogger, C. J. Nelson.
Nature, Vol. 226, 1124 - 1125 (1970).

Travelling ionospheric disturbances in Athens during the March 7 solar eclipse.
M. Anastassiadis, G. Moraitis.
Nature, Vol. 226, 1125 - 1126 (1970).

VLF radio signals observed in Newfoundland during the solar eclipse of March 7, 1970.
M. J. Rycroft, C. D. Reeve.
Nature, Vol. 226, 1126 - 1127 (1970).

VLF propagation effects produced by the eclipse.
R. E. Schaal, A. M. Mendes, S. Ananthakrishnan, P. Kaufmann.
Nature, Vol. 226, 1127 - 1129 (1970).

Comparison of 100 kHz pulse propagation during two solar eclipses.   R. H. Doherty.
Nature, Vol. 226, 1129 (1970).

Preliminary comparison of predicted and observed structure of the solar corona at the eclipse of March 7, 1970.   See Abstr. 074.041.

Prediction and observation of the coronal structure for the solar eclipse of March 7, 1970.
See Abstr. 074.042.

Solar UV flash spectrum 1400 Å – 1960 Å.
See Abstr. 076.020.

Rocket observations of the corona on March 7, 1970.   See Abstr. 074.043.

The Lockheed-aerospace eclipse expedition.
D. C. Martin, S. F. Smith, G. A. Chapman.
Nature, Vol. 226, 1138 - 1139 (1970).

Swiss solar eclipse expedition to Mexico.
M. Waldmeier.

Nature, Vol. 226, 1139 - 1140 (1970).

Los Alamos airborne observations.
A. N. Cox.
Nature, Vol. 226, 1140 - 1141 (1970).

Photometric study of the solar corona.
See Abstr. 074.044.

Doppler temperature of the solar corona.
See Abstr. 074.045.

The outer corona at the eclipse of March 7, 1970.
See Abstr. 074.046.

Variations in line profiles from photosphere to chromosphere.   See Abstr. 071.035.

Association of coronal structures with chromospheric structure.   See Abstr. 074.047.

Photospheric magnetic fields on March 7, 1970.
See Abstr. 071.036.

Hα filtergrams of the sun on March 7, 1970.
See Abstr. 073.044.

Observations at the Sagamore Hill Solar Radio Observatory.   See Abstr. 077.031.

16 GHz observations of the eclipse.
See Abstr. 077.032.

Observations with a 7 GHz polarimeter.
See Abstr. 077.033.

Radio observations of the distribution of chromospheric brightness temperature.   See Abstr. 073.045.

Zur totalen Sonnenfinsternis am 7. März 1970 in Florida.   G. Schindler.
Orion Schaffhausen, 28. Jahrgang, p. 17 - 18 (1970).

Total solar eclipse of March 7, 1970.
J. Klepešta.
Říše hvězd, Vol. 51, 111 - 115 (1970).   In Czech.

Travel and eclipse site guide for Mexico.
L. J. Robinson.
Sky Telescope, Vol. 39, 19 - 20 (1970).

Eclipse survey for the Astronomical League.
R. C. Maag.
Sky Telescope, Vol. 39, 28 - 30 (1970).

Partial phases of the March eclipse.
Sky Telescope, Vol. 39, 88 - 89 (1970).

Recording shadow bands at the March eclipse.
E. M. Paulton.
Sky Telescope, Vol. 39, 132 - 133 (1970).

Scientists' eclipse goals.   L. J. Robinson.
Sky Telescope, Vol. 39, 167 - 171 (1970).

First eclipse reports.
Sky Telescope, Vol. 39, 211 - 214, 270 (1970).

With the eclipse expeditions in Mexico.
R. Little, L. J. Robinson.
Sky Telescope, Vol. 39, 280 - 285 (1970).

Total eclipse along the eastern seaboard.
Sky Telescope, Vol. 39, 285 - 289 (1970).

A colorful total eclipse.
Sky Telescope, Vol. 39, 308 - 309 (1970).

Partial-eclipse observations on March 7th.
Sky Telescope, Vol. 39, 324 - 328 (1970).

The March eclipse rocket program at Wallops Island.   C. A. Accardo.
Sky Telescope, Vol. 39, 344 - 349 (1970).

The High Altitude Observatory's 1970 eclipse expedition.   R. A. Kopp.
Sky Telescope, Vol. 39, 359 - 362 (1970).

The March 7, 1970 total eclipse in Mexico.
T. E. Bell, J. Laborde, W. B. Lindley.
Strolling Astronomer, Vol. 22, 101 - 105 (1970).

Sonnenfinsternis über Mexiko.
M. Waldmeier.
Umschau, 70. Jahrgang, p. 391 (1970).

The F-region during an eclipse – A theoretical study.
See Abstr. 083.057.

079.103   Solar eclipse, 1896 August 9

Observations of the August 9, 1896 total solar eclipse in Hokkaido, Japan.
K. Saito, S. Shinozawa.
Tokyo Astron. Obs. Rep., Vol. 15, No. 1 (No. 56), p. 12 - 54 (1970).   In Japanese.

079.104   Solar eclipse, 1968 September 22

A method of measuring the slope of the solar radio emission spectrum and its application to observations during the eclipse of September 22, 1968 in Gorki.
See Abstr. 077.010.

079.105   Solar eclipse, 1967 May 9

Brightness distribution in the polar zone of the solar limb at 4.0 cm as determined from observations of the solar eclipse on May 9, 1967.   See Abstr. 077.011

079.106   Solar eclipse, 1966 May 20

The influence of the solar eclipse of May 20, 1966 on the ionospheric E-region.
K. Serafimov, J. Taubenheim.
Geomagn. Aeronom., Vol. 10, 451 - 460 (1970).   In Russian.

An attempt to measure the solar limb darkening using the lunar profile during the eclipse of May 20, 1966.
See Abstr. 071.033.

The determination of the two-dimensional structure of the noise storm on May 20, 1966 from data of eclipse observations at 4 stations.   See Abstr. 077.015.

Electron densities measured by the partial reflection method compared with simultaneous rocket measurements.
See Abstr. 083.058.

079.107   Solar eclipse, 1965 May 30

Observations of the brightness and polarization of the outer corona during the 1966 November 12 total eclipse of the sun. Comparisions with the eclipse 1965 May 30.
See Abstr. 074.024.

079.108   Solar eclipse, 1962 February 4

Coronagraph observations of the coronal condensation of 4 February 1962.   See Abstr. 074.049.

079.109   Solar eclipse, 1961 February 15

Observation of the polarization of the solar corona during the solar eclipse at Hvar on 15 II 1961.
J. Arsenijević.
Bull. Obs. Astron. Beograd, Vol. 27, No. 2, p. 24 - 28 (1969). In French.
Based on the obtained 8 second exposures with suitable calibrations, the results for the smaller part of the corona were obtained. The values for the degree and direction of the polarization agree favorably with the results of previous measurements of other authors.

080 Solar Figure, Internal Constitution, Rotation, Miscellanea

**080.001 The solar oblateness and the gravitational quadru-
pole moment.** R. H. Dicke.
Astrophys. Journ., Vol. 159, 1 - 24 (1970).

The connection between the oblateness and the gravitational quadrupole moment of the sun is developed under
the assumption of the absence of magnetic and velocity fields
in the surface layers. This discussion is generalized to include
also the effects of surface rotation for rotation on cylinders.
The theory of the generation by surface fields of oblateness
and equatorial temperature excess is developed. The severe
constraint on allowed surface fields introduced by the observed latitude independence of solar brightness is analyzed. It
is concluded that an acceptable explanation of the solar oblateness, other than the effects of a gravitational quadrupole
and surface rotation, has yet to be found.

**080.002 Modeling of the distribution of the sun's radiative
energy in the spectral region between 500 and
1500 Å by means of a gas-beam source.**
Eh. T. Verkhovtseva, A. V. Kravchenko, V. S. Osyka,
Ya. M. Fogel'.
Kosmich. Issled.,Vol. 8, 140 - 145 (1970). In Russian.

080.003 On large-scale solar convection.
R. P. Davies-Jones, P. A. Gilman.
Solar Physics, Vol. 12, 3 - 22 (1970).

Quantitatively realistic models of large-scale solar convection, which predict the dominant mode, its amplitude and
structure, are still in the future, due to the difficulties in
handling simultaneously such effects as compressibility,
partial ionization, spherical geometry, differential rotation,
eddy viscosity and conductivity, variation of gravity, etc.
We show here that solutions from even the simplest models
of convection in a rotating system contain several properties
of relevance to large-scale solar phenomena.

080.004 Spectroscopic determinations of solar rotation.
R. Howard, J. Harvey.
Solar Physics, Vol. 12, 23 - 51 (1970).

Spectral line shift data obtained from full-disk magnetograms recorded at Mt. Wilson are analyzed for differential
rotation. The method of analysis is discussed and the results
from the data for 1966 through 1968 are presented. The
average equatorial velocity over this period is found to be
1.93 km/sec or 13.76 deg/day (sideral). This corresponds to
a sideral period of 26.16 days.

**080.005 Some observed characteristics of solar radar echoes
and their implications.** J. C. James.
Solar Physics, Vol. 12, 143 - 162 (1970).

The results of a series of radar studies of the sun at
38.2 MHz are presented. The echoes imply ever-present
compressional waves in the corona, and these waves are
likely associated with coronal heating. Some echoes are
refracted by plasma clouds high in the corona. Other echoes
are reflected by dense plasma irregularities moving outward
very slowly at 0 to 20 km/sec.

**080.006 Neutrino astronomy of the sun. Part I. Neutrino
production.** B. Kuchowicz.
Postępy Astron., Vol. 18, 149 - 174 (1970). In Polish.

In a general introduction properties of the neutrinos are
summarized together with a short historical outline of development of the neutrino research. Various neutrino sources
(occurring in nature and artificial) are presented and a detailed discussion of the thermonuclear reactions as neutrino
sources in the main sequence stars is given. The extremely

weak interaction of the neutrino with matter makes it an
especially suited probe for stellar interiors.

080.007 A line-blanketed solar model.
R. G. Athay.
Bull. American Astron. Soc., Vol. 2, 181 (1970). – Abstr.
AAS.

**080.008 A model of the solar temperature-minimum
region.** E. H. Avrett, J. L. Linsky.
Bull. American Astron. Soc., Vol. 2, 181 (1970). – Abstr.
AAS.

**080.009 Composition uncertainty and the solar-core
opacity.** W. D. Watson.
Bull. American Astron. Soc., Vol. 2, 224 (1970). – Abstr.
AAS.

**080.010 Zum Einfluß eines Quadrupolmoments der Sonne
auf die Bahnlage der Planeten.**
S. Böhme.
Astron. Nachr., Vol. 292, 35 - 36 = Mitt. Astron. Rechen-
Inst. Heidelberg, Ser. B, No. 21 (1970).

Formulae are derived for the transformation of the
secular perturbations of the elements giving the position
of the orbit of a planet, produced by a solar quadrupole
moment and related to the equator of the sun, to perturbations related to the ecliptic. For Mercury, Venus, and
earth the numerical values of the coefficients of the
transforming equations are given.

**080.011 Thermally driven motions in a gravitational atmo-
sphere.** R. J. Bessey, M. Kuperus.
Solar Physics, Vol. 12, 216 - 228 (1970).

Numerical solutions of the non-linear equations of fluid
dynamics for a compressible inviscid initially isothermal atmosphere are given using Lax' method for the integration of the
equations when discontinuities occur in the flow. The motion
of the atmosphere is studied following the heating of a thin
layer in the atmosphere. It is found that for a sufficiently large
heat input the atmosphere strongly expands towards the
regions of lower densities. In most cases a shock wave is formed
which precedes the expanding region. The possible occurrence
of thermally generated motions in the solar chromosphere is
discussed.

**080.012 Determination of the spread function for solar
stray light.** L. Staveland.
Solar Physics, Vol. 12, 328 - 331 (1970). – Research note.

**080.013 Solar-neutrino fluxes with recent corrections to
opacity.** J. N. Bahcall, R. K. Ulrich.
Astrophys. Journ. *(Letters)*, Vol. 160, L57 - L60 (1970).

Improved calculations of the rate of solar-neutrino
capture for experiments with ^{37}Cl and ^7Li are presented. There
is a sizable discrepancy between the calculated values and the
experimental upper limit of Davis et al. The primordial
helium abundance inferred from the relevant solar models
ranges from $Y = 0.22$ to $Y = 0.26$. The expected neutrino
capture rates in the ^{37}Cl and ^7Li experiments and the inferred
primordial helium abundance are sensitive to the present-day
photospheric abundance ratio of iron to hydrogen.

080.014 The ultraviolet solar opacity. O. Gingerich.
IAU Symposium No. 36, (see 012.014), p. 140 -
142 (1970).

Ultraviolet solar observations are compared with predictions from a new solar model. From 3600 to 1700 Å there is heavy line blanketing; probably one or more major sources of opacity are missing from the theoretical calculation in this region.

080.015 Image motion of the solar limb measured at the Norikura Corona Station. M. Irie.
Univ. Tokyo, Tokyo Astron. Obs. Report, Vol. 15, (No. 2), (No. 57), p. 317 - 320 (1970). In Japanese.

080.016 Electron correlations and solar neutrino counts. M. O. Diesendorf.
Nature, Vol. 227, 266 - 267 (1970).

This report shows that corrections to the Thomson cross-section produced by classical and quantum electron correlations reduce the electron-scattering opacity in the core of the sun by approximately one-third, and that this leads to a significant reduction in the theoretical neutrino flux.

080.017 The course of the C-N-O cycle in the sun's interior in the process of evolution and neutrino generation.
V. A. Kuzmin, V. I. Chelpanov.
Cosmic Rays No. 11, Moscow, p. 139 - 141 (1969). In Russian.

080.018 On the spectrum of solar B^8 -neutrinos. Yu. S. Kopysov, V. A. Kuzmin.
Cosmic Rays No. 11, Moscow, p. 142 - 145 (1969). In Russian.

The energy spectrum of neutrinos from the fissure B^8

(e^+, ν) Be^{8*} in the interior of the sun is calculated with consideration of transitions to highly excited level of the nucleus Be^8.

080.019 A summary of recent measurements of the solar constant outside the earth's troposphere.
E. A. Makarova, A. V. Kharitonov.
Astron. Tsirk., No. 543, p. 3 - 5 (1970). In Russian.

080.020 Influence of a horizontal magnetic field on the criterion of convective instability.
N. S. Petrukhin.
Solnechnye Dannye 1970 Byull., No. 3, p. 107 - 110 (1970). In Russian.

Solare Neutrinos.
See Abstr. 061.004.

Toroidal magnetic fields and differentially rotating sun. See Abstr. 062.009.

The energy transfer in the convective envelope of a rotating star and the solar differential rotation.
See Abstr. 064.039.

On thermonuclear reactions in the interior of stars similar to the sun. See Abstr. 065.087.

Limb darkening and solar rotation – II. Observed darkening in 1966. See Abstr. 071.026.

Earth

081 Figure, Composition, and Gravity of the Earth

081.001 Novel correlations between global features of the earth's gravitational and magnetic fields.
R. Hide, S. R. C. Malin.
Nature, Vol. 225, 605 - 609 (1970).
There is a significant correlation between the earth's gravitational field and the non-dipole part of the geomagnetic field, provided the latter is displaced in longitude.

081.002 Imperfections of elasticity and continental drift.
H. Jeffreys.
Nature, Vol. 225, 1007 - 1008 (1970).
Imperfection of elasticity in the earth follows a law that forbids convections and continental drift.

081.003 The contribution of satellite geodesy to the determination of the figure and the internal constitution of the earth. E. Groten.
Journ. British Interplanet. Soc., Vol. 23, 325 - 328 (1970).
A general review of the present and future contribution of satellite geodesy to precise mapping of the earth's surface and to terrestrial geophysics is presented.

081.004 Seasonal variations of the geopotential inferred from satellite observations. Y. Kozai.
SAO, *Cambridge, Mass.,* Special Rep., No. 312, 3 + 6 pp. (1970).
Annual variations of J_2, the coefficient of the dominant term in the geopotential, is identified from the variations in the longitudes of the ascending nodes for two satellites. It is concluded that about half the seasonal variation of the length of day is due to a change in the principal moment of inertia.

081.005 A Taylor series approach to the rotation and evaluation of spherical harmonics. R. W. James.
Geophys. Journ. Roy. Astron. Soc., Vol. 19, 203 - 205 (1970).
Letter.

081.006 Datations terrestres et cosmiques.
O. Godart.
Ciel et Terre, Vol. 86, 1 - 22 (1970). – Review article.

081.007 Astro-geo project Spitsbergen 1968 - 1970.
P. Melchior, M. Bonatz, J. Blankenburgh.
Ciel et Terre, Vol. 86, 85 - 99 (1970).
The aims endeavoured to obtain by this mission are: the study of the characteristics of the earth tides in the neighbourhood of the north pole, the slow movements of the earth's crust and an attempt to make a junction of this region to Europe by satellite triangulation.

081.008 La nouvelle conception de la dérive des continents.
2. L'enregistrement paléomagnétique de l'expansion des fonds océaniques.** A. de Vuyst.
Ciel et Terre, Vol. 86, 100 - 124 (1970).

081.009 La gravimétrie à l'Observatoire Royal de Belgique.
B. Ducarme.
Ciel et Terre, Vol. 86, 125 - 144 (1970).
The earth's gravity field is examined according to three main topics: absolute measurements, variations at the surface

and periodic changes of *g* under the action of the sun and of the moon. The gravimetric tides are studied at the Royal Observatory of Belgium since the I.G.Y. in connection with the internal constitution of the earth.

081.010 The earth as a planet.
A. H. Cook.
Nature, Vol. 226, 18 - 20 (1970).
A comparison of the earth with the moon and other planets can help considerably in the solving of geophysical problems. The author deals with some of these problems in this article, which is a summary of the inaugural lecture he gave at Edinburgh on January 15.

081.011 Aeromagnetic tests for continental drift in Africa and South America.
D. W. Strangway, P. R. Vogt.
Earth Planet. Sci. Letters, Vol. 7, 429 - 435 (1970).

081.012 The recurrent mesozoic drift of South America and Africa. J. F. Vilas, D. A. Valencio.
Earth Planet. Sci. Letters, Vol. 7, 441 - 444 (1970).
Recent palaeomagnetic investigations on Mesozoic lavas from Argentina have indicated that the relative drift between South America and Africa started later than upper Triassic but before middle Cretaceous.

081.013 The electrical conductivity of the earth.
A. T. Price.
Quarterly Journ. Roy. Astron. Soc., Vol. 11, 23 - 42 (1970).
Harold Jeffreys lecture delivered at Burlington House on 1969 October 24.

081.014 Turbulence and mixing in stably stratified environments.
Quarterly Journ. Roy. Astron. Soc., Vol. 11, 49 - 51 (1970).
Report of a geophysical discussion held in Burlington House on 1969 February 28.

081.015 Fit between Africa and Antarctica: A continental drift reconstruction.
R. S. Dietz, W. P. Sproll.
Science, Vol. 167, 1612 - 1614 (1970).
A computerized best fit position is obtained for the juxtaposition of Africa and Antarctica in a continental drift reconstruction.

081.016 A simple layer model of the geopotential from a combination of satellite and gravity data.
K.-R. Koch, F. Morrison.
Journ. Geophys. Res., Vol. 75, 1483 - 1492 (1970).
Instead of the expansion in spherical harmonics currently used in satellite geodesy, the geopotential in this analysis is represented by the potential of a simple layer distributed over the surface of the earth. Density values of this layer for 48 surface elements have been determined from Baker-Nunn camera observations of 4 satellites for 5 weeks. Existing gravity anomalies are combined with the solution obtained from the satellite observations. The combination does not markedly differ from the satellite solution because the de-

termination of the geopotential from satellite data is stronger.

081.017 Geophysical implications of the interrelationship of the geogravity and geomagnetic fields.
M. A. Khan, G. P. Woollard.
Nature, Vol. 226, 340 - 343 (1970).

081.018 Comments on paper by Milo Wolff, 'Direct measurements of the earth's gravitational potential using a satellite pair'. F. Morrison.
Journ. Geophys. Res., Vol. 75, 2142 - 2143 (1970).

081.019 ˙Evaporites and continental drift.
E. Bonatti, M. Ball, C. Schubert.
Naturwissenschaften, Vol. 57, 107 - 108 (1970).

081.020 Magnetohydrodynamic twisting oscillations in the core of the earth and variations of the length of the day. S. I. Braginsky.
Geomagn. Aeronom., Vol. 10, 3 - 12 (1970). In Russian.

081.021 Développement de l'aplatissement et de la pesenteur de la terre jusqu'aux termes du troisième ordre inclusivement. V. K. Hristov.
Izv. Tsentr. labor. geod. B'lg. AN, Vol. 9, 113 - 128 (1969).

081.022 On the selection of a standard ellipsoid.
O. M. Ostach.
Geod., kartogr. i aehrofotos"emka. Mezhved. resp. nauchnotekhn. sb., 1969 vyp. (No.) 9, p. 47 - 49. In Russian.
Abstr. in Referativ. Zhurn. 52.Geod. Aehros"emka, 3.52.108 (1970).

081.023 Precession in a disk dynamo model of the earth's dipole field. G. C. J. Suffolk.
Nature, Vol. 226, 628 - 629 (1970).
It is suggested that several polarity intervals of the dipole field of the earth may be connected with the effects of precession acting on the earth's dynamo process.

081.024 Earth models consistent with geophysical data.
F. Press.
Phys. Earth Planet. Interiors, Vol. 3, 3 - 22 (1970). − Conference paper (see 012.004).

081.025 Phase transformations within the earth's mantle as a cause of crustal movement and a source of crustal material. S. I. Subbotin.
Phys. Earth Planet. Interiors, Vol. 3, 499 - 502 (1970). − Conference paper (see 012.004).

081.026 The evolution of the earth's core and its magnetic field. J. A. Jacobs.
Phys. Earth Planet. Interiors, Vol. 3, 513 - 518 (1970). − Conference paper (see 012.004).

081.027 Großschollentektonik und Kontinentaldrift.
D. P. McKenzie.
Endeavour, Vol. 29, (No. 106), 39 - 44 (1970).

081.028 On the modified Lomnitz law of damping.
H. Jeffreys, S. Crampin.
Monthly Notices, Roy. Astron. Soc., Vol. 147, 295 - 301 (1970).
Calculations based on the modified Lomnitz law of damping show that the effect on a unit pulse can be conveniently expressed in terms of two variables, and results are found based on the new estimate of damping of the 14-month period. Separate calculations are made for the most probable value and values increased and decreased by the standard error. Previous results on the effect on rotations, the moon's libra-

tions, and the rates of subsidence of its dynamical ellipticities and of gravity anomalies on the earth are revised; the results are consistent with the facts.

081.029 Convection in the mantle and its effect upon the earth's gravitational field. D. G. Ashworth.
The Moon, Vol. 1, 339 - 346 (1970).
The failure of an equilibrium model to provide an adequate representation of the earth's external gravitational field suggests that one should consider a more general hydrodynamical model for the interior of the terrestrial globe, and the most probable cause of motion, which may significantly effect the distribution of density inside the earth, is convection throughout the mantle. In the present paper we investigate the effects of convection in the mantle on the gravitational field of the earth and calculate the velocity of convection necessary to account for the observed characteristics of the external gravitational field.

081.030 On the spherical harmonics series expansion of the geopotential. E. Groten.
Bull. Géod., Nouvelle Série, No. 96, p. 169 - 181 (1970).
The spherical harmonics development of the gravitational potential at the earth's surface and on the geoid is discussed. First, the effect of Kelvin transformations is studied; secondly, numerical information as obtained from satellite and other data is investigated.

081.031 East Canary Islands as a microcontinent within the Africa–North America continental drift fit.
R. S. Dietz, W. P. Sproll.
Nature, Vol. 226, 1043 - 1045 (1970).

081.032 Anorthosites in the early crust of the earth and on the moon. B. F. Windley.
Nature, Vol. 226, 333 - 335 (1970).
Archaean (pre-3000 m.y.) calcic chromiferous anorthosites occur as layers up to 6 km thick in high grade gneisses in several parts of the world especially west Greenland. They are fragments of stratiform igneous cumulates separated from their ultramafic differentiates at the base of the crust early in the evolution of the earth. They are compositionally similar to the calcic chromiferous lunar anorthosites. This suggests that both planetary bodies went through parallel stages of anorthositic development.

081.033 Expansion of the geopotential in ellipsoidal harmonics. H. G. Walter.
Scientific Report, ESRO SR-11 (ESOC), 6 + 12 pp. (1969).
The potential of the earth is developed in ellipsoidal harmonics, and the mathematical tools required, which are the generation of Lamé's functions and the relationship between rectangular and ellipsoidal coordinates, are compiled. With the aim of carrying out a numerical integration of the Lagrangian equations of planetary motion, the functional dependence of the disturbing earth potential on the orbital elements for elliptic motion is given. In particular, formulae for the partial derivatives of the disturbing potential with respect to the orbital elements are derived, thus making possible the numerical calculation of these partial derivatives from orbital elements.

081.034 Timing of the breakup of the continents around the Atlantic as determined by paleomagnetism.
E. E. Larson, L. LaFountain.
Earth Planet. Sci. Letters, Vol. 8, 341 - 351 (1970).

081.035 The earth's thermal regime.
B. G. Poliak, Ia. B. Smirnov.
Priroda, No. 5.70, p. 12 - 18 (1970). In Russian.

081.036 **On temporal variations of the second derivatives of the gravitational potential.** V. A. Kazinskii.
Dokl. Akad. Nauk SSSR, Ser. Mat. Fiz., Vol. 192, 790 - 792 (1970). In Russian.

081.037 **The cause of the ice ages.**
G. Bacsák.
Bull. Obs. Astron. Beograd, Vol. 27, No. 2, p. 12 - 23 (1969). In German.

081.038 **New data on the figure and dimensions of the earth.**
B. A. Volynskij.
Uch. Zap. Yaroslavsk. gos. ped. in-t, 1969 vyp. (No.) 75, p. 175 - 181. In Russian. – Abstr. in Referativ. Zhurn. 52. Geod. Aehros"emka, 5.52.88 (1970).

081.039 **On the external gravitational field of the earth.**
V. I. Vajnrot.
Izv. vyssh. uchebn. zavedenij. Geod. i aehrofotos"emka, 1968 No. 6, p. 71 - 74. In Russian. – Abstr. in Referativ. Zhurn. 52 Geod. Aehros'emka, 5.52.90 (1970).

081.040 **Auf das Bessel-Ellipsoid bezogene astrogeodätische Geoide für die Bundesrepublik Deutschland.**
S. Heitz.
Nachr. Karten–, Vermessungswesen, Reihe I, Heft No. 48, 33 pp. (1970).
For the purpose of practical geodetic surveying in the Federal Republic of Germany the author performed determinations of the geoid on the basis of deflections of the vertical in the system of the "Reichsdreiecksnetz" basing on Bessel's ellipsoid. The computations being made by interpolation according to the least squares method. The author describes the fundamentals of this work and gives the results of his geoid determinations.

081.041 **The role of the vertical gravitational gradient in the theory of the figure of the earth.** M. I. Marych.
Geod., kartogr. i aehrofotos"emka. Mezhved. resp. nauchno-tekhn. sb., 1969, vyp. (No.) 8, p. 125 - 128, with comments by N. K. Migal'. In Russian. – Abstr. in Referativ. Zhurn. 52. Geod. Aehros"emka, 6.52.88; 6.52.89 (1970).

081.042 **Method of common determination of the reduction constants of the theory of the figure of the earth.**
A. T. Dul'tsev.
Geod., kartogr. i aehrofotos"emka. Mezhved. resp. nauchno-tekhn. sb., 1969, vyp. (No.) 8, p. 3 - 8. In Russian. – Abstr. in Referativ. Zhurn. 52. Geod. Aehros"emka, 6.52.91 (1970).

081.043 **Continental drift and the movement of India.**
A. R. Crawford.
Naturwissenschaften, Vol. 57, 344 - 348 (1970).

081.044 **Phase delay of the solid earth tide.**
S. W. Smith, P. Jungels.
Phys. Earth Planet. Interiors, Vol. 2, 233 - 238 (1970).
Tidal strain data from Isabella, California and Nana, Peru have been analyzed to determine the phase shift of the M_2 solid tide and the Love number combination ($h - 3l$). The phase advance of the tidal bulge is $3.0 \pm 1°$, and ($h - 3l$) = 0.475. Indirect tidal effects due to ocean and atmosphere loading of the crust are minimized by performing the tidal analysis on the areal dilatation rather than on individual strain components. The lunar retarding torque calculated from this result is $(5.7 \pm 1.7) \times 10^{+23}$ dyne cm which is in agreement with the value obtained from discrepancies in the lunar orbit.

081.045 **1969 Smithsonian Standard Earth (II).**
E. M. Gaposchkin, K. Lambeck.
SAO, *Cambridge, Mass.,* Special Rep., No. 315, 104 pp. (1970).

081.046 **Continental drift before 1900.** N. A. Rupke.
Nature, Vol. 227, 349 - 350 (1970).
The idea that Francis Bacon and other seventeenth and eighteenth century thinkers first conceived the notion of continental drift does not stand up to close scrutiny. The few authors who expressed the idea viewed the process as a catastrophic event.

081.047 **Global photography of the earth.**
B. V. Vinogradov.
Zemlya i Vselennaya, No. 1, p. 31 - 38 (1970). In Russian.

081.048 **On M. S. Molodensky's second approximation of the disturbing potential.** M. I. Marych.
Geod., kartogr. i aehrofotos"emka. Mezhved. resp. nauchno-tekhn. sb., 1969, vyp. (No.) 10, p. 17 - 27. In Russian. – Abstr. in Referativ. Zhurn. 52. Geod. Aehros"emka, 7.52.71 (1970).

081.049 **Mareas terrestres.** C. Machin.
Las Ciencias, Vol. 34, No. 4, p. 3 - 9 = Publ. Seminario Astron. Geod. Univ. Madrid, No. 62 (1969). – Review article.

081.050 **Analyse des enregistrements de marée terrestre par la méthode des moindres carrés.**
J. C. Usandivaras, B. Ducarme.
Bull. Cl. Sci. Acad. Roy. Belgique, 5ᵉ Sér., Vol. 55, 560 - 569 = Obs. Royal Belgique, *Uccle,* Commun., Sér. B, No. 45 = Sér. Géophys., No. 95 (1969).

081.051 **Testrechnungen zur Bestimmung des zonalen Anteils des Gravitationsfeldes der Erde aus Koordinatenstörungen von Satelliten.** C. Reigber.
Bundesministerium für Bildung und Wissenschaft, Forschungsber. W 70-37, (Weltraumforschung), 57 pp. (1970).
The use of a generalized Fourier-analysis of satellite coordinate perturbations for the determination of force field parameters is investigated. To test the applicability conditions the method has been applied to the determination of zonal harmonics by analyzing the deviations of the simulated satellite motion from an arbitrarily chosen reference orbit.

The Earth: Its Origin, History and Physical Constitution. See Abstr. 003.055.

La précession–nutation et la structure de la Terre. See Abstr. 043.003.

Colour photography of the moon and of the earth from outer space. See Abstr. 094.194.

Âge des roches lunaires et terrestres. See Abstr. 094.241.

082 The Earth's Atmosphere including Refraction, Scintillation, Extinction, Airglow, Site Testing

082.001 **Identification of the ν_3 NO_2 band in the solar spectrum observed from a balloon borne spectrometer.** A. Goldman, D. G. Murcray, F. H. Murcray, W. J. Williams, F. S. Bonomo.
Nature, Vol. 225, 443 - 444 (1970).
Since the detection of HNO_3 in the atmosphere, it has been expected that oxides of nitrogen, such as NO_2, should also be present. A previous tentation identification of NO_2 absorption was confirmed.

082.002 **Observations of atmospheric turbulence with a radio telescope at 5 GHz.**
R. A. Hinder.
Nature, Vol. 225, 614 - 617 (1970).
Variations of the radio refractive index of the atmosphere are important to radio astronomers and in satellite communications. This article describes the situation at 5 GHz, based on work with the one-mile telescope at Cambridge.

082.003 **Retro-reflexion.** M. Minnaert.
Nature, Vol. 225, 718 (1970).
Comment on J. A. Howard: Increased luminance in the direction of reflex reflexion − a recently observed natural phenomenon, Nature, Vol. 224, 1102 (1969).

082.004 **Time delays of electromagnetic pulses due to molecular resonances in the atmosphere and the interstellar medium.** D. B. Trizna, T. A. Weber.
Astrophys. Journ., Vol. 159, 309 - 317 (1970).
Pulsed radio signals in a medium will travel at less than the speed of light in a vacuum when the main frequency components of the pulse lie in a region of anomalous dispersion. We estimate time delays due to anomalous dispersion by molecular radiofrequency resonances for signals propagating through the earth's atmosphere and the interstellar medium. For resonances below microwave frequencies, it appears that nominal time delays may occur for little measurable absorption. This effect may allow identification, and perhaps column-density estimates, of neutral constituents of these media. We also indicate how it may be pertinent in radar observations of other planets.

082.005 **Rare gases on the earth and in meteorites.**
M. N. Rao.
Astrophys. Space Sci., Vol. 6, 315 - 320 (1970).
Analysis of abundance patterns of rare gases Ne^{22}, Ar^{36}, Kr^{84} and Xe^{130} on earth and in ordinary and carbonaceous chondrites is presented. A mechanism of chemical adsorption of rare gases at the planetesimal stage during their accretion is proposed to generate the abundance pattern of the heavy rare gases on the earth. The calculated values for Xe and Kr agree well with the observed values whereas for Ar, the agreement is poor.

082.006 **The measurement of ozone from a satellite.**
D. E. Miller.
Journ. British Interplanet. Soc., Vol. 23, 291 - 296 (1970).
Two techniques are described to make determinations of ozone from a satellite, both of which depend on measuring the attenuation of ultraviolet sunlight which it produces.

082.007 **On the measurement of molecular oxygen concentration by absorption spectroscopy.**
J.-A. Quessette.
Journ. Geophys. Res., Vol. 75, 839 - 844 (1970).
Two methods are compared to derive molecular oxygen concentration by absorption spectroscopy of the solar hydrogen Lyman-alpha line: a classical one, using a constant absorption cross section for O_2 in the vicinity of Lyman α and, another one, using a wavelength-dependent cross section. It is shown that the results differ slightly but that this difference cannot explain the abnormal absorption observed by the author between the altitudes 90 and 100 km.

082.008 **Discussion of paper by R. C. Haymes, S. W. Glenn, G. J. Fishman, and F. R. Harnden, Jr., 'Low-energy gamma radiation in the atmosphere at midlatitudes'.**
E. L. Chupp, D. J. Forrest.
Journ. Geophys. Res., Vol. 75, 871 - 872, with a reply by R. C. Haymes, S. W. Glenn, G. J. Fishman, F. R. Harnden, Jr., p. 873 - 874 (1970). − Letter.

082.009 **The determination of lunar daily geophysical variations by the Chapman–Miller method.**
S. R. C. Malin, S. Chapman.
Geophys. Journ. Roy. Astron. Soc., Vol. 19, 15 - 35 (1970).
The Chapman–Miller method is the one most widely used for the determination of lunar daily variations in geophysical data (atmosphere, geomagnetic field, ionosphere). The method is described with minimum mathematical detail, the method of determination of probable errors is improved, and a computer program is appended.

082.010 **Ein neuer Prozeß in der Atmosphäre: Kontinuum-Absorption.** H. Quenzel.
Umschau, 70. Jahrgang, p. 113 - 114 (1970).

082.011 **Atmospheric photochemistry.**
R. D. Cadle, E. R. Allen.
Science, Vol. 167, 243 - 249 (1970).
The photochemistry of the lower atmosphere is dominated by atoms, molecules, and free radicals.

082.012 **Variations of the atmospherical parameters during geomagnetic storms.**
V. V. Mikhnevich, T. A. Solonenko.
Kosmich. Issled.,Vol. 8, 85 - 97 (1970). In Russian.

082.013 **Scattering by high cirrus: Its effect on submillimetre wave determinations of atmospheric water vapour.**
H. A. Gebbie, P. M. Kuhn, R. A. Bohlander.
Nature, Vol. 226, 71 - 72 (1970).

082.014 **Far-infrared observations of the night sky: Different data.** D. P. McNutt, P. D. Feldman, J. R. Houck, M. Harwit.
Science, Vol. 167, 1277 (1970).

082.015 **Atomic oxygen infrared emission in the earth's upper atmosphere.**
G. Kockarts, W. Peetermans.
Planet. Space Sci., Vol. 18, 271 - 285 (1970).
The atomic oxygen emission at 63 μ is analyzed between 50 km and 250 km for several atmospheric models. The frequency integrated intensities and the volume emission rates are computed and discussed in connection with experimental and theoretical investigations of the atomic oxygen distribution.

082.016 **Air density at a height of 120 km, from visual ob-**

servations of 1967-95A.
D. M. Brierley.
Planet. Space Sci., Vol. 18, 309 - 314 (1970).

During the last few weeks of its life the Molniya satellite 1967-95A attained an orbit with the exceptionally low perigee height of 116 km. Visual observations made by the author have been analysed to give values of air density at a height of 120 km during the period 1969 February 10 - 17.

082.017 Variations in exospheric density during 1967–68, as revealed by Echo 2.
G. E. Cook.
Planet. Space Sci., Vol. 18, 387 - 394 (1970).

Values of atmospheric density at heights near 1000 km have been obtained from the orbit of Echo 2 during 1967 and 1968. The results clearly show the existence in the exosphere of 27-day variations associated with the rotation of the sun and also correlation with geomagnetic disturbances. The semi-annual variation in the exosphere is again apparent; the magnitude of the semi-annual density variation shows no definite dependence on the level of solar activity. The irregular year-to-year variations in the effect are similar to those at lower heights.

082.018 Meteorological circumstances at the Norikura Corona Station. H. Morishita.
Tokyo Astron. Obs. Rep., Vol. 15, No. 1 (No. 56), p. 1 - 11 (1970). In Japanese.

082.019 Absorption measurements of earth's hydrogen atmosphere from solar hydrogen Lyman alpha rocket data. R. A. Jones, E. C. Bruner, Jr., W. A. Rense.
Journ. Geophys. Res., Vol. 75, 1849 - 1853 (1970).

The high-resolution profile of the solar H Ly α line has been measured for five altitudes from spectrograms taken during an Aerobee flight. The absorption due to the earth's hydrogen atmosphere was analyzed. These data are compared with the absorption predicted by two current models of the earth's hydrogen atmosphere.

082.020 A Raman scattering method for precise measurement of atmospheric oxygen balance.
R. L. Schwiesow, V. E. Derr.
Journ. Geophys. Res., Vol. 75, 1629 - 1632 (1970).

The purpose of this paper is to report recent quantitative results on Raman spectra of N_2, O_2, and CO_2 taken under experimental conditions similar to those suited to a precise measurement of $O_2/CO_2/N_2$ ratios in the atmosphere.

082.021 Vertical distribution of ozone in the winter subpolar region. J. S. Randhawa.
Journ. Geophys. Res., Vol. 75, 1693 - 1696 (1970). – Brief report.

082.022 Atmospheric ozone – A short review.
H. U. Dütsch.
Journ. Geophys. Res., Vol. 75, 1707 - 1712 (1970).

082.023 "Super-rotation" of the upper atmosphere at heights of 150 - 170 km.
D. G. King-Hele.
Nature, Vol. 226, 439 - 440 (1970).

Orbit analysis of satellite 1968-59A indicates that the atmosphere is rotating faster than the earth not only at heights above 200 km, but also down to heights as low as 170 km or possibly 150 km.

082.024 Solar-wind dependence of the diurnal temperature variation in the thermosphere.
L. G. Jacchia.
SAO, *Cambridge, Mass.,* Special Rep., No. 311, 4 + 8 pp.

(1970).

The ratio $r = T_{max}/T_{min}$ of the maximum to the minimum exospheric temperature considered on a global scale was derived by means of static models using the densities obtained from the atmospheric drag of three low-inclination satellites. It is found that r varies with the solar cycle, lagging more than a year behind the variations of the 10.7-cm solar flux; on the other hand, the variations of r are in phase with K_p, which is obtained by averaging the planetary geomagnetic index K_p over the whole year. The obvious conclusion is that in the mechanism of the diurnal temperature variation in the thermosphere, the solar wind is at least as important as, if not more important than, solar EUV.

082.025 Sul clima di St. Barthélemy. M. G. Fracastoro.
Mem. Soc. Astron. Italiana, Nuova Serie, Vol. 41, 125 - 127 (1970). – Letter.

082.026 Interpretation of airglow emissions – OH emissions.
R. L. Gattinger.
Ann. Géophys., Vol. 25, 825 - 830 (1969).

A time-dependent solution of the equations describing the variations of the important constituents in an oxygen-hydrogen atmosphere has been obtained with the aid of a digital computer.

082.027 Some peculiarities of the heating of the upper atmosphere during geomagnetic storms and aurorae. Yu. L. Truttse.
Kosm. Issled., Vol. 8, 298 - 305 (1970). In Russian.

082.028 Intensity measurement of protons of low energy in the upper atmosphere.
V. F. Tulinov, V. A. Lipovetskij.
Kosm. Issled., Vol. 8, 306 - 307 (1970). In Russian.
Brief information.

082.029 Dependence of the intensity of the corpuscular radiation in the upper atmosphere on solar activity. V. F. Tulinov, Yu. N. Moiseev, I. G. Shapiro, L. A. Ulanova.
Kosm. Issled., Vol. 8, 307 - 309 (1970). In Russian.
Brief information.

082.030 A possibility of investigating the composition of the upper atmosphere from effects of resonance scattering. E. G. Shvidkovskij, O. K. Kostko, Eh. A. Chayanova.
Kosm. Issled., Vol. 8, 310 - 311 (1970). In Russian.
Brief information.

082.031 Atmospheric density determined from artificial earth satellite observations. W. Góral.
Artificial Satellites Polish Acad. Sci., Vol. 4, No. 1, p. 16 - 20 (1968).

The paper presents the method of evaluating the density of the atmosphere from observations of artificial earth satellites. The considerations concern such satellites for which the main factor altering the value of the total energy is represented by the atmospheric drag.

082.032 Surface albedo and the filling-in of Fraunhofer lines in the day sky. D. M. Hunten.
Astrophys. Journ., Vol. 159, 1107 - 1110 = Contr. Kitt Peak National Obs.,No. 514 (1970).

The observed partial filling-in (Ring effect) of solar lines in light from the sky has been explained by Brinkmann as due to rotational Raman scattering. Analogous effects in the albedo contribution are examined, and are shown to be capable of accounting for a variation with solar zenith angle, and for the marked variability that is observed. The principal effect responsible for smearing of the spectrum is probably the so-called Rayleigh wings, which can be regarded as heavily damped rotational Raman scattering in condensed matter.

082.033 Seasonal changes in the vertical distribution of dust in the lower troposphere.
S. B. Idso, P. C. Kangieser.
Journ. Geophys. Res., Vol. 75, 2179 - 2184 (1970).
Measurements of solar radiation at two different elevations in Phoenix, Arizona, yield a mean transmittance difference between the two levels (Δ elevation $\approx$ 464 meters) of 0.012 for the summer period May through August and 0.066 for the winter period November through February. A mechanism is proposed to explain this seasonal variation in the vertical distribution of surface-generated dust, and local meteorological data are presented to support its validity.

082.034 On the possibility of utilization of lunar eclipses for exploring optical properties of the atmosphere.
V. G. Fessenkov.
Astron. Zhurn. Akad. Nauk SSSR, Vol. 47, 237 - 245 (1970). In Russian. – English translation in Soviet Astron., AJ, Vol. 14, No. 2.
It is shown that in the lunar eclipse phenomenon a refractional dispersion, equivalent to the apparent flattening of the solar disc, occurring without change of its visible brightness, plays the leading part. For the full interpretation of a lunar eclipse it is necessary to take into account ozone absorption and gas-aerosol extinction in the earth's atmosphere, as well as the intensity distribution over the solar disc. Results are given of corresponding calculations, showing the possibility to determine the optical properties of the atmosphere at various heights on the basis of photometric properties of the lunar eclipse at various wavelengths.

082.035 The experience of the aerological investigation of volcanic covers of Kamchatka.
Yu. N. Lipsky, G. S. Shteinberg, M. M. Pospergelis, V. V. Novikov.
Astron. Zhurn. Akad. Nauk SSSR, Vol. 47, 411 - 419 (1970). In Russian. – English translation in Soviet Astron., AJ, Vol. 14, No. 2.

082.036 A comparison of satellite drag measurement techniques. D. B. DeBra.
Astrophys. Space Sci. Library, Vol. 15, 563 - 573 (1970). Conference paper (see 012.007).

082.037 Departures from Jeans' escape rate for H and He in the earth's atmosphere. R. T. Brinkmann.
Planet. Space Sci., Vol. 18, 449 - 478 (1970).
An examination of the theory of thermal or gravitational escape leads to new values for the escape rates of H and He in the earth atmosphere. Escape rates of H and He are 70 - 75 per cent and 97 - 99 per cent of the Jeans rate, respectively.

082.038 A method of determining the atmospheric sodium concentration from twilight observations.
G. I. Kvifte, L. Wallace.
Planet. Space Sci., Vol. 18, 623 - 635 (1970).
Simple formulae are given for computing the simultaneous amount of sodium in the twilight and the daytime sky from one set of observations of the $D1$ and $D2$ line intensities in twilight. The formulae are derived from an approximate solution of the transfer equation for a plane parallel atmosphere and conservative isotropic scattering.

082.039 A balloon-borne observation of the intensity variation of the OH emission in the evening twilight.
G. Moreels, W. F. J. Evans, J. E. Blamont, A. V. Jones.
Planet. Space Sci., Vol. 18, 637 - 640 (1970). – Research note.

082.040 A study of the diurnal variations of the 5577 Å [OI] airglow emission at selected IGY stations.
J. G. Brenton, S. M. Silverman.
Planet. Space Sci., Vol. 18, 641 - 653 (1970).
In the present paper the study of the diurnal variation is extended to 22 stations using data available from the Annals of the International Geophysical Year (Yao, 1962) and data from the AFCRL station at Sacramento Peak, New Mexico. Definite seasonal effects were found for the 13 stations for which sufficient data were available. No sunspot cycle effect was found.

082.041 Excitation and radiative transport of OI 1304 Å resonance radiation – I. The dayglow.
D. J. Strickland, T. M. Donahue.
Planet. Space Sci., Vol. 18, 661 - 689 (1970).
Radiative transfer theory is developed for the 1304 Å triplet of atomic oxygen which includes the effects of pure absorption by molecular oxygen and variability of the temperature governing the relative populations of the ground state ^3P_j levels.

082.042 Variations in the width of the nightglow OI λ6300 line during the magnetic storm of October 30 – November 2, 1968.
E. B. Armstrong, J. A. Bell.
Planet. Space Sci., Vol. 18, 784 - 789 (1970). – Research note.

082.043 Xenon in natural gases.
G. A. Bennett, O. K. Manuel.
Geochim. Cosmochim. Acta, Vol. 34, 593 - 610 (1970).
It is shown that the unusually high yields of Xe^{129}, Xe^{131} and Xe^{132} in well gas could alternately arise from the addition of Xe^{130} to the earth's atmosphere. The addition of xenon from the solar wind to meteorites and to the earth's atmosphere could explain the very low fission yields of Xe^{132} reported in xenon-rich meteorites.

082.044 Six-colour photometry of the night sky at declination – 19° between wavelengths 3900 Å and 7100 Å. J. Dachs.
Astron. Astrophys., Vol. 6, 155 - 164 (1970). In German.
The surface brightness of the night sky at declination – 19.2° has been measured at six wavelengths ranging from 3920 to 7100 Å, in a circular field of view of 4.5 square degrees with a photoelectric zenith photometer by one-year observations at Tsumeb, South West Africa, from July, 1965, until June, 1966. The surface brightness at right ascension $0^h 50^m$, declination – 19° (galactic latitude – 82°) for wavelength 5580 Å, omitting the [O I] 5577 Å airglow emission component, is $240 S_{10}$ (A 0 V). It is shown that the spectrum of the zodiacal light must have essentially the same energy distribution as the sun in the wavelength range from 4900 Å to 6300 Å. A southern extension of the isophotes of the galactic component of the night sky brightness in the constellation Orion is implicated by the observations, the 50 S_{10} isophote at wavelength 5580 Å reaching as far as galactic latitude – 60°.

082.045 The night sky brightness measured from satellites Kosmos 51 and 213.
N. A. Dimov, A. B. Severny, A. M. Zvereva.
IAU Symposium No. 36, (see 012.014), p. 325 - 333 (1970).
(1) The minimum measured brightness of the night sky (in the visual region) is about 100 stars of the 10th magnitude per square degree near the galactic poles. (2) The ratio of fluxes in the ultraviolet (2300 - 3000 Å) and visual regions is approximately in agreement with expected theoretical data based on the models of stellar atmospheres for spectral classes B0 to G5 and on the distribution of stars of the different spectral

classes over the sky. (3) The nature of possible deviations (theory minus observation) is discussed.

082.046 **Some results of visual observations and of a spectrophotometry of the dusk halo of the terrestrial atmosphere made by the Soyuz 5 spaceship.**
K. J. Kondratiev, B. V. Volynov, A. P. Galtsev, V. V. Koltsov, O. I. Smoktii, E. V. Khrunov.
Dokl. Akad. Nauk SSSR, Ser. Mat. Fiz., Vol. 190, 327 - 330 (1970).   In Russian.

The spectral, angular and space evolution of the brightness of the twilight atmosphere have been measured in the range of 400 - 650 mμ at the Soyuz 5 spaceship. The spectral and angular resolution of the hand-spectrograph used was 5mμ and 2'. Spectrophotometry has been complemented by visual observations as well as colour and black-and-white photography. Three series of observations with different cloudiness have been obtained. Near the earth's surface long-wave radiation prevails in the dusk halo. With increasing height the maximum brightness is shifted to shorter wavelengths. The total maximum brightness is at 480 mμ. A local minimum at 600 mμ is caused by ozone. The spectral brightness depends strongly on the azimuth and on the depression of the sun. Aerosol absorptions are observable only with a depression of the sun over 3°. Multiple scattering predominates near the surface, at heights over 30 - 40 km it is negligible. Further experiments have been suggested.
B. Onderlička

082.047 **The atmospheric aerosol structure as defined by the method of spectral transparency.**
N. I. Nikitinskaia, A. J. Perelman, K. S. Shifrin.
Dokl. Akad. Nauk SSSR, Ser. Mat. Fiz., Vol. 190, 331 - 333 (1970).   In Russian.

Observations of atmospheric transparency have been carried out in 1951 by means of a thermoelectric actinometer with interference filters defining 9 narrow bands from 0.37 to 1 μ. Only 15 days with extremely high transparency have been used for the determination of the aerosol optical depth. From these data three frequency maxima of particle radii, 03 - 0.35, 0.6 - 0.7, and 1.0 - 1.3 μ, have been calculated by a method of Shifrin and Perelman. The results are consistent with other independent investigations.
B. Onderlička

082.048 **New static models of the thermosphere and exosphere with empirical temperature profiles.**
L. G. Jacchia.
SAO, *Cambridge, Mass.,* Special Rep. No. 313, 7 + 87 pp. (1970).

The present models are patterned after similar models published by the author. The main differences consist in the lower height (90 km instead of 120 km) of the constant-boundary surface and in a higher ratio of atomic-oxygen to molecular-oxygen density. Mixing is assumed to prevail to a height of 105 km, diffusion above this height. All the recognized variations that can be connected with solar, geomagnetic, temporal, and geographic parameters are represented by empirical equations.

082.049 **Determination of size distribution of atmospheric aerosol particles from spectral solar radiation measurements.**   H. Quenzel.
Journ. Geophys. Res., Vol. 75, 2915 - 2921 (1970).

The extinction coefficient in a vertical column of atmosphere was determined from spectral solar radiation measurements. The corresponding aerosol size distributions were derived by comparison with Mie calculations.

082.050 **Chemical factors regulating free alkali metal atoms in the upper atmosphere.**   E. R. Allen.

Journ. Geophys. Res., Vol. 75, 2947 - 2950 (1970).

The nightglow emission is probably due to chemical reaction of alkali metal compounds. Studies on bulk meteoritic material suggest that, because of the high recombination coefficients observed, a layer of meteoritic dust at about 90 km would greatly perturb the atomic-particle density near this altitude.

082.051 **Die Erforschung der Hochatmosphäre.**
K. Bischoff.
Astron. in der Schule, 7. Jahrgang, p. 39 - 42 (1970).

082.052 **On the Hα radiation in the night glow.**
S. Isobe.
Univ. Tokyo, Tokyo Astron. Obs. Report, Vol. 15, (No. 2), (No. 57), p. 373 - 375 (1970).   In Japanese.

082.053 **Visual colour estimates of twilight sky according to observations from the spacecraft Soyuz 5.**
K. J. Kondratiev, A. P. Galtsev, O. I. Smoktii, E. V. Khrunov.
Dokl. Akad. Nauk SSSR, Ser. Mat. Fiz., Vol. 191, 824 - 825 (1970).   In Russian.

082.054 **Colorimetry of twilight sky based on horizon spectra obtained from the spacecraft Soyuz 5.**
K. J. Kondratiev, A. P. Galtsev, O. I. Smoktii, E. V. Khrunov.
Dokl. Akad. Nauk SSSR, Ser. Mat. Fiz., Vol. 191, 1044 - 1047 (1970).   In Russian.

082.055 **Determination of upper atmosphere temperatures from diffusion of vapour clouds.**
J. N. Desai, M. S. Narayanan.
Journ. Atmosph. Terr. Phys., Vol. 32, 1235 - 1245 (1970).

In this paper we present a more accurate method for eliminating the sky background of brightness measurements of atmospheric vapour clouds. We also present here a new procedure for calculating the upper atmosphere temperatures from the measured diffusion profile which requires knowledge of atmospheric parameters at a single height only.

082.056 **Night airglow observations during the IQSY.**
F. E. Roach, L. L. Smith, J. R. McKennan.
Ann. IQSY, (see 012.016), Vol. 4, 375 - 387 (1969).

082.057 **Airglow research during and since the IQSY.**
G. Weill.
Ann. IQSY, (see 012.016), Vol. 4, 388 - 399 (1969).

082.058 **Hydrogen and hydroxyl emissions in the nightglow.**   N. N. Shefov, Yu. L. Truttse.
Ann. IQSY, (see 012.016), Vol. 4, 400 - 406 (1969).

082.059 **Noctilucent cloud observations and deductions: Water vapour in the stratosphere and mesosphere.**
E. Hesstvedt.
Ann. IQSY, (see 012.017), Vol. 5, 23 - 31 (1969).

082.060 **Concentrations of hydrogen and helium in the outer atmosphere: Geocorona.**   N. N. Shefov.
Ann. IQSY, (see 012.017), Vol. 5, 215 - 228 (1969).

082.061 **Effects of solar XUV radiation on the earth's atmosphere.**   H. E. Hinteregger.
Ann. IQSY, (see 012.017), Vol. 5, 305 - 321 (1969).

The absolute intensity distribution in the spectrum of solar XUV fluxes (EUV and X-rays) incident upon the upper atmosphere, required for any quantitative physical model of this atmospheric region, represents the main topic of the present review.

082.062 **Atmospheric density variations during solar maxi-**

mum and minimum.   L. G. Jacchia.
Ann. IQSY, (see 012.017), Vol. 5, 323 - 339 (1969).

**082.063   Geocoronal hydrogen: An analysis of the Lyman-
alpha airglow observed from OGO-4.**
R. R. Meier, P. Mange.
Planet. Space Sci., Vol. 18, 803 - 821 (1970).
    Observations of the hydrogen Lyman-alpha glow, sur-
rounding the earth, have been carried out from the OGO-4
spacecraft. The dependence of this emission feature on
solar zenith angle was measured in the altitude range of
about 400 - 900 km.

**082.064   Ballistical transport phenomena in a collision-free
exosphere.   H. J. Fahr.**
Planet. Space Sci., Vol. 18, 823 - 834 = Mitt. Astron. Inst.
Bonn, No. 117 (1970).

**082.065   Aircraft airglow intensity measurements: Variations
in OH and OI (5577).**
K. A. Dick, G. G. Sivjee, H. M. Crosswhite.
Planet. Space Sci., Vol. 18, 887 - 894 (1970).

**082.066   Low energy gamma radiation in the atmosphere
during active and quiet periods on the sun.**
E. L. Chupp, D. J. Forrest, A. A. Sarkady, P. J. Lavakare.
Planet. Space Sci., Vol. 18, 939 - 943 (1970).
    In the present paper, we give a progress report on the
results from a series of five mid latitude balloon flights
covering the time period December 1966 to April 1968.
A search for the solar emission of the 0.51 MeV annihilation
line and the 2.2 MeV neutron-proton capture line was made
during different flights in correlation with the occurrence
of solar flares.

**082.067   Metastable oxygen: Origin of atmospheric
absorption near 50 kilometers.**
J. F. Noxon.
Science, Vol. 168, 1120 - 1121 (1970).

**082.068   Attenuation on an earth-space path measured in the
wavelength range of 8 to 14 micrometers.**
R. W. Wilson.
Science, Vol. 168, 1456 - 1457 (1970).
    The measurements reported here show that, in a climate
such as New Jersey's, the attenuation of the lower atmosphere
can be quite high in the 8- to 14-μm window for an appreciable fraction of the time.

**082.069   On the mechanism of the infrared radiation of the
upper atmosphere.   B. F. Gordiets, N. M. Markov,**
L. A. Shelepin.
Kosmich. Issled., Vol. 8, 437 - 448 (1970).   In Russian.

**082.070   On the density of the sources of charges at the
mutual interaction between charged particles and**
the upper atmosphere.   N. K. Osipov, V. G. Pivovarov,
N. B. Pivovarova.
Kosmich. Issled., Vol. 8, 460 - 462 (1970).   In Russian.
Brief information.

**082.071   Investigation of the sign of the spatial charge of
the atmosphere up to 86 km height.**
Yu. A. Bragin.
Kosmich. Issled., Vol. 8, 465 - 467 (1970).   In Russian.
Brief information.

**082.072   Determination of atmospheric parameters in an
intermediate region of 80 - 120 km height.**
E. N. Golubev, V. V. Mikhnevich, Yu. M. Trishina.
Kosmich. Issled., Vol. 8, 467 - 470 (1970).   In Russian.

Brief information.

**082.073   A possible method for the measurement of the
temperature of the neutral component of the
upper atmosphere above 150 km height.**
M. B. Belotserkovskij, V. A. Sokolov.
Kosmich. Issled., Vol. 8, 470 - 473 (1970).   In Russian.
Brief information.

082.074   Upper atmosphere dynamics.
M. P. Friedman.
SAO *Cambridge*, Mass., Special Rep. No. 316, 8 + 51 pp.
(1970).
    We are able to construct three - dimensional global
models of the upper atmosphere (120 to 800 km) by solving
the conservation equations of mass, momentum, and energy.
We show that variation of conditions at the lower boundary,
120 km, affects the atmosphere only to about 200 - km
altitude. Differing results obtained by in situ satellite or
rocket measurements in the 120 - to 200 - km range can
therefore be explained as resulting from atmospheric behavior
below 120 km.

**082.075   Surface temperature of the early earth and the
nature of the terrestrial atmosphere.**
A. J. Meadows.
Nature, Vol. 226, 927 - 928 (1970).
    The evolution of the surface temperature of the earth
since its formation is examined on the assumption that the
terrestrial atmosphere accumulated by degassing. It is pointed
out that either an oxidising ($CO_2 + H_2O$) or a reducing ($NH_3 +$
H_2O) atmosphere would produce an appreciable greenhouse
effect. These two combinations act differently in the presence
of liquid water, since NH_3 is more soluble than CO_2. Hence,
a reducing atmosphere should produce a lower average surface
temperature, with most of the water present as ice. The
existence of metamorphosed sediments some 3.5×10^9 years
old suggests that liquid water, and therefore a CO_2 atmosphere,
appeared on earth prior to that time.

**082.076   The coefficient of the transparency of the atmo-
sphere for some sites in Tadjikistan.**
A. M. Bakharev.
Dokl. AN Tadzh. SSR, Vol. 12, No. 8, p. 16 - 18 (1969).
In Russian. – Abstr. in Referativ. Zhurn. 51. Astron., 4.51.102
(1970).

082.077   Scintillazione e jet-streams.   C. Blanco.
Mem. Soc. Astron. Italiana, Nuova Serie, Vol. 41,
227 - 229 (1970).

082.078   Far infrared emission of the night sky.
M. Harwit.
Říše hvězd, Vol. 51, 121 - 124 (1970).   In Czech.

082.079   The astroclimate at Naugarzan.
V. I. Kardopolov, V. E. Slutskij.
Tsirk. Astron. Inst., *Tashkent,* No. 18 (365), p. 18 - 22
(1970). In Russian.

**082.080   Observations of clear air turbulence by high power
radar.   K. A. Browning, C. D. Watkins.**
Nature, Vol. 227, 260 - 263 (1970).
    Clear air turbulence is a hazard to aviation and is thought
to have important effects on atmospheric dynamics. This article describes the structure and evolution of clear air turbulence at high altitudes as revealed by a high power radar and
vertical soundings of wind and temperature.

**082.081   On the emission near 5610 Å in the twilight
spectrum.   T. I. Toroshelidze.**

Astron Tsirk., No. 550, p. 7 - 8 (1970). In Russian.

082.082 **Light of the night-sky.** M. Huruhata.
Trans. IAU, Vol. 14A, (see 003.028), 193 - 206 (1970). – Report of Commission 21.

082.083 **Clear weather over Nashville.** A. M. Heiser.
Sky Telescope, Vol. 40, 62 (1970).

082.084 **On site testing methods.**
S. B. Novikov, A. A. Ovtchinnikov, P. V. Sheglov.
Astron. Tsirk., No. 554, p. 1 - 2 (1970). In Russian.

082.085 **Study of the disturbing air layer at the mountain station of the Sternberg Institute.** V. V. Rodionov.
Astron. Tsirk., No. 554, p. 3 - 5 (1970). In Russian.

082.086 **Double-beam site testing at the mountain station of the Sternberg Institute.** V. V. Rodionov.
Astron. Tsirk., No. 554, p. 5 - 6 (1970). In Russian.

082.087 **Real diurnal variations of H_a-nightglow.**
L. M. Fishkova, N. M. Martsvaladze, P. V. Sheglov.
Astron. Tsirk., No. 555, p. 7 - 8 (1970). In Russian.

082.088 **Results of one year double-beam site testing at Mount Sanglock.**
A. V. Bagrov, Yu. F. Nikitin, G. V. Novikov, P. V. Sheglov.
Astron. Tsirk., No. 556, p. 7 - 8 (1970). In Russian.

082.089 **On observations of temperature inhomogeneities at Mt. Sanglock.** A. V. Bagrov, A. A. Ovchinnikov.
Astron. Tsirk., No. 558, p. 7 - 8 (1970). In Russian.

082.090 **Determination of short-periodic atmospheric density variations from quasi-simultaneous visual observations of artificial satellites.** M. Ill.
Byull. Stantsij Optichesk. Nablyud. Iskusstv. Sputnikov Zemli, No. 53 (1), p. 9 - 22 (1969). In Russian.
Density variations with the amplitude of the order of 64.3% in two days interval were discovered for the satellite Cosmos 53.

082.091 **Electron impact excitation of the dayglow.**
A. Dalgarno, M. B. McElroy, A. I. Stewart.
Journ. Atmosph. Sci., Vol. 26, 753 - 762 = Contr. Kitt Peak National Obs., No. 446 (1969).
Calculations are described of the equilibrium velocity distributions of the photoelectrons produced in the F region by solar ionizing radiation. Detailed estimates are presented of the intensities and altitude profiles of emission features of atomic oxygen, molecular nitrogen, and molecular oxygen, appearing in the dayglow as a result of photoelectron impacts.

082.092 **Clear Weather over Nashville.** A. M. Heiser.
Journ. Tennessee Acad. Sci., Vol. 45, 19 - 20 = Repr. Arthur J. Dyer Obs., Vanderbilt Univ., No. 52 (1970).

082.093 **An atlas of air absorptions in the infrared.**
D. J. Lovell.
Contr. Four College Obs., *Amherst,* No. 49, 82 pp. (1969).
An atlas of the spectral absorption characteristic of air is presented in the spectral region from 4000 to 250 cm^{-1}. Spectra were observed over a 92 meter path under two conditions: in a near vacuum and at ambient pressure and temperature.

082.094 **Condizioni meteorologiche a Serra La Nave nel 1968. Prima indagine sulla distribuzione del vento (1965–68).**
F. Affronti, C. Blanco.
Oss. Astrofis. Catania, Pubbl., Nuova Ser., No. 140, 17 pp. (1969).

082.095 **Report on the Pachino solar site testing.**
C. Blanco, G. Godoli, L. Paternò, A. Righini, M. Rodonò.
Oss. Astrofis. Catania, Pubbl., Nuova Ser., No. 142, 77 pp. (1969).

082.096 **Kvalitet slike u funkciji meteoroloških uslova u Beogradu.** P. M. Djurković, G. Popović, D. Zulević.
Publ. Obs. Astron. Beograd, No. 16, (see 012.021), p. 48 - 54 (1969). In Serbo-Croatian.

082.097 **Astronomical refraction and its problems.**
G. Teleki.
Publ. Obs. Astron. Beograd, No. 16, (see 012.021), p. 113 - 117 (1969). In Serbo-Croatian.

Seasonal changes in thermospheric composition.
See Abstr. 083.022.

Interstellar and atmospheric clouds.
See Abstr. 131.027.

Measurements of absolute sky brightness temperatures at 320 and 707 MHz. See Abstr. 157.005.

083 Ionosphere

083.001 Satellite observations of equatorial phenomena and defocusing of VLF electromagnetic waves.
R. R. Scarabucci.
Journ. Geophys. Res., Vol. 75, 69 - 84 (1970).

The purpose of this paper is to present and discuss two new ionospheric phenomena involving the propagation and absorption of VLF waves near the magnetic equator.

083.002 Backscatter observations of F-region field-aligned irregularities during the I.Q.S.Y.
W. J. Baggaley.
Journ. Geophys. Res., Vol. 75, 152 - 158 (1970).

Field-aligned irregularities have been observed in the F region by using a 17-MHz swept-azimuth sounder operating in the United Kingdom during the period 1964–1966.

083.003 Seasonal variation of the F_1-region ion composition.
J. V. Evans, L. P. Cox.
Journ. Geophys. Res., Vol. 75, 159 - 164 (1970).

Thomson scatter observations at a wavelength of 23 cm are described that supplement earlier ionospheric studies made at the Millstone Hill Radar of the region 130–230 km. By making certain reasonable assumptions about the altitude dependence of the electron and ion temperatures, it has been possible to recover from the observations the ratio of atomic ions (O^+) to molecular ions (O_2^+ or NO^+) in this region.

083.004 Diurnal model of the E region.
T. J. Keneshea, R. S. Narcisi, W. Swider, Jr.
Journ. Geophys. Res., Vol. 75, 845 - 854 (1970).

Detailed computer calculations of the electron, O_2^+, and NO^+ concentrations are provided for the E region on a diurnal basis. The present work contains the first continuous solution of these concentrations during sunrise and sunset. Comparison is made between this model and observations, particularly new results at sunrise and sunset. The model is compatible with the experimental data within a factor of 2, although some larger discrepancies occur.

083.005 The day-to-day variability of the equatorial electrojet in Peru. K. Burrows.
Journ. Geophys. Res., Vol. 75, 1319 - 1323 (1970).

By means of a semi-automatic procedure, scalings taken from the H records of a latitude spread of seven magnetometers, located close to the dip equator in Peru, have been used to investigate the variability of the electrojet on a day-to-day basis for the period from the December solstice to the equinox during a solar minimum.

083.006 Photoelectron flux and protonospheric heating during the conjugate point sunrise.
B. C. N. Rao, E. J. R. Maier.
Journ. Geophys. Res., Vol. 75, 816 - 822 (1970).

083.007 The exospheric plasma during the International Years of the Quiet Sun.
J. O. Thomas, M. J. Rycroft.
Planet. Space Sci. Vol. 18, 41 - 63 (1970).

It is the purpose of this paper to provide information about a number of surveys made as an extensive programme of analysis carried out in the Space Sciences Division of the NASA Ames Research Center, the detailed results of which, presented in a series of NASA Special Publications, provide data on the overall morphology of the topside ionosphere during, and close to, the International Years of the Quiet Sun. Attention is mainly concentrated on data derived for the highest altitudes possible (near 1000 km). The scale height distribution can be understood in terms of diffusive equilibrium theory, using which the mean ionic mass of the plasma can be derived.

083.008 Model of the polar ion-exosphere.
J. Lemaire, M. Scherer.
Planet. Space Sci., Vol. 18, 103 - 120 (1970).

A model of a polar ion-exosphere in which the geomagnetic field lines are open is developed.

083.009 The use of multiple receivers to measure the wave characteristics of very-low-frequency noise in space.
S. D. Shawhan.
Space Sci. Rev., Vol. 10, 689 - 736 (1970). – Review article.

083.010 The electron density distributions in the D-region during the night and pre-sunrise period.
L. Thomas, M. D. Harrison.
Journ. Atmosph. Terr. Phys., Vol. 32, 1 - 14 (1970).

Experimental measurements of reflection and conversion coefficients over a range of frequencies from 16 to 71 kHz are used to deduce electron density distributions for night-time and the pre-sunrise period.

083.011 A theoretical investigation of corpuscular radiation effects on the F-region of the ionosphere.
D. G. Torr, M. R. Torr.
Journ. Atmosph. Terr. Phys., Vol. 32, 15 - 34 (1970).

The aim of this paper is to investigate the importance of corpuscular radiation as a source of ionization in the ionosphere. In order to carry out this investigation quantitatively, the time dependent ionospheric continuity equation is solved taking into account production of ionization by solar extreme ultraviolet and corpuscular radiation, loss by non-linear recombination type processes, transport by plasma diffusion in an atmosphere which is in thermal non-equilibrium and atmospheric winds.

083.012 Oscillation of electrostatically trapped particles.
V. C. Liu, R. J. Hung.
Journ. Atmosph. Terr. Phys., Vol. 32, 69 - 82 (1970).

083.013 Solar cosmic ray ionization in the low ionosphere.
P. Velinov.
Journ. Atmosph. Terr. Phys., Vol. 32, 139 - 147 (1970).

The enhanced electron production rate produced by solar cosmic rays in the lower ionosphere is theoretically considered in this paper.

083.014 Saturation and focusing effects in radio-star and satellite scintillations. D. G. Singleton.
Journ. Atmosph. Terr. Phys., Vol. 32, 187 - 208 (1970).

The diffraction and refraction theories of radio-star and satellite scintillations are re-examined and successfully reconciled to produce a composite result. This result gives new emphasis to the importance of focusing and saturation effects in scintillation studies.

083.015 On the theory of ionospheric scintillations of lunar radar echoes. G. Perona.
Journ. Atmosph. Terr. Phys., Vol. 32, 277 - 291 (1970).

In the present paper some experiments that can be performed using lunar echoes to study ionospheric scintillations are described, together with a brief account of the experimental procedures that should be followed in order to collect useful information about ionospheric parameters.

083.016   **The effect of irregularity shape on radio star and satellite scintillations.**   D. G. Singleton.
Journ. Atmosph. Terr. Phys., Vol. 32, 315 - 343 (1970).
     An irregularity model of F-region irregularities is put forward which includes the shapes as special cases. Using this model and diffraction theory, a theoretical investigation is made of the effect these shapes have on (a) the nature of the variation of scintillation index with the zenith angle and azimuth of the direction of propagation at a point on the ground and (b) the shape and orientation of the characteristic ellipse of the amplitude diffraction pattern observed at the ground.

083.017   **The diurnal variations of the concentrations of NO^+, O_2^+, NO and N in the ionospheric E-region.**
P. E. Monro.
Journ. Atmosph. Terr. Phys., Vol. 32, 373 - 382 (1970).

083.018   **A two-ion model of electron-ion recombination in the D-region.**   A. Haug, B. Landmark.
Journ. Atmosph. Terr. Phys., Vol. 32, 405 - 407 (1970).
Short paper.

083.019   **Interpretation of loss rates observed during auroral absorption in terms of a two-ion model of recombination.**   K. Folkestad, R. J. Armstrong.
Journ. Atmosph. Terr. Phys., Vol. 32, 409 - 412 (1970).
Short paper.

083.020   **Electric fields and F-region electron densities over Peru.**   E. Dunford.
Journ. Atmosph. Terr. Phys., Vol. 32, 421 - 425 (1970).
Short paper.

083.021   **The correlation between sudden phase anomalies and solar microwave radio bursts.**
P. Kaufmann, A. Mendes.
Journ. Atmosph. Terr. Phys., Vol. 32, 427 - 432 (1970).
     More than two hundred SIDs were observed as VLF phase anomalies (SPA) at times when measurements were being made of solar radio noise at 7 GHz. 99 per cent of the SPAs were correlated with microwave solar events.

083.022   **Seasonal changes in thermospheric composition.**
G. A. M. King.
Journ. Atmosph. Terr. Phys., Vol. 32, 433 - 437 (1970).
Short paper.

083.023   **Measurements of total electron content at Huancayo, Peru.**   P. Bandyopadhyay.
Planet. Space Sci., Vol. 18, 129 - 135 (1970).
     Measurements of total electron content of the ionosphere above Huancayo, Peru by observation of Faraday fading of satellite signals during January through March 1967 are reported. The data are examined to study principally the diurnal variation of total electron content and of equivalent slab thickness as well as the latitudinal variation of the electron content.

083.024   **Ion densities in the night ionosphere.**
R. A. Howard, J. T. Vanderslice, S. G. Tilford.
Planet. Space Sci., Vol. 18, 145 - 153 (1970).
     The nighttime concentration profiles of the major ions in the ionosphere between 120 and 450 km have been calculated from the known daytime concentrations. Agreement between the observed and calculated nighttime profiles of the O^+, NO^+, O_2^+ ions is quite good if one allows normal Chapman-Enskog diffusion.

083.025   **Determination of the columnar electron content and the layer shape factor of the plasmasphere up to the**

plasmapause.   O. G. Almeida, O. K. Garriott, A. V. Da Rosa.
Planet. Space Sci., Vol. 18, 159 - 170 (1970).
     Measurements of the total columnar electron content of the plasmasphere up to the plasmapause have been made using the beacon transmitters aboard the geostationary satellite ATS-III. This paper describes the analysis used to obtain both the absolute value of content and the shape factor.

083.026   **New type of ionospheric disturbance.**
R. G. Rastogi.
Nature, Vol. 225, 258 - 259 (1970).
     A sharp discontinuity or kink noticed first on the F_1 trace of the ionograms at Thumba (dip $0.6°$ S) is found on the subsequent records to progress upward on the height or the frequency scale. Real height calculations of these kinks with time indicate an upward velocity of about 20 m/sec. This seems to be a direct evidence of the uplifting of the equatorial ionosphere during the day-time due to the electro-dynamic interaction between the equatorial electrojet current and geomagnetic field.

083.027   **Some ionospheric properties at 1000 kilometers altitude within and near the auroral zone.**
P. L. Dyson, A. J. Zmuda.
Journ. Geophys. Res., Vol. 75, 1893 - 1901 (1970).
     The authors report and discuss measurements of electron density and temperature of the ionosphere made with Explorer 22 between April 26 and August 30, 1967, at the Applied Physics Laboratory tracking station in Maryland.

083.028   **Dependence of satellite scintillations on zenith angle and azimuth.**   D. G. Singleton.
Journ. Atmosph. Terr. Phys., Vol. 32, 789 - 803 (1970).
     The purpose of the present paper is to investigate experimentally the manner in which amplitude scintillations of satellites depend on the geometry of observation.

083.029   **Seasonal variations of electron densities below 100 km at mid-latitudes – III. Stratospheric-ionospheric coupling.**   J. B. Gregory, A. H. Manson.
Journ. Atmosph. Terr. Phys., Vol. 32, 837 - 852 (1970).

083.030   **Simultaneous solution of the time dependent coupled continuity equations, heat conduction equations, and equations of motion for a system consisting of a neutral gas, an electron gas, and a four component ion gas.**   P. Stubbe.
Journ. Atmosph. Terr. Phys., Vol. 32, 865 - 903 (1970).

083.031   **"Lunar tide in D-region" of the ionosphere near the magnetic equator.**
S. C. Chakravarty, R. G. Rastogi.
Journ. Atmosph. Terr. Phys., Vol. 32, 945 - 948 (1970).
Short paper.

083.032   **Attenuation of energetic electrons and impact ionization in the ionosphere.**
R. Y. Prasad, R. N. Singh.
Earth Planet. Sci. Letters, Vol. 8, 169 - 172 (1970).
     Attenuation of energetic electrons in the energy range 1–300 keV precipitating in the ionosphere has been calculated and the resulting height profile in the range 50–300 km has been shown. The impact ionization produced by precipitating electrons of energy 3–300 keV has been calculated and its height profile is shown. It is seen that impact ionization increases with increasing energy of precipitating electrons. High energy electrons penetrate deeper into the ionosphere and produce high degree of ionization.

083.033   **The gradient of annual variation of the characteris-**

tic of the ionization state of the F2 layer.
M. A. Likhachev.
Geomagn. Aeronom., Vol. 10, 52 - 55 (1970).   In Russian.

083.034 Investigation of the connection between the rio-
           meter absorption and the ionospheric-magnetic
activity in the auroral zone.
Yu. N. Gorshkov, N. M. Denisenko.
Geomagn. Aeronom., Vol. 10, 73 - 76 (1970).   In Russian.

083.035 A comparison of the variability of auroral absorp-
           tion of radio waves in the ionosphere in Loparskaya
and College.   Z. Tz. Rapoport.
Geomagn. Aeronom., Vol. 10, 77 - 83 (1970).   In Russian.

083.036 Obtaining on a computer approximate dependences
           of the mean square declinations of critical frequen-
cies of the F2 layer from latitude and solar activity.
V. S. Gubenko, Ya. S. Rodionov.
Geomagn. Aeronom., Vol. 10, 144 - 146 (1970).   In Russian.
Brief information.

083.037 The ionization function for minimum solar
           activity.   B. N. Velichansky, M. K. Ivelskaya,
N. N. Klimov.
Geomagn. Aeronom., Vol. 10, 149 - 151 (1970).   In Russian.
Brief information.

083.038 Ionospheric electron content at temperate latitu-
           des during the increasing phase of the solar cycle.
N. N. Rao, K. C. Yeh, M. Y. Youakim.
Australian Journ. Phys., Vol. 23, 37 - 43 (1970).
     Ionospheric electron content results obtained from the
radio beacon satellite Beacon Explorer–B during the increasing phase of the solar cycle 1964–8 are presented. The
diurnal, seasonal, solar cycle, latitudinal, and day-to-day
variations of electron content are discussed.

083.039 Measurement of the electron concentration of the
           ionosphere with a high-frequency impedance probe.
G. P. Komrakov, V. P. Ivanov, I. V. Popkov, V. N. Tyukin.
Kosm. Issled., Vol. 8, 278 - 283 (1970).   In Russian.

083.040 Measurement of the total electron concentration
           in the near-polar ionosphere by registrations of
the signals from the 3rd A.E.S. (artificial earth satellite).
G. K. Solodovnikov, V. A. Misyura, V. M. Migunov,
I. I. Gorbatchev.
Geomagn. Aeronom., Vol. 10, 334 - 336 (1970).
In Russian. – Brief information.

083.041 The connection of the earth's magnetic field with
           the E-layer.   T. V. Kolesnikova, A. A. Staro-
vatov, L. D. Filonova.
Geomagn. Aeronom., Vol. 10, 358 - 359 (1970).
In Russian. – Brief information.

083.042 Investigation of the lower ionosphere during the
           solar eclipse of September 22, 1968.
V. S. Yampolsky.
Geomagn. Aeronom., Vol. 10, 360 - 362 (1970).
In Russian. – Brief information.

083.043 Introductory lecture on ion waves.
           D. Pfirsch.
Astrophys. Space Sci. Library, Vol. 14, 1 - 11 (1969).
Conference paper (see 012.006).

083.044 General features and satellite observations of
           magnetoionic and magnetohydrodynamic waves
in the outer ionosphere.   H. Kikuchi.

Astrophys. Space Sci. Library, Vol. 14, 12 - 60 (1969).
Conference paper (see 012.006).

083.045 Effects of field-aligned currents on the structure
           of the ionosphere.
L. P. Block, C.-G. Fälthammar.
Astrophys. Space Sci. Library, Vol. 14, 69 - 77 (1969).
Conference paper (see 012.006).

083.046 Effects of ion-neutral collisions on ion-acoustic
           instabilities in the auroral ionosphere.
N. D'Angelo.
Astrophys. Space Sci. Library, Vol. 14, 78 - 86 (1969).
Conference paper (see 012.006).

083.047 Scintillations of satellite signals.
           L. Liszka.
Astrophys. Space Sci. Library, Vol. 14, 192 - 206 (1969).
Conference paper (see 012.006).

083.048 Inhomogeneities in the ionosphere measured by
           radio signals from the beacon satellite Explorer-22,
emphasizing satellite scintillations.   G. K. Hartmann.
Astrophys. Space Sci. Library, Vol. 14, 207 - 215 (1969).
Conference paper (see 012.006).

083.049 Production and prediction of sporadic E.
           J. D. Whitehead.
Rev. Geophys. Space Phys., Vol. 8, 65 - 144 (1970).
     This is a survey of the experimental observations and
theories of formation of all types of sporadic E.

083.050 A study of scintillations at low latitudes during a
           period from sunspot minimum to sunspot maximum.
G. O. Walker, T. Chan.
Journ. Geophys. Res., Vol. 75, 2517 - 2528 (1970).
     Records taken of the 20 and 40 MHz transmissions from
the satellite S 66 during the period October 1964 to July 1968
have been analysed for scintillations both north and south
of Hong Kong. Scintillation activity has been found to be
greatest (1) around midnight, (2) in the summer and (3) at
sunspot minimum.

083.051 Production and loss of electrons in the quiet day-
           time D region of the ionosphere.   G. C. Reid.
Journ. Geophys. Res., Vol. 75, 2551 - 2562 (1970).
     Rocket probe measurements of the concentration of free
electrons in the undisturbed daytime D region typically show
the presence of a steep ledge in electron density at an altitude
that lies between 80 and 90 km, in the vicinity of the meso-
pause. This paper is chiefly concerned with the interpretation
of this feature.

083.052 Thermal diffusion in the $F2$-region of the iono-
           sphere.   R. W. Schunk, J. C. G. Walker.
Planet. Space Sci., Vol. 18, 535 - 557 (1970).

083.053 The density of O^{2+} ions in the topside ionosphere.
           J. C. G. Walker.
Planet. Space Sci., Vol. 18, 559 - 564 (1970).
     Illustrative calculations show that the O^{2+} density
increases slowly with altitude and that the ratio of O^{2+} to O^+
is about 1 per cent at 500 km.

083.054 The calculation of ionospheric profiles from data
           given on oblique incidence ionograms.
M. S. Smith.
Journ. Atmosph. Terr. Phys., Vol. 32, 1047 - 1056 (1970).
     A method is described for calculating the electron density
distribution in the ionosphere, between two locations, from
the information contained in an oblique incidence ionogram.

083.055 **Lunar tidal variations in the equatorial sporadic-E-layer.** B. K. Joshi, K. M. Kotadia.
Journ. Atmosph. Terr. Phys., Vol. 32, 1057 - 1066 (1970).
Results of the study on the lunar variation of f_0E_s and f_bE_s at Kodaikanal are given and compared with the lunar variations in the H-field.

083.056 **The seasonal anomaly in the electron density of the topside F2-region.** M. N. Fatkullin.
Journ. Atmosph. Terr. Phys., Vol. 32, 1067 - 1075 (1970).
In the northern hemisphere over the American continent the anomaly extends to heights near 500 km at geographic latitudes between 46 and 53°N. The seasonal anomaly in the F2-layer appears to be mainly due to anomalous behavior in summer. The results obtained refer to a period near sunspot minimum.

083.057 **The F-region during an eclipse – A theoretical study.** P. Stubbe.
Journ. Atmosph. Terr. Phys., Vol. 32, 1109 - 1116 (1970).
Theoretical solutions are presented for the electron and ion densities, neutral, electron, and ion temperatures, and plasma velocities in the height range from 120 to 1500 km for a total eclipse. The computations were carried out for the coming eclipse at Wallops Island on March 7, 1970.

083.058 **Electron densities measured by the partial reflection method compared with simultaneous rocket measurements.**
A. Haug, M. Jespersen, J. A. Kane, E. V. Thrane.
Journ. Atmosph. Terr. Phys., Vol. 32, 1139 - 1142 (1970).
Electron densities measured by the partial reflection method and by rocket techniques during the solar eclipse of 20 May 1966, are in satisfactory agreement below about 82 km height. A discrepancy above 82 km is almost certainly due to a systematic error in the partial reflection experiment caused by an increasing importance of oblique echoes at high altitudes.

083.059 **Auroral enhancement of ionospheric electron density.** H. F. Bates.
Journ. Atmosph. Terr. Phys., Vol. 32, 1153 - 1157 (1970).
The purpose of this paper is to combine a number of previously reported results as a basis for suggesting that the same flux that produces the aurora causes the polar peak.

083.060 **Harmonic ion cyclotron resonances observed by the OGO-4 satellite.** H. Kikuchi.
Nature, Vol. 225, 257 - 258 (1970).
A new phenomenon associated with proton and electron whistlers observed from the OGO-4 satellite appears on a frequency-time spectrogram as several separated, narrow, steady bands of noise whose center lines are located at multiples of the ion cyclotron frequency (0.475 kHz). These spectral lines are identified to be a manifestation of the 2nd, 4th, and 6th harmonic ion cyclotron resonances. It is suggested that moderate to low energy ions in addition to an ambient thermal plasma are responsible for this excitation and that incoming electron whistlers provide a triggering effect.

083.061 **The altitude dependence of the quiettime cosmic ray ionization over the polar regions at solar minimum.** M. J. George.
Journ. Geophys. Res., Vol. 75, 3154 - 3158 (1970).
The integrating ionization chamber on the OGO 2 satellite has measured the quiettime cosmic ray ionization from 430 to 1540 km over the polar regions. The ionization in this region, called the polar plateau, can have five possible sources: (1) primary galactic cosmic rays, (2) splash-albedo particles, (3) solar cosmic rays, (4) magnetospheric particles, and (5) spacecraft radioactivity. The last three of these must be eliminated in the analysis in order to find the effect of the first two.

083.062 **Observations of the cosmic ray knee with a polar orbiting ionization chamber.** M. J. George.
Journ. Geophys. Res., Vol. 75, 3159 - 3166 (1970).
The cosmic ray knee has been observed with the ion chambers on the OGO 2 (October 1965 – February 1966) and OGO 4 (August 1967) polar orbiting satellites, using the same graphical definition that applies to balloon observations. Here we will discuss knee observations with the ion chambers of the OGO 2 and 4 satellites.

083.063 **In situ measurements of neutral and electron density wave structure from the Explorer 32 satellite.**
P. L. Dyson, G. P. Newton, L. H. Brace.
Journ. Geophys. Res., Vol. 75, 3200 - 3210 (1970).
In situ measurements of neutral particle and electron density from the Explorer 32 satellite have provided direct evidence that gravity waves in the thermosphere are associated with wave-like structure in the F region electron density.

083.064 **Underside morphology of the F2-ionospheric layer.** W. C. Knudsen, G. W. Sharp, K. K. Harris.
Journ. Atmosph. Terr. Phys., Vol. 32, 1183 - 1190 (1970).

083.065 **Evaluation of electron density from VLF Doppler measurements in a rocket.**
A. Egeland, G. Bjøntegård, T. L. Aggson.
Journ. Atmosph. Terr. Phys., Vol. 32, 1191 - 1204 (1970).

083.066 **Solar X-ray control of the E-layer of the ionosphere.** P. R. Sengupta.
Journ. Atmosph. Terr. Phys., Vol. 32, 1273 - 1282 (1970).
In this paper we have evaluated the relative importance of X-ray and ultraviolet ionization from a correlation between f_0E, the ordinary ray critical frequency of the E-layer and integrated solar X-ray flux in 8 – 20 Å and 44 – 60 Å bands recorded by N. R. L. detectors on Explorer 30, OGO-4 and OSO-4 satellites.

083.067 **Synoptic ionospheric observations, including absorption, drifts and special programmes.**
K. Rawer.
Ann. IQSY, (see 012.017), Vol. 5, 97 - 130 (1969).

083.068 **A review of the large-scale structure of the ionospheric F layer.** J. W. King.
Ann. IQSY, (see 012.017), Vol. 5, 131 - 165 (1969).
The major features of the ionospheric F layer between 200 and 2000 km in the equatorial, mid-latitude, and high-latitude regions are reviewed.

083.069 **The ionosphere as the base of the magnetosphere.** S.-I. Akasofu.
Ann. IQSY, (see 012.017), Vol. 5, 167 - 180 (1969).

083.070 **Ion and neutral composition of the ionosphere.** C. Y. Johnson.
Ann. IQSY, (see 012.017), Vol. 5, 197 - 213 (1969).

083.071 **Radio star and satellite scintillations.** G. G. Getmantsev, L. M. Eroukhimov.
Ann. IQSY, (see 012.017), Vol. 5, 229 - 259 (1969).
The paper presents the observational data of radio star and satellite scintillations for the period 1958 – 1965. The diurnal, seasonal, and latitude variations of scintillations are considered, as well as the association of scintillations with magnetic and solar acitivity.

083.072   Ionospheric plasma disturbances due to a moving
space vehicle.   H. Oya.
Planet. Space Sci., Vol. 18, 793 - 802 (1970).

083.073   F2-region disturbances associated with major
magnetic storms.   L. Thomas.
Planet. Space Sci., Vol. 18, 917 - 928 (1970).

083.074   The ion composition of the upper atmosphere in
130 - 155 km height during the activity of the
Orionids meteor stream.   A. D. Danilov, V. G. Istomin,
V. K. Semenov.
Kosmich. Issled., Vol. 8, 473 - 475 (1970).   In Russian.
Brief information.

083.075   The electric field in the ionosphere and in the magnetosphere of the earth at the presence of an inhomogeneity of fast particles.   E. E. Tzedilina.
Geomagn. Aeronom., Vol. 10, 408 - 416 (1970).   In Russian.

083.076   The influence of the atmosphere on the longitudinal
dependence of the electron intensity in the region
of the anomaly.   E. V. Gorchakov, G. A. Timofeev.
Geomagn. Aeronom., Vol. 10, 423 - 427 (1970).   In Russian.

083.077   Altitude distribution of the effect of storms in the
outer ionosphere.
M. N. Fatkullin, A. D. Legen'ka.
Geomagn. Aeronom., Vol. 10, 435 - 442 (1970).   In Russian.

083.078   Cases when the F-layer is absent in the earth's
ionosphere at high latitudes.   M. N. Fatkullin.
Geomagn. Aeronom., Vol. 10, 443 - 446 (1970).   In Russian.

083.079 NO_2^+ and O_2^+ ions in the lower ionosphere and
corpuscular ionizing radiation.   V. F. Tulinov.
Geomagn. Aeronom., Vol. 10, 538 - 540 (1970).
In Russian. – Brief information.

083.080   The seasonal anomaly of the F2-layer and corpus-
cular radiation.   N. M. Boenkova.
Geomagn. Aeronom., Vol. 10, 541 - 543 (1970).
In Russian. – Brief information.

083.081   The change of the N(h)-profiles during positive
ionospheric disturbances in the years of the maxi-
mum and minimum of solar activity.
E. E. Goncharova, R. A. Zevakina, E. V. Lavrova, L. A.
Yudovitch.
Geomagn. Aeronom., Vol. 10, 547 - 549 (1970).
In Russian. – Brief information.

083.082   The method of the definition of the distribution of
the electron concentration in the ionosphere.
I. S. Kutiev.
Geomagn. Aeronom., Vol. 10, 549 - 551 (1970).
In Russian. – Brief information.

083.083   On the effective frequency of electron encounters
in the auroral ionosphere.
N. K. Osipov, N. B. Pivovarova.
Geomagn. Aeronom., Vol. 10, 551 - 552 (1970).
In Russian. - Brief information.

083.084   On the structure and parameters of sporadic forma-
tions in the auroral ionosphere.
N. K. Osipov, N. B. Pivovarova, A. G. Chiryaev.
Geomagn. Aeronom., Vol. 10, 553 - 554 (1970).
In Russian. – Brief information.

083.085   Additional calculations of the rate of electron

production due to solar cosmic rays in the lower
ionosphere.   L. I. Dorman, T. M. Krupitskaya.
Izv. AN SSSR. Ser. fiz., Vol. 33, 1933 - 1939 (1969).
In Russian. – Abstr. in Referativ. Zhurn. 51. Astron.,
4.51.477 (1970).

083.086   Influence of the characteristics of solar corpuscular
streams on the rate of the origin of electrons in the
lower ionosphere.   P. Velinov.
Izv. AN SSSR. Ser. fiz., Vol. 33, 1918 - 1920 (1969).
In Russian. – Abstr. in Referativ. Zhurn. 51. Astron.,
4.51.480 (1970).

083.087   The influence of solar corpuscular streams on the
nocturnal lower ionosphere at mean latitudes.
G. N. Nesterov.
Izv. Geofiz. in-t. B"lg. AN, Vol. 14, 53 - 61 (1969).
In Bulgarian.

083.088   Nocturnal absorption in the ionosphere at mean
solar activity.   G. Nesterov.
Izv. Geofiz. in-t. B"lg. AN, Vol. 14, 63 - 73 (1969).
In Bulgarian. – Abstr. in Referativ. Zhurn. 51. Astron.,
6.51.462 (1970).

083.089   Mesures ionosphériques.
Edited by "Division des Prévisions Ionosphériques
du Centre National d'études des Télécommunications, Issy-les-
Moulineaux.
Bull. Mesures Ionosph. (BMI), Vol. 8, Nos. 10 - 12 [Dakar,
Djibuti, Paris-Saclay, Tahiti, Tananarive] (1969). – 1966
Octobre - Decembre.

083.090   Penetration of solar particles to ionospheric heights
at low latitudes.   S. Ganguly, M. Rao.
Nature, Vol. 225, 169 - 170 (1970).
This article reports what is believed to be evidence for
the penetration of solar particles to lower ionospheric heights
at a low-latitude station.

083.091   Considérations sur la haute atmosphère pendant
une aurore polaire.   J. C. Gérard.
Bull. Soc. Roy. Sci. Liège, Vol. 38, 61 - 71 = Univ. Liège,
Inst. d'Astrophys., Coll. 8°, No. 577 (1969).
Description is given of the processes of ionisation,
recombination and charge transfer affecting the upper
atmosphere during an aurora.

Ionospheric effects of X-ray emission from an
active region with a proton flare (30 June - 11 July 1966).
See Abstr. 073.076.

The X-ray and extreme ultraviolet radiation of the
7 July 1966 proton flare as deduced from sudden ionospheric
disturbance data.   See Abstr. 073.089.

The geomagnetic crochet of 7 July 1966.
See Abstr. 073.090.

Ionospheric conditions following the proton flare
of 7 July 1966 as deduced from topside soundings.
See Abstr. 073.115.

Ionospheric conditions following the proton flare
event of 7 July 1966 as measured at ground-based stations.
I. Low-energy particle effects in the lower ionosphere at
medium latitudes.   See Abstr. 073.116.

Ionospheric conditions following the proton flare
event of 7 July 1966 as measured at ground-based stations:
II. F-region effects.   See Abstr. 073.117.

Ionospheric disturbances after the proton flare of 7 July 1966.   See Abstr. 073.118.

Changes in the electron content of the ionosphere. See Abstr. 079.102.

Changes in the lower ionosphere during the eclipse– A preliminary report of the Canadian programme. See Abstr. 079.102.

Comparison of changes in total electron content along three paths.   See Abstr. 079.102.

Eclipse effects in the ionosphere. See Abstr. 079.102.

Eclipse observations at Arecibo, Puerto Rico, on March 7, 1970.   See Abstr. 079.102.

Observations of ionospheric electron content during the March 7, 1970, solar eclipse. See Abstr. 079.102.

Synoptic preview of ionospheric data taken at Fort Monmouth, New Jersey, during the eclipse. See Abstr. 079.102.

Variation of electron density in the D-region. See Abstr. 079.102.

Mathematical models of magnetospheric convection and its coupling to the ionosphere.   See Abstr. 084.251.

Decametric variability of Cas A and 3C84. See Abstr. 141.048.

The influence of Forbush effects on the state of the cosmic ray layer in the lower ionosphere. See Abstr. 143.033.

084 Aurorae, Geomagnetic Field, Radiation Belts

Aurorae

084.001 **Satellite observations of soft particle fluxes in the auroral zone.** W. J. Heikkila.
Nature, Vol. 225, 369 - 370 (1970).

This is a preliminary report of measurements of electron and proton precipitation in the auroral zone by a soft particle spectrometer (SPS) in the ISIS-1 satellite.

084.002 **Temporal behavior of energetic particle precipitation during an auroral substorm.**
B. A. Whalen, I. B. McDiarmid.
Journ. Geophys. Res., Vol. 75, 123 - 132 (1970).

Results from two rocket experiments launched from Fort Churchill Research Range into the breakup and post-breakup phases of an auroral substorm are described. Unusual characteristics were observed in energy spectra and pitch-angle distributions of energetic electrons $(E > 23$ kev$)$ and protons $(E > 28$ kev$)$ simultaneously with the arrival of the northern edge of a series of northward-propagating auroral arcs.

084.003 **Rapid fluctuations of energetic auroral particles.**
R. L. Arnoldy.
Journ. Geophys. Res., Vol. 75, 228 - 232 (1970). – Letter.

084.004 **Rocket-borne measurements of Hβ emissions and energetic hydrogen fluxes during an auroral breakup.**
R. L. Wax, W. Bernstein.
Journ. Geophys. Res., Vol. 75, 783 - 787 (1970).

Simultaneous rocket-borne measurements of the intensity and altitude profile of Hβ light emission and the flux and energy spectrum of 0.5 – 22 kev precipitated hydrogen have been performed during an auroral breakup.

084.005 **Far-ultraviolet altitude profiles and molecular oxygen densities in an aurora.**
C. B. Opal, H. W. Moos, W. G. Fastie.
Journ. Geophys. Res., Vol. 75, 788 - 796 (1970).

The altitude profiles and absolute intensities of several far-ultraviolet emission features in a steady IBC I$^+$ aurora were measured with a filterwheel photometer carried in an Aerobee rocket over Churchill, Manitoba, at 2241 CST on February 8, 1968.

084.006 **Observations of the aurora in the far ultraviolet from OGO 4.** T. A. Chubb, G. T. Hicks.
Journ. Geophys. Res., Vol. 75, 1290 - 1311 (1970).

This paper is concerned with observations of the earth's far ultraviolet polar aurora as seen from the OGO 4 spacecraft. We present here a number of small studies on the shape and location of the auroral oval, the relative positions of the proton and electron auroras, the variation in auroral intensity with magnetic activity, and the relative intensities of day and night auroras.

084.007 **Behavior of hydroxyl emission during aurora.**
A. W. Harrison.
Journ. Geophys. Res., Vol. 75, 1330 - 1333 (1970).

Analysis of auroral spectra (1.02 to 1.13 μ) and photometric records of the O I (5577) auroral emission obtained during 6 hours of continuous IBC II-III aurora on November 1-2, 1968, leads to the conclusion that on this occasion the hydroxyl emission normally present in the near infrared night airglow spectrum was not significantly correlated with the O I (5577) brightness fluctuations.

084.008 **An observation in situ of an auroral pulsation.**
T. D. Parkinson, E. C. Zipf, K. A. Dick.
Journ. Geophys. Res., Vol. 75, 1334 - 1338 (1970).

An auroral pulsation was observed by all instruments on board an Aerobee rocket that was launched into a diffuse IBC class I aurora above Fort Churchill, Canada. This event was also observed from the ground by a multichannel photometer. The data suggest that the apparent downward propagation speed for the incoming electron flux pulse was only 5 km/sec.

084.009 **Horizontal gradients in auroral radio absorption and their effect on the heights determined by multifrequency riometry.** J. K. Hargreaves.
Journ. Atmosph. Terr. Phys., Vol. 32, 123 - 126 (1970). Short paper.

084.010 **Simultaneous forward-scatter, riometer, and bremsstrahlung observations of a daytime electron precipitation event in the auroral zone.**
D. K. Bailey, R. R. Brown, M. H. Rees.
Journ. Atmosph. Terr. Phys., Vol. 32, 149 - 169 (1970).

084.011 **The lunar influence on radio-aurora.**
P. A. Forsyth.
Journ. Atmosph. Terr. Phys., Vol. 32, 251 - 255 (1970).

A series of radio-auroral observations extending over an 11-yr period are analyzed in an effort to detect the 29.5 day lunar period. While such a periodicity is present in the data it appears to arise as a result of the chance interaction of other well-known periodicities and not as a result of an interaction between the moon and the magnetosphere.

084.012 **Investigation of geoactive particles and photoelectrons with the satellite Cosmos 261. 1. Description of the experiment.**
A. D. Bolyunova, M. L. Bragin, Yu. I. Gal'perin, V. A. Gladyshev, N. V. Dzhordzhio, G. N. Zlotin, I. N. Kiknadze, R. A. Kovrazhkin, T. M. Mulyarchik, Yu. N. Ponomarev, V. V. Temnyj, N. I. Fedorova, Yu. P. Shilyaev, F. K. Shujskaya, R. V. Shulenina.
Kosmich. Issled.,Vol. 8, 104 - 107 (1970). In Russian.

084.013 **Investigation of geoactive particles and photoelectrons with the satellite Cosmos 261. 2. Measurements of low energy electrons.**
Yu. I. Gal'perin, N. V. Dzhordzhio, I. D. Ivanov, I. P. Karpinskij, Eh. L. Lein, T. M. Mulyarchik, B. V. Polenov, V. V. Temnyj, N. I. Fedorova, B. I. Khazanov, A. V. Shifrin, F. K. Shujskaya.
Kosmich. Issled.,Vol. 8, 108 - 119 (1970). In Russian.

084.014 **Investigation of geoactive particles and photoelectrons with the satellite Cosmos 261. 3. Measurements of low energy ions.** Yu. l. Gal'perin, V. A. Gladyshev, I. D. Ivanov, I. P. Karpinskij, T. M. Mulyarchik, B. V. Polenov, V. V. Temnyj, B. I. Khazanov, A. V. Shifrin, F. K. Shujskaya.
Kosmich. Issled.,Vol. 8, 120 - 126 (1970). In Russian.

084.015 **Investigation of geoactive particles and photoelectrons with the satellite Cosmos 261. 4. Measurements of charged particles of mean and high energy.**
A. D. Bolyunova, A. D. Verevkin, Yu. I. Gal'perin, L. S. Gorn, L. S. Zhurina, I. D. Ivanov, R. N. Isaeva, I. P. Karpinskij, R. A. Kovrazhkin, V. V. Temnyj, B. I. Khazanov, A. V. Shifrin, F. K. Shujskaya.
Kosmich. Issled.,Vol. 8, 126 - 135 (1970). In Russian.

084.016 **Ion composition and ion chemistry in an aurora.**
T. M. Donahue, E. C. Zipf, Jr., T. D. Parkinson.
Planet. Space Sci., Vol. 18, 171 - 186 (1970).

Ion composition measurements are reported as obtained with a rocket borne mass spectrometer flown during an aurora. The results appear to demand effective ion temperatures greater than $1000°K$ and electron temperatures even larger. They also require efficient channels for conversion of O_2^+ to NO^+. The O_2^+ densities measured are too low to permit dissociative recombination to contribute significantly to OI green line excitation except at high altitude.

084.017 **Rocket investigation of the auroral green line.**
T. D. Parkinson, E. C. Zipf, Jr., T. M. Donahue.
Planet. Space Sci., Vol. 18, 187 - 198 (1970).

Results obtained from rocket borne photometers, electron energy analyzers and ion mass spectrometers show that direct excitation of atomic oxygen by auroral electrons can excite only a small fraction of the auroral green line observed.

084.018 **On the ionic constitution of class I auroras.**
W. Swider, Jr., R. S. Narcisi.
Planet. Space Sci., Vol. 18, 379 - 385 (1970).

The purpose of this paper is to construct an ion composition model for a class I aurora which agrees with the experimental results.

084.019 **Photometric and interferometric observations of a mid-latitude stable auroral red arc.**
R. G. Roble, P. B. Hays, A. F. Nagy.
Planet. Space Sci., Vol. 18, 431 - 439 (1970).

Observational results of the mid-latitude stable auroral red arcs of October 30/31 and October 31/November 1, 1968 are presented. The structure, intensity, and position of the red arc were determined from photometer scan measurements. The results of the Doppler temperature measurements made with the Fabry-Pérot interferometer show no measurable neutral gas temperature increase within the red arc.

084.020 **Auroral micropulsation instability.**
F. V. Coroniti, C. F. Kennel.
Journ. Geophys. Res., Vol. 75, 1863 - 1878 (1970).

In this paper we describe a drift instability of Alfvén waves driven by the sharp electron thermal gradient at the inner edge of the electron plasma sheet.

084.021 **Auroral emission from O_2 ($^1\Delta_g$).**
J. F. Noxon.
Journ. Geophys. Res., Vol. 75, 1879 - 1891 (1970).

We describe a large enhancement of the infrared oxygen band at $1.27\,\mu$ in <IBC II auroras. A comparison with N_2^+ band intensity suggests that the O_2 emission is too strong to be sustained by the electron flux responsible for N_2^+. Alternative sources of energy, such as chemical reactions and electric fields, are considered.

084.022 **Auroral activity near Banff, Alberta.**
R. J. Hoch, F. I. Carver, L. L. Smith, K. C. Clark.
Journ. Geophys. Res., Vol. 75, 1935 - 1936 (1970).

Auroral activity near Banff during the period April 1, 1968 to April 1, 1969, is summarized. The analysis is based on meridian spectrograph records. Frequency distributions of auroral activity with respect to Ap magnetic index are given.

084.023 **About the zone of pulsating aurorae.**
V. K. Roldugin, G. V. Starkov.
Geomagn. Aeronom., Vol. 10, 97 - 100 (1970). In Russian.

084.024 **Ring current and oval of aurorae in morning and evening hours.** G. V. Starkov, Ya. I. Feldstein.
Geomagn. Aeronom., Vol. 10, 162 - 164 (1970). In Russian.

084.025 **Statistical regularities of the hydrogen emission of aurorae.** V. G. Sobolev.
Geomagn. Aeronom., Vol. 10, 291 - 294 (1970).
In Russian.

084.026 **On the meridional movement of bays of auroral absorption.** V. M. Driatzky, O. I. Shumilov.
Geomagn. Aeronom., Vol. 10, 305 - 311 (1970).
In Russian.

084.027 **On pulsating aurorae in conjugated points.**
M. B. Gokhberg, B. N. Kazak, O. M. Raspopov,
V. K. Roldugin, V. A. Troitzkaya, V. I. Fedoseev.
Geomagn. Aeronom., Vol. 10, 367 - 370 (1970).
In Russian. – Brief information.

084.028 **Role of the universal instability in auroral phenomena.** N. D'Angelo.
Astrophys. Space Sci. Library, Vol. 14, 87 - 93 (1969).
Conference paper (see 012.006).

084.029 **A survey of low-energy ($E \gtrsim 5$ kev) electron energy fluxes over the northern auroral regions with satellite Injun 4.** J. D. Craven.
Journ. Geophys. Res., Vol. 75, 2468 - 2480 (1970).

084.030 **Recent occurrence of stable auroral red arcs.**
R. J. Hoch, K. C. Clark.
Journ. Geophys. Res., Vol. 75, 2511 - 2515 (1970).

Stable auroral red (SAR) arcs were definitely recorded on nine occasions and possibly recorded on six other occasions at Richland, Washington, between September 1967 and mid-May 1969. New spectrographic evidence verifies that [O I] 6300 - 6364 Å is the predominant radiation of the SAR arc.

084.031 **Local time behavior of the alignment and position of a stable auroral red arc.**
N. W. Glass, J. H. Wolcott, L. W. Miller, M. M. Robertson.
Journ. Geophys. Res., Vol. 75, 2579 - 2582 (1970).
Brief report.

084.032 **Detection of geomagnetically aligned currents associated with an auroral arc.**
P. A. Cloutier, H. R. Anderson, R. J. Park, R. R. Vondrak,
R. J. Spiger, B. R. Sandel.
Journ. Geophys. Res., Vol. 75, 2595 - 2600 (1970). – Letter.

084.033 **Motions associated with auroral zone electron precipitation.** G. Maral.
Journ. Geophys. Res., Vol. 75, 2601 - 2605 (1970). – Letter.

084.034 **Spatial separation of $\lambda 3914$- and $\lambda 5577$-Å emission in an aurora.** R. Duysinx, A. Monfils.
Journ. Geophys. Res., Vol. 75, 2606 - 2607 (1970). – Letter.

084.035 **Auroral pulsations – television image and X-ray correlations.** M. W. J. Scourfield, G. R. Cresswell, G. R. Pilkington, N. R. Parsons.
Planet. Space Sci., Vol. 18, 495 - 499 (1970).

The association between auroral X-rays and luminosity has been studied for pulsating auroras by coordinated observations with a balloon borne X-ray detector and a ground based image intensifier-television system. The degree of correlation between X-ray and luminosity variations was found to depend upon the number of pulsating forms common to both detector fields of view.

084.036 **The auroral oval and the boundary of closed field lines of geomagnetic field.**
Y. I. Feldstein, G. V. Starkov.
Planet. Space Sci., Vol. 18, 501 - 508 (1970).

A comparison of the position of the auroral oval with the boundary of the stable trapping region and the limit of closed geomagnetic field lines has been carried out; Alouette 2 data are used to obtain the trapping boundary.

084.037 Characteristics of a visual aurora following a solar flare. J. W. Meriwether, Jr., W. M. Benesch, S. G. Tilford.
Planet. Space Sci., Vol. 18, 525 - 534 (1970).

The present work concerns ground-based auroral data collected after the flare on February 13, 1967. An unusual amount of λ6300 intensity and an increase in the apparent vibrational temperature of the molecular nitrogen are stated.

084.038 Polarization of auroral echoes.
L. Harang, J. Tröim.
Planet. Space Sci., Vol. 18, 655 - 660 (1970).

084.039 Excitation and radiative transport of OI 1304 Å resonance radiation − II. The aurora.
T. M. Donahue, D. J. Strickland.
Planet. Space Sci., Vol. 18, 691 - 697 (1970).

Radiative transfer calculations appropriate to optically thick media in the presence of molecular absorption have been performed which derive the rate of excitation of the $O(3^3S)$ term in an aurora from observed intensity profiles.

084.040 Energy deposition by auroral electrons in the atmosphere.
M. J. Berger, S. M. Seltzer, K. Maeda.
Journ. Atmosph. Terr. Phys., Vol. 32, 1015 - 1045 (1970).

This paper describes transport calculations pertinent to the analysis of the auroral phenomena that occur when the upper atmosphere is bombarded with electrons.

084.041 On the heights of aurorae in the period of a solar minimum. D. A. Andrienko, N. N. Bliznyuk.
Vestn. Kiev. Univ., Ser. Astron., No. 11, p. 136 - 137 (1969).
In Russian.

084.042 The magnetospheric substorm.
K. A. Anderson.
Ann. IQSY, (see 012.016), Vol. 4, 62 - 64 (1969).

084.043 The phenomenology and morphology of aurorae.
Ya. I. Feldstein, S. I. Isaev, A. I. Lebedinsky.
Ann. IQSY, (see 012.016), Vol. 4, 311 - 348 (1969).

The space-time distribution of the discrete aurorae in the polar cap and of the zone aurorae is discussed.

084.044 Spectroscopic morphology of aurora.
A. V. Jones.
Ann. IQSY, (see 012.016), Vol. 4, 349 - 363 (1969).

The problem of determining spectroscopically the global excitation patterns of aurora excited by incident electrons and protons is reviewed.

084.045 Incoming lower energy particles and their association with airglow and aurorae. B. J. O'Brien.
Ann. IQSY, (see 012.016), Vol. 4, 364 - 374 (1969).

Numerous measurements have now been made of the electrons and protons precipitated into the atmosphere to cause auroras. A review is given of the measurements from ground-based devices as well as those carried by balloons, rockets, and satellites.

084.046 Energy transfer from $N_2(A^3\Sigma_u^+)$ as a source of $O(^1S)$ in the aurora.
T. D. Parkinson, E. C. Zipf.
Planet. Space Sci., Vol. 18, 895 - 900 (1970).

084.047 Conjugate riometer studies of auroral zone cosmic noise absorption. L. A. Hajkowicz.
Australian Journ. Phys., Vol. 23, 187 - 196 (1970).

084.048 Protons of aurorae and the "resonance" conception of a sub-storm. Yu. I. Gal'perin, V. A. Gladyshev, A. V. Gurevich, A. K. Kuz'min, Yu. N. Ponomarev.
Kosmich. Issled., Vol. 8, 457 - 460 (1970). In Russian.
Brief information.

084.049 Observation of polar aurora on March 8, 1970 at the observatory Lomnický Štít.
M. Rybanský.
Říše hvězd, Vol. 51, 129 - 131 (1970). In Slovakian.

084.050 Mid-latitude aurorae on March 23 - 24, 1969.
Yu. L. Truttse, N. N. Shefov, O. T. Yurchenko.
Astron. Tsirk., No. 562, p. 3 - 5 (1970). In Russian.

084.051 Étude spectroscopique dans le domaine visible et ultra-violet proche des aurores du 19 novembre 1966 et 4 février 1967.
R. Malbrouck, R. Duysinx.
Bull. Soc. Roy. Sci. Liège, Vol. 38, 35 - 42 = Univ. Liège, Inst. d'Astrophys., Coll. 8°, No. 579 (1969).

Notes on some recent auroras.
Sky Telescope, Vol. 39, 400 - 401 (1970).

The polar-cap absorption on 7 - 10 July 1966.
See Abstr. 073.103.

The 7 July 1966 solar cosmic-ray event.
See Abstr. 073.104.

Very low radio frequency observations of the solar flare and polar-cap absorption event of 7 July 1966.
See Abstr. 073.105.

Solar particle observations over the polar caps.
See Abstr. 078.006.

Geomagnetic Field

084.201 **On the electric field in the earth's distant magnetotail.** J. A. Van Allen.
Journ. Geophys. Res., Vol. 75, 29 - 38 (1970).

Solar electrons of energy $E_e \gtrsim 50$ kev are used as test
particles for studying electric and magnetic fields in the distant magnetotail of the earth. During the prolonged solar
electron event November 10–22, 1967, simultaneous observations were made with the earth-orbiting satellite Explorer 33
in interplanetary space and with the moon-orbiting satellite
Explorer 35 as the latter crossed the magnetotail.

084.202 **Local-time dependence of geomagnetic cutoffs for
solar protons, $0.52 \leqslant E_p \leqslant 4$ Mev.**
H. R. Flindt.
Journ. Geophys. Res., Vol. 75, 39 - 49 (1970).

Low-altitude, high-altitude observations of $0.52 \leqslant E_p \leqslant$
4 Mev protons by the University of Iowa/NASA satellite Injun
4 are presented for three solar proton events having maximum
intensities on January 20, May 30, and June 23, 1966. Only
weak magnetic storms were associated with these events. A
substantial body of new data is given on the magnetic local
time (MLT) dependence of the geomagnetic cutoffs and the
counting rate versus invariant latitude profiles from 0600
to 1300 hours and 1800 to 2200 hours for low-energy protons
moving approximately normal to the local magnetic field.

084.203 **A study of the influence of magnetic activity on the
location of the plasmapause as measured by OGO 5.**
C. R. Chappell, K. K. Harris, G. W. Sharp.
Journ. Geophys. Res., Vol. 75, 50 - 56 (1970).

The plasmapause position has been measured quite well
by the light ion mass spectrometer aboard OGO 5, which
measures the concentrations of H^+, He^+, and O^+ ions as a
function of L and local time. The influence of magnetic
activity on this plasmapause position has been studied for the
local-time regions at 1000 ± 2 hours and 0200 ± 2 hours.

084.204 **A three-dimensional model current system for
polar magnetic substorms.**
B. Bonnevier, R. Boström, G. Rostoker.
Journ. Geophys. Res., Vol. 75, 107 - 122 (1970).

A model current system, in which magnetospheric and
ionospheric sections are connected by currents flowing along
the geomagnetic field lines, is proposed to represent the
current system responsible for polar magnetic substorms. The
magnetic perturbations from model current systems of this
type are studied in terms of elementary loops, whose magnetic
effects are evaluated numerically.

084.205 **Initial deceleration of solar wind positive ions in
the earth's bow shock.** M. Neugebauer.
Journ. Geophys. Res., Vol. 75, 717 - 733 (1970).

High time resolution ($\Delta t = 0.288$ to 9.5 seconds) plasma
measurements have been made on the upstream edge of the
earth's bow shock by the combination of a Faraday cup with
modulation grid and a curved-plate analyzer on the satellite
OGO 5. These observations show that the solar wind positive
ions often undergo a substantial deceleration just upstream
of the shock's steep gradient of magnetic-field strength.

084.206 **Geometry of the geomagnetic tail.**
K. W. Behannon.
Journ. Geophys. Res.,Vol. 75, 743 - 753 (1970).

This investigation combines the Explorer 35 tail field
measurements with those of Explorer 33 for the period July
1967 to August 1968. Evidence is presented for the existence
of (1) a broad region of depressed field magnitude and increased B_z component surrounding the neutral sheet; (2) a

slight divergence of tail field lines from the axis of the tail
with distance from the earth; and (3) an aberration of the
tail axis. The location of the neutral sheet relative to the
solar magnetospheric XY plane is discussed, and the contributions of the expansion of the tail and the reconnection at
the neutral sheet to the observed gradient are considered.

084.207 **OGO 3 observations of ELF noise in the magneto-
sphere. 2. The nature of the equatorial noise.**
C. T. Russell, R. E. Holzer, E. J. Smith.
Journ. Geophys. Res., Vol. 75, 755 - 768 (1970).

An examination of the noise present between 1 and
1000 Hz at the magnetic equator with the OGO 3 search coil
magnetometer has revealed a previously unobserved class of
signals existing only in the outer plasmasphere.

084.208 **Plasma temperatures in the magnetosphere.**
S. Sanatani, W. B. Hanson.
Journ. Geophys. Res., Vol. 75, 769 - 775 (1970).

Rocket measurements of ion temperature are found to be
in good agreement with the calculated values. It is shown that
for low electron concentrations the ion temperature may be
seriously underestimated when the boundary condition on
ion heat transport is imposed at too low an altitude. The
opacity of the magnetosphere to photoelectrons is also examined.

084.209 **A magnetosphere model based on two zones of
precipitating energetic particles.**
V. M. Mishin, T. I. Saifudinova, I. A. Zhulin.
Journ. Geophys. Res., Vol. 75, 797 - 806 (1970).

In this paper the main features of two quasi-circular
zones precipitating energetic particles are presented on the
basis of the instantaneous patterns of the distribution of geomagnetic activity as well as on the basis of the dynamics of
these patterns. A magnetosphere model reflecting these features is presented. The suggested model links the lower-latitude
zone with processes occurring in the magnetospheric tail,
whereas the higher-latitude zone is connected with particle
injection through the neutral points.

084.210 **Magnetic measurements in the earth's magneto-
sphere and magnetosheath: Mariner 5.**
E. J. Smith, L. Davis, Jr.
Journ. Geophys. Res., Vol. 75, 1233 - 1245 (1970).

Magnetic field and charged particle measurements were
obtained by Mariner 5 near Venus, in interplanetary space,
and near earth. In this paper we present an analysis of the
magnetic fields detected near earth on June 14, 1967, by
the low-field vector helium magnetometer.

084.211 **Hydromagnetic wave interaction with the magnetopause and the bow shock.** J. F. McKenzie.
Planet. Space Sci., Vol. 18, 1 - 23 (1970).

The results of analyses of hydromagnetic reflection and
refraction at a shear layer and at a shock are applied to situations representative of the magnetopause and the earth's bow
shock. It is noted that both longitudinal and transverse waves
can be amplified on passage through a strong shock. Thus the
amplification of the turbulent spectrum of hydromagnetic
waves in the solar wind on passage through the earth's bow
shock may account for the (at least) order of magnitude
increase of the noise spectrum in the magnetosheath over that
in the unshocked solar wind.

084.212 **The magnetospheric plasmapause and the electron
density trough at the Alouette I orbit.**
M. J. Rycroft, J. O. Thomas.
Planet. Space Sci., Vol. 18, 65 - 80 (1970).

The location of the magnetospheric plasmapause or
'knee' and the position of the electron density minimum or

'trough' at the orbit of the Alouette I satellite near 1000 km have been derived from whistler data and from topside ionograms respectively. It is inferred that the plasmapause in the magnetosphere and the trough near the exospheric base are related phenomena, and that a sharp decrease of plasma density occurs beyond a particular field line, the position of which depends on local time and on the degree of geomagnetic activity.

084.213 **The solar wind interaction with the geomagnetic field.** J. H. Wolfe, D. S. Intriligator.
Space Sci. Rev., Vol. 10, 511 - 596 (1970).

A review is presented of the interaction of the solar wind with the magnetic field of the earth. The material is developed primarily from an observational point of view. The early observations are covered through late 1963, with primary emphasis on the sunward interaction region. The historical review of the early results is discussed in terms of the significant contributions of each satellite observation and in the light of our present concept of the solar wind-geomagnetic field interaction.

084.214 **Ultra low frequency waves in the magnetosphere.** J. W. Dungey, D. J. Southwood.
Space Sci. Rev., Vol. 10, 672 - 688 (1970).

A brief summary of properties of ultra low frequency waves in the magnetosphere is given; the remainder of the paper is devoted to observations from satellites and recent developments in the theory, which have shown considerable points of agreement with satellite observations.

084.215 **The secular variation of the magnetic field and its cyclic components.** B. N. Bhargava, A. Yacob.
Journ. Atmosph. Terr. Phys., Vol. 32, 365 - 372 (1970).

084.216 **Record of observations at Victoria Magnetic Observatory 1967.** D. R. Auld, D. G. Holmes.
Publ. Dominion Obs. Ottawa, Vol. 38, (No. 6), 399 - 455 (1969).

084.217 **Record of observations at Fort Churchill Magnetic Variometer Station 1966.** G. Jansen van Beek.
Publ. Dominion Obs. Ottawa, Vol. 38, (No. 5), 323 - 397 (1969).

084.218 **Record of observations at Resolute Bay Magnetic Observatory 1967.** A. E. Evans.
Publ. Dominion Obs. Ottawa, Vol. 38, (No. 1), 1 - 85 (1969).

084.219 **Record of observations at Baker Lake Magnetic Observatory 1967.** G. Jansen van Beek.
Publ. Dominion Obs. Ottawa, Vol. 38, (No. 2), 87 - 164 (1969).

084.220 **Record of observations at Mould Bay Magnetic Observatory 1967.** A. E. Evans.
Publ. Dominion Obs. Ottawa, Vol. 38, (No. 3), 165 - 243 (1969).

084.221 **The magnetic reciprocal effect between the magnetosphere and the filamentary inhomogeneity of the solar wind.** N. V. Mikerina, K. G. Ivanov.
Kosmich. Issled.,Vol. 8, 149 - 151 (1970). In Russian.

084.222 **Record of observations at Alert Magnetic Observatory 1967.** A. E. Evans.
Publ. Dominion Obs. Ottawa, Vol. 38, (No. 4), 245 - 322 (1969).

084.223 **Observations of magnetopause geometry and waves at the lunar distance.**

J. D. Mihalov, D. S. Colburn, C. P. Sonett.
Planet. Space Sci., Vol. 18, 239 - 258 (1970).

This paper analyzes magnetopause crossings observed by the Ames magnetometer on Explorer 35 in orbit around the moon. The observations presented and discussed in this paper provide gross features and some details of boundary structure downstream from earth at the lunar orbit (~ 60 earth radii). These results are contrasted with earlier observations and differences in boundary orientation and thickness are noted. Models for oscillations of boundary location and boundary normal orientation are developed.

084.224 **Equatorial variation of the 27-day oscillation in the horizontal intensity.**
B. N. Bhargava, D. R. K. Rao.
Planet. Space Sci., Vol. 18, 287 - 290 (1970).

A 27-day oscillation of 4 – 6 gamma amplitude in the horizontal intensity in the low latitudes is well known and has been associated with solar synodic rotation. The present paper reports the results of a study in the frequency domain of amplitude of this oscillation as a function of dip for periods of high, moderate and low solar activity.

084.225 **The viscous magnetopause.** P. Cassen, J. Szabo.
Planet. Space Sci., Vol. 18, 349 - 366 (1970).

Under the assumption that wave-particle interactions cause a hydromagnetic viscosity, we represent the magnetopause by the mixing region of two streams of plasma with different velocities and different magnetic fields (the magnetosheath flow and the magnetospheric convection). The plasma is considered to be compressible, with non-zero viscosity, thermal conductivity, and resistivity. Solutions are found for the variations of velocity, magnetic field, and temperature through the boundary layer.

084.226 **Luni-solar tides in H at stations within the equatorial electrojet.** R. G. Rastogi, N. B. Trivedi.
Planet. Space Sci., Vol. 18, 367 - 377 (1970).

The paper describes lunar daily (L) variation at fixed lunar ages and lunar monthly (M) variation at individual solar hours of the horizontal magnetic field, H, at stations close to the magnetic equator for the IGY–IGC period. These results indicate an intimate relation between the lunar tides at the equatorial stations with the electrojet currents in respect to the diurnal as well as longitudinal variations.

084.227 **Magnetospheric convections and damped-type geomagnetic pulsations associated with storms.**
T. Namikawa, S. Matsushita.
Planet. Space Sci., Vol. 18, 407 - 415 (1970).

It is shown that the steady magnetospheric motion generated by the solar wind is current free in the first approximation, regardless of the driving mechanism. Also the sufficient condition for the hydromagnetic stability of cold plasma in the presence of an external magnetic field and a steady motion is obtained. Based on these theoretical estimations the damped-type geomagnetic pulsations associated with geomagnetic storms are interpreted as an interaction between hydromagnetic oscillations and magnetospheric motions caused by solar winds.

084.228 **VLF wave generation by the Čerenkov process in the inner magnetosphere.**
R. P. Singh, R. N. Singh.
Nature, Vol. 225, 49 - 50 (1970).

The Čerenkov radiation as a possible source of VLF wave generation in the inner magnetosphere (L = 1.1) has been considered. The calculated Čerenkov power generated by low energy electrons in the VLF range is shown to agree well with the measured VLF power.

084.229 **On the structure of the inner magnetosphere.**
C. T. Russell, R. M. Thorne.
Cosmic Electrodynamics, Vol. 1, 67 - 89 (1970).

This paper investigates the morphology of the thermal and energetic particles in the inner magnetosphere. The motion and relative position of several well defined features of the particle distribution are followed during geomagnetic disturbances and over the solar cycle.

084.230 **Geomagnetic intensity: Changes during the past 3000 years in the western hemisphere.**
V. Bucha, R. E. Taylor, R. Berger, E. W. Haury.
Science, Vol. 168, 111 - 114 (1970).

A series of archeomagnetic measurements have been carried out on archeologic materials from Arizona and Mexico which can be compared with results from Europe and Asia. This comparison shows a westward drift of geomagnetic intensity at a rate of about 0.24 degree per year. Furthermore, an apparent coincidence between changes in the earth's magnetic moment and changes in the production rate of radiocarbon is observed.

084.231 **A layer of energetic electrons (>40 kev) near the magnetopause.** C. I. Meng, K. A. Anderson.
Journ. Geophys. Res., Vol. 75, 1827 - 1836 (1970).

The purpose of the present work is to report the observation of energetic electron fluxes at the magnetopause by the IMP 3 satellite, which penetrated the magnetosphere at rather high latitudes over a wide range of local times. The data from IMP 1 and Explorer 35 were also examined.

084.232 **Low-frequency noise observed in the distant magnetosphere with OGO 1.**
N. Dunckel, B. Ficklin, L. Rorden, R. A. Helliwell.
Journ. Geophys. Res., Vol. 75, 1854 - 1862 (1970).

084.233 **Permanent aspects of the earth's non-dipole magnetic field over upper tertiary times.**
R. L. Wilson.
Geophys. Journ. Roy. Astron. Soc., Vol. 19, 417 - 437 (1970).

Investigation of collected palaeomagnetic results from continental igneous rocks and from oceanic sediment-cores, shows a persistent off-centre displacement of the effective dipole source about the past two million years. This dipole displacement is 191 ± 38 km northward along the rotational axis.

084.234 **Mathematical model for the magnetosphere.**
R. C. Hewson-Browne, D. N. Burghes.
Journ. Atmosph. Terr. Phys., Vol. 32, 757 - 765 (1970).

In this paper, a possible mathematical model for the interaction of the solar wind with the geomagnetic field is considered. This model includes a neutral point at the front of the magnetosphere and a neutral sheet downstream. The cavity boundary and the field lines inside the cavity are found for various values of the tail flux parameter.

084.235 **Methods of calculation of the earth's magnetic field upward in the near-earth space.**
B. D. Vintz, V. I. Pochtarev, R. Sh. Rakhmatulin.
Geomagn. Aeronom., Vol. 10, 119 - 128 (1970). In Russian.

084.236 **Investigation of the motion of charged particles in the model of the earth's magnetosphere.**
E. V. Kolomeetz, L. A. Mirkin.
Geomagn. Aeronom., Vol. 10, 137 - 138 (1970). In Russian.
Brief information.

084.237 **Fast time-resolved spectra of electrostatic turbulence in the earth's bow shock.**

R. W. Fredricks, F. V. Coroniti, C. F. Kennel, F. L. Scarf.
Phys. Rev. Letters, Vol. 24, 994 - 998 (1970).

We present time-resolved spectra of electrostatic turbulence in the earth's bow-shock structure. Spectral details on scales for a few Debye lengths indicate that single modes or groups of single modes dominate the turbulent spectrum. These modes are probably ion-acoustic or Buneman instabilities of short wavelength ($k\lambda_D \sim 1$) which are generated in parts of the shock microstructure containing diamagnetic drift currents.

084.238 **Magnetosphäre.** M. Siebert.
Plenarvorträge, 34. Physikertagung 1969 Salzburg (B. G. Teubner, Stuttgart), p. 52 - 77 (1969). – Conference paper.

084.239 **Magnetfeld der Erde.** J. Untiedt.
Plenarvorträge, 34. Physikertagung 1969 Salzburg (B. G. Teubner, Stuttgart), p. 490 - 509 (1969). – Conference paper.

084.240 **Daily variation of the geomagnetic field at the dip equator.** K. N. Nair, R. G. Rastogi, V. Sarabhai.
Nature, Vol. 226, 740 - 741 (1970).

084.241 **Spatial distribution and temporal variations of weak electron currents in the earth's magnetosphere from data of a trap of charged particles on board of Elektron 2 and their connection with the orientation of the earth's magnetic dipole.** M. Z. Khokhlov.
Kosm. Issled., Vol. 8, 261 - 272 (1970). In Russian.

084.242 **The spectrum of Alfvén oscillations of the magnetosphere.** A. V. Gulielmi.
Geomagn. Aeronom., Vol. 10, 234 - 239 (1970).
In Russian.

084.243 **On a possible cause of the dependence of the distance to the boundary of the magnetosphere on geographical longitude.** N. M. Rudneva.
Geomagn. Aeronom., Vol. 10, 312 - 315 (1970).
In Russian.

084.244 **On the magnetic field of the neutral layer in the tail of the magnetosphere.**
K. G. Ivanov.
Geomagn. Aeronom., Vol. 10, 333 - 334 (1970).
In Russian. – Brief information.

084.245 **On some peculiarities of the connection between geomagnetic and solar activity.**
G. P. Berishvili.
Geomagn. Aeronom., Vol. 10, 365 - 367 (1970).
In Russian. – Brief information.

084.246 **Diagnostics of the parameters of the magnetosphere and of the interplanetary space by means of micropulsations.** V. A. Troitskaya, A. V. Gul'elmi.
Astrophys. Space Sci. Library, Vol. 14, 120 - 136 (1969).
Conference paper (see 012.006).

084.247 **Entry of solar cosmic rays into the earth's magnetosphere.** K. A. Anderson.
Astrophys. Space Sci. Library, Vol. 17, 3 - 17 (1970). – Conference paper (see 012.008).

084.248 **Formation and geometry of geomagnetic tail.**
A. J. Dessler.
Astrophys. Space Sci. Library, Vol. 17, 18 - 23 (1970).
Conference paper (see 012.008).

084.249 **Magnetotail plasma and magnetospheric substorms.**
E. W. Hones, Jr.
Astrophys. Space Sci. Library, Vol. 17, 24 - 33 (1970).
Conference paper (see 012.008).

084.250 **A model current system for the magnetospheric substorm.** S.-I. Akasofu.
Astrophys. Space Sci. Library, Vol. 17, 34 - 45 (1970).
Conference paper (see 012.008).

084.251 **Mathematical models of magnetospheric convection and its coupling to the ionosphere.**
V. M. Vasyliunas.
Astrophys. Space Sci. Library, Vol. 17, 60 - 71 (1970).
Conference paper (see 012.008).

084.252 **Plasma measurements near earth's bow shock: Vela 4.** M. D. Montgomery.
Astrophys. Space Sci. Library, Vol. 17, 95 - 101 (1970).
Conference paper (see 012.008).

084.253 **AC electric and magnetic fields and collisionless shock structures.**
F. L. Scarf, R. W. Fredricks, C. F. Kennel.
Astrophys. Space Sci. Library, Vol. 17, 102 - 108 (1970).
Conference paper (see 012.008).

084.254 **Energetic particle phenomena in the earth's magnetospheric tail.** J. A. van Allen.
Astrophys. Space Sci. Library, Vol. 17, 111 - 121 (1970).
Conference paper (see 012.008).

084.255 **Anisotropic distributions of energetic electrons in the earth's magnetotail and magnetosheath.**
S. Singer, S. J. Bame.
Astrophys. Space Sci. Library, Vol. 17, 122 - 131 (1970).
Conference paper (see 012.008).

084.256 **Trapped and polar particles during the June 9, 1968 magnetic storm.**
P. L. Rothwell, V. H. Webb, L. Katz.
Astrophys. Space Sci. Library, Vol. 17, 132 - 140 (1970).
Conference paper (see 012.008).

084.257 **The reaction of the plasmapause to varying magnetic activity.**
C. R. Chappell, K. K. Harris, G. W. Sharp.
Astrophys. Space Sci. Library, Vol. 17, 148 - 153 (1970).
Conference paper (see 012.008).

084.258 **Magnetic fields in the earth's tail.**
K. W. Behannon.
Astrophys. Space Sci. Library, Vol. 17, 157 - 164 (1970).
Conference paper (see 012.008).

084.259 **Magnetic field observations in high β regions of the magnetosphere.**
M. Sugiura, T. L. Skillman, B. G. Ledley, J. P. Heppner.
Astrophys. Space Sci. Library, Vol. 17, 165 - 170 (1970).
Conference paper (see 012.008).

084.260 **Electric fields in the ionosphere and magnetosphere.**
G. Haerendel, R. Lüst.
Astrophys. Space Sci. Library, Vol. 17, 213 - 228 (1970).
Conference paper (see 012.008).

084.261 **Auroral and polar cap electric fields from barium releases.**
E. M. Wescott, J. D. Stolarik, J. P. Heppner.
Astrophys. Space Sci. Library, Vol. 17, 229 - 238 (1970).
Conference paper (see 012.008).

084.262 **Hydromagnetic waves and instabilities in the magnetosphere.** A. Hasegawa.
Astrophys. Space Sci. Library, Vol. 17, 284 - 291 (1970).
Conference paper (see 012.008).

084.263 **Pioneer 6 plasma measurements in the magnetosheath.** H. C. Howe, Jr.
Journ. Geophys. Res., Vol. 75, 2429 - 2437 (1970).
Measurements of the magnetosheath plasma made by the MIT plasma experiment during the outbound passage of Pioneer 6 in the dusk meridian (December 16, 1965) are presented and compared with theoretical predictions and other simultaneous experimental measurements for the same region.

084.264 **On the equilibrium of the magnetopause.**
E. T. Karlson.
Journ. Geophys. Res., Vol. 75, 2438 - 2448 (1970).
In the present paper we show that a layer without any potential difference cannot be part of a self-consistent model. It is shown that at the outer border of the layer there must be a potential barrier, retarding the ions and accelerating the electrons in such a way that there is an overall charge neutrality of the stream particles within the layer.

084.265 **Transverse structure of the earth's magnetotail and fluctuations of the tail magnetic field.**
A. Hruška, J. Hrušková.
Journ. Geophys. Res., Vol. 75, 2449 - 2457 (1970).

084.266 **Penetration of low-energy protons deep into the magnetosphere.** A. J. Chen.
Journ. Geophys. Res., Vol. 75, 2458 - 2467 (1970).

084.267 **Warum hat die Erde ein Magnetfeld?**
M. Steenbeck.
Phys. Blätter, 26. Jahrgang, p. 158 - 168 (1970).

084.268 **Zur Deutung der Neigung und der Westdrift des erdmagnetischen Hauptfeldes.**
M. Steenbeck, G. Helmis.
Monatsber. Deutsch. Akad. Wiss. Berlin, Vol. 11, 723 - 734 (1969).

084.269 **Vela 4 plasma observations near the earth's bow shock.** M. D. Montgomery, J. R. Asbridge, S. J. Bame.
Journ. Geophys. Res., Vol. 75, 1217 - 1231 (1970).
This paper presents results of recent Vela 4 plasma measurements near the bow shock. Examples of jumps in density, bulk speed, and electron and proton temperature are given.

084.270 **Record of observations at Meanook Magnetic Observatory 1967.** A. B. Cook, S. J. Sprysak.
Publ. Dominion Obs. Ottawa, Vol. 39, (No. 1), 1 - 55 (1969).

084.271 **Magnetic anomaly maps of the nordic countries and the Greenland and Norwegian seas.**
G. V. Haines, W. Hannaford, P. H. Serson.
Publ. Dominion Obs. Ottawa, Vol. 39, (No. 5), 119 - 149 (1970).

084.272 **Record of observations at Meanook Magnetic Observatory, 1951 - 1952.**
H. E. Cook, A. B. Cook.
Publ. Dominion Obs. Ottawa, Vol. 17C, (No. 3), 169 - 276 (1970).

084.273 **Pioneer 8 electric field measurements in the distant geomagnetic tail.** F. L. Scarf, I. M. Green, G. L. Siscoe, D. S. Intriligator, D. D. McKibbin, J. H. Wolfe.

Journ. Geophys. Res., Vol. 75, 3167 - 3179 (1970).

In this report, we correlate the Pioneer 8 wave and particle observations in detail and show that during the extended tail crossings the average broadband wave levels were reduced. Enhanced 400-Hz activitiy was frequently detected near the tail boundaries, however, and the observations suggest that tail breakup and field-line reconnection phenomena begin to be important within 500 R_E.

084.274 **Deformation of a magnetic dipole field by trapped particles.** K. Lackner.
Journ. Geophys. Res., Vol. 75, 3180 - 3192 (1970).

In this paper we present a model for stationary symmetric ring currents based on a solution of Vlasov's equation for a quasi-neutral plasma.

084.275 **Variation of the magnetopause position with substorm activity.** C. I. Meng.
Journ. Geophys. Res., Vol. 75, 3252 - 3254 (1970). – Brief report.

084.276 **Rocket and satellite studies of the geomagnetic field during the IQSY.** L. J. Cahill, Jr.
Ann. IQSY, (see 012.017), Vol. 5, 349 - 368 (1969).

084.277 **Geomagnetic conjugacy variations.**
D. A. Mendis, W. I. Axford.
Comments Astrophys. Space Phys., Vol. 2, 99 - 108 (1970).

The authors find that it is possible to understand qualitatively why there should be geomagnetic conjugacy variations, which effects are likely to be most important, and where the most pronounced variations are likely to occur.

084.278 **The structure of the magnetosphere in high latitudes.** E. G. Eroshenko, A. E. Antonova,
Kosmich. Issled., Vol. 8, 397 - 407 (1970). In Russian.

084.279 **A test of the earth's magnetic field during Permian time.** F. G. Stehli.
Journ. Geophys. Res., Vol. 75, 3325 - 3342 (1970).

Quantitative paleontologic data sensitive to the planetary temperature gradient are used with similar data for living organisms to test two possible Permian latitude models for the northern hemisphere.

084.280 **Surface integral formulae for geomagnetic studies.** B. A. Hobbs, A. T. Price.
Geophys. Journ. Royal Astron. Soc., Vol. 20, 49 - 63 (1970).

Surface integral formulae are derived expressing any one of certain field quantities, namely current functions, magnetic potentials and normal components of magnetic fields, in terms of any one other, for current systems flowing in concentric spherical surfaces. In all, 36 such formulae are obtained, which should prove useful in many geomagnetic studies, escpecially in geomagnetic induction problems.

084.281 **An interpretation of the behaviour of the geomagnetic field during polarity transitions.**
K. M. Creer, Y. Ispir.
Phys. Earth Planet. Interiors, Vol. 2, 283 - 293 (1970).

084.282 **Statistical characteristics of anomalous gravitational and magnetic fields.**
V. N. Lugovenko, A. I. Soroka.
Geomagn. Aeronom., Vol. 10, 513 - 518 (1970). In Russian.

084.283 **Polarization splitting of the spectrum of the Alfvén oscillations of the magnetosphere.**
A. V. Gulielmi.
Geomagn. Aeronom., Vol. 10, 524 - 526 (1970).

In Russian. – Brief information.

084.284 **Determination of the plasma concentration in the magnetosphere from the periods of geomagnetic pulsations.** A. A. Namgaladze.
Geomagn. Aeronom., Vol. 10, 526 - 528 (1970).
In Russian. – Brief information.

084.285 **The theoretical form of the magnetosphere of the non-dipole geomagnetic field.**
N. M. Rudneva, S. I. Rogovaya.
Geomagn. Aeronom., Vol. 10, 530 - 531 (1970).
In Russian. – Brief information.

084.286 **Record of observations at Meanook Magnetic Observatory, 1949 - 1950.**
H. E. Cook, A. B. Cook, R. G. Madill.
Publ. Dominion Obs. Ottawa, Vol. 17C (No. 4), 277 - 384 (1970).

084.287 **Computation of solar and lunar geomagnetic variations. II.**
S. Chapman, J. C. Gupta, S. R. C. Malin.
Gerlands Beiträge Geophys., Vol. 79, 5 - 10 (1970). In German.

084.288 **Solar wind and semi-annual variation of geomagnetic activity.** M. Siebert.
Zeitschr. Geophys., Vol. 36, 41 - 56 (1970).

Using the monthly means of the linear activity measure of Ap of the years 1932 to 1968, is shown that the semi-annual maxima of activity occur exactly at those times when the mean direction of the incident solar wind, as observed from the earth, is perpendicular to the rotational axis of the earth.

084.289 **Configuration of the magnetic field of the earth in the modified double-dipole model of the magnetosphere.** A. E. Antonova, V. P. Shabansky.
Cosmic Rays No. 11, Moscow, p. 115 - 118 (1969).
In Russian.

084.290 **Some phenomena preceding geomagnetic storms.**
Yu. L. Truttse, N. N. Shefov.
Astron. Tsirk., No. 553, p. 7 - 8 (1970). In Russian.

084.291 **Geophysical numerology.** J. A. Jacobs.
Nature, Vol. 227, 161 - 162 (1970).

Perturbations décelées dans la magnétosphère, au moyen des sifflements radioélectriques, à la suite de l'événement à proton du 7 juillet 1966. See Abstr. 073.110.

Evidence of contraction of the earth's thermal plasmaphere subsequent to the solar flare events of 7 and 9 July 1966. See Abstr. 073.111.

Solar particle observations inside the magnetosphere during the 7 July 1966 proton flare event. See Abstr. 073.112.

Preliminary analysis of micropulsations of the earth's electromagnetic field in connection with the proton flare of 7 July 1966. See Abstr. 073.113.

The variations of the geomagnetic field in middle and high latitudes during the proton flare event of July 1966. See Abstr. 073.114.

Solar wind stimulation of the magnetosphere. See Abstr. 074.025.

Geophysical implications of the interrelationship of the geogravity and geomagnetic fields.
See Abstr. 081.017.

The electric field in the ionosphere and in the magnetosphere of the earth at the presence of an inhomogeneity

of fast particles.   See Abstr. 083.075.

The flux of the interplanetary magnetic field and the K_p-index of geomagnetic activity.
See Abstr. 106.010.

Radiation Belts

084.401   Time history of the inner radiation zone, October 1963 to December 1968.
C. O. Bostrom, D. S. Beall, J. C. Armstrong.
Journ. Geophys. Res., Vol. 75, 1246 - 1256 (1970).
We have presented data on trapped electrons and protons between $L = 1.2$ and $L = 3.0$ over a time period of 5 years and 3 months from September 1963 through December 1968. Statements following from these data are given.

084.402   Injection of protons into the radiation belt by solar neutron decay.
E. S. Claflin, R. S. White.
Journ. Geophys. Res., Vol. 75, 1257 - 1262 (1970).
In this paper we examine the strength of the solar neutron decay source. We calculate the coefficient χ for injecting protons into the radiation belt and compare this source to the cosmic-ray albedo neutron decay source.

084.403   Direct detection of asymmetric increases of extraterrestrial 'ring current' proton intensities in the outer radiation zone.   L. A. Frank.
Journ. Geophys. Res., Vol. 75, 1263 - 1268 (1970).
Measurements of the spatial distributions and temporal variations of the extraterrestrial 'ring current' proton intensities near the magnetic equator during selected phases of two moderate magnetic storms on July 9 and September 8, 1966, provide direct evidence of symmetric enhancement of these proton intensities deep in the outer radiation zone, during the early development of the latter magnetic storm.

084.404   Omnidirectional intensity contours of low-energy protons ($0.5 \leqslant E \leqslant 50$ kev) in the earth's outer radiation zone at the magnetic equator.
L. A. Frank, H. D. Owens.
Journ. Geophys. Res., Vol. 75, 1269 - 1278 (1970).
Our goal for the present analysis is to construct contours of proton intensities as functions of the magnetic shell parameter L and time, similar to those that have been published for outer-zone electron intensities with energies >40 kev, >230 kev, and >1.6 Mev observed with Explorer 14.

084.405   About the instability connected with the anisotropy of the distribution function of ultrarelativistic electrons in the earth's radiation belts.
B. N. Breizman, V. V. Mirnov.
Geomagn. Aeronom., Vol. 10, 34 - 37 (1970).   In Russian.

084.406   Results of the measurements of the concentration of charged particles in the plasma cloud of the earth carried out on board of Elektron 2 and Elektron 4.
V. V. Bezrukikh.
Kosm. Issled., Vol. 8, 273 - 277 (1970).   In Russian.

084.407   The connection of variations of the electron intensity of the outer belt with the indices of

geomagnetic activity.   S. N. Vernov, S. N. Kuznetzov, G. B. Lopatina, V. G. Stolpovsky.
Geomagn. Aeronom., Vol. 10, 214 - 220 (1970).
In Russian.

084.408   Observation of the van Allen radiation zone with the radio sounding technique from the earth's surface in Antarctic Continent.   J. Molski.
Artificial Satellites Polish Acad. Sci., Vol. 5, No. 1, p. 13 -25 (1969).

084.409   On the origin of radiation belt and auroral primary ions.   W. I. Axford.
Astrophys. Space Sci. Library, Vol. 17, 46 - 59 (1970).
Conference paper (see 012.008).

084.410   Summary of particle populations in the magnetosphere.   J. I. Vette.
Astrophys. Space Sci. Library, Vol. 17, 305 - 318 (1970).
Conference paper (see 012.008).

084.411   Further comments concerning low energy charged particle distributions within the earth's magnetosphere and its environs.   L. A. Frank.
Astrophys. Space Sci. Library, Vol. 17, 319 - 331 (1970).
Conference paper (see 012.008).

084.412   The origin and distribution of energetic electrons in the van Allen radiation belts.   J. R. Winckler.
Astrophys. Space Sci. Library, Vol. 17, 332 - 352 (1970).
Conference paper (see 012.008).

084.413   Recent measurements of inner belt protons.
H. Elliot, R. J. Hynds.
Astrophys. Space Sci. Library, Vol. 17, 353 - 363 (1970).
Conference paper (see 012.008).

084.414   Alpha particles trapped in the earth's magnetic field.   S. M. Krimigis.
Astrophys. Space Sci. Library, Vol. 17, 364 - 379 (1970).
Conference paper (see 012.008).

084.415   Measurements of trapped α-particles: $2 \leqslant L \leqslant 4.5$.
J. B. Blake, G. A. Paulikas.
Astrophys. Space Sci. Library, Vol. 17, 380 - 384 (1970).
Conference paper (see 012.008).

084.416   Introductory survey of radiation belt diffusion.
C.-G. Fälthammar.
Astrophys. Space Sci. Library, Vol. 17, 387 - 395 (1970).
Conference paper (see 012.008).

084.417   Trapped protons $\geqslant 100$ keV and possible sources.
D. J. Williams.
Astrophys. Space Sci. Library, Vol. 17, 396 - 409 (1970).
Conference paper (see 012.008).

084.418   Radial diffusion of trapped particles.

M. Walt.
Astrophys. Space Sci. Library, Vol. 17, 410 - 415 (1970).
Conference paper (see 012.008).

084.419 The effect of a large ring current on the topology of the magnetosphere.
C. Sozou, D. W. Windle.
Planet. Space Sci., Vol. 18, 699 - 707 (1970).

084.420 The earth's radiation belts.
S. N. Vernov.
Ann. IQSY, (see 012.016), Vol. 4, 281 - 301 (1969).

084.421 Comments and speculations concerning the radiation belts. C. E. McIlwain.

Ann. IQSY, (see 012.016), Vol. 4, 302 - 306 (1969).

084.422 Some quantitative aspects of electron precipitation in and near the auroral zone. D. K. Bailey.
Ann. IQSY, (see 012.016), Vol. 4, 307 - 308 (1969).

084.423 High energy electrons in nearby cosmic space.
N. L. Grigorov, L. F. Kalinkin, E. I. Kogan-Laskina, I. A. Savenko.
Kosmich. Issled., Vol. 8, 418 - 422 (1970). In Russian.

Investigation of the cosmic radiation and the radiation belts on board a vertical cosmic probe.
See Abstr. 143.043.

085 Solar-Terrestrial Relations

085.001 Secular perturbations in the elements of the earth's orbit and the astronomical theory of climate variations. S. G. Sharaf, N. A. Boudnikova.
Trudy Inst. Teoret. Astron., *Leningrad,* Vyp. (No.) 14, p. 48 - 84 + 28 pp. (1969). In Russian.
The paper deals with the astronomical theory of climate variations. Tables of the secular perturbations in the earth's orbital elements for 3×10^6 years before and 1×10^6 years after the epoch 1950.0 are given. The secular course of the solar radiation (in "canonical" units) at 65° north and south is illustrated by means of tables and diagrams for the same interval of time, and in form of diagrams for the past 3×10^7 and the future 1×10^6 years. The problem of change of temperature due to variations of solar radiation arriving at the earth's surface is discussed, whereby only dynamical influences on the sun's radiation are considered.

085.002 Astronomical theory of climatic change: Support from New Guinea.
H. H. Veeh, J. Chappell.
Science, Vol. 167, 862 - 865 (1970).
Radiocarbon and thorium-230 dates if uplifted coral reef terraces on New Guinea appear to support theories of glaciation which utilize Milankovitch cycles as a controlling trigger mechanism. In addition to high sealevel stands recognized by other workers, the New Guinea data clearly indicate a marine transgression between 50.000 and 35.000 years before the present. A eustatic sea level curve reconstructed from field observations and radiometric dates shows a close correlation with temperature fluctuations in high latitudes as predicted by astronomical data.

085.003 Zusammenhänge zwischen Meteorerscheinungen und Sonnenfleckenzyklus. R. Müller.
SuW, Vol. 9, 130, 132 (1970).

085.004 Geoactivity of solar flares depending on brightness of metal luminescence in a flare.
I. N. Odintzova, N. S. Shilova.
Geomagn. Aeronom., Vol. 10, 326 - 328 (1970).
In Russian. – Brief information.

085.005 Measurement of the phase of ULW-signals during the eclipse of September 22, 1968.
L. A. Protopopov, B. A. Khadzhi.
Geomagn. Aeronom., Vol. 10, 363 - 364 (1970).
In Russian. – Brief information.

085.006 Solar activity index: Validity supported by oxygen isotope dating. J. R. Bray.
Science, Vol. 168, 571 - 572 (1970).
A significant correlation between oxygen-18 concentration in a Greenland ice core and an index of solar activity supports the validity of the solar index. The correlation may result from an apparent control of temperature by solar acitivity.

085.007 Korpuskularstrahlung und Wetter.
H. Toelle.
Sterne, 46. Jahrgang, p. 69 - 76 (1970).

085.008 Ritmica diurna nei sistemi sensibili ed attività solare. G. Piccardi.
Coelum, Vol. 38, 117 - 123 (1970).

085.009 Temporal and spatial covariation of high-latitude geophysical phenomena.
R. Y. Prasad, R. N. Singh.
Journ. Phys. A, Gneral Phys., Ser. 2, Vol. 3, 400 - 403 (1970).
A possible mechanism for the unification of regular and irregular fluctuations in high-latitude geophysical phenomena has been given. In view of this mechanism the simultaneity of various geophysical fluctuations has been discussed.

085.010 On the suitability of solar data in solar-terrestrial physics. G. Godoli, G. L. Tagliaferri.
Oss. Astrofis. Catania, Pubbl., Nuova Ser., No. 135, 6 pp. (1969).

Search for an effect of the sun on the frequency of 18-centimeter radiation.
See Abstr. 131.083.

Short-period fluctuations of cosmic ray intensity at the geomagnetic equator and their solar and terrestrial relationship. See Abstr. 143.032.

Planetary System

091 Physics of the Planetary System (Planetary Atmospheres, Figure, Interior, Magnetic Fields, Rotation, etc.)

091.001 **Equatorial jets in planetary atmospheres.**
R. Hide.
Nature, Vol. 225, 254 - 255 (1970). −Note.

091.002 **Multiple scattering in a plane-parallel atmosphere. I. Successive scattering in a semi-infinite medium.**
A. Uesugi, W. M. Irvine.
Astrophys. Journ., Vol. 159, 127 - 135 = Contr. Four College Obs., Amherst, No. 29 (1970).

The reflection function for a plane-parallel semi-infinite conservative atmosphere with an isotropic phase function is computed by the method of successive scattering.

091.003 **Behavior of absorption lines in a hazy planetary atmosphere.** J. W. Chamberlain.
Astrophys. Journ., Vol. 159, 137 - 158 = Contr. Kitt Peak National Obs. No. 502 (1970).

An approximate, analytic theory is developed for the formation of spectral absorption lines in a hazy atmosphere with isotropic scattering and a homogeneous mixture of scattering and absorbing matter. The behavior−i.e., the curve of growth and the dependence of equivalent width on incident and emergent angles−is examined systematically for various possible situations, with emphasis on the physical reasons for a particular behavior. Limiting cases derived in earlier papers are extracted from the more general solution.

091.004 **On obtaining electron-density profiles of planetary ionospheres.** A. C. Kak.
Journ. Atmosph. Terr. Phys., Vol. 32, 233 - 236 (1970). Short paper.

091.005 **The dependence of convection in planetary mantles upon the magnitude of the Rayleigh number.**
D. G. Ashworth.
The Moon, Vol. 1, 253 - 263 (1970).

The conservation equations of mass, energy and momentum when applied to the problem of convection produced by isothermal compression as well as by thermal expansion in a self-gravitating sphere of uniform density, consisting of a core extending to a fraction η of the radius of the sphere and with a viscous mantle overlying the core, lead to a relationship between two characteristic numbers. In the present paper these characteristic numbers are evaluated, for the case when both boundaries of the mantle are free, for spherical harmonic disturbances of order $1 \rightarrow 6$.

091.006 **Chemical homogeneity of groups of terrestrial planets.** J. C. Jamieson, J. V. Smith.
Nature, Vol. 225, 354 - 355 (1970).

McCrea argued that Lyttleton's theory of planetary origin by break-up of primary bodies is strongly supported by near-equality of densities (implying chemical homogeneity) of Mercury−Venus and Earth−Moon−Mars aggregates. Jamieson and Smith argued that actual densities should not be added, and that chemical comparison requires uncompressed densities. Use of uncompressed densities leads to inconclusive implications.

091.007 **Groups of terrestrial planets − a reply.**
W. H. McCrea.
Nature, Vol. 225, 355 - 356 (1970).

McCrea mentions several ways in which the comments made by Jamieson and Smith appear to be not wholly relevant to his work. In particular, their discussion of densities depends upon values of the "expanded densities" of the earth and Venus; these can scarcely be estimated to the degree of accuracy implied by them. On the other hand, their discussion appears to show that a small differential effect ascribed by McCrea to compressibility is of the order of magnitude to be expected.

091.008 **The atmospheres of the planets of the solar system.**
A. P. Vinogradov.
Vestn. Mosk. un-ta. Geologiya, 1969 No. 4, p. 3 - 14.
In Russian. − Abstr. in Referativ. Zhurn. 51. Astron., 3.51.259 (1970).

091.009 **Sur les marées exercées par les planètes sur le soleil et la prévision de l'activité solaire.**
A. Dauvillier.
Comptes Rendus Acad. Sci. Paris, Sér. B, Vol. 270, 1119 - 1121 (1970).

L'auteur montre que l'amplitude du cycle de l'activité solaire, entre 1910 et 1968, est proportionnelle à l'écart existant entre le centre du soleil et le centre de gravité du système solaire.

091.010 **On the determination of the factor of roughness of planetary surfaces.** N. P. Barabashov.
Astron. Tsirk., No. 540, p. 1 - 2 (1969). In Russian.

091.011 **Diffuse reflection and transmission of light by a planetary atmosphere according to the three-term scattering indicatrix.** A. K. Kolesov, O. I. Smokty.
Astron. Zhurn. Akad. Nauk SSSR, Vol. 47, 397 - 406 (1970). In Russian. − English translation in Soviet Astron., AJ, Vol. 14, No. 2.

The rigorous theory of anisotropic light scattering developed recently by V. V. Sobolev is used for the solution of the problem of diffuse reflection and transmission of light by a planetary atmosphere. The case of the three-term scattering indicatrix is considered. Exact formulae for the reflection and transmission coefficients (for arbitrary particle albedo λ) and for the albedo of a planetary atmosphere, the spherical albedo and the illumination of the planetary surface (for $\lambda = 1$) are derived.

091.012 **Diffuse reflection of solar rays by a spherical shell atmosphere.** R. E. Bellman, H. H. Kagiwada, R. E. Kalaba, S. Ueno.
Icarus, Vol. 11, 417 - 423 (1969).

In this paper is given an invariant imbedding formulation for the angular distribution of diffusely reflected radiation over the surface of an inhomogeneous, anisotropically scattering, planetary atmosphere illuminated by uniform parallel rays.

091.013 Characteristic γ - and X-radiation in the planetary system. P. Gorenstein, H. Gursky.
Space Sci. Rev., Vol. 10, 770 - 829 (1970).

In this review we address ourselves to the question of what can be expected in the way of determining chemical abundances from a vehicle in orbit around a planet that carries presently existing instrumentation for the detection of characteristic X- and gamma-rays. Remote sensing and passive observation is the framework of the measurements we consider. Mechanisms for producing these characteristic lines are discussed and we describe various estimates for the flux plus some experimental results that have been reported so far apply to the earth's atmosphere and the moon; in addition, there are measurements that represent upper limits for some of the other planets.

091.014 Atmospheric composition of the Jovian planets. M. B. McElroy.
Journ. Atmosph. Sci., Vol. 26, 798 - 812 (1969). − Conference paper (see 012.009).

This paper reviews the present status of research on the composition of the atmospheres of Jupiter, Saturn, Uranus and Neptune.

091.015 The structure of the atmospheres of the major planets. L. M. Trafton, G. Münch.
Journ. Atmosph. Sci., Vol. 26, 813 - 825 (1969). − Conference paper (see 012.009).

We review the current knowledge of the atmospheric structures of the major planets emphasizing the limits of completeness and present weaknesses. In addition, we recommend specific planetary observations and laboratory investigations for significantly improving the state of this knowledge.

091.016 The pressure-induced infrared spectrum of hydrogen and its application to the study of planetary atmospheres. H. L. Welsh.
Journ. Atmosph. Sci., Vol. 26, 835 - 840 (1969). − Conference paper (see 012.009).

The main results of experimental and theoretical investigations on the pressure-induced infrared absorption of nonpolar molecules are outlined, with particular emphasis on those properties which may be of importance in studies of planetary atmospheres containing hydrogen and helium. The structures of the fundamental and overtone absorptions of hydrogen are explained in some detail, and a possible explanation for the frequency discrepancy of the planetary S_3 (0) and S_4 (0) features is given. Our present knowledge of the shape of the high frequency wing in induced absorption is reviewed.

091.017 Dynamics of the atmospheres of the major planets with an appendix on the viscous boundary layer
at the rigid bounding surface of an electrically-conducting rotating fluid in the presence of a magnetic field.
R. Hide.
Journ. Atmosph. Sci., Vol. 26, 841 - 853 (1969). − Conference paper (see 012.009).

091.018 Mutual diffusion coefficient of hydrogen atoms and molecules. R. J. W. Henry.
Journ. Atmosph. Sci., Vol. 26, 918 - 919 = Contr. Kitt Peak National Obs., No. 464 (1969). − Conference paper (see 012.009).

091.019 Phase equilibria in planetary atmospheres. W. B. Streett.
Journ. Atmosph. Sci., Vol. 26, 924 - 931 (1969). − Conference paper (see 012.009).

Experimental phase diagrams for binary mixtures, extrapolated to very high pressures, suggest that layered structures may exist deep in the atmospheres of Jupiter and Saturn, as a result of solid-fluid and fluid-fluid phase separations in the primary H_2-He mixture. Consideration of these phase separations, together with the barotropic phenomenon, leads to a model in which discontinuous changes in fluid density and composition may occur.

091.020 Discrete space theory of radiative transfer and its application to problems in planetary atmospheres.
G. E. Hunt, I. P. Grant.
Journ. Atmosph. Sci., Vol. 26, 963 - 972 (1969). − Conference paper (see 012.009).

We briefly describe a method of computation that uses discrete space techniques depending on concepts of invariance. The solution algorithms compute internal and external light fields for inhomogeneous plane parallel atmospheres with arbitrary internal, external source distributions and scattering diagrams. To illustrate our techniques, we briefly discuss the practical problem of making theoretical predictions of the spectral properties of ice clouds at selected spectral intervals between $2.6 - 150 \mu$ in the infrared. The predictions are consistent with recent measurements.

091.021 Computation of synthetic spectra for a semi-infinite atmosphere.
A. Uesugi, W. M. Irvine.
Journ. Atmosph. Sci., Vol. 26, 973 - 978 (1969). − Conference paper (see 012.009).

The computation of absorption spectra in a model planetary atmosphere is shown to be feasible using the Neumann series (successive orders of scattering) solution to the equation of radiative transfer for semi-infinite atmospheres. The method may be applied for arbitrary single scattering albedo and phase function.

091.022 Preliminary results of experiments with symmetric baroclinic instabilities.
P. H. Stone, S. Hess, R. Hadlock, P. Ray.
Journ. Atmosph. Sci., Vol. 26, 991 - 996 (1969). − Conference paper (see 012.009).

An experiment has been designed to test the predictions of nongeostrophic baroclinic stability theory. The first qualitative observations derived from the experiment are described and are found to agree well with the theory.

091.023 The atmospheres of the major planets. R. Goody.
Journ. Atmosph. Sci., Vol. 26, 997 - 1001 (1969). − Conference paper (see 012.009).

091.024 Computer simulations of systems of strongly interacting planets and Bode's law.
J. G. Hills.
Bull. American Astron. Soc., Vol. 2, 199 (1970). − Abstr. AAS.

091.025 Distribution and covariance function of elevations on a cratered planetary surface. Part I: Theory.
A. H. Marcus.
The Moon, Vol. 1, 297 - 337 (1970).

We derive the distribution and covariance function of elevations on a cratered planetary surface from a representation of the surface as the 'moving average' of a random point process.

091.026 On the evolution of commensurabilities between natural satellites. R. R. Allan.
Symposia Mathematica, Vol. 3, 75 - 96 (1970). − Conference paper (see 012.010).

091.027 Generalized Cassini's laws. S. J. Peale.

Symposia Mathematica, Vol. 3, 175 - 191 (1970).
Conference paper (see 012.010).

091.028 The rotation of the planets and their interiors.
S. K. Runcorn.
Symposia Mathematica, Vol. 3, 193 - 202 (1970). - Conference
paper (see 012.010).

**091.029 Imperfect Rayleigh scattering in a semi-infinite
atmosphere.** K. D. Abhyankar, A. L. Fymat.
Astron. Astrophys., Vol. 4, 101 - 110 (1970).

Planetary atmospheres exhibit imperfect scattering
irrespective of the assumed law of scattering. The reflection
matrix for a semi-infinite plane-parallel stratified homogeneous
atmosphere, scattering in accordance with the conservative
Rayleigh phase matrix, was obtained by Chandrasekhar (1950).
The corresponding solutions for a nonconservative Rayleigh
atmosphere in which the albedo for single scattering Ω is
constant, but different from unity, are presented for some
representative values of Ω.

091.030 Axial rotation of the planets. V. Mitra.
Astron. Astrophys., Vol. 4, 263 - 267 (1970).

A simple method based upon particle accretion is developed for obtaining the spin periods of the planets. It is argued
that the cloud of particles as considered by Giuli (1968) was
a gaseous volume before condensation. The dimensions of
the gaseous cloud and hence its angular momentum has been
obtained in the orbital plane of the planet. Using the observed
orbital angular momentum of the planet, the spin angular
momentum and hence its spin period is obtained. The present
treatment gives values in better agreement with the observations than the other investigations.

**091.031 The Eddington approximation for planetary atmo-
spheres.** Y. Kawata, W. M. Irvine.
Astrophys. Journ., Vol. 160, 787 - 790 = Contr. Four College
Obs., *Amherst*, No. 65 (1970).

The accuracy of the Eddington approximation in determining the albedo of a planetary atmosphere is discussed as a
function of optical thickness and angle of incidence for
anisotropic phase functions which are typical of clouds and
hazes. An error in an earlier paper by Irvine is pointed out. It
is found that the Eddington approximation gives more accurate results for large angles of incidence than was previously
indicated, but that the approximation still fails for thin layers
and single-scattering albedos $a \neq 1$ when the phase function
is very forward directed.

**091.032 Importance de la polarisation dans le rayonnement
diffuse par une atmosphère planétaire.**
J. Lenoble.
Journ. Quant. Spectrosc. Radiat. Transfer, Vol. 10, 533 -
556 (1970).

We have studied the variation of the radiance and of
the degree of polarization in an absorption line, as observed
in the radiation scattered at the center of a planetary disc,
for a semi-infinite atmosphere with Rayleigh scattering.
We have shown that it is possible to compute the radiance
with a good accuracy when the polarizing effect of scattering
is neglected.

**091.033 External radiation fields for isotropically scattering
finite atmospheres bounded by a Lambert law
reflector.** J. Casti, H. Kagiwada, R. Kalaba.
Journ. Quant. Spectrosc. Radiat. Transfer, Vol. 10, 637 -
651 (1970).

Formulas for obtaining the diffusely transmitted and
reflected radiation fields for a planetary, isotropically
scattering atmosphere of finite thickness in terms of the
solution to the problem with no planetary surface are

provided. Numerical results then show that these reflected
and transmitted fluxes are essentially the same whether isotropic or Rayleigh laws are assumed.

**091.034 The similarity theory for the large-scale motions of
planetary atmospheres.** G. S. Golitsyn.
Dokl. Akad. Nauk SSSR, Ser. Mat. Fiz., Vol. 190, 323 -
326 (1970). In Russian.

The intensity of atmospheric circulation and the temperature differences are correlated and determined by the influx
of heat, the mass of the atmosphere, and its thermal properties. Thermal balance is established by emission of long-wave
radiation, a gray atmosphere being supposed. By means of
similarity and dimensional theories relations between characteristic atmospheric parameters and dimensionless constants
are derived. The rotation of a planet is allowed for. The following values of the mean velocity of atmospheric motions
and characteristic atmospheric temperature differences are
derived: for the earth 10 m/s and 25°K (owing to the zonal
pattern of winds the real values are twice that), for Mars 50
m/s and 100°K (consistent with the results of numerical
modelling of atmospheric circulation by Loewy and Mintz),
and for Venus 0.7 m/s and 2°K. Strong absorption of solar
radiation in the Venus atmosphere would reduce the velocities and increase the temperature differences.

B. Onderlička

**091.035 On the temperature of the exospheres of the earth,
Mercury, Venus, Mars, Jupiter, and of the solar
corona.** I. M. Matora.
Dokl. Akad. Nauk SSSR, Ser. Mat. Fiz., Vol. 190, 1303 -
1304 (1970). In Russian.

Temperatures and gravitational potentials are compared
for the earth's exosphere and the solar corona at minimum
heights, where uncollisional orbital motion of hydrogen atoms
is possible (800 km for the earth, 1.4×10^6 km for the sun).
In spite of very different physical conditions the ratio
$3/2$ kT/m_H U (T, m_H, U being absolute temperature, mass
of the hydrogen atom, and gravitational potential) is in
both cases about the same, nearly 0.5. Assuming the same
value of this ratio for other planets the following maximum
temperatures of the exospheres are calculated: Mercury
400°K, Venus 800°K, Mars 300°K, and Jupiter 25000°K.
The measured exosphere temperature for Venus (650 -
700°K) is near the maximum calculated value.

B. Onderlička

**091.036 New optical measurements of the diameters of
Jupiter, Saturn, Uranus, and Neptune.**
A. Dollfus.
Icarus, Vol. 12, 101 - 117 (1970).

A range of new measurements and discussed means of
older measurements are presented for the polar and equatorial
diameters of the Jovian planets and for the lateral dimensions
of the rings of Saturn.

**091.037 La photographie planétaire à haute résolution au
Pic du Midi.** P. Guérin.
L'Astronomie, 84e année, p. 241 - 256 (1970).

**091.038 Note on application of Emden's equation to plane-
tary interiors.** K. E. Bullen.
Monthly Notices, Roy. Astron. Soc., Vol. 149, 51 - 52 (1970).

Although the Emden equation commonly used in the
past in determining density-depth relations inside the terrestrial planets is now known to be seriously unreliable in certain
contexts, a modified Emden equation may be safely used to
good approximation. The modified equation differs from the
older form only in respect of the definition of certain
constants. It has the cardinal advantage that it can be applied
in regions where there are significant deviations from uniform

chemical composition and phase and from adiabatic temperature gradients.

091.039 Line formation in planetary atmospheres. I. The simple reflecting model. J. Regas, C. Sagan.
Comments Astrophys. Space Phys., Vol. 2, 116 - 120 (1970).

In this paper we attempt, not a comprehensive and rigorous treatment of the subject, but rather a simple physical picture of how absorption lines are formed in a scattering planetary atmosphere like that on Venus. Because of its simplicity this picture, while reproducing many of the earlier results, permits us also to draw several new conclusions.

091.040 Intermediary orbits of planetary satellites. E. L. Lukashevich.
Kosmich. Issled., Vol. 8, 449 - 451 (1970). In Russian.
Brief information.

091.041 Principles of symmetry for polarization studies of planets. J. W. Hovenier.
Astron. Astrophys., Vol. 7, 86 - 90 (1970).

Two general principles of symmetry, which have a very wide range of validity, are applied to the light reflected and transmitted by a planetary atmosphere. It is shown how inhomogeneities on the surface of a planet can be detected by the use of polarimetry. The assumption that light which is scattered more than once is unpolarized can be directly tested.

091.042 The Titius-Bode rule and inclinations of the axes of rotation of the earth and planets.
V. A. Aliev.
Dokl. AN Azerb. SSR, Vol. 25, No. 7, p. 3 - 5 (1969).
In Russian. — Abstr. in Referativ. Zhurn. 51. Astron., 4.51.242 (1970).

091.043 An experiment to determine the structure of a planetary atmosphere.
S. C. Sommer, L. Yee.
Journ. Spacecraft Rockets, Vol. 6, (No. 6), 704 - 710 (1969).

091.044 Physical study of planets and satellites. J. S. Hall.
Trans. IAU, Vol. 14A, (see 003.028), 153 - 168 (1970).
Report of Commission 16.

091.045 Television observation of the planets.

P. R. Lichtman.
Optical Spectra, Vol. 3, No. 5, p. 83 - 85 (1969).

Formulas originally derived to relate focal plane image intensity to planetary parameters have now been successfully used to determine optimum intensity in connection with vidicon and image orthicon observations.

091.046 Geologie der Planeten. J. Chodak.
Ideen exakt. Wissens, (Wiss. Technik Sowjetunion), No. 10.69, p. 627 - 636 (1969).

091.047 Exploration of the outer solar system. D. D. Meisel.
Ann. New York Acad. Sci., Vol. 163, No. 1, p. 476 - 492 (1969).

091.048 Exploration of the solar system and of interstellar space. K. A. Ehricke.
Ann. New York Acad. Sci., Vol. 163, No. 1, p. 493 - 553 (1969).

Superdense water ice.
See Abstr. 022.009.

Vision in astronomical instruments and the observation of planetary surfaces. See Abstr. 031.029.

Determinación de la presencia de compuestos quimicos en atmosferas estelares y planetarias, cometas y espacio interestelar. See Abstr. 064.012.

Vibration of a Coulomb crystal at high pressure.
See Abstr. 065.033.

Positions and motions of minor planets, comets and satellites. See Abstr. 098.034.

The outer planets: Some early history.
See Abstr. 107.007.

On the evolution of the angular velocity of bodies in the solar system. See Abstr. 107.009.

Stellar and planetary spectra in the infrared from 1.35 to 4.10 microns. See Abstr. 114.011.

092 Mercury

092.001 **Le passage de Mercure du 9 mai 1970.**
J. Meeus.
Ciel et Terre, Vol. 86, 161 - 163 (1970).

092.002 **Passage de Mercure sur le disque du soleil le 9 mai 1970.** H. Marichal.
L'Astronomie, 84ᵉ année, p. 43 - 45 (1970).

092.003 **Transits of Mercury and Venus.**
J. G. Porter.
Journ. British Astron. Ass., Vol. 80, 182 - 189 (1970).

092.004 **The coming transit of Mercury.**
Sky Telescope, Vol. 39, 232 - 233 (1970).

092.005 **Mercury: Surface features observed during radar studies.** R. M. Goldstein.
Science, Vol. 168, 467 - 469 (1970).
Radar studies of Mercury have shown the presence of several large, rough surface features and of one smooth area.

092.006 **Beobachtung des Merkurdurchganges vom 9. Mai 1970.** H. Haupt.
Sternenbote, 13. Jahrgang, p. 92 - 93 (1970).

092.007 **Thermal and tidal effect on the rotation of Mercury.**
H.-S. Liu.
Celestial Mechanics, Vol. 2, 4 - 8 (1970).
It is shown that the influence of the thermal and tidal effects on Mercury's libration are in equilibrium with the periods of rotation and revolution of Mercury locked in the 3 : 2 resonant state. The suggestion by Liu that the solar gravitational couple on the thermal bulges accelerates Mercury's rotation is investigated and the production of mechanical energy to balance the dissipation of the bodily tides is discussed. It is possible for Mercury to rotate with two bulges as a solar thermal engine; the tidal effect causes this engine to function and its maximum power is close to 10^{16} ergs per sec.

092.008 **A note on the instantaneous rotational velocity of Mercury.** H.-S. Liu.
Celestial Mechanics, Vol. 2, 123 - 126 (1970).
The fluctuation in the angular velocity of the present rotation of Mercury is investigated. The instantaneous rotational rate in terms of orbital mean motion at different positions along Mercury's orbit is given. It is found that the difference between the rotational periods derived from the motions at perihelion and aphelion is 4.68 min and that the maximum rate of rotation occurs at $f = \pi/4$ and $f = 7\pi/4$, where f is the true anomaly.

092.009 **Spin-orbit resonance of Mercury.**
C. C. Counselman III, I. I. Shapiro.
Symposia Mathematica, Vol. 3, 121 - 169 (1970). – Conference paper (see 012.010).

092.010 **The microwave spectrum of Mercury.**
D. Morrison, M. J. Klein.
Astrophys. Journ., Vol. 160, 325 - 332 (1970).
We have measured the disk temperature of Mercury at wavelengths of 1.95 and 6.0 cm; the phase-invariant components of the temperature are $350° \pm 30°$K and $385° \pm 20°$K, respectively. These data, combined with previous measurements, indicate that the time-averaged disk temperature of Mercury increases by 25 percent from short millimeter wavelengths to 6 cm. This departure from a thermal spectrum is probably due to an increase in temperature with depth. This temperature gradient could be maintained by a subsurface "greenhouse effect" if radiative heat conduction, which is strongly temperature-dependent, supports a large fraction of the diurnal flow of heat in Mercury's epilith. The observations are consistent with a model in which the ratio of the electrical to thermal skin depths is near unity at a wavelength of 1 cm.

092.011 **The phase anomaly of Mercury and Venus.**
G. J. Kirby.
Journ. British Astron. Ass., Vol. 80, 293 - 295 (1970).

092.012 **De Mercurius-overgang van 9 mei 1970.**
J. Meeus.
Hemel en Dampkring, Vol. 68, 65 - 68 (1970).

092.013 **Merkurdurchgang 1970 Mai 9.**
W. Malsch.
VdS Nachrichtenblatt, 19. Jahrgang, p. 64 (1970).

092.014 **The transit of Mercury on 1970 May 9.**
R. G. Hodgson.
Strolling Astronomer, Vol. 22, 63 - 64 (1970).

092.015 **Il transito di Mercurio sul disco del Sole del 9 maggio 1970.** A. Betti.
Coelum, Vol. 38, 69 - 71 (1970).

092.016 **Photometric studies of the planet Mercury.**
K. A. Hämeen-Anttila, T. Pikkarainen, H. Camichel.
The Moon, Vol. 1, 440 - 448 (1970).
Mercury's surface brightness has been measured from 16 photographic plates representing an equal amount of different phases. The resulting photometric function turns out to be almost identical with the lunar photometric function.

092.017 **Findings from Mercury's transit.**
J. Ashbrook.
Sky Telescope, Vol. 40, 20 - 24 (1970).

092.018 **Planet Mercury.**
C. Sagan, D. Morrison.
Zemlya i Vselennaya, No. 1, p. 9 - 17 (1970). In Russian.

Generalized Cassini's laws. See Abstr. 091.027.

093 Venus

093.001 Origin and dynamical behavior of thermal protons in the Venusian ionosphere.
P. M. Banks, W. I. Axford.
Nature, Vol. 225, 924 - 926 (1970).

An importance source of thermal H^+ and/or D^+ has been overlooked in the ionosphere of Venus which supports the choice of H^+ and/or D^+ as the dominant light ion at high altitudes.

093.002 Venus clouds: Test for carbon suboxide.
W. T. Plummer, R. K. Carson.
Astrophys. Journ., Vol. 159, 159 - 163 (1970).

Carbon suboxide has previously been suggested as the principal material of the Venus clouds. New laboratory reflection spectra of carbon suboxide frost, which must closely resemble a carbon suboxide cloud spectrum, show narrow absorption minima at 2.27 and 2.67 μ and a broad, deep absorption from 2.9 to 3.4 μ.

093.003 A radar determination of the rotation of Venus.
R. L. Carpenter.
Astron. Journ., Vol. 75, 61 - 66 (1970).

The rotation period of Venus is determined by comparing the positions of surface features observed by radar on four successive inferior conjunctions from 1962 through 1967. The best estimated sidereal period is 242.982 ± 0.04 d retrograde with the spin vector pointing toward R.A. = 94.1 ± 3°, Dec. = –71.4 ± 1° (1950.0). This period suggests that Venus is not in synodic resonance with the earth.

093.004 Evidence for severe microwave signal fading in the atmosphere of Venus due to turbulence.
D. A. de Wolf.
Journ. Geophys. Res., Vol. 75, 1202 - 1208 (1970).

The goal of this study is to estimate the possibility of atmospheric turbulence on Venus and its effect on 1- to 10-GHz radio wave propagation.

093.005 Some problems posed by the planet Venus.
J. Nikander.
Spaceflight, Vol. 12, 180 - 183 (1970).

093.006 Die untere Venuskonjunktion 1969, Beobachtungs-methoden und Fehlerquellen. P. Ahnert.
Sterne, 45. Jahrgang, p. 238 - 244 (1969).

093.007 Topside ionosphere of Venus and its interaction with the solar wind.
S. J. Bauer, R. E. Hartle, J. R. Herman.
Nature, Vol. 225, 533 - 534 (1970).

The pressure balance between the solar wind and the ionospheric plasma of Venus is used as boundary condition to develop a self-consistent model of the Venus ionosphere in agreement with Mariner V occultation data. The topside ionosphere is proposed to consist of He^+ and H^+ ions, with CO_2^+ ions being predominant near the maximum. The temperature of the ionospheric electrons is found to be 4100°K, that of the ions 1900°K and that of the neutral gas, 550°K.

093.008 The ionospheres of Venus and Mars. II. Mariner V and the Venus ionosphere.
R. Goody.
Comments Astrophys. Space Phys., Vol. 2, 7 - 11 (1970).

Electron-density profiles were obtained for both daytime and nighttime, when Mariner V flew by Venus in October 1967. The results are compared with theory and with a discussion of the Martian upper atmosphere.

093.009 The obliquity of Venus.
P. Goldreich, S. J. Peale.
Astron. Journ., Vol. 75, 273 - 284 (1970).

Solar gravitational tides tend to reduce the obliquity of Venus from its present value near 180° whether or not Venus is locked in the synodic spin resonance with the earth. Both a thermally driven atmospheric tidal torque and dissipation of energy at the boundary between a rigid mantle and a differentially rotating liquid core are possible mechanisms for maintaining the retrograde spin. The latter mechanism, where precession of the spin vector about the orbit normal induces the differential rotation, is almost certainly capable of driving the obliquity to 180° from values greater than 90° for a wide range of reasonable core viscosities and spin angular velocities.

093.010 The dark ultra-violet formation on Venus.
O. M. Starodubtzeva.
Astron. Tsirk., No. 540, p. 5 - 6 (1969).   In Russian.

093.011 Etude de la rotation rétrograde, en 4 jours, de la couche extérieure nuageuse de Vénus.
C. Boyer, P. Guérin.
Icarus, Vol. 11, 338 - 355 (1969).

We review briefly photographic as well as spectrographic observations which have allowed us to prove and confirm the retrograde rotation of about 4 days of the external cloud layer of Venus. Certain configurations visible in ultraviolet light – in particular two spots having the shape of horizontal Y – seem permanent, and we can establish their cartography. But the observations suggest that highlying clouds cover these configurations from time to time. An examination of the displacement of the equatorial spots during time intervals of several hours furnishes an instantaneous synodic rotation period somewhat larger than 4 days for the permanent spots. The synodic period is shorter for the variable spots. Ephemerides have been calculated for the years to come.

093.012 Atmospheric tides and resonant rotation of Venus.
T. Gold, S. Soter.
Icarus, Vol. 11, 356 - 366 (1969).

If a solar atmospheric tide exists, partly thermally induced and similar to that known on the earth, its torque may counteract that due to the solar solid body tide at a particular rotation period. The small interaction with the earth is then sufficient to look the period to one of the resonances in the vicinity of that angular velocity.

093.013 Geochemistry of the volatile elements on Venus.
J. S. Lewis.
Icarus, Vol. 11, 367 - 385 (1969).

A detailed study is made of the chemistry of the volatile elements on Venus. It is concluded that the dominant cloud-forming species in the atmosphere are compounds of mercury: Hg_2Br_2, Hg_2I_2, HgS, Hg, and Hg_2Cl_2 clouds are predicted. A number of other proposed cloud constituents are critically discussed. The implications of such a cloud structure with respect to the composition of the upper and lower atmosphere are described, and it is shown that COS, HBr, NH_3, and HI must be essentially absent from the observable portion of the atmosphere. HBr may be abundant in the lower atmosphere, and it is suggested that species other than CO_2 may be responsible for the attenuation of radar signals of $\lambda \simeq 3$ cm by the atmosphere.

093.014 High-dispersion spectroscopic observations of Venus. V. The carbon dioxide band at 8689 Å.
L. D. G. Young, R. A. Schorn, E. S. Barker, M. MacFarlane.

Icarus, Vol. 11, 390 - 407 (1969)

An average rotational temperature of 242° ± 2°K is found. The variation of the equivalent width of the 8689-Å band with Venus phase is seen to agree generally with the observations of Kuiper; the equivalent width decreases with increasing phase angles.

093.015 Spectra of Venus and Jupiter from 1800 to 3200 Å. R. C. Anderson, J. G. Pipes, A. L. Broadfoot, L. Wallace.
Journ. Atmosph. Sci., Vol. 26, 874 - 888 = Contr. Kitt Peak National Obs., No. 463 (1969). – Conference paper (see 012.009).

The results of two Aerobee rocket flights are reported. One obtained spectra of Venus from 3200 to 1900 Å at 16.5 Å resolution and the other spectra of Jupiter from 3200 to 1800 Å at 28 Å resolution. The spectra of both planets are of much higher statistical accuracy than those that have been obtained previously. The peculiarities indicated by the previous observations are not confirmed. In particular, there does not appear to be an absorption feature in the Jupiter spectrum at 2600 Å or an ozone absorption in the Venus spectrum. Two extreme models are used to interpret the data: the reflecting layer model and the cloud model.

093.016 Interpretation of the Venus CO_2 absorption bands. J. W. Chamberlain, C. R. Smith.
Astrophys. Journ., Vol. 160, 755 - 765 = Contr. Kitt Peak National Obs., No. 533 (1970).

The available information on the various CO_2 absorption bands in Venus's spectra is examined to ascertain whether a scattering model for the atmosphere can fit all the data without additional ad hoc assumptions.

093.017 The runaway greenhouse and the accumulation of CO_2 in the Venus atmosphere. S. I. Rasool, C. de Bergh.
Nature, Vol. 226, 1037 - 1039 (1970).

Conditions on earth would be as hostile as on Venus if the earth were closer to the sun by only 6 – 10 million miles.

093.018 Radar backscattering properties of Venus at 70 cm. R. F. Jurgens, R. B. Dyce.
Astron. Journ., Vol. 75, 297 - 314, 509, 510 (1970).

Radar echoes from the surface of Venus are analyzed with a view to describing the mean behavior by suitable adjustment of available theories. By shortening the pulse width to give better resolution near the front portion of Venus, we find that a theory by Muhleman fits better than other theories. We show variability of the echoes near normal incidence as the subradar point moves. We find that difficulty arises in fitting a spherically symmetric model to Venus because the echoes exhibit patchiness or structure which move as the planet rotates.

093.019 Investigation of the composition of the Venus atmosphere at the automatic space probes Venus 5 and Venus 6. A. P. Vinogradov, Iu. A. Surkov, B. M. Andreichikov.
Dokl. Akad. Nauk SSSR, Ser. Mat. Fiz., Vol. 190, 552 - 554 (1970). In Russian.

The composition of the Venus atmosphere was determined by the space probe Venus 5 at pressure levels (temperatures) 0.6 kg/cm² (25°C) and 5 kg/cm² (150°C), by Venus 6 at 2 kg/cm² (85°C) and 10 kg/cm² (225°C). The results are the following: CO_2 97 ± 4% (by weight), N_2 (including inert gases) ≤ 2%, O_2 ≤ 0.1%, H_2O (at 0.6 atm) about 11 mg/l. The H_2O content decreases with decreasing height. The change of temperature with height is nearly adiabatic. Adiabatic extrapolation leads to surface values about 100 atm and 500°C for pressure and temperature.   *B. Onderlička*

093.020 On the phase anomaly of the inner planets. D. C. Heggie.
Journ. British Astron. Ass., Vol. 80, 288 - 292 (1970).

093.021 The problem of the rotation of Venus. V. A. Firsoff.
Journ. British Astron. Ass., Vol. 80, 303 - 304 (1970).

093.022 Carbon atoms in the upper atmosphere of Venus. F. F. Marmo, A. Engelman.
Icarus, Vol. 12, 128 - 130 (1970).

The detection of carbon atoms in the upper atmosphere of Venus is reported on the basis of an extension of a previously reported analysis of low-resolution spectra of the solar-illuminated atmosphere of Venus. It is demonstrated that the previously reported anomalously high albedo for $\lambda\lambda > 1500$ Å can be explained by invoking the presence of a 14 kR (kilorayleigh) CI signal at 1657 Å.

093.023 The analysis of the brightness temperature spectrum and the structure of the lower Venus atmosphere. O. N. Rzhiga.
Astron. Zhurn. Akad. Nauk SSSR, Vol. 47, 566 - 576 (1970). In Russian. English translation in Soviet Astron. AJ, Vol. 14, No. 3.

The analysis of the brightness temperature spectrum indicates that a near-surface layer of the Venus atmosphere with thickness of about 10 km can be in the isothermal state or close to it. This allows to draw important conclusion on the physical conditions in the lower atmosphere of the planet.

093.024 Interpretation of radar and radio observations of Venus. M. A. Slade, I. I. Shapiro.
Journ. Geophys. Res., Vol. 75, 3301 - 3317 (1970).

Constraints on the atmospheric conditions at the surface of Venus are determined from the radar angular scattering law, radar cross section, and microwave brightness spectrum. A rigorous derivation, considering refraction effects, is given for the relationship between delay-Doppler coordinates and spherical coordinates on the planet's surface.

093.025 An estimate of the present-day deep-mantle degassing rate from data on the atmosphere of Venus. J. C. G. Walker, K. K. Turekian, D. M. Hunten.
Journ. Geophys. Res., Vol. 75, 3558 - 3561 (1970).

We argue that the present rate of escape of hydrogen from the upper atmosphere of Venus establishes an upper limit for the rate of degassing of water from the interior of the planet. This upper limit is 2.4×10^9 grams of water per year. If we assume that this value holds also for deep-mantle degassing from the earth, we find that the rate is insufficient, by five orders of magnitude, to supply the water at the earth's surface in 4.5×10^9 years. An early episode of greatly enhanced degassing appears to be required.

093.026 Peculiarities in the propagation of radio waves of the millimetre and centimetre regions in the atmosphere of Venus. N. N. Krupenio.
In-t kosmich. issled. AN SSSR, Moskau. 10 pp. (1970). In Russian. – Abstr. in Referativ. Zhurn. 62. Issled. kosm. prostranstv., 6.62.157 (1970).

093.027 Dehydrogenation of Venus. R. F. Mueller.
Nature, Vol. 227, 363 - 365 (1970).

The purpose of this paper is to show that the water content of the atmosphere of Venus may be established independently of the mechanism of transport if the time scale of the heterogeneous reactions involving water in the lithosphere is short compared with that governing the mechanism of escape. In such conditions the atmospheric water would be a direct function of the lithospheric composition and the temperature.

093.028 **Preliminary results of interferometric observations of Venus at 11.1-cm wavelength.**
A. C. E. Sinclair, J. P. Basart, D. Buhl, W. A. Gale, M. Liwshitz.
Radio Science, Vol. 5, 347 - 354 = National Radio Astron.
Obs., *Green Bank*, Repr. Ser. A, No. 159 (1970).

Radio interferometry measurements have been made of the temperature variation over Venus. The results imply that the temperature has no significant poleward variation. A limit of 12° K can be placed on the surface temperature difference between poles and equator.

093.029 **A study of the average and anomalous radar scattering from the surface of Venus at 70 cm wavelength.**   R. F. Jurgens.
Thesis, Cornell. Univ., Ithaca, N. Y. [Available from Univ. Microfilms, Ann Arbor, Mi.], 241 pp. (1968). — See Phys. Abstr., Vol. 73, No. 15879 (1970).

The right ascensions of the sun and Venus obtained from observations with the Tashkent meridian circle.
See Abstr. 041.024.

Transits of Mercury and Venus.
See Abstr. 092.003.

The phase anomaly of Mercury and Venus.
See Abstr. 092.011.

The atmospheres of Mars and Venus.
See Abstr. 097.007.

Observations of Mars and Venus at 11.1 cm.
See Abstr. 097.022.

Comparative characteristics of the ionosphere of the planets of the terrestrial group: Mars, Venus, and the earth.
See Abstr. 097.038.

Disturbances of the interplanetary plasma near Venus according to data obtained with the space probes Venus 4 and Venus 6.   See Abstr. 106.025.

094 Moon

094.001 **Carbon compounds in Apollo 11 lunar samples.**
B. Nagy, W. M. Scott, V. Modzeleski, L. A. Nagy,
C. M. Drew, W. S. McEwan, J. E. Thomas, P. B. Hamilton,
H. C. Urey.
Nature, Vol. 225, 1028 - 1032 (1970).

Attempts are made to answer the question whether
carbon compounds are entrapped solar wind particles,
meteoritic or cometary contributions, remnants of a primordial atmosphere or products of the degassing of the moon.

094.002 **Wavelength dependence of polarization. XX. The
integrated disk of the moon.**
G. V. Coyne, S. F. Pellicori.
Astron. Journ., Vol. 75, 54 - 60 (1970).

Photoelectric polarimetry of the whole lunar disk in eight
wavelength intervals between 0.3 and 0.6μ at a range of phase
angles between 0° and |140°| is presented. The polarization is
maximum near + and −100° phase angle, and it increases progressively with decreasing wavelength. The polarization maximum occurs at smaller phase angles for the longer wavelengths.
For phase angles greater than |40°| the electric-vector maxi-
mum is perpendicular to the scattering plane and there is no
rotation of the plane of polarization within an accuracy of
±2°. Some excess polarization with the electric-vector maximum in the scattering plane has been found during certain
eclipses.

094.003 **Star-calibrated lunar photography.**
M. Moutsoulas.
The Moon, Vol. 1, 173 - 189 (1970).

Accuracy in the determination of the absolute coordinates of lunar features or the physical libration of the moon can
be greatly improved by a technique providing lunar photographs on star-calibrated plates. An outline of the photographic technique and the computer programme for the redcution
of the plates is given. In the appendix the technical details
of the double slide plate carrier which, adjusted to the Data
Corporation Camera, secures high quality stellar images,
are presented.

094.004 **Lunar Orbiter gravity analysis.**
J. Lorell.
The Moon, Vol. 1, 190 - 231 (1970).

This paper presents the results to date of the Jet Propulsion Laboratory (JPL) effort at analyzing the tracking data
from the five Lunar Orbiter spacecraft. Emphasis is placed
on the long-arc evaluation, to which most of the work was
directed, rather than on the mascon analysis, which will be
reported separately. The model of the moon's gravity
field that has been derived from the long-arc analysis is interpreted in terms of the moon's mass distribution. The results
are evaluated from the point of view of consistency and of
statistical significance. In particular, the various factors
contributing to the uncertainty in the estimation process
are discussed.

094.005 **The impact-theory interpretation of the distribution
of maria on the lunar surface.**
R. Metcalfe, N. A. Barricelli.
The Moon, Vol. 1, 232 - 236 (1970).

The satellite impact interpretation of the surface distribution of lunar maria is presented according to Barricelli and
Metcalfe (1969). It is emphasized that the formation of
molten rock (lava) which, according to the Apollo 11 findings,
seems to have been the origin of the material of which maria
are composed, can be the result of heat developed by the
impacts which created the respective maria (Gilbert 1893)

and does not necessarily imply a volcanic or internal origin
of this material. The distribution of mascons and some of its
possible interpretations are discussed.

094.006 **Terrestrial analogs to lunar dimple (drainage)
craters.** R. Greeley.
The Moon, Vol. 1, 237 - 252 (1970).

In addition to modes of formation offered by previous
investigators, terrestrial drainage craters formed over lava
tubes in Oregon are presented as analogs to lunar drainage
craters. Craters associated with lava tubes result from:
(1) drainage of surface material through roof fractures;
(2) plastic collapse of the partially cooled lava tube roof; and
(3) drainage of surface material into roof collapses. The former
two categories result in shallow, often elongate craters; the
latter category form classic 'dimple' shaped craters. Elongate
dimple craters formed over volcanic fissures in southern Idaho
are also discussed and presented in support of one mode of
formation proposed by previous investigators for lunar
drainage craters.

094.007 **Adsorptive mechanism for loss of the moon's
atmosphere.** A. M. Gutkin, M. S. Markov,
C. M. Raitburd, M. V. Slonimskaya.
Journ. British Interplanet. Soc., Vol. 23, 165 - 167 (1970).
Translated by E. R. Hope from a paper in Russian, Dokl.
Akad. Nauk, Vol. 182, 1294 - 1295 (1968).

094.008 **On the origin of the large lunar craters and circular
maria.** G. S. Shteinberg.
Journ. British Interplanet. Soc., Vol. 23, 237 - 242 (1970).
Translated by E. R. Hope from a paper in Russian, Dokl. Akad.
Nauk, Vol. 184, 566 - 569 (1969).

094.009 **Method for estimating the electrical conductivity
of the lunar interior.**
W. R. Sill, J. L. Blank.
Journ. Geophys. Res., Vol. 75, 201 - 210 (1970).

A theoretical investigation of the lunar interaction with
the solar wind and its electric and magnetic fields reveals a
method for estimating the electrical conductivity of the lunar
interior. Data from a single lunar surface magnetometer, in
conjunction with simultaneous measurements of the interplanetary magnetic field and solar wind velocity, should provide
sufficient information to distinguish between various classes
of conductivity models. Such models are based on different
thermal histories and assumed lunar materials.

094.010 **Mascons as structural relief on a lunar 'Moho'.**
D. U. Wise, M. T. Yates.
Journ. Geophys. Res., Vol. 75, 261 - 268 (1970).

A mechanism for the creation of lunar mascons is proposed that requires no abnormal density materials or major
density inversions. The mascons are produced by mantle plugs
upwelling into giant impact basins punched through the lunar
crust followed by volcanic filling of the remainder of the
crater above the plug. It is explicitly shown that continued
volcanic filling is not inhibited by the attainment of isostatic
equilibrium.

094.011 **Lunar conducting islands and formation of a lunar
limb shock wave.** J. V. Hollweg.
Journ. Geophys. Res., Vol. 75, 1209 - 1216 (1970).

We propose a mechanism for the creation of a weak
solar-wind shock at the lunar limb, in which the fluctuating
interplanetary magnetic field interacts with a thin highly
conducting crust. The crust is assumed to be fractures and

describable in terms of what we call 'conducting islands'. The fluctuating interplanetary field produces currents inside the conducting islands, which in turn produce magnetic fields capable of deflecting the solar wind.

094.012 **The lunar laser reflector.**
J. E. Faller, E. J. Wampler.
Sci. American, Vol. 222, No. 3, p. 38 - 50 (1970).
A reflector array placed on the moon by its first visitors returns pulses of light emitted by lasers on the earth. The round-trip travel time yields the distance to the moon with an accuracy of six inches.

094.013 **Der Mond nach Apollo 11.** H. Wänke.
Umschau, 70. Jahrgang, p. 138 - 143 (1970).
Popular article.

094.014 **Keine Erzlagerstätten auf dem Mond.**
H. Borchert.
Umschau, 70. Jahrgang, p. 180 - 181 (1970). — Popular article.

094.015 **The latest Apollo's experiments.**
R. N. Watts, Jr.
Sky Telescope, Vol. 39, 11 - 13 (1970).

094.016 **The Apollo 12 explorers on the moon.**
R. Hillenbrand.
Sky Telescope, Vol. 39, 95 - 98 (1970).

094.017 **Findings from a sample of lunar material.**
R. N. Watts, Jr.
Sky Telescope, Vol. 39, 144 - 147 (1970).

094.018 **The Apollo 12 landing site.**
R. J. Fryer.
Spaceflight, Vol. 12, 23 - 25 (1970).

094.019 **A guide to the moon's farside.**
C. A. Cross.
Spaceflight, Vol. 12, 50 - 55 (1970).

094.020 **Lunar igneous intrusions.**
F. El-Baz.
Science, Vol. 167, 49 - 50 (1970).
Photographs taken from Apollo 10 and 11 reveal a number of probable igneous intrusions, including three probable dikes that crosscut the wall and floor of an unnamed 75-kilometer crater on the lunar farside. These intrusions are distinguished by their setting, textures, structures, and brightness relative to the surrounding materials. Recognition of these probable igneous intrusion in the lunar highlands supports the indications of the heterogeneity of lunar materials.

094.021 **Lunar surface rocks and fines: Chemical composition.** L. H. Ahrens, R. V. Danchin.
Science, Vol. 167, 87 - 88 (1970).
A critical examination of the chemical abundance data on 12 specimens of lunar surface material has led to several conclusions of remarkable geochemical and cosmochemical interest.

094.022 **Iron in synthetic quartz: Heat and radiation induced changes.** A. J. Cohen, F. Hassan.
Science, Vol. 167, 176 - 177 (1970).
As part of a study of the possible changes in silicate material on the lunar surface produced by solar radiation, effects of ionizing radiation, of heating, and of heating followed by ionizing radiation on iron-doped synthetic quartzes have been studied as models. The optical changes are mainly related to valence changes in ferric or ferrous ions.

094.023 **Apollo 11 Laser Ranging Retro-Reflector: Initial measurements from the McDonald Observatory.**
C. O. Alley, R. F. Chang, D. G. Currie, J. Mullendore, S. K. Poultney, J. D. Rayner, E. C. Silverberg, C. A. Steggerda, H. H. Plotkin, W. Williams, B. Warner, H. Richardson, B. Bopp.
Science, Vol. 167, 368 - 370 (1970).
Acquisition measurements of the round-trip travel time of light, from the McDonald Observatory to the Laser Ranging Retro-Reflector deployed on the moon by the Apollo 11 astronauts, were made on 20 August and on 3, 4, and 22 September 1969. The uncertainty in the round-trip travel time was ± 15 nanoseconds, with the pulsed ruby laser and timing system used for the acquisition.

094.024 **Chemical composition of the lunar surface in Sinus Medii.** E. J. Franzgrote, J. H. Patterson, A. L. Turkevich, T. E. Economou, K. P. Sowinski.
Science, Vol. 167, 376 - 379 (1970).
More precise and comprehensive analytical results for lunar material in Sinus Medii have been derived from the alpha-scattering experiment on Surveyor VI. The amounts of the principal constituents at this mare are approximately the same as those of constituents at Mare Tranquillitatis.

094.025 **Lunar libration tables.**
D. H. Eckhardt.
The Moon, Vol. 1, 264 - 275 (1970).
The selenographic direction cosines of the earth and of the pole of the ecliptic are developed in trigonometric series; the time dependence of each term enters through the Delaunay arguments. These and other series are tabulated for different libration parameters to provide means for calculating at any epoch the lunar librations and their partial derivatives with respect to the parameters.

094.026 **Lunar regolith at Tranquillity Base.**
E. M. Shoemaker, M. H. Hait, G. A. Swann, D. L. Schleicher, D. H. Dahlem, G. G. Schaber, R. L. Sutton.
Science, Vol. 167, 452 - 455 (1970). — Abstr. in 'The Moon', Vol. 1, 353 (1970).

094.027 **Passive seismic experiment.**
G. V. Latham, M. Ewing, F. Press, G. Sutton, J. Dorman, Y. Nakamura, N. Toksöz, R. Wiggins, J. Derr, F. Duennebier.
Science, Vol. 167, 455 - 457 (1970). — Abstr. in 'The Moon', Vol. 1, 353 (1970).

094.028 **Laser ranging retro-reflector: Continuing measurements and expected results.**
C. O. Alley, R. F. Chang, D. G. Currie, S. K. Poultney, P. L. Bender, R. H. Dicke, D. T. Wilkinson, J. E. Faller, W. M. Kaula, G. J. F. MacDonald, J. D. Mulholland, H. H. Plotkin, W. Carrion, E. J. Wampler.
Science, Vol. 167, 458 - 460 (1970). — Abstr. in 'The Moon', Vol. 1, 353 - 354 (1970).

094.029 **Age of the moon: An isotopic study of uranium-thorium-lead systematics of lunar samples.**
M. Tatsumoto, J. N. Rosholt.
Science, Vol. 167, 461 - 463 (1970). — Abstr. in 'The Moon', Vol. 1, 354 (1970).

094.030 **Ages, irradiation history, and chemical composition of lunar rocks from the Sea of Tranquillity.**
A. L. Albee, D. S. Burnett, A. A. Chodos, O. J. Eugster, J. C. Huneke, D. A. Papanastassiou, F. A. Podosek, G. P. Russ, II, H. G. Sanz, F. Tera, G. J. Wasserburg.
Science, Vol. 167, 463 - 466 (1970). — Abstr. in 'The Moon', Vol. 1, 354 - 355 (1970).

094.031 Argon-40/argon-39 dating of lunar rock samples.
G. Turner.
Science, Vol. 167, 466 - 468, with a correction, p. 1759
(1970). – Abstr. in 'The Moon', Vol. 1, 355 (1970).

094.032 Uranium-thorium-lead isotope relations in lunar
materials. L. T. Silver.
Science, Vol. 167, 468 - 471 (1970). – Abstr. in 'The Moon',
Vol. 1, 355 (1970).

094.033 Rubidium-strontium, uranium, and thorium-lead
dating of lunar material.
K. Gopalan, S. Kaushal, C. Lee-Hu, G. W. Wetherill.
Science, Vol. 167, 471 - 473 (1970). – Abstr. in 'The Moon',
Vol. 1, 355 - 356 (1970).

094.034 Rubidium-strontium relations in Tranquillity Base
samples. P. M. Hurley, W. H. Pinson, Jr.
Science, Vol. 167, 473 - 474 (1970). – Abstr. in 'The Moon',
Vol. 1, 356 (1970).

094.035 Rubidium-strontium chronology and chemistry of
lunar material.
W. Compston, P. A. Arriens, M. J. Vernon, B. W. Chappell.
Science, Vol. 167, 474 - 476 (1970). – Abstr. in 'The Moon',
Vol. 1, 356 (1970).

094.036 Rubidium-strontium age and elemental and isotopic
abundances of some trace elements in lunar samples.
V. R. Murthy, R. A. Schmitt, P. Rey.
Science, Vol. 167, 476 - 479 (1970). – Abstr. in 'The Moon',
Vol. 1, 356 - 357 (1970).

094.037 Age determinations and isotopic abundance mea-
surements on lunar samples.
R. K. Wanless, W. D. Loveridge, R. D. Stevens.
Science, Vol. 167, 479 - 480 (1970). – Abstr. in 'The Moon',
Vol. 1, 357 (1970).

094.038 Lead and thallium isotopes in Mare Tranquillitatis
surface material.
T. P. Kohman, L. P. Black, L. Ihochi, J. M. Huey.
Science, Vol. 167, 481 - 483 (1970). – Abstr. in 'The Moon',
Vol. 1, 357 (1970).

094.039 Abundance of alkali metals, alkaline and rare earths,
and strontium-87/strontium-86 ratios in lunar
samples. P. W. Gast, N. J. Hubbard.
Science, Vol. 167, 485 - 487 (1970). – Abstr. in 'The Moon',
Vol. 1, 357 (1970).

094.040 Rare earth elements in returned lunar samples.
L. A. Haskin, P. A. Helmke, R. O. Allen.
Science, Vol. 167, 487 - 490 (1970). – Abstr. in 'The Moon',
Vol. 1, 357 - 358 (1970).

094.041 Trace elements and radioactivity in lunar rocks:
Implications for meteorite infall, solar-wind flux,
and formation conditions of moon.
R. R. Keays, R. Ganapathy, J. C. Laul, E. Anders, G. F. Her-
zog, P. M. Jeffery.
Science, Vol. 167, 490 - 493 (1970). – Abstr. in 'The Moon',
Vol. 1, 358 (1970).

094.042 Potassium, rubidium, strontium, barium, and rare-
earth concentrations in lunar rocks and separated
phases. J. A. Philpotts, C. C. Schnetzler.
Science, Vol. 167, 493 - 495 (1970). – Abstr. in 'The Moon',
Vol. 1, 358 (1970).

094.043 Total carbon and nitrogen abundances in lunar
samples. C. B. Moore, C. F. Lewis, E. K. Gibson, W. Nichiporuk.
Science, Vol. 167, 495 - 497 (1970). – Abstr. in 'The Moon',
Vol. 1, 359 (1970).

094.044 Instrumental neutron activation analyses of lunar
specimens. G. G. Goleš, M. Osawa, K. Randle,
R. L. Beyer, D. Y. Jérome, D. J. Lindstrom, M. R. Martin,
S. M. McKay, T. L. Steinborn.
Science, Vol. 167, 497 - 499 (1970). – Abstr. in 'The Moon',
Vol. 1, 359 (1970).

094.045 Isotopic abundances of actinide elements in lunar
material. P. R. Fields, H. Diamond,
D. N. Metta, C. M. Stevens, D. J. Rokop, P. E. Moreland.
Science, Vol. 167, 499 - 501 (1970). – Abstr. in 'The Moon',
Vol. 1, 359 - 360 (1970).

094.046 Trace elements and accessory minerals in lunar
samples. G. W. Reed, Jr., S. Jovanovic,
L. H. Fuchs.
Science, Vol. 167, 501 - 503 (1970). – Abstr. in 'The Moon',
Vol. 1, 360 (1970).

094.047 Gallium, germanium, indium, and iridium in lunar
samples. P. A. Baedecker, J. T. Wasson.
Science, Vol. 167, 503 - 505 (1970). – Abstr. in 'The Moon',
Vol. 1, 360 (1970).

094.048 Multielement analysis of lunar soil and rocks.
G. H. Morrison, J. T. Gerard, A. T. Kashuba,
E. V. Gangadharam, A. M. Rothenberg, N. M. Potter,
G. B. Miller.
Science, Vol. 167, 505 - 507 (1970). – Abstr. in 'The Moon',
Vol. 1, 361 (1970).

094.049 Neutron activation analysis of milligram quantities
of lunar rocks and soils.
K. K. Turekian, D. P. Kharkar.
Science, Vol. 167, 507 - 509 (1970). – Abstr. in 'The Moon',
Vol. 1, 361 (1970).

094.050 Elemental composition of lunar surface material.
A. A. Smales, D. Mapper, M. S. W. Webb, R. K.
Webster, J. D. Wilson.
Science, Vol. 167, 509 - 512 (1970). – Abstr. in 'The Moon',
Vol. 1, 361 (1970).

094.051 Abundances of 30 elements in lunar rocks, soil,
and core samples.
R. A. Schmitt, H. Wakita, P. Rey.
Science, Vol. 167, 512 - 515 (1970). – Abstr. in 'The Moon',
Vol. 1, 361 - 362 (1970).

094.052 Alpha-particle activity of Apollo 11 samples.
K. A. Richardson, D. S. McKay, W. R. Greenwood,
T. H. Foss.
Science, Vol. 167, 516 - 517 (1970). – Abstr. in 'The Moon',
Vol. 1, 362 (1970).

094.053 Electron-microprobe analyses of phases in lunar
samples. N. G. Ware, J. F. Lovering.
Science, Vol. 167, 517 - 520 (1970). – Abstr. in 'The Moon',
Vol. 1, 362 - 363 (1970).

094.054 Semimicro chemical and X-ray fluorescence analysis
of lunar samples. H. J. Rose, Jr., F. Cuttitta,
E. J. Dwornik, M. K. Carron, R. P. Christian, J. R. Lindsay,
D. T. Ligon, R. R. Larson.
Science, Vol. 167, 520 - 521 (1970). – Abstr. in 'The Moon',
Vol. 1, 363 (1970).

094.055   Emission spectrographic determination of trace elements in lunar samples.
C. Annell, A. Helz.
Science, Vol. 167, 521 - 523 (1970). – Abstr. in 'The Moon', Vol. 1, 363 (1970).

094.056   Major and trace elements and cosmic-ray produced radioisotopes in lunar samples.
H. Wänke, F. Begemann, E. Vilcsek, R. Rieder, F. Teschke, W. Born, M. Quijano-Rico, H. Voshage, F. Wlotzka.
Science, Vol. 167, 523 - 525 (1970). – Abstr. in 'The Moon', Vol. 1, 363 - 364 (1970).

094.057   Lunar rock compositions and some interpretations.
A. E. J. Engel, C. G. Engel.
Science, Vol. 167, 527 - 528 (1970). – Abstr. in 'The Moon', Vol. 1, 364 (1970).

094.058   Oxygen, silicon, and aluminum in lunar samples by 14 MeV neutron activation.
W. D. Ehmann, J. W. Morgan.
Science, Vol. 167, 528 - 530 (1970). – Abstr. in 'The Moon', Vol. 1, 364 (1970).

094.059   Chemical composition of lunar material.
J. A. Maxwell, S. Abbey, W. H. Champ.
Science, Vol. 167, 530 - 531 (1970). – Abstr. in 'The Moon', Vol. 1, 364 (1970).

094.060   Chemical analyses of lunar samples 10017, 10072, and 10084.   H. B. Wiik, P. Ojanpera.
Science, Vol. 167, 531 - 532 (1970). – Abstr. in 'The Moon', Vol. 1, 365 (1970).

094.061   Quantitative chemical analysis of lunar samples.
L. C. Peck, V. C. Smith.
Science, Vol. 167, 532 (1970). – Abstr. in 'The Moon', Vol. 1, 365 (1970).

094.062 $^{18}O/^{16}O$, $^{30}Si/^{28}Si$, D/H, and $^{13}C/^{12}C$ studies of lunar rocks and minerals.
S. Epstein, H. P. Taylor, Jr.
Science, Vol. 167, 533 - 535 (1970). – Abstr. in 'The Moon', Vol. 1, 365 (1970).

094.063   Oxygen isotope fractionation between minerals and an estimate of the temperature of formation.
N. Onuma, R. N. Clayton, T. K. Mayeda.
Science, Vol. 167, 536 - 538 (1970). – Abstr. in 'The Moon', Vol. 1, 365 - 366 (1970).

094.064   Water, hydrogen, deuterium, carbon, carbon-13, and oxygen-18 content of selected lunar material.
I. Friedman, J. R. O'Neil, L. H. Adami, J. D. Gleason, K. Hardcastle.
Science, Vol. 167, 538 - 540 (1970). – Abstr. in 'The Moon', Vol. 1, 366 (1970).

094.065   Concentration and isotopic composition of carbon and sulfur in Apollo 11 lunar samples.
I. R. Kaplan, J. W. Smith.
Science, Vol. 167, 541 - 543 (1970). – Abstr. in 'The Moon', Vol. 1, 366 (1970).

094.066   Rare gases, hydrogen, and nitrogen: Concentrations and isotopic composition in lunar material.
H. Hintenberger, H. W. Weber, H. Voshage, H. Wänke, F. Begemann, E. Vilcsek, F. Wlotzka.
Science, Vol. 167, 543 - 545 (1970). – Abstr. in 'The Moon', Vol. 1, 366 - 367 (1970).

094.067   Isotopic analysis of rare gases from stepwise heating of lunar fines and rocks.
J. H. Reynolds, C. M. Hohenberg, R. S. Lewis, P. K. Davis, W. A. Kaiser.
Science, Vol. 167, 545 - 548 (1970). – Abstr. in 'The Moon', Vol. 1, 367 (1970).

094.068   Solar wind gases, cosmic ray spallation products, and the irradiation history.
K. Marti, G. W. Lugmair, H. C. Urey.
Science, Vol. 167, 548 - 550 (1970). – Abstr. in 'The Moon', Vol. 1, 367 (1970).

094.069   Isotopic composition of rare gases in lunar samples.
R. O. Pepin, L. E. Nyquist, D. Phinney, D. C. Black.
Science, Vol. 167, 550 - 553 (1970). – Abstr. in 'The Moon', Vol. 1, 367 - 368 (1970).

094.070   Cosmic ray production of rare gas radioactivities and tritium in lunar material.
R. W. Stoenner, W. J. Lyman, R. Davis, Jr.
Science, Vol. 167, 553 - 555 (1970). – Abstr. in 'The Moon', Vol. 1, 368 (1970).

094.071   Inert gases in lunar samples.
D. Heymann, A. Yaniv, J. A. S. Adams, G. E. Fryer.
Science, Vol. 167, 555 - 558 (1970). – Abstr. in 'The Moon', Vol. 1, 368 (1970).

094.072   Trapped solar wind noble gases, Kr^{81}/Kr exposure ages and K/Ar ages in Apollo 11 lunar material.
P. Eberhardt, J. Geiss, H. Graf, N. Grögler, U. Krähenbühl, H. Schwaller, J. Schwarzmüller, A. Stettler.
Science, Vol. 167, 558 - 560 (1970). – Abstr. in 'The Moon', Vol. 1, 368 (1970).

094.073   Gas analysis of the lunar surface.
J. G. Funkhouser, O. A. Schaeffer, D. D. Bogard, J. Zähringer.
Science, Vol. 167, 561 - 563 (1970). – Abstr. in 'The Moon', Vol. 1, 369 (1970).

094.074   Solid state studies of the radiation history of lunar samples.
G. Crozaz, U. Haack, M. Hair, H. Hoyt, J. Kardos, M. Maurette, M. Miyajima, M. Seitz, S. Sun, R. Walker, M. Wittels, D. Woolum.
Science, Vol. 167, 563 - 566 (1970). – Abstr. in 'The Moon', Vol. 1, 369 (1970).

094.075   Tritium and argon radioactivities in lunar material.
E. L. Fireman, J. C. D'Amico, J. C. DeFelice.
Science, Vol. 167, 566 - 568, with a correction, p. 1759 (1970). – Abstr. in 'The Moon', Vol. 1, 369 - 370 (1970).

094.076   Particle track, X-ray, thermal, and mass spectrometric studies of lunar material.
R. L. Fleischer, E. L. Haines, R. E. Hanneman, H. R. Hart, Jr., J. S. Kasper, E. Lifshin, R. T. Woods, P. B. Price.
Science, Vol. 167, 568 - 571 (1970). – Abstr. in 'The Moon', Vol. 1, 370 (1970).

094.077   Rare gases in lunar samples: Study of distribution and variations by a microprobe technique.
T. Kirsten, F. Steinbrunn, J. Zähringer.
Science, Vol. 167, 571 - 574 (1970). – Abstr. in 'The Moon', Vol. 1, 370 (1970).

094.078   Pattern of bombardment-produced radionuclides in rock 10017 and in lunar soil.

J. P. Shedlovsky, M. Honda, R. C. Reedy, J. C. Evans, Jr.,
D. Lal, R. M. Lindstrom, A. C. Delany, J. R. Arnold,
H.-H. Loosli, J. S. Fruchter, R. C. Finkel.
Science, Vol. 167, 574 - 576, with a correction, p. 1759
(1970). – Abstr. in 'The Moon', Vol. 1, 370 - 371 (1970).

094.079 **Cosmogenic and primordial radionuclides in lunar
samples by nondestructive gamma-ray spectrometry.**
R. W. Perkins, L. A. Rancitelli, J. A. Cooper, J. H. Kaye,
N. A. Wogman.
Science, Vol. 167, 577 - 580 (1970). – Abstr. in 'The Moon',
Vol. 1, 371 (1970).

094.080 **Elemental compositions and ages of lunar samples
by nondestructive gamma-ray spectrometry.**
G. D. O'Kelley, J. S. Eldridge, E. Schonfeld, P. R. Bell.
Science, Vol. 167, 580 - 582 (1970). – Abstr. in 'The Moon',
Vol. 1, 371 - 372 (1970).

094.081 **Mineralogy and petrology of some lunar samples.**
S. O. Agrell, J. H. Scoon, I. D. Muir, J. V. P. Long,
J. D. C. McConnell, A. Peckett.
Science, Vol. 167, 583 - 586 (1970). – Abstr. in 'The Moon',
Vol. 1, 372 (1970).

094.082 **Petrologic history of moon suggested by petro-
graphy, mineralogy, and crystallography.**
A. T. Anderson, Jr., A. V. Crewe, J. R. Goldsmith, P. B. Moore,
J. C. Newton, E. J. Olsen, J. V. Smith, P. J. Wyllie.
Science, Vol. 167, 587 - 590 (1970). – Abstr. in 'The Moon',
Vol. 1, 372 (1970).

094.083 **Electron microprobe analysis of lunar samples.**
I. Adler, L. S. Walter, P. D. Lowman, B. P. Glass,
B. M. French, J. A. Philpotts, K. J. F. Heinrich, J. I. Gold-
stein.
Science, Vol. 167, 590 - 592 (1970). – Abstr. in 'The Moon',
Vol. 1, 372 - 373 (1970).

094.084 **Mineralogical and petrological investigations of
lunar samples.**
J. C. Bailey, P. E. Champness, A. C. Dunham, J. Esson,
W. S. Fyfe, W. S. MacKenzie, E. F. Stumpfl, J. Zussman.
Science, Vol. 167, 592 - 594 (1970). – Abstr. in 'The Moon',
Vol. 1, 373 (1970).

094.085 **Mineralogy and deformation in some lunar samples.**
J. A. V. Douglas, M. R. Dence, A. G. Plant,
R. J. Traill.
Science, Vol. 167, 594 - 597 (1970). – Abstr. in 'The Moon',
Vol. 1, 373 (1970).

094.086 **Mineral chemistry of lunar samples.**
K. Keil, M. Prinz, T. E. Bunch.
Science, Vol. 167, 597 - 599 (1970). – Abstr. in 'The Moon',
Vol. 1, 374 (1970).

094.087 **Petrographic, mineralogic, and X-ray fluorescence
analysis of lunar igneous-type rocks and spherules.**
G. M. Brown, C. H. Emeleus, J. G. Holland, R. Phillips.
Science, Vol. 167, 599 - 601 (1970). – Abstr. in 'The Moon',
Vol. 1, 374 (1970).

094.088 **Lunar anorthosites.**
J. A. Wood, J. S. Dickey, Jr., U. B. Marvin,
B. N. Powell.
Science, Vol. 167, 602 - 604 (1970). – Abstr. in 'The Moon',
Vol. 1, 374 (1970).

094.089 **Experimental petrology of lunar material: The
nature of mascons, seas, and the lunar interior.**

M. J. O'Hara, G. M. Biggar, S. W. Richardson.
Science, Vol. 167, 605 - 607 (1970). – Abstr. in 'The Moon',
Vol. 1, 375 (1970).

094.090 **Petrogenesis of lunar basalts and the internal
constitution and origin of the moon.**
A. E. Ringwood, E. Essene.
Science, Vol. 167, 607 - 610 (1970). – Abstr. in 'The Moon',
Vol. 1, 375 (1970).

094.091 **Crystallization of some lunar mafic magmas and
generation of rhyolitic liquid.**
I. Kushiro, Y. Nakamura, H. Haramura, S.-I. Akimoto.
Science, Vol. 167, 610 - 612, with a correction, p. 1759
(1970). – Abstr. in 'The Moon', Vol. 1, 375 (1970).

094.092 **Iron-titanium oxides and olivine from 10020 and
10071.**   S. E. Haggerty, F. R. Boyd, P. M. Bell,
L. W. Finger, W. B. Bryan.
Science, Vol. 167, 613 - 615 (1970). – Abstr. in 'The Moon',
Vol. 1, 376 (1970).

094.093 **Opaque minerals of the lunar rocks and dust from
Mare Tranquillitatis.**
P. Ramdohr, A. El Goresy.
Science, Vol. 167, 615 - 618 (1970). – Abstr. in 'The Moon',
Vol. 1, 376 (1970).

094.094 **Search for magnetite in lunar rocks and fines.**
J. Jedwab, A. Herbosch, R. Wollast, G. Naessens,
N. van Geen-Peers
Science, Vol. 167, 618 - 619 (1970). – Abstr. in 'The Moon',
Vol. 1, 376 (1970).

094.095 **Quantitative optical and electron-probe studies of
the opaque phases.**   P. R. Simpson, S. H. U.
Bowie.
Science, Vol. 167, 619 - 621 (1970). – Abstr. in 'The Moon',
Vol. 1, 377 (1970).

094.096 **Lunar troilite: Crystallography.**
H. T. Evans, Jr.
Science, Vol. 167, 621 - 623 (1970). – Abstr. in 'The Moon',
Vol. 1, 377 (1970).

094.097 **Opaque minerals in lunar samples.**
E. N. Cameron.
Science, Vol. 167, 623 - 625 (1970). – Abstr. in 'The Moon',
Vol. 1, 377 (1970).

094.098 **Diffraction and Mössbauer studies of minerals from
lunar soils and rocks.**
P. Gay, G. M. Bancroft, M. G. Bown.
Science, Vol. 167, 626 - 628 (1970). – Abstr. in 'The Moon',
Vol. 1, 377 - 378 (1970).

094.099 **Lunar clinopyroxenes: Chemical composition,
structural state, and texture.**
M. Ross, A. E. Bence, E. J. Dwornik, J. R. Clark, J. J. Papike.
Science, Vol. 167, 628 - 630 (1970). – Abstr. in 'The Moon',
Vol. 1, 378 (1970).

094.100 **Compositional zoning and its significance in pyro-
xenes from three coarse-grained lunar samples.**
R. B. Hargraves, L. S. Hollister, G. Otalora.
Science, Vol. 167, 631 - 633 (1970). – Abstr. in 'The Moon',
Vol. 1, 378 (1970).

094.101 **Crystallography of some lunar plagioclases.**
D. B. Stewart, D. E. Appleman, J. S. Huebner,
J. R. Clark.

Science, Vol. 167, 634 - 635 (1970). – Abstr. in 'The Moon', Vol. 1, 378 - 379 (1970).

094.102 **Petrology of a fine-grained igneous rock from the Sea of Tranquillity.**
D. F. Weill, I. S. McCallum, Y. Bottinga, M. J. Drake, G. A. McKay.
Science, Vol. 167, 635 - 638, with a correction, p. 1758, 1759 (1970). – Abstr. in 'The Moon', Vol. 1, 379 (1970).

094.103 **High-voltage transmission electron microscopy study of lunar surface material.**
S. V. Radcliffe, A. H. Heuer, R. M. Fisher, J. M. Christie, D. T. Griggs.
Science, Vol. 167, 638 - 640 (1970). – Abstr. in 'The Moon', Vol. 1, 379 (1970).

094.104 **Silicate liquid immiscibility in lunar magmas, evidenced by melt inclusions in lunar rocks.**
E. Roedder, P. W. Weiblen.
Science, Vol. 167, 641 - 644 (1970). – Abstr. in 'The Moon', Vol. 1, 379 - 380 (1970).

094.105 **Petrology of unshocked crystalline rocks and shock effects in lunar rocks and minerals.**
E. C. T. Chao, O. B. James, J. A. Minkin, J. A. Boreman, E. D. Jackson, C. B. Raleigh.
Science, Vol. 167, 644 - 647 (1970). – Abstr. in 'The Moon', Vol. 1, 380 (1970).

094.106 **Lunar soil: Size distribution and mineralogical constituents.** M. B. Duke, C. C. Woo, M. L. Bird, G. A. Sellers, R. B. Finkelman.
Science, Vol. 167, 648 - 650 (1970). – Abstr. in 'The Moon', Vol. 1, 380 (1970).

094.107 **Mineralogy and petrology of coarse particulate material from lunar surface at Tranquillity Base.**
E. A. King, Jr., M. F. Carman, J. C. Butler.
Science, Vol. 167, 650 - 652 (1970). – Abstr. in 'The Moon', Vol. 1, 380 - 381 (1970).

094.108 **High crystallization temperatures indicated for igneous rocks from Tranquillity Base.**
B. J. Skinner.
Science, Vol. 167, 652 - 654 (1970). – Abstr. in 'The Moon', Vol. 1, 381 (1970).

094.109 **Morphology and related chemistry of small lunar particles from Tranquillity Base.**
D. S. McKay, W. R. Greenwood, D. A. Morrison.
Science, Vol. 167, 654 - 656 (1970). – Abstr. in 'The Moon', Vol. 1, 381 (1970).

094.110 **Mineralogy and petrography of lunar samples.**
B. Mason, K. Fredriksson, E. P. Henderson, E. Jarosewich, W. G. Melson, K. M. Towe, J. S. White, Jr.
Science, Vol. 167, 656 - 659 (1970). – Abstr. in 'The Moon', Vol. 1, 381 - 382 (1970).

094.111 **Phase chemistry, structure, and radiation effects in lunar samples.** G. Arrhenius, S. Asunmaa, J. I, Drever, J. Everson, R. W. Fitzgerald, J. Z. Frazer, H. Fujita, J. S. Hanor, D. Lal, S. S. Liang, D. MacDougall, A. M. Reid, J. Sinkankas, L. Wilkening.
Science, Vol. 167, 659 - 661 (1970). – Abstr. in 'The Moon', Vol. 1, 382 (1970).

094.112 **Mineralogy, petrology, and surface features of lunar samples 10062,35, 10067,9, 10069,30, and 10085,16.** J. L. Carter, I. D. MacGregor.
Science, Vol. 167, 661 - 663 (1970). – Abstr. in 'The Moon', Vol. 1, 382 (1970).

094.113 **Lunar glasses and micro-breccias: Properties and origin.** K. Fredriksson, J. Nelen, W. G. Melson, E. P. Henderson, C. A. Andersen.
Science, Vol. 167, 664 - 666 (1970). – Abstr. in 'The Moon', Vol. 1, 382 - 383 (1970).

094.114 **Deformation of silicates from the Sea of Tranquillity.** N. L. Carter, I. S. Leung, H. G. Ave'Lallemant.
Science, Vol. 167, 666 - 669 (1970). – Abstr. in 'The Moon', Vol. 1, 383 (1970).

094.115 **Shock metamorphism in lunar samples.**
W. von Engelhardt, J. Arndt, W. F. Müller, D. Stöffler.
Science, Vol. 167, 669 - 670 (1970). – Abstr. in 'The Moon', Vol. 1, 383 (1970).

094.116 **Impact metamorphism of lunar surface materials.**
W. Quaide, T. Bunch, R. Wrigley.
Science, Vol. 167, 671 - 672 (1970). – Abstr. in 'The Moon', Vol. 1, 383 - 384 (1970).

094.117 **Evidence and implications of shock metamorphism in lunar samples.** N. M. Short.
Science, Vol. 167, 673 - 675 (1970). – Abstr. in 'The Moon', Vol. 1, 384 (1970).

094.118 **Shock-wave damage in minerals of lunar rocks.**
C. B. Sclar.
Science, Vol. 167, 675 - 677 (1970). – Abstr. in 'The Moon', Vol. 1, 384 (1970).

094.119 **Cathodoluminescence properties of lunar rocks.**
R. F. Sippel, A. B. Spencer.
Science, Vol. 167, 677 - 679 (1970). – Abstr. in 'The Moon', Vol. 1, 384 - 385 (1970).

094.120 **Mineralogy and composition of lunar fines and selected rocks.**
C. Frondel, C. Klein, Jr., J. Ito, J. C. Drake.
Science, Vol. 167, 681 - 683 (1970). – Abstr. in 'The Moon', Vol. 1, 385 (1970).

094.121 **Mössbauer spectrometry of lunar samples.**
C. L. Herzenberg, D. L. Riley.
Science, Vol. 167, 683 - 686 (1970). – Abstr. in 'The Moon', Vol. 1, 385 (1970).

094.122 **Mössbauer effect and high-voltage electron microscopy of pyroxenes in type B samples.**
H. Fernández-Morán, S. S. Hafner, M. Ohtsuki, D. Virgo.
Science, Vol. 167, 686 - 688 (1970). – Abstr. in 'The Moon', Vol. 1, 385 - 386 (1970).

094.123 **Mössbauer spectroscopy of moon samples.**
A. H. Muir, Jr., R. M. Housley, R. W. Grant, M. Abdel-Gawad, M. Blander.
Science, Vol. 167, 688 - 690 (1970). – Abstr. in 'The Moon', Vol. 1, 386 (1970).

094.124 **Magnetic properties of lunar samples.**
D. W. Strangway, E. E. Larson, G. W. Pearce.
Science, Vol. 167, 691 - 693 (1970). – Abstr. in 'The Moon', Vol. 1, 386 (1970).

094.125 **Magnetic properties of lunar dust and rock samples.**
C. E. Helsley.
Science, Vol. 167, 693 - 695 (1970). – Abstr. in 'The Moon',
Vol. 1, 387 (1970).

094.126 **Magnetic studies of lunar samples.**
R. R. Doell, C. S. Grommé, A. N. Thorpe,
F. E. Senftle.
Science, Vol. 167, 695 - 697 (1970). – Abstr. in 'The Moon',
Vol. 1, 387 (1970).

094.127 **Magnetic properties of lunar samples.**
S. K. Runcorn, D. W. Collinson, W. O'Reilly,
A. Stephenson, N. N. Greenwood, M. H. Battey.
Science, Vol. 167, 697 - 699 (1970). – Abstr. in 'The Moon',
Vol. 1, 387 (1970).

094.128 **Magnetic properties of lunar sample 10048-22.**
A. Larochelle, E. J. Schwarz.
Science, Vol. 167, 700 - 701 (1970). – Abstr. in 'The Moon',
Vol. 1, 387 - 388 (1970).

094.129 **Search for magnetic monopoles in the lunar sample.**
L. W. Alvarez, P. H. Eberhard, R. R. Ross,
R. D. Watt.
Science, Vol. 167, 701 - 703 (1970). – Abstr. in 'The Moon',
Vol. 1, 388 (1970).

094.130 **Magnetic properties of the lunar crystalline rock
and fines.**   T. Nagata, Y. Ishikawa,
H. Kinoshita, M. Kono, Y. Syono, R. M. Fisher.
Science, Vol. 167, 703 - 704 (1970). – Abstr. in 'The Moon',
Vol. 1, 388 (1970).

094.131 **Magnetic resonance properties of some lunar
material.**   R. A. Weeks, A. Chatelain,
J. L. Kolopus, D. Kline, J. G. Castle.
Science, Vol. 167, 704 - 707 (1970). – Abstr. in 'The Moon',
Vol. 1, 388 - 389 (1970).

094.132 **Optical and high-frequency electrical properties
of the lunar sample.**
T. Gold, M. J. Campbell, B. T. O'Leary.
Science, Vol. 167, 707 - 709 (1970). – Abstr. in 'The Moon',
Vol. 1, 389 (1970).

094.133 **Magnetic resonance studies of lunar samples.**
S. L. Manatt, D. D. Elleman, R. W. Vaughan,
S. I. Chan, F.-D. Tsay, W. T. Huntress, Jr.
Science, Vol. 167, 709 - 711 (1970). – Abstr. in 'The Moon',
Vol. 1, 389 (1970).

094.134 **Thermoluminescence of lunar samples.**
G. B. Dalrymple, R. R. Doell.
Science, Vol. 167, 713 - 715 (1970). – Abstr. in 'The Moon',
Vol. 1, 389 - 390 (1970).

094.135 **Luminescence and thermoluminescence induced
by bombardment with protons of 159 million
electron volts.**   J. A. Edgington, I. M. Blair.
Science, Vol. 167, 715 - 717 (1970). – Abstr. in 'The Moon',
Vol. 1, 390 (1970).

094.136 **Luminescence, electron paramagnetic resonance,
and optical properties of lunar material.**
J. E. Geake, A. Dollfus, G. F. J. Garlick, W. Lamb,
C. Walker, G. A. Steigmann, C. Titulaer.
Science, Vol. 167, 717 - 720 (1970). – Abstr. in 'The Moon',
Vol. 1, 390 (1970).

094.137 **Luminescence of Apollo 11 lunar samples.**
N. N. Greenman, H. G. Gross.
Science, Vol. 167, 720 - 721 (1970). – Abstr. in 'The Moon',
Vol. 1, 391 (1970).

094.138 **Luminescence and reflectance of Tranquillity
samples: Effects of irradiation and vitrification.**
D. B. Nash, J. E. Conel, R. T. Greer.
Science, Vol. 167, 721 - 724 (1970). – Abstr. in 'The Moon',
Vol. 1, 391 (1970).

094.139 **Thermal radiation properties and thermal con-
ductivity of lunar material.**
R. C. Birkebak, C. J. Cremers, J. P. Dawson.
Science, Vol. 167, 724 - 726 (1970). – Abstr. in 'The Moon',
Vol. 1, 391 (1970).

094.140 **Elastic wave velocities of lunar samples at high
pressures and their geophysical implications.**
H. Kanamori, A. Nur, D. Chung, D. Wones, G. Simmons.
Science, Vol. 167, 726 - 728 (1970). – Abstr. in 'The Moon',
Vol. 1, 391 - 392 (1970).

094.141 **Infrared and thermal properties of lunar rock.**
J. A. Bastin, P. E. Clegg, G. Fielder.
Science, Vol. 167, 728 - 730 (1970). – Abstr. in 'The Moon',
Vol. 1, 392 (1970).

094.142 **Thermal diffusivity and conductivity of lunar
material.**   K.-I. Horai, G. Simmons, H. Kanamori, D. Wones.
Science, Vol. 167, 730 - 731 (1970). – Abstr. in 'The Moon',
Vol. 1, 392 (1970).

094.143 **Compressibilities of lunar crystalline rock, micro-
breccia, and fines to 40 kilobars.**
D. R. Stephens, E. M. Lilley.
Science, Vol. 167, 731 - 732 (1970). – Abstr. in 'The Moon',
Vol. 1, 392 - 393 (1970).

094.144 **Sound velocity and compressibility for lunar rocks
17 and 46 and for glass spheres from the lunar soil.**
E. Schreiber, O. L. Anderson, N. Soga, N. Warren, C. Scholz.
Science, Vol. 167, 732 - 734 (1970). – Abstr. in 'The Moon',
Vol. 1, 393 (1970).

094.145 **Apollo 11 drive-tube core samples: An initial
physical analysis of lunar surface sediment.**
R. Fryxell, D. Anderson, D. Carrier, W. Greenwood,
G. Heiken.
Science, Vol. 167, 734 - 737 (1970). – Abstr. in 'The Moon',
Vol. 1, 393 (1970).

094.146 **Spectral reflectivity of lunar samples.**
J. B. Adams, R. L. Jones.
Science, Vol. 167, 737 - 739 (1970). – Abstr. in 'The Moon',
Vol. 1, 393 - 394 (1970).

094.147 **Apollo 11 soil mechanics investigation.**
N. C. Costes, W. D. Carrier, J. K. Mitchell,
R. F. Scott.
Science, Vol. 167, 739 - 741 (1970). – Abstr. in 'The Moon',
Vol. 1, 394 (1970).

094.148 **Interferometric examination of small glassy
spherules and related objects in a 5-gram lunar
dust sample.**   S. Tolansky.
Science, Vol. 167, 742 - 743 (1970). – Abstr. in 'The Moon',
Vol. 1, 394 (1970).

094.149 **Surface properties of lunar samples.**
J. J. Grossman, J. A. Ryan, N. R. Mukherjee,

M. W. Wegner.
Science, Vol. 167, 743 - 745 (1970). – Abstr. in 'The Moon',
Vol. 1, 394 - 395 (1970).

094.150 **Solar radiation effects in lunar samples.**
B. W. Hapke, A. J. Cohen, W. A. Cassidy,
E. N. Wells.
Science, Vol. 167, 745 - 747 (1970). – Abstr. in 'The Moon',
Vol. 1, 395 (1970).

094.151 **Determination of manganese-53 by neutron activa-**
tion and other miscellaneous studies on lunar dust.
W. Herr, U. Herpers, B. Hess, B. Skerra, R. Woelfle.
Science, Vol. 167, 747 - 749 (1970). – Abstr. in 'The Moon',
Vol. 1, 395 (1970).

094.152 **Specific heats of lunar surface materials from 90**
to 350 degrees Kelvin.
R. A. Robie, B. S. Hemingway, W. H. Wilson.
Science, Vol. 167, 749 - 750, with a correction, p. 1759
(1970). – Abstr. in 'The Moon', Vol. 1, 395 - 396 (1970).

094.153 **Lunar organic compounds: Search and characteri-**
zation.   A. L. Burlingame, M. Calvin,
J. Han, W. Henderson, W. Reed, B. R. Simoneit.
Science, Vol. 167, 751 - 752 (1970). – Abstr. in 'The Moon',
Vol. 1, 396 (1970).

094.154 **Search for alkanes of 15 to 30 carbon atom length.**
W. G. Meinschein, E. Cordes, V. J. Shiner, Jr.
Science, Vol. 167, 753 - 754 (1970). – Abstr. in 'The Moon',
Vol. 1, 396 (1970).

094.155 **Fluorometric examination of a lunar sample.**
J. H. Rho, A. J. Bauman, T. F. Yen, J. Bonner.
Science, Vol. 167, 754 - 755 (1970). – Abstr. in 'The Moon',
Vol. 1, 396 (1970).

094.156 **Search for organic material in lunar fines by mass**
spectrometry.   R. C. Murphy, G. Preti,
M. M. Nafissi-V., K. Biemann.
Science, Vol. 167, 755 - 757 (1970). – Abstr. in 'The Moon',
Vol. 1, 397 (1970).

094.157 **Organic analysis of the returned lunar sample.**
P. I. Abell, G. H. Draffan, G. Eglinton, J. M. Hayes,
J. R. Maxwell, C. T. Pillinger.
Science, Vol. 167, 757 - 759, with a correction, p. 1759
(1970). – Abstr. in 'The Moon', Vol. 1, 397 (1970).

094.158 **Pyrolysis-hydrogen flame ionization detection of**
organic carbon in a lunar sample.
R. D. Johnson, C. C. Davis.
Science, Vol. 167, 759 - 760 (1970). – Abstr. in 'The Moon',
Vol. 1, 397 (1970).

094.159 **Search for organic compounds in the lunar dust**
from the Sea of Tranquillity.
C. Ponnamperuma, K. Kvenvolden, S. Chang, R. Johnson,
G. Pollock, D. Philpott, I. Kaplan, J. W. Schopf,
C. Gehrke, G. Hodgson, I. A. Breger, B. Halpern, A. Duffield,
K. Krauskopf, E. Barghoorn, H. Holland, K. Keil.
Science, Vol. 167, 760 - 762 (1970). – Abstr. in 'The Moon',
Vol. 1, 397 - 398 (1970).

094.160 **Search for porphyrins in lunar dust.**
G. W. Hodgson, E. Peterson, K. A. Kvenvolden,
E. Bunnenberg, B. Halpern, C. Ponnamperuma.
Science, Vol. 167, 763 - 765 (1970). – Abstr. in 'The Moon',
Vol. 1, 398 (1970).

094.161 **Organogenic elements and compounds in surface**
samples from the Sea of Tranquillity.
J. Oró, W. S. Updegrove, J. Gibert, J. McReynolds, E. Gil-av,
J. Ibanez, A. Zlatkis, D. A. Flory, R. L. Levy, C. Wolf.
Science, Vol. 167, 765 - 767 (1970). – Abstr. in 'The Moon',
Vol. 1, 398 - 399 (1970).

094.162 **Bio-organic compounds and glassy microparticles**
in lunar fines and other materials.
S. W. Fox, K. Harada, P. E. Hare, G. Hinsch, G. Mueller.
Science, Vol. 167, 767 - 770 (1970). – Abstr. in 'The Moon',
Vol. 1, 399 (1970).

094.163 **Organic compounds in lunar samples: Pyrolysis**
products, hydrocarbons, amino acids.
B. Nagy, C. M. Drew, P. B. Hamilton, V. E. Modzeleski,
M. E. Murphys, W. M. Scott, H. C. Urey, M. Young.
Science, Vol. 167, 770 - 773 (1970). – Abstr. in 'The Moon',
Vol. 1, 399 (1970).

094.164 **A search for viable organisms in a lunar sample.**
V. I. Oyama, E. L. Merek, M. P. Silverman.
Science, Vol. 167, 773 - 775 (1970). – Abstr. in 'The Moon',
Vol. 1, 400 (1970).

094.165 **Micropaleontological study of lunar material.**
E. S. Barghoorn, D. Phillpott, C. Turnbill.
Science, Vol. 167, 775 (1970). – Abstr. in 'The Moon', Vol. 1,
400 (1970).

094.166 **Micromorphology and surface characteristics of**
lunar dust and breccia.
P. Cloud, S. V. Margolis, M. Moorman, J. M. Barker,
G. R. Licari, D. Krinsley, V. E. Barnes.
Science, Vol. 167, 776 - 778 (1970). – Abstr. in 'The Moon',
Vol. 1, 400 (1970).

094.167 **Analysis of lunar material for organic compounds.**
S. R. Lipsky, R. J. Cushley, C. G. Horvath,
W. J. McMurray.
Science, Vol. 167, 778 - 779 (1970). – Abstr. in 'The Moon',
Vol. 1, 400 - 401 (1970).

094.168 **Micropaleontological studies of lunar samples.**
J. W. Schopf.
Science, Vol. 167, 779 - 780 (1970). – Abstr. in 'The Moon',
Vol. 1, 401 (1970).

094.169 **Physics and chemistry of the moon (I), (II).**
F. Zwicky.
Urania Kraków, Vol. 41, 66 - 74, 104 - 113 (1970).
In Polish.

094.170 **The distribution of lunar domes.**
K. J. Delano.
Strolling Astronomer, Vol. 22, 8 - 13 (1970).

094.171 **The A.L.P.O. Lunar Section Aristarchus-Herodotus**
mapping project: Final report. VI. The Aristarchus-
Herodotus region.   J. E. Westfall.
Strolling Astronomer, Vol. 22, 13 - 24 (1970).

094.172 **Tables pour le calcul des périgées et des apogées**
de la Lune.   J. Meeus.
Ciel et Terre, Vol. 86, 61 - 68 (1970).

094.173 **Desert stream channels resembling lunar sinuous**
rilles.   J. D. Burke, R. G. Brereton, P. M. Muller.
Nature, Vol. 225, 1234 - 1236 (1970).
Some of the sinuous rilles on the moon seem to have

been formed by fluid flow, but what the fluid was is unknown. Lava, ash, water, or combinations such as mud, or water under or over permafrost have been suggested. We have examined two of these rilles in the Deadman Lake Quadrangle approximately 12 miles north of Twentynine Palms, to find how they are formed.

094.174 Indigenous lunar methane and ethane.
P. I. Abell, G. Eglinton, J. R. Maxwell, C. T. Pillinger, J. M. Hayes.
Nature, Vol. 226, 251 - 252 (1970).

Recent analyses of Apollo 11 lunar samples have shown that methane and other gaseous hydrocarbons, ranging from C_2 to C_4, are released on treatment of the fines with aqueous HCl, H_3PO_4 or HF. It was uncertain whether these hydrocarbons were formed by acid hydrolysis or were present as such. The authors show that methane and ethane arise from both sources.

094.175 The lunar controversy – II, III.
G. J. H. McCall.
Journ. British Astron. Ass., Vol. 80, 100 - 106, 190 - 199 (1970).

094.176 Some geological interpretations of the structure of the Orientale region of the moon.
I. Ridpath, J. Murray.
Journ. British Astron. Ass., Vol. 80, 115 - 124 (1970).

Mare Orientale is the largest and most complicated basin structure on the moon. The report is based on a study of Lunar Orbiter IV photograph M-187.

094.177 Results of Apollo 11 research.
R. N. Watts, Jr.
Sky Telescope, Vol. 39, 226 - 227 (1970).

094.178 Preliminary examination of lunar samples from Apollo 12.
Science, Vol. 167, 1325 - 1339 (1970).

This is the first scientific report on the examination of the samples returned from the Apollo 12 landing at a site south-southwest of Copernicus in Oceanus Procellarum on 19 November 1969. A physical, chemical, mineralogical, and biological analysis of 34 kilograms of lunar rocks and fines is presented.

094.179 Viscosity of lunar lavas.
T. Murase, A. R. McBirney.
Science, Vol. 167, 1491 - 1493 (1970).

The viscosity of a synthetic silicate liquid with the composition of a lunar rock has been determined experimentally and found to be lower than that of any previously studied volcanic rock on earth. This fact suggests that lunar lava flows will be very thin and extensive unless they are ponded, and that lava tubes should be common and of larger dimensions than those on earth. Coarse crystallinity can be a feature of rapidly cooled surface lavas.

094.180 Bed forms in base-surge deposits: Lunar implications.
D. R. Grine, R. V. Fisher, A. C. Waters.
Science, Vol. 167, 1637 - 1638 (1970).

094.181 Alpha radioactivity of the lunar surface at the landing sites of Surveyors 5, 6, and 7.
A. L. Turkevich, J. H. Patterson, E. J. Franzgrote, K. P. Sowinski, T. E. Economou.
Science, Vol. 167, 1722 - 1724 (1970).

Evidence has been obtained for a radioactive deposit on the lunar surface at Mare Tranquillitatis with a total intensity of 0.09 ± 0.03 alpha disintegration per second per square centimeter. The presence of polonium-210 in amounts that are close to equilibrium indicates a continuous turnover rate of lunar material at this site of less than 0.1 micrometer per year. The lack of such a deposit at two other lunar sites suggests lower local concentrations of uranium there.

094.182 Rb-Sr ages of lunar rocks from the Sea of Tranquillity.
D. A. Papanastassiou, G. J. Wasserburg, D. S. Burnett.
Earth Planet. Sci. Letters, Vol. 8, 1 - 19 (1970).

This study reports the detailed results of Rb and Sr analyses on Apollo 11 samples. The approach used is that of analyzing total rock samples and separated mineral phases from each rock. This procedure avoids the necessity of assuming a genetic relationship between different rock types or between lunar rocks and meteorites.

094.183 The isotopic composition of Gd and the neutron capture effects in samples from Apollo 11.
O. Eugster, F. Tera, D. S. Burnett, G. J. Wasserburg.
Earth Planet. Sci. Letters, Vol. 8, 20 - 30 (1970).

We report here the isotopic composition and concentration of Gd extracted from ten samples of lunar material collected at Tranquillity Base during the Apollo 11 mission. The large cross-sections of ^{157}Gd and ^{155}Gd for thermal neutrons allow a determination of the total thermal neutron exposure of the samples.

094.184 Impact phenomena of micrometeorites on lunar surface material.
G. Neukum, A. Mehl, H. Fechtig, J. Zähringer.
Earth Planet. Sci. Letters, Vol. 8, 31 - 35 (1970).

The NASA lunar surface samples from the Apollo 11 mission were searched for micrometeorite impact craters by means of an optical reflecting microscope and a Stereoscan electron microscope. Craters with diameters between 2 microns and 0.8 mm were observed. Approx. 1 crater larger than 0.15 mm in diameter per 1 mm² and 1 crater larger than 2μm per 500μm² could be found. The time of exposure is calculated from crater number densities and cosmic dust influx values to be 10000 yr and 100 yr for layers of respective thicknesses.

094.185 Lunar cratering and erosion from Orbiter 5 photographs.
C. R. Chapman, J. A. Mosher, G. Simmons.
Journ. Geophys. Res., Vol. 75, 1445 - 1466 (1970).

Incremental diameter-frequency relations for twelve lunar regions are presented for four classes of craters (sharp to soft), 0.02 to 2 km diameter. Our data are consistent with this plausible (but not unique) interpretation:(1) Subsequent to extensive early blanketing, impact cratering has occurred on *highlands and old crater walls,* with a probable admixture of endogenic and secondary cratering; (2) The *maria* underwent similar processes, but have many additional endogenic craters, probably collapse features formed in lava flows. Our counts of craters by class should also be useful for refining alternative models for cratering and erosion.

094.186 On the wavelength dependence of radar echoes from the moon.
A. A. Burns.
Journ. Geophys. Res., Vol. 75, 1467 - 1482 (1970).

High delay resolution (20 μsec) radar studies of the moon at 6-, 7.5-, and 12-meter wavelengths show that the echo strength at the leading edge peak and the total radar cross section are very sensitive to the position of the subradar point. It is observed that the long wavelength normalized angular power laws for subradar points in the rough highland region are virtually identical to the short wavelength ones, as well as the previously noted 68-cm-7.84-meter agreement for those subradar points. The present and previous observations show that this is not true at other subradar points.

094.187 Lunar luminescence and neutral particles.
T. W. Snouse, D. L. Anderson.
Journ. Geophys. Res., Vol. 75, 1573 - 1578 (1970).

Samples of powdered silica and basalt have been bombarded by 3- to 17-kev atomic hydrogen beams, and the resultant luminescence has been measured. Spectral distributions of luminescence show a dependence on the energy of the bombarding ion as well as a dependence on dose. The results are discussed relative to lunar luminescence.

094.188 Lunar rilles and a possible terrestrial analogue.
G. J. H. McCall.
Nature, Vol. 225, 714 - 716 (1970).

The author proposes an alternative to the popular surface channel scouring explanation for lunar rilles: he illustrates his argument with a terrestrial analogue from Silali, Kenya, proposing that rilles are reflections of gas escape up complex fracture aligned conduits where transverse systems interfere, sinuous rilles develop.

094.189 Gaseous species in equilibrium with the Apollo 11 holocrystalline rocks during their crystallization.
T. R. Wellman.
Nature, Vol. 225, 716 - 717 (1970).

The vesicle gas of the crystallizing Apollo 11 holocrystalline rocks had an extremely low oxygen partial pressure. Hydrogen was more abundant than water and carbon monoxide more abundant than carbon dioxide. Graphite was probably not a stable phase during crystallization of these rocks.

094.190 Mineralogy and petrology of the Apollo 11 lunar sample. J. A. Wood, U. B. Marvin, B. N. Powell, J. S. Dickey, Jr.
SAO, *Cambridge, Mass.,* Special Rep., No. 307, 16 + 99 pp. (1970).

We prepared and studied thin sections of 1676 rock fragments (diameter range 1 to 5 mm) from the Apollo 11 bulk sample. In almost all cases, the rocks are fine-grained enough for fragments of this size to constitute representative samples. We found the following proportions of rock types, evidently a mixture from many sources: soil breccias, 52.4%; basalts, 37.4%; glasses, 5.1%; anorthositic rocks, 3.6%; others (including less than 0.1% recognizable meteoritic debris), 1.5%. A lunar structural model is proposed in which a 25-km anorthosite crust, produced by magmatic fractionation, floats on denser gabbro. Where early major impacts punched through the crust, basaltic lava welled up to equilibrium surface levels and solidified, forming the maria. An explanation for mascons is offered.

094.191 Positions of lunar features from Apollo 8.
R. Nance.
Journ. Geophys. Res., Vol. 75, 2029 - 2034 (1970).

From Apollo 8 data, positions were obtained for three lunar far-side features. The positions may be used to improve the accuracy of lunar far-side maps. Based on these positions, the best lunar far-side maps are estimated to be accurate to within 60 km in latitude and longitude.

094.192 Mikroskopische Untersuchung der Mondproben des Apollo 11-Unternehmens.
P. Ramdohr, A. El Goresy.
Naturwissenschaften, Vol. 57, 98 - 106 (1970).

094.193 Altersbestimmungen am Mondgestein.
J. Zähringer.
SuW, Vol. 9, 117 - 121 (1970).

094.194 Colour photography of the moon and of the earth from outer space. V. D. Bolshakov, N. P. Lavrova.

Priroda, No. 1.70, p. 66 - 69 (1970).   In Russian.

094.195 The nature of the seismic echo on the moon.
A. K. Mukhamedzhanov.
Priroda, No. 3.70, p. 74 - 75 (1970).   In Russian.

094.196 The mascons of the depressions of lunar seas.
M. S. Markov, A. L. Sukhanov.
Priroda, No. 3.70, p. 84 - 87 (1970).   In Russian.

094.197 A red spot inside the lunar crater Aristarchus observed on April 1, 1969. N. A. Kozyrev.
Astron. Zhurn. Akad. Nauk SSSR, Vol. 47, 179 - 181 (1970). In Russian. English translation in Soviet Astron.,AJ, Vol. 14, No. 1.

In the evening on April 1, 1969 a series of spectrograms of Aristarchus was obtained at the Crimean Astrophysical Observatory by means of a spectrograph with small dispersion (500 Å at Hα) on a scale of 34″ per mm on the plate. The spectrum of a red spot, located on the internal western slope of the crater, has been detected. The measurements show that this spectrum consists mainly of broad emission which can be identified reliably with the red system of CN molecules. Narrow emission features which correspond, apparently, to the border of bands of the first positive group of nitrogen molecules N_2 are noticed on this spectrogram as well.

094.198 Anelasticity and lunar seismology.
E. Strick.
Earth Planet. Sci. Letters, Vol. 8, 229 - 233 (1970).

Derived relations from the author's power-law (PL) model for the anelastic behavior of sedimentary earth rock are presented for possible use in understanding the long-time response behavior of lunar seismograms obtained in the Apollo 12 experiment.

094.199 Anelasticity and the Apollo 12 LP lunar seismogram. E. Strick.
Earth Planet. Sci. Letters, Vol. 8, 234 - 236 (1970).

Further refined arguments are given to support the author's low-Q regolith layer model as deduced from the Apollo 12 LP seismogram. It is also shown that this model can yield the same response envelope as does a high-Q model based upon the diffusion equation that has been suggested by Latham et al.

094.200 On the difference in the laws of light reflection from moon and Mars.
N. P. Barabashov, L. A. Akimov.
Astron. Tsirk., No. 540, p. 2 - 4 (1969).   In Russian.

094.201 Apollo 13 lunar heat flow experiment.
M. G. Langseth, Jr., A. E. Wechsler, E. M. Drake, G. Simmons, S. P. Clark, Jr., J. Chute, Jr.
Science, Vol. 168, 211 - 217 (1970).

Direct measurement of the heat escaping from the lunar interior should be made during Apollo 13.

094.202 Missile impacts as sources of seismic energy on the moon. G. V. Latham, W. G. McDonald, H. J. Moore.
Science, Vol. 168, 242 - 245 (1970).

Seismic signals recorded from impacts of missiles at the White Sands Missile Range are radically different from the signal recorded from the Apollo 12 lunar module impact. This implies that lunar structure to depths of at least 10 to 20 kilometers is quite different from the typical structure of the earth's crust. Results obtained from this study can be used to predict seismic wave amplitudes from future man-made lunar impacts. Seismic energy and crater dimensions from impacts are compared with measurements from chemical explosions.

094.203 Lunar gravity over large craters from Apollo 12 tracking data. P. Gottlieb, P. M. Muller, W. L. Sjogren, W. R. Wollenhaupt.
Science, Vol. 168, 477 - 479 (1970).

The Doppler residuals from the Apollo 12 lunar module radio tracking data indicate large negative accelerations over the craters Ptolemaeus and Albategnius. The mass deficiencies required to produce these accelerations are approximately equivalent to the removal of the surface material to a depth of 1 kilometer over the entire area of these craters. Several other features of the gravity fine structure can also be correlated with topography.

094.204 Radiation effects and oxygen vacancies in silicates. A. Chatelain, J. L. Kolopus, R. A. Weeks.
Science, Vol. 168, 570 - 571 (1970).

Processes are described which may contribute to the erosion and transport of lunar surface materials.

094.205 Radioactivity induced in Apollo 11 lunar surface material by solar flare protons.
H. R. Heydegger, A. Turkevich.
Science, Vol. 168, 575 - 576 (1970).

Comparison of values of the specific radioactivities reported for lunar surface material from the Apollo 11 mission with analogous data for stone meteorites suggests that energetic particles from the solar flare of 12 April 1969 may have produced most of the cobalt-56 observed.

094.206 Origin of glass deposits in lunar craters.
J. Green, E. D. Dietz, P. J. Vergano, W. R. Greenwood, G. Heiken, T. Gold.
Science, Vol. 168, 608 - 611 (1970).

094.207 Problems of astronomical navigation on the lunar surface. J. Mietelski.
Artificial Satellites Polish Acad. Sci., Vol. 5, No. 1, p. 3 - 11 (1969).

A method of determination of the observer's position on the lunar surface is proposed and suggestions are made pertaining to the nautical ephemeris for the moon. Also the question of the number of navigation stars is discussed.

094.208 Construction of a hypsometric map of the visible side of the moon. Yu. N. Lipsky, V. A. Nikonov.
Astron. Zhurn. Akad. Nauk SSSR, Vol. 47, 407 - 410 (1970). In Russian. – English translation in Soviet Astron., AJ, Vol. 14, No. 2.

On the basis of Hopmann's catalogue (1052 points) and the topographic map compiled by the Army Map Service a contour map of the visible lunar side is worked out.

094.209 On the problem of the translational-rotational motion of the moon under the action of the earth.
G. F. Osipov.
Astron. Zhurn. Akad. Nauk SSSR, Vol. 47, 420 - 425 (1970). In Russian. – English translation in Soviet Astron., AJ, Vol. 14, No. 2.

The translational-rotational motion of the axisymmetric moon in the spherical field of the earth is investigated. The solution close to the regular "fluid" solution is obtained by Poincaré's method of variation of conditions. The initial conditions of the disturbed motion correspond to Cassini's laws and the classical formulas of the elliptic motion.

094.210 Atmospheric pressure on the moon from the data of twilight photography with Surveyor VII.
G. V. Rozenberg.

Astron. Zhurn. Akad. Nauk SSSR, Vol. 47, 449 - 452 (1970). In Russian. – English translation in Soviet Astron., AJ, Vol. 14, No. 2.

On the basis of the analysis of the image of the twilight horizon, transmitted by Surveyor VII from the lunar surface, an upper estimation of the air density at the height of about 10 m above the lunar surface is obtained.

094.211 Intermediate scale lunar roughness.
J. A. Bastin, D. O. Gough.
Icarus, Vol. 11, 289 - 319 (1969).

A model surface consisting of parallel troughs has been examined in order to assess the importance of roughness in accounting for the properties of lunar thermal radiation. Insolation, emission from the surface, reabsorption of emitted radiation, and conduction are all considered. Brightness temperatures both in the midinfrared and microwave region are computed for eclipse and lunation conditions, not only as a function of lunar phase, latitude, longitude, and direction of observation, but also for a variety of trough dimensions. All those features of the observed thermal radiation which cannot be accounted for on the basis of a plane homogeneous model are listed and the extent to which they can be accounted for by the proposed model is considered.

094.212 Absolute ages of the lunar maria and large craters.
R. B. Baldwin.
Icarus, Vol. 11, 320 - 331 (1969).

By an analysis of the changes in rim heights of lunar craters through isostatic settling, combined with inferences from counts of small impact craters contained within and subsequent to large craters, it is concluded that: (1) the maria are not older than about 640 million years; and (2) the flux of masses which produced craters larger than 160 km in diameter reached a peak rate at a maximum age of about 2.5 billion years ago, and not very early in the moon's history as has usually been thought.

094.213 The effect of pulverization on the albedo of lunar rocks. M. G. J. Minnaert.
Icarus, Vol. 11, 332 - 337 (1969).

Measures of the albedo under full-moon conditions have been made on two samples of very dark rocks, pulverized and sieved so as to obtain powders of different grain size. Below a size of 0.05 mm the albedo suddenly increases, obviously because the individual grains become transparent. By a rough calculation of the radiation transfer this is made understandable and a relation is found between the grain size and the absorption coefficient necessary to produce in combination an albedo of 0.10.

094.214 Stoßwelleneffekte in den Mondproben.
W. F. Müller.
Umschau, 70. Jahrgang, p. 331 - 335 (1970).

094.215 Problems of selenodesy. K. Kozieł.
Postępy Astron., Vol. 18, 139 - 147 (1970).
In Polish.

The article contains a review of investigations on the determination of the libration constants of the moon from the times of Bessel up to present. The libration constants determined by the author and the coordinates of the crater Mösting A are given.

094.216 A new homogeneous reduction of four libration series of the moon. K. Kozieł.
Postępy Astron., Vol. 18, 189 - 191 (1970). In Polish.
Abstr. Polish Astron. Soc.

094.217 Method to determine sense and magnitude of electric field from lunar particle shadows.

K. A. Anderson.
Journ. Geophys. Res., Vol. 75, 2591 - 2594 (1970). – Letter.

094.218 **Determination of lunar gravitational harmonics from secular and long-period effects of lunar satellites.** P. A. Laing, A. Liu.
Bull. American Astron. Soc., Vol. 2, 204 (1970). – Abstr. AAS.

094.219 **Isotopenanalysen an Mondproben.** T. Kirsten.
Naturwissenschaften, Vol. 57, 236 - 239 (1970).

094.220 **Apollo 11 evidence for the differentiation of lunar materials.** R. F. Mueller.
Nature, Vol. 226, 925 - 927 (1970).

The compositions of the surface rock at Tranquillity Base seem to be consistent with a differentiation mechanism, perhaps of a complex character embodying igneous and other processes, but which probably contained elements similar to that which produced the crust of the earth.

094.221 **Étude de la luminescence lunaire sur la lune pendant une mission Apollo.** F. Link.
The Moon, Vol. 1, 338 (1970). – Research note.

094.222 **Physical libration of the moon in longitude.** M. Moutsoulas.
The Moon, Vol. 1, 347 - 351 (1970).

The application of modern computing techniques in the study of the physical libration of the moon in longitude brings into a new perspective this problem that has been debated so much in the past.

094.223 **The moon after satellites Luna 10 and Lunar Orbiter I, III, IV.** M. Caputo, G. F. Panza.
Symposia Mathematica, Vol. 3, 97 - 119 (1970). – Conference paper (see 012.010).

094.224 **Bisherige Ergebnisse der Mondlandungen I. 1. Mondoberfläche und Mondgestein.**
E. Stuhlinger.
Phys. Blätter, 26. Jahrgang, p. 151 - 157 (1970).

094.225 **Apollo 11 photography. 70-mm, 16-mm, and 35-mm frame index.**
Prepared by the Mapping Sciences Laboratory at the NASA Manned Spacecraft Center, Houston, Texas.
National Space Science Data Center, NSSDC 70-02, NASA, Greenbelt, Md. 131 pp. (1970).

094.226 **Apollo 11 70-mm photographic catalog.**
National Space Science Data Center, NSSDC 70-07, NASA, Greenbelt, Md. 230 pp. (1970).

094.227 **Apollo 11 lunar photography (NSSDC ID No. 69-059A-01).**
Prepared by: A. T. Anderson, C. K. Michlovitz, K. Hug.
Data Users' Note, NSSDC 70-06, National Space Science Data Center, NASA, Goddard Space Flight Center, Greenbelt, Md. 5 + 11 + A8 pp. (1970).

094.228 **Apollo mission 11. Photography indexes.**
Prepared under the direction of the Department of Defense by the Aeronautical Chart and Information Center, United States Air Force for the NASA, Greenbelt, Maryland. 4 sheets (1969).

094.229 **Geodetic measurements with lunar laser-ranging.**
O. Calame, B. Guinot, J. Kovalevsky, A. Orszag.
Astron. Astrophys., Vol. 4, 18 - 30 (1970). In French.
An analytic work has been undertaken in order to eva-

luate the possibility of precise determination of the various parameters involved in ranging a fiducial point of the moon. Ranging is to be effected by measuring the round trip time of a light pulse reflected from a lunar cube corners-panel. Using an accurate lunar ephemeris and estimated values for pertinent parameters will yield a theoretical value of observatory to retroreflectors distance. The analytical expression giving the distance have been computed and derived.

094.230 **The position of the moon.** J. S. Griffith.
Journ. Roy. Astron. Soc. Canada, Vol. 64, 5 - 7 (1970).

Report on the new analytic lunar theory at the present time in preparation (Eckert and Smith).

094.231 **The morphology, distribution, and origin of lunar sinuous rilles.**
G. Schubert, R. E. Lingenfelter, S. J. Peale.
Rev. Geophys. Space Phys., Vol. 8, 199 - 224 (1970).

The morphology of the lunar sinuous rilles requires that they be features of surface water erosion. We have shown that such erosion can take place in the present lunar environment and have proposed that water outgassed from the lunar interior and trapped beneath a permafrost could have been released by meteoritic impact, faulting, or volcanism to erode a rille. The distribution of these rilles thus indicates the past and perhaps present distribution of subsurface water and other volatiles. The concentration of these rilles on the margins of the circular mare basins suggests that the large impacts that formed these mare basins directly caused or strongly influenced the outgassing of volatiles from the lunar interior.

094.232 **Rb–Sr ages from the Ocean of Storms.**
D. A. Papanastassiou, G. J. Wasserburg.
Earth Planet. Sci. Letters, Vol. 8, 269 - 278 (1970).

Rb-Sr internal isochrons for two texturally and mineralogically distinct crystalline rocks returned by Apollo 12 yield similar ages of 3.36 ± 0.10 and $3.26 \pm 0.10 \times 10^9$ years. The initial Sr compositions determined are relatively primitive but distinct so that these samples represent two different rock bodies. These ages are slightly but distinctly younger than the Mare Tranquillitatis Rb-Sr ages and indicate that the flooding of the maria may have occurred only during a short interval of $\sim 300 \times 10^6$ years duration 3.6×10^9 years ago. The soil at the Ocean of Storms yields a well determined model age of 4.44×10^9 years and therefore the special nature and older age of the lunar soil determined at the Apollo 11 site is found to be a widespread phenomenon.

094.233 **A model for small-impact erosion applied to the lunar surface.** L. A. Soderblom.
Journ. Geophys. Res., Vol. 75, 2655 - 2661 (1970).

A model for erosion of the lunar surface by impact of small projectiles is developed that provides an analytic representation of the change of crater shape as a function of time. The model is applied to the erosion of craters approximately 1 meter to 1 km in diameter. The lifetime of a crater in this size range, which is steadily eroded by impact, is approximately proportional to its radius. The model predicts the observed steady-state size frequency distribution of small lunar craters and the dependence of this distribution on the crater-production curve.

094.234 **Infrared emissivity of lunar surface features. 1. Balloon-borne observations.**
F. H. Murcray, D. G. Murcray, W. J. Williams.
Journ. Geophys. Res., Vol. 75, 2662 - 2669 (1970).

The thermal emission spectrums (7.0 to 13.5 μ) of six selected areas on the lunar surface were measured from an altitude of 32 km and their spectral emissivities calculated. All spectrums show departures from black- or gray-body emission.

Emissivities in the 8.0- to 9.0-μ region of the spectrum are significantly higher than those for wavelengths greater than 10 μ. Differences are noted in the wavelengths of peak emissivity, particularly between the highland areas and the maria.

094.235 Infrared emissivity of lunar surface features.
2. Interpretation. J. W. Salisbury, R. K. Vincent,
L. M. Logan, G. R. Hunt.
Journ. Geophys. Res., Vol. 75, 2671 - 2682 (1970).
The purpose of this paper is to interpret the lunar spectra obtained by the University of Denver in the light of our laboratory studies, to compare these conclusions with those drawn from other compositional experiments, and to explore the geological implications of the results.

094.236 Chemical composition of the lunar surface in a terra
region near the crater Tycho.
J. H. Patterson, A. L. Turkevich, E. J. Franzgrote, T. E. Economou, K. P. Sowinski.
Science, Vol. 168, 825 - 828 (1970).
More precise and comprehensive analytical results for lunar surface material in a terra region have been derived from the data of the alpha-scattering experiment on Surveyor 7. The silicon content and the low sodium abundance are close to that of mare material. The abundances of titanium and iron are at least a factor of 2 lower, whereas the abundances of aluminum and calcium are significantly higher.

094.237 One-way radar range to the moon.
G. C. McVittie, P. L. Bender.
Science, Vol. 168, 1012 (1970).

094.238 Transient lunar phenomena and electrostatic
glow discharges. A. A. Mills.
Nature, Vol. 225, 929 - 930 (1970).
The observed properties of transient lunar phenomena (particularly those on the dark side of the moon) do not appear to be satisfactorily explained by previously suggested mechanisms such as proton luminescence or thermoluminescence. It is therefore proposed that obscurations result from the disturbance of fine dust by gases emitted from the interior of the moon. Occasionally, tidal stresses promote the release of sufficient accumulated gas to generate electrostatic glow discharges within the transient gas phase.

094.239 Reaction of carbon and sulphur isotopes in
Apollo 11 samples with solar hydrogen atoms.
R. Berger.
Nature, Vol. 226, 738 - 739 (1970).
Analyses of carbon and sulfur isotopes in lunar material returned from the Apollo 11 landing site in the Mare Tranquillitatis suggest that the solar wind reacts with exposed carbon and sulfur isotopes such that preferentially the heavier isotopes are retained. The mechanisms of carbon deposition on the moon and the effect of solar wind erosion are discussed.

094.240 Measurements of lunar radiation in the wavelength
range centred at 1.2 mm.
P. E. Clegg, B. S. Carter.
Monthly Notices, Roy. Astron. Soc., Vol. 148, 261 - 274 (1970).
The results of several investigations of the lunar surface thermal properties are reported including observational and theoretical considerations of the effect of roughness on the polarization of this radiation. An effective dielectric constant for the surface material is deduced and a tentative model for the centimetre scale roughness of the surface advanced.

094.241 Âge des roches lunaires et terrestres.
J. Dulemba.

L'Astronomie, 84e année, p. 223 - 224 (1970).

094.242 Bisherige Ergebnisse der Mondlandungen II.
Instrumente und Messungen.
E. Stuhlinger.
Phys. Blätter, 26. Jahrgang, p. 200 - 207 (1970).

094.243 The lunar controversy – IV.
G. J. H. McCall.
Journ. British Astron. Ass., Vol. 80, 263 - 269 (1970).

094.244 Preliminary note on lunar cratering rates and
absolute time-scales. W. K. Hartmann.
Icarus, Vol. 12, 131 - 133 (1970).
Solidification ages from Mare Tranquillitatis (Apollo 11) are used to compute the lunar cratering rate over the last 3.9 aeons. Results agree with estimates from more recent terrestrial cratering.

094.245 Excerpts from NASA description of Apollo 12
through 20. J. E. Naugle.
Icarus, Vol. 12, 134 - 139 (1970).

094.246 Lunar radar mapping: Correlation between radar
reflectivity and stratigraphy in north-western Mare
Imbrium.
G. G. Schaber, R. E. Eggleton, T. W. Thompson.
Nature, Vol. 226, 1236 - 1239 (1970).
In this letter, the "polarized" echo represents the power in circular polarization which is opposite in direction to the transmitted signal. These radar echoes are interpreted in a simple, two-component model.

094.247 Premiers échos lumineux sur la lune obtenus par
le télémètre laser du Pic du Midi.
O. Calame, M.-J. Fillol, G. Guérault, R. Muller, A. Orszag,
J.-C. Pourny, J. Rösch, Y. de Valence.
Comptes Rendus Acad. Sci. Paris, Sér. B, Vol. 270, 1637 - 1640 (1970).
Nous relatons ici les premiers échos obtenus avec l'ensemble de télémétrie laser de l'Observatoire du Pic du Midi, visant les cataphotes déposés sur la lune par Apollo XI.

094.248 Anomalous propagation of elastic energy within
the moon. B. I. Pandit, D. C. Tozer.
Nature, Vol. 226, 335 (1970).
The anomalous seismic data from the moon may imply a high Q and intense wave scattering for its near surface region. Some support for a 'welded' aggregate which might show these characteristics is found in our experiments in which Q of porous terrestrial rocks consistently increases (~5 fold) when placed under vacuum (10^{-2} torr); and, in those reported earlier which show that surfaces of lunar material cleft under ultra high vacuum adhere when placed in contact.

094.249 The A.L.P.O. Lunar Section Aristarchus-Herodotus
mapping project: Final report. J. E. Westfall.
Strolling Astronomer, Vol. 22, 46 - 63 (1970).

094.250 La surface lunaire – Origine de la lune.
M. Fluckiger.
Orion Schaffhausen, 28. Jahrgang, p. 74 - 77 (1970).

094.251 The principles of the physical mapping of the
lunar surface. Yu. N. Lipsky, V. V. Shevchenko.
Astron. Zhurn. Akad. Nauk SSSR, Vol. 47, 586 - 598 (1970).
In Russian. English translation in Soviet Astron., AJ, Vol. 14, No. 3.
The construction of cartographic pictures of the whole lunar surface or separate regions, containing information on

the presence and distribution of various physical characteristics of the lunar cover can be most effective for analyzing diverse informations.

094.252 A physical mapping of the moon from photometric data.   V. V. Shevchenko.
Astron. Zhurn. Akad. Nauk SSSR, Vol. 47, 599 - 609 (1970). In Russian. English translation in Soviet Astron.,AJ, Vol. 14, No. 3.
Problems of compiling and the interpretation of photometric maps of the lunar surface are stated. On the basis of B. Hapke's "improved" formula systems of equations for 403 parts of the surface are compiled. As results of the solution of the system of equations, maps of the distribution of photometric parameters are compiled and the typological division into regions of the surface is carried out.

094.253 Systematic and accidental differences of selenodetic base networks.   V. S. Kisljuk.
Astron. Zhurn. Akad. Nauk SSSR, Vol. 47, 610 - 612 (1970). In Russian. English translation in Soviet Astron.,AJ, Vol. 14, No. 3.
The differences of the system of modern catalogues of basic points on the moon's surface are given. The dispersion of the positions of these points are estimated for different pairs of catalogues. The correlation between values of absolute heights of common points has been studied.

094.254 On the orientation of C.B. Watts' charts of the marginal zone of the moon.   L. V. Morrison.
Monthly Notices, Roy. Astron. Soc., Vol. 149, 81 - 90 (1970).
It is shown, from an analysis of occultations of stars by the moon, that limb profiles from Watts' charts accord better with observation when the tabular position angles of the charts are amended systematically by $-0\overset{\prime}{.}25$ ($\pm$ $0\overset{\prime}{.}01$).

094.255 Tektite glass in Apollo 12 sample.   J. A. O'Keefe.
Science, Vol. 168, 1209 - 1210 (1970).
The glassy portion of lunar sample 12013 from Apollo 12 is chemically more like some tektites from Java than like any terrestrial igneous rock. It satisfies all the chemical criteria for a tektite. Tektites are relatively recent and acid rocks, whereas the moon is chiefly ancient and basaltic; hence, tektites are probably ejected volcanically, rather than by impact, from the moon.

094.256 Chemical individuality of lunar, meteoritic, and terrestrial silicate rocks.   M. C. Ulbrich.
Science, Vol. 168, 1375 - 1376 (1970).

094.257 Lunar ephemeris: Delaunay's theory revisited.   A. Deprit, J. Henrard, A. Rom.
Science, Vol. 168, 1569 - 1570 (1970).
Delaunay's reduced Hamiltonian of the main problem in lunar theory is checked against a new analytical theory based on Lie transforms. It is found to be correct up to order 9 with the exception of one error in addition at order 7.

094.258 Properties and composition of lunar materials: Earth analogies.   E. Schreiber, O. L. Anderson.
Science, Vol. 168, 1579 - 1580 (1970).
The sound velocity data for the lunar rocks were compared to numerous terrestrial rock types and were found to deviate widely from them. A group of terrestrial materials were found which have velocities comparable to those of the lunar rocks, but they do obey velocity-density relations proposed for earth rocks.

094.259 Massaknobbels in de maan.

M. Minnaert.
Hemel en Dampkring, Vol. 68, 127 - 130 (1970).

094.260 The secular acceleration of the moon.   T. C.Van Flandern.
Astron. Journ., Vol. 75, 657 - 658 (1970).
The secular acceleration of the moon over the period 1955–1969, as determined from occultation observations, differs substantially from the value assumed in current lunar theories.

094.261 Low-resolution differential drift scans of the moon at 22 microns.   W. W. Mendell, F. J. Low.
Journ. Geophys. Res., Vol. 75, 3319 - 3324 (1970).
Differential drift scans were made across the 3-day-old moon at wavelength of 22 μ with a beamwidth of 2.4 arc min. Certain scans were processed to reveal the temperature distribution from the cold limb to the sunrise terminator.

094.262 On the fissions of a solid body under influence of tidal force; with application to the problem of twin craters on the moon.   N. Sekiguchi.
The Moon, Vol. 1, 429 - 439 (1970).
The internal strain due to the tidal force in the proximity of a tide-generating body (in the present case, the moon) is calculated according to the Lord Kelvin theory of earth tides. The conditions for which a uniform elastic sphere possessing a definite tensile strength is crushed near the surface of the moon is investigated. Many lunar features, such as twin craters, craterous walled plains of irregular forms, compound craters, may be explained by fission of the meteoritic material before impact.

094.263 Localisation of lunar acoustic energy.   J. A. Bastin.
The Moon, Vol. 1, 449 - 450 (1970).
The author gives an explanation for the long reverberation time measured at the Apollo 12 site following the closeby explosion of the lunar module.

094.264 On the depth of the lunar regolith.   Z. Kopal.
The Moon, Vol. 1, 451 - 461 (1970).
The aim of this paper is to point out that if the sinuous rilles on the moon represent trenches in the mare ground in which they meander, the existence of a great number of individual boulders on their slopes – as discovered on the high-resolution photographs taken by US Lunar Orbiters 4 and 5 in 1967 – suggests that the solid substrate of the lunar globe is covered by broken-up debris produced by cosmic abrasion – and hereafter referred to as lunar regolith – of thickness comparable with the depth of the respective rilles – at least of those lacking flat floors; which is generally in the order of 200 - 300 m.

094.265 The role of accretionally and electrically inverted thermal profiles in lunar evolution.
C. P. Sonett, D. S. Colburn.
The Moon, Vol. 1, 483 - 484 (1970). – Abstract of a conference paper (see 011.031).

094.266 Moments of inertia of the moon.   W. H. Michael, Jr.
The Moon, Vol. 1, 484 - 485 (1970). – Abstract of a conference paper (see 011.031).

094.267 Lunar mass concentrations.   W. M. Kaula.
The Moon, Vol. 1, 485 - 486 (1970).– Abstract of a conference paper (see 011.031).

094.268 Results from Apollo passive seismic experiment.   G. V. Latham.

The Moon, Vol. 1, 486 - 487 (1970). – Abstract of a conference paper (see 011.031).

094.269 **Implications of the lunar meteoroid environment derived from Apollo age dating results.**
D. E. Gault.
The Moon, Vol. 1, 493 - 494 (1970). – Abstract of a conference paper (see 011.031).

094.270 **Is the moon a pre-planetary object?** H. C. Urey.
The Moon, Vol. 1, 494 - 495 (1970). – Abstract of a conference paper (see 011.031).

094.271 **The handling and analysis of the Apollo lunar samples.** R. D. Bell.
The Moon, Vol. 1, 497 (1970). – Abstract of a conference paper (see 011.032).

094.272 **The petrology and mineralogy of Apollo 11 moon rocks.** G. M. Brown.
The Moon, Vol. 1, 497 - 498 (1970). – Abstract of a conference paper (see 011.032).

094.273 **Some speculations on lunar chemistry and composition.**
K. Randle, G. G. Goles, D. J. Lindstrom.
The Moon, Vol. 1, 498 (1970). – Abstract of a conference paper (see 011.032).

094.274 **Magnetic properties of lunar samples.**
D. W. Collinson, W. O'Reilly, S. K. Runcorn, A. Stephenson.
The Moon, Vol. 1, 498 (1970). - Abstract of a conference paper (see 011.032).

094.275 **Age determination and isotope effects in lunar material.** G. Turner.
The Moon, Vol. 1, 499 (1970). – Abstract of a conference paper (see 011.032).

094.276 **Lunar petrogenesis.** B. J. Wood, R. G. J. Strens.
The Moon, Vol. 1, 499 (1970). – Abstract of a conference paper (see 011.032).

094.277 **Shock effects in lunar rocks and minerals.**
W. v. Engelhardt, H. Arndt, W. Müller, D. Stöffler.
The Moon, Vol. 1, 499 - 500 (1970). – Abstract of a conference paper (see 011.032).

094.278 **The nature of seas, mascons and the lunar interior in the light of experimental studies.**
M. J. O'Hara, G. M. Biggar, S. W. Richardson, C. E. Ford, B. G. Jamieson.
The Moon, Vol. 1, 500 (1970). – Abstract of a conference paper (see 011.032).

094.279 **Nuclear track studies of ancient solar radiations and dynamic lunar surface processes.**
G. Crozaz, U. Haack, M. Hair, M. Maurette, R. Walker, D. Woolum.
The Moon, Vol. 1, 501 (1970). – Abstract of a conference paper (see 011.032).

094.280 **Solid state properties of lunar materials.**
R. Walker, H. Hoyt.
The Moon, Vol. 1, 501 (1970). – Abstract of a conference paper (see 011.032).

094.281 **Elements of the moon.** A. A. Smales.
The Moon, Vol. 1, 501 (1970). – Abstract of a conference paper (see 011.032).

094.282 **Analysis of first returned lunar samples by techniques based on Mössbauer studies of Apollo 11 lunar samples.** C. L. Herzenberg, D. L. Riley.
The Moon, Vol. 1, 501 - 502 (1970). – Abstract of a conference paper (see 011.032).

094.283 **Mössbauer studies of Apollo 11 lunar samples.**
N. N. Greenwood, A. T. Howe.
The Moon, Vol. 1, 502 (1970). – Abstract of a conference paper (see 011.032).

094.284 **Methods and potentialities of the study of lunar samples by magnetic resonance.** E. E. Schneider.
The Moon, Vol. 1, 502 (1970). – Abstract of a conference paper (see 011.032).

094.285 **Gravity measured at the Apollo 12 landing site.**
R. L. Sjogren.
The Moon, Vol. 1, 502 - 503 (1970). – Abstract of a conference paper (see 011.032).

094.286 **The lunar gravity field.**
W. L. Sjogren.
The Moon, Vol. 1, 503 (1970). – Abstract of a conference paper (see 011.032).

094.287 **The moon's physical librations.** H. Jeffreys.
The Moon, Vol. 1, 503 (1970). – Abstract of a conference paper (see 011.032).

094.288 **The physical libration of the moon.**
M. Moutsoulas.
The Moon, Vol. 1, 503 - 504 (1970). – Abstract of a conference paper (see 011.032).

094.289 **Lunar data from occultation observations.**
L. V. Morrison.
The Moon, Vol. 1, 504 (1970). – Abstract of a conference paper (see 011.032).

094.290 **The question of isostasy on the moon.**
R. B. Baldwin.
The Moon, Vol. 1, 504 (1970). – Abstract of a conference paper (see 011.032).

094.291 **Interpretation of the lunar gravitational field.**
W. M. Kaula.
The Moon, Vol. 1, 504 - 505 (1970). – Abstract of a conference paper (see 011.032).

094.292 **The figure of the moon.** S. K. Runcorn.
The Moon, Vol. 1, 505 (1970). – Abstract of a conference paper (see 011.032).

094.293 **Mascons and the moon's spherical harmonic gravity coefficients.** J. R. Booker, R. L. Kovach.
The Moon, Vol. 1, 505 - 506 (1970). – Abstract of a conference paper (see 011.032).

094.294 **Volcanism and internal activity in the moon and Mars.** G. Fielder.
The Moon, Vol. 1, 506 (1970). – Abstract of a conference paper (see 011.032).

094.295 **A fluidization hypothesis for the origin of craters.**
A. A. Mills.
The Moon, Vol. 1, 506 (1970). – Abstract of a conference paper (see 011.032).

094.296 **Impact fluidization: Further modes of crater formation.** I. F. Ferguson.

The Moon, Vol. 1, 506 - 507 (1970). – Abstract of a conference paper (see 011.032).

094.297 Large boulders and rock strata in the bright-rayed lunar craters. D. Buhl.
The Moon, Vol. 1, 507 (1970). – Abstract of a conference paper (see 011.032).

094.298 Statistical aspects of the formation and evolution of lunar rocks. A. H. Marcus.
The Moon, Vol. 1, 508 (1970). – Abstract of a conference paper (see 011.032).

094.299 Meteoroid environment and age relationships for the lunar and Martian surfaces. D. E. Gault.
The Moon, Vol. 1, 508 - 509 (1970). – Abstract of a conference paper (see 011.032).

094.300 Size of rocks on the lunar surface. M. J. Gross.
The Moon, Vol. 1, 509 (1970). – Abstract of a conference paper (see 011.032).

094.301 Localisation of lunar seismic energy. J. A. Bastin.
The Moon, Vol. 1, 509 (1970). – Abstract of a conference paper (see 011.032).

094.302 The early history of the moon. H. Gerstenkorn.
The Moon, Vol. 1, 509 (1970). – Abstract of a conference paper (see 011.032).

094.303 Orbit-orbit coupling and the history of the earth-moon system. R. G. Hipkin.
The Moon, Vol. 1, 509 - 510 (1970). – Abstract of a conference paper (see 011.032).

094.304 An origin of the moon compatible with its present condition. P. R. Bell.
The Moon, Vol. 1, 510 (1970). – Abstract of a conference paper (see 011.032).

094.305 Electrical conductivity of the moon: Recent results. C. P. Sonett.
The Moon, Vol. 1, 510 (1970). – Abstract of a conference paper (see 011.032).

094.306 Solar wind flow past moon and lunar electrical conductivity. N. F. Ness.
The Moon, Vol. 1, 510 - 511 (1970). – Abstract of a conference paper (see 011.032).

094.307 Laboratory simulation of solar wind interaction with the moon. L. J. Srnka.
The Moon, Vol. 1, 511 (1970). – Abstract of a conference paper (see 011.032).

094.308 Mascons, mare rock and isostasy. K. A. Howard.
Nature, Vol. 226, 924 - 925 (1970).
Lunar mascons, which represent nonisostatic conditions in large impact basins, may be thick fills of dense mare volcanic rock. On earth, imbalences similarly caused by volcanic flooding eventually rectify themselves through isostatic creep, but the moon has been rigid at least since the time of mare flooding. To explain why some old basins have no mascons, it is postulated that the lunar rigidity was lower early in its history, so that the oldest impact basins had rebounded isostatically before flooding, but the younger ones were still deep and so were deeply flooded to excess of isostasy.

094.309 On the methods of determination of the figure and gravitational field of the moon.
N. A. Chujkova.

Vestn. Mosk. un-ta. Fiz., astron., 1969, No. 5, p. 81 - 88. In Russian. – Abstr. in Referativ. Zhurn. 51. Astron., 4.51.294 (1970).

094.310 Technique of mapping and a legend for geologic-morphological maps of the moon on a scale of 1 : 1000000. A. L. Sukhanov, V. G. Trifonov.
(Trudy) Geol. in-t AN SSSR, 1969, vyp. (No.) 204, p. 11 - 36. In Russian. – Abstr. in Referativ. Zhurn. 62. Issled. kosm. prostranstv., 5.62.188; 51. Astron., 6.51.303 (1970).

094.311 Old maria of the moon. A. L. Sukhanov.
(Trudy) Geol. in-t AN SSSR, 1969, vyp. (No.) 204, p. 147 - 164. In Russian. – Abstr. in Referativ. Zhurn. 62. Issled. kosm. prostranstv., 5.62.193; 51. Astron., 6.51.304 (1970).

094.312 Some peculiarities of small craters of the moon. A. L. Sukhanov.
(Trudy) Geol. in-t AN SSSR, 1969, vyp. (No.) 204, p. 188 - 205. In Russian. – Abstr. in Referativ. Zhurn. 62. Issled. kosm. prostranstv., 5.62.195; 51. Astron., 6.51.305 (1970).

094.313 A comparison of the lunar topography with terrestrial volcanic formations. V. G. Trifonov.
(Trudy) Geol. in-t AN SSSR, 1969, vyp. (No.) 204, p. 229 - 243. In Russian. – Abstr. in Referativ. Zhurn. 62. Issled. kosm. prostranstv., 5.62.197; 51. Astron., 6.51.306 (1970).

094.314 Volcanic ridges on the moon. A. L. Sukhanov.
(Trudy) Geol. in-t AN SSSR, 1969, vyp. (No.) 204, p. 244 - 261. In Russian. – Abstr. in Referativ. Zhurn. 62. Issled. kosm. prostranstv., 5.62.198; 51. Astron., 6.51.307 (1970).

094.315 Some general regularities in the development of lunar structures and peculiarities of its geological history. M. S. Markov, A. L. Sukhanov.
(Trudy) Geol. in-t AN SSSR, 1969, vyp. (No.) 204, p. 262 - 273. In Russian. – Abstr. in Referativ. Zhurn. 62. Issled. kosm. prostranstv., 5.62.199; 51. Astron., 6.51.308 (1970).

094.316 Geological comparison of the moon and the earth. V. G. Trifonov, P. V. Florenskij.
(Trudy) Geol. in-t AN SSSR, 1969, vyp. (No.) 204, p. 274 - 285. In Russian. – Abstr. in Referativ. Zhurn. 51. Astron., 6.51.309 (1970).

094.317 Comment on the geologic-morphological maps of the southern part of Mare Imbrium, crater Copernicus and its surroundings (charts LAC-40, 58). P. V. Florenskij.
(Trudy) Geol. in-t AN SSSR, 1969, vyp. (No.) 204, p. 46 - 56. In Russian. – Abstr. in Referativ. Zhurn. 62. Issled. kosm. prostranstv., 5.62.190 (1970).

094.318 Comment on the geologic-morphological map of the northern part of Mare Nubium (chart LAC-76). M. S. Markov.
(Trudy) Geol. in-t AN SSSR, 1969, vyp. (No.) 204, p. 57 - 70. In Russian. – Abstr. in Referativ. Zhurn. 62. Issled. kosm. prostranstv., 5.62.191 (1970).

094.319 Comment on the geologic-morphological maps of the region of the craters Theophilus and Ptolemaeus (charts LAC-77, 78). A. L. Sukhanov.
(Trudy) Geol. in-t AN SSSR, 1969, vyp. (No.) 204, p. 71 - 90. In Russian. – Abstr. in Referativ. Zhurn. 62. Issled. kosm. prostranstv., 5.62.192 (1970).

094.320 Peculiarities of the formation of lava covers on the

moon.
A. M. Gutkin, Ts. M. Rajtburd, M. V. Slonimskaya, N. G.
Sushkin, I. A. Fomina.
(Trudy) Geol. in-t AN SSSR, 1969, vyp. (No.) 204, p. 165 -
187. In Russian. – Abstr. in Referativ. Zhurn. 62. Issled.
kosm. prostranstv., 5.62.194 (1970).

094.321 **Unequal distribution of the ring structures of the**
moon according to their diameters.
P. V. Florenskij, E. I. Zabelin, S. V. Mochalov, Yu. G.
Pimenov.
(Trudy) Geol. in-t AN SSSR, 1969, vyp. (No.) 204, p. 206 -
228. In Russian. – Abstr. in Referativ. Zhurn. 62. Issled. kosm.
prostranstv., 5.62.196 (1970).

094.322 **Mondbodenproben unter dem Mikroskop.**
D. Stöffler.
Sternenbote, 13. Jahrgang, p. 86 - 91 (1970).

094.323 **Preliminary results of the analysis of Apollo 11**
lunar samples. M. Grün.
Pokroky, Vol. 15, 85 - 88 (1970). In Czech.

094.324 **Preliminary investigations of Apollo 11 lunar**
samples. J. Sadil.
Říše hvězd, Vol. 51, 25 - 29 (1970). In Czech.

094.325 **Petrology and mineralogy of Apollo 11 lunar**
samples. V. Bouška.
Vesmír, Vol. 49, 165 (1970). In Czech.

094.326 **First analysis of lunar samples.** V. Bouška.
Vesmír, Vol. 49, 18 - 19 (1970). In Czech.

094.327 **Mare Orientale gravity anomaly.**
J. R. Booker.
Nature, Vol. 227, 56 (1970).
Sjogren has reported that there is an annular negative
gravity anomaly surrounding a small positive anomaly over
the central basin of Mare Orientale. It is pointed out here
that outside this negative ring, there is a positive ring and perhaps further alternating rings.

094.328 **Hypsometric levels of lunar maria.**
J. F. Rodionova.
Astron. Tsirk., No. 545, p. 6 - 7 (1970). In Russian.

094.329 **The moon.** A. Dollfus.
Trans. IAU, Vol. 14A, (see 003.028), 169 - 175
(1970). – Report of Commission 17.

094.330 **On the origin of lunar rocks.**
A. P. Vinogradov.
Zemlya i Vselennaya, No. 3, p. 3 - 11 (1970). In Russian.

094.331 **Lunar mascons.** V. S. Safronov.
Zemlya i Vselennaya, No. 3, p. 32 - 38 (1970).
In Russian.

094.332 **On the moon's free libration and Eulerian motion**
of its poles. Sh. T. Khabibullin, Yu. A. Chikanov.
Trudy Kazansk. gor. astron. observ. 1969, No. 36, p. 49 - 60.
In Russian. – Abstr. in Referativ. Zhurn. 51. Astron.,7.51.127
(1970).

094.333 **On peculiarities of lava-cover formation on the**
moon. A. M. Gutkin, Ts. M. Rajtburd, M. V.
Slonimskaya, N. G. Sushkin, I. A. Fomina.
(Trudy) Geol. in-t AN SSSR, 1969, vyp. (No.) 204, p. 165 -
187. In Russian. – Abstr. in Referativ. Zhurn. 51. Astron.,
7.51.302 (1970).

094.334 **On the thickness of the Procellarian complex.**
A. L. Sukhanov, L. M. Shkerin.
(Trudy) Geol. in-t AN SSSR, 1969, vyp. (No.) 204, p. 37 - 45.
In Russian. – Abstr. in Referativ. Zhurn. 51. Astron.,7.51.312
(1970).

094.335 **Explanatory note to geologic-morphologic maps of**
the southern part of Mare Imbrium, Copernicus and
its vicinity (sheets LAC-40, 58). P. V. Florenskij.
(Trudy) Geol. in-t AN SSSR, 1969, vyp. (No.) 204, p. 46 - 56.
In Russian. – Abstr. in Referativ. Zhurn. 51. Astron., 7.51.313
(1970).

094.336 **Explanatory note to the geologic-morphologic map**
of the northern part of Mare Nubium (sheet LAC-
76). M. S. Markov.
(Trudy)Geol. in-t AN SSSR, 1969, vyp. (No). 204, p. 57 - 70.
In Russian. – Abstr. in Referativ. Zhurn. 51. Astron., 7.51.314
(1970).

094.337 **Explanatory note to the geologic-morphologic maps**
of the Theophilus and Ptolemaeus regions (sheets
LAC-77, 78). A. L. Sukhanov.
(Trudy) Geol. in-t AN SSSR, 1969, vyp. (No.) 204, p. 71 - 90.
In Russian. – Abstr. in Referativ. Zhurn. 51. Astron.,7.51.315
(1970).

094.338 **Irregular distribution of ring structures of the moon**
in regard of their diameters. P. V. Florenskij,
E. I. Zabelin, S. V. Mochalov, Yu. G. Pimenov.
(Trudy) Geol. in-t AN SSSR, 1969, vyp. (No.) 204, p. 206 -
228. In Russian. – Abstr. in Referativ. Zhurn. 51. Astron.,
7.51.316 (1970).

094.339 **Thermal history of the moon.**
J. J. Gilvarry.
Nature, Vol. 225, 623 - 625 (1970).
The purpose of this article is to use the approximation to
the central temperature set by the result of Ness in order to
correct previous solutions of the heat-conduction equation,
to infer possible thermal histories consistent with the present
thermal regime.

094.340 **Lunar continental migration and maria spreading.**
J. W. Elder.
Nature, Vol. 225, 842 - 844 (1970).
This report considers the possibility that the lunar maria
were formed by internal geological processes similar to those
which formed the ocean floors of the earth.

094.341 **On the origin of lunar sinuous rilles.**
V. R. Oberbeck, W. L. Quaide, R. Greeley.
Modern Geol., Vol. 1, (No. 1), 75 - 80 (1969).
Sinuous rilles on the lunar surface have morphologic
features indicative of an origin through fluid flow. Evidence
is presented which suggests that the fluid can be lava and
that some smaller sinuous rilles may have been formed by
collapse of lava tubes.

094.342 **Shock metamorphism of basalt.**
N. M. Short.
Modern. Geol., Vol. 1, (No. 1), 81 - 95 (1969).
Technique for recognizing possible shock damage in
lunar basalts and other fine crystalline basic rocks is suggested.

094.343 **Travel times of body waves in the moon.**
Y. Nakamura, G. V. Latham.
Bull. Seismol. Soc. America, Vol. 60, No. 1, p.63 - 78 (1970).
Travel times and amplitudes of body waves in lunar
models have been computed.

094.344 Lunar thermal anomalies and internal heating.
J. M. Saari.
Report D1-82-0809, Boeing Sci. Res. Labs., Washington.
[Available from CFSTI, Springfield, Va.], 25 pp. (1969).

The evidence for the existence of the so-called linear thermal anomaly recently reported on the western margin of Mare Humorum is examined critically and rejected.

094.345 Apollo 12 70-mm Photographic Catalog.
R. Musgrove, G. Gutschewski, A. Patteson,
A. T. Anderson.
National Space Science Data Center, National Aeronautics and Space Administration, Greenbelt, Maryland. NSSDC 70-10, 291 pp. (1970).

This catalog contains proof prints of the 70-mm photography exposed during the Apollo 12 mission.

**094.346 Apollo 12 Photography, 70-mm, 16-mm, and
35-mm frame index.**   A. T. Anderson.
National Space Science Data Center, National Aeronautics and Space Administration, Greenbelt, Maryland.
NSSDC 70-11, 5 + 149 pp. (1970).

094.347 Origin of the moon: The precipitation hypothesis.
A. E. Ringwood.
Earth Planet. Sci. Letters, Vol. 8, 131 - 140 (1970).

The "single stage" hypothesis of the origin of the earth proposed by the author in 1960 (AJB 60, 7154) and extended to apply to the origin of the moon in 1966 (AJB 66, 5604; "Advances in Earth Science", P. Hurely (Editor), MIT Press, Boston (1966), p. 287 - 356) is developed further in the present paper.

**094.348 Additions to the A.L.P.O. lunar photograph library:
Amateur, JPL, Orbiter-IV and -V, and Apollo-8
photographs.**   J. E. Westfall.
Strolling Astronomer, Vol. 22, 84 - 88 (1970).

094.349 Lunar notes.
J. E. Westfall, C. L. Ricker, H. W. Kelsey, K. J. De-
lano, H. D. Jamieson.
Strolling Astronomer, Vol. 22, 88 - 90 (1970).

**094.350 Non-diffuse infrared emission from the lunar
surface.**   J. K. Harrison.
International Journ. Heat Mass Transfer, Vol. 12, (No. 6), 689 - 697 (1969).

Presents the results of calculations of the thermal radiation from the lunar surface incident onto a flat surface of unit area located a small distance above the moon.

094.351 Apollo mission 12: Lunar photography indexes.
Prepared under the direction of the Department of Defense by the Aeronautical Chart and Information Center, United States Air Force for the National Aeronautics and Space Administration. 4 sheets (1970).

**094.352 Är Apollo 11-bergarterna differentiationsproduk-
ter?**   H. B. Wiik.
Astron. Tidssk., Vol. 3, 69 - 71 (1970).

094.353 Lunar gravimetry and mascons.
P. M. Muller, W. L. Sjogren.
Applied Mech. Rev., Vol. 22, 955 - 959 (1969).

Landeplatz von Apollo 11.
Naturwissenschaften, Vol. 57, 97 (1970).

The lunar rocks.
Spaceflight, Vol. 12, 183 (1970).

The Moon as Viewed by Lunar Orbiter.
See Abstr. 003.059.

New details about lunar specimens.
See Abstr. 011.028.

Secular changes in the lunar elements.
See Abstr. 043.001.

Flutreibung und Akzeleration des Mondes.
See Abstr. 044.015.

**Hydromagnetic aspects of solar wind flow past the
moon.**   See Abstr. 074.015.

**Long-wave cosmic radio radiation in the space
close to moon.**   See Abstr. 077.017.

On the modified Lomnitz law of damping.
See Abstr. 081.028.

**Anorthosites in the early crust of the earth and
on the moon.**   See Abstr. 081.032.

**On the theory of ionospheric scintillations of lunar
radar echoes.**   See Abstr. 083.015.

**Characteristic γ- and X-radiation in the planetary
system.**   See Abstr. 091.013.

Generalized Cassini's laws.   See Abstr. 091.027.

**A comparison of the frequency–size distribution of
Martian "oases" with those of lunar craters and of Martian
craters.**   See Abstr. 097.003.

Lunar glass: Origin and effects.
See Abstr. 105.076.

095 Lunar Eclipses

095.001 **Enlargement of the earth's shadow during the lunar eclipse of April 13, 1968.** J. Bouška.
Bull. Astron. Inst. Czechoslovakia, Vol. 21, 61 - 64 (1970).

The enlargement of the umbral shadow has been determined from the observed times when the lunar craters entered or left the earth's shadow during the lunar eclipse of April 13, 1968. During this total eclipse the enlargement of the umbra was 1/40, somewhat larger than usual.

095.002 **De verduistering van de maan op 25 september 1969.** G. W. E. Beekman.
Hemel en Dampkring, Vol. 68, 42 - 43 (1970).

095.003 **Partial lunar eclipse observed.**
Sky Telescope, Vol. 39, 267 (1970).

095.004 **L'éclipse partielle du Lune du 17 août 1970.**
J. Meeus.
Ciel et Terre, Vol. 86, 257 - 258 (1970).

095.005 **Prediction of the total eclipse of the moon and its local prediction in Japan, February 10, 1971 and August 6, 1971.** M. Takahashi, K. Imai.
Univ. Tokyo, Tokyo Astron. Obs. Report, Vol. 15, (No. 2), (No. 57), p. 321 - 329 (1970). In Japanese.

095.006 **Osservazioni dell'eclisse di luna del 25 settembre 1969.** F. Ficarrotta.
ᴸCoelum, Vol. 38, 36 - 38 (1970).

095.007 **Eclisse lunare di penombra del 25 settembre 1969.**
W. Ferreri.
Coelum, Vol. 38, 39 - 40 (1970).

095.008 **Analyse de l'ombre extérieure pendant les éclipses de lune.** J. Dubois, F. Link.
The Moon, Vol. 1, 462 - 475 (1970).

Photometric analysis of the peripheric umbra during 20 eclipses between 1921 and 1968 based upon the homogeneous observational material reveals the existence of the lunar luminescence excited by solar corpuscular radiations. The influence of the terrestrial upper atmosphere at about 25 km height is detectable on the border of the umbra.

095.009 **Photometric observations of the total lunar eclipse of April 12/13, 1968.**
A. Feinstein, J. C. Muzzio, O. Ferrer, L. García, H. Levato.
Asoc. Argentina Astron. Bol., No. 14 (see 012.020), p. 50 - 51 (1968).

On the possibility of utilization of lunar eclipses for exploring optical properties of the atmosphere.
See Abstr. 082.034.

096 Lunar Occultations

096.001 **Die Plejadenbedeckung durch den Mond am 21.12.1969.** B. Wedel.
VdS Nachrichtenblatt, 19. Jahrgang, p. 16 (1970).

096.002 **Rakende sterbedekkingen zichtbaar in Nederlanden en België, januari – september 1970.**
J. Meeus.
Hemel en Dampkring, Vol. 68, 16 - 17 (1970).

096.003 **1970 occultaties van zwakke sterren.**
Hemel en Dampkring, Vol. 68, 17 - 18 (1970).

096.004 **Occultations of bright stars by the moon in 1970.**
M. M. Dagaev.
Priroda, No. 3.70, p. 125 - 126 (1970). In Russian.

096.005 **Beobachtung einer streifenden Sternbedeckung durch den Mond von sechs verschiedenen Standorten aus.** K. Locher.
Orion Schaffhausen, 28. Jahrgang, p. 19 - 20 (1970).

096.006 **Grazing occultation of ZC 2298, observed at Zeerust, South Africa, on 1969 October 14.**
M. D. Overbeck.
Monthly Notes Astron. Soc. Southern Africa, Vol. 29, 42 - 43 (1970).

096.007 **Vorausberechnete Sternbedeckungen durch den Mond 1970.**
Astron. Nachr., Vol. 292, 39 - 43 (1970).

096.008 **Observations d'occultations d'étoiles par la lune.**
G. Hilaire-Henry.
Astron. Astrophys., Suppl. Series, Vol. 1, 259 - 261 (1970).

On présente les résultats de 50 observations d'occultations faites à l'Observatoire de Besançon de janvier 1967 à décembre 1968.

096.009 **Photoelectric observations of lunar occultations.**
C. de Vegt, E. Pansch.
Astron. Astrophys., Vol. 5, 328 - 329 (1970).

Photoelectric observations of lunar occultations have been obtained at the Hamburg Observatory from April 1969 until December 1969.

096.010 **Quelques occultations rasantes visibles en France (juillet – décembre 1970).** J. Meeus.
L'Astronomie, 84ᵉ année, p. 293 - 295 (1970).

096.011 **Rakende sterbedekkingen zichtbaar in Nederland en België.** J. Meeus.
Hemel en Dampkring, Vol. 68, 111 - 112 (1970).

096.012 **Photoelectric measurement of lunar occultations. I. The process.** R. E. Nather, D. S. Evans.
Astron. Journ., Vol. 75, 575 - 582 (1970).

096.013 **Photoelectric measurement of lunar occultations. II. Instrumentation.** R. E. Nather.
Astron. Journ., Vol. 75, 583 - 588 (1970).

096.014 **Photoelectric measurement of lunar occultations. III. Lunar limb effects.** D. S. Evans.
Astron. Journ., Vol. 75, 589 - 599 (1970).

As part of a program of high-speed electronic astronomy at McDonald Observatory, appropriate apparatus was constructed and a regular program of occultation observations was begun in December 1968. It is the purpose of this series of papers to describe the occultation process, the electronic equipment used, the data analysis and reduction procedures, and to report on the results of the observations so far obtained.

096.015 **BD +0°2902.**
H. Povenmire, W. Sander, M. Seslar.
IAU Circ. No. 2242 (1970).

096.016 **Lunar occultations of infrared objects.**
G. Wallerstein.
IAU Circ. No. 2251 (1970).

096.017 **Lunar occultations 1968 - 1971 visible at Belgrade.**
Bull. Obs. Astron. Beograd, Vol. 27, No. 2, p. 133 - 141 (1969).

096.018 **Streifende Sternbedeckungen.**
W. Jaschek.
Sternenbote, 13. Jahrgang, p. 2 - 6 (1970).

096.019 **Observations of occultations of stars by the moon in Tashkent 1967 - 1968.**
M. R. Ehshmanov, S. Z. Sadykov.
Tsirk. Astron., Inst., *Tashkent,* No. 14 (361), p. 24 - 26 (1969). In Russian.

096.020 **Lunar occultations.** H. de Souza, L. E. da Silva Machado.
Contr. Obs. Valongo, Univ. Federal Rio de Janeiro, Sér. III, Nos. 10 - 16 (1970).

The results obtained in the observations of lunar occulta tions are reported for 1969 September to 1970 March.

096.021 **Observations of occultations made the Štefánik Observatory Prague—Petřín in the years 1967 - 1969.**
Contr. Observations People's Obs. Prague, Vol. 7, Ser. 2, No. 1, 4 pp. (1970).

096.022 **Observations of lunar occultations.**
Yamamoto Circ., Nos. 1716, 1720 (1970).
In Japanese.

On the orientation of C. B. Watts' charts of the marginal zone of the moon. See Abstr. 094.254.

Additional occultation studies of weak radio sources at Arecibo Observatory: List 4.
See Abstr. 141.119.

Occultation observations of the Pleiades.
See Abstr. 153.007.

097 Mars

097.001 **L'atmosphère primitive de la planète Mars d'après les photographies transmises par les sondes Mariner VI et VII.**
A. Dollfus, R. Fryer, C. Titulaer.
Comptes Rendus Acad. Sci. Paris, Sér. B, Vol. 270, 424 - 426 (1970).

La morphologie des cratères martiens, la statistique de leur nombre en fonction de leur diamètre et l'étude des processus d'érosion indiquent que la planète aurait possédé une atmosphère originelle relativement dense qui se serait ultérieurement résorbée pour laisser place au résidu actuel.

097.002 **Les anomalies de la surface du sol de Mars dans la région «Hellas».** A. Dollfus.
Comptes Rendus Acad. Sci. Paris, Sér. B, Vol. 270, 641 - 644 (1970).

Les observations télescopiques de la planète Mars révèlent de fréquents changements de l'aspect de la région «Hellas». La sonde spatiale Mariner 7 a montré que cette région est dépourvue de tous cratères, pourtant nombreux dans les régions voisines. Il est montré par l'analyse polarimétrique de la lumière, que ces deux propriétés ne sont pas attribuables à des brumes mais qu'elles appartiennent réellement à la surface du sol.

097.003 **A comparison of the frequency–size distribution of Martian "oases" with those of lunar craters and of Martian craters.** J. A. Russell, M. J. Mayo.
Meteoritics, Vol. 4, 293 (1969). – Abstract.

097.004 **Spectral reflectance of CO_2 – H_2O frosts.**
H. Kieffer.
Journ. Geophys. Res.,Vol. 75, 501 - 509 (1970).

The spectral reflectance of CO_2 and H_2O frosts grown individually, simultaneously, and sequentially have been measured from 0.8 to 3.2μ. Contamination and metamorphism were minimized by growing the frosts from high-purity gases in a cold vacuum chamber.

097.005 **Interpretation of the Martian polar cap spectra.**
H. Kieffer.
Journ. Geophys. Res., Vol. 75, 510 - 514 (1970).

The purpose of this work is to compare the telescope observations with laboratory spectra for simulated Martian frosts and to examine the possible role of H_2O in a predominantly CO_2 polar cap model. This study includes only CO_2 and H_2O as they are the only likely chemicals with the required physical properties.

097.006 **Photochemistry of CO_2 in the atmosphere of Mars.** M. B. McElroy, D. M. Hunten.
Journ. Geophys. Res., Vol. 75, 1188 - 1201 (1970).

We first discuss CO_2 photochemistry in general terms. Then in sequential sections we apply our treatment to the upper and lower atmospheres of Mars. Some brief comments are offered on possible ambiguities in the interpretation of dayglow emissions from the Mars atmosphere.

097.007 **The atmospheres of Mars and Venus.**
R. Goody.
Naturwissenschaften, Vol. 57, 10 - 16 (1970). – Review article.

097.008 **Mars in focus.**
Spaceflight, Vol. 12, 98 - 103 (1970).

097.009 **Mars: The changing picture.**
E. J. Öpik.

Irish Astron. Journ., Vol. 9, 136 - 148 (1969).

097.010 **The frequency-size distribution of Martian oases.**
J. A. Russell, M. J. Mayo.
Publ. Astron. Soc. Pacific, Vol. 82, 138 - 142 (1970).

Comparison of the diameter distribution of Martian oases with those of lunar and Martian craters supports the hypothesis that the oases are large craters. Oasis counts fall below those for Martian craters, but an incompleteness factor should be expected. Distribution curves for craters and oases have very similar slopes, suggesting a common mode of origin.

097.011 **Evidence for solid carbon dioxide in the upper atmosphere of Mars.**
K. C. Herr, G. C. Pimentel.
Science, Vol. 167, 47 - 49 (1970).

The infrared spectra recorded by Mariner 6 and 7 show reflections at 4.3 microns, which suggest the presence of solid carbon dioxide in the upper atmosphere of Mars.

097.012 **Martian mass and earth-moon mass ratio from coherent S-band tracking of Mariners 6 and 7.**
J. D. Anderson, L. Efron, S. K. Wong.
Science, Vol. 167, 277 - 279 (1970).

Range and Doppler tracking data from Mariners 6 and 7 have been used to obtain values for the ratio of the mass of the earth to that of the moon which are in substantial agreement with those determined from other Mariner and Pioneer spacecraft. There is an inconsistency of about 0.004 percent in values for the mass of the moon determined from lunar trajectories. A gravitational constant for Mars of 42,828.48 ± 1.38 cubic kilometers per second per second has been obtained.

097.013 **Report on the 1967 apparition of Mars.**
K. R. Brasch.
Strolling Astronomer, Vol. 22, 29 - 35 (1970).

097.014 **Mars vu par Mariner 6 et 7.** J. Kovalevsky.
L'Astronomie, 84e année, p. 57 - 67 (1970).

097.015 **Mars, aus der Nähe besehen.** H. Zimmer.
Weltraumfahrt, 21. Jahrgang, p. 5 - 10 (1970).

097.016 **Mariners 6 and 7 – Mars encounter.**
H. Miles.
Journ. British Astron. Ass., Vol. 80, 107 - 114 (1970).

097.017 **Blue haze and Mariner 6 pictures of Mars.**
S. L. Hess, P. B. Boyce, B. A. Smith, A. T. Young, C. B. Leovy.
Science, Vol. 167, 906 - 908 (1970).

097.018 **Mariner 6: Origin of Mars ionized carbon dioxide ultraviolet spectrum.**
A. Dalgarno, T. C. Degges, A. I. Stewart.
Science, Vol. 167, 1490 - 1491 (1970).

The calculations show that most of the excitations arise from photo-ionization of CO_2 and fluorescent scattering by CO_2^+ and that comparatively few excitations arise from electron impacts.

097.019 **Neue Argumente für aktive Vulkane auf Mars.**
H. Heuseler.
Sterne, 46. Jahrgang, p. 18 - 23 (1970).

097.020 Petrology of Mars.   Yu. A. Khodak.
Byull. Mosk. o-va ispyt. prirody. Otd. geol., Vol.
44, No. 5, p. 144 - 145 (1969).   In Russian.

**097.021 Halo phenomena in the atmosphere of Mars due to
ice and unknown crystals.**   V. D. Davydov.
Astron. Zhurn. Akad. Nauk SSSR, Vol. 47, 172 - 278 (1970).
In Russian. English translation in Soviet Astron. AJ, Vol. 14,
No. 1.

On two independent graphs of the activity distribution of
unusually bright formations on Mars from phase angles, cases
of conformity between some earth and Martian halo rings
are found. Such a conformity can be considered as an indication of the presence of crystals of ice in the Martian atmosphere. Besides that, maxima, which have no analogy between
well-known halo rings, are discovered on graphs. The question,
whether these maxima can belong to solid carbon dioxide demands a special investigation.

097.022 Observations of Mars and Venus at 11.1 cm.
K. S. Stankevich.
Australian Journ. Phys., Vol. 23, 111 - 112 = Separate print
Division Radiophys. C.S.I.R.O. Sydney (1970).

The radio temperature of Mars was measured at 11.1 cm
using the 210 ft radio telescope at Parkes during June 10 and
11, 1969. During June 11 and 12 radio emission from Venus
was also measured.

097.023 Observations of Mars in 1969.
V. I. Garazha.
Astron. Tsirk., No. 540, p. 4 - 5 (1969).   In Russian.

**097.024 On possibilities of the investigation of Martian
matter properties from radio radiation of Mars.**
V. S. Troitsky.
Astron. Zhurn. Akad. Nauk SSSR, Vol. 47, 384 - 391 (1970).
In Russian. – English translation in Soviet Astron., AJ, Vol.
14, No. 2.

A theoretical analysis of Martian radio radiation shows
the possibility of the determination of damping of electromagnetic waves in the matter of the Martian upper cover.

**097.025 Thermal conditions in the region of the south pole
of Mars.**   V. I. Aleshin, O. B. Shchuko.
Astron. Zhurn. Akad. Nauk SSSR, Vol. 47, 392 - 396 (1970).
In Russian. – English translation in Soviet Astron., AJ, Vol.
14, No. 2.

Thermal conditions of the soil and annual variations of
the polar cap thickness in the vicinity of the south pole of
Mars are considered, depending on properties of the underlying
soil and processes of condensation and sublimation of CO_2.

**097.026 The opposition effect of Mars and some possible
peculiarities of the structure of crystals in the**
Martian atmosphere.   V. D. Davydov.
Astron. Zhurn. Akad. Nauk SSSR, Vol. 47, 446 - 449 (1970).
In Russian. – English translation in Soviet Astron., AJ, Vol.
14, No. 2.

**097.027 High-dispersion spectroscopic studies of Mars. III.
Preliminary results of 1968–1969 water-vapor**
studies.   R. A. Schorn, C. B. Farmer, S. J. Little.
Icarus, Vol. 11, 283 - 288 (1969).

Recent high-resolution spectra of Mars have confirmed
the existence of water vapor in the atmosphere of that planet.
New laboratory measurements and improved temperature
corrections have resulted in more accurate abundances than
previously. Our abundance and temperature estimates lead
to a surprisingly high relative humidity (over 50%), and
we feel that the possibility of small amounts of liquid water
on the planet cannot be dismissed out of hand.

**097.028 Interpretation of high-resolution spectra of Mars.
I. CO_2 abundance and surface pressure derived
from the curve of growth.**   L. D. G. Young.
Icarus, Vol. 11, 386 - 389 (1969).

High-resolution spectra of Mars were obtained by Pierre
and Janine Connes. These have been used to obtain a curve
of growth for carbon dioxide and carbon monoxide lines
formed in the Martian atmosphere. A rotational temperature
of $201° ± 6°K$ is found for the CO_2 lines while the CO lines
indicate a temperature of $203° ± 8°K$. By a nonlinear least-
squares fit to the curve of growth we find, for an air mass
$\eta = 3.5$, an approximate CO_2 abundance of 70 m-atm and a
surface pressure of 5 mb. This surface pressure is approximately equal to the CO_2 partial pressure.

097.029 Further Mariner 6 and 7 photos.
Icarus, Vol. 11, 424 - 431 (1969).

097.030 Marsaufnahmen von Mariner 6 und 7.
H. Müller.
Orion Schaffhausen, 28. Jahrgang, p. 37 - 41 (1970).

097.031 Mars 1969.   S. Cortesi.
Orion Schaffhausen, 28. Jahrgang, p. 49 - 50
(1970). – Rapport No. 20 du «Groupement planétaire SAS».

**097.032 Goethite on Mars: A laboratory study of physically
and chemically bound water in ferric oxides.**
J. B. Pollack, D. Pitman, B. N. Khare, C. Sagan.
SAO, *Cambridge, Mass.,* Special Rep. No. 314, 6 + 32 pp.
(1970).

A thermogravimetric analysis of the decomposition of
geothite-rich samples of limonite and measurement of the
equilibrium vapor pressure of the water physically bound in
the sample are performed. The results imply that physical
sorption will tend to take place preferentially over condensation. This tends on Mars to inhibit the formation of
a water polar cap.

**097.033 The figure of Mars from Mariner radio occultation
measurements.**   A. Kliore.
Bull. American Astron. Soc., Vol. 2, 202 - 203 (1970).
Abstr. AAS.

**097.034 Ephemeris, radar radius, and radar topography of
Mars.**   D. A. O'Handley, W. G. Melbourne,
R. M. Goldstein, G. A. Morris, G. S. Downs.
Bull. American Astron. Soc., Vol. 2, 211 - 212 (1970).
Abstr. AAS.

**097.035 Observations of Mars and Deimos with the astro-
graph of the Kiev University Observatory.**
N. D. Kovalenko, V. V. Telnyuk-Adamchuk.
Vestn. Kiev. Univ., Ser. Astron., No. 11, p. 129 - 130 (1969).
In Russian.

097.036 The problems of the satellites of Mars.
G. A. Wilkins.
Symposia Mathematica, Vol. 3, 29 - 43 (1970). – Conference
paper (see 012.010).

097.037 The surface of Mars.   R. B. Leighton.
Sci. American, Vol. 222, No. 5, p. 26 - 41 (1970).

The remarkable pictures taken by Mariners 6 and 7
reveal a planet with features unlike those seen on the moon
and also unlike those on the earth, including a polar cap
evidently composed of dry ice.

**097.038 Comparative characteristics of the ionospheres of
the planets of the terrestrial group: Mars, Venus,
and the earth.**   K. I. Gringauz, T. K. Breus.

Space Sci. Rev., Vol. 10, 743 - 769 (1970). – Review article.

097.039 Phobos: Preliminary results from Mariner 7.
B. A. Smith.
Science, Vol. 168, 828 - 830 (1970).

Analysis of an image of Phobos on Mariner 7 frame 7F91 indicates that the martian satellite is larger and has a darker surface than had previously been thought. The limb profile measures 18 by 22 kilometers and is elongated along the orbital plane. Phobos has an average visual geometric albedo of 0.065 lower than that known for any other body in the solar system. It seems probable that Phobos did not form by accretion around primordial Mars, but was captured at some later time.

097.040 Mars: Occurrence of liquid water.
A. P. Ingersoll.
Science, Vol. 168, 972 - 973 (1970).

In the absence of juvenile liquid water, condensation of water vapor to ice and subsequent melting of ice are the only means of producing liquid water on the martian surface. However, the evaporation rate is so high that the available heat sources cannot melt pure ice. Liquid water is therefore limited to concentrated solutions of strongly deliquescent salts.

097.041 The Mariner Mars 1971 experiments. Introduction.
R. H. Steinbacher, S. Z. Gunter.
Icarus, Vol. 12, 3 - 9 (1970).

The Mariner Mars 1971 scientific experiments and Principal Investigators for each experiment are listed. The general objectives of the Mariner Mars 1971 science effort are (1) to provide broad topographic and thermal coverage, (2) to study seasonal variations in the atmosphere and on the surface, and (3) to obtain other long-term dynamic observations. To accomplish these objectives, JPL is developing two spacecraft to orbit Mars in late 1971.

097.042 Television experiment for Mariner Mars 1971.
H. Masursky, R. Batson, W. Borgeson, M. Carr, J. McCauley, D. Milton, R. Wildey, D. Wilhelms, B. Murray, N. Horowitz, R. Leighton, R. Sharp, W. Thompson, G. Briggs, P. Chandeysson, E. Shipley, C. Sagan, J. Pollack, J. Lederberg, E. Levinthal, W. Hartmann, T. McCord, B. Smith, M. Davies, G. de Vaucouleurs, C. Leovy.
Icarus, Vol. 12, 10 - 45 (1970).

The primary objective of the television experiment on Mariner Mars 1971 is to provide imaging data that will increase scientific knowledge of Mars and the solar system. This objective is defined in detail.

097.043 Infrared radiometry experiment for Mariner Mars 1971.
S. Chase, Jr., E. Miner, G. Münch, G. Neugebauer.
Icarus, Vol. 12, 46 - 47 (1970).

The infrared radiometer is designed to provide brightness temperatures of the surface of Mars by measuring the energy radiated in the 8 to 12 and 18 to 25 μ wavelength bands.

097.044 Infrared spectroscopy experiment for Mariner Mars 1971.
R. A. Hanel, B. J. Conrath, W. A. Hovis, V. Kunde, P. D. Lowman, C. Prabhakara, B. Schlachman, G. V. Levin.
Icarus, Vol. 12, 48 - 62 (1970).

The infrared spectroscopy (IRIS M) experiment for Mariner Mars 1971 is designed to provide spectral measurements of the thermal emission of the Martian surface and atmosphere. The orbital mission allows the Martian atmospheric and surface properties to be studied with respect to geographic location and time variation.

097.045 Ultraviolet spectroscopy experiment for Mariner Mars 1971.
C. W. Hord, C. A. Barth, J. B. Pearce.
Icarus, Vol. 12, 63 - 77 (1970).

The Mariner Mars 1971 ultraviolet spectrometer will obtain spectra in the 1100 to 3400 Å wavelength range. The spectral features in this range will help to characterize the Martian atmosphere in a quantitative way. Ultraviolet spectroscopy will contribute to a biological understanding of the planet.

097.046 Celestial mechanics experiment for Mariner Mars 1971.
J. Lorell, J. D. Anderson, I. I. Shapiro.
Icarus, Vol. 12, 78 - 81 (1970).

A spacecraft experiment is described. This experiment is expected to test the theory of general relativity, improve the Martian ephemeris, and provide new gravimetric data for Mars. The anticipated measurements will be obtained from the radio tracking system used to navigate the Mariner Mars 1971 orbiter.

097.047 S-band occultation experiment for Mariner Mars 1971.
A. Kliore, D. L. Cain, B. L. Seidel, G. Fjeldbo.
Icarus, Vol. 12, 82 - 90 (1970).

The S-band occultation experiment, performed with the Mariner Mars 1971 orbiters, will provide approximately 90 occultations of the spacecraft in orbit about Mars. These occultations will be well dispersed in latitude and longitude, providing atmospheric and ionospheric profiles as well as the radius of Mars at about 180 locations on its surface. Because of the large amount of data that will be gathered in a relatively short time of 30 days, more sophisticated data handling and processing methods will have to be used.

097.048 Martian blue-clearing during 1967 apparition.
C. F. Capen.
Icarus, Vol. 12, 118 - 127 (1970).

The occurrence of Martian blue-clearing during a 20-month observational patrol is tabulated and summarized. Periods of clearing were more numerous and stronger during the 1967 apparition than during 1965. The percentage of nights in which some form of clearing was detected was 48.5% of the total nights of quality violet-blue observations. Blue-clearings were noted 175 days before opposition and 349 days after opposition.

097.049 Visuelle Marsbeobachtungen während der Opposition 1969. F. Kimberger.
SuW, Vol. 9, 154 - 156 (1970).

097.050 La planète Mars en 1969. J. Dragesco.
L'Astronomie, 84e année, p. 281 - 292 (1970).

097.051 Additional A.L.P.O. 1967 Mars drawings.
Strolling Astronomer, Vol. 22, 43 - 46 (1970).

097.052 Photographic observations of the satellites of Mars at the Pulkovo Observatory. A. N. Deutsch.
Izv. Glav. Astron. Obs. v Pulkove, No. 185, p. 103 - 109 (1970).
In Russian.

The measurements of Deimos and Phobos positions relative to Mars were made from observations at Pulkovo with the normal astrograph during the oppositions 1956 and 1967 and also the 26″(65 cm) refractor during 1967. The results of these measurements and the comparison of observations with the satellite motion ephemeris are given.

097.053 Photographic observations of Deimos made with the Pulkovo normal astrograph during the opposi-

tion of Mars in 1967.   N. V. Fatchikhin.
Izv. Glav. Astron. Obs. v Pulkove, No. 185, p. 110 - 116
(1970).   In Russian.
  Altogether 34 plates with the image of Mars were obtained by all the observers. On these plates the Phobos images
were absent and the Deimos images were found on 15 plates
taken by the author being among them. From all these observations 26 positions of Deimos relative to Mars were determined.

**097.054   Observations of the satellites of Mars made with
a special light reducing slit.**   I. I. Kanaev.
Izv. Glav. Astron. Obs. v Pulkove, No. 185, p. 117 - 118
(1970).   In Russian.
  The design of a special light reducing slit is described. The
method of observations of the Mars satellites by means of this
shutter is described. The results of observations are given.

097.055   Mariners report on Mars.
I. Halliday.
Journ. Royal Astron. Soc. Canada, Vol. 64, 191 - 192 (1970).

097.056   On properties of the Martian surface.
N. P. Barabashov.
Astron. Tsirk., No. 557, p. 2 - 4 (1970).   In Russian.

**097.057   On the possibility of determining the parameters
of the aerosol distribution in dependence on height**
in the Martian atmosphere.   A. V. Morozhenko.
Astron. Tsirk., No. 557, p. 4 - 6 (1970).   In Russian.

097.058   The structure of the Martian atmosphere.
J. S. Hogan.
Thesis, New York Univ. [Available from Univ. Microfilms,
Ann Arbor, Mi.], 193 pp. (1969).

**097.059   An estimate of the solar cyclic variation of the
Martian upper atmosphere.**
D. N. Vachon, D. Weidner, K. Lichtenfeld.
Ann. New York Acad. Sci., Vol. 163, No. 1, p. 69 - 80
(1969). − See Bull. Signalét., Vol. 31, Section 120, No. 7069
(1970).

The hills and dales of Mars.
Sci. American, Vol. 222, No. 3, p. 60, 62 (1970).

**Photographic positions of Mars obtained with the
short-focus double astrograph of the Pulkovo Observatory
during 1962 - 1965.**   See Abstr. 041.014.

**The ionospheres of Venus and Mars. II. Mariner V
and the Venus ionosphere.**   See Abstr. 093.008.

**On the difference in the laws of light reflection
from moon and Mars.**   See Abstr. 094.200.

**Volcanism and internal activity in the moon and
Mars.**   See Abstr. 094.294.

**Meteoroid environment and age relationships for
the lunar and Martian surfaces.**   See Abstr. 094.299.

098 Minor Planets

098.001 **Spatial density of the known asteroids.**
D. J. Kessler.
Meteoritics, Vol. 4, 279 - 280 (1969). – Abstract.

098.002 **Minor planets and related objects. IV. Asteroid (1566) Icarus.**
T. Gehrels, E. Roemer, R. C. Taylor, B. H. Zellner.
Astron. Journ., Vol. 75, 186 - 195 (1970).
With various telescopes at Kitt Peak and in the Catalina Mountains, observations were made of positions, brightness, colors, lightcurves, and polarization of asteroid (1566) Icarus. The absolute magnitude of Icarus on the UBV system is $B(1,0) = 17.55$, and the colors are $B - V = +0.80$ and $U - B = +0.66$. The sidereal or true period of rotation is $2^h 16^m 23^s \pm 3^s$ (p.e.). The wavelength dependence of polarization shows a minimum, and the wavelength dependence of brightness a maximum, near 0.6μ. By indirect methods the geometric albedo is determined to be 26% at 0.4μ and the diameter then is 1.08 km. Icarus appears to be a rough stony-iron body, nearly spherical, with nonuniform reflectivity over the surface. The possibility of Icarus being an extinct cometary nucleus is discussed.

098.003 **The axial rotation of asteroids.** Z. Kopal.
Astrophys. Space Sci., Vol. 6, 33 - 35 (1970).
The observed fact that light changes of the asteroids exhibit no beat periods is interpreted as an indication that they do not wobble in space like spinning tops, but spin about only one axis. Since, moreover, the damping of three-dimensional rotation by jovi-solar attraction would require a time which is long in comparison with the age of the solar system, it is concluded that the present uni-axial rotation must represent a property preserved from the time when the asteroids were formed.

098.004 **Small bodies of the solar system.** V. Rijves.
Astron. Geod. Eston. SSR, Tartu, (see 003.004), p. 92 - 98 (1969). In Russian.

098.005 **Waarnemingen van Bamberga.**
J. Meeus.
Hemel en Dampkring, Vol. 68, 43 - 44 (1970).

098.006 **Positionsbestimmungen an Planetoiden.**
T. Düring.
SuW, Vol. 9, 132 - 133 (1970).

098.007 **Erfolgreiche Beobachtungen des Planetoiden (1620) Geographos während seiner großen Annäherung an die Erde im August/September 1969.**
R. A. Naef.
Orion Schaffhausen, 28. Jahrgang, p. 41 - 42 (1970).

098.008 **Commensurability cases among the minor planets.**
J. Schubart.
Symposia Mathematica, Vol. 3, 23 - 27 (1970). – Conference paper (see 012.010).

098.009 **Positions de petites planètes.**
G. Soulié, R. Dumont.
Astron. Astrophys., Suppl. Series, Vol. 1, 233 - 241 (1970).
On donne des positions des petites planètes 1, 2, 3, 4, 6, 7, 11, 18, 39, 40, 51, 354, 1566 et 1736 observées à l'équatorial photographique de 33 cm de l'Observatoire de Bordeaux.

098.010 **Observations photographiques de petites planètes.**
G. Hilaire-Henry.

Astron. Astrophys., Suppl. Series, Vol. 1, 243 - 257 (1970).
On donne des positions des petites planètes 1, 2, 3, 4, 5, 6, 7, 11, 14, 16, 18, 39, 40 et 51 observées à l'Observatoire de Besançon de novembre 1966 à janvier 1969.

098.011 **Index to minor planet elements.**
P. Herget.
Minor Planet Circ., Nos. 3001 - 3007 (1969).

098.012 **The fragmentation of the asteroids.**
B. Hellyer.
Monthly Notices, Roy. Astron. Soc., Vol. 148, 383 - 390 (1970).
The mathematical problem of the evolution of the asteroidal and meteoritic mass distributions under collisional fragmentation is re-examined. A power law of index $-5/3$ is shown to be a good approximation to the asymptotic solution for the most massive bodies (i.e. the larger asteroids). For less massive bodies a power law of index about -1.8 is indicated, the precise value of the index depending on the exact mode of break-up assumed.

098.013 **Die Identifikation Kleiner Planeten.**
J. W. Ekrutt.
SuW, Vol. 9, 145 - 148 (1970).

098.014 **The dynamical problems of small bodies in the solar system.** G. A. Chebotarev.
Byull. Inst. Teoret. Astron., *Leningrad*, Vol. 12, 1 - 7 (1970). In Russian.
The present state of the dynamical problems of small bodies (asteroids, comets and meteors) in the solar system is considered. The urgent problems are formulated.

098.015 **Minor planets 1968.** N. S. Samoilova-Yakhontova.
Byull. Inst. Teoret. Astron., *Leningrad*, Vol. 12, 8 - 15 (1970). In Russian.

098.016 **The evolution of orbits of the Hilda group planets and of the planet Thule.**
G. A. Chebotarev, N. A. Beljaev, R. P. Yeremenko.
Byull. Inst. Teoret. Astron., *Leningrad*, Vol. 12, 82 - 103 (1970). In Russian.
The motion of the Hilda group planets and of Thule has been investigated by the method of special perturbations for the time interval of about 400 years (1660 - 2059). It turns out that some well-known statistical distributions of the orbital elements of planets are temporary. The relation between Schwarzschild's periodic orbits and the real orbits has been studied in detail.

098.017 **Structure of the central part of the asteroid belt.**
G. A. Chebotarev, M. Ja. Shmakova.
Byull. Inst. Teoret. Astron., *Leningrad*, Vol. 12, 104 - 118 (1970). In Russian.
The central part of asteroid belt ($610''- 1110''$) is divided by two gaps (the commensurabilities 2/5 and 1/3) into three secondary rings ($610''- 740''$, $750''- 890''$ and $910''- 1110''$). This article contains the detailed statistical analysis of these three secondary rings (each of them having its specific physical properties).

098.018 **Observations of minor planets made at the Crimean Astrophysical Observatory.** (12-th report).
L. I. Chernykh.
Byull. Inst. Teoret. Astron., *Leningrad*, Vol. 12, 234 - 242 (1970). In Russian.

098.019 **Asteroid Vesta: Spectral reflectivity and compositional implications.** T. B. McCord, J. B. Adams, T. V. Johnson.
Science, Vol. 168, 1445 - 1447 (1970).

The reflection spectrum for Vesta contains a strong absorption band centered near 0.9 micron and a weaker absorption feature between 0.5 and 0.6 micron. The reflectivity decreases strongly in the ultraviolet. Vesta shows the strongest and best-defined absorption bands yet seen in the reflection spectrum for the solid surface of an object in the solar system. Comparison with laboratory measurements on meteorites and Apollo 11 samples indicates that the surface of Vesta has a composition very similar to that of certain basaltic achondrites.

098.020 **Minor planets and related objects. V. The density of Trojans near the preceding Lagrangian point.**
C. J. Van Houten, I. Van Houten-Groeneveld, T. Gehrels.
Astron. Journ., Vol. 75, 659 - 662 (1970).

Two fields near the Lagrangian point L_5 in the sun-Jupiter system were searched for Trojan asteroids. Combinations of this investigation with the Palomar-Leiden asteroid survey yielded the results that there are about 700 Trojans brighter than opposition magnitude 20.9 associated with this libration point.

098.021 **Observations of 1620 Geographos.**
B. F. Mintz.
Astron. Journ., Vol. 75, 663 - 664 (1970).

During its recent close approach to the earth, the minor planet 1620 Geographos was observed. Seventy-four positions were obtained, spanning an arc length of approximately 100 deg.

098.022 **(1566) Icarus.**
V. I. Vlăsceanu.
Stud. Cerc. Astron., Vol. 15, 59 - 60 (1970).
25 photographic positions during 1968 are given.

098.023 **(1566) Icarus.**
S. G. Brauenfeld, Z. N. Grigor'eva, E. K. Denisjuk, E. S. Erošević, V. F. Kartašov, L. N. Kondrat'eva, V. S. Matjagin, L. P. Sorokina, L. A. Usoltseva, A. A. Ščipenstein.
IAU Circ. No. 2201, 2208 (1970).

098.024 **(1620) Geographos.**
M. Antal.
IAU Circ. No. 2210 (1970).

098.025 **Rapidly moving asteroid.**
C. Kowal.
IAU Circ. No. 2220 (1970).

098.026 **(887) Alinda.** A. Mrkos, Maršálková.
IAU Circ. No. 2226 (1970).

098.027 **(1620) Geographos.** H. Ross, S. Naftilan.
IAU Circ. No. 2226 (1970).

098.028 **(887) Alinda.** P. Wild.
IAU Circ. No. 2228 (1970).

098.029 **Object Kowal.** K. Aksnes, E. F. Helin.
IAU Circ. No. 2233 (1970).

098.030 **(1620) Geographos.**
S. Vasilevskis, K. M. Cudworth, E. A. Harlan.
IAU Circ. No. 2237 (1970).

098.031 **Second object Kowal.** C. Kowal.
IAU Circ. No. 2236 (1970).

098.032 **1968 AA.**
E. Roemer, R. McCallister, B. Schreur, R. C. Elliott.
IAU Circ. No. 2255 (1970).

098.033 **Heliocentric coordinates of Ceres, Pallas, Juno, Vesta, 1928 – 2000.** R. L. Duncombe.
Astron. Papers, prepared for the use of the American Ephemeris and Nautical Almanac, [published by the Nautical Almanac Office, U. S. Naval Observatory – United States Government Printing Office, Washington], Vol. 20, Part II, p. 133 - 309 (1969).

098.034 **Positions and motions of minor planets, comets and satellites.** G. A. Chebotarev.
Trans. IAU, Vol. 14A, (see 003.028), 187 - 192 (1970).
Report of Commission 20.

098.035 **The minor planet named in honour of V. I. Lenin.**
N. S. Yakhontova.
Zemlya i Vselennaya, No. 2, p. 60 - 61 (1970). In Russian.

098.036 **Infrared diameter of Vesta.** D. A. Allen.
Nature, Vol. 227, 158 - 159 (1970).

Observations of Vesta were made during the past winter with the 30-inch telescope of the O'Brien observatory, University of Minnesota, at effective wavelengths of 8·5, 11·8 and 21·3 μm. The weighted mean infrared diameter is 573±6 km and the corresponding Bond albedo 0·119. Making allowance for roughness and rotation, the linear diameter must be about 600 km.

098.037 **Determination of 20 asteroid orbits with a IBM 360 computer.** R. Peralta, C. Torres, H. Wroblewski.
Asoc. Argentina Astron. Bol., No. 14 (see 012.020), p. 58 - 62 (1968). In Spanish.

On the non-linear equations of secular perturbations of minor planets. See Abstr. 042.049.

A comparison of perturbations from Neptune in the motion of Ceres as determined by the methods of Laplace – Newton and Hill. See Abstr. 042.054.

Mission to an asteroid.
See Abstr. 051.009.

On the relationship between comets and minor planets. See Abstr. 102.003.

Dynamics of meteor streams and new asteroid–meteor and comet–meteor associations.
See Abstr. 104.022.

Radioactivity caused by cosmic rays in bodies of asteroid size. See Abstr. 143.064.

099 Jupiter

099.001 The infrared spectra of Jupiter and Saturn at 1.2 - 4.2 microns. H. L. Johnson.
Astrophys. Journ. (*Letters*), Vol. 159, L1 - L5 (1970).

The infrared spectra of Jupiter and Saturn have been observed with a rapid-scanning Michelson interferometer-spectrometer. The spectral resolution is about 8 cm⁻¹. These spectra cover the range from 1.2μ (8300 cm⁻¹) to 4.2μ (2400 cm⁻¹). All spectra have been corrected completely for atmospheric extinction and have been divided by an outside-atmosphere lunar spectrum; they are, therefore, freed of the effects of terrestrial atmospheric absorption, solar spectral lines, and the wavelength dependences of detector and interferometer efficiencies. Thus, they represent the Jovian and Saturnian reflection spectra.

099.002 Wasserpflanze überlebt in simulierter Jupiter-Atmosphäre. E. Koch.
Umschau, 70. Jahrgang, p. 216 - 217 (1970).

099.003 Periodische Schwankungen in der Längenbewegung des GRF. H. Heuseler.
SuW, Vol. 9, 50 (1970).

099.004 Spectral observations of Jupiter in the frequency interval 18.5 – 24.0 GHz: 1968.
D. E. Jones, B. L. Meredith, E. Arthur.
Publ. Astron. Soc. Pacific, Vol. 82, 122 - 125 (1970).

For 13 days between January 16 and April 10, 1968, Jupiter's brightness temperatures were measured at seven discrete local oscillator frequencies between 18.5 and 24.0 GHz with a Dicke radiometer on the 30-ft Cassegrain radio telescope at Goldstone. The results are in excellent agreement with those of S. E. Law and D. H. Staelin (Astrophys. Journ., Vol. 154, 1077 - 1086, 1968).

099.005 The 1964–65 apparition of Jupiter.
P. K. Mackal.
Strolling Astronomer, Vol. 22, 1 - 8 (1970).

099.006 A note on Jupiter.
P. K. Mackal.
Strolling Astronomer, Vol. 22, 8 (1970).

099.007 The atmosphere of Jupiter.
T. Owen.
Science, Vol. 167, 1675 - 1681 (1970).

This giant planet appears to represent an early stage in the history of the solar system.

099.008 A theory of Jovian dekametric emission.
R. A. Duncan.
Planet. Space Sci., Vol. 18, 217 - 228 (1970).

This paper describes new characteristics of Jovian dekametric emission and of its modulation by Jovian rotation and Io, and suggests an explanation of these phenomena.

099.009 Interferometry of Jupiter at 18 MHz with a 52800λ baseline. W. F. Block, M. P. Paul, T. D. Carr, G. R. Lebo, V. M. Robinson, N. F. Six.
Astrophys. Letters, Vol. 5, 133 - 136 (1970).

Millisecond bursts received at 18 MHz during the 1967 – 1968 apparition of Jupiter were analyzed with a tape-recording fringe interferometer having a baseline of 880 km (52800λ). The measurements indicated source diameters as small as 0.72 arc sec. No significant difference in the time of arrival of the bursts at the two stations was observed, other than that due to geometry.

099.010 Characteristics of Jupiter's decametric radio source measured with arc-second resolution.
G. A. Dulk.
Astrophys. Journ., Vol. 159, 671 - 684 (1970).

Long-baseline measurements of Jupiter's decametric radiation at 34 MHz were made with an intensity interferometer between Boulder, Colorado, and Clark Lake, California (120000λ), and with a phase interferometer between Boulder and the Arecibo Ionospheric Observatory (487000λ). On time scales of less than 0.1 second, the radiation received at the two ends of the baselines was highly correlated, giving an upper limit to the size of an incoherent source of 0.″1, or 400 km, at Jupiter. The observations also imply an upper limit to a possible coherent source of 1″, or 4000 km. The stability of the fringe rate shows that the source position is stable to within 700 km on time scales of at least 10 seconds, and apparent position shifts caused by interplanetary scintillations are less than 0.″2.

099.011 Morphology of the fine structure in the dynamic spectra of Jupiter's decametric radiation.
J. J. Riihimaa, G. A. Dulk, J. W. Warwick.
Astrophys. Journ., Suppl. Series, Vol. 19, 175 - 192 (1970).

High-resolution spectral recordings of Jupiter's decametric emission during the years 1963–1968 exhibit several different spectral types. We classify the spectra into three components: the radiation envelope, which has a duration imposed by the interplanetary medium; the substructure within the envelope, which is inherent to the source, and the superimposed, interference-fringelike modulation lanes, which have an unknown cause but seem to be related to conditions at Jupiter. We show how the substructure relates to Jupiter's longitude and Io's position and how the sense of drift of the modulation lanes is characteristic of the longitude facing earth. Of several suggested mechanisms to produce the modulation lanes, the most likely employ an interference phenomenon.

099.012 Search for visual aurorae on Jupiter. G. A. Dulk, J. A. Eddy, J. P. Emerson.
Astrophys. Journ., Vol. 159, 1123 - 1124 = Contr. Kitt Peak National Obs.,No. 482 (1970).

A new search for predicted Jupiter aurorae with high sensitivity but limited space and time coverage revealed no emission with intensity greater than 10 kilorayleighs.

099.013 Hydrogen abundance in Jupiter's atmosphere.
J. P. Emerson, J. A. Eddy, G. A. Dulk.
Icarus, Vol. 11, 413 - 416 = Contr. Kitt Peak National Obs., No. 516 (1969).

Spectra of Jupiter showing the (4–0)S(1) hydrogen quadrupole line at 6367.80 Å have been used to derive the amount of hydrogen contributing to the formation of the line: 75 ± 15 km-atm, neglecting scattering. This and measurements by other authors are used to test whether the C/H ratio for Jupiter's atmosphere is equal to the solar value.

099.014 Jupiter: Présentation 1969.
S. Cortesi.
Orion Schaffhausen, 28. Jahrgang, p. 8 - 12 (1970). – Rapport No. 18 du «Groupement planétaires SAS».

099.015 On the invisible Jovian satellites. A. Michalec.
Postępy Astron., Vol. 18, 215 - 217 (1970).
In Polish. – Short report.

099.016 **The upper atmosphere of Jupiter.**
D. M. Hunten.
Journ. Atmosph. Sci., Vol. 26, 826 - 834 = Contr. Kitt Peak National Obs., No. 462 (1969). – Conference paper (see 012.009).
The aeronomy of Jupiter's atmosphere at pressures <25 mb is surveyed. Attention is drawn to those areas most in need of further work, including both planetary and laboratory studies.

099.017 **Molecular absorption and the possible structure of the cloud layers of Jupiter and Saturn.**
V. G. Teifel.
Journ. Atmosph. Sci., Vol. 26, 854 - 859 (1969). – Conference paper (see 012.009).
This paper deals with spectrographic and spectrophotoelectric observations of the absorption bands of CH_4 at 6190 and 7250 Å, and NH_3 at 6450 Å, in the spectra of Jupiter and Saturn carried out in recent years.

099.018 **An estimate of line width and pressure from the high resolution spectrum of Jupiter at 11,000 Å.**
C. B. Farmer.
Journ. Atmosph. Sci., Vol. 26, 860 - 861 (1969). – Conference paper (see 012.009).
A preliminary analysis of high resolution image tube spectra of Jupiter in the region of the R branch of the $3\nu_3$ band of methane is reported. The results indicate, for a simple reflecting layer model, that the effective pressure for the formation of lines in this band is 2.3 ± 0.5 atm.

099.019 **Extension of calculations of rotational temperature and abundance of methane in the Jovian atmosphere.**
J. S. Margolis, K. Fox.
Journ. Atmosph. Sci., Vol. 26, 862 - 864 (1969). – Conference paper (see 012.009).
Rotational temperatures and abundances of methane in the Jovian atmosphere have been calculated for Lorentz half-widths $\gamma = 0.15, 0.20$ and 0.30 cm^{-1}. The results are within ~10 to 20% of those previously determined for $\gamma = 0.10$ cm^{-1}, and are not far from the values corresponding to no saturation.

099.020 **Distribution of ammonia on Jupiter.**
V. I. Moroz, D. P. Cruikshank.
Journ. Atmosph. Sci., Vol. 26, 865 - 869 (1969). – Conference paper (see 012.009).
Observations of the variation in intensity of an NH_3 band at 1.53μ in different positions on the disk of Jupiter are presented with a simple physical interpretation model.

099.021 **New studies of Jupiter's atmosphere.**
T. Owen, H. P. Mason.
Journ. Atmosph. Sci., Vol. 26, 870 - 873 = Contr. McDonald Obs. No. 447 (1969). – Conference paper (see 012.009).
Assuming that the shapes of weak methane lines are determined by hydrogen and helium collisional broadening, the value of H/He in the Jovian atmosphere is found to be much greater than unity. A preliminary interpretation of differences in the amount of absorption by a strong methane band over the disk of the planet leads to an estimation of height differences of 18 km over the Great Red Spot and 12 km over the equatorial zone. The yellow coloration observed on Jupiter and Saturn is compared with the properties of ammonium sulfide $(NH_4)_2 S$, a compound that may occur in the atmospheres of these planets.

099.022 **A two-layer model of the Jovian clouds.**
R. E. Danielson, M. G. Tomasko.
Journ. Atmosph. Sci., Vol. 26, 889 - 897 (1969). – Conference paper (see 012.009).

A two-layer model of the Jovian clouds consisting of an upper cloud layer and a lower semi-infinite cloud separated by a clear space is investigated.

099.023 **The thermal structure of the Jovian atmosphere.**
J. S. Hogan, S. I. Rasool, T. Encrenaz.
Journ. Atmosph. Sci., Vol. 26, 898 - 905 (1969). – Conference paper (see 012.009).
Thermal structure calculations have been carried out for the atmosphere of Jupiter above the level of the dense clouds, for boundary conditions suggested by recent observations. The resulting models are characterized by an extensive region of dynamical control above the cloud level, and a thermal inversion in the mesosphere, produced by absorption of solar IR energy in the 3020 cm^{-1} band of methane.

099.024 **The photochemistry of methane in the Jovian atmosphere.** D. F. Strobel.
Journ. Atmosph. Sci., Vol. 26, 906 - 911 = Contr. Kitt Peak National Obs., No. 460 (1969). – Conference paper (see 012.009).
The photochemistry of methane in the Jovian atmosphere is reviewed and the photolysis of acetylene, ethylene and ethane is discussed. Approximate photochemical calculations are made for a mixed atmosphere of CH_4 and H_2 with a CH_4 mixing ratio of 10^{-3}. In regions where diffusion effects are important, the solutions are modified to include these effects.

099.025 **The absorption of extreme ultraviolet solar radiation by Jupiter's upper atmosphere.**
R. J. W. Henry, M. B. McElroy.
Journ. Atmosph. Sci., Vol. 26, 912 - 917 = Contr. Kitt Peak National Obs., No. 465 (1969). – Conference paper (see 012.009).
The absorption of extreme ultraviolet (EUV) sunlight by Jupiter's upper atmosphere is investigated and a detailed description is offered for the collision processes by which photoelectrons lose energy. It is argued that thermal equilibrium between electrons, ions and neutrals is probable at least with current ideas on the structure of Jupiter's ionosphere. An estimate is given for the fraction of absorbed EUV energy converted locally into heat.

099.026 **Polarization measures of Jupiter and Saturn.**
J. S. Hall, L. A. Riley.
Journ. Atmosph. Sci., Vol. 26, 920 - 923 (1969). – Conference paper (see 012.009).

099.027 **The spectrum and transmission of ammonia under Jovian conditions.** J. C. Gille, T.-H. Lee.
Journ. Atmosph. Sci., Vol. 26, 932 - 940 (1969). – Conference paper (see 012.009).
The theory and data for calculating the positions, intensities, and half-widths of the spectral lines of ammonia with frequencies < 1400 cm^{-1} are described. The ground state rotation, ν_2, $2\nu_2 - \nu_2$, and the rotation-inversion bands of ν_2 and $2\nu_2$ were included.

099.028 **Laboratory studies of the visible NH_3 bands with applications to Jupiter.**
L. P. Giver, R. W. Boese, J. H. Miller.
Journ. Atmosph. Sci., Vol. 26, 941 - 942 (1969). – Conference paper (see 012.009).

099.029 **Collision-narrowed curves of growth for H_2 applied to new photoelectric observations of Jupiter.**
U. Fink, M. J. S. Belton.
Journ. Atmosph. Sci., Vol. 26, 952 - 962 = Contr. Kitt Peak National Obs., No. 472 (1969). – Conference paper (see 012.009).

Curves of growth suitable for the interpretation of hydrogen quadrupole lines in planetary atmospheres are calculated. Collisional narrowing of the line profile is included in the theory and it is shown how the curves of growth are affected by the phenomenon.

099.030 Radiative time constants in the atmosphere of Jupiter. P. J. Gierasch, R. M. Goody.
Journ. Atmosph. Sci., Vol. 26, 979 - 980 (1969). – Conference paper (see 012.009).

We have computed radiative decay times for thermal disturbances near the cloud tops of Jupiter and conclude that they are much larger than probable dynamical time constants. Under these circumstances radiative equilibrium calculations are of little significance.

099.031 Dynamics of Jupiter's cloud bands. A. P. Ingersoll, J. N. Cuzzi.
Journ. Atmosph. Sci., Vol. 26, 981 - 985 (1969). – Conference paper (see 012.009).

The observed zonal motion of Jupiter's atmosphere near the cloud tops is investigated assuming geostrophic balance and a systematic temperature difference between light and dark bands.

099.032 Jupiter's zonal winds: Variation with latitude. C. R. Chapman.
Journ. Atmosph. Sci., Vol. 26, 986 - 990 (1969). – Conference paper (see 012.009).

Visual data on Jovian spot motions for 1897 – 1966 are assembled to show the variation of zonal velocity with Jovian latitude. No physical interpretation of these observations is attempted.

099.033 Twelve-year periodicities in Jupiter's decametric radiation. T. D. Carr, A. G. Smith, F. F. Donivan, H. I. Register.
Bull. American Astron. Soc., Vol. 2, 186 (1970). – Abstr. AAS.

099.034 Broadband observations of Jovian S-burst activity. R. S. Flagg, D. L. Smoleny, G. R. Lebo.
Bull. American Astron. Soc., Vol. 2, 192 (1970). – Abstr. AAS.

099.035 Rocket UV observations of Jupiter. Y. Kondo.
Bull. American Astron. Soc., Vol. 2, 203 (1970). – Abstr. AAS.

099.036 Modulation lanes in the dynamic spectra of Jovian L bursts. J. J. Riihimaa.
Astron. Astrophys., Vol. 4, 180 - 188 (1970).

The fine structure of the dynamic spectra of Jovian L bursts displays repeated, tilted lanes. A spaced-spectrograph experiment in which two identical radio spectrographs at 21 to 23 MHz were operated over a 160-km baseline indicated that the lanes originate at Jupiter. The interplanetary scintillation only affects the group appearance of the lanes. The sign and magnitude of the drift of these lanes is a strong function of System III central meridian longitude of Jupiter. The lanes are an intensity effect. *

099.037 The abundance of ammonia in the atmosphere of Jupiter. H. P. Mason.
Astrophys. Space Sci., Vol. 7, 424 - 436 = Contr. Univ. Illinois, Chicago Circle Phys. Dep., No. 12 (1970).

A laboratory curve of growth analysis was made on the lines in two ammonia bands located at 6450 Å and 10800 Å and the abundance of ammonia in the atmosphere of Jupiter was determined. Strengths for some of the lines in the 6450 Å band were determined and half-widths of some strong lines were also measured.

099.038 Jupiters GRF und die solare Aktivität. H. Heuseler.
SuW, Vol. 9, 158 (1970).

099.039 An attempt of the determination of Jupiter's mass from observations of the minor planet 10 Hygiea. N. S. Chernykh.
Byull. Inst. Teoret. Astron., *Leningrad,* Vol. 12, 127 - 154 (1970). In Russian.

On the basis of the observations of the minor planet 10 Hygiea 1932 - 1967 Jupiter's mass is determined. About 300 observations were used. The conditional equations of 28 normal places were formed and solved for six versions with 6 to 13 unknowns. The correction to the reciprocal mass of Jupiter $1/m = 1047.355$ is $\Delta (1/m) = -0.041 \pm 0.031$ (mean error).

099.040 Procedures for recording central meridian transits on Jupiter and their reduction. P. W. Budine.
Strolling Astronomer, Vol. 22, 37 - 38 (1970).

099.041 La radioemissione decametrica di Giove. U. Dall'Olmo.
Coelum, Vol. 38, 59 - 68 (1970).

099.042 La radioemissione decametrica di Giove. U. Dall'Olmo.
Coelum, Vol. 38, 107 - 116 (1970).

099.043 On results of spectrophotometry of absorption bands of methane (6190 Å) and ammonia (6441 and 6478 Å) on Jupiter's disk. V. V. Avramchuk.
Astron. Zhurn. Akad. Nauk SSSR, Vol. 47, 577 - 585 (1970). In Russian. English translation in Soviet Astron. AJ, Vol. 14, No. 3.

It is established that intensities of absorption bands CH_4 6190 Å and NH_3 6441 and 6478 Å decrease to the edge of Jupiter's disk. It is shown that time variations of the intensity of bands of methane and ammonia have a real character and make up 40 - 45% for the disk center. The observed course of absorption bands of methane and ammonia over Jupiter's disk is explained satisfactorily within the limits of a model of the semi-infinite atmosphere.

099.044 Modulation of Jupiter's decametric radio emission by Io. S. F. Dermott.
Monthly Notices, Roy. Astron. Soc., Vol. 149, 35 - 44 (1970).

A d.c. circuit model has been proposed by Goldreich and Lynden-Bell to account for the modulation of Jupiter's decametric radio emission by Io. The model accounts for the observations only if the electrical resistance of Io is less than some critical resistance. These resistances are estimated and it is shown that Io's is probably the greater. It is also shown that the model probably cannot be applied to Amalthea.

099.045 Jovian decimetre radiation. J. E. Wilkinson.
Australian Journ. Phys., Vol. 23, 197 - 202 (1970).

The drift in longitude of electrons in Jupiter's magnetic field is considered. The characteristic time of synchrotron radiation yields a magnetic flux density of some 30 G at the equator of Jupiter.

099.046 Jupiter: His limb darkening and the magnitude of his internal energy source. L. M. Trafton, R. L. Wildey.
Science, Vol. 168, 1214 - 1215 (1970).

The most accurate infrared photometric observations

(8 to 14 microns) to date of the average limb darkening of Jupiter have been combined with the most refined deduction of jovian model atmospheres in which flux constancy has been closely maintained in the upper regime of radiative equilibrium and a much more accurate approximation of the 10- and 16-micron vibration-rotation bands of ammonia has been incorporated. The resulting comparison indicates that Jupiter is radiating from three to four times as much power as the planet is receiving from the sun.

099.047   Jupiter's convection and its Red Spot.
R. Smoluchowski.
Science, Vol. 168, 1340 - 1342 (1970).

Physical properties of the liquid hydrogen-helium layer of Jupiter are calculated and used in evaluating convection and in interpreting the approximately constant rate of longitudinal motion of the Red Spot on the basis of the Hide-Streett model.

099.048   Numerical analysis of orbits of hypothetical comets ejected from the Jovian surface.
M. A. Mamedov.
Izv. AN Azerb. SSR. Ser. fiz-tekhn. i matem. n., 1969, No. 3, p. 83 - 95.   In Russian. – Abstr. in Referativ. Zhurn. 51. Astron., 4.51.124 (1970).

099.049   Jupiter IX and XII.
S. Herrick, G. E. Newnam II.
IAU Circ. No. 2211 (1970).

099.050   On the nature of light and dark formations of Jupiter.   T. Sato.
Říše hvězd, Vol. 51, 7 - 11 (1970).   In Czech.

099.051   On the radio frequency spectrum of Jupiter.
L. J. Gleeson, M. P. C. Legg, K. C. Westfold.
Proc. Astron. Soc. Australia, Vol. 1, 320 - 322 (1970).

099.052   Spectrophotometry of the CH_4 absorption band in the titan spectrum.   Yu. D. Davudov.
Astron. Tsirk., No. 553, p. 1 - 2 (1970).   In Russian.

099.053   Lunar occultation observations of Jupiter at 74 cm and 128 cm.   S. Gulkis.
Radio Science, Vol. 5, 505 - 511 = National Radio Astron. Obs., *Green Bank*, Repr. Ser. A, No. 149 (1970).

This paper reports the results of a recent lunar-occulta - tion observation of Jupiter at 74 cm and 128 cm from which we have been able to derive the one-dimensional strip bright - ness distributions for strips nearly perpendicular to the equa - torial plane. These results support those earlier studies that show that the source size increases with increasing wavelength. The results of these observations are used in conjunction with observations at the higher frequencies to determine broad con - straints on two model magnetospheres.

099.054   Thermal radio emission from the major planets.
K. I. Kellermann.
Radio Science, Vol. 5, 487 - 493 = National Radio Astron. Obs., *Green Bank*, Repr. Ser. A, No. 151 (1970).

Measurements of the thermal radio emission from all major planets have been reported at millimeter and centime - ter wavelengths, and from Jupiter and Saturn at decimeter wavelengths as well. The measured brightness temperatures deduced from these measurements generally exceeds the ex - pected equilibrium temperature calculated from solar heating.

099.055   Satellites of Jupiter.   L. E. da Silva Machado, D. P. Pinto Filho.
Contr. Obs. Valongo, Univ. Federal Rio de Janeiro, Sér. II, No. 9 (1970).

099.056   The 1968–69 apparition of Jupiter.
P. K. Mackal.
Strolling Astronomer, Vol. 22, 73 - 84 (1970).

099.057   A procedure for drawing and notating strip sketches of Jupiter.   P. W. Budine.
Strolling Astronomer, Vol. 22, 97 - 101 (1970).

099.058   Structural differences in the Jupiter millisecond pulses.   H. Torgersen.
Phys. Norvegica, Vol. 3, (No. 3), 195 - 202 (1969).

The profile of millisecond pulses in the Jovian deca-metric radio emission (16 and 18 MHz) has been studied. A technique giving a time resolution of approx. 0.2 msec was developed. It seems possible to distinguish between two types of pulses with different characteristic profiles and occurrences.

099.059   Jupiter's Red Spot in 1967 – 1968.
H. G. Solberg, Jr.
Report NASA-CR-97610. [Available from CFSTI, Springfield, Va.]. 10 pp. (1968).

Photographic observations of Jupiter's Red Spot between 12 September 1967 and 17 July 1968 are reported. As the activity in the SEBs decreased, the Red Spot became larger and darker, finally regaining the prominence it had during early 1966.

099.060   Jupiter's conductive molten core of moderate tem - perature.   N. Kawai, Y. Inokuti.
Journ. Phys. Soc. Japan,Vol. 27, 1686 - 1689 (1969). – See Phys. Abstr., Vol. 73, No. 26312 (1970).

099.061   Determination of the mass of Jupiter from a study of the motion of (57) Mnemosyne.
A. D. Fiala.
Thesis, Yale Univ., New Haven, Conn. [Available from Univ. Microfilms, Ann Arbor, Mi.], 190 pp. (1969).

Mnemosyne has now been observed through one period of its long-period inequality since its discovery in 1859. A definitive set of osculating elliptic elements for the orbit was derived. An investigation was made to determine which other parameters of the model, principally the mass of Jupiter, might be corrected to obtain a better fit of theory to observation.

099.062   Decametric radiation from Jupiter.   R. G. Wilson.
Thesis, Univ. California, Los Angeles. [Available from Univ. Microfilms, Ann Arbor, Mi.], 115 pp. (1969).

099.063   Pulse shape of the millisecond pulses in Jupiter's decametric radio emission.   H. Torgersen.
Phys. Norvegica, Vol. 3, (No. 3), 210 (1969).

Abstract only given with material recorded on magnetic tape during Jupiter's 1966/67 and 1968/69 apparition as source, pulse shapes of millisecond pulses in the decametric emission are studied.

Spectra of Venus and Jupiter from 1800 to 3200 Å.   See Abstr. 093.015.

100 Saturn

100.001 **Sur la mise en évidence d'un quatrième anneau et d'une nouvelle division obscure dans le système des anneaux de Saturne.** P. Guérin.
Comptes Rendus Acad. Sci. Paris, Sér. B, Vol. 270, 125 - 128 (1970).
Au cours de quatre nuits, les 16, 26, 27 et 28 octobre 1969, nous avons pris environ 200 photographies de Saturne à l'opposition, au foyer Cassegrain de 17 m du télescope de 105 cm du Pic du Midi. L'examen des négatifs à la loupe nous révéla, sur chaque image, la présence de deux petites zones faiblement noircies au voisinage immédiat des bords équatoriaux oriental et occidental du globe, ces zones étant séparées de la limite intérieure de l'anneau C par une division obscure (transparente sur les négatifs) semblable à celle de Cassini et obéissant comme elle aux lois de la perspective.

100.002 **Observations of Saturn at λ 3.75 cm.**
T. V. Seling.
Astron. Journ., Vol. 75, 67 - 68 (1970).
Observations of Saturn were made at $\lambda 3.75$ cm and a disk temperature of $(168 \pm 11)°K$ was obtained. This value is consistent with measurements made at other wavelengths. The spectrum of Saturn shows an upturn at longer wavelengths, suggesting the possibility of nonthermal radiation from Saturn.

100.003 **The effect of meteoroidal bombardment on Saturn's rings.** A. F. Cook, F. A. Franklin.
Astron. Journ., Vol. 75, 195 - 205 (1970).
This study tries to establish whether Saturn's rings are undergoing a loss or a gain of material as a result of impacts with interplanetary particles. Cometary meteoroids probably dominate over interstellar dust at Saturn, and meteoroids from comets like P/Comet Halley probably constitute the bulk of the interplanetary dust near Saturn. There is some doubt as to whether the rings are slowly accreting or losing matter, although our results somewhat favor the latter.

100.004 **Observations of Saturn's rings at the moments of the earth's transit through their plane (1966).**
R. I. Kiladze.
Byull. Abastumansk. Astrofiz. Obs. No. 37, p. 151 - 164 (1969). In Russian.
Photographic photometry of the rings of Saturn in the 10-m focus of the 70-cm meniscus telescope of Abastumani Astrophysical Observatory was fulfilled. The moments of the transit of the earth through the plane of the rings, the surface brightness of their shadow side and the thickness of the rings (equal to 1.42 ± 0.49 km) were determined.

100.005 **The composition of Saturn's rings.**
G. P. Kuiper, D. P. Cruikshank, U. Fink.
Sky Telescope, Vol. 39, 14 (1970), with a correction in Sky Telescope, Vol. 39, 80 (1970).

100.006 **Saturnreport. Ein zusammenfassender Bericht über die Oppositionen 1965, 1966 und 1967/68.**
H. Haug.

SuW, Vol. 9, 76 - 78 (1970).

100.007 **Saturn 1968 - 1969.**
A. W. Heath.
Journ. British Astron. Ass., Vol. 80, 133 - 145 (1970). – Report Saturn Section British Astron. Ass.

100.008 **Saturn's rings: Identification of water frost.**
C. B. Pilcher, C. R. Chapman, L. A. Lebofsky, H. H. Kieffer.
Science, Vol. 167, 1372 - 1373 (1970).
A recently published infrared spectrum of Saturn's rings resembles our laboratory spectra of water frosts. Furthermore, there are discrepancies between the ring spectrum and ammonia frost spectra in the 2- to $2.5-\mu$ region. These discrepancies render unlikely a reported identification of ammonia frost in the ring spectrum.

100.009 **The thickness of Saturn's rings.** M. S. Bobrov.
Priroda, No. 3.70, p. 66 - 69 (1970). In Russian.

100.010 **The millimeter-wave spectrum of Saturn.**
G. T. Wrixon, W. J. Welch.
Bull. American Astron. Soc., Vol. 2, 226 (1970). – Abstr. AAS.

100.011 **A model of radial structure of the rings of Saturn.**
G. Colombo, F. A. Franklin.
Symposia Mathematica, Vol. 3, 65 - 74 (1970). – Conference paper (see 012.010).

100.012 **Study of the planet Saturn by the Hamilton Centre.**
K. E. Chilton.
Journ. Roy. Astron. Soc. Canada, Vol. 64, 53 - 54 (1970).

100.013 **Study of the intensity distribution in the methane band 6190 Å on Saturn's disk.**
V. G. Teifel, G. A. Kharitonova.
Astron. Tsirk., No. 549, p. 5 - 8 (1970). In Russian.

New white spot on Saturn.
Sky Telescope, Vol. 39, 56 (1970).

Semi-numerical method for solving Hill's problem. - Application to Phoebe. See Abstr. 042.039.

The infrared spectra of Jupiter and Saturn at 1.2 - 4.2 microns. See Abstr. 099.001.

Molecular absorption and the possible structure of the cloud layers of Jupiter and Saturn. See Abstr. 099.017.

Polarization measures of Jupiter and Saturn. See Abstr. 099.026.

Thermal radio emission from the major planets. See Abstr. 099.054.

101 Uranus, Neptune, Pluto, Transplutonian Planet

101.001 The occultation of B.D. -17° 4388 by Neptune on 1968 April 7. G. E. Taylor.
Monthly Notices, Roy. Astron. Soc., Vol. 147, 27 - 33 (1970).

All the available observations of the occultation of the star B.D. -17° 4388 by Neptune on 1968 April 7 are discussed. The photoelectric observations are used to provide (a) an extremely accurate determination of the relative positions of the star and the planet and (b) a new value for the diameter of the planet. The 'occultation' diameter (equatorial) thus determined, which presumably corresponds to the upper reaches of the atmosphere of Neptune, is 50830 km ± (s.d.) 140 km and this is increased to 50940 km ± (s.d.) 140 km if the correction for the deflection of light in a gravitational field is applied.

101.002 The diameter and structure of the atmosphere of Neptune from the occultation of BD -17° 4388.
E. F. Guinan, J. S. Shaw.
Bull. American Astron. Soc., Vol. 2, 195 - 196 (1970).
Abstr. AAS.

101.003 Data for Neptune from occultation observations. K. C. Freeman, G. Lynga.
Astrophys. Journ., Vol. 160, 767 - 780 (1970).

The present paper aims at a full discussion of the information to be derived from the photoelectric light curves of the occultation of BD -17°4388 by Neptune on 1968 April 7 with consistent time determination and with due regard to the refraction effects in the atmosphere of Neptune. We first derive the dimensions of Neptune and its position relative to the occulted star, then a model of the atmosphere, and finally the mean density of the planet itself.

101.004 Investigation on meridian observation of Neptune from the comparison with photoelectric observation of occultation. H. Yasuda.
Ann. Tokyo Astron. Obs., Second Series, Vol. 11, 202 - 207 (1969).

The angular distances between Neptune and BD -17°4388 were measured with meridian circle at Tokyo for a month around the occultation by Neptune on April 7 1968. To coincide our observational results with the photoelectric observations of the occultation, the position of Neptune obtained by meridian observations should be revised by +0."30 in right ascension. Various sources of the discrepancy are searched in both meridian and photoelectric observations. Its origin cannot be explained clearly.

101.005 De diameter van Neptunus nauwkeurig bepaald. H. J. Lamers.
Hemel en Dampkring, Vol. 68, 130 - 133 (1970).

101.006 Utflykt till Neptunus. G. Lyngå.
Astron. Tidssk., Vol. 3, 1 - 6 (1970).

Thermal radio emission from the major planets.
See Abstr. 099.054.

102 Comets

102.001 Resonance fluorescence of the CN free radical.
J. A. Myer, R. W. Nicholls.
Nature, Vol. 225, 928 - 929 (1970).

Cometary emission spectra exhibit features of some diatomic radicals. In this report we record some recent work on the fluorescent excitation of CN in a static system.

102.002 Comets and nongravitational forces. III.
B. G. Marsden.
Astron. Journ., Vol. 75, 75 - 84 (1970).

Three particular problems encountered in paper II of this series are explored further. The first concerns the difference between the nongravitational effects detected in the motions of the short- and the long-period comets. A further study has been made of the second problem, that of the systematic residuals in the orbit of P/Encke. The longest section of the paper deals with the difficulties that seem to arise when a comet strongly affected by nongravitational forces makes a relatively close approach to Jupiter. We have not succeeded in uncovering precisely what went amiss in the cases of P/Schaumasse and P/Perrine-Mrkos. On the other hand, we have established that the secular variation in the motion of P/Pons-Winnecke changed sign around the turn of the century, a fact that is presumably connected with this comet's repeated encounters with Jupiter.

102.003 On the relationship between comets and minor planets. B. G. Marsden.
Astron. Journ., Vol. 75, 206 - 217 (1970).

Earlier results on the librations of minor planets are confirmed here by means of numerical integrations in which allowance is also made for the perturbations by planets other than Jupiter. With the exception of Hidalgo, all known minor planets are able to avoid encounters with Jupiter. Since the short-period comets characteristically have frequent encounters with Jupiter, it follows that although cometary remnants may look like minor planets, they have subsequent lifetimes that are several orders of magnitude shorter than those of conventional minor planets. There is little real evidence that comets turn into truly planetary objects, and it is still questionable whether the Apollo objects, for example, can be ex-comets. It is reasonable that a comet that avoids Jupiter or temporarily librates could indeed be approaching the end of its cometary life.

102.004 Simulation experiment on the tail of type 1 comets.
H. Kubo, N. Kawashima, T. Itoh.
Journ. Geophys. Res., Vol. 75, 1937 - 1939 (1970).

The ionized tail of a comet is usually not a simple tail, but has a ray structure, composed of several thin filaments. An experiment has been conducted in the laboratory to simulate the situation by using a plasma from a coaxial plasma gun.

102.005 Changes in total energy for 392 long-period comets, 1800-1970. E. Everhart, N. Raghavan.
Astron. Journ., Vol. 75, 258 - 272 (1970).

All the long-period comets since 1800 are studied to find the changes $\Delta(1/a)$ in the reciprocal major axes of their orbits, both before and after their passage. The effects of all nine planets are included, the contribution of each planet to each comet's change in energy being tabulated. The results are compared with previous calculations and with the predicted statistical distributions of these changes in energy. The distribution of net changes (before plus after) is found to be symmetrical about zero.

102.006 Type I tail structures of comets within the inner coma region. K. Wurm, J. Rahe.
Icarus, Vol. 11, 408 - 412 (1969).

Several photographs of comets are reproduced and described which give information about the Type I tail structures within the inner coma region.

102.007 Physico-chemical phenomena in comets — I. Experimental study of snows in a cometary environment.
A. H. Delsemme, A. Wenger.
Planet. Space Sci., Vol. 18, 709 - 715 (1970).

This paper gives some experimental information on the behavior of snows of water and of the clathrate hydrate of methane at low pressure in the $100°K$ temperature range, which is the meaningful range for the onset of activity of the cometary nucleus.

102.008 Physico-chemical phenomena in comets — II. Gas adsorption in the snows of the nucleus.
A. H. Delsemme, D. C. Miller.
Planet. Space Sci., Vol. 18, 717 - 730 (1970).

Circumstantial evidence shows that water is likely to be abundant in the cometary nucleus. Therefore adsorption of gases on water snows must be taken into consideration. It is first shown that for an incoming comet the regulating mechanism in gas production is not the desorption process, but remains the sublimation rate of water snow. A model is proposed where some of the observed lifetimes are nothing else than lifetimes of ice grains in the evaporating halo, releasing molecules and radicals within an extended source.

102.009 The interaction of gravitational and nongravitational forces affecting the motions of comets.
B. G. Marsden.
Bull. American Astron. Soc., Vol. 2, 208 (1970). — Abstr. AAS.

102.010 Comets and nongravitational forces.
B. G. Marsden.
Symposia Mathematica, Vol. 3, 3 - 11 (1970). — Conference paper (see 012.010).

102.011 Collisional effects in cometary atmospheres. Part I. Model atmospheres and synthetic spectra.
D. J. Malaise.
Astron. Astrophys., Vol. 5, 209 - 227 (1970).

A model of cometary atmospheres has been developed in order to compute theoretical spectra of the head. The model takes into account collisional effects through a distribution of the total gas density in the atmosphere. It also comprises the differential Doppler shifts due to the gas expansion. Computations have been made for the CN violet band, and 4 high and 4 medium dispersion spectra have been taken for the comparison. As a result, total densities 4 to 5 orders of magnitude higher than currently admitted have been found.

102.012 Dust from cometary nuclei. W. F. Huebner.
Astron. Astrophys., Vol. 5, 286 - 297 (1970).

The influence of the sun's radiation on the evaporation of dust constituted from materials with high latent heats of vaporization, between 20 and 100 kcal/mole, is investigated. The life time and motion of evaporating dust particles, which become important in the coma of a comet at heliocentric distances $r_h < 0.7$ a.u., are discussed.

102.013 **Ionic comet tails and the direction of the solar wind.**
J. C. Brandt, J. Hardorp.
Astron. Astrophys., Vol. 5, 322 - 324 (1970).
Schlosser and Hardorp have interpreted the waviness of tail-rays in comet Morehouse (1908 III) in terms of fluctuations in the direction of the solar wind, which they found much smaller than expected from satellite measurements in the ecliptic plane. Their conclusion that this is due to Morehouse's high heliographic latitude conflicts with measurements of the gross orientations of ionic comet tails. Two alternative explanations are offered.

102.014 **Comets, interplanetary space, and the problems of the solar system.** S. K. Vsekhsvyatskii.
Ann. IQSY, (see 012.016), Vol. 4, 110 - 118 (1969).
The number of comets now observed and the distribution of their paths in the solar system are sufficient to indicate the conditions in a wide region of interplanetary space. Comets are also important to the understanding of the origin and evolution of the solar system.

102.015 **Surface temperature distribution over the rotating nucleus of a comet deficient in volatiles.**
Z. Sekanina.
Astron. Astrophys., Vol. 7, 109 - 119 (1970).
The surface temperature distribution along the equator of a rotating spherical comet nucleus is calculated on the assumption that the absorbed solar flux is fully spent on heating the nucleus, the losses on both the gas vaporization and the radiation flux outward from the nucleus being neglected.

102.016 **Analysis of the observed density distribution in cometary atmospheres.** P. Egibekov.
Dokl. AN Tadzh. SSR, Vol. 12, No. 9, p. 21 - 23 (1969). In Russian. – Abstr. in Referativ. Zhurn. 51. Astron., 5.51.315 (1970).

102.017 **A new method for the solution of the reverse problem of the mechanical theory of cometary forms.** O. V. Dobrovolsky, Kh. Ibadinov.
Byull. Inst. Astrofiz., *Dushanbe,*No. 50, p. 20 - 21 (1968). In Russian.
It was found a new formula for the solution of the reverse problem of the mechanical theory of cometary forms which simplifies the solution of this problem very much.

102.018 **Physical study of comets.** L. Biermann.
Trans. IAU, Vol. 14A, (see 003.028), 141 - 152 (1970). – Report of Commission 15.

102.019 **The radiants of the comets 1951 - 1961 and their identification with meteor showers.**
I. N. Zentsev.
Astron. Tsirk., No. 559, p. 7 - 8 (1970). In Russian.

102.020 **Numerical analysis of the distribution of orbits for hypothetical comets generated by eruption from the surface of a Jupiter satellite.** M. A. Mamedov.
Dokl. AN Azerb. SSR, Vol. 25, No. 9, p. 15 - 17 (1969). In Russian. – Abstr. in Referativ. Zhurn. 51. Astron., 7.51.124 (1970).

102.021 **On plasma envelopes of comets.** O. Mamadov.
Izv. AN Tadzh. SSR. Otd. fiz.-matem. i geol.-khim. n., 1969, No. 2 (32), p. 3 - 6. In Russian. – Abstr. in Referativ. Zhurn. 51. Astron., 7.51.325 (1970).

102.022 **The sun-grazing comets.**
D. Conger.
Strolling Astronomer, Vol. 22, 90 - 95 (1970).

Atlas of Cometary Forms. See Abstr. 003.099.

Superdense water ice.
See Abstr. 022.009.

The dynamical problems of small bodies in the solar system. See Abstr. 098.014.

Positions and motions of minor planets, comets and satellites. See Abstr. 098.034.

Dynamics of meteor streams and new asteroid–meteor and comet–meteor associations.
See Abstr. 104.022.

Comets and the formation of planets.
See Abstr. 107.014.

103 Comets: Listed Objects

103.001 **Comet notes.** E. Roemer.
Publ. Astron. Soc. Pacific, Vol. 82, 151 - 157
(1970). – Concerning the comets 1963 VI, VII; 1967 n;
1968 g, j; 1969 a, b, d, e, f, g, h.

103.002 **Report on comets.** J. C. Bennett.
Monthly Notes Astron. Soc. Southern Africa, Vol.
29, 3 - 5 (1970). – Concerning 1969g, i; 1970a.

103.003 **Comet notes.**
Journ. Astron. Soc. Victoria, Vol. 23, 17 - 22,
48 - 51 (1970).

103.004 **Comet notes.** E. Roemer.
Publ. Astron. Soc. Pacific, Vol. 82, 356 - 366
(1970). – Concerning 1968g, j; 1969 a, b, d, g, h, i; 1970a,
b, c.

103.005 **Die periodischen Kometen des Jahres 1970.**
K. Mayrhofer.
Sternenbote, 13. Jahrgang, p. 22 - 28 (1970).

103.006 **Kometer 1969.**
H. Q. Rasmusen.
Astron. Tidssk., Vol. 3, 92 - 93 (1970).

103.100 **Comet 1682 Halley**

Halley's comet in 1682. J. Classen.
Sky Telescope, Vol. 39, 102 (1970).

103.101 **Comet 1969g Tago-Sato-Kosaka**

**Ephemeris of the comet Tago-Sato-Kosaka
(1969g).**
Astron. Tsirk., No. 550, p. 1 (1970). In Russian.

Comet Tago-Sato-Kosaka 1969 g.
British Astron. Ass. Circ., No. 515 (1970).

Comet Tago-Sato-Kosaka 1969 g.
S. W. Milbourn.
British Astron. Ass. Circ., No. 519 (1970).

Osservazione della cometa Tago-Sato-Kosaka.
M. Lavelli.
Coelum, Vol. 38, 89 (1970).

Comet Tago-Sato-Kosaka (1969g).
Z. M. Pereyra, B. Oviedo, J. J. Rodriguez, F. Dossin, B. Mintz.
IAU Circ. No. 2197 (1970).

Comet Tago-Sato-Kosaka (1969g).
Z. M. Pereyra.
IAU Circ. No. 2200 (1970).

Comet Tago-Sato-Kosaka (1969g).
A. D. Code, T. E. Houck, C. F. Lillie.
IAU Circ. No. 2201 (1970).

Comet Tago-Sato-Kosaka (1969g).
B. Milet.
IAU Circ. No. 2202 (1970).

Comet Tago-Sato-Kosaka (1969g).
V. M. Blanco, B. Mintz, A. Gomez, Z. M. Pereyra, J. J. Rodriguez, B. Oviedo, T. Seki, M. Miranian.
IAU Circ. No. 2204 (1970).

Comet Tago-Sato-Kosaka (1969g).
I. Andruszkiw, B. J. Harris, M. P. Candy, I. Nikoloff, D. Gans, D. Harwood, T. Seki, B. Milet.
IAU Circ. No. 2208 (1970).

Comet Tago-Sato-Kosaka (1969g).
Z. M. Pereyra, J. J. Rodriguez, B. Oviedo, B. G. Jørgensen, B. Reipurth, A. Kizilirmak, N. Güdür, A. Heiser, S. Furia.
IAU Circ. No. 2212 (1970).

Comet Tago-Sato-Kosaka (1969g).
E. Roemer, G. E. D. Alcock.
IAU Circ. No. 2214 (1970).

Comet Tago-Sato-Kosaka (1969g).
B. Reipurth, B. G. Jørgensen, B. Milet, J. E. Bortle, M. Sugano, B. G. Marsden.
IAU Circ. No. 2218 (1970).

Comet Tago-Sato-Kosaka (1969g).
B. Milet.
IAU Circ. No. 2223 (1970).

Comet Tago-Sato-Kosaka (1969g).
M. Antal, A. N. Argue, H. A. Couper, T. Seki, A. Mrkos, H. Debehogne, G. Roland.
IAU Circ. No. 2224 (1970).

Comet Tago-Sato-Kosaka (1969g).
B. G. Jørgensen, B. Reipurth, T. Seki, A. Mrkos.
IAU Circ. No. 2230 (1970).

Comet Tago-Sato-Kosaka (1969g).
D. Gans, D. Harwood, P. Birch, I. Andruszkiw.
IAU Circ. No. 2235 (1970).

Comet Tago-Sato-Kosaka (1969g).
H. Potter, A. Lokalov, C. Torres, M. Wischniewsky, J. Petit, H. Wroblewski, G. Soulié, T. Seki, A. Mrkos, I. Hasegawa.
IAU Circ. No. 2242 (1970).

Comet Tago-Sato-Kosaka (1969g).
C. Torres, M. Wischnjewsky, J. Petit, H. Wroblewski.
IAU Circ. No. 2249 (1970).

Comet Tago-Sato-Kosaka, 1969g.
Kometn. Tsirk., *Kiev,* No. 95 (1970). In Russian.

Comet Tago-Sato-Kosaka, 1969g.
Kometn. Tsirk., *Kiev,* No. 96 (1970). In Russian.

Comet Tago-Sato-Kosaka, 1969g.
Kometn. Tsirk., *Kiev,* No. 97 (1970). In Russian.

Comet Tago-Sato-Kosaka, 1969g.
Kometn. Tsirk., *Kiev,* No. 98 (1970). In Russian.

Comet Tago-Sato-Kosaka, 1969g.
Kometn. Tsirk., *Kiev,* No. 99 (1970). In Russian.

Spectrum and explosion of the comet Tago-Sato-Kosaka, 1969g.

M. B. Babaev, M. S. Gadzhiev, E. B. Gusev.
Kometn. Tsirk., *Kiev,* No. 99 (1970). In Russian.

Comet Tago-Sato-Kosaka, 1969g.
Kometn. Tsirk., *Kiev,* No. 100 (1970). In Russian.

Comet Tago-Sato-Kosaka, 1969g.
Kometn. Tsirk., *Kiev,* No. 101 (1970). In Russian.

Comet Tago-Sato-Kosaka, 1969g.
Kometn. Tsirk., *Kiev,* No. 102 (1970). In Russian.

Comet Tago-Sato-Kosaka, 1969g.
Kometn. Tsirk., *Kiev,* No. 103 (1970). In Russian.

Comet Tago-Sato-Kosaka 1969g. J. Bouška.
Říše hvězd, Vol. 51, 85 - 88 (1970). In Czech.

Spectra of comet Tago-Sato-Kosaka.
F. Dossin.
Sky Telescope, Vol. 39, 152 - 153 (1970).

Comet Tago-Sato-Kosaka (1969g).
Sky Telescope, Vol. 39, 196 - 198 (1970).

Photos from comet 1969 g from Cerro Tololo.
A. Gomez, V. Blanco.
Sky Telescope, Vol. 39, 228-231 (1970).

Comet 1969 g observed around the world.
Sky Telescope, Vol. 39, 262 - 266 (1970).

Ephemeris of comet Tago-Sato-Kosaka 1969g.
Strolling Astronomer, Vol. 22, 71 - 72 (1970).

Comet Tago-Sato-Kosaka (1969g).
I. Janos.
Urania Kraków, Vol. 41, 153 - 155 (1970). In Polish.

Komet 1969g Tago-Sato-Kosaka.
C. Kowalec.
VdS Nachrichtenblatt, 19. Jahrgang, p. 29 - 30 (1970).

Komet 1969g – Tago-Sato-Kosaka. C. Kowalec.
VdS Nachrichtenblatt, 19. Jahrgang, p. 41 (1970).

Comet Tago-Sato-Kosaka, 1969 g.
Yamamoto Circ., Nos. 1711, 1712, 1713, 1715, 1716, 1718 (1970). In Japanese.

Dunkelkammerarbeit an einer Kometenaufnahme.
See Abstr. 036.005.

103.102 Comet 1969i Bennett

Comet Bennett (1969i).
Astron. Tsirk., No. 557, p. 1 (1970). In Russian.

Comet Bennett 1969i.
Astron. Tsirk., No. 558, p. 1 (1970). In Russian.

The 10-micron emission peak of comet Bennett
1969i. R. W. Maas, E. P. Ney, N. J. Woolf.
Astrophys. Journ. *(Letters),* Vol. 160, L101 - L104 (1970).
Photometric observations of comet Bennett from 2 to 20 microns show a blackbody-like continuum at short wavelengths, and a sharp peak at 10 microns identified as due to silicate grains.

Spectroscopic observations of comet Bennett near
perihelion. M. E. Méndez.
Bol. Obs. Tonantzintla y Tacubaya, Vol. 5 (No. 34), 239 - 240 (1970).
Several spectrograms of comet Bennett were obtained, which cover the spectral range from $\lambda\lambda 3100$ to 8700 Å. Molecules of CN, C_2, CH, C_3, CH^+ and OH are present.

New comet Bennett 1969 i.
J. C. Bennett.
British Astron. Ass. Circ., No. 515 (1970).

Comet Bennett 1969 i.
M. P. Candy.
British Astron. Ass. Circ., No. 519 (1970).

Comet Bennett 1969 i.
British Astron. Ass. Circ., No. 520 (1970).

Comet Bennett 1969 i.
British Astron. Ass. Circ., No. 523 (1970).

Cometa Bennett (1969i). A. Bernasconi.
Coelum, Vol. 38, 87 - 88 (1970).

Observation de l'émission d'hydrogène atomique
de la comète Bennett. J.-L. Bertaux, J. Blamont.
Comptes Rendus Acad. Sci. Paris, Sér. B, Vol. 270, 1581 - 1584 (1970).

Komeet Bennett 1969i.
Hemel en Dampkring, Vol. 68, 151 - 156 (1970).

Comet Bennett (1969i).
B. J. Harris, M. P. Candy.
IAU Circ. No. 2198 (1970).

Comet Bennett (1969i).
B. J. Harris, M. P. Candy, D. Gans.
IAU Circ. No. 2199 (1970).

Comet Bennett (1969i).
M. P. Candy.
IAU Circ. No. 2202 (1970).

Comet Bennett (1969i).
M. P. Candy, Z. M. Pereyra, J. J. Rodriguez, B. Oviedo.
IAU Circ. No. 2203, 2208 (1970).

Comet Bennett (1969i).
Z. M. Pereyra, B. Oviedo, J. J. Rodriguez.
IAU Circ. No. 2205 (1970).

Comet Bennett (1969i).
C. Torres, H. Wroblewski, M. Wischniewsky, S. Barros, H. Potter, A. Lokalov, M. P. Candy, I. Nikoloff, D. Gans, D. Harwood.
IAU Circ. No. 2207 (1970).

Comet Bennett (1969i).
F. de J. Bateman, Z. M. Pereyra, J. J. Rodriguez, B. Oviedo.
IAU Circ. No. 2213 (1970).

Comet Bennett (1969i).
J. A. Bruwer.
IAU Circ. No. 2214 (1970).

Comet Bennett (1969i).
Z. M. Pereyra, J. J. Rodriguez.
IAU Circ. No. 2217 (1970).

Comet Bennett (1969i).
B. G. Marsden.
IAU Circ. No. 2219 (1970).

Comet Bennett (1969i).
J. J. Rodriguez, B. Oviedo, S. C. McMillan, V. L. Matchett.
IAU Circ. No. 2223 (1970).

Comet Bennet (1969i).
IAU Circ. No. 2226 (1970).

Comet Bennett (1969i).
B. Mintz, A. Gomez, F. Dossin, M. V. Jones, V. L. Matchett,
B. G. Marsden.
IAU Circ. No. 2228 (1970).

Comet Bennett (1969i).
H. Potter, A. Lokalov, C. Torres, H. Wroblewski, J. J. Rodri-
guez, Z. M. Pereyra, T. Seki, J. C. Bennett, K. Simmons,
D. Li, J. E. Bortle.
IAU Circ. No. 2229 (1970).

Comet Bennett (1969i).   B. Milet.
IAU Circ. No. 2231 (1970).

Comet Bennett (1969i).
B. J. Harris, M. P. Candy, I. Nikoloff, D. Gans, I. Andruszkiw,
P. Muller, B. Milet, P. L. Bernacca, A. Mammano.
IAU Circ. No. 2232 (1970).

Comet Bennett (1969i).
B. Milet, Z. M. Pereyra, F. W. Gerber, L. González, J. E.
Bortle.
IAU Circ. No. 2234 (1970).

Comet Bennett (1969i).
H. Potter, A. Lokalov, C. Torres, M. Wischniewsky, J. Petit,
H. Wroblewski, M. Bielicki, B. Milet.
IAU Circ. No. 2238 (1970).

Comet Bennett (1969i).
C. Torres, H. Wroblewski, M. Wischnjewsky, J. Petit,
Heudier-Helmer, P. Muller, A. Mrkos.
IAU Circ. No. 2247 (1970).

Comet Bennett (1969i).   J. Codina, A. Mrkos.
IAU Circ. No. 2250 (1970).

Comet Bennett (1969i).
R. L. Waterfield, M. J. Hendrie, H. Morgan, R. South.
IAU Circ. No. 2252 (1970).

Comet Bennett (1969i).
H. Debehogne, G. Roland, T. Urata.
IAU Circ. No. 2255 (1970).

Comet Bennett (1969i).
H. Soper, S. W. Milbourn, B. Milet, A. Mrkos.
IAU Circ. No. 2258 (1970).

New comet Bennett, 1969i.
Kometn. Tsirk., *Kiev*, No. 95 (1970).   In Russian.

Comet Bennett, 1969i.
Kometn. Tsirk., *Kiev*, No. 96 (1970).   In Russian.

Comet Bennett, 1969i.
Kometn. Tsirk., *Kiev*, No. 99 (1970).   In Russian.

Comet Bennett, 1969i.
Kometn. Tsirk., *Kiev*, No. 100 (1970).   In Russian.

Comet Bennett, 1969i.
Kometn. Tsirk., *Kiev*, No. 101 (1970).   In Russian.

Comet Bennett, 1969i.
Kometn. Tsirk., *Kiev*, No. 102 (1970).   In Russian.

Comet Bennett, 1969i.
N. Chernykh, M. Shmakova.
Kometn. Tsirk., *Kiev*, No. 103 (1970).   In Russian.

Comet Bennett, 1969i.
Kometn. Tsirk., *Kiev*, No. 104 (1970).   In Russian.

Comet Bennett, 1969i.
Kometn. Tsirk., *Kiev*, No. 105 (1970).   In Russian.

Observations de la comète Bennett (1969i) à
l'Observatoire de Haute-Provence.   C. Fehrenbach.
L'Astronomie, 84e année, p. 257 - 260 (1970).

La comète Bennett (1969i).
Orion Schaffhausen, 28. Jahrgang, p. 96 (1970).

Ephemeris of comet Bennett 1969i.
Strolling Astronomer, Vol. 22, 72 (1970).

Bright comet Bennett (1969i).
Sky Telescope, Vol. 39, 199 (1970).

Comet Bennett's fine show.
Sky Telescope, Vol. 39, 330 - 333 (1970).

The great comet of 1970.
Sky Telescope, Vol. 39, 351 - 356 (1970).

Aufnahmen des Kometen 1969i (Bennett).
SuW, Vol. 9, 156 - 157 (1970).

Komet 1969i Bennett.
E. Mädlow.
VdS Nachrichtenblatt, 19. Jahrgang, p. 30 (1970).

Der Komet Bennett (1969i).
B. Wedel, P. Völker, J. Herrmann, H. B. Brenske.
VdS Nachrichtenblatt, 19. Jahrgang, p. 50 - 54 (1970).

Komet Bennett (1969i).
C. Kowalec.
VdS Nachrichtenblatt, 19. Jahrgang, p. 66 (1970).

Comet Bennett, 1969 i.
Yamamoto Circ., Nos. 1711, 1712, 1713, 1715, 1717,
1718, 1719, 1720 (1970).   In Japanese.

103.103   Comet 1970a Daido-Fujikawa

New comet Daido-Fujikawa 1970 a.
G. E. D. Alcock.
British Astron. Ass. Circ., No. 516 (1970).

Comet Daido-Fujikawa 1970 a.
B. G. Marsden.
British Astron. Ass. Circ., No. 517 (1970).

Comet Daido-Fujikawa 1970 a.
B. G. Marsden.
British Astron. Ass. Circ., No. 519 (1970).

Comet Daido-Fujikawa (1970a).

Daido, Fujikawa, H. Kosai.
IAU Circ. No. 2203 (1970).

Comet Daido-Fujikawa (1970a).
Kanai, T. Urata, H. Kosai, K. Locher.
IAU Circ. No. 2205 (1970).

Comet Daido-Fujikawa (1970a).
M. Honda, H. Kosai, T. Seki, K. Tomita, B. Milet, M. Mattei.
IAU Circ. No. 2206 (1970).

Comet Daido-Fujikawa (1970a).
T. Seki, B. Milet, H. Kosai, K. Tomita.
IAU Circ. No. 2208 (1970).

Comet Daido-Fujikawa (1970a).
T. Urata, T. Yumoto, H. L. Giclas, E. Roemer.
IAU Circ. No. 2210 (1970).

Comet Daido-Fujikawa (1970a).
T. Seki.
IAU Circ. No. 2215 (1970).

Comet Daido-Fujikawa (1970a).
B. Milet, T. Kurosaki, T. Urata, T. Seki.
IAU Circ. No. 2216 (1970).

New comet Daido-Fujikawa, 1970a.
Kometn. Tsirk., *Kiev*, No. 97 (1970).   In Russian.

Elements and ephemeris of the comet Daido-Fujikawa, 1970a.
Kometn. Tsirk., *Kiev*, No. 97 (1970).   In Russian.

Comet Daido-Fujikawa, 1970a.
Kometn. Tsirk., *Kiev*, No. 98 (1970).   In Russian.

Comet Daido-Fujikawa, 1970a.
Kometn. Tsirk., *Kiev*, No. 99 (1970).   In Russian.

Komet 1970a Daido-Fujikawa.
E. Mädlow.
VdS Nachrichtenblatt, 19. Jahrgang, p. 30 (1970).

Comet Daido-Fujikawa, 1970 a.
Yamamoto Circ., Nos. 1713, 1714, 1716, (1970).
In Japanese.

103.104   Comet 1967 VII Mitchell-Jones-Gerber

An investigation of the brightness of two comets.
D. D. Meisel.
Astron. Journ., Vol. 75, 252 - 257 (1970).

Major sources of systematic errors in comet magnitude estimates and their relation to standard reduction formulas are discussed. Empirical regressions using over 90 visual magnitude estimates for two recent comets – 1967f (Mitchell-Jones-Gerber) and 1967d (P/Tempel 2) – are given. The results are discussed.

103.105   Comet 1967 X Tempel 2

An investigation of the brightness of two comets.
D. D. Meisel.
Astron. Journ., Vol. 75, 252 - 257 (1970).
See Abstr. 103.104.

103.106   Comet 1963 III Alcock

Observations of comet Alcock (1963b).
G. Chincarini.
Mem. Soc. Astron. Italiana, Nuova Serie, Vol. 41, 9 - 20 (1970).

During the outburst of comet Alcock 1963b some halos with an expansion velocity of about 0.6 km/sec were visible. After a few days a rather detailed type I tail appeared which lasted about one day. Rays were visible in the coma to a distance $\leqslant 1.8 \times 10^4$ km from the nucleus. A particular and very intense jet developed afterwards, the observed velocity of the dust grains was 45 m/sec.

103.107   Comet 1908 III Morehouse

Lebensdauer und Dichte der CO⁺-Ionen im Kometen Morehouse.   J. Rahe.
Astron. Nachr., Vol. 292, 31 - 33 (1970).

A photometry of five photographs of comet Morehouse 1908 III gives a decay constant for CO^+ ions of $\gamma = 3.5 \times 10^{-6}$ s^{-1} (heliocentric distance of the comet at the time of observation 1.4 A. U.). In a distance from the nucleus of 1.2×10^6 km the average CO^+ ion density amounts to 130/cm^3.

103.108   Comet 1968 VI Honda

Polarization measurements of the comet Honda 1968c.   R. S. Osherov.
Dokl. AN Tadzh. SSR, Vol. 13, No. 1, p. 15 - 18 (1970). In Russian. – Abstr. in Referativ. Zhurn. 51. Astron., 7.51.328 (1970).

Comète Honda (1968c).
Gaz. astron., No. 2, p. 5 - 6 (1969).

Observations of comets.
H. Debehogne, G. Roland.
IAU Circ. No. 2236 (1970).

Comet Honda, 1968c.   V. 1. Voronenko.
Kometn. Tsirk., *Kiev*, No. 102 (1970).   In Russian.

103.109   Comet 1853 III Klinkerfues

The thesis by I. N. Uljanov "Olbers' method for the determination of parabolic orbits and its application to the comet 1853 Klinkerfues".   G. A. Chebotarev.
Byull. Inst. Teoret. Astron., *Leningrad*, Vol. 12, 119 - 126 (1970).   In Russian.

I. N. Uljanov (1831 - 1886), a famous Russian teacher and democrat, father of V. I. Uljanov-Lenin, in 1854 supported the thesis on the subject: "Olbers' method for the determination of parabolic orbits and its application to the comet 1853 Klinkerfues" at the University in Kazan. The manuscript has not been published up to now. In the present paper the detailed commentary on the work of I. N. Uljanov and additional information about the comet 1853 III are given.

103.110 Comet 1969b Kohoutek

Comet Kohoutek 1969 b.
B. G. Marsden.
British Astron. Ass. Circ., No. 516 (1970).

Comet Kohoutek 1969 b.
R. L. Waterfield.
British Astron. Ass. Circ., No. 517 (1970).

Comet Kohoutek (1969b).
S. I. Gerasimenko, B. A. Burnaševa, A. Mrkos.
IAU Circ. No. 2197 (1970).

Comet Kohoutek (1969b).
B. G. Marsden.
IAU Circ. No. 2199 (1970).

Comet Kohoutek (1969b).
R. L. Waterfield.
IAU Circ. No. 2204 (1970).

Comet Kohoutek (1969b).
T. Seki, B. Milet.
IAU Circ. No. 2210 (1970).

Comet Kohoutek (1969b).
A. Mrkos.
IAU Circ. No. 2217 (1970).

Comet Kohoutek (1969b).
A. Mrkos, B. Milet.
IAU Circ. No. 2225 (1970).

Comet Kohoutek (1969b).
T. Seki, A. Mrkos.
IAU Circ. No. 2233 (1970).

Comet Kohoutek (1969b).   A. Mrkos.
IAU Circ. No. 2246 (1970).

Comet Kohoutek (1969b).
L. Kohoutek, B. Milet.
IAU Circ. No. 2256 (1970).

Observations of comet Kohoutek, 1969b.
Kometn. Tsirk., *Kiev*, No. 95 (1970).   In Russian.

Comet Kohoutek, 1969b.
Kometn. Tsirk., *Kiev*, No. 96 (1970).   In Russian.

Comet Kohoutek, 1969b.
Kometn. Tsirk., *Kiev*, No. 97 (1970).   In Russian.

Comet Kohoutek, 1969b.
Kometn. Tsirk., *Kiev*, No. 98 (1970).   In Russian.

Comet Kohoutek, 1969b.
Kometn. Tsirk., *Kiev*, No. 100 (1970).   In Russian.

Comet Kohoutek, 1969b.
Kometn. Tsirk., *Kiev*, No. 101 (1970).   In Russian.

Comet Kohoutek, 1969b.
Kometn. Tsirk., *Kiev*, No. 103 (1970).   In Russian.

Comet Kohoutek, 1969b.
Kometn. Tsirk., *Kiev*, No. 104 (1970).   In Russian.

Comet Kohoutek, 1969b.
Kometn. Tsirk., *Kiev*, No. 105 (1970).   In Russian.

Comet Kohoutek 1969b.
Strolling Astronomer, Vol. 22, 72 (1970).

Comet Kohoutek, 1969 b.
Yamamoto Circ., Nos. 1712, 1720 (1970).   In Japanese.

103.111 Comet 1965 III Wolf-Harrington

Ephemeris of the periodic comet Wolf-Harrington (1952 II) for its reappearance in 1970–72.
G. Sitarski.
Acta Astron., Vol. 20, 155 - 157 (1970).
The ephemeris is based on the linkage of three apparitions of the comet using 81 observations made in 1951 – 52, 1957 – 59 and 1964 – 65. The perturbations from Mercury to Neptune as well as the nongravitational changes in the motion were included. The conditions of observation of the comet in 1971/72 are presented graphically.

Periodic comet Wolf-Harrington.
G. Sitarski.
IAU Circ. No. 2257 (1970).

103.112 Comet 1969f Slaughter-Burnham

Improved ephemeris of the periodic comet Slaughter-Burnham (1958 VI) for 1970 – 71.
G. Sitarski.
Acta Astron., Vol. 20, 159 - 161 (1970).
Using 14 observations made in 1958/59 and three made in 1969 the orbital elements for 1959 Feb. 1.0 E.T. were improved, representing all the observations with a mean residual of $\pm 0\overset{.}{''}87$. Then including the perturbations from Mercury to Neptune and integrating the equations of motion of the comet by Cowell's method, the elements for 1970 April 4.0 E.T. were obtained and the improved ephemeris of the comet was computed.

Other comets.
British Astron. Ass. Circ., No. 515 (1970).

Periodic comet Slaughter-Burnham (1969f).
G. Sitarski.
IAU Circ. No. 2253 (1970).

103.113 Comet 1969h Čurjumov-Gerasimenko

Ephemeris of comet Čurjumov-Gerasimenko (1969h).
Astron. Tsirk., No. 543, p. 2 (1970).   In Russian.

Periodic comet Čurjumov-Gerasimenko (1969h).
B. A. Burnaševa, S. I. Gerasimenko, V. L. Afanas'ev,
S. K. Vsehsvjatskij.
IAU Circ. No. 2201 (1970).

Periodic comet Čurjumov-Gerasimenko (1969h).
M. P. Candy, T. Seki, B. Milet.
IAU Circ. No. 2209 (1970).

Periodic comet Čurjumov-Gerasimenko (1969h).
E. Roemer, B. Schreur, A. Mrkos, B. Milet.
IAU Circ. No. 2224 (1970).

Periodic comet Čurjumov-Gerasimenko (1969h).

A. Mrkos.
IAU Circ. No. 2237 (1970).

**Short-period comet Čurjumov-Gerasimenko,
1969h.**
Kometn. Tsirk., *Kiev,* No. 95 (1970).   In Russian.

Comet Čurjumov-Gerasimenko, 1969h.
Kometn. Tsirk., *Kiev,* No. 98 (1970).   In Russian.

Periodic comet Čurjumov-Gerasimenko, 1969h.
Kometn. Tsirk., *Kiev,* No. 100 (1970).   In Russian.

Periodic comet Čurjumov-Gerasimenko, 1969h.
Kometn. Tsirk., *Kiev,* No. 101 (1970).   In Russian.

Periodic comet Čurjumov-Gerasimenko, 1969h.
V. Afanas'ev, M. Shmakova.
Kometn. Tsirk., *Kiev,* No. 103 (1970).   In Russian.

The comet Čurjumov-Gerasimenko 1969th.
K. I. Churiumov, S. I. Gerasimenko.
Priroda, No. 4.70, p. 88 - 89 (1970).   In Russian.

Observations of comets.
Yamamoto Circ., No. 1711 (1970).   In Japanese.

103.114   Comet 1969d Fujikawa

Comet Fujikawa (1969d).
Z. M. Pereyra, B. Oviedo, J. J. Rodriguez.
IAU Circ. No. 2198 (1970).

Comet Fujikawa (1969d).
I. Nikoloff, M. P. Candy.
IAU Circ. No. 2215 (1970).

Comet Fujikawa, 1969d.
Kometn. Tsirk., *Kiev,* No. 97 (1970).   In Russian.

Comet Fujikawa, 1969d.
N. N. Kiselev, A. G. Krylov.
Kometn. Tsirk., *Kiev,* No. 100 (1970).   In Russian.

103.115   Comet 1969a Faye

Other comets.
British Astron. Ass. Circ., No. 515 (1970).

Periodic comet Faye (1969a).
Z. M. Pereyra, B. Oviedo, J. J. Rodriguez.
IAU Circ. No. 2200 (1970).

Periodic comet Faye (1969a).
H. L. Giclas.
IAU Circ. No. 2202 (1970).

Periodic comet Faye (1969a).
M. Antal, L. Pajdušáková, D. Harwood, B. J. Harris,
Petrovičová, A. Mrkos, B. Milet.
IAU Circ. No. 2215 (1970).

Periodic comet Faye (1969a).
M. P. Candy.
IAU Circ. No. 2240 (1970).

Periodic comet Faye (1969a).

D. Ferguson, T. Griess, S. Kanagy, J. Levine.
IAU Circ. No. 2249 (1970).

Continued ephemeris of comet Faye, 1969a.
Kometn. Tsirk., *Kiev,* No. 95 (1970).   In Russian.

Short-periodic comet Faye, 1969a.
Kometn. Tsirk., *Kiev,* No. 96 (1970).   In Russian.

Comet Faye, 1969a.
Kometn. Tsirk., *Kiev,* No. 98 (1970).   In Russian.

Periodic comet Faye, 1969a.
Kometn. Tsirk., *Kiev,* No. 100 (1970).   In Russian.

103.116   Comet 1970b Pons-Winnecke

Other comets.
British Astron. Ass. Circ., No. 519 (1970).

Periodic comet Pons-Winnecke (1970b).
E. Roemer.
IAU Circ. No. 2206 (1970).

Recovery of the comet Pons-Winnecke, 1970b.
Kometn. Tsirk., *Kiev,* No. 97 (1970).   In Russian.

Comet P/Pons-Winnecke, 1970 b.
Yamamoto Circ., No. 1714 (1970).   In Japanese.

103.117   Comet 1969e Honda-Mrkos-Pajdušáková

Other comets.
British Astron. Ass. Circ., No. 515 (1970).

Periodic comet Honda-Mrkos-Pajdušáková (1969e).
B. Milet.
IAU Circ. No. 2211 (1970).

103.118   Comet 1968g Comas Solá

Periodic comet Comas Solá (1968g).
B. J. Harris, I. Nikoloff, B. Milet, T. Seki.
IAU Circ. No. 2212 (1970).

Periodic comet Comas Solá (1968g).
T. Seki, A. Mrkos, Petrovičová, B. Milet.
IAU Circ. No. 2227 (1970).

Periodic comet Comas Solá (1968g).
D. Gans, A. Mrkos, R. Petrovičová.
IAU Circ. No. 2235 (1970).

Periodic comet Comas Solá (1968g).
D. Ferguson, T. Griess, S. Kanagy, J. Levine, A. Mrkos,
R. Petrovičová.
IAU Circ. No. 2243 (1970).

Periodic comet Comas Solá, 1968g.
Kometn. Tsirk., *Kiev,* No. 98 (1970).   In Russian.

Periodic comet Comas Solá, 1968g.
Kometn. Tsirk., *Kiev,* No. 99 (1970).   In Russian.

Periodic comet Comas Solá, 1968g.

Kometn. Tsirk., *Kiev*, No. 101 (1970). In Russian.

Periodic comet Comas Solá, 1968g.
Kometn. Tsirk., *Kiev*, No. 103 (1970). In Russian.

Observations of comets.
Yamamoto Circ., No. 1711 (1970). In Japanese.

103.119 Comet 1970c Kopff

Other comets.
British Astron. Ass. Circ., No. 519 (1970).

Periodic comet Kopff (1970c).
E. Roemer.
IAU Circ. No. 2213 (1970).

Periodic comet Kopff (1970c).
E. Roemer, B. Schreur.
IAU Circ. No. 2223 (1970).

Recovery of the comet Kopff, 1970c.
Kometn. Tsirk., *Kiev*, No. 99 (1970). In Russian.

Short-periodic comet Kopff, 1970c.
Kometn. Tsirk., *Kiev*, No. 105 (1970). In Russian.

Comet P/Kopff, 1970 c.
Yamamoto Circ., No. 1716 (1970). In Japanese.

103.120 Comet 1969c Whipple

Periodic comet Whipple (1969c).
B. G. Marsden.
IAU Circ. No. 2220 (1970).

Ephemeris of comet Whipple, 1969c.
Kometn. Tsirk., *Kiev*, No. 95 (1970). In Russian.

103.121 Comet 1941 VII Du Toit-Neujmin-Delporte

**Periodic comet Du Toit-Neujmin-Delporte
(1941 VII).** B. G. Marsden.
IAU Circ. No. 2222 (1970).

103.122 Comet 1970d D'Arrest

Comet P/D'Arrest 1970 d.
British Astron. Ass. Circ., No. 520 (1970).

Periodic comet D'Arrest (1970d). E. Roemer.
IAU Circ. No. 2227 (1970).

Periodic comet D'Arrest (1970d).
E. Roemer, R. Elliott, B. Schreur.
IAU Circ. No. 2249 (1970).

Periodic comet D'Arrest (1970d).
T. Urata.
IAU Circ. No. 2254 (1970).

**Recovery of the short-periodic comet D'Arrest,
1970d.**

Kometn. Tsirk., *Kiev*, No. 101 (1970). In Russian.

Short-periodic comet D'Arrest, 1970d.
Kometn. Tsirk., *Kiev*, No. 105 (1970). In Russian.

Comet P/D'Arrest (1970 d).
Yamamoto Circ., Nos. 1717, 1721 (1970).
In Japanese.

103.123 Comet 1963 IV Johnson

Periodic comet Johnson.
N. A. Beljaev, E. A. Vorob'ev.
IAU Circ. No. 2236 (1970).

**Elements and ephemeris of the comet Johnson
(1949 II).** E. A. Vorob'ev.
Kometn. Tsirk., *Kiev*, No. 102 (1970). In Russian.

103.124 Comet 1968 V Whitaker-Thomas

Observations of comets.
H. Debehogne, G. Roland.
IAU Circ. No. 2236 (1970).

103.125 Comet 1936 IV Jackson-Neujmin

Periodic comet Jackson-Neujmin (1936 IV).
B. G. Marsden.
IAU Circ. No. 2240 (1970).

Periodic comet Johnson-Neujmin (1936 IV).
Kometn. Tsirk., *Kiev*, No. 104 (1970). In Russian.

103.126 Comet 1970e Ashbrook-Jackson

Comet P/Ashbrook-Jackson 1970 e.
British Astron. Ass. Circ., No. 522 (1970).

Periodic comet Ashbrook-Jackson (1970e).
Z. M. Pereyra.
IAU Circ. No. 2241 (1970).

Periodic comet Ashbrook-Jackson (1970e).
Z. M. Pereyra, B. Oviedo.
IAU Circ. No. 2244 (1970).

Short-periodic comet Ashbrook-Jackson, 1970e.
Kometn. Tsirk., *Kiev*, No. 104 (1970). In Russian.

Comet P/Ashbrook-Jackson, 1970 e.
Yamamoto Circ., No. 1719 (1970). In Japanese.

103.127 Comet 1967 XIII Encke

Periodic comet Encke. B. G. Marsden.
IAU Circ. No. 2244 (1970).

**Elements and ephemeris of the short-periodic
comet Encke.** N. Bokhan.
Kometn. Tsirk., *Kiev*, No. 103 (1970). In Russian.

103.128 Comet 1970f White-Ortiz-Bolelli

New bright comet White-Ortiz-Bolelli 1970 f.
British Astron. Ass. Circ., No. 521 (1970).

Comet White-Ortiz-Bolelli 1970 f.
British Astron. Ass. Circ., No. 522 (1970).

Comet White-Ortiz-Bolelli 1970 f.
British Astron. Ass. Circ., No. 523 (1970).

Comet White-Ortiz-Bolelli (1970f).
G. White, E. Ortiz, C. Bolelli, V. M. Blanco, Z. M. Pereyra, H. M. Maitzen, A. Moffat, H. E. Schuster, H. Potter, B. J. Harris, M. P. Candy, G. N. Sprott.
IAU Circ. No. 2246 (1970).

Comet White-Ortiz-Bolelli (1970f).
V. L. Matchett, S. C. McMillan, Z. M. Pereyra.
IAU Circ. No. 2248 (1970).

Comet White-Ortiz-Bolelli (1970f).
E. Ortiz, H. M. Maitzen, A. Moffat, H. E. Schuster, J. B. Savio, C. Bolelli, R. Gonzáles, V. M. Blanco, B. Mintz, M. V. Jones, S. C. McMillan, V. L. Matchett, C. Torres.
IAU Circ. No. 2250 (1970).

Comet White-Ortiz-Bolelli (1970f).
G. L. White, F. W. Gerber, L. González, J. B. Savio, V. L. Matchett, M. V. Jones, S. C. McMillan.
IAU Circ. No. 2251 (1970).

Comet White-Ortiz-Bolelli (1970f).
E. Ortiz, Stewart, A. Jones, C. Bolelli, B. Mintz, M. V. Jones, S. C. McMillan.
IAU Circ. No. 2253 (1970).

New bright comet White-Ortiz-Bolelli, 1970f.
Kometn. Tsirk., *Kiev,* No. 104 (1970). In Russian.

Comet White-Ortiz-Bolelli, 1970f.
Kometn. Tsirk., *Kiev,* No. 105 (1970). In Russian.

Comet White-Ortiz-Bolelli, 1970 f.
Yamamoto Circ., Nos. 1719, 1720, 1721 (1970). In Japanese.

103.129 Comet 1964 V Arend-Rigaux

Periodic comet Arend-Rigaux.
B. G. Marsden.

IAU Circ. No. 2248 (1970).

103.130 Comet 1938 I Gale

Periodic comet Gale.
IAU Circ. No. 2256 (1970).

103.131 Comet 1960 III Schaumasse

Definitive orbit of the comet Schaumasse from observations of 1951 - 52. K. P. Matsukov.
Uch. zap. Kemerovsk.gos. ped. in-t, 1969, vyp. (No.) 10, p. 3 - 24. In Russian. — Abstr. in Referativ. Zhurn. 51. Astron., 6.51.149 (1970).

103.132 Comet 1966 V Kilston

Photographic observations of the comet Kilston 1966 b and Barbon 1966 c in Tashkent.
N. R. Alieva, Yu. M. Ivanov.
Tsirk. Astron. Inst., *Tashkent,* No. 6 (353), p. 13 - 17 (1968). In Russian.

103.133 Comet 1966 II Barbon

Photographic observations of the comet Kilston 1966 b and Barbon 1966 c in Tashkent.
See Abstr. 103.132.

103.134 Comet 1965 VIII Ikeya-Seki

On the main tail of the comet Ikeya- Seki 1965 f.
O. V. Dobrovolsky, I. Kh. Ibadinov.
Byull. Inst. Astrofiz., *Dushanbe,* No. 49, p. 28 - 31 (1969). In Russian.
   The main tail of the comet 1965 f was investigated on the basis of pre-perihelion photographs.

103.135 Comet 1970g Abe

New comet Abe, 1970g.
Kometn. Tsirk., *Kiev,* No. 105 (1970). In Russian.

104 Meteors, Meteor Streams

104.001 On the orbits of bright fireballs.
L'. Kresák.
Bull. Astron. Inst. Czechoslovakia, Vol. 21, 1 - 9 (1970).

The differences of the orbits of bright sporadic fireballs from those of faint photographic meteors are critically examined. The clustering of meteors in the diagram of geocentric velocity vs. inclination, interpreted by Rajchl as evidence of two classes of fireball orbits, is explained by the geometrical conditions of encounter combined with a higher resistivity of certain types of orbits to planetary perturbations and collisional destruction.

104.002 On the problem of hyperbolic meteors.
J. Štohl.
Bull. Astron. Inst. Czechoslovakia, Vol. 21, 10 - 17 (1970).

The generally accepted explanation of the observed hyperbolicity of meteors is studied in somewhat greater detail on the basis of the statistics of meteors and their radiant distributions. The effect of perturbations on meteor orbits is investigated, and the conditions are derived under which a meteor with a hyperbolic orbit caused by perturbations could be observed at the earth. The question of hyperbolic interstellar meteors is reexamined on the basis of a model for which the annual variation of interstellar meteors is derived.

104.003 Comparison of radar and optical meteor observations. Results of the meteor expedition Ondřejov 1962.
L. Kohoutek, J. Grygar, Z. Plavcová, J. Kvizová.
Bull. Astron. Inst. Czechoslovakia, Vol. 21, 18 - 28 (1970).

Simultaneous radar and optical observations were performed during the meteor expedition in 1962 at Ondřejov. Within 8 nights in August—September a total of 1792 visual, 5611 telescopic, and 2893 radar observations was obtained. The diurnal variation of the hourly rates for telescopic meteors is very small; it is more conspicuous and mutually similar for visual and radar meteors. The luminosity function was studied.

104.004 Physiological distortion of telescopic meteor observations. M. Šulc.
Bull. Astron. Inst. Czechoslovakia, Vol. 21, 29 - 37 (1970).

In this paper, we made an attempt to broaden the knowledge achieved so far by results of the research of subjective influence on the determination of position angles of meteors and to solve the problem of the correctness of one assumption of the independent-calculation method.

104.005 Structure of the meteor stream associated with comet Halley. A. Hajduk.
Bull. Astron. Inst. Czechoslovakia, Vol. 21, 37 - 45 (1970).

The structure of the Orionid and the Eta Aquarid meteor streams along and across their orbits has been studied. Based on an extensive observational material of the Orionids, the following results have been obtained. 1. the density of the stream varies semi-regularly with a period of 5 to 15 years; 2. the density and the particle size distribution in a cross-section of the stream are different for different sections of the orbit as met at annual apparitions; 3. stream filaments with diameter of the order of 10^6 km are identified along the orbit. The Eta Aquarids show very similar structural features. The association of both these streams with comet Halley appears confirmed.

104.006 Faint meteor classification.
B. Baldwin, Y. Sheaffer.
Meteoritics, Vol. 4, 258 - 259 (1969). – Abstract.

104.007 The Leonids by radar – 1957 to 1968.
B. A. McIntosh, P. M. Millman.
Meteoritics, Vol. 4, 283 - 284 (1969). – Abstract.

104.008 Faint meteor ablation processes.
B. Baldwin, Y. Sheaffer.
Journ. Geophys. Res., Vol. 75, 495 - 498 (1970).

The purpose of this note is to present additional evidence in support of the hypothesis that most meteoroids are solid prior to entry into the atmosphere. A model is proposed for the sequence of events leading to eventual vaporization of the meteoroid. Explanations are given for the phenomena of wake blending (persistent luminosity in the wake) and meteor trails that appear to begin abruptly (Jacchia et al., 1965).

104.009 Microparticle collection experiments during the 1966 Orionid and Leonid meteor showers.
G. V. Ferry, M. B. Blanchard, N. H. Farlow.
Journ. Geophys. Res., Vol. 75, 859 - 870 (1970).

Results are presented from two microparticle collection experiments carried out in 1966 during Orionid and Leonid meteor showers. No collection of particles in either size range was detected above the background contamination levels for either flight. Particles from the Luster flight were analyzed for elemental composition using an electron microprobe. Certain elemental groups were found in increased concentrations on flight surfaces in comparison with similar surfaces that were not exposed. These results are in general agreement with the results of the 1965 Luster flight into the Leonid meteor shower. If the increased concentrations of particles with certain compositions reflect true collections, then either the Orionid and the Leonid meteor showers are similar in composition, or a collection of sporadic meteoric material has been made.

104.010 Über die Beobachtung von Meteoren.
R. Lukas.
SuW, Vol. 9, 50, 52 (1970).

104.011 Meteoric matter in the earth's neighbourhood.
T. N. Nazarova.
Kosmich. Issled., Vol. 8, 154 - 156 (1970). In Russian. Brief information.

104.012 The 11 Canis Minorids – a new meteor stream probably associated with comet Mellish 1917 I.
K. B. Hindley, M. A. Houlden.
Nature, Vol. 225, 1232 - 1233 (1970).

We have detected a new meteor stream complex, active between solar longitudes $\lambda_\odot$ = 257° and 262°, with a true radiant some 3° north-west of 11 Canis Minor. The stream was first recorded on the morning of December 10, 1964. The true radiant and preliminary orbit are given. The orbit is similar to that of the comet Mellish (1917 I).

104.013 Quadrantid meteors in 1970.
K. B. Hindley.
Sky Telescope, Vol. 39, 269 (1970).

104.014 On the determination of the appearance time of meteors. T. Takenouchi, K. Tomita, M. Nukariya.
Tokyo Astron. Obs. Rep., Vol. 15, No. 1 (No. 56), p. 195 - 201 (1970). In Japanese.

104.015 Fireball and meteorite of April 25, 1969.
K. B. Hindley, H. G. Miles.
Nature, Vol. 225, 255 - 257 (1970).

An accurate atmospheric trajectory has been determined from 300 visual observations of the British fireball of 1969 April 25. At φ = +51.9°, λ = 4.2°W, the path was towards 332° ± 2°, descending at 9° ± 2°. Two associated meteorite falls lie close to this track. The corrected radiant has been used to determine the most likely orbit, an ellipse inclined at 13°, with perihelion at 0.68AU. Attention is drawn to the great similarity shown by this orbit and that of the New Jersey (USA) fireball of 1962 April 24.

104.016 **Definition of the angular sizes of a meteor train by the method of frequency variation.**
A. F. Yakovetz.
Geomagn. Aeronom., Vol. 10, 164 - 166 (1970).   In Russian.
Brief information.

104.017 **Selected records of astronomical phenomena (meteor events) from old Armenian chronicles.**
I. S. Astapovich, B. E. Tumanyan.
Uch. zap. Erevansk. un-t. Estestv. n., 1969 No. 2 (111), p. 40 - 47.   In Armenian. – Abstr. in Referativ. Zhurn. 51. Astron., 3.51.6 (1970).

104.018 **Measurements of radiants and orbits of weak meteors.**   B. L. Kashcheev, I. A. Delov,
B. S. Dudnik, N. V. Novoselova, A. A. Tkachuk.
Vestn. Khar'kovsk. politekhn. in-ta, 1969 No. 36 (84), p. 15 - 21.   In Russian. – Abstr. in Referativ. Zhurn. 51. Astron., 3.51.348 (1970).

104.019 **Magnetic fields of meteor streams as a possible mechanism of diffusion of cosmic rays in the** circumsolar space.   S. A. Belsky.
Astron. Zhurn. Akad. Nauk SSSR, Vol. 47, 201 - 205 (1970). In Russian. English translation in Soviet Astron. AJ, Vol. 14, No. 1.
On the basis of an earlier estimation of the possible value of magnetic fields in meteor streams, formed as a result of passing of charged micrometeors through the magnetic plasma, it is supposed that magnetic fields of meteor streams can yield a valuable contribution to diffusion of cosmic rays in the circumsolar space.

104.020 **The determination of the initial radii of meteor trains.**   W. J. Baggaley.
Monthly Notices, Roy. Astron. Soc., Vol. 147, 231 - 243 (1970).
Estimates of the spatial density, mass distribution and velocity characteristics of interplanetary matter derived from observations of radio-meteors may contain uncertainties if neglect is made of the initial transverse dimensions of the ionized meteor trains. Observations of radio-meteors employing a dual wavelength system are described and values of train initial radius together with their variation with height and meteoroid velocity are presented.

104.021 **On the physical theory of Super-Schmidt meteors.**
A. F. Cook.
Bull. American Astron. Soc., Vol. 2, 188 - 189 (1970).
Abstr. AAS.

104.022 **Dynamics of meteor streams and new asteroid— meteor and comet—meteor associations.**
Z. Sekanina.
Bull. American Astron. Soc., Vol. 2, 217 - 218 (1970).
Abstr. AAS.

104.023 **Results of photographic meteor observations in Kiev in 1962 - 1964.**
V. G. Kruchinenko, V. V. Benyuch, A. A. Demenko, S. S. Tryashin, N. A. Hinkulova, L. M. Sherbaum.

Vestn. Kiev. Univ., Ser. Astron., No. 11, p. 59 - 90 (1969). In Russian.
The results of photographic meteor observations are given. A table contains the corrected and true radiants, initial and final heights, the velocities and initial masses, the elements of the orbit. Besides, "new" values of the meteor masses are calculated, taking into consideration the dependence of the luminosity factor on mass and structure of the meteor body.

104.024 **Absolute photometry of meteor flares.**
V. V. Benyuch, E. V. Sandakova.
Vestn. Kiev. Univ., Ser. Astron., No. 11, p. 91 - 95 (1969). In Russian.
A method for absolute photometry of meteor flares is given. The influence of development effects and light scattering in a photographic emulsion is considered. The light curve of a meteor obtained by the usual method and by that described in this paper is given.

104.025 **The meteor train of November 18, 1966.**
E. V. Sandakova, A. A. Demenko.
Vestn. Kiev. Univ., Ser. Astron., No. 11, p. 96 - 98 (1969). In Russian.
The diffusion coefficient and luminescence of the meteor train are discussed. The photographs of the train were obtained near Ashkhabad.

104.026 **Some questions of the physical theory of meteors.**
A. R. Kolomiets.
Vestn. Kiev. Univ., Ser. Astron., No. 11, p. 99 - 104 (1969). In Russian.
It is noted that the usual definition of the ionization coefficient does not correspond to its true physical meaning. Having assumed simple ionization of meteor atoms, we obtained a new equation for the linear electron density in the trail. Some considerations on the form of the evaporation equation are suggested.

104.027 **Radar observations of meteors at the epoch of the 1967 Leonids.**
E. I. Fialko, I. V. Bairachenko, R. I. Moisya, G. I. Kolomiets, V. I. Melnik, Yu. V. Bitsenko, V. N. Doniy, V. F. Romanyuk.
Vestn. Kiev. Univ., Ser. Astron., No. 11, p. 105 - 109 (1969). In Russian.

104.028 **The distribution of orbit parameters and the changes in incident meteor particle flux density.**
N. S. Andrianov, U. A. Pupysev, V. V. Sidorov.
Monthly Notices, Roy. Astron. Soc., Vol. 148, 227 - 237 (1970).
This paper presents the results of measurements, made at the Kazan Radio Astronomy Observatory (U.S.S.R.), of the orbits of meteors and the incident flux of meteoroids. The elimination of velocity selection has not resulted in considerable changes in the distributions of orbit parameters. Observations for many years have shown that seasonal peculiarities in the distributions of sporadic meteor radiants over the celestial sphere and the average incident meteor particle flux value remain unchanged from year to year.

104.029 **The Leonids by radar – 1957 to 1968.**
B. A. McIntosh, P. M. Millman.
Meteoritics, Vol. 5, 1 - 18 (1970).
Radar observations of the Leonid meteor shower, made near Ottawa during the years from 1957 to 1968 inclusive, are analyzed and reduced to give comparative flux rates. A strength classification has been made in terms of the ratio of shower rates to background rates. The relative strengths found by radar, showing marked variability from year to year, are confirmed by analysis of available visual observations. There is also great variation in the distribution of particle sizes.

104.030   **The mechanics of meteor showers.**   A. C. Parkin.
Journ. Astron. Soc. Victoria, Vol. 23, 46 (1970).

104.031   **The Leonids - 1969.**   P. M. Millman.
Journ. Roy. Astron. Soc. Canada, Vol. 64, 55 -
57 (1970).

104.032   **The fireball and meteorite of 1969 April 25.**
K. B. Hindley, H. G. Miles.
Journ. British Astron. Ass., Vol. 80, 313 - 322 (1970).

104.033   **The luminosity function of telescopic meteors.**
Z. Kvíz, J. Mikušek.
Contr. Public Obs. and Planetarium Brno 4, p. 7 - 15 (1967).
In Czech.
The present paper deals with observations made by
means of the independent counting method in the course of
telescopic meteor expedition to Hlavačky. The frequency of
sporadic meteors has been found to increase by 5.6 times per
unit increase in magnitude for meteors of brightness 7^m.

104.034   **Observation of the α-Lyrid meteor stream.**
J. Grygar, L. Kohoutek, Z. Kvíz, J. Mikušek.
Contr. Public Obs. and Planetarium Brno 4, 17 - 28 (1967).
In Czech.
In the course of our Mt. Bezovec meteor expedition an
exceptionally high hourly rate of telescopic meteors was recorded. These records were fully confirmed by an observation of an extraordinal α-Lyrid stream at the Simferopol
department of the VAGO dating from the same time.

104.035   **Subjective distortions of telescopically observed
meteors.**   J. Grygar, L. Kohoutek.
Contr. Public Obs. and Planetarium Brno 4, p. 29 - 51 (1967).
In Czech.
Results of the Mount-Bezovec meteor expedition 1958
are communicated.

104.036   **Atmospheric trajectories of telescopic meteors.**
L. Kohoutek, J. Grygar.
Contr. Public Obs. and Planetarium Brno 4, p. 51 - 102 (1967).
In Czech.
Observations of telescopic meteors were performed from
a 2.5 km long base line by a group of eight observers during
the Mt. Bezovec expedition. In eight nights – from 10th till
27th July, 1958 – 123 pairs were registered, that is 62 per
cent of all cases recorded.

104.037   **Observations of telescopic meteors in the years
1958 - 1962.**   M. Šulc.
Contr. Public Obs. and Planetarium Brno 7, p. 1 - 31 (1969).
In Czech.
Distribution of meteors according to the position angle
and according to the hourly rates for individual days of observation were obtained from the material observed in the
years 1958 to 1962.

104.038   **Meteoritic flux determined from visual observations.**
P. M. Millman.
Journ. Royal Astron. Soc. Canada, Vol. 64, 187 - 190 (1970).
From 155 visual meteor observations a new mean hourly
flux rate on a monthly basis is determined.

104.039   **On the origin and distribution of meteoroids.**
J. S. Dohnanyi.
Journ. Geophys. Res., Vol. 75, 3468 - 3493 (1970).
In this paper we examine the mass distribution of
sporadic meteors. Using reasonable assumptions regarding
the physics of hypervelocity impact we show that the mass
distribution of sporadic meteoroids is not stable under the
influence of mutual inelastic collisions between individual

particles. Theory and observation of the mass distribution
of several major showers are compared. The present theory
is found to be self-consistent and in reasonable agreement with
observations.

104.040   **Methods of investigating multipath propagation
on a meteor radiotrace.**   A. F. Yakovetz.
Geomagn. Aeronom., Vol. 10, 473 - 477 (1970).   In Russian.

104.041   **Fireballs and the physical theory of meteors.**
R. E. McCrosky, Z. Ceplecha.
SAO *Cambridge*, Mass., Special Rep. No. 305, 10 + 65 pp.
(1969).
The purpose of this paper is to investigate the new observational material of very bright meteors provided by the
Prairie Network and to derive additional constraints on the
meteor theory or on the structure of the meteoroid. Although
we cannot resolve the question of the origin of meteorites
with the new information, we can increase our confidence in
the low-density interpretation of a significant part of the faint-
meteor data.

104.042   **Drifts of meteor trains in Ashkhabad 1965.
Results of telescopic observations.**
A. P. Savrukhin.
Izv. AN Turkm. SSR. Ser. fiz.-tekhn., khim. i geol. n., 1969,
No. 5, p. 79 - 84.   In Russian. – Abstr. in Referativ. Zhurn.
51. Astron., 4.51.343 (1970).

104.043   **Statistical analysis of observations of meteoric
phenomena.**   L. V. Chernysheva.
Zhurn. tekhn. fiz., Vol. 39, 1689 - 1693 (1969).   In Russian.
Abstr. in Referativ. Zhurn. 51. Astron., 5.51.320 (1970).

104.044   **On the theory of ablation of meteors.**
V. S. Getman.
Dokl. AN Tadzh. SSR, Vol. 12, No. 11, p. 10 - 11 (1969).
In Russian. – Abstr. in Referativ. Zhurn. 51. Astron.,
5.51.322 (1970).

104.045   **Orbits of 77 photographic meteors 1964.**
P. B. Babadzhanov, T. I. Getman, A. F. Zausayev,
S. A. Karaselnikova.
Byull. Inst. Astrofiz., *Dushanbe*, No. 49, p. 3 - 12 (1969).
In Russian.

104.046   **Results of measurements of drifts of meteor trails
over Tadzhikistan during the IQSY period.**
Sh. O. Isamutdinov, R. P. Chebotarev.
Byull. Inst. Astrofiz., *Dushanbe*, No. 49, p. 13 - 16 (1969).
In Russian.
From more than 85 000 measurements made during the
IQSY, data on the wind at heights of 70 - 105 km are presen-
ted.

104.047   **Measurement of the angular coordinates of meteor
trails with the Doppler method.**
V. M. Kolmakov.
Byull. Inst. Astrofiz., *Dushanbe*, No. 49, p. 17 - 20 (1969).
In Russian.
A method for measuring the coordinates of meteor
trails is described. The theoretical basis for the method is
presented and the possibility of a control of the measurements
with meteors is shown.

104.048   **On the scales of turbulence from photoobservations
of meteor trains.**   U. Shodiev.
Byull. Inst. Astrofiz., *Dushanbe*, No. 49, p. 21 - 23 (1969).
In Russian.
The dimensions of small-scale and large-scale turbulences
for heights of 82 - 94 km are given. The dissipation energy

according to the scales of turbulence is determined.

104.049 **The Scorpionids.**   A. M. Bakharev.
 Byull. Inst. Astrofiz., *Dushanbe,* No. 49, p. 24 - 27
(1969).   In Russian.
 The history of investigation of the Scorpionid shower,
its radiants and its connection with the comet 1770 I are
described.

104.050 **The dispersion of meteors in meteor streams.**
    I. The size of the radiant areas.
L'. Kresák, V. Porubcan.
Bull. Astron. Inst. Czechoslovakia, Vol. 21, 153 - 170 (1970).
 The dispersion of individual radiants in meteor showers
is investigated using all available double-station photographic
observations of high accuracy. The variability of the radiant
pattern with the position within the stream and with the semi-
major axis of meteor orbits is examined.

104.051 **The deceleration of faint meteors.**
    N. V. Novoselova.
Astron. Tsirk., No. 546, p. 1 - 3 (1970).   In Russian.

104.052 **Dependence of the coefficient of ambipolar dif-**
    fusion of a meteor trail on height.
V. V. Zhukov.
Astron. Tsirk., No. 546, p. 3 - 4 (1970).   In Russian.

104.053 **Meteors and meteorites.**   Z. Ceplecha, E. Anders.
    Trans. IAU, Vol. 14A, (see 003.028), 207 - 223
(1970). – Report of Commission 22.

104.054 **Addition to the Commission 22 report of the IAU.**
    P. B. Babadžanov.
Bull. Astron. Inst. Czechoslovakia, Vol. 21, 192 - 193 (1970).

104.055 **Spectrum of a Leonid meteor.**
    M. S. Rao, P. V. S. Ramarao.
Current Sci. India, Vol. 38, (No. 11), 262 - 263 (1969).
 Describes the third meteor spectrum taken from India
at Waltair. The meteor appeared on 17/18 November 1965,
and was identified to be a Leonid. Its apparent visual
magnitude was visually estimated to be minus one.

104.056 **Fluid mechanics of meteor trails.**
    V. C. Liu.
Phys. Fluids, Vol. 13, 62 - 65 (1970).

104.057 **Meteors and radar.**   E. J. Maanders.
    Philips Electronic Meas. & Microwave Notes,
Vol. 69, No. 1, p. 1 - 10 (1969).
 This paper comprises a general description of the radar
equipment at Eindhoven Univ., a short account of some of
the components and a survey of the properties of ionised
meteor trails which can be measured.

104.058 **Long enduring meteor trains. Fourth paper.**
    C. P. Olivier.
Proc. American Phil. Soc., Vol. 113, No. 2, p. 127 - 139 =
Flower and Cook Obs., *Philadelphia,* Repr. No. 189 (1969).

104.059 **Meteor reports. American Meteor Society for 1968.**
    C. P. Olivier.
Flower and Cook Obs., *Philadelphia,* Repr. No. 190, 17 pp.
(1969).

104.060 **Long-base observation of telescopic meteors.**
    V. Znojil.
Contr. Obs. and Planetarium, Brno, No. 8/I, 12 pp. (1969).
In Czech.

104.061 **Meteor sighting probability and the problem of the**
    actual number of meteors.   V. Znojil.
Contr. Obs. and Planetarium, Brno, No. 8/II, 19 pp. (1969).
In Czech.

104.062 **Frequency occurrence of small particles in meteor**
    showers. I. Aurigids, Lyrids, Scorpio-Sagittarids,
Taurids.   V. Znojil.
Contr. Obs. and Planetarium, Brno, No. 8/III, 17 pp. (1969).
In Czech.

104.063 **Frequency occurrence of small particles in meteor**
    showers. II. Orionids, ϵ Geminids.
V. Znojil.
Contr. Obs. and Planetarium, Brno, No. 8/IV, 28 pp. (1969).
In Czech.

November Leonid meteors observed.
Sky Telescope, Vol. 39, 62 - 63 (1970).

Geminid meteors in 1969.
Sky Telescope, Vol. 39, 202 - 203 (1970).

The ion composition of the upper atmosphere in
130 - 155 km height during the activity of the Orionids
meteor stream.   See Abstr. 083.074.

The fragmentation of the asteroids.
See Abstr. 098.012.

The dynamical problems of small bodies in the
solar system.   See Abstr. 098.014.

The radiants of the comets 1951 - 1961 and their
identification with meteor showers.
See Abstr. 102.019.

Atmospheric freezing nuclei and the structure of the
interplanetary matter.   See Abstr. 106.021.

Influence of meteor streams on the level of cosmic
radiation on the earth.   See Abstr. 143.063.

105 Meteorites, Meteorite Craters

105.001 **Twin terrestrial impact craters.**
W. D. Ehmann, J. W. Morgan.
Nature, Vol. 225, 255 (1970).
We wish to point out that at least one terrestrial example of multiple impact craters with dividing walls essentially retained does exist. It belongs to the Henbury meteorite crater group in Central Australia.

105.002 ^{81}**Kr radiation ages of stone meteorites.**
M. W. Rowe.
Nature, Vol. 225, 368 (1970).
The purpose of this report is to suggest that the relative production rate, $P(^{81}$Kr$)/P(^{78}$Kr$)$ may have been overestimated by Marti and Eugster et al.

105.003 **Impact craters and the relative ages of earth and moon.** C. S. Beals.
Nature, Vol. 225, 368 - 369 (1970).
Recent aerial photographic and ground studies of islands in the Lake of the Woods area of the Canadian Shield have revealed a series of arcuate ridges which together make up a conspicuous circular feature 16 km in diameter. The appearance of these ridges suggests that they are igneous intrusions which have been constrained to conform to a circular pattern with predominantly outward dipping layers by distortions in the earth's crust produced by an ancient meteorite impact.

105.004 **The Borgo San Donnino meteorite: Mineralogy and chemistry.** B. Baldanza, G. R. Levi-Donati, C. F. Lewis.
Meteoritics, Vol. 4, 258 (1969). – Abstract.

105.005 **Chromium and silicon in iron meteorites.**
P. L. Beaulieu, C. B. Moore.
Meteoritics, Vol. 4, 259 - 260 (1969). – Abstract.

105.006 **Isotopic variations in trapped meteoritic argon.**
D. C. Black.
Meteoritics, Vol. 4, 260 (1969). – Abstract.

105.007 **A project to observe meteorite falls.**
A. T. Blackwell, I. Halliday.
Meteoritics, Vol. 4, 260 - 261 (1969). – Abstract.

105.008 **Preliminary results of artificial meteor ablation.**
M. B. Blanchard.
Meteoritics, Vol. 4, 261 (1969). – Abstract.

105.009 **The constrained equilibrium theory: Sulfide phases.**
M. Blander.
Meteoritics, Vol. 4, 261 - 262 (1969). – Abstract.

105.010 **Heating and vaporization of basalts with a carbon dioxide laser.**
M. Blander, K. Keil, L. S. Nelson, S. R. Skaggs.
Meteoritics, Vol. 4, 262 (1969). – Abstract.

105.011 **Possible origin of chondrules and meteorites.**
T. E. Bridge.
Meteoritics, Vol. 4, 262 - 263 (1969). – Abstract.

105.012 **Investigation of Winkler crater, Kansas.**
D. G. Brookins.
Meteoritics, Vol. 4, 263 - 264 (1969). – Abstract.

105.013 **Results of a large volume micrometeorite collection at an altitude of 115.000 feet.**
D. E. Brownlee, P. W. Hodge.
Meteoritics, Vol. 4, 264 (1969). – Abstract.

105.014 **The Gibeon meteorites.** V. F. Buchwald.
Meteoritics, Vol. 4, 264 - 265 (1969). – Abstract.

105.015 **"Bushman Land" and "Karasburg", two new iron meteorites from SW Africa.**
V. F. Buchwald.
Meteoritics, Vol. 4, 265 - 266 (1969). – Abstract.

105.016 **The Eaton copper meteorite—brass from space?**
P. R. Buseck, E. F. Holdsworth, G. R. Scott.
Meteoritics, Vol. 4, 267 (1969). – Abstract.

105.017 **Serra de Magé: An odd meteorite.**
E. A. Carver, E. Anders.
Meteoritics, Vol. 4, 267 - 268 (1969). – Abstract.

105.018 **The Iron River meteorite.**
V. D. Chamberlain.
Meteoritics, Vol. 4, 268 (1969). – Abstract.

105.019 **Magnetic surveys at West Hawk Lake, Manitoba, Canada.** J. F. Clark.
Meteoritics, Vol. 4, 268 (1969). – Abstract.

105.020 **The characteristics of very high speed small particle impacts in rocks and glasses.**
B. G. Cour-Palais.
Meteoritics, Vol. 4, 268 - 269 (1969). – Abstract.

105.021 **Meteorite craters and astroblemes: New information.**
R. S. Dietz.
Meteoritics, Vol. 4, 269 (1969). – Abstract.

105.022 **Barringerite: How it formed in the ollague pallasite.**
A. S. Doan, Jr.
Meteoritics, Vol. 4, 269 - 270 (1969). – Abstract.

105.023 **Experimental formation of artificial chondrules.**
D. L. Fernald, W. W. Salisbury.
Meteoritics, Vol. 4, 270 (1969). – Abstract.

105.024 **Uranium measurements in hypersthene chondrites and the reality of the 700 million year "event".**
D. E. Fisher.
Meteoritics, Vol. 4, 270 (1969). – Abstract.

105.025 **Meteoroid impact study on spacecraft windows.**
R. E. Flaherty.
Meteoritics, Vol. 4, 271 (1969). – Abstract.

105.026 **Carbon distribution in chondrites.**
K. Fredriksson, J. Nelen.
Meteoritics, Vol. 4, 271 - 272 (1969). – Abstract.

105.027 **A mineralogical and chemical study of the Burdett, Kansas, chondrite.**
R. V. Fodor, K. Keil, E. Jarosewich, G. I. Huss.
Meteoritics, Vol. 4, 272 (1969). – Abstract.

105.028 **Secondary craters as a clue to primary crater origin on the moon.**
R. F. Fudali, W. G. Melson.
Meteoritics, Vol. 4, 273 (1969). – Abstract.

105.029 **Total nitrogen abundances in chondritic meteorites.**
E. K. Gibson, Jr., C. B. Moore.
Meteoritics, Vol. 4, 273 - 274 (1969). – Abstract.

105.030 **The effect of phosphorus on the development of the Widmanstatten pattern in iron meteorites.**
J. I. Goldstein, A. S. Doan, Jr.
Meteoritics, Vol. 4, 274 - 275 (1969). – Abstract.

105.031 **Rubidium-strontium studies of the black chondrite, Orvinio and its mineral phases: Evidence of recent equilibration due to shock and reheating.**
K. Gopalan, G. W. Wetherill.
Meteoritics, Vol. 4, 275 (1969). – Abstract.

105.032 **Rubidium-strontium studies of enstatite chondrites: Whole meteorite and mineral ages.**
K. Gopalan, G. W. Wetherill.
Meteoritics, Vol. 4, 275 - 276 (1969). – Abstract.

105.033 **Terrestrial analogs to lunar dimple (drainage) craters.** R. Greeley.
Meteoritcs, Vol. 4, 276 (1969). – Abstract.

105.034 **Determination of spallogenic ^{53}Mn etc. and a group of nobler trace elements in meteorites by neutron activation.** W. Herr, U. Herpers.
Meteoritics, Vol. 4, 277 (1969). – Abstract.

105.035 **Electron microprobe study of metal in the Kingfisher meteorite.** D. Heymann, G. J. Taylor.
Meteoritics, Vol. 4, 277 (1969). – Abstract.

105.036 **Spallation yields of xenon from irradiation of Cs, Ce, Nd, Dy and a rare earth mixture with 730 MeV protons.** C. M. Hohenberg, M. W. Rowe.
Meteoritics, Vol. 4, 278 (1969). – Abstract.

105.037 **Shock pressure calibration of rock forming minerals using Debye–Scherrer X-ray techniques.**
F. Hörz, W. L. Quaide.
Meteoritics, Vol. 4, 278 - 279 (1969). – Abstract.

105.038 **The near-earth micrometeorite environment.**
J. F. Kerridge.
Meteoritics, Vol. 4, 279 (1969). – Abstract.

105.039 **The chemical composition of carbonaceous material from the Tieschitz chondrite.**
G. Kurat.
Meteoritics, Vol. 4, 280 - 281 (1969). – Abstract.

105.040 **Significance of some very high temperature mineral associations from the Lancé carbonaceous chondrite.**
G. Kurat.
Meteoritics, Vol. 4, 281 (1969). – Abstract.

105.041 **Elemental abundances in meteorites and the earth.**
J. Larimer.
Meteoritics, Vol. 4, 282 (1969). – Abstract.

105.042 **Bismuth and thallium contents of chondrites.**
J. C. Laul, I. Pelly, M. E. Lipschutz.
Meteoritics, Vol. 4, 282 (1969). – Abstract.

105.043 **Three new meteorite finds from Niger.**
M. Christophe-Michel-Levy.
Meteoritics, Vol. 4, 283 (1969). – Abstract.

105.044 **Chemical fractionation in stony meteorites.**
H. von Michaelis.
Meteoritics, Vol. 4, 284 - 285 (1969). – Abstract.

105.045 **Sample preparation of stony meteorites.**
H. von Michaelis.
Meteoritics, Vol. 4, 285 (1969). – Abstract.

105.046 **A comparison of various meteorite-orbit studies.**
P. M. Millman.
Meteoritics, Vol. 4, 286 (1969). – Abstract.

105.047 **Lithium: Its abundance and distribution in chondritic meteorites.**
W. Nichiporuk, C. B. Moore.
Meteoritics, Vol. 4, 286 - 287 (1969). – Abstract.

105.048 **Organic analysis of the Pueblito de Allende meteorite.** J. Oro, E. Gelpi.
Meteoritics, Vol. 4, 287 (1969). – Abstract.

105.049 **Manganese and sodium homogeneity in chondritic meteorites and terrestrial obsidians.**
T. W. Osborn, R. A. Schmitt.
Meteoritics, Vol. 4, 287 - 289 (1969). – Abstract.

105.050 **Silicates in mesosiderites and their genetic implications.** B. N. Powell.
Meteoritics, Vol. 4, 289 (1969). – Abstract.

105.051 **"Cañon Diablo – a transmission electron microscopy study".** S. V. Radcliffe.
Meteoritics, Vol. 4, 290 (1969). – Abstract.

105.052 **"Peckelsheim" a new bronzite achondrite from Westfalia, Germany.**
P. Ramdohr, A. El Goresy.
Meteoritics, Vol. 4, 291 (1969). – Abstract.

105.053 **Significance of the breccia dikes of the Charlevoix structure.** J. Rondot.
Meteoritics, Vol. 4, 291 - 292 (1969). – Abstract.

105.054 **Refraction of a shock wave passing through a cross folded crystalline basement.**
D. W. Roy.
Meteoritics, Vol. 4, 292 - 293 (1969). – Abstract.

105.055 **Use of lunar data to determine space erosion of meteorites.** S. F. Singer.
Meteoritics, Vol. 4, 294 (1969). – Abstract.

105.056 **Electrolytic separation of kamacite and taenite in iron meteorites.**
S. L. Tackett, R. A. Tucker, F. R. Duncan.
Meteoritics, Vol. 4, 294 (1969). – Abstract.

105.057 **Two types of taenite in ordinary chondrites.**
G. J. Taylor, D. Heymann.
Meteoritics, Vol. 4, 294 - 295 (1969). – Abstract.

105.058 **The Knoop microhardness of the metallic phases in the Holbrook chondrite.**
W. M. Walker.
Meteoritics, Vol. 4, 295 (1969). – Abstract.

105.059 **Determination of the P_{O_2}-T equilibrium of indochinite tektites.**
L. S. Walker, A. S. Doan.
Meteoritics, Vol. 4, 295 - 296 (1969). – Abstract.

105.060 **Ni, Ga, Ge, and Ir in the metal of iron meteorites with silicate inclusions.** J. T. Wasson.

Meteoritics, Vol. 4, 296 (1969). – Abstract.

105.061 Iron meteorites of chemical group I.
J. T. Wasson.
Meteoritics, Vol. 4, 297 (1969). – Abstract.

105.062 Inhomogeneities in moldavites.
W. Weiskirchner.
Meteoritics, Vol. 4, 297 (1969). – Abstract.

105.063 Compositions of 'irradiated' pyroxenes and feldspars in the Kapoeta Howardite.
L. Wilkening, D. Lal, A. M. Reid.
Meteoritics, Vol. 4, 298 (1969). – Abstract.

105.064 The mineralogical and chemical composition of silicate inclusions in the El Taco (Campo del Cielo) iron meteorite. F. Wlotzka, E. Jarosewich.
Meteoritics, Vol. 4, 298 - 299 (1969). – Abstract.

105.065 Microprobe analyses of spherules surrounding meteorite craters.
F. W. Wright, P. W. Hodge.
Meteoritics, Vol. 4, 299 (1969). – Abstract.

105.066 Total nitrogen abundances in iron meteorites.
E. K. Gibson, C. B. Moore.
Meteoritics, Vol. 4, 304 - 305 (1969). – Abstract.

105.067 A note on the meteoroid hazard.
N. H. Langton.
Journ. British Interplanet. Soc., Vol. 23, 79 - 82 (1970).
The probability of a space vehicle or lunar base of given target area being hit by a meteoroid depends upon the frequency and distribution of the particles in space. Data from terrestrial and satellite measurements are compared, and it is suggested that the apparent discrepancy supports Whipple's theory of the low density meteoroid.

105.068 Some paradoxes in Australasian microtektite compositional trends. B. E. Leake.
Journ. Geophys. Res., Vol. 75, 349 - 356 (1970).
Critical compositional plots chosen to enable a distinction to be made between rocks formed by igneous processes, sedimentary processes, and vapor fractionation processes are presented. Niggli numbers are used in order to circumvent apparent correlations resulting from the swamping influence of major variations in silica. None of these processes seems to explain the Australasian microtektite compositional trends at present.

105.069 Earth accretion in 'black spherules'.
O. Vittori A.
Journ. Geophys. Res., Vol. 75, 371 - 374 (1970).
Airborne ferromagnetic particles were collected at 2000 meters altitude and analyzed under an optical microscope to determine the earth accretion rate in 'black spherules'. Concentrations and ratios of iron, nickel, and cobalt were determined in the samples and compared with data from other stations. The results suggest that most ferromagnetic fallout is independent of sampling locality and may therefore be of extraterrestrial origin. An improved method for determining particle counts is presented.

105.070 Mechanical properties of iron meteorites and the structure of their parent planets.
R. B. Gordon.
Journ. Geophys. Res., Vol. 75, 439 - 447 (1970).
The strength and ductility of metal from the meteorite Gibeon have been determined over a range of temperatures and strain rates. The absence of structural evidence of

ductile fracturing strongly suggests that the Gibeon irons are not fragments of some larger metal mass, such as a planetary core, but existed as separate metal pieces surrounded by stony material in the parent planet. The available evidence indicates that a similar conclusion probably can be drawn for most iron meteorites.

105.071 Aluminum 26–magnesium 26 dating of feldspar in meteorites. W. B. Clarke, J. R. de Laeter, H. P. Schwarcz, K. C. Shane.
Journ. Geophys. Res., Vol. 75, 448 - 462 (1970).
A search for 'fossil' Mg^{26} from the decay of 0.72 m.y. Al^{26} in an aluminum-rich (plagioclase feldspar) phase of several stone meteorites has been carried out. An electron impact ion source in a tandem mass spectrometer was used for a series of measurements of the relative abundances of table magnesium isotopes in meteoritic and terrestrial feldspar, as well as in unseparated ('common') magnesium from meteorites, from meteoritic chondrules, and from laboratory reagents.

105.072 Rubidium 87–strontium 87 age of carbonaceous chondrites. S. K. Kaushal, G. W. Wetherill.
Journ. Geophys. Res., Vol. 75, 463 - 468 (1970).
Nine carbonaceous chondrites have been analyzed for potassium, rubidium, and strontium elemental concentrations and Sr isotopic composition. Two additional carbonaceous chondrites were analyzed for K and Rb concentrations only.

105.073 Thermoluminescence in tektites.
S. A. Durrani, C. Christodoulides, K. V. Ettinger.
Journ. Geophys. Res., Vol. 75, 983 - 995 (1970).
Thermoluminescence (TL) in tektites has been studied with a view to dating tektites. Powdered samples, both natural and after irradiation by Co^{60} γ rays, are heated in a nitrogen atmosphere at a rate of 30°C/sec, the glow being examined as a function of temperature by a specially selected photomultiplier. The structure of glow curves is found to be constant and reproducible, showing several distinct peaks besides some unresolved ones. Exposure to UV light results in a severe bleaching of the TL glow within a few hours. Isothermal annealing of various glow peaks is studied in detail, and an attempt is made to derive from these observations the half-lives and energy depths of electron traps responsible for the glow. Some tentative age estimates are made (a few times 10^5 years for an indochinite) subject to certain assumptions discussed in the text.

105.074 Age of australite fall. E. D. Gill.
Journ. Geophys. Res., Vol. 75, 996 - 1002 (1970).
If (1) the K/Ar age 0.7 m.y. for australite glass is correct, and (2) the age of formation and the age of fall are not very different, then (3) there is a major discrepancy between the K/Ar result and the stratigraphic information now available. The idea that age of material and age of fall are much the same should perhaps be reexamined. This investigation suggests a young age for australites.

105.075 Ureilites. G. P. Vdovykin.
Space Sci. Rev., Vol. 10, 483 - 510 (1970).
The review article contains the following sections: Introduction; Classification of ureilites; General data about ureilites; Chemical composition; Mineral composition and structure; Diamonds; Age and preatmospheric sizes; Origin.

105.076 Lunar glass: Origin and effects.
K. J. K. Buettner, M. LeMone.
Naturwissenschaften, Vol. 57, 87 - 88 (1970).

105.077 The Lost City meteorite fall. R. E. McCrosky.
Sky Telescope, Vol. 39, 154 - 158 (1970).

105.078 The Lost City meteorite – A deep-space probe for cosmic rays. E. L. Fireman.
Sky Telescope, Vol. 39, 158 (1970).

105.079 Zur Genese des kohligen Materials im Meteoriten von Tieschitz. G. Kurat.
Earth Planet. Sci. Letters, Vol. 7, 317 - 324 (1970).

A carbonaceous in inclusion in a chondrule and a carbonaceous fragment from the Tieschitz meteorite are described. Electronmicroprobe analyses of that material and of the carbonaceous matrix are compared with analyses of carbonaceous chondrites. This comparison shows that the carbonaceous material from the Tieschitz meteorite probably is a residual material of the chondrule-forming process.

105.080 Ca-Al rich phases in the Allende meteorite. U. B. Marvin, J. A. Wood, J. S. Dickey, Jr.
Earth Planet. Sci. Letters, Vol. 7, 346 - 350 (1970).

This newly fallen type III carbonaceous chondrite contains abundant, irregular bodies of unusual bulk chemistry: rich in Ca and Al, poor in Si. Phases identified in these bodies include spinel, hercynite, gehlenite, anorthite, nepheline, diopside, fassaite, ferroaugite, perovskite and Ca-Al-rich glass. High temperatures and an effective mechanism of chemical fractionation were required to form these assemblages. We suggest that they may represent early, high temperature condensates from the solar nebula.

105.081 Australite distribution pattern in southern central Australia. D. H. McColl, G. E. Williams.
Nature, Vol. 226, 154 - 155 (1970).

Reports of australite discoveries at many new localities, together with the better documented of those previously recorded, have been incorporated in a revised location map for southern central Australia.

105.082 Neutron capture effects in Gd from the Norton County meteorite.
O. Eugster, F. Tera, D. S. Burnett, G. J. Wasserburg.
Earth Planet. Sci. Letters, Vol. 7, 436 - 440 (1970).

The isotopic composition of Gd in one chondrite, two achondrites and the silicate inclusions of two meteorites have been determined. When corrected for mass discrinimation, Gd in all samples except the Norton County achondrite shows the same relative isotopic abundances as terrestrial Gd.

105.083 Magnetic particles extracted from manganese nodules: Suggested origin from stony and iron meteorites. R. B. Finkelman.
Science, Vol. 167, 982 - 984 (1970).

On the basis of X-ray diffraction and electron microprobe data, spherical and ellipsoidal particles extracted from manganese nodules were divided into three groups. Group I particles are believed to be derived from iron meteorites, and group II particles from stony meteorites. Group III particles are believed to be volcanic in origin.

105.084 Endogenous carbon in carbonaceous meteorites. J. W. Smith, I. R. Kaplan.
Science, Vol. 167, 1367 - 1370 (1970).

Seven carbonaceous chondrites have been analyzed for soluble organic compounds, carbonate, and residual carbon.

105.085 Meteoritic spherules in the soil surrounding terrestrial impact craters. P. W. Hodge, F. W. Wright.
Nature, Vol. 225, 717 - 718 (1970).

Soil-sampling surveys in the vicinity of three different terrestrial impact crater areas are compared. They are the Henbury, Sikhote-Alin, and Canyon Diablo craters. A major meteoritic component of the soil samples consists of spherules in the decamicron range of size and with nickel-iron compo-

sition ranging from ~ 60 percent to ~ 10 percent. These particles are of interest because of the possible similarities in their modes of formation with particles on the lunar surface.

105.086 Comparison of microstructures in carbonaceous chondrites with organic microspheres in terrestrial hydrothermal minerals. G. Mueller.
Meteoritics, Vol. 4, 286 (1969). – Abstract.

105.087 Extinct radioactivities.
A. G. W. Cameron.
Comments Astrophys. Space Phys., Vol. 2, 18 - 24 (1970).

105.088 The isotopic composition of lead in iron meteorites. V. M. Oversby.
Geochim. Cosmochim. Acta, Vol. 34, 65 - 75 = Lamont-Doherty Contr. No. 1359 (1970).

Lead isotopic composition and lead concentration were measured in four troilite samples from Canyon Diablo, two troilites from Toluca, and single troilites from Bogou, Baquedano, Gibeon, Grant and Staunton. Uranium concentrations were measured in four samples; they ranged from 2.6 to 25 ppb. Observed isotopic compositions were corrected for fractionation by the double spike technique. The lead compositions ranged from primitive compositions to modern radiogenic compositions.

105.089 Bismuth contents of chondrites.
J. C. Laul, D. R. Case, F. Schmidt-Bleek, M. E. Lipschutz.
Geochim. Cosmochim. Acta, Vol. 34, 89 - 103 (1970).

A neutron activation technique has been developed for determination of the bismuth concentrations in C-, H-, L- and LL-group chondrites. The analyses of 14 carbonaceous and 5 equilibrated ordinary chondrite samples have confirmed that Bi is indeed a "strongly-depleted" element in meteorites.

105.090 Isotopic composition of primordial helium in carbonaceous chondrites.
E. Anders, D. Heymann, E. Mazor.
Geochim. Cosmochim. Acta, Vol. 34, 127 - 132 (1970).

The He^3/He^4 ratio of primordial He in carbonaceous chondrites varies with the Ne^{20}/Ne^{22} ratio, according to the relation: $He^3/He^4 = (6.87 \pm 0.72) \times 10^{-5} (Ne^{20}/Ne^{22}) - (4.38 \pm 0.76) \times 10^{-4}$. Apparently primordial He, like Ne, consists of at least two distinct components, of $He^3/He^4 = (1.25 \pm 0.76) \times 10^{-4}$ and $(4.20 \pm 0.76) \times 10^{-4}$.

105.091 Trapped helium-neon isotopic correlations in gas-rich meteorites and carbonaceous chondrites.
D. C. Black.
Geochim. Cosmochim. Acta, Vol. 34, 132 - 140 (1970).

Stepwise heating experiments have been carried out on the dark portions of some gas-rich meteorites, to determine more precisely the march of trapped helium neon and argon. Isotopic variations of up to a factor of 7 exist in trapped-helium and correlate with variations in trapped neon.

105.092 Composition of the metal, schreibersite and perryite of enstatite achondrites and the origin of enstatite chondrites and achondrites. J. T. Wasson, C. M. Wai.
Geochim. Cosmochim. Acta, Vol. 34, 169 - 184 (1970).

In this paper we report investigations of the composition of the metal, phosphide and silicide phases found in enstatite achondrites. We have evaluated the possibility that the enstatite achondrites also form a systematic sequence, and have considered the possibility of a relationship between the enstatite chondrites and achondrites.

105.093 Xenon from the Angra dos Reis meteorite.

C. M. Hohenberg.
Geochim. Cosmochim. Acta, Vol. 34, 185 - 191 (1970).

Xenon released in stepwise heating of the Angra dos Reis stone is found to be largely cosmogenic and fissiogenic. This meteorite contains the largest proportion of these two components observed to date with some temperature fractions containing nearly pure spallation and fission-produced xenon.

105.094 **Light hydrocarbon gases, C^{13}, and origin of organic matter in carbonaceous chondrites.**
T. Belsky, I. R. Kaplan.
Geochim. Cosmochim. Acta, Vol. 34, 257 - 278 (1970).

Compounds ranging from C_1 to C_6 were identified in 11 meteorites, including 8 carbonaceous chondrites. The distribution of compounds was different in the different meteorites studied, both saturated and unsaturated hydrocarbons were detected, and benzene was generally a relatively minor component.

105.095 **Dating of meteorites by the high-temperature release of iodine-correlated Xe^{129}.** F. A. Podosek.
Geochim. Cosmochim. Acta, Vol. 34, 341 - 365 (1970).

Correlations between the amounts of Xe^{129} and Xe^{128} released in stepwise heating of neutron-irradiated meteorites are used to determine the initial ratio I^{129}/I^{127} and hence a relative formation time for the various samples. In such an experiment, we obtain the formation times of 9 specimens. The reliability of iodine-xenon ages of individual meteorites is considered; in particular, the ages of Bishopville and Lafayette are less reliable than those of most other meteorites studied, especially in view of the anomalous ages reported. The relevance of iodine-xenon dating to theories of nucleosynthesis, early solar system chronology, and theories of meteorite parent-body formation is discussed.

105.096 **Chemical fractionations in meteorites – III. Major element fractionations in chondrites.**
J. W. Larimer, E. Anders.
Geochim. Cosmochim. Acta, Vol. 34, 367 - 387 (1970).

Some 20 elements, including the major constituents of chondritic matter, are fractionated among the several chondrite classes. We have tried to explain these fractionations on the assumption that they occurred in the solar nebula, starting from material of carbonaceous chondrite composition.

105.097 **Silicon in the Nedagolla ataxite and the relationship between Si and Cr in reduced iron meteorites.**
C. M. Wai, J. T. Wasson.
Geochim. Cosmochim. Acta, Vol. 34, 408 - 410 (1970).

The Nedagolla ataxite contains 0.14% Si in the metal. The high Cr contents of this meteorite and the previously studied high-Si iron, Tucson, suggest an association between Cr and Si. A study of irons with contents of Ni, Ga, Ge and Ir similar to those of Nedagolla and Tucson resulted in the discovery of 2 ataxites with slightly elevated Cr contents, but Si concentrations less than 30 ppm.

105.098 **Argon 37 and argon 39 in recently fallen meteorites and cosmic-ray variations.**
E. L. Fireman, R. Goebel.
Journ. Geophys. Res., Vol. 75, 2115 - 2124 (1970).

The 35-day Ar^{37} and the 270-year Ar^{39} isotopes were measured in the freshly fallen meteorites Alandroal, Allende, and Sprucefield to 5% accuracy. The argon radioactivity results were used to obtain the cosmic-ray fluxes at different times. Other radioactivities were also discussed.

105.099 **Chemical and mineralogical compositions of cosmic and terrestrial spherules from a marine sediment.**
H. T. Millard, Jr., R. B. Finkelman.
Journ. Geophys. Res., Vol. 75, 2125 - 2134 (1970).

The mineralogical and chemical compositions and the specific gravities were determined for particles in a collection of large (149 – 351 μ diameter), dense, magnetic spherules separated from 750 kg of Pacific red clay sediment. Two groups of spherules were found.

105.100 **Extraterrestial magnetic spherules: Their association with meteor showers and rainfall frequency.**
J. Rosinski.
Journ. Atmosph. Terr. Phys., Vol. 32, 805 - 827 (1970).

The relationship between meteoric activity and extraterrestrial particle concentration in the troposphere is examined experimentally using magnetic spherules as an indicator of the larger flux. Frequency of rainfall and periods of maximum meteor shower activity correlate over a 52-yr period where synoptic conditions favor rain initiation through particle flux into high cirrus levels. The study concludes that the experimental evidence, together with proposed revisions in the Bowen hypothesis, restore credibility to the concept of meteoric influences on rainfall.

105.101 **Uranium and thorium in tektites: An additional comment.** J. W. Morgan.
Earth Planet. Sci. Letters, Vol. 8, 141 - 142 (1970).

The relationship between uranium and thorium in tektites and achondrites is reconsidered in the light of the oxygen isotopic composition of lunar rocks. A correlation is found between the uranium and thorium abundances in three high ^{18}O achondrites and those in tektites which is particularly strong in the case of the Australasian group.

105.102 **Effect of strain on fission-track ages of tektites.**
S. A. Durrani, D. A. Hancock.
Earth Planet. Sci. Letters, Vol. 8, 157 - 162 (1970).

Spurious tracks in unirradiated tektite sections can seriously affect the ages determined by the fission-track analysis method. It is shown that strain can cause spurious tracks which resemble the genuine fission tracks under certain circumstances. Methods of eliminating such errors are described.

105.103 **The abundance of ^{244}Pu in the early solar system.**
F. A. Podosek.
Earth Planet. Sci. Letters, Vol. 8, 183 - 187 (1970).

The abundance ratio $^{244}Pu/^{238}U$ at the time the St. Severin meteorite began to retain xenon has been determined to be 0.0127 ± 0.0026 by a method involving only xenon isotope ratio measurements.

105.104 **Les sphérules cosmiques dans les nodules de manganèse.** J. Jedwab.
Geochim. Cosmochim. Acta, Vol. 34, 447 - 457 (1970).

A definition is given of an objective criterion for the identification of natural spherules in pelagic sediments, offering the possibility of eliminating technologic spherules from the observational field. This criterion is based on the formation process and rate of formation of manganese nodules.

105.105 **C^{13}/C^{12} abundances in components of carbonaceous chondrites and terrestrial samples.**
H. R. Krouse, V. E. Modzeleski.
Geochim. Cosmochim. Acta, Vol. 34, 459 - 474 (1970).

The C^{13}/C^{12} compositions of components from nine carbonaceous chondrites were measured mass spectrometrically.

105.106 **Cooling rates and thermal histories of iron and stony-iron meteorites.**
P. E. Fricker, J. I. Goldstein, A. L. Summers.
Geochim. Cosmochim. Acta, Vol. 34, 475 - 491 (1970).

It is the purpose of this study to investigate the cooling

histories of parent meteorite bodies on the basis of thermal models, which include the effects of melting, redistribution of radioactive heat sources and surface heating, and to compare the results of those calculations with meteoritic cooling rates.

105.107 Gold and iridium in meteorites and some selected rocks.   W. D. Ehmann, P. A. Baedecker, D. M. McKown.
Geochim. Cosmochim. Acta, Vol. 34, 493 - 507 (1970).

Gold and iridium abundances in stony meteorites and various meteoritic and terrestrial materials have been determined by the technique of neutron activation. Gold abundances in chondrites are reported. Iridium data were obtained by use of an improved radiochemical separation technique and also a non-destructive technique employing gamma-gamma coincidence spectrometry. The significance of the meteorite data to the life history of these objects is discussed.

105.108 Magnetic properties of meteorites from the collection of the Institute of Geology, Academy of Sciences, Estonian SSR.   E. Gus'kova.
ENSV Tead. Akad. toimetised. Keemia, geol., Izv. AN Ehst. SSR, Khimiya, geol., Vol. 18, No. 3, p. 259 - 269 (1969). In Russian. – Abstr. in Referativ. Zhurn. 51. Astron., 3.51.367 (1970).

105.109 A new hexagonal modification of carbon in meteorites.   G. P. Vdovykin.
Geokhimiya, 1969 No. 9, p. 1145 - 1148.   In Russian. Abstr. in Referativ. Zhurn. 51. Astron., 3.51.371 (1970).

105.110 $(Mg, Fe)_2 SiO_4$ spinel in a meteorite.   R. A. Binns.
Phys. Earth Planet. Interiors, Vol. 3, 156 - 160 (1970). – Conference paper (see 012.004).

105.111 The mass distribution of the Sikhote Alin' meteorite shower.   B. Hellyer.
Observatory, Vol. 90, 55 - 57 (1970).

The total mass of fallen material has been estimated at 70 tons, of which 23 tons has been recovered. The missing material is accounted for by craters as yet unexcavated and those smaller particles which have escaped detection.

105.112 Serra de Magé: A meteorite with an unusual history.   E. A. Carver, E. Anders.
Earth Planet. Sci. Letters, Vol. 8, 214 - 220 (1970).

Serra de Magé has a fission-track age of 540 ± 90 Myr, and may hence be genetically related to three other meteorite classes that suffered major collisions at about that time: hypersthene chondrites, group III irons, and shergottites.

105.113 Rare gases from Serra de Magé feldspar.   R. Ganapathy.
Earth Planet. Sci. Letters, Vol. 8, 221 - 222 (1970).

The noble gases He, Ne, Ar, and Xe were measured in a plagioclase concentrate separated from Serra de Magé. A mean apparent radiation age of 2.5 Myr was obtained by combining earlier and the present work. The concentration of excess ^{136}Xe, presumably from the decay of ^{244}Pu, was found to be 2.34×10^{-12} cm^3 STP/g.

105.114 The chemical composition of the silicate inclusions in the Weekeroo Station iron meteorite.   E. Olsen, E. Jarosewich.
Earth Planet. Sci. Letters, Vol. 8, 261 - 266 (1970).

A chemical analysis of the silicate portion of the iron meteorite, Weekeroo Station, is compared with those of other meteorite types. The composition is not similar to that of any chondrite type, however, it is similar to the residue of

a bulk chondritic composition from which 50% olivine (^{85}Fo) has been subtracted, with only the minor elements Cr and P showing significant anomalies.

105.115 Fission track ages and ages of deposition of deep-sea microtektites.
W. Gentner, B. P. Glass, D. Storzer, G. A. Wagner.
Science, Vol. 168, 359 - 361 (1970).

The Australasian and Ivory Coast deep-sea microtektites have fission track ages of 0.71 and 1.09 million years, respectively. These ages are in good agreement with the ages of deposition of the microtektites determined from paleomagnetic data. Both the fission track ages and ages of deposition of the microtektites agree with the potassium/argon and fission track ages of tektites from the respective tektite strewn fields.

105.116 Surfaces for micrometeoroid impact crater detection.   D. E. Brownlee, P. W. Hodge.
SAO, *Cambridge, Mass.,* Special Rep. No. 308, 5 + 16 pp. (1970).

Because of the importance of increasing the sensitivity of micrometeoroid-detection experiments, a Van de Graaff microparticle accelerator was used to investigate the general properties of micron-sized impact craters in various surface types. Of the 22 surfaces studied, glass, Au film on lucite, and Al film on glass were found highly efficient in displaying impacts and could be used for statistically meaningful determinations of flux and species in spite of the low particle flux implied by many recent rocket and satellite measurements.

105.117 Central Australia's meteorite craters.   P. W. Hodge.
Astron. Soc. Pacific, Leaflet No. 487, 8 pp. (1970).

105.118 Mineralogical relationships in Seymour, a coarse octahedrite.   J. C. Drake.
Meteoritics, Vol. 5, 19 - 31 (1970).

Seymour is a coarse octahedrite weighing 24.5 kg. It contains two types of troilite-graphite nodules, four morphologically distinct types of phosphide and three types of cohenite.

105.119 Micro-chondrules.   G. R. Levi-Donati.
Meteoritics, Vol. 5, 33 - 42 (1970).

Very small chondrules, which are less than 250μ in size and are referred to as microchondrules, were discovered in twelve chondritic meteorites. The morphology, composition, and structure of these microchondritic bodies are investigated.

105.120 Electrolytic corrosion of iron meteorites.   S. L. Tackett, R. A. Tucker, F. R. Duncan.
Meteoritics, Vol. 5, 43 - 55 (1970).

Six iron meteorites were electrolyzed and the resulting corrosion was studied by a potentiostatic technique. It was found that both iron and nickel in the kamacite phase dissolve, and that neither iron nor nickel dissolve from taenite. The rate of corrosion was shown to be inversely proportional to the nickel content.

105.121 Discovery of another meteorite specimen from the 1912 Holbrook, Arizona fall site.
E. K. Gibson, Jr.
Meteoritics, Vol. 5, 57 - 60 (1970).

A 1.5 kilogram specimen of the Holbrook chondrite was recovered during a visit to the 1912 fall site. The heavily-weathered stone was complete with fusion crust still intact.

105.122 Carbonaceous chondrites. – The Murchison meteorite.   G. M. Pitt.
Journ. Astron. Soc. Victoria, Vol. 23, 41 - 45 (1970).

105.123 A tape recording of the Belfast meteorite.
P. M. Millman.
Journ. Roy. Astron. Soc. Canada, Vol. 64, 57 - 59 (1970).

105.124 Meteorite orbits. P. M. Millman.
Journ. Roy. Astron. Soc. Canada, Vol. 64, 114 - 116 (1970).

105.125 Isotopic composition of gadolinium and neutron-capture effects in some meteorites.
O. Eugster, F. Tera, D. S. Burnett, G. J. Wasserburg.
Journ. Geophys. Res., Vol. 75, 2753 - 2768 (1970).
The isotopic composition of Gd in one chondrite, two achondrites, and the silicate inclusions of two iron meteorites has been determined.

105.126 Pyroxene-garnet transformation in Coorara meteorite.
J. V. Smith, B. Mason.
Science, Vol. 168, 832 - 833 (1970).
Majorite is a new garnet in a veinlet of the Coorara meteorite. Its chemical composition is compatible with derivation mostly from original pyroxene, not from olivine as originally reported. Silicon is partly in sixfold coordination. Ringwoodite, a spinel of olivine composition, occurs as purple grains set in a matrix of fine-grained garnet. The similar mineralogy and texture of the Coorara and Tenham meteorites suggest a common parent body.

105.127 On the deposition of sediments in craters.
C. S. Beals, A. Hitchen.
Publ. Dominion Obs. Ottawa, Vol. 39, (No. 4), 101 - 118 (1970).
Profile studies of a model crater made during the process of sedimentary deposition, have revealed interesting analogies with the Holleford crater (known to be an ancient meteorite crater filled with Paleozoic sediments) and with the Mecatina crater, the origin and subsequent history of which are still uncertain.

105.128 Description d'une météorite ferrique.
C. Biemans.
Gaz. astron., No. 2, p. 6 - 8 (1969).

105.129 Thermoluminescence induite dans des cibles épaisses par des protons de 3 GeV.
C. Lalou, U, Brito, D. Nordemann, M. Mary.
Comptes Rendus Acad. Sci. Paris, Sér. B, Vol. 270, 1706 - 1708 (1970).
La thermoluminescence artificielle de poudre de météorite irradiée par des protons de 3 GeV a été mesurée. Les courbes donnant la variation de l'intensité de cette thermoluminescence en fonction de la profondeur dans une cible de 1 m de long ont été établies et montrent le rôle prépondérant joué par les particules secondaires crées dans la cible par interactions nucléaires.

105.130 Der Meteorit von Bovedy.
F. Schmeidler.
Sterne, 46. Jahrgang, p. 81 - 83 (1970).

105.131 Zirconium and hafnium in meteorites by activation analysis. W. D. Ehmann, T. V. Rebagay.
Geochim. Cosmochim. Acta, Vol. 34, 649 - 658 (1970).
Zirconium and hafnium have been determined in 28 chondrites and 7 achondrites by the use of thermal and 14 MeV neutron activation analyses. The average Zr/Hf ratio on a weight basis for all the stony meteorites analyzed in this work was 43, a value which is in good agreement with that reported in the literature for the earth's crust.

105.132 Electron microprobe study of metal particles in the Kingfisher meteorite. G. J. Taylor, D. Heymann.
Geochim. Cosmochim. Acta, Vol. 34, 677 - 687 (1970).
Ni, Fe and P contents were determined with an electron microprobe in five metal particles of the Kingfisher chondrite. These particles show microstructures which can best be described as very coarse plessites. The results indicate that the microstructures were formed by the transformation of preexisting α_2-phase. In the case of Kingfisher the transformation occurred when the meteorite was reheated by shock, some 500 m.y. ago.

105.133 Atmospheric collection of debris from the Revelstoke and Allende fireballs. M. H. Carr.
Geochim. Cosmochim. Acta, Vol. 34, 689 - 700 (1970).
In two separate events, Revelstoke and Allende, the air through which a fireball had been observed to pass was sampled for meteoritic debris. In the Revelstoke type, large amounts of meteoritic debris is left in the atmosphere and little reaches the ground in large coherent fragments; in the Allende type, little material remains in the atmosphere but large fragments reach the ground in the fall area.

105.134 Iron meteorites picked up by primordial people?
J. Classen.
Priroda, No. 5.70, p. 95 - 97 (1970). In Russian.
Translated from English by I. T. Zotkin.

105.135 Rubidium-strontium studies on enstatite chondrites: Whole meteorite and mineral isochrons.
K. Gopalan, G. W. Wetherill.
Journ. Geophys. Res., Vol. 75, 3457 - 3467 (1970).
Eight of the eleven known enstatite chondrite falls have been analyzed for K, Rb, and Sr concentrations and Sr isotopic composition. Several density fractions having a grain size of less than 37 microns have also been separated from one of them, Indarch, and Rb-Sr analyses made.

105.136 The Meteoritical Bulletin.
Edited by "Postoyan. komis. po meteoritam.
Mezhdunar. soyuza geol. nauk; No. 47, 4 pp. (1969).
In Russian. — Abstr. in Referativ. Zhurn. 51. Astron., 4.51.370 (1970). — Report on meteorites in U.S.A.

105.137 La radioactivité de la météorite Dosso (chute du 19 février 1962) mesurée par spectrométrie γ.
J. Tobailem, D. Nordemann.
Comptes Rendus Acad. Sci. Paris, Sér. B, Vol. 271, 262 - 264 (1970).
Les activité d'émetteurs γ (^{54}Mn, ^{22}Na, ^{60}Co, ^{26}Al et ^{40}K) présents dans un fragment de 0.805 kg de la météorite de pierre Dosso ont été mesurées par spectrométrie γ à faible mouvement propre.

105.138 The Meteoritical Bulletin.
Edited by "Postoyan. komis. po meteoritam.
Mezhdunar. soyuza geol. nauk, No. 48, 3 pp. (1969). In Russian. — Abstr. in Referativ. Zhurn. 51. Astron., 6.51.340 (1970). — Report on four new meteorites.

105.139 Rare gas anomalies and intense muon fluxes in the past. J. Takagi.
Nature, Vol. 227, 362 - 363 (1970).
In this report I shall point out that the anomalies in the rare gases from underground sources can be explained by a muon flux similar to that assumed for the explanation of the xenon anomaly in tellurides.

105.140 Organic analysis of the Pueblito de Allende meteorite. R. L. Levy, C. J. Wolf, M. A. Grayson,

J. Gilbert, E. Gelpi, W. S. Updegrove, A. Zlatkis, J. Oro'.
Nature, Vol. 227, 148 - 150 (1970).

It appears unlikely that the organic material detected in the meteorite that fell in Mexico last year can have been introduced by contamination.

105.141   On the existence of super-heavy elements near
Z = 114 and N = 184 in meteorites.
M. N. Rao.
Nuclear Phys., Ser. A, Vol. 140, No. 1, p. 69 - 73 (1970).

The observed excess abundance of ^{136}Xe and ^{86}Kr in primitive chondritic meteorites can be attributed to spontaneous fission of the super-heavy elements (near $Z = 114$ and $N = 184$) decaying through symmetric and asymmetric fission modes respectively.

105.142   Magnetic profiles at Holleford crater, Eastern
Ontario.   J. F. Clark.
Proc. Geol. Ass. Canada, Vol. 20, 24 - 29 = Contr. Dominion Obs., Ottawa, Vol. 8, No. 11 (1969).

The attenuation of magnetic anomalies observed at an altitude of 500 feet as compared to ground observations is studied along four profiles across the crater.

105.143   Die Bestimmung der Spallationsnuklide ^{26}Al, ^{53}Mn,
^{45}Sc und verschiedener Spurenelemente in Meteoriten mit Hilfe der Neutronenaktivierung und spezifischer "Low-Level" Meßverfahren.   U. Herpers.
Thesis, Univ. Köln, 102 pp. (1969).

105.144   Anwendung der quantitativen ^{57}Fe-Mössbauerspektroskopie auf Probleme der Meteoritik.
B. Skerra.
Thesis, Univ. Köln, 78 pp. (1968).

Oklahoma asteroid.
Sci. American, Vol. 222, No. 3, p. 59 - 60 (1970).

Rare gases on the earth and in meteorites.
See Abstr. 082.005.

Chemical factors regulating free alkali metal atoms
in the upper atmosphere.   See Abstr. 082.050.

Terrestrial analogs to lunar dimple (drainage)
craters.   See Abstr. 094.006.

The lunar controversy – II, III.
See Abstr. 094.175.

Chemical individuality of lunar, meteoritic, and
terrestrial silicate rocks.   See Abstr. 094.256.

Microparticle collection experiments during the
1966 Orionid and Leonid meteor showers.
See Abstr. 104.009.

Fireball and meteorite of April 25, 1969.
See Abstr. 104.015.

Meteors and meteorites.   See Abstr. 104.053.

Interstellar dust and aromatic carbon.
See Abstr. 131.002.

Radioactivity caused by cosmic rays in bodies of
asteroid size.   See Abstr. 143.064.

106 Interplanetary Matter, Interplanetary Magnetic Field, Zodiacal Light

106.001 **The thermal emission of the F corona.**
C. B. Kaiser.
Astrophys. Journ., Vol. 159, 77 - 92 (1970).
A model of the heliocentric dust cloud is constructed which allows for the variation of temperature with particle size and optical properties, and the thermal emission is computed. The results show that a stony dust material produces two emission peaks at near-infrared wavelengths. Comparison with recent observations at wavelength 2.2 μ indicates that there are two dust materials present near the sun. The thermal emission predicted by the model is computed at longer wavelengths for comparison with future observations, and the variation with wavelength is found to reflect the reststrahlen band structure of the dust materials.

106.002 **Consequences of the meteoroid penetration measurements with respect to zodiacal light observations.** H. A. Zook, D. J. Kessler.
Meteoritics, Vol. 4, 300 (1969). – Abstract.

106.003 **Observation of a solar flare induced interplanetary shock and helium-enriched driver gas.**
J. Hirshberg, A. Alksne, D. S. Colburn, S. J. Bame, A. J. Hundhausen.
Journ. Geophys. Res., Vol. 75, 1 - 15 (1970).
On February 13, 1967, a class 3B solar flare occurred at 20°N, 10°W. The resultant disturbance in the solar wind was observed by the Los Alamos plasma probe on Vela 3A and the Ames Research Center magnetometer on Explorer 33. The initial discontinuity in the solar wind was identified as a shock. Nine hours after the shock passed, plasma containing 22% helium was observed.

106.004 **On the presence of low-energy protons ($5 \lesssim E \lesssim$ 50 kev) in the interplanetary medium.**
L. A. Frank.
Journ. Geophys. Res., Vol. 75, 707 - 716 (1970).
A fortuitous set of observational conditions has allowed first measurements of low-energy proton ($5 \lesssim E \lesssim$ 50 kev) distributions heretofore unobserved in the interplanetary medium. Salient features of the two major enhancements of interplanetary proton ($5 \lesssim E \lesssim$ 50 kev) intensities observed during the period July 25 through August 13, 1967, are described in detail. The presence of these interplanetary proton intensities may be one of the important conditions for the development of main-phase geomagnetic storms.

106.005 **Dispersion analysis of interplanetary scintillation.**
M. G. Golley, P. A. Dennison.
Planet. Space Sci., Vol. 18, 95 - 101 (1970).
By the application of dispersion analysis to records of interplanetary scintillation taken from a triangular arrangement of receivers, it is found that the velocity of drift of the diffraction pattern increases with temporal frequency of different Fourier components of the pattern.

106.006 **Theory of generation of bow-shock-associated hydromagnetic waves in the upstream interplanetary medium.** A. Barnes.
Cosmic Electrodynamics, Vol. 1, 90 - 114 (1970).
This paper discusses a model of the generation of upstream hydromagnetic-wave turbulence by the streaming-instability mechanism, and shows that it agrees reasonably well with spacecraft observations of magnetic fluctuations and proton streams.

106.007 **Interplanetary shock waves. II. Shock structure.**
C. P. Sonett.
Comments Astrophys. Space Phys., Vol. 2, 59 - 65 (1970).

106.008 **OGO 3 observations of the Lyman alpha intensity and the hydrogen concentration beyond 5 R_E.**
P. Mange, R. R. Meier.
Journ. Geophys. Res., Vol. 75, 1837 - 1847 (1970).
The intensity of Lyman α was measured from the OGO 3 spacecraft at altitudes from 5 to 19 R_E. The variation of intensity with distance reveals a mean hydrogen density at 50.000 km of about 20 atoms cm^{-3} for the summer 1966 epoch and negligible contribution to the signal from geocoronal hydrogen beyond 12 radii. An extraterrestrial background of some 750 rayleighs was observed at apogee. A correlation of the background with solar activity over a 40-day period suggests that a portion of the background is , solar-related.

106.009 **Comparison of measurements of the interplanetary magnetic field with the space probes Venus 4 and Mariner 5.** Sh. Sh. Dolginov, E. G. Eroshenko,
L. N. Zhuzgov.
Kosm. Issled., Vol. 8, 290 - 297 (1970). In Russian.

106.010 **The flux of the interplanetary magnetic field and the K_p-index of geomagnetic activity.**
K. G. Ivanov, N. V. Mikerina.
Geomagn. Aeronom., Vol. 10, 331 - 333 (1970).
In Russian. – Brief information.

106.011 **Analysis of observations of interplanetary scintillations.**
R. V. E. Lovelace, E. E. Salpeter, L. E. Sharp, D. E. Harris.
Astrophys. Journ., Vol. 159, 1047 - 1055 (1970).
It is shown that under suitable conditions it may be possible to derive the speed of the solar wind from observations of interplanetary scintillations taken with a single antenna. Theory suggests that the characteristic diffraction pattern of the interplanetary medium may be identified as a sequence of minima in a Bessel transform of the time autocorrelation function of the observed variations in intensity; the location of the minima gives the speed of the solar wind. The necessary analysis has been done on scintillation data of CTA 21 taken at 430 MHz.

106.012 **Interplanetary gas. XV. Nonradial plasma motions from the orientations of ionic comet tails.**
J. C. Brandt, J. Heise.
Astrophys. Journ., Vol. 159, 1057 - 1066 (1970).
Orientations of type I comet tails show a dispersion apparently attributable to effects of waves and/or discontinuities in the interplanetary medium. The mean azimuthal plasma velocity is calculated with an improved analysis, and a value of about 9 km sec^{-1} is derived, which confirms Brandt's result; in addition, the azimuthal velocity is found to increase with increasing geomagnetic activity. These results allow a computation of the total loss rate of angular momentum from the sun, and an e-folding time of 3–4 × 10^9 years is found for a constant solar wind. The initial sun could have had a much higher equatorial rotational velocity, in agreement with Kraft's results, if the flux of the solar wind has a comparable time scale.

106.013 **Observations of scatter-free propagation of**

~40-kev solar electrons in the interplanetary medium. R. P. Lin.
Journ. Geophys. Res., Vol. 75, 2583 - 2586 (1970).

We present observations of low-energy (~40 kev) solar electron events that appear to propagate from the sun to earth through the interplanetary medium without undergoing any noticeable scattering. These events are characterized by their short duration, especially short decay times, and the rapid transit times of the particles from the sun to the earth.

106.014 Statistical significance of the proposed heliographic latitude dependence of the dominant polarity of the interplanetary magnetic field. J. M. Wilcox.
Journ. Geophys. Res., Vol. 75, 2587 - 2590 (1970). – Letter.

106.015 Zodiacal light models based on Apollo 11 lunar sample analysis. N. S. Kovar, R. P. Kovar.
Bull. American Astron. Soc., Vol. 2, 203 (1970). – Abstr. AAS.

106.016 The source of the Gegenschein. R. G. Roosen.
Bull. American Astron. Soc., Vol. 2, 216 (1970). – Abstr. AAS.

106.017 Interplanetary scintillations and the structure of solar-wind fluctuations.
J. R. Jokipii, J. V. Hollweg.
Astrophys. Journ., Vol. 160, 745 - 753 (1970).

It is demonstrated that the observed correlation scale of the interplanetary scintillation of radio sources is consistent with a plasma-density correlation length of 10^6 km or more. This result is in sharp contrast to previous analyses which inferred a correlation length of 100 - 200 km. Fluctuations in plasma density may therefore have a structure similar to that observed for the interplanetary magnetic field and plasma velocity. We find that in this case of a long correlation length the rms phase fluctuation produced in the radio wave by the solar plasma is very large (~10^4 radians) and that the 100 - 200-km scale inferred in previous work is then closely related to the "inner scale" of the fluctuations, i.e., that wavelength below which there is little power.

106.018 Anisotropic atomic hydrogen distribution in interplanetary space.
W. H. Chambers, P. E. Fehlau, J. C. Fuller, W. E. Kunz.
Nature, Vol. 225, 713 - 714 (1970).

Observations with satellite-borne Lyman-α detectors aboard the two Vela 4 spacecraft in nearly circular orbits at 110.000 km are described and compared to current theories. The predominant feature of the data is a repetitive and marked asymmetry in the Lyman-α flux seen looking radially away from the earth. Typically, the flux rises from 60 Rayleighs to a broad maximum of 160 Rayleighs which is centered at RA 265° and δ + 32°.

106.019 Possibility of a terrestrial component in the Doppler shifted zodiacal light.
V. Vanýsek, M. Harwit.
Nature, Vol. 225, 1231 - 1232 (1970).

The Fraunhofer Hβ line shifts observed by Reay and Ring in the zodiacal scattered light are discussed. One can show that a better approximation in the scattering theory of grains gives no better fit to the low velocity shifts found at large elongation angles. If much of the dust at large elongations is in circumterrestrial orbits, a reasonable fit to the data can be obtained.

106.020 Interstellar hydrogen densities in the surroundings of the solar system. H. J. Fahr.
Nature, Vol. 226, 435 - 436 (1970).

The Lyman-α background intensity has been measured by several satellite-borne photometers. Assuming that this background intensity is due to scattering of solar Lyman-α by interplanetary hydrogen, one is able to deduce some properties of this interplanetary medium. By the use of a theoretical distribution of interstellar hydrogen within interplanetary space it was possible to fit the intensity measurements of Chambers et al. with the Vela 7 spacecraft. The Lyman-α intensity in the direction of the solar apex could be reproduced if a density of 0.06 cm^{-3} of interstellar hydrogen in the vicinity of the sun is taken.

106.021 Atmospheric freezing nuclei and the structure of the interplanetary matter. Z. Kvíz.
Contr. Public Obs. and Planetarium Brno 4, p. 103 - 112 (1967). In Czech.

106.022 Beobachtungen des Morgen-Zodiakallichtes im November/Dezember 1969.
W. Sandner.
Sterne, 46. Jahrgang, p. 91 (1970).

106.023 Direct measurements of interplanetary magnetic fields and plasmas. N. F. Ness.
Ann. IQSY, (see 012.016), Vol. 4, 88 - 109 (1969).

This paper reviews the quantitative results and the synoptic picture of interplanetary plasma which has emerged of the extended solar corona as observed from 0.7 to 1.5 AU during the last six years by the Mariner, IMP, Pioneer, and Vela spacecraft.

106.024 Studies of magnetic imhomogeneities in interplanetary space using solar and galactic cosmic-ray measurements. A. N. Charakhchyan, T. N. Charakhchyan.
Ann. IQSY, (see 012.016), Vol. 4, 119 - 130 (1969).

The results obtained from stratospheric measurements of galactic cosmic-ray modulation are discussed in the light of theoretical considerations.

106.025 Disturbances of the interplanetary plasma near Venus according to data obtained with the space probes Venus 4 and Venus 6. K. I. Gringauz, V. V. Bezrukikh, G. I. Volkov, L. S. Musatov, T. K. Breus.
Kosmich. Issled., Vol. 8, 431 - 436 (1970). In Russian.

106.026 One of the possible mechanisms of formation of inhomogeneities in the interplanetary plasma.
M. Ya. Kotsarenko, S. V. Koshova, A. K. Yukhimuk.
Ukr. fiz. zhurn., Vol. 14, 2048 - 2051 (1969). In Russian. Abstr. in Referativ. Zhurn. 62. Issled. kosm. prostranstv., 4.62.187 (1970).

106.027 Large-scale structure of the interplanetary magnetic field according to the data of the annual variations of cosmic-ray intensity during the cycle of solar activity.
L. I. Dorman, A. A. Luzov, V. P. Mamrukova.
Cosmic Rays No. 11, Moscow, p. 5 - 22 (1969). In Russian.

The available experimental data on the structure of solar corpuscular streams indicate comparatively large angle dimensions of the streams in the plane of the ecliptic. At the same time some geophysical phenomena reveal annual changes which are connected with the heliolatitude of the earth. It is shown, that the region, covered by the corpuscular streams from the sun in the direction perpendicular to the plane of the ecliptic, is small. The position of the region changes from year to year.

106.028 Observations of fine structure in the interplanetary medium. P. A. Dennison.
Report ADP-62, Adelaide Univ., Australia. 17 pp. (1968).
Observations of the interplanetary scintillation of radio

sources using spaced receivers enable the size and motion of plasma irregularities in the solar wind to be studied.

106.029 **The dynamics of charged interplanetary grains.**
         L. V. Standeford.
Thesis, Univ. Illinois, Urbana. [Available from Univ. Microfilms, Ann Arbor, Mi.], 140 pp. (1969).
    The cumulative effects of the Lorentz force arising from the interplanetary magnetic field as the orbital inclinations of interplanetary grains are evaluated.

        Observations of the interplanetary magnetic field,
4 - 12 July 1966.   See Abstr. 073.106.

        Observed particle effects of an interplanetary shock
wave on 8 July 1966.   See Abstr. 073.107.

        Observations of the interplanetary plasma subse-
quent to the 7 July 1966 proton flare.
See Abstr. 073.109.

        Solar wind interactions − hypersonic analogue.
See Abstr. 074.017.

        Solar wind structure determined by corotating

coronal inhomogeneities. 1. Velocity-driven perturbations.
See Abstr. 074.019.

        Solar corona, interplanetary plasma and prominen-
ces.   See Abstr. 074.036.

        Deformation velocity of the solar wind, large-scale
variations of the interplanetary magnetic field and the K_p-index of magnetic activity.   See Abstr. 074.064.

        Comets, interplanetary space, and the problems of
the solar system.   See Abstr. 102.014.

        Magnetic fields of meteor streams as a possible
mechanism of diffusion of cosmic rays in the circumsolar space.   See Abstr. 104.019.

        Radial gradients and anisotropies of cosmic rays in
the interplanetary medium.   See Abstr. 143.027.

        Variations of cosmic rays and the interplanetary
medium.   See Abstr. 143.045.

        Optimal methods of calculation of the primary
streams of cosmic rays in interaction with the earth's magnetic field.   See Abstr. 143.065.

107 Cosmogony of the Planetary System

107.001 **Jet streams in space.**   H. Alfvén.
         Astrophys. Space Sci., Vol. 6, 161 - 174 (1970).
    If viscosity is taken into account, Keplerian motion of a large number of grains in a gravitational field has a tendency to lead to the formation of jet streams. In order to treat problems of this kind it is advantageous to present celestial mechanics by a simple perturbation approach. Inelastic collisions between a number of grains will tend to make their orbits similar. This leads to the formation of jet streams. Finally, we discuss the possible application of the jet stream theory to meteor streams, to asteroidal jet streams, and to the cosmogonic accretion process.

107.002 **Comments on cosmic physics. III. Tidal evolution**
         of the earth-moon system; corals and molluscan
shells.   E. J. Öpik.
Irish Astron. Journ., Vol. 9, 120 - 135 (1969).

107.003 **Planetary fission events: Dynamical constraints.**
         J. M. Bailey.
Nature, Vol. 225, 48 - 49 (1970).
    Mass distribution in the solar system is subject to certain constraints. The limitations of this places on any history of planetary interaction and fission were evaluated. Results support the idea that the moon is a former planet (Luna), displaced from an orbit between Venus and Mercury by interaction with the latter. In only two situations (asteroids and Neptune–Pluto) it is possible that separate masses now orbiting the sun were once part of the same body. It is thus unlikely that fission of primary planetary bodies gave rise to the planetary configurations observed today.

107.004 **Planetary fission events.− Reply to J. M. Bailey.**
         W. H. McCrea.
Nature, Vol. 225, 49 (1970).

    The fission events envisaged by McCrea would necessarily have to occur early in the history of the solar system. Consequently even if, say, Mars originated by fission from the earth, its orbit would be subjected to approximately the same "circularization" processes as the orbit of, say, Jupiter or of the earth.

107.005 **Dynamic relaxation of planetary systems and**
         Bode's law.   J. G. Hills.
Nature, Vol. 225, 840 - 842 (1970).
    Eleven different planetary systems with a central star of $1 M_\odot$ but with widely different planetary mass functions were evolved from arbitrary initial orbits on an IBM 360/67 electronic computer by numerically integrating the equations of motion. The systems evolved towards stationary states in which adjacent orbits are commensurate. Bode's law results from some commensurabilities being more favored that others. The time necessary for the Jovian planets to evolve from their initial orbits to their present-day orbits is estimated to be $10^5 - 10^6$ years.

107.006 **Acerca de la formación de los planetas.**
         V. Mitra, P. K. Sharma.
Revista Astron., Vol. 41, No. 171, p. 17 - 19 (1969).

107.007 **The outer planets: Some early history.**
         R. Wildt.
Journ. Atmosph. Sci., Vol. 26, 795 - 797 (1969). − Conference paper (see 012.009).

107.008 **Magnetohydrodynamics and angular momentum**
         transfer.   D. W. Allan.
Symposia Mathematica, Vol. 3, 171 - 174 (1970). − Conference paper (see 012.010).

107.009 **On the evolution of the angular velocity of bodies in the solar system.** G. Colombo.
Symposia Mathematica, Vol. 3, 203 - 214 (1970). — Conference paper (see 012.010).

107.010 **Possible clues to the early history of the solar system.** H. Mabuchi, A. Masuda.
Nature, Vol. 226, 338 - 339 (1970).
^{138}La and ^{180}Ta isotopic abundances in meteorites were proposed as possible clues to the early solar irradiation assumed by Hayakawa and FGH. This proposal is based on the nuclear reactions ^{138}Ba (p, n) ^{138}La and ^{180}Hf (p, n) ^{180}Ta induced by low energy solar protons. The sensitivity of these nuclides to the irradiation was compared with already cited Li, S, K, Cr and V isotopes.

107.011 **Recent theories on the origin of the planets.** G. E. Satterthwaite.
Journ. British Astron. Ass., Vol. 80, 299 - 302 (1970).

107.012 **On the formation of planets.** V. Mitra.
Astron. Astrophys., Vol. 6, 491 - 493 (1970).
On the basis of the accretional theory for the formation of the planets as recently developed by the author (Mitra, 1970), the dimensions of the gaseous disc formed in the equatorial plane of the solar nebula have been determined. It is concluded that the particle accretion commenced only after the gaseous disc split into separate gaseous rings and that this splitting took place in two distinct groups.

107.013 **The melting of asteroidal-sized bodies by unipolar dynamo induction from a primordial T Tauri sun.**
C. P. Sonett, D. S. Colburn, K. Schwartz, K. Keil.
Astrophys. Space Sci., Vol. 7, 446 - 488 (1970).
This paper examines the heating of asteroidal parent bodies by electrical induction during early solar evolution and prior to positioning of the sun onto the main sequence. Calculations also include cases of joint heating by fossil radionuclides and electrical induction. Finally, some consequences of the mechanism applied to planets in the presence of an intense solar wind are considered.

107.014 **Comets and the formation of planets.** E. J. Öpik.
The Moon, Vol. 1, 487 - 493 (1970). — Abstract of a conference paper (see 011.031).

107.015 **Development of the primitive solar nebula.** A. G. W. Cameron.
The Moon, Vol. 1, 495 - 496 (1970). — Abstract of a conference paper (see 011.031).

107.016 **Catastrophes in the early history of the earth—moon system.** E. L. Ruskol.
Zemlya i Vselennaya, No. 3, p. 73 - 74 (1970). In Russian.

107.017 **Cosmic rays and the origin of the solar system.** A. J. R. Prentice, D. ter Haar.
Acta Phys. Acad. Sci. Hungaricae, Vol. 27, 231 - 235 (1969).
It is shown that cosmic rays may have caused sufficient ionisation in a solar disc in a Kant–von Weizsäcker type of theory about the origin of the solar system to produce a conductivity high enough for magnetohydrodynamic effects to be important.

107.018 **Energy regularity of the solar system and its possible origin.** C. J. Lavagnino.
Asoc. Argentina Astron. Bol., No. 14 (see 012.020), p. 76 (1968). In Spanish. — Abstract.

Chemical homogeneity of groups of terrestrial planets. See Abstr. 091.006.

Axial rotation of the planets. See Abstr. 091.030.

Comets, interplanetary space, and the problems of the solar system. See Abstr. 102.014.

Extinct radioactivities. See Abstr. 105.087.

The abundance of ^{244}Pu in the early solar system. See Abstr. 105.103.

Stars

111 Stellar Parallaxes

111.001 Spektroskopische Parallaxen.
H. Schmidt.
SuW, Vol. 9, 62 - 64 (1970).

111.002 Comparison of trigonometric parallaxes in right
ascension and declination determined with the
Sproul 24-inch refractor. S. L. Lippincott.
Bull. American Astron. Soc., Vol. 2, 206 (1970). − Abstr.
AAS.

111.003 Parallaxes and proper motions of 22 stars.
A. R. Upgren, W. S. Mesrobian, R. M. Nelson,
R. Grossenbacher, S. J. Kerridge.
Astron. Journ., Vol. 75, 319 - 320 (1970).

The relative parallaxes and proper motions of 22 stars
are given, ten of which have no previous trigonometric
parallax determination. All are faint red dwarfs found
spectrophotometrically by Vyssotsky. Nine of the stars
were measured on the automatic measuring machine of the
U. S. Naval Observatory. A substantial improvement in
accuracy in both coordinates was found for the stars with
automatic measures.

111.004 The parallax, proper motion and position of the
star LP 9−231 obtained from observations with the
26″-refractor at Pulkovo. A. A. Kiselev, N. K. Sumzina.
Izv. Glav. Astron. Obs. v Pulkove, No. 185, p. 119 - 123
(1970). In Russian.

The results are given of determinations of the parallax,
proper motion and position of the 14.3 magnitude star
LP9−231 ("Zwicky's Blue Pygmy"), obtained with the 65-cm
refractor at Pulkovo. The results obtained shows that the star
LP9−231 is a white dwarf of +13.7 absolute magnitude.

111.005 Stellar parallaxes and proper motions.
W. J. Luyten.
Trans. IAU, Vol. 14A, (see 003.028), 227 - 229 (1970).
Report of Commission 24.

Probleme der langbrennweitigen Astrometrie.
See Abstr. 041.001.

On the distance scale of the near-by classical
Cepheids. See Abstr. 122.057.

112 Proper Motions, Radial Velocities, Space Motions

112.001 The radial velocities of 335 late B-type stars.
D. P. Hube.
Mem. Roy. Astron. Soc., Vol. 72, 233 - 280 (1970).

Newly determined radial velocities are presented for
335 stars north and south of the equator. Most of these stars
have HD spectral type B8 or B9, and all are listed in the
Catalogue of Bright Stars. Individual plate velocities are
given along with MK spectral types where available. Previously published radial velocities are listed for comparison
with these new results. The spectrograms were obtained and
measured at the David Dunlap, Radcliffe, and Dominion
Astrophysical Observatories.

112.002 Proper Motion Survey with the forty-eight inch
Schmidt telescope. The zone +70 to +75.
W. J. Luyten.
Separate print Univ. Minnesota, Minneapolis, Minnesota.
Rep. Contract N onr 5060 (00), NR 046752.144 pp. (1970).

In continuation of the General Catalogue of Proper
Motions in the North Polar Cap − the area north of declination
+75 − data are here given for the proper motions of 7041
stars with declinations between +70 and +75 (1950).

112.003 Radial velocities of field horizontal-branch stars.
II. 1 HLF 2. A. G. D. Philip.
Astron. Journ., Vol. 75, 246 = Contr. Kitt Peak National Obs.
No. 524 (1970).

A table is presented giving the radial velocities of six

horizontal-branch stars in the 1 HLF 2 area ($l^{II} = 76°$, $b^{II} = -30°$).
Their velocity dispersion is ±109 km/sec, confirming their
assignment to membership in population II.

112.004 Estrellas de gran movimiento propio.
J. C. Muzzio.
Revista Astron., Vol. 41, No. 170, p. 55 - 57 (1969).

112.005 Photoelectric radial velocities of 87 seventh-magni-
tude K stars previously observed by Redman.
R. F. Griffin.
Monthly Notices, Roy. Astron. Soc., Vol. 148, 211 - 225
(1970).

A photoelectric instrument has been used to give very
accurate radial velocities for a sample of 87 seventh-magnitude
late-type stars. The more recent of two series of measurements
yields a standard error of 0.64 km s^{-1} for each observation.
Redman, who observed the same stars photographically 40
years ago, correctly estimated the errors of his observations.
In comparison with his work, the photoelectric instrument
offers a gain in information rate of about 4000 times.

112.006 Rigorous computation of proper motions and their
effects on star positions.
H. Eichhorn, A. Rust.
Astron. Nachr., Vol. 292, 37 - 38 (1970).

New rigorous formulas are given for the computation of
the effects of proper motions and radial velocities on star

positions, and for the transformation of proper motion components and radial velocities from one epoch to another. These expressions depend explicitly only on the values of the star's coordinates and distance, proper motion components and radial velocity, at the initial epoch.

112.007   A catalogue of proper motions for 437 A stars.
H. J. Fogh Olsen.
Astron. Astrophys., Suppl. Series, Vol. 1, 189 - 197 (1970).
Proper motions and radial velocities are given for 437 A stars observed photometrically by Johansen and Gyldenkerne. Most of the proper motions are improved GC motions transformed to the FK4 system.

112.008   Erratum: Radial velocities of field horizontal-branch stars. I. [Astrophys. Journ. *(Letters)*, Vol. 158, L113 - L115 (1969)].   A. G. D. Philip.
Astrophys. Journ. *(Letters)*, Vol. 160, L61 (1970).

112.009   The radial velocity of Arcturus determined from interferometric spectra.
B. W. Bopp, F. N. Edmonds, Jr.
Publ. Astron. Soc. Pacific, Vol. 82, 299 - 305 (1970).
From high-resolution, infrared, interferometric spectra, the radial velocity of Arcturus is determined by simple measurements between superposed spectral tracings of the star and the sun. A velocity of -5.47 ± 0.07 km/sec obtained from metal lines is comparable to a highly accurate result obtained by Petrie and Fletcher (1967) using standard methods. A velocity of -5.82 ± 0.18 obtained from CO lines is interpreted in terms of a differentially expanding atmosphere.

112.010   Radial velocity of EN (16) Lacertae.
M. Bloch, N. Morguleff, A. Terzan.
Astron. Astrophys., Vol. 6, 322 - 324 (1970).   In French.
The variations of radial velocity of EN (16) Lacertae have been studied on spectrograms taken at Haute Provence Observatory on the 26th of September 1968. The results confirm the period of 0.169.166 day of the primary pulsation.

112.011   Radial velocities of early-type stars.
J. P. Kaufmann.
IAU Symposium No. 38, (see 012.013), p. 290 (1970). Abstract.

112.012   The velocity dispersions of O- and B-stars within a few kpc.   R. B. Shatsova.
IAU Symposium No. 38, (see 012.013), p. 291 - 294 (1970).
The results of a study of the dispersion of velocities of O-B5 stars up to a distance of 3 kpc are described.

112.013   On the distribution of space velocities of OB stars.
L. V. Mirzoyan, M. A. Mnatsakanian.
IAU Symposium No. 38, (see 012.013), p. 295 - 296 (1970).
In this report we present briefly the results of a study of the space velocities of O-B1 stars in stellar associations which can give some information on the internal motions in the spiral arms.

112.014   Some results of the determination of absolute proper motions of stars referred to galaxies.
N. V. Fatchikhin.
Astron. Zhurn. Akad. Nauk SSSR, Vol. 47, 619 - 632 (1970). In Russian.   English translation in Soviet Astron. AJ, Vol. 14, No. 3.
From the analysis of differences in the proper motions of 779 stars of the AGK3 and our catalogue, corrected for magnitude equation, the correction to Newcomb's precessional constant and the motion of the equinox have been derived. For the stars of photographic magnitude 15.0 coordinates of solar apex, the secular parallax, and Oort's constants were determined.

112.015   Catalog of individual radial velocities, $0^h - 12^h$, measured by astronomers of the Mount Wilson Observatory.   H. A. Abt.
Astrophys. Journ., Suppl. Series, Vol. 19, 387 - 505 = Contr. Kitt Peak National Obs., No. 518 (1970).
For those stars for which only undated mean velocities have been published by Mount Wilson astronomers, the individual velocities, times of observation, and other pertinent data are given. Included in this catalog are approximately 11000 velocities of 3500 stars, observed and measured between 1909 and 1951.

112.016   Preliminary results of determinations of the proper motions of stars with reference to galaxies.
N. V. Fatchikhin.
Izv. Glav. Astron. Obs. Pulkove, No. 185, p. 93 - 97 (1970). In Russian.
The absolute proper motions of 14000 stars (including 1283 AGK 3 stars) relative to 271 galaxies were obtained in 82 sky areas. The coordinates of the solar apex, the secular parallax and Oort's constants were determined relative to the stars of 14.5 - 15.0 photographic magnitude. From an analysis of the differences of proper motions of 742 AGK 3 stars which were obtained at Pulkovo and also deduced by W. Dieckvoss in the FK 4 system and sent kindly to Pulkovo, the correction to Newcomb's precessional constant and the motion of the equinox have been computed.

112.017   Proper motions of 1160 late-type stars.
H. J. Fogh Olsen.
Astron. Astrophys., Suppl. Series, Vol. 2, 69 - 95 (1970).
Proper motions and radial velocities are given for 1160 late-type stars observed photoelectrically by Dickow et al. Most of the proper motions are improved GC motions transformed to the FK4 system.

112.018   Proper motions of long-period variable stars. II.
G. Alfieri, N. Missana.
Mem.Soc. Astron. Italiana, Nuova Serie, Vol. 41, 151 - 165 (1970).
Formulae are deduced for obtaining precise stellar positions from micrometric measurements on plates of Schmidt telescopes. The measurements made on 19 plates taken with the Schmidt of Asiago, scale 1 mm = 96", are compared with the standard coordinates of the Carte du Ciel catalogues and the proper motions of 13 stars are deduced. Considering the small number of cases available, the results appear satisfactory.

112.019   Proper motions of two U Gem type variables and of stars of their surroundings.
Sh. Primkulov.
Tsirk. Astron. Inst., *Tashkent*, No. 10 (357), p. 12 - 32 (1968). In Russian.

112.020   Radial velocities.   D. S. Evans.
Trans. IAU, Vol. 14A, (see 003.028), 335 - 342 (1970). – Report of Commission 30.

Catalogue of the positions and proper motions of stars between declinations $-40°$ and $-50°$, reduced to the equinox of 1950 without applying proper motions. See Abstr. 041.005.

Parallaxes and proper motions of 22 stars. See Abstr. 111.003.

The parallax, proper motion and position of the star LP 9-231 obtained from observations with the 26"-refractor at Pulkovo.   See Abstr. 111.004.

Stellar parallaxes and proper motions.
See Abstr. 111.005.

Radial velocities of companions to Mira variables.
See Abstr. 122.025.

Runaway stars and the pulsars near the Crab nebula.
See Abstr. 141.130.

7 Sextantis, a possible run-away star from upper Centaurus Lupus.   See Abstr. 152.011.

Large Magellanic Cloud. List of LMC members and list of galactic stars.   See Abstr. 159.015.

113 Stellar Magnitudes, Colors, Photometry

113.001 The use of *UBVr* photometry for the discovery of peculiar stars.
E. J. Mannery, G. Wallerstein.
Astron. Journ., Vol. 75, 169 - 170 = Contr. Kitt Peak National Obs. No. 526 (1970).

By plotting the ultraviolet excess, $\delta\,(U - B)$ against the index ΔM_4, it is easy to isolate barium, CH, and carbon stars. These peculiar stars can be distinguished from metal-poor, metal-rich, or composite stars.

113.002 Erratum: Photoelectric *UBV* sequences in Taurus.
[Astron. Journ., Vol. 72, 1012 - 1018 (1967)].
A. U. Landolt.
Astron. Journ., Vol. 75, 218 (1970).

113.003 Be stars which recently showed large photometric changes. A. Feinstein.
Publ. Astron. Soc. Pacific, Vol. 82, 132 - 135 = Contr. Cerro Tololo Inter-American Obs. No. 99 (1970).

113.004 On the application of a fiber-optic image intensifier to stellar photographic photometry.
J. B. de Veny.
Publ. Astron. Soc. Pacific, Vol. 82, 142 - 146 = Contr. Kitt Peak National Obs. No. 513 (1970).

The purpose of this note is to report the results of stellar iris photometry of direct image-tube plates and to discuss the limitations on photographic photometry with this type of intensifier.

113.005 Catalogue of magnitudes, color-indices, spectral and luminosity classes of stars in the Milky Way in Taurus (the area III of the P. P. Parenago plan).
M. D. Metreveli.
Byull. Abastumansk. Astrofiz. Obs. No. 38, p. 93 - 125 (1969).
In Russian.

113.006 Astrophysique en lumière infrarouge. F. J. Low.
L'Astronomie, 84e année, p. 11 - 22 (1970). –
Translated from Science, Vol. 164, 501 - 505 (1969).

113.007 Les magnitudes bolométriques. J. Breysacher.
L'Astronomie, 84e année, p. 110 - 114 (1970).

113.008 Quantitative analytic composite photography.
J. D. Wray.
Astron. Journ., Vol. 75, 247 - 251, 285 (1970).

The purpose of composite photography, as discussed in this paper, is to photographically perform a point-by-point subtraction of the image density of one negative from the image density of another. The result is an image of the density differences between the two negatives. In principle this subtraction is accomplished by reversing the sign of the densities of one negative (making it positive) and adding these densities to the other negative by superimposing the positive and negative.

113.009 Photoelectric photometry of stars in the system UPXYZVS with glass filters. II.
K. Zdanavičius, J. Sūdžius, Z. Sviderskienė, V. Straižys, V. Burnašov, R. Drazdys, A. Bartkevičius, G. Kakaras, G. Kavaliauskaitė, V. Jasevičius.
Bull. Vilnius Astron. Obs., No. 26, p. 3 - 12 (1969).
In Russian.

The results of photoelectric photometry of 315 stars in the intermediate band photometric system UPXYZVS are given. The standard system is the system of the 48-cm reflector

of Vilnius Observatory near Simeis. All color-indices are corrected for atmospheric extinction and interstellar reddening.

113.010 Photoelectric photometry of stars in the system UPXYZVS with glass filters. III.Metal – deficient stars. A.Bartkevičius, L. P. Metik.
Bull. Vilnius Astron. Obs., No. 26, p. 13 - 22 (1969).
In Russian.

The catalogue contains the color-indices of 58 stars with metal deficiency, i.e. subdwarfs and late-type giants of population II, observed with the 70-cm reflector of Crimean Astrophysical Observatory and the 48-cm reflector of Vilnius Astronomical Observatory in Simeis. All observations are reduced to the standard system of the 48-cm reflector.

113.011 Photometry of classical cepheids η Aql, SU Cas and ζ Gem in the system UPXYZVS.
J. Sūdžius.
Bull. Vilnius Astron. Obs., No. 26, p. 23 - 27 (1969).
In Russian.

Three classical cepheids η Aql, SU Cas and ζ Gem were observed in the UPXYZVS system. The variations of all color-indices can be explained by temperature effects. The dependence of amplitude on $1/\lambda$ is not linear and is similar to that in the six-color Stebbins–Whitford system. The color-excesses and MK spectra of the cepheids determined by two color diagrams are in good agreement with those determined by other methods.

113.012 The band-width effect in infrared photometry.
V. Straižys.
Bull. Vilnius Astron. Obs., No. 26, p. 29 - 35 (1969).
In Russian.

113.013 Photographic photometry of stars in fields with variable background density. M. E. Dixon.
Observatory, Vol. 90, 57 - 63 (1970).

Some numerical examples show that Ross's method of photographic photometry may be used to determine accurate stellar magnitudes in the presence of variable background density.

113.014 Delta Scorpii, an infrared deficient star and the value of R for the Scorpius region.
B. Iriarte Erro.
Bol. Obs. Tonantzintla y Tacubaya, Vol. 5 (No. 32), 101 - 103 (1969).

δ Sco, a B0 V star, was used among others in a previous determination of R for the Scorpius region. In this investigation it is shown that δ Sco is a peculiar star with a conspicuous infrared deficiency. A new value of R for Scorpius is suggested.

113.015 BVR, Potsdam, and Harvard photometries.
E. E. Mendoza V., T. Gómez.
Bol. Obs. Tonantzintla y Tacubaya, Vol. 5 (No. 33), 111 - 179 (1969).

Multiple regression analysis has been used to find equations to transform Potsdam and Harvard visual photometries into the BVR photoelectric system. The results indicate that the transformations are good. They strongly depend on the colors (B–V or V–R).

113.016 The calibration of narrow band photometry – I. Cambridge observations of G and K giants.
R. A. Bell.
Monthly Notices, Roy. Astron. Soc., Vol. 148, 25 - 52 (1970).
It has been shown that model stellar atmospheres and

laboratory data for atomic and molecular lines may be used to compute narrow band indices for G and K giant stars. These indices are in reasonable agreement with the observed indices. The dependence of the indices on the model parameters has been examined. On the basis of the present calculations and their observed indices, a number of K giants are found to be metal rich whilst others are metal poor.

113.017 **Some characteristics of the peculiar B stars with color-spectrum discrepancies.**
R. F. Garrison.
Bull. American Astron. Soc., Vol. 2, 194 (1970). — Abstr. AAS.

113.018 **The intrinsic colours of stars and two-colour reddening lines.** M. P. FitzGerald.
Astron. Astrophys., Vol. 4, 234 - 243 (1970).

Mean intrinsic colours, on the Johnson UBV and Cape U_cBV system, are obtained for stars of all MK spectral classes by analysis of the data in the Photoelectric Catalogue (Blanco et al., 1968). The intrinsic colours of stars of a given MK class are determined from: (1) the average colours of unreddened stars, for A to M giants, sub-giants, and dwarfs; (2) blue-most envelopes in two-colour and spectral-colour diagrams, for A to M bright giants and supergiants, and also for O and B type stars; and (3) two-colour reddening lines, for O and B type stars. The coefficients of the two-colour reddening line equation are obtained for the early type stars. The mean slope of the reddening line on the UBV system for O9 to B1 main sequence stars is found to be 0.75 ± 0.01 in Cygnus, and 0.70 ± 0.01 in the rest of the sky.

113.019 **R – I colour index for 330 late type stars.**
P. - U. Jacobsen.
Astron. Astrophys., Vol. 4, 302 - 308 (1970).

The infra-red colour index $R – I$ is given for 330 stars, the majority of which are G and K type giants brighter than 7th apparent visual magnitude. An ITT FW 118 photomultiplier with an end-on S 1 cathode cooled with dry-ice was employed. The results have been transformed to $R – I$ in Johnson's system.

113.020 **Etude entre 3600 et 6000 Å des étoiles G et K par photométrie photoélectrique à 11 bands passantes.**
N. Morguleff, M. P. Véron.
Astron. Astrophys., Vol. 4, 391 - 403 (1970).

By means of narrow-band photometry (photoelectric techniques), we have studied the classification (spectral type and luminosity class) of two hundred G and K stars, brighter than 8th magnitude. For the determination of the intensities in the bands centered on Na I, Mg I, Ca I, K of Ca II, CN lines, the method of Öhman was used and the measured intensities compared with the results obtained by other authors. Finally, the correlation is shown between some of our values and the fundamental parameters of classification.

113.021 **Photométrie photoélectrique UBV d'étoiles chaudes des "Selected Areas" 40 et 49, situées au voisinage du plan galactique.**
J. H. Bigay, R. Garnier.
Astron. Astrophys., Suppl. Series, Vol. 1, 15 - 28 (1970).

On donne les résultats de mesures photoélectriques UBV de 174 étoiles B et A0 dans la S.A.40 et de 112 étoiles B, A0 et A1 dans la S.A. 49.

113.022 **UBV-Messungen in den "Selected Areas" 100 und 112.** U. Haug, K. Walter.
Astron. Astrophys., Suppl. Series, Vol. 1, 29 - 33 = Mitt. Astron. Inst. Tübingen, No. 110 (1970).

Auf der Nordhalbkugel gemessene UBV-Helligkeiten von Sternen der Äquatorfelder S. A. 100 und 112 werden mitgeteilt und mit den in Südafrika erhaltenen Messungen verglichen.

113.023 **UBV observations of luminous stars in three Milky Way fields (Cassiopeia, Camelopardalis and Gemini).**
U. Haug.
Astron. Astrophys., Suppl. Series, Vol. 1, 35 - 104 = Mitt. Astron. Inst. Tübingen, No. 100 = Contr. Kitt Peak National Obs., No. 390 (1970).

All stars contained in three fields of the "Luminous Stars in the Northern Milky Way" catalogues have been observed in UBV with a 16-inch telescope at Kitt Peak National Observatory in 1964–1965. The results are assembled in a catalogue together with preliminary intrinsic colours, reddening values and distances. H II regions and other nebulae which may be connected with program stars are listed.

113.024 **Photoelectric observations of early A stars.**
K. T. Johansen, K. Gyldenkerne.
Astron. Astrophys., Suppl. Series, Vol. 1, 165 - 188 (1970).

Photoelectric observations of the β index and of indices similar to those of the Strömgren $uvby$ system have been made for 437 field stars and for a number of stars in the Coma Berenices, Praesepe, NGC 6633 and NGC 1662 clusters. Of the field stars, 377 are A0-A2 stars north of declination −5° with $V \leqslant 6.0$; in the clusters mostly A stars were observed. The night-to-night correction method is described and the transformation of the observed indices to the $uvby$ and β standard systems is discussed in detail.

113.025 **Narrow-band photometry of late-type stars.**
L. Häggkvist, T. Oja.
Astron. Astrophys., Suppl. Series, Vol. 1, 199 - 232 (1970).

The paper is part of an investigation of the distribution of stars in the direction perpendicular to the galactic plane. It is limited to the late-type stars, which are studied by means of interference filters; the break at the G band and the cyanogen absorption are measured. The catalogue includes, *i.a.*, all late-type stars brighter than $V = 5$ north of declination −10°, and those brighter than $V = 6$ north of galactic latitude +60°. The relation between the two-dimensional classification established for the G and K stars and the MK classification is studied. The absolute magnitudes of the late-type giants are discussed.

113.026 **Near-infrared photometry of two extremely red objects.** G. W. Lockwood.
Astrophys. Journ. *(Letters),* Vol. 160, L47 - L50 = Contr. Kitt Peak National Obs. No. 537 (1970).

The infrared objects IRC + 10216 and IRC + 40004 (CIT 1) have been observed on a five-color narrow-band photometric system capable of detecting near-infrared TiO and VO features in M-type stars. IRC + 40004 appears to be about M9 on this system, but IRC + 10216 is featureless.

113.027 **Photometric and spectroscopic observations of infrared stars.**
F. J. Low, H. L. Johnson, D. E. Kleinmann, A. S. Latham, S. L. Geisel.
Astrophys. Journ., Vol. 160, 531 - 543 (1970).

Photometric observations from U to Q (0.36 to $22\,\mu$) are reported for six stars that are bright in the infrared: Becklin's Object, HD 45677, R Mon, T Tau, NML Cyg, and VY CMa. Spectroscopic observations from 2500 to 7000 cm^{-1} have been obtained for three of these objects: NML Cyg, VY CMa, and HD 45677. These data suggest that all six systems contain protostars at various stages of evolution ranging in bolometric luminosity from 15 $L_\odot$ to 30000 $L_\odot$.

113.028 **Magnitudes, colours and coordinates of 175 ultraviolet excess objects in the field 13^h, +36°.**
A. Braccesi, L. Formiggini, E. Gandolfi.
Astron. Astrophys., Vol. 5, 264 - 279 (1970).

Here reported are the positions and magnitudes in a four colour (u, b, v, i) system of 175 ultraviolet excess objects down to b = 19.5 in a 6° × 6° degree field centered at 13ʰ, +36°. Most of these objects appear to be Q.S.O.'s (Braccesi and Formiggini, 1969). The paper is divided in two sections: the first gives an exhaustive description of the photometric system employed and how it has been calibrated against the classical U, B, V system. The second section describes the catalogue itself.

113.029 The blanketing effect in Am stars.
O. Ferrer, M. Jaschek, C. Jaschek.
Astron. Astrophys., Vol. 5, 318 - 321 (1970).

It is shown that blanketing corrections for Am stars can be deduced from $UBVRI$ photometry alone and that the metallicity, defined spectroscopically, is related to the modulus of the blanketing vector, defined photometrically.

113.030 Schilt's method: Application to stellar photometry with the electronic camera.
A. Bijaoui, M. Dantel.
Astron. Astrophys., Vol. 6, 51 - 59 (1970).

The application of the electronic camera to stellar photometry allows us to retain almost all the advantages of both classical photographic plate and those of photomultiplier characteristics. We determined by Schilt's method the characteristic curves for different diaphragms. We applied a two diaphragm-method to remove the background density. The linearity of the characteristic curves is better in this particular case. We studied the different effects which can introduce systematic errors: star-setting error, field curvature and distortion. In order to remove the possible errors due to these effects we suggest a method with three diaphragms. In this case the measurement discrepancies are slightly amplified by the systematic effects are reduced significantly.

113.031 Recent absolute calibration work at Palomar Mountain.
J. B. Oke, R. Schild.
IAU Symposium No. 36, (see 012.014), p. 13 - 17 (1970).

Bright stars such as α Lyr are compared directly with light sources under identical instrumental conditions.

113.032 Far-ultraviolet intensities of Orion stars.
G. R. Carruthers.
IAU Symposium No. 36, (see 012.014), p. 100 - 108 (1970).

This paper covers some of the results of an Aerobee rocket flight from White Sands Missile Range, New Mexico, on January 30, 1969, in which an electronographic objective spectrograph and ultraviolet photon-counter photometers were used to obtain spectra in the 1000 - 1600 Å wavelength range, and photometric data in the 1050 - 1180 Å and 1230 - 1350 Å wavelength ranges, for early-type stars in Orion.

113.033 Ultraviolet photometry of stars obtained with the Celescope experiment in the Orbiting Astronomical Observatory.
R. J. Davis.
IAU Symposium No. 36, (see 012.014), p. 109 - 119 (1970).

We have used the television photometers in the Celescope OAO experiment to measure the far ultraviolet brightness of several thousand stars, including parts of the constellations Draco, Lyra, Puppis, Vela, Taurus, and Orion; and the Moon.

113.034 Photographic magnitudes of 201 stars at 2600 Å.
J.-P. Sivan, M. Viton.
IAU Symposium No. 36, (see 012.014), p. 120 - 129 (1970).

The magnitudes of 201 stars at 2600 Å (1000 Å passband) were derived from two plates of the winter Milky Way obtained with a large field camera. A preliminary investigation of the interstellar reddening allowed us to plot a color-spectral type diagram. Stars of type O seem to be brighter than predicted.

113.035 Absolute stellar photometry in the region 1200 - 3000 Å.
J. W. Campbell.
IAU Symposium No. 36, (see 012.014), p. 135 - 137 (1970).

113.036 Preliminary note on the astronomical satellite Kosmos 215.
N. Dimov.
IAU Symposium No. 36, (see 012.014), p. 138 (1970).

113.037 Ultraviolet photometry of stars from OSO II.
K. L. Hallam.
IAU Symposium No. 36, (see 012.014), p. 139 (1970).
Abstract.

113.038 UBV photoelectric photometry in four southern Milky Way fields.
W. H. Wooden II.
Astron. Journ., Vol. 75, 324 - 336, 513 - 516 (1970).

Three-color observations in the UBV system define photoelectric sequences in four southern Milky Way fields. Each of the sequences consists of predominantly main-sequence A, main-sequence F, and giant K stars.

113.039 An attempt to define luminosity criteria in O stars via narrow-band photoelectric photometry.
A. U. Landolt.
Astron. Journ., Vol. 75, 337 - 344 = Contr. Louisiana State Univ. Obs., *Baton Rouge*, No. 37 (1970).

Narrow-band photoelectric indices were defined at the N III triplet $\lambda\lambda$4634, 4640, 4641 and at He II λ4686 for 156 O and Of stars. There is some slight indication that the indices defined are sensitive to luminosity. If true, the VI Cygni association is much nearer than previously thought. Comments are made concerning variability of these spectral features.

113.040 Calibration of 4-color and Hβ photometry for B- and A-type stars.
D. L. Crawford.
IAU Symposium No. 38, (see 012.013), p. 283 (1970).

113.041 Questions concerning the usefulness and feasibility of large-scale stellar statistics.
W. Seitter.
IAU Symposium No. 38, (see 012.013), p. 297 (1970).
Abstract.

113.042 Mesures de l'ultra-violet stellaire à partir de ballons stratosphériques.
A. Gaide.
Arch Sci. Genève, Vol. 22, Fasc. 1, p. 5 - 47 (1969) = Publ. Obs. Genève, Sér. A, Fasc. 77/I (1970).

This work is based on the study of the evolutionary criteria of early type stars and consists in carrying out photometric observations of the stellar near ultra-violet from a balloon-borne observatory. In the first three chapters we analyse the observational possibilities, describe the astronomical gondola, and deal with the sky slow-scanning method, the telescope used and the method of analysing the recordings. In the two last chapters the results are discussed.

113.043 Dépouillement et résultats astronomiques d'observations stellaires dans le proche ultraviolet.
C. Navach.
Arch. Sci. Genève, Vol. 23, Fasc. 1, p. 41 - 59 (1970) = Publ. Obs. Genève, Sér. A, Fasc. 77/II (1970).

Une caméra Maksutov équipée d'un diapositif photométrique à 2 couleurs a volé à 33 km d'altitude à bord d'une nacelle de ballon stratosphérique. Un indice de couleur a pu être établi pour 27 étoiles. La comparaison entre les résultats expérimentaux et les modèles d'atmosphère stellaires révèle l'insuffisance des modèles sans blanketing. Une loi d'extinction interstellaire est établie pour $\lambda \geq 2900$ Å.

113.044 Groupes physiques très jeunes. III. L'influence du

«coudé» dans les lois d'extinction. G. Goy.
Arch. Sci. Genève, Vol. 23, Fasc. 1, p. 61 - 66 (1970) = Publ.
Obs. Genève, Sér. A, Fasc. 77/VII (1970).

On montre que la position du «coudé» autour de 4300 Å
dans les lois d'extinction interstellaire de Nandy influence con-
sidérablement l'indice (B1–B2) de la photométrie en 7 couleurs de l'Observatoire de Genève, ainsi que le paramètre [d]
qui le contient.

113.045 **Photométrie des étoiles A. II. Métallicité et rotation des étoiles Am.** B. Hauck.
Bull. Soc. Vaudoise Sci. nat., Vol. 70, Fasc. 7 (No. 332), p.
309 - 312 (1970) = Publ. Obs. Genève, Sér. A, Fasc. 77/VIII
= Obs. Univ. Lausanne Commun., No. 19 (1970).

The meausres of Am stars in the photometric system of
Geneva Observatory show that the distribution of these stars
in the plane $\Delta m_2/B_2$-V_1 is not aleatory. The coldest Am stars
have the greatest values of the parameter of metallicity, Δm_2.

113.046 **Four-color and Hβ photometry for bright stars in
the southern hemisphere.**
D. L. Crawford, J. V. Barnes, J. C. Golson.
Astron. Journ., Vol. 75, 624 - 635 = Contr. Cerro Tololo
Inter-American Obs., No. 109 (1970).

Photoelectric data on the *uvby* and β systems for 392
B-, A-, and F-type stars south of declination -10° are given
in a table. Many of the stars can be used as standards for
southern hemisphere photometry with these systems.

113.047 *UBV* **observations of Mira stars.**
N. R. Evans.
Astron. Journ., Vol. 75, 636 - 640 (1970).

This paper presents new observations of 40 Mira stars at
or near maximum light. Photoelectric magnitudes and colors
for Mira stars with $|b^{II}| > 15°$ have been compiled. A color-
color diagram for Mira stars at maximum light is included.

113.048 **Erratum: Photometric standards for the southern
hemisphere.** [Astron. Journ., Vol. 74, 1125 -
1130 (1969), see 02.113.035]. B. J. Bok, P. F. Bok.
Astron. Journ., Vol. 75, 665 (1970).

113.049 **Some remarks regarding photoelectric photometry.**
E. E. Mendoza V.
Bol. Obs. Tonantzintla y Tacubaya, Vol. 5 (No. 34), 209 -
212 (1970).

An investigation of photoelectric data of 8 flare stars in
the Orion nebula leads to the following results: The photometric errors are not large; internal agreement of the photometry is good; the observed flare stars have infrared excesses,
unexplained by interstellar extinction alone.

113.050 **Photoelectric photometry of faint blue stars at
high galactic latitudes.** B. Iriarte Erro.
Bol. Obs. Tonantzintla y Tacubaya, Vol. 5 (No. 34), 213 -
217 (1970).

In this note thirty three stars from Chavira's (1958, 1959)
lists were observed in the course of other photometric programs carried out by the author with the 40″ reflector at the
Tonantzintla Observatory. The results of the observations are
given in a table.

113.051 **Erratum: Delta Scorpii, an infrared deficient star
and the value of R for the Scorpius region.**
[Bol. Obs. Tonantzintla y Tacubaya, Vol. 5 (No. 32), 101 -
103 (1969), see 03.113.014] B. Iriarte Erro.
Bol. Obs. Tonantzintla y Tacubaya, Vol. 5 (No. 34), 218
(1970).

113.052 **Photoelectric photometry of 1160 late-type stars.**
P. Dickow, K. Gyldenkerne, L. Hansen, P.-U. Jacob-
sen, K. T. Johansen, P. Kjaergaard, E. H. Olsen.
Astron. Astrophys., Suppl. Series, Vol. 2, 1 - 67 (1970).

Photoelectric narrow-band observations have been carried
out for 1160 late-type stars north of declination -30°. Most of
the stars are of spectral type G and K, and a number of high-
velocitiy stars has been included in the programme. Measured
are: the G-band break (index g), the cyanogen absorption with
band head at 4216 Å (index n), the break at 4000 Å including
the K line of Ca II (index k), the metallic-line index m, and
two colour indices f and u.

113.053 **Effects of reddening on colour transformations.**
A. Gutiérrez-Moreno, H. Moreno.
Astron. Astrophys., Vol. 7, 35 - 48 (1970).

By means of numerical integrations, it is shown that
colour-system transformation equations, obtained from unreddened stars, are not satisfied by reddened stars. Correction
formulae are determined. These corrections are found to
depend on the colour of the star, the amount of reddening
and the transformation coefficients to the standard system.

113.054 **Sur le raccordement des systèmes photométriques.**
I. Todoran.
Stud. Cerc. Astron., Vol. 15, 99 - 104 (1970).

113.055 **Neues Verfahren zur visuellen Photometrie.**
G. Apflauer.
Sternenbote, 13. Jahrgang, p. 38 - 43 (1970).

113.056 **Stellar photometry.** A. W. J. Cousins.
Trans. IAU, Vol. 14A, (see 003.028), 231 - 247
(1970). – Report of Commission 25.

113.057 **Narrow-band photoelectric photometry of the two
WR stars HD 192765 and HD 192103 with no
detectable signs of spectral duplicity.**
A. M. Tsherepashuk, V. A. Erizhokov.
Astron. Tsirk., No. 561, p. 1 - 4 (1970). In Russian.

113.058 **Spectrophotometry of cool stars in the near infrared.
Results for a region in Cygnus.**
K. Nandy, F. Smriglio.
Publ. Roy. Obs., Edinburgh, Vol. 7, (No.1), 1 - 18 (1970).

The distribution of relatively cool stars has been investi -
gated from objective prism spectra in the near infrared in a
region of about 12 square degrees in Cygnus. Spectral types
and apparent infrared magnitudes of 442 M and C type stars
have been obtained. The limiting infrared magnitude is 12^m0.
From the available material and the published data, estimates
of the interstellar extinction in the infrared and of infrared
absolute magnitudes have been obtained. The spatial distribu -
tion of M stars has been tentatively determined. In the galactic
plane in the direction of Cygnus N stars are more abundant
than R stars.

113.059 **Spectrophotometric investigation of the star MWC
334.** Ya. N. Chkhikvadze.
Soobshch. AN Gruz. SSR, Vol. 57, 321 - 324 (1970). In
Russian. – Abstr. in Referativ. Zhurn. 51. Astron., 7.51.502
(1970).

113.060 **UBV photoelectric photometry of SX Phoenicis.**
S. Tapia.
Asoc. Argentina Astron. Bol., No. 12, (see 012.019), p. 32
(1968). In Spanish. – Abstract.

113.061 **Catalogue of photoelectric measurements.**
C. Jaschek, A. Sierra, E. Hernández, A. Gerhardt.
Asoc. Argentina Astron. Bol., No. 12, (see 012.019), p. 38
(1968). – Abstract.

113.062 **Four-colour photometry.** H. Moreno.
Asoc. Argentina Astron. Bol., No. 14 (see 012.020), p. 54 - 55 (1968). In Spanish. – Abstract.

113.063 **Reddening effects in the transformation of colour systems.** A. Gutierrez-Moreno, H. Moreno.
Asoc. Argentina Astron. Bol., No. 14 (see 012.020), p. 63 - 64 (1968). In Spanish.

Some methodic problems of astrophotography.
See Abstr. 031.017.

A photometric investigation of strong-cyanogen stars. See Abstr. 114.010.

Stellar spectrophotometry from a pointed rocket.
See Abstr. 114.027.

Spectral classifications and multi-band colour indices. See Abstr. 114.133.

Nyere resultater fra den infrarøde astronomi.
See Abstr. 114.141.

Studies of A stars. I: Catalogue of spectral types and colours. See Abstr. 114.150.

Studies of bright southern stars.
See Abstr. 114.154.

Two subluminous stars near the galactic plane.
See Abstr. 115.005.

Photometric and spectroscopic data for some distant O and B stars. See Abstr. 153.017.

A photometric search for distant OB stars in Norma and an investigation of the Norma cloud.
See Abstr. 155.018.

114 Stellar Spectra, Temperatures, Spectroscopy

114.001 **ν Indi: a weak-lined halo star with excess deficiency of nitrogen.** D. L. Harmer, B. E. J. Pagel.
Nature, Vol. 225, 349 - 351 (1970).
The weakness of ultraviolet CN bands in the spectrum of ν Indi implies that, while carbon and metals are deficient (compared with the sun) by a factor of about fifteen, nitrogen is still more deficient by a factor between four and eight.

114.002 **Rotation and chemical abundances in the peculiar A stars – I.** C. J. Durrant.
Monthly Notices, Roy. Astron. Soc., Vol. 147, 59 - 73 (1970).
The classifications and atmospheric structures displayed by early type peculiar stellar spectra are compared empirically with those of normal stars. It is concluded that the structures of the metal line forming regions of the atmospheres of the peculiar A and B stars are identical with those of the normal stars of the same hydrogen type. A consistent temperature sequence can be established for both types later than spectral type B5.

114.003 **Rotation and chemical abundances in the peculiar A stars – II.** C. J. Durrant.
Monthly Notices, Roy. Astron. Soc., Vol. 147, 75 - 93 (1970).
New measurements of the distribution of silicon line stengths amongst early type stars having similar atmospheric structures do not allow a clear-cut separation into normal and anomalous types. It is shown that those stars having line strengths much larger than the mean tend to have rotational velocities less than 200 km s^{-1}.

114.004 **Expanding atmospheres in OB supergiants – IV. A mass-loss survey.** J. B. Hutchings.
Monthly Notices, Roy. Astron. Soc., Vol. 147, 161 - 176 (1970).
A study of medium-dispersion spectra of seven very bright early-type stars has indicated the presence, to a varying degree, of expansion in the outer layers of their envelopes. A brief survey of other early-type supergiants has indicated that a similar situation arises in eleven of them. Combined with the results of detailed studies of four stars with large mass loss, these conclusions indicate that mass loss is common among the brightest hot stars. The rate of mass loss increases with luminosity and effective temperature amongst the OB stars and may well be an important consideration in the evolution of massive stars.

114.005 **An absolute spectrophotometric calibration of the energy distribution of twelve standard stars.** D. S. Hayes.
Astrophys. Journ., Vol. 159, 165 - 176 (1970).
Energy distributions of 12 standard stars have been calibrated with two Phillips standard-lamps having a tungsten-ribbon filament which were calibrated in Heidelberg in 1958 and 1966. The observations were made with a reflection-grating photoelectric spectrum scanner on the 36-inch Crossley reflector of Lick Observatory and cover the range 3200–10870 Å. The results for the twelve stars are tabulated.

114.006 **Lines of neutral helium in O- and B-type stars.** H. L. Shipman, S. E. Strom.
Astrophys. Journ., Vol. 159, 183 - 193 (1970).
The effect of ultraviolet line blanketing on predicted departures from LTE is investigated. Previous conclusions suggesting that departures are small for stars later than B3 are not altered when ultraviolet blanketing is included in model calculations. New descriptions of line broadening for the He I diffuse triplet λ4471 and the diffuse singlet λ4922 have been used in conjunction with model atmospheres to investigate the behavior of these lines for O and B stars. Qualitative agreement with the observed behavior of these lines is found for a wide range of spectral types and luminosities. It now seems possible to obtain reliable He abundances from observation of λ4471.

114.007 **Silicon monoxide bands in some low-temperature stars.** J. H. Fertel.
Astrophys. Journ. (Letters), Vol. 159, L7 - L8 (1970).
Several bands which appear in the near-infrared spectra of cool stars may be due to SiO.

114.008 **The spectrum and a quantitative analysis of the Ap star Alpha Draconis.** J. Zverko.
Bull. Astron. Inst. Czechoslovakia, Vol. 21, 56 - 61 (1970).
Three spectrograms, obtained by the 193-cm telescope's coudé spectrograph at Obsérvatoire du Haute Provence, have been used for measuring the equivalent line widths in the spectrum of the peculiar star α Draconis. A comparison of the observed contours of H_γ with the theoretical ones computed by D. Mihalas was made. The excitation temperature increases with increasing ionization potential of the ionized elements. The abundance anomalies for Si, Cr, and Sr are discussed.

114.009 **Observations of water-vapor emission associated with infrared stars.** P. R. Schwartz, A. H. Barrett.
Astrophys. Journ. (Letters), Vol. 159, L123 - L127 (1970).
A survey of 134 infrared stars for microwave H_2O emission has resulted in the detection of emission from three new sources: NML Cygni, the Mira variable U Herculis, and the semiregular variable W Hydrae. In all cases the radial velocities of the microwave lines lie between radial velocities of the optical emission and absorption lines of H_2O.

114.010 **A photometric investigation of strong-cyanogen stars.** R. D. McClure.
Astron. Journ., Vol. 75, 41 - 52 = Contr. Kitt Peak National Obs. No. 512 (1970).
A cyanogen-band index which is sensitive to metallicity for late-type giant stars, but is virtually independent of surface gravity, has been formed using the David Dunlap Observatory intermediate-bandpass photoelectric system. The index is used to obtain a quantitative measure of the cyanogen anomaly for a large group of strong-cyanogen stars.

114.011 **Stellar and planetary spectra in the infrared from 1.35 to 4.10 microns.** F. F. Forbes, W. F. Stonaker, H. L. Johnson.
Astron. Journ., Vol. 75, 158 - 164 (1970).
Infrared spectra of late-type stars and some planets have been obtained through the use of an infrared spectrometer using a rotating, circular, variable-thickness interference filter. The resolution of these spectra is of the order of 100 ($=\Delta\lambda/\lambda$) over the interval 1.35–4.10μ. The spectra of normal M stars indicate a positive luminosity effect based on the first and second overtones of CO. All three Mira variables observed exhibit stellar steam absorption. NML Cygnus and NML Taurus show extremely red energy distributions as expected. Features in the spectrum of NML Cygnus, which may be interpreted as the first overtone of CO, would seem to confirm its identity as a supergiant provided it is of normal composition.

114.012 **MK classification for F- and G-type stars. II.** E. A. Harlan, D. C. Taylor.

Astron. Journ., Vol. 75, 165 - 166 = Lick Obs. Bull. No. 607 (1970).

MK spectral classifications are given for 166 stars of HD types F2-G5, and having m_V = 7.5 or brighter. The classifications were made on slit spectrograms of dispersion 75 Å/mm.

114.013 The helium-rich stars of early spectral type.
A. S. Dinger.
Astrophys. Space Sci., Vol. 6, 118 - 130 (1970).

The properties of the known helium-rich stars of early type are reviewed and the elemental abundances are analyzed. Current evolutionary theories of mixing and mass loss may explain the origin of these stars.

114.014 Correlations between statistical population indices and spectral characteristics of carbon stars.
J. Krempeć.
Astrophys. Space Sci., Vol. 6, 131 - 140 (1970).

The coefficients of correlation between spectroscopic data published by Yamashita (1967) and others for carbon stars and the statistical population indices calculated for these stars at the Toruń Observatory are calculated.

114.015 Two-dimensional quantitative spectral classification of F0–G5 stars by means of objective prism spectra.
M. A. Shiukashvili.
Byull. Abastumansk. Astrofiz. Obs. No. 37, p. 43 - 67 (1969). In Russian.

Quantitative spectral classes and absolute magnitudes were determined for nearly 210 F0–G5 stars. The mean square error of spectral class amounts to ± 0.6 of the spectral subclass, of absolute magnitude to ± 0.8 and ± 0.5 for F0–F5 and F5–G5 stars respectively. The diagram spectrum—absolute magnitude confirms the presence of Hertzsprung gap for relatively faint stars in the interval of F6–G5.

114.016 Spectral classification of stars in the near ultraviolet with moderate- and low-dispersion spectra.
M. V. Dolidze, G. N. Jimsheleishvili.
Byull. Abastumansk. Astrofiz. Obs. No. 37, p. 68 - 88 (1969). In Russian.

The results of the spectral classification of stars in the near ultraviolet region are given. They are compared with the data of existing catalogues and lists of stellar spectra.

114.017 Some problems of spectral classification and spectrophotometry of late type stars.
G. N. Jimsheleishvili.
Byull. Abastumansk. Astrofiz. Obs. No. 37, p. 89 - 116 (1969).   In Russian.

An investigation concerning the spectral classification of late stars and the photometry of their continuous spectrum has been carried out. The classification method of stars in the C-system has been developed for our conditions. Forty new and twenty-four standard stars have been re-classified. The possibility of selection of peculiar C-stars with cyanogen red band intensity has been revealed. The ultraviolet anomaly of C-stars may be used as second criterion of the abundance parameter. A blue excess has been revealed in CH Cyg.

114.018 Observations of interstellar sodium lines in stars in the direction of the galactic center.
G. Wallerstein.
Publ. Astron. Soc. Pacific, Vol. 82, 5 - 9 (1970).

Radial velocities and equivalent widths are presented for components of interstellar sodium in the spectra of nine stars within 20° of the galactic center. The data are compared with observations of neutral hydrogen in the direction of each star. A component at –61 km/sec in HD 160529 is discussed in some detail because it may be the first evidence from optical interstellar lines of the gas flowing out from the galactic center.

114.019 Faint new Wolf-Rayet stars in Carina.
D. J. MacConnell, N. Sanduleak.
Publ. Astron. Soc. Pacific, Vol. 82, 80 - 85 = Contr. Cerro Tololo Inter-American Obs. No. 96 (1970).

Nine new faint Wolf-Rayet stars were found on objective-prism plates of the Carina region taken with the Curtis-Schmidt telescope at Cerro Tololo. Ultimately, these stars may help to define spiral structure in that direction at distances perhaps in excess of 10 kpcs from the sun.

114.020 Catalogue of stellar spectra in the Orion region.
E. B. Kostyakova.
Byull. Abastumansk. Astrofiz. Obs. No. 38, p. 19 - 92 (1969). In Russian.

The results of the one-dimensional spectral classification of about 7000 stars (m ≤ 12^m) in the Orion region are given.

114.021 Observations spectroscopiques de l'étoile η Arietis.
A. Baranne, F. Spite, M. Spite.
Comptes Rendus Acad. Sci. Paris, Sér. B, Vol. 270, 916 - 917 (1970).

De l'étude de deux spectres de l'étoile η Arietis, il semble ressortir que cette étoile n'est pas déficiente en fer comme le laissait présager son indice de métallicité. Par contre, η Ari semble être très déficiente en vanadium.

114.022 Wolf-Rayet sterren.   R. van Helden.
Hemel en Dampkring, Vol. 68, 5 - 6 (1970).

114.023 The lithium content in late-type giants.
A. M. Boesgaard.
Astrophys. Letters, Vol. 5, 145 - 147 (1970).

It is shown that Warner's recent re-analysis of Bonsack's Li abundances in G and K stars leads to a consistent picture of the Li content in all late-type giant stars. The Li content decreases monotonically with temperature from F5 to M4, with no pronounced dip for the late K stars as found earlier.

114.024 The spectrum of a^2 CVn. II.   J. G. Cohen.
Astrophys. Journ., Vol. 159, 473 - 484 (1970).

The behavior of the continuum of a^2 CVn has been studied. The Balmer discontinuity does not vary but the slope of the Paschen continuum changes over the cycle. The line-blanketing coefficients have been measured at several phases and do not change sufficiently to produce the observed variations of the continuum. The effective temperature (12000°K) and surface gravity (log g = 4.0) are determined from the scans and hydrogen-line profiles, and the behavior of equivalent width as a function of phase is discussed. A crude abundance analysis is performed.

114.025 The unusual composition of +39°4926.
K. Kodaira, J. L. Greenstein, J. B. Oke.
Astrophys. Journ., Vol. 159, 485 - 512 (1970).

The extremely metal-poor star has T_e + 7500°K, log g = 1, and a very large Balmer jump (1.7 mag). Model atmospheres explain the hydrogen spectrum, and over plausible ranges of T_e the abundances are insensitive to errors in T_e. Weak lines of He I are present, but the He/H ratio is temperature-dependent. Strong lines of C I and O I are observed and yield abundance ratios C/H and O/H near that in the sun and insensitive to T_e. The metal abundances average 1 percent of their solar values, differ from element to element, and show an excessively large odd-even alternation. The velocity seems variable in a long period. The absolute magnitude is near –3, the mass less than the sun.

114.026 The chemical composition of Iota Herculis.
G. J. Peters, L. H. Aller.

Astrophys. Journ., Vol. 159, 525 - 542 (1970).

The chemical composition of the sharp-lined B3 V star ι Her has been determined by a model-atmosphere technique. All elements considered in this analysis seem to have solar-type abundances. The effective temperature determined for this star by the criterion of ionization balance is dependent upon whether blanketed or unblanketed model atmospheres are utilized.

114.027 Stellar spectrophotometry from a pointed rocket.
T. P. Stecher.
Astrophys. Journ., Vol. 159, 543 - 550 (1970).

A 13-inch telescope with a three-channel photoelectric scanner was successfully flown on an Aerobee rocket. The whole rocket assembly was pointed at four program stars with an absolute accuracy exceeding 20 seconds of arc. Two photometric scans (of 20-sec duration) of each program star were made with a resolution of 10 Å. The spectra are discussed.

114.028 Ultraviolet fluxes and bolometric corrections for late B to F main-sequence stars.
J. Davis, R. J. Webb.
Astrophys. Journ., Vol. 159, 551 - 568 (1970).

A comparison of observed ultraviolet fluxes with the predictions of line-blanketed model atmospheres for stars of spectral type B5 and later has revealed significant discrepancies that increase toward later spectral types. Possible sources of ultraviolet opacity are discussed which are not included in the models but which might explain the discrepancies. Bolometric corrections for existing model atmospheres, based on a clear definition that uses solar data to establish the zero point, are presented. The theoretical values are compared with empirical bolometric corrections.

114.029 Zeeman measures of sharp-lined early B stars.
P. S. Conti.
Astrophys. Journ., Vol. 159, 723 - 725 = Contr. Lick Obs., No. 298 (1970).

Measures of Zeeman-analyzed spectrograms of early B stars in Orion are presented. There is no evidence that any of the stars studied have detectable fields. These results do not confirm the presence of magnetic fields in these stars reported by Sargent, Sargent, and Strittmatter.

114.030 Spectre de l'étoile particulière à émission HBV 475 dans le proche infrarouge. Y. Andrillat.
Comptes Rendus Acad. Sci. Paris, Sér. B, Vol. 270, 1066 - 1069 (1970).

Dans le proche infrarouge, le spectre de l'étoile variable HBV 745 montre de fortes émissions dues à H, He I, O I et Ca II. On identifie TiO en absorption. Le continu est comparable à celui d'une étoile de type M4. Cette étude semble donc confirmer le caractère symbiotique de cette étoile, qui est masqué dans les autres domaines spectraux par de nombreuses émissions.

114.031 The spectra of the two metallic-line stars ν_1 and ν_2 Draconis. S. Islik Engin.
Mem. Soc. Astron. Italiana, Nuova Serie, Vol. 41, 33 - 45 (1970).

The spectra of the two members of the visual binary ν_1 and ν_2 Dra are studied. The two companions are metallic-line stars. They are compared with normal stars of about the same temperature.

114.032 Spectroscopy and astrophysics.
S. L. Mandelshtam.
Priroda, No. 2.70, p. 8 - 19 (1970). In Russian.

114.033 The continuous radiation of bright B3 V stars.
K. Kodaira.

Astrophys. Journ., Vol. 159, 931 - 943 (1970).

Spectrophotometry from $\lambda 3297$ to $\lambda 7850$ is presented for twenty-seven B3 V stars brighter than 5.6 mag in the northern sky. No serious discrepancy is found between the observed and theoretical continua when a model atmosphere of $T_e = 9600°K$ and $\log g = 4.0$, calculated by Mihalas, is adopted for the primary standard α Lyr, according to the system of Oke. Apparent effective temperatures are determined for the stars in this program by comparing the colors and Balmer jumps with those of model atmospheres. Most of the observed B3 V stars are found to have an effective temperature between 16000° and 18000°K. We discuss an excessively large Balmer jump observed for some rapidly rotating stars, with special attention to the effect of rotation on the fluxes.

114.034 Spectral observations, 1965 - 1968, of the peculiar emission object V1016 Cygni (MHα 328-116).
M. P. FitzGerald, N. Houk.
Astrophys. Journ., Vol. 159, 963 - 972 (1970).

Identifications and intensities are given for approximately 130 emission lines between 3130 and 5030 Å measured on fifteen spectrograms, with dispersions ranging from 12 to 130 Å mm^{-1}, taken during the 1965 - 1968 seasons. Available photometric data, measurements of radial velocity, and line structure are also discussed. The radial velocity is -60 ± 1.5 km/sec.

114.035 Infrared CN bands in M supergiants and carbon stars. R. F. Wing, H. Spinrad.
Astrophys. Journ., Vol. 159, 973 - 983 (1970).

Bands of the red system of CN in the region $1.1 - 2.5\mu$ have been identified in published spectra of M supergiants and carbon stars; their strengths are consistent with our observations of red CN bands in the $1-\mu$ region. The CN features longward of 1.1μ in M supergiants had previously been attributed to H_2O; the revised identifications indicate that these stars have oxygen/carbon ratios very close to unity. In carbon stars the CN bands produce broad and deep depressions which cause an apparent weakening of the rotation-vibration bands of CO; the observed differences in CO strength among carbon stars are caused primarily by differences in the CN abundance.

114.036 Inhomogeneities of the chemical composition and physical conditions at the surface of the Si II Ap stars CU Vir and 56 Ari. V. L. Khokhlova.
Astron. Zhurn. Akad. Nauk SSSR, Vol. 47, 132 - 138 (1970). In Russian. English translation in Soviet Astron. AJ, Vol. 14, No. 1.

The variation of Si II intensities in the Ap stars CU Vir and 56 Ari are considered assuming the oblique rotator model. The curves of intensity variations of Si II lines have two maxima, which correspond to two spots on the surface of the star. The relative intensities of the maxima depend on the excitation potential of the lines. This phenomenon is explained by the temperature difference in Si II spots.

114.037 Relative abundances of magnesium isotopes in Arcturus. R. A. Bell, D. Branch.
Astrophys. Letters, Vol. 5, 203 - 206 (1970).

Rotational lines of the $A^2\Pi - X^2\Sigma$ (0,0) band of MgH have been used to determine the relative abundances of the magnesium isotopes in the atmosphere of Arcturus. It is found that $Mg^{24} : Mg^{25} : Mg^{26} = 80 : 10 : 10$, in agreement with the terrestrial and meteoritic ratios.

114.038 Reddening of η Carinae from permitted Fe II lines. P. Ade, B. E. J. Pagel.
Observatory, Vol. 90, 6 - 9 (1970).

A spectrum of η Car was taken on 1967 Dec. 12 using the Mount Stromlo 74-inch coudé spectrograph at a dispersion of 6.7 Å/mm. From a study of the equivalent widths of permitted Fe II lines, adopting the method of common upper levels, the colour index was found considerably larger than that due to interstellar reddening in the surrounding H II region. It is supposed that there is a dust cloud around the star.

114.039 Identification of oxygen in the helium star HD 168476. P. W. Hill.
Observatory, Vol. 90, 10 - 12 (1970).

The upper limit to the oxygen abundance in the helium star HD 168476 was previously estimated at one-quarter to one-third the normal abundance from the non-detectability of O II lines in the blue. This has been confirmed by the observed strength of the O I blend at 7774 Å.

114.040 On H_2O^- and cool stars. M. S. Vardya.
Observatory, Vol. 90, 30 - 31 (1970). — Letter.

114.041 Forbidden oxygen in a Serpentis and other stars. R. Griffin, R. Griffin.
Observatory, Vol. 90, 70 - 71 (1970). — Letter.

114.042 The spectrum of R Coronae Australis.
E. E. Mendoza V., M. Jaschek, C. Jaschek.
Bol. Obs. Tonantzintla y Tacubaya, Vol. 5 (No. 32), 107 - 109 (1969).

The spectrum of R CrA cannot be classified uniquely in the MK system. However, the K-line of Ca II and some hydrogen lines indicate a spectral type earlier than A7, perhaps A5. Shell characteristics, with extremely strong Ti II (specially $\lambda\lambda 3759-61$) are observed. A radial velocity of –28 km/sec is derived from all well-identified absorption lines.

114.043 The intensity of Sc I $\lambda 6305$ Å in late-type stars. R. F. Griffin.
Monthly Notices, Roy. Astron. Soc., Vol. 147, 303 - 321 (1970).

The possibility of using the Cambridge technique of narrow-band spectrometry to make direct measurements of the equivalent widths of stellar absorption lines is discussed. An attempt to measure the resonance line of Sc I at $\lambda 6305$ Å in some 300 late-type stars has been made. As far as equivalent widths are concerned, the attempt was not successful. The Sc I ratios show a rapid variation with spectral type; at a given type, high-luminosity stars show the strongest scandium lines. The relationship between Sc I and $(B-V)$ is almost independent of luminosity class.

114.044 Influence of noncoherent electron scattering on a line profile in Wolf-Rayet stars.
J. I. Castor, L. F. Smith, D. van Blerkom.
Astrophys. Journ., Vol. 159, 1119 - 1121 (1970).

The $\lambda 3483$ line of N IV in the WN 6 star HD 192163 is observed to have a P Cygni-type profile, except for the presence of very extensive emission wings. It is shown that noncoherent scattering by free electrons in the W-R envelope can account for this profile.

114.045 Line strengths for southern OB stars — III. Balmer profiles for slow rotators. W. Buscombe.
Monthly Notices, Roy. Astron. Soc., Vol. 148, 75 - 78 (1970).

From intensity tracings of high-dispersion spectra of 23 stars with small rotational velocities, detailed measures of the profiles of the first four hydrogen lines in the Balmer series are listed.

114.046 Abundance analysis of cool carbon stars. K. Utsumi.
Publ. Astron. Soc. Japan, Vol. 22, 93 - 112 (1970).

Curve-of-growth analyses were made for 22 cool carbon stars, and the excitation temperatures, turbulent velocities,

damping constants, electron pressures, and relative abundances of the following elements were determined: Ca, Sc, Ti, V, Cr, Mn, Fe, Sr, Y, Zr, Ba, La, Ce, Pr, Nd, and Sm.

114.047 Spectral variability of HD 217050. R. A. Gobros.
Astron. Zhurn. Akad. Nauk SSSR, Vol. 47, 445 - 446 (1970). In Russian. — English translation in Soviet Astron., AJ, Vol. 14, No. 2.

114.048 Ultraviolet spectrophotometry of Canopus from Gemini XI. Y. Kondo, K. G. Henize, C. L. Kotila.
Astrophys. Journ., Vol. 159, 927 - 930 (1970).

The energy curve of Canopus in the wavelength range 2400 – 4000 Å has been obtained from spectrograms having a dispersion of 183 Å mm^{-1}. This curve is in good agreement with previous observations and with a model atmosphere. The equivalent width of the Mg II doublet at 2795 and 2802 Å is found to be 22 Å.

114.049 Line strengths for southern OB stars — IV. Emission-line profiles. W. Buscombe.
Monthly Notices, Roy. Astron. Soc., Vol. 148, 79 - 85 (1970).

Emission features seen on spectrograms of intermediate and high dispersion for 57 stars of early type, exposed at Mount Stromlo between 1955 and 1967, are listed. Detailed profiles of Hβ and Hγ from tracings of coudé plates of six stars are illustrated. Hα is also shown for the shell stars o Aqr and ϵ Cap. A list of sharp absorption lines due to metallic ions in the shell of ϵ Cap is presented.

114.050 Equivalent widths of the first three Balmer lines in Be stars. J. C. Boone.
Bull. American Astron. Soc., Vol. 2, 183 (1970). — Abstr. AAS.

114.051 Cosmic sources of infrared radiation.
G. R. Burbidge, W. A. Stein.
Bull. American Astron. Soc., Vol. 2, 185 - 186 (1970). Abstr. AAS.

114.052 Summary of the Smithsonian Observatory's results from the Orbiting Astronomical Observatory.
R. J. Davis.
Bull. American Astron. Soc., Vol. 2, 190 (1970). — Abstr. AAS.

114.053 Short-period variability in G 44 – 32.
B. M. Lasker, J. E. Hesser.
Bull. American Astron. Soc., Vol. 2, 205 (1970). — Abstr. AAS.

114.054 The helium content of population I B stars.
D. S. Leckrone.
Bull. American Astron. Soc., Vol. 2, 205 (1970). — Abstr. AAS.

114.055 Spectrophotometric comparison of the 3883- and 4215-Å CN bands.
T. E. Lutz, K. M. Yoss.
Bull. American Astron. Soc., Vol. 2, 207 (1970). — Abstr. AAS.

114.056 Emission-star studies at the Vatican Observatory.
M. F. McCarthy.
Bull. American Astron. Soc., Vol. 2, 208 (1970). — Abstr. AAS.

114.057 The absolute spectral-energy distribution of α Lyrae. J. B. Oke, R. E. Schild.
Bull. American Astron. Soc., Vol. 2, 212 (1970). — Abstr. AAS.

114.058 1 – 4-μ spectra of several C, M, and K stars.
R. I. Thompson, H. W. Schnopper, R. L. Mitchell,
H. L. Johnson.
Bull. American Astron. Soc., Vol. 2, 221 - 222 (1970).
Abstr. AAS.

114.059 Spectrum variability in β Carinae.
H. J. Wood.
Bull. American Astron. Soc., Vol. 2, 226 (1970). – Abstr.
AAS.

114.060 A spectrophotometric investigation of the star
MWC 84.   I. N. Chkhikvadze.
Astrofizika, Vol. 6, 65 - 76 (1970).   In Russian. – English
translation in Astrophysics, Vol. 6, No. 1.

The lines in the spectrum of the star MWC 84 obtained
at the Abastumani Astrophysical Observatory have been
identified and the relative intensities of the emission lines
and the energy distribution in the continuous spectrum
have been established. The observed Balmer decrement
differs but little from that calculated for a gaseous nebula.
The temperature of the star has been estimated to be
$40000°K$. It has been found that the distance of the star
is of the order of 1 kpc.

114.061 A search for classification criteria in spectra of K-
and M-type stars in the region $\lambda\lambda$ 5700 - 6800 Å.
G. F. Gahm.
Astron. Astrophys., Vol. 4, 268 - 279 (1970).

Slit-spectra of 33 stars of spectral types K 2 – M 6, dispersion 185 Å/mm at $\lambda6150$ Å, have been investigated in
order to obtain classification criteria with respect to spectral
type and luminosity class and to study spectral features dependent on other specific properties of the stars. Several criteria
for luminosity classification have been derived from the depths
of the following depressions; Ba II at λ 6497 Å; Fe, V and
Y0 at λ 6135 Å, Hα, Na D and the TiO bands. The derived
indices operate in limited intervals of spectral type where
separation between dwarfs, giants and supergiants is obtained.

114.062 A study of high-dispersion spectrograms of the shell
star ζ Tauri in 1964 and 1966.
T. van der Wel.
Astron. Astrophys., Vol. 4, 341 - 356 (1970).

By means of very high-dispersion spectrograms of the
shell star ζ Tauri obtained in 1964 and 1966 it has been
possible to get a good set of line profiles and equivalent
widths from lines of the Si II, Mg II, Ni II and Fe II spectra.
Some qualitative conclusions about the movements and the
structure of the shell are drawn from the measured radial
velocities and line profiles.

114.063 The rotational temperature of a TiO band in the
spectrum of R Hydrae.   C. F. Keller, B. V.
Jackson, A. I. Poland, B. F. Peery, Jr.
Astron. Astrophys., Vol. 4, 415 - 418 (1970).

A procedure is presented for the determination of a
rotational temperature from observations of the (1.0) γ_3 TiO
band. The method is based upon comparison between accurately measured wavelengths of blended absorption features in
the band and corresponding wavelengths in the computed
band profile. The rotational temperature found from McDonald coudé spectrograms of R Hydrae is $T_{rot} = 1430 \pm 290 °K$.

114.064 Observations d'étoiles Ap et Am.
M. Floquet.
Astron. Astrophys., Suppl. Series, Vol. 1, 1 - 5 (1970).

Les spectrogrammes de 27 étoiles A à spectre particulier
et à raies métalliques, et ceux de trois étoiles soupçonnées
Am, ont été étudiés et décrits, ce qui a permis de donner
une classification ou de confirmer celle existant déjà.

114.065 Classification des spectres de 112 étoiles A et F
dont 89 étoiles Am.   C. Bertaud.
Astron. Astrophys., Suppl. Series, Vol. 1, 7 - 13 (1970).

Les spectres de 112 étoiles ont été obtenus à l'Observatoire de Haute-Provence, en février et septembre 1968, avec
le télescope de 120 cm d'ouverture et le spectrographe à un
prisme donnant une dispersion de 60 Å/mm à Hδ. Quatre-
vingt-neuf étoiles Am, dont trois sont nouvelles, ont été
classées suivant l'intensité de la raie K et suivant l'intensité
des raies métalliques. Enfin une nouvelle étoile Ap, classée
Si-Cr, est signalée.

114.066 The effective temperatures of six Wolf-Rayet stars.
D. C. Morton.
Astrophys. Journ., Vol. 160, 215 - 219 (1970).

The ratio of Lyman-continuum flux to visual flux for
six WN-type Wolf-Rayet stars is derived from the free-free
radio emission of surrounding nebulae and the apparent magnitudes of the stars. Comparison with the theoretical ratios
obtained from model atmospheres which omit the emission
lines, but include ultraviolet line blanketing where appropriate,
gives estimates of the effective temperatures. Values around
$50000°K$ are found for WN 4.5 and WN 5, $33000°$ for WN 6,
and $23000°$ for WN 8.

114.067 Dependence of chromospheric emission upon
bolometric luminosity for the Hyades.
O. C. Wilson.
Astrophys. Journ., Vol. 160, 225 - 231 (1970).

Fluxes in the centers of the H and K lines were measured
for sixty-five Hyades stars between F4 and K5. Analysis of
these measures shows that, in terms of bolometric luminosity,
the radiation in the chromospheric H and K emission lines
increases by a factor of 2 between $B - V = 0.45$ and $B - V =
1.25$. Other results give the emission as a function of the local
continuum, as well as the true emission ratios for stars in this
color range.

114.068 Identification of infrared CN bands in the spectra
of several carbon stars.   R. I. Thompson,
H. W. Schnopper.
Astrophys. Journ. (Letters), Vol. 160, L97 - L100 (1970).

Identification of the 2–4 and 3–5 bands of the CN red
system is made for W Ori, 19 Psc, U Hya, and X Cnc. These
bands occur in the infrared spectral region between 2 and
2.5 μ. A possible identification has been made of the
$\Delta_v = -1$ bands in the $(1.8–1.4)$-μ region. This identification
is consistent with the expected high concentration of the
CN molecule in the atmospheres of carbon stars.

114.069 Spectral classification of A and F stars.
D. C. Barry.
Astrophys. Journ., Suppl. Series, Vol. 19, 281 - 304 (1970).

The ultraviolet features in the 3300-4000 Å region of
117 Å mm^{-1} grating spectra of A and F stars are studied for
the purpose of improving the relation between MK spectral
type and $b - y$ color. More than 160 stars are classified
according to temperature, luminosity, and line strength. The
resulting relation between spectral type and color is an
improvement over that of the MK standards and the Hyades
main-sequence stars classified by Morgan and Hiltner. The
new luminosity criteria permit the use of more luminosity
subclasses than are possible with the standard MK criteria.
Photometric methods for identifying Am and λ Bootis
stars are suggested. While the photometry of spectroscopically identified weak-line F stars indicates low luminosity
and low metallic-line strength, the photometry of the fast
rotating spectroscopically identified weak-line A stars
indicates luminosity slightly higher than normal and low
metallic-line strength. Evidence is presented indicating the
need for improvement in the definition of the Am stars.

114.070 Erratum: Observations of water-vapor emission associated with infrared stars. [Astrophys. Journ. (*Letters*), Vol. 159, L123 - L127 (1970)].

P. R. Schwartz, A. H. Barrett.

Astrophys. Journ. (*Letters*), Vol. 160, L61 (1970).

See Abstr. 114.009.

114.071 Spectrum variations in 56 Arietis. II. New observations of silicon and helium lines.

W. K. Bonsack, W. A. Wallace.

Publ. Astron. Soc. Pacific, Vol. 82, 249 - 273 (1970).

Spectrum line intensity and wavelength measurements have been made on 95 spectrograms of this Ap-type spectrum variable, sampling 1100 cycles of variation. Measurements of the He I lines at 4471 and 4026 Å, and the Si II lines at 3954, 4076, 4128 - 31, and 4201 Å support a model in which the stellar surface contains two distinct regions in which the He I lines are strong, and three to five region in which the Si II lines are strong. The spectrum variation occurs as these are carried through the visible hemisphere by rotation with a period of 0^d728. No evidence is found for phase shifts or other changes in the 0^d728 period.

114.072 New peculiar stars noted on objective-prism plates.

H. E. Bond.

Publ. Astron. Soc. Pacific, Vol. 82, 321 - 328 = Contr. Louisiana State Univ. Obs., Baton Rouge, No. 34 (1970).

A list of previously unknown peculiar stars noted during an objective-prism survey is given. Included are 51 Ap, 33 Am, two Bp, and two Ba II stars, two G-K stars showing strong Ca II emission, and a new, relatively bright CH star. The distribution of galactic latitudes of the Ap Si stars in the survey region suggests that the silicon stars are related to the late B-type stars, and have not evolved from the early B stars.

114.073 OH radio emission associated with infrared stars.

W. J. Wilson, A. H. Barrett, J. M. Moran.

Astrophys. Journ., Vol. 160, 545 - 571 (1970).

A search of sixty infrared stars for OH emission at 1612 MHz has been made. OH emission was detected from seven of these objects, and their spectra and polarization properties have been measured. These OH sources are the first to be identified with stellar objects. Also, seven T Tauri and eight known red-giant stars were observed, with negative results. The OH source in NML Cygnus was studied with a very-long-baseline interferometer and found to be coincident in position with the infrared star within 5″ and to have individual spectral features within 2″ of one another with characteristic sizes of 0.″08. The characteristics of the OH emission are best explained by a partially saturated maser pumped by near-infrared radiation. A model of an evolved, very luminous red-giant star with an expanding atmosphere is proposed to explain the observed properties.

114.074 Cosmic sources of infrared radiation.

G. R. Burbidge, W. A. Stein.

Astrophys. Journ., Vol. 160, 573 - 593 (1970).

The physics of radiation processes is applied to an analysis of the observations of infrared sources. It is shown that in the case of most galactic infrared sources small mass fractions ($10^{-4} - 10^{-6} M_\odot$) of solid particles surrounding luminous stars can explain the data. An analysis is made of the observations of the galactic center, but is does not seem possible at this time to decide whether the infrared radiation is due to nonthermal processes or an energy source surrounded by dust. Calculations applied to the dust hypothesis of the origin of infrared radiation in the nuclei of Seyfert galaxies and quasistellar objects lead to results that appear difficult to reconcile with infrared and optical observations. It is shown that the synchrotron models are able to explain the infrared flux from the nuclei of galaxies, provided that the magnetic field

strengths are ~1 - 100 gauss. Nuclei made up of a number of small components are indicated in some cases.

114.075 Rocket spectroscopy of Zeta Puppis.

A. M. Smith.

Astrophys. Journ., Vol. 160, 595 - 608 (1970).

A spectrum of ζ Pup extending from 920 to 1360 Å with approximately 0.8 Å resolution has been recorded at rocket altitudes. Tentative identification of 102 multiplets of both stellar and interstellar origin has been made, from which it is concluded that all lines included in existing model atmospheres have been detected with the exception of those masked by telluric N_2 or strong P Cygni-type profiles. Additional weak absorption lines indicate a wide range of ionization and excitation entirely consistent with observations in the visible spectral region of stars of similar type; they also appear to affect sensibly the energy distribution within the spectrum.

114.076 A study of the shell spectrum 48 Librae.

H. G. Geuverink.

Astron. Astrophys., Vol. 5, 341 - 354 (1970).

This paper contains the complete material. The main points of the observations in 1967 and 1968 are presented and a comparison is made between the spectrum in 1957 and in 1967. A thorough study has been made of the Hβ profiles in the years 1957, 1958, 1967 and 1968 and the component arising from the shell has been separated from the stellar line.

114.077 Calcium emission intensities as indicators of stellar age.

O. Wilson, R. Woolley.

Monthly Notices, Roy. Astron. Soc., Vol. 148, 463 - 475 (1970).

Fresh data are presented concerning the calcium emission intensities of 325 main-sequence late-type stars estimated by one of the authors (O. W.). These are compared with parameters of the galactic orbits of the stars according to methods worked out at Royal Greenwich Obs. A very clear correlation is found between the calcium emission intensities and the eccentricities and box angles (inclinations) of the orbits.

114.078 The chemical composition of twelve late F dwarfs.

A. L. T. Powell.

Monthly Notices, Roy. Astron. Soc., Vol. 148, 477 - 488 (1970).

The chemical composition of 12 late F dwarfs has been determined by differential curve-of-growth analysis using blue coudé plates. For each star fourteen elemental abundances have been determined from ninety-two spectral lines. In addition the electron pressure, damping parameter and total Doppler velocity have been found spectroscopically.

114.079 Lithium in F and G type stars–I. Isotope abundances.

M. W. Feast.

Monthly Notices, Roy. Astron. Soc., Vol. 148, 489 - 499 (1970).

The wavelength of the lithium 6708 Å line has been determined from 65 spectra at 13.7 Å mm^{-1} for 12 lithium-rich F and G type stars. Li^7 is always the predominant isotope. The results are consistent with an upper limit of about 0.5 for R', the Li^6/Li^7 ratio. A review of the available data indicates that the few stars with high $R'(\sim 0.5)$ are all subgiants. Radial velocity determinations for these and some other stars are given in the appendix.

114.080 Spectroscopic investigation of B 3 V stars: ι Herculis, η Hydrae, HD 58343.

K. Kodaira, M. Scholz.

Astron. Astrophys., Vol. 6, 93 - 113 (1970).

New observations of ι Her, η Hya, and HD 58343 are

reported and interpreted by means of LTE model-atmosphere techniques. Fine analyses are carried out for models with T_{eff} = 20200° K, log g = 3.75, ξ_t = 5 km/s (ι Her and HD 58343), and 21900° K, 4.0, 5 km/s (η Hya), respectively; Mihalas' unblanketed models and the KG-theory for Balmer lines are used.

114.081 The hydrogen lines in a B type supergiant.
A. B. Underhill.
Astron. Astrophys., Vol. 6, 114 - 123 (1970).

Observations of the Hγ profile in the supergiants β Orionis, B8 Ia, η Canis Majoris, B5 Ia, o^2 Canis Majoris, B3 Ia, and ϵ Orionis, B0 Ia, are presented as well as observations of the confluence of the Balmer series. It is shown that the break off of the Balmer and of the Paschen series must be due chiefly to Stark effect. A simple theory based on the one-layer hypothesis is developed to investigate the effects of temperature, electron density and relative helium content on the extent of the wings of the Balmer lines. It is shown that the extent of the wings is a sensitive function of temperature at least through the range 12000° < T < 18000 °K, the wings decreasing significantly as T is increased. No secure conclusion can be made about the helium content in the Ia supergiants of type B until the temperature in the wing-forming layers of the atmosphere is known accurately.

114.082 The relative metal-to-hydrogen ratio for the sun, Hyades and 53 F 5 - G 2 stars.
P. E. Nissen.
Astron. Astrophys., Vol. 6, 138 - 150 = Contr. Kitt Peak National Obs., No. 545 (1970).

A new method to determine the metal-to-hydrogen ratio in stars based on photoelectric observations of the strength of a group of weak metal lines in a band of the width of 3.5 Å near 4800 Å is described. The method has been applied at the 60-inch Robert McMath solar telescope and through a model-atmosphere analysis of the observations [Fe/H] values between – 0.3 and 0.6 for 53 F 5 – G 2 stars have been derived. Nine Hyades stars have been observed and the average value of [Fe/H] is found to be 0.38 ± 0.02.

114.083 Fine analysis of five stars of the γ Leonis group.
G. Zielke.
Astron. Astrophys., Vol. 6, 206 - 218 (1970).   In German.

Applying the method of fine analysis to five stars of the γ Leonis group, we determine model parameters from spectra and colors. HD 136202, for which we have the most extensive material, is analysed relative to the sun, the other stars relative to HD 136202. Combining their effective temperatures and gravities with calculations of stellar evolution by Iben, we obtain "spectroscopic evolution-parallaxes". They are compared with the spectroscopic, trigonometric, and group parallaxes. We remeasure the radial velocities and discuss the space motion.

114.084 The effective temperatures of the O stars.
D. C. Morton.
IAU Symposium No. 36, (see 012.014), p. 59 - 63 (1970).

Effective temperatures of O-type stars imbedded in diffuse nebulae are derived from measurements of Hα and radio fluxes from the nebulae and the apparent magnitudes of the stars.

114.085 The effect of silicon and carbon opacity on ultraviolet stellar spectra.
O. Gingerich, D. Latham.
IAU Symposium No. 36, (see 012.014), p. 64 - 70 (1970).

114.086 The stellar temperature scale from O5 to A0.
D. S. Hayes.
IAU Symposium No. 36, (see 012.014), p. 83 - 89 (1970).

A stellar temperature scale has been determined by fitting the measured sizes of the Balmer discontinuity of 43 stars to blanketed model atmospheres.

114.087 Low resolution stellar spectrophotometric observations in the region 1500 Å - 3000 Å.
G. C. Sudbury.
IAU Symposium No. 36, (see 012.014), p. 134 (1970).

114.088 Observations of ultraviolet stellar spectra.
R. Wilson.
IAU Symposium No. 36, (see 012.014), p. 147 - 162 (1970).

114.089 Photoelectric rocket spectra at 10 Å resolution.
T. P. Stecher.
IAU Symposium No. 36, (see 012.014), p. 163 (1970).
Abstract.

114.090 Rocket spectroscopy of ζ Puppis below 1100 Å.
A. M. Smith.
IAU Symposium No. 36, (see 012.014), p. 164 - 172 (1970).

A spectrum of ζ Pup extending from 920 Å to 1360 Å with approximately 0.8 Å resolution has been recorded at rocket altitudes. Several heretofore unobserved P Cygni-like profiles have been identified as have a number of weak, photospheric lines arising from excited levels in both abundant and relatively rare ions. In this report these data will be presented and discussed.

114.091 Observations of strong stellar lines with the OAO.
A. D. Code, R. C. Bless.
IAU Symposium No. 36, (see 012.014), p. 173 - 177 (1970).

This paper reports on preliminary analysis of spectral scans of early-type stars obtained with the Orbiting Astronomical Observatory. The discussion is confined to the spectra of 50 stars observed with a resolution of approximately 10 Å over the spectral interval from 1050 Å to 2000 Å.

114.092 The far-ultraviolet spectrum of γ Cassiopeiae.
D. C. Morton, E. B. Jenkins, R. C. Bohlin.
IAU Symposium No. 36, (see 012.014), p. 178 - 179 (1970).

114.093 UV spectrophotometry of Canopus from Gemini XI.
Y. Kondo, K. G. Henize, C. L. Kotila.
IAU Symposium No. 36, (see 012.014), p. 180 - 184 (1970).

The UV energy curve of Canopus in the 2400 - 4000 Å wavelength range has been obtained from spectrograms having a dispersion of 183 Å mm^{-1}. This curve is in good agreement with previous observations and with a model atmosphere.

114.094 Review of ground-based observations of spectra relevant to the ultraviolet.   M. W. Feast.
IAU Symposium No. 36, (see 012.014), p. 187 - 198 (1970).

114.095 Chromospheric activity in red giants, and related phenomena.   A. J. Deutsch.
IAU Symposium No. 36, (see 012.014), p. 199 - 208 (1970).

Normal red giants of a given spectral type are shown to be heterogeneous with respect to the following chromospheric features: the Balmer absorption lines, the emission line at Hϵ, and the double-reversed emission lines at Ca II H and K. These chromospheric lines are also shown to be strongly time variable, in at least some red giants, on a time scale of a few months or years.

114.096 Mass loss from early-type stars.   J. B. Hutchings.
IAU Symposium No. 36, (see 012.014), p. 209 - 212 (1970).

Following the detailed study of four very high luminosity OB stars, a survey has been made for spectroscopic evidence of mass loss in a number of early-type supergiants. A list of spectroscopic criteria is given and the mass loss estimates for 24 stars plotted on the HR diagram. The dependence of the phenomenon on spectral type and luminosity is discussed as well as its significance in terms of stellar evolution.

114.097 A discussion of the theory for interpreting ultra-violet stellar spectra.   A. B. Underhill.
IAU Symposium No. 36, (see 012.014), p. 215 - 225 (1970).

Some numerical examples are presented demonstrating that with the UV resonance lines the opacity in the centre of a line may exceed the continuous opacity by a factor $10^6 - 10^8$. A summary is given of the chief factors which should be taken into account in any theory of line formation when the hypothesis of LTE is not valid. Some examples of the distribution in energy of the lower energy levels of the ion are presented for typical ions of interest.

114.098 Possibility of fluorescence phenomena in the ultra-violet spectrum of symbiotic stars and long period variables.   J. P. Swings, P. Swings.
IAU Symposium No. 36, (see 012.014), p. 226 - 231 (1970).

We discuss essentially the cases where molecular fluorescences may be excited by Lyman α and other strong discrete ultraviolet emissions, including lines beyond the Lyman limit. The stars involved are the symbiotic objects and the long period variables. The molecules are H_2, N_2, O_2, NO and CO which have their resonance systems in the ultraviolet.

114.099 Radiative acceleration and ultraviolet resonance line profiles in OB supergiants.   J. B. Hutchings.
IAU Symposium No. 36, (see 012.014), p. 232 - 235 (1970).

114.100 Observations of interstellar Lyman-α absorption.   E. B. Jenkins.
IAU Symposium No. 36, (see 012.014), p. 281 - 301 (1970).

Absorption at the Lyman-α transition from interstellar neutral hydrogen has been observed in the ultraviolet spectra of 18 nearby O and B stars. Several effects which might introduce uncertainties into the Lyman-α measurements are considered, but none seems to be able to produce enough error to explain the disagreement with the 21-cm data. The possibility that small-scale irregularities in the interstellar gas could give significantly lower values at Lyman-α is explored.

114.101 Observations of interstellar Lyman-α with the Orbiting Astronomical Observatory.
B. D. Savage, A. D. Code.
IAU Symposium No. 36, (see 012.014), p. 302 - 314 (1970).

The equivalent width of the blended line at Lyman α is given for 48 stars measured with the OAO-A2 scanning spectrometer. The correlation between the OAO blended equivalent widths and color excess, 4430 Å absorption, and interstellar sodium absorption are examined. Excellent correlation between sodium and hydrogen column densities is found.

114.102 Interstellar lines other than hydrogen.   G. H. Herbig.
IAU Symposium No. 36, (see 012.014), p. 315 - 319 (1970).

114.103 Space and ground-based stellar spectrophotometry; a summary.   C. de Jager.
IAU Symposium No. 36, (see 012.014), p. 355 - 361 (1970).

114.104 An atlas of low-dispersion spectra of S stars in the blue region.   C. B. Stephenson, H. E. Ross.
Astron. Journ., Vol. 75, 321 - 323, 511 (1970).

Published reproductions of S-star spectra in the blue region are rare; to remedy this defect for objective-prism workers, we reproduce a number of typical S-type spectra. We suggest that AlH be considered as a possible major contributor to the violet opacity in the S and C-S stars, and perhaps in N stars.

114.105 Erratum: MK classifications for F- and G-type stars. II. [Astron. Journ., Vol. 75, 165 - 166 (1970)].
E. A. Harlan, D. C. Taylor.
Astron. Journ., Vol. 75, 507 - 508 = Lick Obs. Bull., No. 607 (1970). – See Abstr. 114.012.

114.106 Progress report on the Cleveland-Chile survey for southern OB stars.   C. B. Stephenson, N. Sanduleak.
IAU Symposium No. 38, (see 012.013), p. 276 - 277 (1970).

114.107 Wolf-Rayet stars, ring-type H II regions, and spiral structure.   T. Schmidt-Kaler.
IAU Symposium No. 38, (see 012. 013), p. 284 - 286 (1970).

114.108 Stellar abundances and the origin of the elements, Part I, II.   A. O. J. Unsöld.
Astron. Soc. Pacific, Leaflet Nos. 491, 492, 8 + 8 pp. (1970). – This article is a condensed version of an article in Science, Vol. 163, 1015 - 1025 (1969).

114.109 MK types for southern supergiants.   G. F. Benedict, W. Buscombe.
Inform. Bull. Southern Hemisphere, No. 15, p. 39 (1969).

114.110 The continuum of MHα 328-116.   F. Caputo, N. Panagia, H. Gerola.
Astrophys. Letters, Vol. 5, 275 - 277 (1970).

The peculiar continuum of MHα 328-116 is found to correspond to that of a nebula where thermal collisions are responsible for the excitation, at $T_e = 1.3 \times 10^4$ °K and $N_e = 5 \times 10^6$ cm^{-3}.

114.111 The energy distribution in spectra of 8 stars.   N. S. Komarov, V. A. Pozigun.
Astron. Zhurn. Akad. Nauk SSSR, Vol. 47, 551 - 556 (1970). In Russian. English translation in Soviet Astron. AJ, Vol. 14, No. 3.

The energy distribution in spectra of 8 stars are determined. The results are compared with Willstrop's data.

114.112 Rectification of P Cygni line profiles.   D. Van Blerkom.
Monthly Notices, Roy. Astron. Soc., Vol. 149, 53 - 57 (1970).

In order to separate emission and absorption components of P Cygni type line profiles, a rectification procedure has been used by some investigators. This consists of reflecting the observed red wing about line centre and determining the amount of absorption necessary to give the observed violet wing. It is shown that this procedure spuriously introduces the occultation effect into the violet wing and may yield an underestimate of the absorption component and distort its profile.

114.113 Spectral line formation in Wolf-Rayet envelopes.   J. I. Castor.
Monthly Notices, Roy. Astron. Soc., Vol. 149, 111 - 127 (1970).

The escape probability method for treating the transfer of line radiation in a stellar envelope which is in rapid radial expansion is developed, including the effect of radiation from the stellar core. This method is applied to a line formed by a two-level atom, yielding an explicit expression for the line source function. These results are used to calculate three representative line profiles assuming different distributions of absorbing atom density but with a fixed velocity distribution.

114.114 R. M. Petrie and the B-star program of the Dominion Astrophysical Observatory.
J. K. Petrie.

Journ. Royal Astron. Soc. Canada, Vol. 64, 163 - 172 (1970). Conference paper.

114.115 Instabilities in the envelope of the Be star HD 37202 ζ Tauri. A. M. Delplace.
Astron. Astrophys., Vol. 7, 68 - 85 (1970). In French.

The profiles of the underlying star were determined in order to estimate the contribution in emission and absorption of the envelope of HD 37202. These have been compared 1) to the theoretical profiles of Mihalas and 2) to the observed profiles of B stars of the same spectral type and of similar class of luminosity. We have compared the value of the radial velocity of the star obtained from our measurements with those found in the literature.

114.116 The barium abundance in ε Pegasi. P. R. Warren.
Observatory, Vol. 90, 101 - 103 (1970).

By analyzing the spectrum of the K2 supergiant ε Peg with respect α Boo in the wavelength range 5700 Å to 6330 Å in a manner similar to that used by Cayrel and Cayrel we have estimated the logarithmic [Ba/Fe] ratio 1.0 ± 0.5 of ε Peg with respect to the sun.

114.117 Photometric calibration of objective prism spectra using a calcite-polaroid filter.
K. Nandy, F. Smriglio.
Observatory, Vol. 90, 114 - 115 (1970).

This note presents a modification of the calcite-polaroid method described by Brück et al. and its results. The method may provide a reliable photometric calibration of objective prism plates.

114.118 The barium abundance of 56 Pegasi.
P. R. Warren, P. M. Williams.
Observatory, Vol. 90, 115 - 118 (1970).

A [Ba/Fe] ratio of +0.9 ± 0.2 was found for the Ba II star 56 Peg.

114.119 The peculiar high-velocity star HD 204613.
H. E. Bond.
Astrophys. Journ., Vol. 160, 1127 - 1128 = Contr. Louisiana State Univ. Obs., No. 36 = Contr. Kitt Peak National Obs., No. 542 (1970).

The early G-type star HD 204613 is kinematically a member of the galactic halo, yet it is not extremely weak-lined and in fact appears to show greatly enhanced abundances of certain elements. Photometrically it is similar to the giants in M92 that show only a small ultraviolet excess.

114.120 M supergiants in the Perseus arm.
R. W. Humphreys.
Astrophys. Journ., Vol. 160, 1149 - 1159 = Contr. Kitt Peak National Obs., No. 549 (1970).

Radial velocities measured from infrared spectra are given for the M supergiants in the Perseus arm. Thirteen M supergiants were confirmed from the MK classification of a number of suspected M supergiants. The space distribution and kinematics of these stars agree well with the other population I objects in the Perseus arm. Seventy percent of the M supergiants are probable members of stellar associations and open clusters. A study of the supergiants in Perseus OB1 gives an internal velocity dispersion of 7.2 km sec^{-1}. Evidence for structure in the association is also discussed.

114.121 On the possibilities of spectral classification of stars from recordings of unbroadened spectra.
V. I. Kuznetsov.
Astrometriya i Astrofiz., Kiev, No. 8, (see 003.024), p. 73 - 76 (1969). In Russian.

A spectral classification was carried out on unbroadened spectra of stars obtained with a 8° objective prism (dispersion 165 Å/mm near Hγ). The classification was made using the registrograms of the spectra.

114.122 Z Andromedae. Variation de l'excitation de 1923 à 1968; spectre en 1967 et 1968.
M. Bloch, N. Jousten, J. P. Swings.
Publ. Obs. Haute Provence, Vol. 10, No. 33, 11 pp. (1969).

There appears to be no periodicity in the presence of lines of different ionization stages and degrees of excitation in the spectrum of Z Andromedae. The variation of the excitation from 1923 to 1968 is discussed. The spectrum of Z Andromedae in 1967 - 1968 is described.

114.123 Z Andromedae, variation de l'excitation de 1923 à 1968; spectre en 1967 et 1968.
M. Bloch, N. Jousten, J. P. Swings.
Bull. Soc. Roy. Sci. Liège, 38ᵉ année, p. 245 - 254 (1969).

114.124 The corbon isotope ratio in some cool carbon stars. II. Y. Fujita, T. Tsuji, H. Maehara.
Proc. Japan Acad., Vol. 45, (No. 6), 484 - 489 = Contr. Dep. Astron. Univ. Tokyo, No. 114 (1969).

114.125 Heavy elements in 73 Draconis.
M. Jaschek, S. Malroda.
Nature, Vol. 225, 246 - 247 (1970).

The star 73 Draconis shows the gross characteristics of an object belonging to the Cr-Eu-Sr group, a strengthening of all iron-peak elements (such as Cr, Mn, Fe, Co, and Ni) and of some heavier elements like Sr. But there are two outstanding features not shared by the other objects of the group, one being the presence of very heavy elements and the other the presence of molecular compounds.

114.126 Computed and observed O I line profiles in the spectra of B3–A0 stars. L. Houziaux.
Commun. Dep. Astrophys., Fac. Sci. Mons, Mons Astrophys. Papers, No. 11, 11 pp. (1970).

In the spectral type range B3 to A0, oxygen can be detected only through the O I lines. Although there are numerous O I transitions in the range λ 950 to λ 10,000, only a few of them are suitable for a study of the oxygen spectrum. In this paper we study the sensitivity of computed profiles for the triplet at λ 7772 to various atmospheric parameters, and the comparison of observed profiles (in bright stars) with computed ones.

114.127 A comparison of the spectra of C and M giants at one micron. T. D. Fay, Jr.
Thesis, Indiana Univ., Bloomington. [Available from Univ. Microfilms, Ann Arbor, Mi., Order No. 68–13686]. 288 pp. (1968).

Spectra of two M supergiants and five carbon stars have been photographed from 1.00 to 1.09 micron using image tubes. The spectra were compared with the one micron spectra of Y CVn and α Ori obtained by Wyller and Spinrad on I–Z plates.

114.128 Estudio del espectro de la estrella 31 Aql en la región λλ 4000 – 6600 Å.
R. Fernández.
Revista Real Acad. Ciencias Exact., Fis., Nat. Madrid, Vol. 63, (No. 3), 391 - 466 (1969).

114.129 The He II λ 4686 emission line in the spectrum of MH$_α$ 276–52. I. N. Chkhikvadze.
Astron. Tsirk., No. 545, p. 8 (1970). In Russian.

114.130 On the stars HD 45677 and HD 4174.
I. N. Chkhikvadze.
Astron. Tsirk., No. 546, p. 4 - 6 (1970). In Russian.

114.131 **On the spectra of three red stars.** M. V. Dolidze.
Astron. Tsirk., No. 552, p. 8 (1970). In Russian.

114.132 **Stellar spectra.**
M. W. Feast, J. B. Oke, G. Cayrel de Strobel,
R. Herman.
Trans. IAU, Vol. 14A, (see 003.028), 319 - 333 (1970).
Report of Commission 29.

114.133 **Spectral classifications and multi-band colour indices.** C. Fehrenbach.
Trans. IAU, Vol. 14A, (see 003.028), 547 - 558 (1970).
Report of Commission 45.

114.134 **Problems in the formation of neutral helium lines in early type stars.** R. W. Simpson.
Proc. Astron. Soc. Australia, Vol. 1, 326 - 328 (1970).

114.135 **Search for plutonium lines in 73 Dra and other Ap stars.** B. Kuchowicz.
Nature, Vol. 227, 156 (1970).
Some remarks on the production of Pu during the r-process are given. The most intense Pu lines for which a search can be made are at 3000.4, 3907.2, 3985.5, and 4273.3 Å.

114.136 **Studies of cool stars in the one-micron region.**
G. W. Lockwood.
Thesis, Univ. Virginia, Charlottesville. [Available from Univ. Microfilms, Ann Arbor, Mi.], 129 pp. (1969).

114.137 **Estudio del espectro de la estrella 31 Aql. en la región λλ 4000 – 6600 Å.**
M. R. Fernández.
Rev. Real Acad. Ciencias Exactas, Fis., Nat. Madrid, Vol. 63, (No. 3), 76 pp. = Publ. Seminario Astron. Geod. Univ. Madrid, No. 61 (1969).
About two thousand six hundred lines have been identified in the spectrum of the G8, IV subdwarf 31 Aql. Relative intensities have been assigned to each line. The BaII 4554 Å line is stronger than in normal stars. The G-band appears also very intense while CN lines around 4215 Å are stronger than in other stars of similar spectral type.

114.138 **Etude à résolution moyenne du spectre U.V. des étoiles chaudes entre 1050 et 3200 Å.**
L. Houziaux, C. Jamar, A. Monfils.
Centre Univ. Mons, Fac. Sci., Dép. Astrophys., Commun., No. 14, 25 pp. (1970).

114.139 **S 2/S 68 UV sky scanning project data handling and analysis group. Computed stellar background radiation around 2800 and 2000 Å.**
A. Blondelot.
Centre Univ. Mons, Fac. Sci., Dép. Astrophys., Commun., No. 16, 8 pp. (1970).

114.140 **Twenty-four southern peculiar emission-line stars.**
E. D. Carlson.
Thesis, Northwestern Univ., Evanston, Ill. [Available from Univ. Microfilms, Ann Arbor, Mi.], 232 pp. (1969).
Observational data are presented for 24 peculiar emission-line stars (and a few associated nebulae) located largely along the fourth quarter of the galactic equator.

114.141 **Nyere resultater fra den infrarøde astronomi.**
H. Stub.
Astron. Tidssk., Vol. 3, 37 - 45 (1970).

114.142 **The spectrum of HD 125823.**
C. Jaschek, M. Jaschek.
Asoc. Argentina Astron. Bol., No. 12, (see 012.019), p. 16 (1968). In Spanish. – Abstract.

114.143 **Spectrophotometry of early type stars.**
A. Gutiérrez-Moreno, H. Moreno, J. Stock.
Asoc. Argentina Astron. Bol., No. 12, (see 012.019), p. 29 - 31 (1968). – Abstract.

114.144 **Early stars with helium deficiency.**
M. Jaschek, C. Jaschek, B. Kucewicz.
Asoc. Argentina Astron. Bol., No. 12, (see 012.019), p. 35 (1968). – Abstract.

114.145 **The spectral atlas in 42 Å/mm: O–F5 types.**
M. Jaschek, C. Jaschek.
Asoc. Argentina Astron. Bol., No. 12, (see 012.019), p. 35 (1968). – Abstract.

114.146 **Identification of lines in peculiar stars.**
M. Jaschek, M. L. Aguilar.
Asoc. Argentina Astron. Bol., No. 12, (see 012.019), p. 39 (1968). In Spanish. – Abstract.

114.147 **The helium variable star HD 125823.**
C. Jaschek, M. Jaschek, W. W. Morgan, A. Slettebak, B. Kucewicz.
Asoc. Argentina Astron. Bol., No. 14 (see 012.020), p. 17 (1968). – Abstract.

114.148 **HR 4817, a new phosphor star.**
M. Jaschek, M. L. Aguilar.
Asoc. Argentina Astron. Bol., No. 14 (see 012.020), p. 36 - 37 (1968). – Abstract.

114.149 **Spectroscopic analysis of peculiar stars. IV: The strontium group.** M. Jaschek, E. Brandi.
Asoc. Argentina Astron. Bol., No. 14, (see 012.020), p. 40 - 42 (1968). – Abstract.

114.150 **Studies of A stars. I: Catalogue of spectral types and colours.** A. Cowley, B. Cowley,
M. Jaschek, C. Jaschek.
Asoc. Argentina Astron. Bol., No. 14 (see 012.020), p. 53 (1968). – Abstract.

114.151 **The spectra of some Be stars. II.**
C. Jaschek, M. Jaschek, S. Malaroda.
Asoc. Argentina Astron. Bol., No. 14 (see 012.020), p. 55 - 57 (1968). In Spanish.

114.152 **The Paschen jump in B stars.**
A. E. Ringuelet-Kaswalder.
Asoc. Argentina Astron. Bol., No. 14 (see 012.020), p. 62 - 63 (1968). In Spanish. – Abstract.

114.153 **The spectrum of γ_2 Velorum.**
V. N. de Monteagudo, J. Sahade.
Asoc. Argentina Astron. Bol., No. 14 (see 012.020), p. 77 - 82 (1968).

114.154 **Studies of bright southern stars.**
M. Jaschek, C. Jaschek, W. A. Hiltner.
Asoc. Argentina Astron. Bol., No. 14 (see 012.020), p. 102 (1968). In Spanish. – Abstract.

114.155 **Catalogue of B stars with Hα emission.**
C. Jaschek, M. Jaschek, L. Ferrer.
Asoc. Argentina Astron. Bol., No. 14 (see 012.020), p. 104 - 107 (1968). In Spanish.

Atlas for Objective Prism Spectra. See Abstr. 003.033.

The Theory of Stellar Spectra.   See Abstr. 003.037.

Experimental oscillator strengths of Fe II lines and the solar iron abundance.   See Abstr. 022.038.

A catalogue of early-type stars whose spectra have shown emission lines.   See Abstr. 041.018.

Neutral-helium line strenghts. III. The singlet-triplet anomaly of Population I stars.   See Abstr. 064.018.

Diffusion processes in peculiar A stars. See Abstr. 064.037.

A model atmosphere analysis of the Ap star κ Cancri. See Abstr. 064.040.

Review of ultraviolet and visual continuum observations and comparisons with models.   See Abstr. 064.041.

A curve-of-growth analysis of the super-metal-rich G dwarf HR 72.   See Abstr. 064.050.

Non-L.T.E. analysis of spectral lines. See Abstr. 064.056.

Microturbulence in main sequence stars. See Abstr. 065.044.

The radial velocity of Arcturus determined from interferometric spectra.   See Abstr. 112.009.

Some characteristics of the peculiar B stars with color-spectrum discrepancies.   See Abstr. 113.017.

Photometric and spectroscopic observations of infrared stars.   See Abstr. 113.027.

Far-ultraviolet intensities of Orion stars. See Abstr. 113.032.

Spectrophotometry of cool stars in the near infrared. Results for a region in Cygnus.   See Abstr. 113.058.

On the dependence of M_v (K) on metal abundance. See Abstr. 115.003.

Quantitative analysis and spectral variation of the Ap star HD 151199.   See Abstr. 116.005.

Harmonic analysis of rigidly rotating Ap stars. See Abstr. 116.007.

Observations of interstellar Lyman-Alpha absorption with the Orbiting Astronomical Observatory. See Abstr. 131.042.

Measurement of interstellar extinction in emission line stars.   See Abstr. 131.089.

The interstellar extinction curve from 4000 Å to 6500 Å.   See Abstr. 131.090.

HBV 475: Evolution stage of a planetary nebula? See Abstr. 133.022.

Photometric and spectroscopic data for some distant O and B stars.   See Abstr. 153.017.

115 Stellar Luminosities, Masses, Diameters, HR-Diagrams and Others

115.001 A new method of stellar radius determinations.
K. Stępień.
Postępy Astron., Vol. 18, 218 - 221 (1970). In Polish.
Short report.

115.002 The derivation of absolute magnitudes from proper motions and radial velocities and the calibration of the H.R. diagram II. J. Jung.
Astron. Astrophys., Vol. 4, 53 - 69 (1970).
We discuss a method of deriving statistical parallaxes based on the principle of maximum likelihood. Checks made with the aid of trigonometric parallax stars and by means of numerical experiments show this method to be more reliable than the usual methods. Its application to stars in the Bright Stars Catalogue results in a new calibration of the H.R. diagram, based on proper motions and radial velocities only. We derive from these data an estimation of both the mean absolute magnitude and the dispersion of the absolute magnitudes for each spectral type. The luminosity of early-type stars derived from cluster parallaxes and distance moduli of associations is confirmed, but late-type giants appear to be brighter than it is generally assumed.

115.003 On the dependence of M_v (K) on metal abundance.
P. Kjaergaard.
Astron. Astrophys., Vol. 5, 165 - 166 (1970).
A comparison of $M_v(K)$, based on the Wilson-Bappu method, and M_v, derived from trigonometric parallaxes has been made for 142 G and K giant stars, which have been divided in four groups according to metal content.

115.004 On ultraviolet fluxes, bolometric corrections and effective temperatures of late B to F stars.
J. Davis, R. J. Webb.
IAU Symposium No. 36, (see 012.014), p. 90 - 99 (1970).
Observed ultraviolet fluxes have been compared with the predictions of line blanketed model atmospheres for stars of spectral type later than B5. Empirical bolometric corrections for the spectral range B8–F5 have been computed from a combination of ultraviolet, visual and infrared observational data. These have been used to derive empirically based effective temperatures for five A and F type stars for which angular diameter measurements are available.

115.005 Two subluminous stars near the galactic plane.
J. L. Greenstein, A. I. Sargent, U. Haug.
Astron. Astrophys., Vol. 7, 1 - 3 (1970).
Photoelectric colors of OB stars in the Hamburg-Case Luminous Star survey revealed that only two out of more than 900 had the colorimetric characteristics of highly sub-

luminous stars. Both are probably members of the halo or old disk population.

115.006 The angular diameter of Lambda Aquarii.
R. E. Nather, M. M. McCants, D. S. Evans.
Astrophys. Journ. (*Letters*), Vol. 160, L181 - L184 (1970).
An occultation observation of λ Aqr gives an angular diameter between 0.″0074 and 0.″0082 according to the assumed law of darkening. The observational uncertainty is of the order of ±0.″0004 in each case. If the cosine law is assumed, the effective temperature lies in the range 3250° –3450°K whatever value of the darkening coefficient is adopted.

115.007 On the absolute magnitudes of U Gem type variables. Sh. Primkulov.
Tsirk. Astron. Inst., *Tashkent*, No. 8 (355), p. 13 - 16 (1968).
In Russian.

115.008 Carbon stars on the two-colour diagram.
Z. K. Alksne.
Astron. Tsirk., No. 544, p. 3 - 6 (1970). In Russian.

115.009 New relation M_v–n_m for O9 – F2 stars.
A. N. Gerashchenko.
Astron. Tsirk., No. 558, p. 2 - 4 (1970). In Russian.

Effective temperatures, gravities, and the mass determination of A and F stars. See Abstr. 064.049.

Les magnitudes bolométriques.
See Abstr. 113.007.

Narrow-band photometry of late-type stars.
See Abstr. 113.025.

An attempt to define luminosity criteria in O stars via narrow-band photoelectric photometry.
See Abstr. 113.039.

Spectrophotometry of cool stars in the near infrared. Results for a region in Cygnus. See Abstr. 113.058

MK classifications for F- and G-type stars. II.
See Abstr. 114.012.

A study of γ² Velorum with a stellar intensity interferometer. See Abstr. 117.009.

The nature of the Beta Cephei phenomenon.
See Abstr. 122.001.

116 Stellar Magnetic Field, Figure, Rotation

116.001 On the possibility of observing interferometrically the surface distortion of rapidly rotating stars.
I. D. Johnston, N. C. Wareing.
Monthly Notices, Roy. Astron. Soc., Vol. 147, 47 - 58 (1970).

The response of an interferometer to a rotating star is analysed, firstly by a perturbation calculation, applicable only to stars which rotate slowly, and then numerically, specifically for the two stars Regulus and Altair. These calculations assume the star is limb-darkened and gravity-darkened according to standard models of stellar atmospheres, and a separate calculation is made for stars with non-spherical gravitational fields. It is concluded that observations of rotational distortion are probably only marginally feasible with present-day equipment, though future interferometers should find them easy.

116.002 Stellar rotation: The anomalous result of Roxburgh, Griffith, and Sweet.
A. D. Sanderson, R. C. Smith, J. Hazlehurst.
Astrophys. Journ. (*Letters*), Vol. 159, L69 - L71 (1970).

Comparison of the luminosity changes in rotating stars predicted by various authors shows that one paper is anomalous. Examination of this paper reveals an internal inconsistency, and recalculation produces a significantly different result.

116.003 The weak magnetic fields of some bright stars.
A. Severny.
Astrophys. Journ. *(Letters)*, Vol. 159, L73 - L76 (1970).

The technique of the solar magnetograph applied at the coudé spectrograph of the 2.6-m reflecting telescope of the Crimean Astrophysical Observatory for the measurement of stellar magnetic fields has recently revealed weak longitudinal magnetic fields (30–300 gauss) of some bright stars.

116.004 Magnetic stars. II. Observations of 56 Ari.
C. Blanco, F. A. Catalano.
Astron. Journ., Vol. 75, 53 - 54 (1970).

From the observed light curves in the *U, B, V* colors, no phase displacement is found in the epoch of primary minimum. Light and spectrum variations may be due to a localized active region.

116.005 Quantitative analysis and spectral variation of the Ap star HD 151199. N. Gökkaya.
Astrophys. Space Sci., Vol. 6, 141 - 153 (1970).

The radial velocity, intensity variations of the Ca II line and chemical composition of the suspected magnetic star HD 151199 have been studied using three 9.6 Å/mm and twenty-one 40 Å/mm dispersion spectrograms which were taken at St. Michel and Asiago Observatories respectively.

116.006 On some theoretical questions on the origin of magnetic fields of stars and nebulae.
A. Ya. Kipper.
Stars, Nebulae, Galaxies, Symposium Byurakan 1968, (see 012.002), p. 173 - 177 (1969). In Russian.

116.007 Harmonic analysis of rigidly rotating Ap stars.
A. J. Deutsch.
Astrophys. Journ., Vol. 159, 985 - 999 (1970).

To facilitate mapping the surfaces of magnetic- and spectrum-variable stars, the requisite formulae are given for the harmonic analysis outlined in an earlier publication.

116.008 The orientation of magnetic axes in the magnetic variables. J. D. Landstreet.

Astrophys. Journ., Vol. 159, 1001 - 1007 (1970).

The distribution of magnetic field over the surface of a star required to reproduce observed magnetic curves of Ap stars on the rigid-rotator model is discussed, and it is shown that the field distribution may be taken to be axisymmetric but is not necessarily antisymmetric to reflections through the magnetic equator. A statistical method for determining the angle β between the rotation and magnetic axes, devised by Preston, is extended to the class of models considered here, and it is shown that most magnetic stars have β near 90° independently of the precise field distribution over the surface which is assumed.

116.009 Comment on a paper of S. Kato and Y. Nakagawa on generation of magnetic fields in rotating stars.
M. Stix.
Astron. Astrophys., Vol. 4, 161 - 162 (1970).

It is pointed out that the effect of momentum transfer between photons and electrons on the generation of a toroidal magnetic field in a rotating star should be taken into account.

116.010 The light-variation of four magnetic variable stars.
A. M. van Genderen.
Astron. Astrophys., Suppl. Series, Vol. 1, 123 - 127 (1970).

The paper deals with photoelectric observations of the magnetic variable stars HD 8441, 21 Per, ι Cas and HD 25354. The first three stars have been observed with a red filter only, the fourth one has also been observed in *UBV*.

116.011 A photoelectric investigation of the magnetic star β CrB. E. S. Brodskaya.
Astron. Zhurn. Akad. Nauk SSSR, Vol. 47, 662 - 664 (1970). In Russian. English translation in Soviet Astron.,AJ, Vol. 14, No. 3.

During two seasons of 1966 and 1968 photoelectric observations of the periodic magnetic variable β CrB were carried out in a system close to UBV. It is found that the brightness variation has the same period as the variation of the magnetic field.

116.012 Stellar rotation: Uniformly rotating stars with hydrogen-line-blanketed model atmospheres. Comparison with observations of A-type stars.
A. Maeder, E. Peytremann.
Astron. Astrophys., Vol. 7, 120 - 132 (1970).

The energy distribution of uniformly rotating stars including hydrogen lines of the Balmer and Lyman series are computed for 5, 2 and 1.4 solar masses, for various rotational velocities and orientations. In order to make a comparison with observations, colours and parameters are computed in *UBV* and Geneva Observatory photometric systems. The predicted magnitudes, colours and Balmer discontinuities show a very good agreement with observational results by Golay (1968) for A-type stars up to types near A7 – F0.

116.013 Spektrographische Untersuchungen des magnetischen Sterns α^2 CVn.
L. Oetken, E. Bartl, R. Orwert.
Astron. Nachr., Vol. 292, 1 - 8 = Mitt. Astrophys. Obs. Potsdam, No. 134 (1970).

The main purpose of this paper was the examination of the Cassegrain-spectrograph of the 2-m-universal-telescope at Tautenburg for investigations of magnetic stars. Therefore from 26 spectrograms of the well unknown magnetically variable star α^2 CVn taken with a reciprocal linear dispersion of 10 Å/mm variations of radial velocity and magnetic field strength for some lines of Eu II, Cr II, Si II, Mg II were determined. The results agree well with those of Babcock and Struve and Swings, derived from spectrograms of higher dispersion. The large variation in the radial velocity of Eu II and Cr II is confirmed. In the oblique rotator model this requires a very strong concentration of Eu II at the poles and of Cr II at the equator.

116.014 The magnetic field of β Coronae Borealis.
S. C. Wolff, R. J. Wolff.
Astrophys. Journ., Vol. 160, 1049 - 1058 (1970).

Measurements of resolved Zeeman patterns in the spectrum of β CrB show that $|H_s|$, the mean surface magnetic field, varies approximately 180° out of phase with the longitudinal component of the field H_e. The maximum observed $|H_s|$ is about 5700 gauss, and the total amplitude of the variation is 800 gauss. The observations are compatible with a field geometry in which one magnetic pole is stronger than the other.

116.015 The large variable magnetic field of HD 126515 and its implications for the rigid-rotator model of magnetic stars. G. W. Preston.
Astrophys. Journ., Vol. 160, 1059 - 1070 (1970).

The mean surface field H_s of the Sr-Cr-Eu star HD 126515 varies approximately sinusoidally between 10 and 17 kilogauss in a period of $130\overset{d}{.}0$. The effective field H_e oscillates between –2 and +2 kilogauss with a single wave in the same period. From the point of view of the rigid-rotator model, these results require that the star have a hemispherically asymmetric magnetic field, the principal features of which are represented by a dipole displaced from the center of the star in the direction opposite to its moment by 0.36 stellar radii.

116.016 On the Li I 6708 line in the spectrum of the magnetic variable β CrB. N. S. Polosukhina.
Astron. Tsirk., No. 549, p. 1 - 3 (1970). In Russian.

116.017 3 Hya = HD 72968.
R. Steinitz, D. M. Pyper.
Inform. Bull. Variable Stars (I.A.U. Commission 27), Konkoly Obs., Budapest, No. 413 (1970).

116.018 Observations photoélectriques de HD 125248.
H. M. Maitzen.
Inform. Bull. Variable Stars (I.A.U. Commission 27), Konkoly Obs., Budapest, No. 421 (1970).

116.019 The magnetic variable Beta CrB.
C. Bartolini, P. Battistini, R. Pecorari.
Inform. Bull. Variable Stars (I.A.U. Commission 27), Konkoly Obs., Budapest, No. 435 (1970).

116.020 Wavelength dependence of polarization in Ap-type magnetic stars. G. V. Coyne.
Ric. Astron. Specola Vaticana, Vol. 8, (No. 7), 117 - 127 (1970).

Polarimetric observations in seven colors for seven Ap-type magnetic stars are presented. Though the average percentage polarization is small, about 0.20%, the wavelength dependence of the polarization gives some indication that the measured polarization is intrinsic to these stars. Six of the seven stars have a larger polarization in the red than in the blue-yellow region of the spectrum. The wavelength dependence of the polarization can be explained by assuming that we are viewing, at nearly pole-on aspects, rapidly rotating A stars with relatively large magnetic fields.

116.021 Variations du champ magnétique d'étoiles Ap périodiques. P. Renson.
Acad. Roy. Belgique, Cl. Sci. Mém. Coll. 8°, 2ᵉ Sér., Vol. 38, Fasc. 5, 102 pp. = Univ. Liège, Inst. d'Astrophys., Coll. 4°, No. 200 (1969).

Magnetic braking by a stellar wind – III. The oblique rotator with a quasi-radial field.
See Abstr. 064.047.

Differential rotation in stellar convection zones.
See Abstr. 065.014.

Rotation in evolving stars.
See Abstr. 065.058.

Non-linear limiting of the Goldreich–Schubert instability. See Abstr. 065.059.

Mass loss from rapidly rotating evolving B-stars.
See Abstr. 065.063.

Nonlinear pulsations and stability of slowly rotating stars. See Abstr. 065.082.

Photométrie des étoiles A. II. Métallicité et rotation des étoiles Am. See Abstr. 113.045.

Rotation and chemical abundances in the peculiar A stars – I. See Abstr. 114.002.

Rotation and chemical abundances in the peculiar A stars – II. See Abstr. 114.003.

The continuous radiation of bright B3 V stars.
See Abstr. 114.033.

Line strengths for southern OB stars – III. Balmer profiles for slow rotators. See Abstr. 114.045.

On the slow rotation of Ap and Am stars.
See Abstr. 117.021.

The very slow spectrum, magnetic, and photometric variations of HD 9996. See Abstr. 119.013.

UBV Observations of the magnetic variable HD 125248. See Abstr. 122.074.

117 Binary and Multiple Stars, Theory

117.001 Evolution of close binaries. V. Stationary models representing the end of the phase of rapid mass loss in case A. J. Horn, S. Kříž, M. Plavec.
Bull. Astron. Inst. Czechoslovakia, Vol. 21, 45 - 54 (1970).

Mass exchange in close binary systems, where the mass-losing component has not yet exhausted hydrogen in its convective core, consists of a rapid and a slow phase. The outcome of the rapid phase can be computed by means of a sequence of stationary models. Such calculations were made for five values of the initial masses of the mass-losing component (9, 7, 6, 5 and 4 solar masses, respectively), for four different evolutionary stages, and for equidistant values of the initial mass ratio.

117.002 One example of mass exchange in case AB in system $5\,M_\odot + 4\,M_\odot$. J. Horn.
Astrophys. Space Sci., Vol. 6, 492 - 496 (1970).

Evolution of a binary system with masses of $5\,M_\odot$ and $4\,M_\odot$, respectively, and with orbital period of 1.41 days is studied by means of non-stationary model calculations under assumptions of conservation of total mass and total angular momentum of the system. As a result of mass exchange between the components we obtain a binary with masses of 8.46 and $0.54\,M_\odot$. Physical parameters of the final product indicate possible connection with shell stars. It is also pointed out that the new secondary component can become rotationally unstable soon after the end of mass exchange.

117.003 Case B of mass exchange in systems $4 + 3.2\,M_\odot$ and $4 + 1.6\,M_\odot$. P. Harmanec.
Astrophys. Space Sci., Vol. 6, 497 - 503 (1970).

Two examples of case B of mass exchange are computed to estimate the effect of basic initial parameters on the course and the results of mass exchange.

117.004 The nearest other solar system?
A. T. Lawton.
Spaceflight, Vol. 12, 170 - 173 (1970).

The article concerns the discovery of a small planetary companion to Barnard's star by P. van de Kamp.

117.005 Measurement of primordial helium abundance from the star μ Cassiopeiae.
D. Hegyi, D. Curott.
Phys. Rev. Letters, Vol. 24, 415 - 418 (1970).

A measurement of the separation between the components of the binary star μ-Cassiopeiae to yield the stellar masses has been made. Applying the mass-luminosity law, we find a helium abundance $Y \sim 0$ where a 1"σ" and 2"σ" random error allows, respectively, $Y \sim 0.05$ and $Y \sim 0.34$. The measured abundance places an upper limit on the helium production in a big bang cosmological model.

117.006 Mass transfer in close binary systems.
S. Piotrowski.
Stars, Nebulae, Galaxies, Symposium Byurakan 1968, (see 012.002), p. 109 - 128 (1969).

117.007 Formation of binary systems in triple encounters.
T. A. Agekian, J. P. Anosova, B. N. Bezgubova.
Astrofizika, Vol. 5, 637 - 644 (1969). In Russian. – English translation in Astrophysics, Vol. 5, No. 4.

The probability of formation of binary systems as a result of triple encounters of stars is investigated. The numerical solutions of equations of motion are obtained using a computer for a sample of 1600 random initial conditions. 656 encounters were accomplished with the formation of binary systems.

117.008 Échange de masse dans les étoiles doubles serrées.
A. Peton.
L'Astronomie, 84e année, p. 23 - 36 (1970).

117.009 A study of γ^2 Velorum with a stellar intensity interferometer.
R. Hanbury Brown, J. Davis, D. Herbison-Evans, L. R. Allen.
Monthly Notices, Roy. Astron. Soc., Vol. 148, 103 - 117 (1970).

In this paper we have reported some of the first measurements of a binary star made with the stellar intensity interferometer. By combining our measurements with spectroscopic and photometric data we have found values for the distance, radius, absolute magnitude, surface gravity and effective temperature of the Wolf-Rayet component of γ^2 Vel; we have also established the size of the region responsible for the emission lines and shown that it is the correct size to fill the critical Roche equipotential lobe around the Wolf-Rayet star.

117.010 Intrinsic polarization in close binary systems.
P. F. Buerger, G. W. Collins II.
Bull. American Astron. Soc., Vol. 2, 185 (1970). – Abstr. AAS.

117.011 On the effects of a supernova explosion in a close binary system. G. E. McCluskey, Jr.,
Y. Kondo.
Bull. American Astron. Soc., Vol. 2, 208 - 209 (1970). Abstr. AAS.

117.012 A study of the structure of rapidly rotating close binary systems. P. G. Martin.
Astrophys. Space Sci., Vol. 7, 119 - 138 (1970).

A polytropic theory for investigating the structure of rapidly rotating close binary systems is developed on the basis of the works of Chandrasekhar (1933) and Monaghan and Roxburgh (1965). Solutions for the interior and exterior potentials and densities are found. The surface, surface gravity and potential are found. These results are used to discuss the critical configurations and contact equipotentials, and the existence of semi-detached and contact binary systems. The theory is compared to previous work, in particular the case of rotation alone and the Roche model.

117.013 On the evolution of close binary systems with a total mass 2.5 $M_\odot$. 2.
P. Giannone, S. Refsdal, A. Weigert.
Astron. Astrophys., Vol. 4, 428 - 436 (1970).

For two close binary systems of total masses $M_1 + M_2 = 2.5\,M_\odot$, the evolution has been calculated from the main sequence through the phase of mass exchange between the components. The systems start with the same mass ratios (1.1/1.4) but with different initial separations. After the mass exchange, these stars, in the order of increasing initial separations, become white dwarfs of $0.366\,M_\odot$ and $0.426\,M_\odot$. For the most massive remnant, the evolution has been followed to the white dwarf stage.

117.014 Forced oscillations in close binaries. The adiabatic approximation. J. P. Zahn.
Astron. Astrophys., Vol. 4, 452 - 461 (1970).

The forced oscillations of a star are discussed, due to a periodic varying gravitational field. The following approximation are made: the star is not rotating, the oscillations are adiabatic and small enough to allow linearization. Olver's method is employed to derive a second-order asymptotic expression for the eigenfunctions. The results predict large amplitudes for the tides near the surface, where however adiabacy is a poor approximation.

117.015 Binary formation by three-body collisions.
P. Mansbach.
Astrophys. Journ., Vol. 160, 135 - 145 (1970).

Keck's "variational method" is used to obtain an expression for the rate of formation of binaries due to stellar three-body collisions, under quite general conditions. In the solar neighborhood this rate is too small to account for any binaries. In globular clusters it could account for a few. The general rate expression is also applicable to other problems, such as comet capture by a star or asteroid capture by a planet. An estimate of the latter, using available data, indicates that the rate is somewhat too low to account for Jupiter's outer moons if the third body is another asteroid. The third body might be a more massive satellite, however; this possibility is promising.

117.016 On the structure of close binary members.
S. Jackson.
Astrophys. Journ., Vol. 160, 685 - 699 (1970).

A double-approximation method, similar to the one described by Roxburgh, Griffith, and Sweet, is used to consider the effect of rotation and the tidal interaction upon the structure of a synchronously rotating binary component. For both the Cowling model and a similar electron-scattering model, this method is used to construct models for two distorted stars —one having a close companion of equal mass and the other being a rapidly rotating single star.

117.017 A potential flow pertaining to binary systems.
Y. Sobouti.
Astron. Astrophys., Vol. 5, 149 - 154 (1970).

Gas flow arising from the orbital motion of stars in a close binary system is a consequence of equations of fluid motion and is inevitable. The velocity of flow is of the order of and less than the orbital velocities. It has the same magnitude and topological structure as most radial velocities derived from observations of spectral lines whose origin lies in the gaseous envelope.

117.018 Models of close and contact binary stars – I. Polytropic models.
B. R. Durney, I. W. Roxburgh.
Monthly Notices, Roy. Astron. Soc., Vol. 148, 239 - 247 (1970).

Polytropic models of close and contact binary stars are constructed using a combination of perturbation techniques and a Laplace approximation previously applied to uniformly rotating stars. Synchronism between orbital and intrinsic angular velocity is assumed.

117.019 Evolution of close binary systems with initial components of 2 and 1.5 $M_\odot$.
P. Giannone, M. A. Giannuzzi.
Astron. Astrophys., Vol. 6, 309 - 317 (1970).

The evolution of the originally more massive stars of three close binary systems has been computed through the phase of mass exchange between the components. The systems have $M_1 + M_2 = 3.5\ M_\odot$, initial mass ratio 0.75 and three different initial separations of $7\ R_\odot$, $10\ R_\odot$ and $15\ R_\odot$. Conservation of total mass and total orbital angular momentum is assumed.

117.020 A theoretical study of the photocentre.
P. J. Morel.
Astron. Astrophys., Vol. 6, 441 - 452 (1970).   In French.

We construct a theoretical model which provides a defi-

nition for the setting position and show how to compute the photocentric correction. Using a convolution product one can calculate the distribution of the scattered light in the photographic emulsion prior to processing. We have taken into account the effects of Airy diffraction, atmospheric turbulence and scattering in the emulsion but have neglected the Eberhard effect. Methods are given for determining the constants which characterize these effects. According to our model the luminosity barycentre does not coincide with the maximum of the photographic density. We show how to determine the photocentric correction if the setting position corresponds to the centre of an isophote.

117.021 On the slow rotation of Ap and Am stars.
J. J. Monaghan.
Astron. Astrophys., Vol. 6, 464 - 467 (1970).

In a long period highly eccentric binary the axial rotation of the secondary is essentially unaffected until the envelope of the primary intersects the orbit of the secondary. Assuming that mass transfer is never fast enough to allow the envelope to adjust to the changing Roche potentials, the effect of the envelope, in turbulent convection, on the dynamics of the secondary is examined.

117.022 Evolution of close binaries. VI. A program for computation of evolution with mass exchange.
J. Ziółkowski.
Acta Astron., Vol. 20, 59 - 74 (1970).

A description of the Henyey-type stellar evolution code allowing to compute the evolution of a single star or a component of a close binary is given.

117.023 The nature of contact binaries.
J. Hazlehurst.
Monthly Notices, Roy. Astron. Soc., Vol. 149, 129 - 146 (1970).

The observed characteristics of the individual components of the W Ursae Majoris systems of low mass are found to be consistent with the primaries being evolved stars, and the secondaries being of age zero. Theoretical models of 'contact' systems in which the primary only is evolved have been used to calculate a period-colour relationship at constant mass-ratio, and this has been compared with the observed period-colour relationship of the W Ursae Majoris systems. The agreement of the slopes is satisfactory. The occurrence in Praesepe of a W Ursae Majoris system which is too red to be of age zero, yet too young to be evolved, remains a paradox.

117.024 The structure of W Ursae Majoris systems.
D. L. Moss, J. A. J. Whelan.
Monthly Notices, Roy. Astron. Soc., Vol. 149, 147 - 165 (1970).

The structure of late-type contact binary systems of the W UMa type is investigated using the Lucy (1968) model with a common convective envelope. However more recent opacities than those adopted by Lucy are used. Some differences are found, but we agree that no zero age systems with unequal components are to be found with a normal population I composition. Zero age systems are found for extreme population I compositions.

117.025 Pre-main-sequence mass transfer in binary systems.
J. A. J. Whelan.
Monthly Notices, Roy. Astron. Soc., Vol. 149, 167 - 177 (1970).

In this paper some aspects of the pre-main-sequence evolution of binary systems are investigated. In particular we are interested in the possibility of and consequences of mass exchange occurring on a dynamical timescale in this stage of evolution. We first discuss the situation of two fully convective stars in contact and then consider the stability of such a system to mass exchange. Finally we consider the results of their instability in conjunction with the observations.

117.026 **Evolution with mass exchange of case C for a binary system of total mass 7 $M_\odot$.** D. Lauterborn.
Astron. Astrophys., Vol. 7, 150 - 159 (1970). − Thesis.

A binary system with a primary of 5 $M_\odot$, a mass ratio of 0.4, and a separation of 302 $R_\odot$ is followed through mass exchange which starts after exhaustion of central helium burning (case C). The final system contains a white dwarf of 1 $M_\odot$ and a main sequence star of 6 $M_\odot$, separated by 840 $R_\odot$; the period is 2.8 years. A few observed case-C systems are briefly discussed.

117.027 **Relative positions and proper motions of the components of ϵ Lyr type triple stars.**
J. P. Anosova.
Vestn. Leningr. un-ta, 1969 No. 19, p. 123 - 133. In Russian. Abstr. in Referativ. Zhurn. 51. Astron., 5.51.635 (1970).

117.028 **The orbital parameters of a binary system in which the more massive component decreases in mass.**
E. P. J. van den Heuvel.
Utrechtse Sterrekundige Overdrukken, No. 102, 16 pp. (1968).

117.029 **Masses and initial chemical composition of early type stars.** G. G. Rodionova.
Nauchn. Informatsii, vyp. (No.) 13, p. 116 - 125 (1969). In Russian.

Results of stellar evolution theory are compared with selected observational data on massive binaries. An estimation of the age of these stars is made.

117.030 **Evolution of close binaries. VI. Case B of mass exchange in systems 4 + 3.2 $m_\odot$ and 4 + 1.6 $m_\odot$.**
P. Harmanec.
Bull. Astron. Inst. Czechoslovakia, Vol. 21, 113 - 131 (1970).

Two examples of case B of mass exchange have been computed to estimate the effect of basic initial parameters of systems on the course and the results of mass exchange. In the case of 4 + 3.2 $m_\odot$ the changes in inner structure of the mass-losing star are discussed in detail.

117.031 **Double stars.** P. Couteau.
Trans. IAU, Vol. 14A, (see 003.028), 249 - 257 (1970). − Report of Commission 26.

117.032 **Significance of the momentum of stars with dark companions.** C. J. Lavagnino.
Asoc. Argentina Astron. Bol., No. 14 (see 012.020), p. 76 - 77 (1968). In Spanish. − Abstract.

117.033 **Statistics of binary stars.** O. Ferrer.
Asoc. Argentina Astron. Bol., No. 14 (see 012.020), p. 103 - 104 (1968). In Spanish.

117.034 **The nature of low-mass companions of stars in the solar neighborhood.** S. S. Kumar.
Ann. New York Acad. Sci., Vol. 163, No. 1, p. 94 - 97 (1969). − See Bull. Signalét., Vol. 31, Section 120, No. 6836 (1970).

Probleme der langbrennweitigen Astrometrie.
See Abstr. 041.001.

The geometry of the Roche coordinates.
See Abstr. 042.046.

L'evoluzione delle stelle. II − Le stelle doppie.
See Abstr. 065.077.

118 Visual Binaries

118.001 **Photometric and astrometric measurements of close double stars with the Kron electronic camera.**
H. D. Ables, R. L. Walker, H. V. Hewitt.
Bull. American Astron. Soc., Vol. 2, 179 (1970). – Abstr. AAS.

118.002 **An intercomparison of three spectroscopic luminosity criteria.** T. E. Lutz.
Bull. American Astron. Soc., Vol. 2, 206 - 207 (1970). Abstr. AAS.

118.003 **Spectroscopic observations of the triple star ADS 14893.** F. R. West.
Bull. American Astron. Soc., Vol. 2, 225 (1970). – Abstr. AAS.

118.004 **Systematic errors in double-star measures.**
C. E. Worley, G. G. Douglass.
Bull. American Astron. Soc., Vol. 2, 226 (1970). – Abstr. AAS.

118.005 **U, B, V, photometry of some visual binaries.**
J. B. Alexander.
Monthly Notes Astron. Soc. Southern Africa, Vol. 29, 44 - 48 (1970).
Photoelectric photometry has been obtained for 24 systems in the southern hemisphere. The observations were all made with the Elizabeth 40-inch reflector at the Cape Observatory between September 1967 and December 1968.

118.006 **Étoiles doubles nouvelles (4ème série), découvertes à Nice avec la lunette de 50 centimètres.**
P. Couteau.
Astron. Astrophys., Suppl. Series, Vol. 1, 105 - 114 (1970).
On donne une liste de 100 étoiles doubles découvertes à la lunette de 50 cm de l'Observatoire de Nice.

118.007 **Orbites de cinq étoiles doubles visuelles.**
P. J. Morel.
Astron. Astrophys., Suppl. Series, Vol. 1, 115 - 121 (1970).
On donne les éléments orbitaux des cinq étoiles doubles visuelles ADS 2980, ADS 7982, ADS 15236, ADS 16314 AB et Rst 4529.

118.008 **Órbita da estrela dupla visual Burnham 524 ≡ A. D. S. 2200 AB e respectivos parâmetros físicos.**
A. Simões da Silva, M. M. Pinheiro.
Comun. Obs. Astron. Univ. Coimbra, No. 6, 28 pp. (1970).

118.009 **Preliminary orbit of the visual binary ADS 9975 = A 1642 = JDS 2417.**
J. Dommanget, E. L. van Dessel.
Astron. Astrophys., Vol. 6, 423 - 425 (1970).
Calculation of the orbital elements and physical characteristics of the visual binary ADS 9975, adopting a change of quadrant for the observations later than ≈ 1950. This assumption is based on the possible variability of one of the components.

118.010 **Photographic measures of double stars.**
P. S. The.
Astron. Astrophys., Suppl. Series, Vol. 1, 357 - 392 (1970).
In this paper the results of photographic measurements and the determinations of the mean positions of 168 mostly southern double stars are reported. The plates used for the measurements were taken with the 60-cm refractor of the Bosscha Observatory at Lembang, Indonesia.

118.011 **Mikrometermessungen von Doppelsternen. VII.**
W. D. Heintz.
Astron. Astrophys., Suppl. Series, Vol. 1, 393 - 398 = Veröff. Sternw. München, Vol. 7, No. 11 (1970).
The paper contains 800 measures of 228 double stars made with the 28 cm refractor in Munich.

118.012 **Mesures d'étoiles doubles à Meudon (2ème série).**
P. Muller.
Astron. Astrophys., Suppl. Series, Vol. 1, 399 - 407 (1970).
Cette publication comprend 289 mesures de 113 couples visuels faites au réfracteur de 83 cm de l'Observatoire de Meudon.

118.013 **Mesures photographiques d'étoiles doubles.**
E. Fossat.
Astron. Astrophys., Suppl. Series, Vol. 1, 409 - 417 (1970).
On donne 246 mesures photographiques de 95 couples, les clichés ayant été faits à la lunette de 50 cm de l'Observatoire de Nice, la plupart avec un agrandisseur de Barlow multipliant par trois la longueur focale de l'instrument, et quelques-uns au foyer direct.

118.014 **Étoiles doubles nouvelles (5ème série), découvertes à Nice avec la lunette de 50 cm.**
P. Couteau.
Astron. Astrophys., Suppl. Series, Vol. 1, 419 - 428 (1970).
On donne un liste de 100 étoiles doubles découvertes à la lunette de 50 cm de l'Observatoire de Nice.

118.015 **Orbites de huit étoiles doubles visuelles.**
P. J. Morel.
Astron. Astrophys., Suppl. Series, Vol. 1, 429 - 439 (1970).
On donne les éléments orbitaux des huit étoiles doubles visuelles ADS 34, ADS 2630, ADS 5726, ADS 7158, ADS 16278, ADS 17030, Kui 102 et Kui 108.

118.016 **Orbites de trois étoiles doubles visuelles.**
P. J. Morel.
Astron. Astrophys., Suppl. Series, Vol. 1, 441 - 445 (1970).
On donne les éléments révisés des orbites des étoiles doubles visuelles ADS 918, ADS 6549 et ADS 10916.

118.017 **De ster van Barnard heeft twee planeten.**
A. G. Jansen.
Hemel en Dampkring, Vol. 68, 53 - 54 (1970).

118.018 **Castor.** J. Meeus.
Hemel en Dampkring, Vol. 68, 56 - 58 (1970).

118.019 **Results of Capella measurements with the Pulkovo stellar interferometer.** E. S. Kulagin.
Astron. Zhurn. Akad. Nauk SSSR, Vol. 47, 557 - 559 (1970).
In Russian. English translation in Soviet Astron.,AJ, Vol. 14, No. 3.
In 1968 - 1969 twelve measurements of Capella were made at Pulkovo with a 6-m stellar interferometer. It is shown that the orbit of Capella determined by Merrill from the first interferometric measurements in 1921 represents the observations well enough up to now.

118.020 **Erratum: Parallax and mass ratio of the visual binary Hu 1176 from photographs taken with the 24-inch Sproul refractor.** [Astron. Journ., Vol. 71, 524 - 526 (1966), see AJB 66, 11226]. P. van de Kamp, S. Moore.
Astron. Journ., Vol. 75, 665 (1970).

118.021 **Measures of photographed binaries with the eye-piece-method.** E. Hock.
Astron. Astrophys., Vol. 7, 160 - 161 (1970). In German.

Measures of photographed binaries ADS 3390, 5514, 6650 AB, AC, 6746 Cc, 10374, 11530, 12515, 14073, 14296, 14412, 14926, 15447, 16057, 16185, 16538, 16638, 16904, and 17178 with the eye-piece-method with additional magnifying optic are given.

118.022 **Photovisual magnitude differences of double stars.** K. A. Strand.
Publ. U. S. Naval Obs., *Washington,* Second Series, Vol. 18, Part V, 31 pp. (1969).

Differences in magnitude between the components of 874 double stars have been derived from visual estimates on several series of multiexposure photographic plates. These plates were obtained with long focus visual refractors. On the basis of the final results, the internal mean error of a single estimate of a magnitude difference is ± 0.^{m}064 which compares favorably with the internal mean error of ± 0.^{m}047 derived from similar material but obtained by a more laborious procedure using the Schilt and Hartmann photometers.

118.023 **Photgraphic measures of double stars.**
V. V. Kallarakal, I. W. Lindenblad, F. J. Josties, R. K. Riddle, M. Miranian, B. F. Mintz, A. P. Klugh.
Publ. U. S. Naval Obs., *Washington,* Second Series, Vol. 18, Part VII, 128 pp. (1969).

From 3418 sets of measurements, with a total of 1,818,217 image bisections (or centering) on 3286 plates, relative positions have been determined for the components of 610 double stars. The difference in scale value of the 26-inch refractor before and after 8 March 1962 is due to a change in configuration of camera and filter which took place at that time.

118.024 **Orbite nouvelle.**
P. Muller.
Circ. Inform. (U. A. I. Commission des Etoiles Doubles), Obs. Meudon, Nos. 50, 51 (1970).

118.025 **Etoiles doubles découvertes à Nice, lunette de 50 cm.** P. Couteau, P.-J. Morel, P. Muller.
Circ. Inform. (U. A. I. Commission des Etoiles Doubles), Obs. Meudon, Nos. 50, 51 (1970).

118.026 **Etoiles doubles découvertes à Beograd, lunette de 65 cm.** G. Popovic, D. Olevic.
Circ. Inform. (U. A. I. Commission des Etoiles Doubles), Obs. Meudon, Nos. 50, 51 (1970).

118.027 **Savremeni problemi izučavanja dvojnih i novi program posmatranja u Beogradu.**
P. M. Djurković.
Publ. Obs. Astron. Beograd, No. 16, (see 012.021), p. 90 - 96 (1969). In Serbo-Croatian.

The spectra of the two metallic-line stars ν_1 and ν_2 Draconis. See Abstr. 114.031.

Eight-colour photometric study of the eclipsing binary 44 i Bootis. See Abstr. 121.026.

Spectra of the Cepheid HR 8157. See Abstr. 122.051.

Search for new flare stars. See Abstr. 122.084.

119 Spectroscopic Binaries

119.001 Analyse de l'étoile à raies métalliques, 15 Vulpeculae, par la méthode différentielle des courbes de croissance. R. Faraggiana, C. van't Veer-Menneret. Comptes Rendus Acad. Sci. Paris, Sér. B, Vol. 270, 765 - 768 (1970).

Les résultats préliminaires de l'analyse par la méthode des courbes de croissance conduisent à une abondance normale du calcium et du fer, mais à une faible surabondance des métaux plus lourds que le fer.

119.002 A mass ratio for 112 Herculis. C. E. Seligman. Publ. Astron. Soc. Pacific, Vol. 82, 128 - 131 (1970).

The mass ratio of the spectroscopic binary 112 Herculis is found to be 2.06 ± 0.17. This is in good agreement with the value expected from the spectral types of the two components (B7 V and A3 V). The sharpness of the line spectra suggests that the stars are rotating synchronously with their orbital motion.

119.003 On the selection in the case of discovery of spectroscopic binaries and interpretation of the B-effect. E. F. Brazhnikova. Astron. Zhurn. Akad. Nauk SSSR, Vol. 47, 149 - 161 (1970). In Russian. English translation in Soviet Astron.,AJ, Vol. 14, No. 1.

The hypothesis is grounded theoretically that if the mass center of a spectroscopic binary moves relative to cosmic dust so that the periastron of the bright component is situated in the direction of the apex of the motion, then near such a star an accumulation of the optically effective cosmic dust takes place. The cosmic dust shields this star most from an observer standing in its apex, and less from an observer standing in its antiapex. The consequences of this hypothesis are discussed.

119.004 CV Ser — a peculiar Wolf-Rayet binary. K. Stępień. Acta Astron., Vol. 20, 13 - 17 (1970).

The two colour photometry of CV Ser made on 46 nights in 1967 shows no evidences of eclipses.

119.005 Orbital elements of six spectroscopic binary stars. B. W. Bopp, D. S. Evans, J. D. Laing, T. J. Deeming. Monthly Notices, Roy. Astron. Soc., Vol. 147, 355 - 366 (1970).

Orbital elements of the spectroscopic binary stars, HD 1273, 148704, 167954, 194215, 202940 and 217792 are presented.

119.006 Orbit of the double-line spectroscopic binary HR 4072 (Ap). K. Nariai. Publ. Astron. Soc. Japan, Vol. 22, 113 - 118 (1970).

Velocities for both components of HR 4072 (= HD 89822) were measured on high dispersion spectrograms. A new spectroscopic orbit was determined. The mass ratio was found to be 1.67 ±0.04, which is very close to the value expected from the mass-luminosity relation, $\log L/L_\odot = 3.3 \log M/M_\odot$ (Guthrie 1966).

119.007 Hβ observations of 44 i Bootis. K. J. Johnston. Bull. American Astron. Soc., Vol. 2, 201 - 202 (1970). Abstr. AAS.

119.008 Anomalous excitation on 57 Cygni. W. McD. Napier, M. W. Ovenden.

Astron. Astrophys., Vol. 4, 129 - 133 (1970).

An attempt is made to interpret the observations of the binary 57 Cygni (Ovenden, 1963) in terms of the thermal reflection effect. It is found that the components cannot heat each other to the degree apparently required by the presence of ions appropriate to an O-type star in the B3 spectrum. A correlation, between the velocity amplitudes of individual absorption lines and their wavelengths, is also shown to be inexplicable in terms of a conventional reflection effect.

119.009 Light-variation of HD 209813. C. Blanco, S. Catalano. Astron. Astrophys., Vol. 4, 482 - 486 (1970).

New light-curve, in U, B, V colours, of the spectroscopic binary HD 209813 are given. It is shown that the light variation cannot be explained in terms of eclipse phenomena or tidal distortion. The photometric period found is different from the spectroscopic one determined by Northcott (1947) from radial velocity observations.

119.010 H.D. 176318: A second epoch spectrographic orbit. D. P. Hube. Journ. Roy. Astron. Soc. Canada, Vol. 64, 98 - 103 (1970).

A new period is derived for the spectrographic binary H.D. 176318 using radial velocities measured in 1914–17, 1944, 1948, and 1963–66. Assuming a constant period, the remaining elements of the orbit in 1963–66 are compared with those in 1948. Possible changes in the elements are suggested but the evidence is not conclusive.

119.011 The double-lined spectroscopic binary ε Lupi (HD 136504). A. D. Thackeray. Monthly Notices, Roy. Astron. Soc., Vol. 149, 75 - 80 (1970).

The first spectroscopic orbit of the 3rd magnitude double-lined B3 binary ε Lupi is derived. The binary is the brighter component of a close visual binary showing signs of orbital motion in 64 years. The triple system probably belongs to the Sco-Cen association. A fourth distant component of A type has also been observed but is probably not physically related.

119.012 α Virginis A as a β CMa type variable. J. Smak. Acta Astron., Vol. 20, 75 - 91 (1970).

Radial velocity data of Baker (1909), Struve and Ebbighausen (1934), and Struve et al. (1958) are used to detect variations connected with the short-period (0.174 day) light variability reported by Shobbrook et al. (1969). The amplitude of radial velocity variations is about 17 km/sec. The basic physical properties of α Vir A are discussed.

119.013 The very slow spectrum, magnetic, and photometric variations of HD 9996. G. W. Preston, S. C. Wolff. Astrophys. Journ., Vol. 160, 1071 - 1076 (1970).

The Cr and Eu line intensities, magnetic field, and UBV photometric properties of the Ap star HD 9996 appear to vary in a period of 22 - 24 years. The phase relations between the various variable parameters are identical with those of well-known Ap stars with shorter periods. HD 9996 is also a single-line spectroscopic binary with a period of 273.2.

119.014 Zeeman measures of four double-lined peculiar A stars. P. S. Conti. Astrophys. Journ., Vol. 160, 1077 - 1082 = Contr. Lick Obs., No. 309 (1970).

Zeeman measures of four Ap stars which are double-lined spectroscopic binaries are presented. There is no evidence for a magnetic field larger than a few hundred gauss in any of the component stars. The absence of strong magnetic fields in these stars, all of the Mn type, is another fundamental difference between this type and the well-known periodic magnetic variables of the Si and Eu-Cr-Sr types. Other established differences concern the abundances of the light elements and the iron-peak group, and the rotational velocities.

119.015 **A programme for the determination of the orbital elements for spectroscopic binaries.**
F. C. Bertiau, J. Grobben.
Ric. Astron. Vaticana, Vol. 8, (No. 1), 1 - 31 (1969).

The programme determines the orbital elements of spectroscopic binaries by using Schlesinger's method in the case of normal eccentricity or Sterne's method in the case of small eccentricity. Provision is made to improve the ori - ginal adopted period, if this is wanted; for determining the systematic errors in the Γ-velocity, if observations made with different spectographs were used; and eventually to determine the elements of the second component either independently of the elements of the first component or not. The programme computes not only the elements but also the probable errors of each element and the theoretical radial velocities corres - ponding to the epoch of each observation.

119.016 **Orbital elements of the spectroscopic binary Zeta Scuti.**   J. Grobben, R. P. Michaelis.
Ric. Astron. Specola Vaticana, Vol. 8, (No. 2), 33 - 39 (1969).

The elements of the spectroscopic binary Zeta Scuti are determined using the new programme (see Abstr. 119.015) for the determination of the orbital elements of spectroscopic binaries.

119.017 **Frequencies of B, A, and K spectroscopic binaries.**   C. Jaschek, A. Gómez.
Asoc. Argentina Astron. Bol., No. 14 (see 012.020), p. 65 - 67 (1968).   In Spanish.

Radial velocity of EN (16) Lacertae.
See Abstr. 112.010.

Spectroscopic observations of the triple star ADS 14893.   See Abstr. 118.003.

Photometry of HD 211853 − a Wolf-Rayet eclips-ing binary.   See Abstr. 121.042.

The frequency of spectroscopic binaries in NGC 6475.   See Abstr. 153.003.

120 Variable Stars: Catalogues, Ephemerides, Miscellanea

120.001 **Reduktion der Beobachtungen veränderlicher Sterne.**   H. Bossen.
SuW, Vol. 9, 19, 21 (1970).

120.002 **Binocular variables.**
P. Moore, M. Ring.
Journ. British Astron. Ass., Vol. 80, 212 - 215 (1970).

120.003 **Het waarnemen van veranderlijke sterren.**
A. Mak.
Hemel en Dampkring, Vol. 68, 10 - 13, 47 - 51 (1970).

120.004 **Zur Methodik der Bestimmung von Entdeckungs-wahrscheinlichkeiten veränderlicher Sterne.**
G. A. Richter.
Astron. Nachr., Vol. 292, 9 - 16 (1970).

The Bernoullian statistics give for rare types of variable stars only inaccurate results concerning the proba-bility of discovery p. In these cases p may be reasonably calcu-lated by means of the "quality function" $Q(\Delta m, m)$, defined by Borgman. $Q(\Delta m, m)$ is the probability of the discovery of a magnitude difference Δm near the apparent magnitude m. Three methods are discussed which allow to find $Q(\Delta m, m)$.

120.005 **Lectures on variable stars. III, IV.**
R. F. Christy.
Journ. Roy. Astron. Soc. Canada, Vol. 64, 8 - 32 (1970).

120.006 **Variable star notes.**   M. W. Mayall.
Journ. Roy. Astron. Soc. Canada, Vol. 64, 69 - 72, 117 - 120 (1970).

120.007 **Flare star programmes 1970.**
F. M. Bateson.

Roy. Astron. Soc. New Zealand, Variable Star Section, Circ. No. 140, 1 pp. (1970).

120.008 **Sequence determination.**
F. M. Bateson, B. Menzies.
Roy. Astron. Soc. New Zealand, Variable Star Section, Circ. No. 148, 2 pp. (1970).

120.009 **Studies of variable stars.**   Yu. N. Efremov.
Vestn. AN SSSR, 1969, No. 11, p. 124 - 125.
In Russian. − Abstr. in Referativ. Zhurn. 51. Astron., 4.51.17 (1970).

120.010 **Variable star notes.**   M. W. Mayall.
Journ. Royal Astron. Soc. Canada, Vol. 64, 205 - 208 (1970).

120.011 **Variable stars.**   L. Detre.
Trans. IAU, Vol. 14A, (see 003.028), 259 - 299 (1970). − Report of Commission 27.

120.012 **Programme of cooperative observations of flare stars for 1970.**
A. D. Andrews, P. F. Chugainov.
Inform. Bull. Variable Stars (I.A.U. Commission 27), Konkoly Obs., Budapest, No. 416 (1970).

General Catalogue of Variable Stars. Volume 1. Constellations Andromeda − Grus.
See Abstr. 003.011.

Statistical Investigation of the Brightness of Irregu-lar and Semiregular Variables.   See Abstr. 003.066.

121 Eclipsing Variables

121.001 **Partial derivatives of eclipse functions α^{oc} and α^{tr}.**
I. Jurkevich.
Astrophys. Space Sci., Vol. 6, 4 - 12 (1970).

Expressions are given for partial derivatives of eclipse functions with respect to geometrical depth, p, and the ratio of radii, k. The derivatives are evaluated for critical combinations of p and k at which indeterminacies occur and the resulting expressions are listed. All expressions are given in a form suitable for numerical evaluation.

121.002 **On light variations of the binary system AX Mon.**
N. L. Magalashvili, I. I. Kumsishvili.
Byull. Abastumansk. Astrofiz. Obs. No. 37, p. 3 - 8 (1969).
In Russian.

Photoelectric observations of the binary star AX Mon were carried out in three colours U, B, V by means of a stellar electrophotometer of the Abastumani Astrophysical Observatory in 1962–1967. Periodic light variations in ultraviolet with an amplitude of about 0.4 mg and a period equal to that of orbital motion (232.5 day) were noticed.

121.003 **Two independent photoelectric light curves and solutions of W Ursae Minoris.**
E. J. Devinney, Jr., D. S. Hall, D. H. Ward.
Publ. Astron. Soc. Pacific, Vol. 82, 10 - 52 = Arthur J. Dyer Obs. Vanderbilt Univ. Repr. No. 49 (1970).

Results of two independent multicolor photoelectric studies of W UMi are presented. Good photometric elements are derived for each and found in very close agreement. Spectroscopic complications are discussed; the range of reasonable absolute dimensions is estimated, and the primary is shown to lie above the main sequence.

121.004 **Application of spectrophotometry to a binary star problem.** R. E. Wilson.
Publ. Astron. Soc. Pacific, Vol. 82, 146 - 150 = Astron. Contr. Univ. South Florida, Tampa, No. 24 (1970).

Photoelectric spectrophotometry may be used to improve the determination of the elements of many partially eclipsing binaries. Requirements and procedures are discussed for using this source of information.

121.005 **Photographic observations of S Equulei.**
N. Tashpulatov.
Peremennye Zvezdy, Vol. 17, 76 - 81 (1969). In Russian.

121.006 **Photographic light curve of the eclipsing binary FK Aquilae and its solution.**
M. E. Kiperman.
Peremennye Zvezdy, Vol. 17, 81 - 86 (1969). In Russian.

121.007 **Elements of BV 443 (DL Virginis).**
G. E. Erleksova, L. M. Vagina.
Peremennye Zvezdy, Vol. 17, 86 - 87 (1969). In Russian.

121.008 **ZZ Delphini.**
V. F. Karamysh.
Peremennye Zvezdy, Vol. 17, 87 - 90 (1969). In Russian.

121.009 **L'importance des minimums secondaires dans l'étude des étoiles variables à éclipses.**
I. Todoran.
L'Astronomie, 84e année, p. 4 - 10 (1970).

121.010 **The lithium isotope ratio in Delta Sagittae.**
A. M. Boesgaard.
Astrophys. Journ., Vol. 159, 727 - 729 (1970).

The apparent Li isotope ratio has been determined from measurements of the shift of the center of gravity of the Li doublet at λ6707 on two 4 Å mm^{-1} spectrograms of δ Sge. The ^{6}Li content is found to be 12 ± 6 percent of the total Li. High-energy spallation reactions predict about 30 percent ^{6}Li. The discrepancy can be understood in terms of post-main-sequence dilution.

121.011 **On the variability of AC Leonis.** G. Sabbatini.
Mem. Soc. Astron. Italiana, Nuova Serie, Vol. 41, 119 (1970). – Letter.

121.012 **Photometric elements of the eclipsing binary system RY Gem.** M. I. Kumsiashvili.
Soobshch. AN Grus. SSR, Vol. 56, No. 1, p. 69 - 78 (1969). In Russian. – Abstr. in Referativ. Zhurn. 51. Astron., 3.51.635 (1970).

121.013 **Determination of the shape and of the limb darkening at λ 4230 of the components of the eclipsing binary SZ Camelopardalis.** J. R. W. Heintze, J. Grygar.
Bull. Astron. Inst. Czechoslovakia, Vol. 21, 77 - 91 (1970).

Revised elements of the early-type binary system SZ Cam have been found from Wesselink's (1941) photographic light curve under the assumption that the stars can be represented by ellipsoids. The average ellipticity seems to be 0.410 instead of 0.305s. It turns out that the derived limb-darkening coefficient depends on the ellipticity of the components.

121.014 **Determination of the spectral and luminosity classes of some eclipsing binaries.**
S. M. Azimov.
Soobshch. Shemakhinsk. Astrofiz. Obs., vyp. (No.) 4, p. 30 - 44 (1969). In Russian.

121.015 **Photoelectric observations of BW Aqr.**
J. M. Kreiner.
Acta Astron., Vol. 20, 19 - 23 (1970).

This paper contains 81 yellow and 78 blue photoelectric observations of the eclipsing binary BW Aqr made with a 64-cm telescope at the Crimean Observatory. The observations were used for the determination of the secondary minima.

121.016 **The light-curves of 10 eclipsing variables.**
R. Szafraniec.
Acta Astron., Vol. 20, 25 - 46 (1970).

The writer's observations of the eclipsing variables: TT, WZ And; CZ Aqr; RY Aur; SY, BO, V687, V698 Cyg; RW, TX Gem, made in Cracow with Argelander's method have been worked out. So far only minima determined from these observations were published.

121.017 **Ergebnisse der Beobachtungen von Bedeckungsveränderlichen.** R. Diethelm, K. Locher.
Orion Schaffhausen, 28. Jahrgang, p. 21 - 22 (1970).

121.018 **Résultats des observations d'étoiles variables à éclipse.** R. Diethelm, K. Locher.
Orion Schaffhausen, 28. Jahrgang, p. 54 (1970).

121.019 **Photoelectric studies of the eclipsing binaries BV 845 and RW PsA.**
C. R. Chambliss.
Bull. American Astron. Soc., Vol. 2, 186 (1970). – Abstr. AAS.

121.020 On the application of nonlinear laws of limb
darkening to the determination of the orbital
elements of select eclipsing binary systems.
M. L. Cooper.
Bull. American Astron. Soc., Vol. 2, 189 (1970). – Abstr.
AAS.

121.021 UBV limb-darkening coefficients of S Cnc.
A. P. Linnell, D. D. Proctor.
Bull. American Astron. Soc., Vol. 2, 205 (1970). – Abstr.
AAS.

121.022 A spectrographic study of R Canis Majoris.
P. Galeotti.
Astrophys. Space Sci., Vol. 7, 87 - 92 (1970).

New spectroscopic elements of R Canis Majoris are
given in this paper, computed from 17 plates secured at the
Astronomical Observatory of Merate. A new orbit has also
been computed with the program by Bertiau, from 81 observations covering a period of about 40 years. Two models
of this binary system are proposed, based on two different
assumptions and on a newly determined mass function. It is
confirmed that the main peculiarity of this binary with
regard to the mass-luminosity relation is real: there is also
an evidence of mass loss from the Lagrangian point L_2 as
suggested by Kitamura. A suspected variation of the velocity
V_0 of the center of mass seems to indicate the presence of a
third body.

121.023 Light curve and photometric elements of AR Aurigae. K. T. Johansen.
Astron. Astrophys., Vol. 4, 1 - 10 (1970).

Photoelectric observations of the eclipsing binary AR
Aur have been carried out in four spectral regions at approximately the same wave-lengths as the Strömgren u, v, b, y
system and in two regions round Hβ. The peak wave-lengths
close to u, v, b, y are all located in the continuum of the spectrum where limb darkening coefficients are available from the
literature, and in these cases the photometric elements have
been determined. From the u, v, b, y and Hβ observations
spectral types and absolute magnitudes have been determined
for the components.

121.024 The multiple system of Algol.
H. Frieboes-Conde, T. Herczeg, E. Høg.
Astron. Astrophys., Vol. 4, 78 - 88 (1970).

An investigation was carried out at the Hamburg Observatory to clarify the structure of the multiple system Algol,
with special respect to possible unseen components. The extended series of astrometric observations by van de Kamp
et al. has been rediscussed by using improved spectroscopic
elements for the triple system, Algol AB-C. We found little
change in the 1$\overset{?}{.}$862 triple star orbit but a combined solution
for this motion and a hypothetical 32-year periodicity revealed
that the latter does not correspond to an orbital motion.
After excluding two hypothetical unseen components and the
possible light time effects caused by them, we put forward a
new description of the observed $O - C$ diagram for the times
of minima.

121.025 Photometric and absolute elements of EE Peg.
S. Catalano, M. Rodonò.
Astron. Astrophys., Vol. 4, 173 - 179 (1970).

The results obtained from our photoelectric observations,
especially concerning the three following points are reported:
1) the photometric period can be considered as constant;
2) the equatorial velocity of the primary component spectroscopically determined (Koch et al., 1963) agrees with that
evaluated from our photometric elements; 3) the absolute
elements lead to the conclusion that EE Peg is one of the few
detached systems with the primary component just coming

out from the main sequence.

121.026 Eight-colour photometric study of the eclipsing
binary 44 i Bootis.
M. Kurpinska, F. van't Veer.
Astron. Astrophys., Vol. 4, 253 - 262 (1970).

Eight-colour photometric observations of the eclipsing
system 44 i Bootis have been analysed. This well known triple
system is composed of an eclipsing contact binary β and a
brighter third companion α. In the integrated light of the
system $a + \beta$ the observations reveal a remarkable behaviour
of the colour-differences between the maxima and the minima.
In order to explain the colour differences as due to the presence of the third companion we computed the colour variation theoretically for a certain number of combinations of the
initial data.

121.027 The binary system χ^2 Hydrae = BV 722.
H. Mauder.
Astron. Astrophys., Vol. 4, 437 - 442 = Veröff. Remeis-
Sternw. Bamberg, Astron. Inst. Univ. Erlangen–Nürnberg,
Band 8, No. 87 (1970).

A radial velocity curve was obtained for the eclipsing
binary χ^2 Hydrae = BV 722 = HD 96314. The pair consists
of a B8 III–IV star of 5.2 solar masses and a B8.5 V star of
2.6 solar masses. The respective radii are 5.12 and 2.87
solar radii. The equatorial rotational velocity of the primary
component due to the bound rotation gives 135 km/s which
is well compatible with the observed broadening.

121.028 Photoelectric observations of 31 Cygni.
K. Gyldenkerne, K. T. Johansen.
Astron. Astrophys., Suppl. Series, Vol. 1, 129 - 148 (1970).

The long-period eclipsing binary 31 Cyg has been observed at the Copenhagen University Observatory in Brorfelde during the 1962 eclipse and in 1965. The observations
were made with UBV standard filters and several narrow-
band filters. A comparison is made with results obtained by
other observers. The atmospheric nature of the eclipse is
demonstrated by the fact that the fraction of the light of the
B-type secondary obscured at a certain phase outside totality
depends on the spectral region of observation. The epoch of
mid-eclipse is J. D. 2 437 685.65 and the period equals
3784 days.

121.029 Photoelectric observations of 32 Cygni.
K. T. Johansen, J. Rudkjobing, K. Gyldenkerne.
Astron. Astrophys., Suppl. Series, Vol. 1, 149 - 164 (1970).

The long-period eclipsing binary 32 Cyg has been observed at the Copenhagen University Observatory in Brorfelde during the 1959, 1962 and 1965 eclipses. The observations were made with UBV standard filters and several narrow-
band filters, and the combined light-curve is discussed. A
comparison is made with results obtained by other observers.
At least the 1965 eclipse seems to be total. The atmospheric
nature of the eclipse is demonstrated by the fact that at a
certain phase outside totality the absorbed fraction of the
B-star depends on the spectral region of observation.

121.030 WY Geminorum – an eclipsing system?
A. Cowley.
Publ. Astron. Soc. Pacific, Vol. 82, 329 - 333 = Contr. Kitt
Peak National Obs. No. 519 (1970).

Spectroscopic evidence suggests that WY Geminorum has
recently suffered an atmospheric eclipse. It is possible that the
star may have also undergone a photometric eclipse. A comparison with the spectroscopic features of VV Cephei is made.

121.031 Differential UBV photometry of β Lyrae, I.
L. P. Lovell, D. S. Hall.
Publ. Astron. Soc. Pacific, Vol. 82, 345 - 349 = Arthur J.

Dyer Obs., Vanderbilt Univ. Repr., No. 50 (1970).

121.032 The K line in Algol.   E. G. Ebbighausen.
Publ. Astron. Soc. Pacific, Vol. 82, 349 - 351 (1970).

During the period 1958-59 the writer obtained a rather large number of spectrograms of Algol at primary minimum at the Dominion Astrophysical Observatory in order to study the sharp metallic lines of Algol C which are so easily visible and numerous for about an hour on either side of the mid-primary minimum.

121.033 The Algol system RV Ophiuchi and the variation of its light curve with time.
K. Walter.
Astron. Astrophys., Vol. 5, 140 - 148 = Mitt. Astron. Inst. Univ. Tübingen, No. 120 (1970).

Photoelectric observations of the Algol variable RV Ophiuchi in B and V are discussed and a photometric solution for this system is given. A model of a gas stream is given by which the chief characteristics of the light curves are represented.

121.034 Photometric elements of the FT Orionis binary system.   S. Cristaldi.
Astron. Astrophys., Vol. 5, 228 - 231 (1970).

The light curve and the photometric elements of FT Orionis are given. Our observations show that the secondary minimum falls at phase $\varphi = 0^d746$ and that the eccentricity is $e = 0.400$. The two components of the system have nearly the same radius. The spectral type of the secondary component, obtained from the photometric data, is A3.

121.035 Origin of the ultra-short period binary WZ Sagittae.
S. C. Vila.
Nature, Vol. 225, 1229 - 1230 (1970).

It is shown that the short period binary WZ Sagittae was originally composed of a main sequence star of several solar masses and an orbiting planet of the size of present day Jupiter. In the subsequent giant phase the star ejected matter that was, in part, captured by the planet. It, then, moved closer to the star until the system acquired its present contact binary configuration.

121.036 Automatic photography of eclipsing variables.
G. O. Rawstron.
Sky Telescope, Vol. 39, 397 - 399 (1970).

121.037 GL Carinae.   G. R. Quast.
Inform. Bull. Southern Hemisphere, No. 15, p. 35 (1969).

121.038 Photoelectric observations of W Serpentis.
J. H. Walraven.
Inform. Bull. Southern Hemisphere, No. 15, p. 40 - 46 (1969).

121.039 V observations of W Serpentis.
A. R. Hogg.
Inform. Bull. Southern Hemisphere, No. 15, p. 47 (1969).

121.040 Photoelectric observations of W Serpentis.
H. Moreno, A. Gutiérrez-Moreno.
Inform. Bull. Southern Hemisphere, No. 15, p. 48 (1969).

121.041 Ergebnisse der Beobachtungen von Bedeckungsveränderlichen.   R. Diethelm, K. Locher.
Orion Schaffhausen, 28. Jahrgang, p. 90 (1970).

121.042 Photometry of HD 211853 – a Wolf-Rayet eclipsing binary.   K. Stępień.
Acta Astron., Vol. 20, 117 - 122 (1970).

An additional two colour photometry of HD 211853 proves that this star is an eclipsing binary. The light curve shows primary minimum 0.06 of a magnitude deep. An estimation of the inclination of its orbit is made and the value 73° is found. Hence, masses of the components are 11.5 and 33 solar masses.

121.043 Photometric observations of BE Vulpeculae.
I. Semeniuk.
Acta Astron., Vol. 20, 123 - 136 (1970).

Photoelectric observations of the 456 V and 443 B differential magnitudes of the eclipsing variable BE Vulpeculae were obtained and used to determine the elements of the photometric orbit of the system.

121.044 The photographic light curve of the supergiant eclipsing variable BL Telescopii.
S. Gaposhkin.
Astron. Journ., Vol. 75, 641 - 642 (1970).

The photometric elements and whole light curve of the unusual variable BL Telescopii are presented, based on observations taken from 1891 to 1953.

121.045 Observations of EL Triangulum Australis.
R. R. D. Austin.
Southern Stars, Vol. 23, 97 - 100 (1970).

The given photometric data confirm Hoffmeister's opinion that the star is of β Lyrae type, showing the classical rounded maxima and the shallow rounded secondary minimum.

121.046 Contributions to the interpretation of the light curves of close binary systems. V. The theoretical light curve and the method of element determination.
V. Ureche.
Stud. Cerc. Astron., Vol. 15, 9 - 16 (1970).

Using the results of previous papers we shall give the necessary formulae for the computation of the theoretical light curve. We shall indicate also the method of element determination, starting from the photometric observations.

121.047 Approximating functions for eclipsing binary solutions.   H. Minţi.
Stud. Cerc. Astron., Vol. 15, 61 - 64 (1970).

The paper advances a complete program for the calculation of the Russell functions α^{oc} and α^{tr} based upon the polynomial approximation functions.

121.048 Détermination des éléments de la binaire à éclipse ϵ Coronae Australis.
R. Dinescu, A. Dumitrescu.
Stud. Cerc. Astron., Vol. 15, 65 - 71 (1970).

Dans ce travail, les auteurs présentent le calcul des éléments de la binaire à éclipse ϵ CrA de type W UMa. On a employé les observations photoélectriques effectuées par S. Tapia, en 1967, à l'Observatoire Inter-American de Cerro Tololo (les observations dans le système V seulement).

121.049 The transit-time effect in Wolf-Rayet binaries.
J. I. Castor.
Astrophys. Journ., Vol. 160, 1187 - 1189 (1970).

It is demonstrated in this note that the transit-time effect in Wolf-Rayet binaries is negligible in all cases of interest. This conclusion is reached by calculating the exact motion of particles ejected from the Wolf-Rayet component of the binary system.

121.050 CV Serpentis has stopped eclipsing!
L. V. Kuhi, F. Schweizer.
Astrophys. Journ. (*Letters*), Vol. 160, L185 - L187 (1970).

The Wolf-Rayet eclipsing binary CV Ser has been observed around the cycle in the light of both continuum and emission lines, and no eclipses have been found.

121.051   The eclipsing triple system VV Orionis.
    G. Beltrami, P. Galeotti.
Mem. Soc. Astron. Italiana, Nuova Serie, Vol. 41, 167 - 175 (1970).

A spectroscopic orbit is computed for the primary couple, deduced from 26 plates (dispersion 34.3 Å/mm) secured at Merate from January 1966. The orbit of the third body was evaluated from all the residuals available, covering about 180 cycles. The masses of the three bodies have been computed from the newly determined mass functions and from a rough estimate of the mass ratio of the primary couple, deduced from the semiamplitude of the secondary component.

121.052   Un atlante di curve di luce di binarie ad eclisse.
    M. G. Fracastoro.
Mem. Soc. Astron. Italiana, Nuova Serie, Vol. 41, 239 - 241 (1970).

121.053   Narrow-band electrophotometry of the eclipsing binary of Wolf-Rayet type CQ Cep.
Ch. Challilulin, A. M. Tsherepashuk.
Astron. Tsirk., No. 551, p. 5 - 7 (1970).   In Russian.

121.054   Photometric double stars.   F. B. Wood.
    Trans. IAU, Vol. 14A, (see 003.028), 491 - 513 (1970). – Report of Commission 42.

121.055   Photoelectric minima of eclipsing variables.
    C. Popovici.
Inform. Bull. Variable Stars (I.A.U. Commission 27), Konkoly Obs., Budapest, No. 419 (1970).

121.056   Photoelectric period for BV 346.   R. C. Tate.
    Inform. Bull. Variable Stars (I.A.U. Commission 27), Konkoly Obs., Budapest, No. 438 (1970).

121.057   New elements for SS Ari.   W. Braune.
    Inform. Bull. Variable Stars (I.A.U. Commission 27), Konkoly Obs., Budapest, No. 440 = BAV Mitt. No. 22 (1970).

121.058   The period of TW Draconis from 1880 to 1970.
    E. Pohl.
Inform. Bull. Variable Stars (I.A.U. Commission 27), Konkoly Obs., Budapest, No. 443 (1970).

121.059   A simple method of determination of photometric orbital elements.   M. I. Lavrov.
Astron. Tsirk., No. 559, p. 5 - 7 (1970).   In Russian.

121.060   Multi-color polarimetry of Beta Lyrae.
    G. V. Coyne.
Ric. Astron. Specola Vaticana, Vol. 8, (No. 5), 85 - 102 (1970).

Polarimetric observations of Beta Lyrae in seven colors are given. The variation in the polarization with the phase of the light curve is explained by scattering from free electrons associated with a flattened disk about the secondary. In order, however, to explain the observations at primary eclipse a tem - porary increase in both the electron density and the opacity of the gas associated with the disk is required. The polariza - tion at primary eclipse is approximately two times greater than that expected from the known decrease in the non-polar - ized light of the system. Huang has proposed a mass exchange mechanism which would satisfy these requirements; namely, a temporary increase in the electron density and in the opaci - ty of the gas.

121.061   Variable polarization in U Ophiuchi.
    G. V. Coyne.
Ric. Astron. Specola Vaticana, Vol. 8, (No. 6), 105 - 116 (1970).

Polarimetric observations of U Ophiuchi show variations as large as about 0.6% in the percentage of polarization in the ultraviolet and red regions of the spectrum, while at interme- diate wavelengths the polarization remains sensibly constant. The largest polarization changes occur near times of eclipse of one or other of the two components. When the variable po- larization is considered in light of information derived from photometric and spectroscopic investigations, it appears that there may be dynamic processes of mass exchange and mass loss taking place in U Oph.

121.062   Photometric observations of RR Cen.
    G. R. Quast.
An. Acad. Brasil. Cienc., Vol. 41, (No. 2), 141 - 147 (1969).
Concerns the determinations of atmospheric extinction and photometric measurements of the eclipsing binary RR Cen. A great number of observations were accumulated, permitting the determination of a precise light curve as well as an epoch of primary minimum.

121.063   Theoretical continuous and line spectra of stars in a close binary system.   P. F. Buerger.
Thesis, Ohio State Univ., Columbus. [Available from Univ. Microfilms, Ann Arbor, Mi.], 89 pp. (1969).

Models have been constructed for rotationally and tidal- ly distorted stars as they appear in synchronous, circular orbits in close binaries. The continuous radiation emitted in the direction of the observer is computed at several wave- lengths. The equivalent width of an absorption line superposed on the gray atmosphere is computed.

121.064   Orbit of Epsilon Coronae Australis.
    C. Hernández, J. Sahade.
Asoc. Argentina Astron. Bol., No. 12, (see 012.019), p. 15 - 16 (1968). – Abstract.

121.065   The problem of the determination of masses in the Algol systems.   A. Ringuelet-Kaswalder,
J. Sahade.
Asoc. Argentina Astron. Bol., No. 12, (see 012.019), p. 25 - 26 (1968).   In Spanish. – Abstract.

121.066   The velocity of the Delta Librae system.
    J. Sahade, C. A. Hernández, T. Fay, H. Cohen.
Asoc. Argentina Astron. Bol., No. 12, (see 012.019), p. 26 (1968). – Abstract.

121.067   R, I photoelectric observations of ε CrA.
    C. A. Hernández.
Asoc. Argentina Astron. Bol., No. 14 (see 012.020), p. 68 (1968). – Abstract.

121.068   Photoelectric photometry of S Velorum.
    R. F. Sisteró.
Asoc. Argentina Astron. Bol., No. 14 (see 012.020), p. 103 (1968).   In Spanish. – Abstract.

The nature of contact binaries. See Abstr. 117.023.

The structure of W Ursae Majoris systems.
See Abstr. 117.024.

CV Ser – a peculiar Wolf-Rayet binary.
See Abstr. 119.004.

Photographische Reihenbeobachtungen von RR-Ly- rae-Sternen und Bedeckungsveränderlichen. See Abstr.122.116.

122 Physical Variables, Flare Stars, Pulsation Theory

122.001   The nature of the Beta Cephei phenomenon.
J. R. Percy.
Astrophys. Journ., Vol. 159, 177 - 182 (1970).
Observational data indicate that the β Cephei instability strip lies about 1.85 mag above the main sequence at its upper end and 1.35 mag above the main sequence at its lower end. It terminates abruptly at or near $M_v = -3.0$. The true instability strip may be less inclined to the main sequence. Evidence suggests that the instability strip lies coincident with the terminal phases of hydrogen burning in the star's evolution. It is hypothesized that the β Cephei phenomenon arises as a result of a semiconvective zone within the star.

122.002   Photometry of BL Lacertae with a time resolution of 15 seconds.   R. Racine.
Astrophys. Journ. (Letters), Vol. 159, L99 - L103 = Commun. David Dunlap Obs. Univ. Toronto, No. 242 (1970).
New photometric observations of the strange object BL Lac = VRO 42.22.01 are reported with the following results: (1) The brightness of BL Lac fluctuates by 0.1 mag over a few hours. (2) Flickers of amplitudes $\Delta V \simeq 0.03$ mag and durations as short as $\Delta t = 2$ min are detected. (3) The $(B - V)$ and $(U - B)$ colors of BL Lac become redder with decreasing brightness. Similarities between BL Lac, N-type galaxies, and Seyfert nuclei are mentioned.

122.003   The three-color photographic photometry of eleven RW Aur type stars.   L. N. Mosidze.
Byull. Abastumansk. Astrofiz. Obs. No. 37, p. 13 - 34 (1969). In Russian.
The results of three-color photometry of eleven RW Aur type stars, based on observations obtained during 1960 – 64 are given. The stars studied are: RW Aur, RY Tau, BP Tau, V_{451} Ori, HK Ori, CO Ori, GW Ori, MHα 265-3, SV Cep, BH Cep, BO Cep.

122.004   Spectral changes of CH Cygni at its outburst.
M. V. Dolidze, G. N. Jimsheleishvili.
Byull. Abastumansk. Astrofiz. Obs. No. 37, p. 35 - 38 (1969). In Russian.
The spectra of the SRa variable CH Cygni at its outburst in autumn 1967 are investigated. The changes in the spectra and in the structure of the radiation are described.

122.005   Spectral classification of some RR Lyrae type stars.
I. F. Alania.
Byull. Abastumansk. Astrofiz. Obs. No. 37, p. 39 - 42 (1969). In Russian.
Spectral classifications of 9 RR Lyrae stars in various light phases have been made on the basis of spectral material obtained with the 70-cm meniscus telescope of Abastumani Observatory. The hydrogen absorption lines and the ionized K line have served as criteria of classification.

122.006   Radioastronomie der Flaresterne.
G. Feix.
SuW, Vol. 9, 40 - 42 (1970).

122.007   The light variation of γ Coronae Borealis.
J. R. Percy.
Publ. Astron. Soc. Pacific, Vol. 82, 126 - 128 = Commun. David Dunlap Obs. Univ. Toronto, No. 250 (1970).
The brightness of γ Coronae Borealis varies in both blue and yellow light by about $0\overset{m}{.}05$ in a time scale of about $0\overset{d}{.}03$. The star appears to be a δ Scuti star. It has the earliest spectral type of any known δ Scuti star, and is the only one for which a direct determination of the mass exists.

122.008   HV 13055 as a symbiotic variable.
P. W. Hodge, F. W. Wright.
Publ. Astron. Soc. Pacific, Vol. 82, 135 - 138 (1970).
HV 13055 in the Large Magellanic Cloud has a light curve, amplitude, cycle length, and color index that are similar to those for symbiotic variables. Its known absolute magnitude agrees with available spectroscopic evidence for galactic symbiotic stars that indicates that they are giants.

122.009   Some questions of the exploration of instationary stars.   V. G. Gorbatskij, L. V. Mirzoyan.
Stars, Nebulae, Galaxies, Symposium Byurakan 1968, (see 012.002), p. 83 - 107 (1969).   In Russian.

122.010   On the continuous radiation of T Tauri stars.
Z. A. Ismailov.
Stars, Nebulae, Galaxies, Symposium Byurakan 1968, (see 012.002), p. 135 - 138 (1969).   In Russian.

122.011   On the statistics of flare objects.
V. A. Ambartsumyan.
Stars, Nebulae, Galaxies, Symposium Byurakan 1968, (see 012.002), p. 283 - 291 (1969).   In Russian.

122.012   Period—radius relation for RR Lyrae stars in the galactic field.   M. S. Frolov.
Peremennye Zvezdy, Vol. 17, 3 - 14 (1969).   In Russian.
A revision is made of relations proposed by the author earlier for RR Lyrae stars. These relations were used to determine radii of 113 RR Lyrae ab stars. The period-radius relation was obtained.

122.013   Application of the dispersion analysis for the search of variable stars.   I. M. Ishchenko.
Peremennye Zvezdy, Vol. 17, 15 - 21 (1969).   In Russian.
An analysis is made of errors in the determination of the brightness of 327 stars measured on 43 plates of the central part of Ori T2. The plates were taken in 1962 - 1963 with the normal astrograph at the Tashkent Observatory and were aimed to verify Parenago's conclusions on multiplicity of microvariables in this region. After systematic errors were taken into account the function of the distribution of the measurement errors was quite satisfactory represented by a normal curve. It was indicated that Parenago's conclusion on multiplicity of microvariables in the Ori T2 region was erroneous.

122.014   Photographic observations of eight cepheids.
E. N. Makarenko.
Peremennye Zvezdy, Vol. 17, 31 - 55 (1969).   In Russian.
Results of photographic observations of the eight cepheids with small amplitudes of light variation X Lac, CR Cep, V 496 Aql, IX Cas, FN Aql, DD Cas, SY Aur, SZ Cas are given. Mean light curves are obtained.

122.015   On three Mira Ceti type stars in Aquila.
V. M. Kovalenko.
Peremennye Zvezdy, Vol. 17, 56 - 63 (1969).   In Russian.
The light of ZZ Aql, BE Aql and CW Aql was estimated on Moscow plates. New elements are obtained for BE Aql; it is found that CW Aql is a Mira Ceti type star. Charts of surroundings, photographic magnitudes of the comparison stars and epochs of light maxima are given. Mean light curves are obtained.

122.016   Observations of RZ Lyrae.
S. I. Belik.

Peremennye Zvezdy, Vol. 17, 93 - 96 (1969).   In Russian.

122.017   Observation of a flare in the blue variable G44-32.
B. Warner, G. W. van Citters, R. E. Nather.
Nature, Vol. 226, 67 - 68 (1970).

We confirmed that G44-32 is a variable star of small amplitude; a full analysis of our results will be reported later. Here we wish to describe the observations made between 10 h 07 m 30 s and 10 h 25 m 30 s UT on the night of February 7, 1970, during which time G44-32 displayed a sharp and unexpected increase in brightness which we believe is due to flare activity.

122.018   A high-dispersion velocity curve of U Sagittarii.
T. S. Jacobsen.
Astrophys. Journ., Vol. 159, 569 - 573 (1970).

A radial-velocity curve based on fourteen high-dispersion coudé spectrograms of the cepheid U Sgr has been constructed. This star has been considered a member of the open cluster M25. Spectra of two cluster members also showed sharp interstellar Na I lines at an average velocity of –9.78 ± 0.33 km sec^{-1}. The radial velocity of the cluster as a whole, found from 30 - 40 members by other observers, shows a remarkably close agreement with the systemic velocity of the cepheid, thus greatly strengthening the case for its cluster membership.

122.019   Pulsation analysis for stars in thermal imbalance.
N. R. Simon.
Astrophys. Journ., Vol. 159, 859 - 870 (1970).

The effects of thermal imbalance on stellar pulsational stability have largely been ignored in the literature. Here we have used the linear, quasi-adiabatic pulsation theory to make a preliminary investigation of such effects.

122.020   Pulsation and the radius of the southern Cepheid S Nor.   M. Breger.
Astron. Journ., Vol. 75, 239 - 243 (1970).

New velocity and light measurements of S Nor in the galactic cluster NGC 6087 are reported. The velocity curve including the secondary hump is very similar to that of β Dor, another 10-d Cepheid. An analysis of all the light data does not support a previous suggestion of an increasing period. A constant period of 9^{d}7542 is derived and a mean radius of 60 R$_\odot$ is suggested.

122.021   On the nature of S Vulpeculae.   J. D. Fernie.
Astron. Journ., Vol. 75, 244 - 245 (1970).

UBV observations of S Vul are presented. On the basis of all the available evidence it is concluded that Wachmann's suggestion that S Vul is a classical Cepheid is most likely correct. In this case it is the Cepheid of longest period known in the Galaxy. Basic data for the star are determined.

122.022   Chi Cyg im Maximum 1969.   E. Heiser.
VdS Nachrichtenblatt, 19. Jahrgang, p. 43 - 44, 45 (1970).

122.023   The beat period of VX Puppis.   R. S. Stobie.
Observatory, Vol. 90, 20 - 23 (1970).

122.024   The nebulosity near the infra-red, OH and H_2O source, VY CMa.   M. W. Feast.
Observatory, Vol. 90, 24 - 25 (1970). – Note.

122.025   Radial velocities of companions to Mira variables.
M. W. Feast.
Observatory, Vol. 90, 25 - 27 (1970). – Concerning U Men and R Car.

122.026   On the variability of V429 Orionis.
H. S. Mahra.

Observatory, Vol. 90, 28 - 30 (1970). – Note.

122.027   HR 4768 – a cepheid variable of small amplitude.
R. S. Stobie, J. B. Alexander.
Observatory, Vol. 90, 66 - 68 (1970). – Note.

122.028   Flare stars in the Orion nebula region.
G. Haro, E. Chavira.
Bol. Obs. Tonantzintla y Tacubaya, Vol. 5 (No. 32), 59 - 78 (1969).

The present results deal mainly with the multiple exposure photographic material obtained during the period comprised between December 1965 up to February 1969. The plates were centered in the Trapezium stars covering a usable area of 16 square degrees, the average ultraviolet limiting magnitude being ~ 17.5. 254 flare stars are marked and numbered.

122.029   On the photoelectric photometry of some Orion flare stars.   G. Haro.
Bol. Obs. Tonantzintla y Tacubaya, Vol. 5 (No. 32), 79 - 88 (1969).

Eleven flare stars have been observed by Mendoza (1968) and Walker (1969) during their photoelectric work, having in common only one flare star. In four of the eleven stars, we detected Hα in emission during the non-flare stage but only one shows in our plates a permanent but variable negative $U-B$ color. This last star seems to lie below the main-sequence.

122.030   On the causes of the spectral change of RW Aur.
Z. A. Ismailov.
Soobshch. Shemakhinsk. Astrofiz. Obs., vyp. (No.) 4, p. 45 - 65 (1969).   In Russian.

122.031   On the variability of O, B stars within nebulae.
V. I. Kardopolov, V. S. Shevchenko.
Astron. Tsirk., No. 540, p. 6 - 8 (1969).   In Russian.

122.032   On the unusual variable HR Aurigae.
P. N. Kholopov.
Astron. Tsirk., No. 542, p. 4 - 5 (1969).   In Russian.

122.033   MU Cas is a new bright ultrashort-period variable star.   G. A. Lange.
Astron. Tsirk., No. 542, p. 5 - 6 (1969).   In Russian.

122.034   Electrophotometric observations of EV Lac and V1216 Sgr.   K. A. Grigorian, R. A. Vardanian.
Astron. Tsirk., No. 542, p. 7 - 8 (1969).   In Russian.

122.035   Photometry of bright southern Cepheids.
R. S. Stobie.
Monthly Notices, Roy. Astron. Soc., Vol. 148, 1 - 15 (1970).

U, B, V. observations of 29 bright southern Cepheids are presented. The new observations enabled existing periods to be checked and revised periods were derived where necessary. Apart from SZ Mon and RY Sco both of which are known to have a variable period, the remaining Cepheids showed no evidence of any variation in the period. A discussion of the effect of a blue companion on the two-colour plot of a Cepheid indicated that three stars showed evidence of such a companion.

122.036   Fluorescent lines of SiH molecule in the spectra of long-period variable stars.   H. Maehara.
Publ. Astron. Soc. Japan, Vol. 22, 119 - 123 (1970).

Two emission lines at λλ 4138 and 4178 which appear in the spectra of long-period variable stars are identified with the fluorescent lines of SiH molecule excited by Hδ emission.

122.037   On the problem of the non-equilibrium character

of the continuous radiation of classical cepheids in the brightness maximum. N. N. Yakimova.
Astron. Zhurn. Akad. Nauk SSSR, Vol. 47, 297 - 307 (1970). In Russian. – English translation in Soviet Astron., AJ, Vol. 14, No. 2.

The value of the possible radiation excess above equilibrium in the brightness maximum of classical cepheids of the Galaxy is estimated statistically, taking into account the existing phase displacement between the moment of the largest light intensity and that of the smallest star radius.

122.038 **UBV photometry of the variable star γ Pegasi.**
M. Jerzykiewicz.
Postępy Astron., Vol. 18, 175 - 177 (1970). In Polish. Abstr. Polish Astron. Soc.

122.039 **UBV photometry of the variable star δ Ceti.**
M. Jerzykiewicz.
Postępy Astron., Vol. 18, 179 - 180 (1970). In Polish. Abstr. Polish Astron. Soc.

122.040 **Pulsation and radius of the Cepheid S Nor in NGC 6087.** M. Breger.
Bull. American Astron. Soc., Vol. 2, 184 (1970). – Abstr. AAS.

122.041 **An incomplete history of Hamburg-Bergedorf variable No. 475.** M. R. Chartrand III, F. M. Stienon.
Bull. American Astron. Soc., Vol. 2, 187 (1970). – Abstr. AAS.

122.042 **Beta Cephei stars: A linear nonadiabatic analysis of radial oscillations.**
W. R. Davey.
Bull. American Astron. Soc., Vol. 2, 190 (1970). – Abstr. AAS.

122.043 **Random variable stars.**
T. J. Deeming.
Bull. American Astron. Soc., Vol. 2, 190 (1970). – Abstr. AAS.

122.044 **The nature of the variations of CH Cygni.**
N. Hamilton, J. Schwarz, J. Veverka, M. Mattei, W. Liller.
Bull. American Astron. Soc., Vol. 2, 196 (1970). – Abstr. AAS.

122.045 **The light variations of AG Peg from 1962 to 1967.**
T. S. Belyakina.
Astrofizika, Vol. 6, 49 - 64 (1970). In Russian. – English translation in Astrophysics, Vol. 6, No. 1.

The results of a three-color photoelectric photometry of AG Peg over the period 1964-67 are given. A combined photoelectric light curve was constructed from the observations of 1962-67. This curve is a sinusoid with a period of 800^d and an amplitude of $0.^m15$. The light variations of AG Peg are synchronous in all the regions of the spectrum, the mean amplitude of light variations is $0.^m3$ in yellow and blue, and $0.^m5$ in ultraviolet. They are assumed to be caused by the orbital movement of a cold component M 3 III.

122.046 **Polarimetric investigation of long-period variable stars.** R. A. Vardanian.
Astrofizika, Vol. 6, 77 - 87 = Byurakan Astrophys. Obs. No. 54 (1970). In Russian. – English translation in Astrophysics, Vol. 6 , No. 1.

The results of a quantitative analysis of polarimetric observations of late type (M, N, R. S) stars are given. The connections between the polarization degree, the brightness

and the wavelength for the stars AB Cyg, AK Peg, V CVn and RX Boo are investigated.

122.047 **Photoelectric observations of the ultra-short period variable DT Velorum.** P. N. J. Wisse.
Astron. Astrophys., Vol. 4, 419 - 422 (1970).

Observations of the ultra-short period RR Lyrae star DT Velorum are presented here. V- and $(V-B)$-curves and mean colours in the Walraven-system are given. A suggestion is made about the interstellar reddening and absolute magnitude. The estimated distance is about 700 pc.

122.048 **Ultrashort-period variables and the masses of blue stragglers in the old disk population.**
O. J. Eggen.
Publ. Astron. Soc. Pacific, Vol. 82, 274 - 292 (1970).

The purpose of the present note is to examine (1) the population distribution of the ultrashort-period cepheids and (2) the evidence that those in the old disk population may contribute to our knowledge of the masses of the "blue stragglers" in that population.

122.049 **Interstellar reddening of RR Lyrae.**
D. H. McNamara, T. M. Helm, S. K. Wilcken.
Publ. Astron. Soc. Pacific, Vol. 82, 293 - 298 (1970).

$uvby\beta$ and (khg) photometric observations of twelve field stars near RR Lyrae are used to derive an accurate value of the interstellar reddening of the variable. The average color excess determined from three different calibrations is $E(b-y) = 0.^m011$ or $E(B-V) = 0.^m016$. The reddening-free $(B-V)_{min}$ value of RR Lyrae is compared with the average $(B-V)_{min}$ value of eleven RR Lyrae stars of similar period at high galactic latitudes. The results indicate that the average reddening near the north galactic pole is extremely small.

122.050 **Photometric and spectroscopic observations of AG and HR Carinae.** H. E. Bond, A. U. Landolt.
Publ. Astron. Soc. Pacific, Vol. 82, 313 - 320 = Contr. Louisiana State Univ. Obs., Baton Rouge, No. 35 = Contr. Cerro-Tololo Inter-American Obs. No. 102 (1970).

Photometric and spectroscopic observations are presented for the two P Cygni-like slow variable stars AG and HR Carinae.

122.051 **Spectra of the Cepheid HR 8157.**
H. A. Abt, S. G. Levy.
Publ. Astron. Soc. Pacific, Vol. 82, 334 - 338 = Contr. Kitt Peak National Obs. No. 525 (1970).

This small-amplitude Cepheid, newly discovered by Millis, is a member of a known, closely spaced visual system with a possible period of 30 years. The velocity curve of the Cepheid is also of small amplitude and with a mean velocity that varies with time, probably due to motion in the visual system. No features in the blue violet or filling in of the absorption lines are definitely attributable to the visual companion.

122.052 **HR 8880: A low-amplitude Delta Scuti star.**
R. L. Millis, D. T. Thompson.
Publ. Astron. Soc. Pacific, Vol. 82, 352 - 355 (1970). Note.

122.053 **Erratum: Observations of the infrared object, VY Canis Majoris.** [Astrophys. Journ., Vol. 158, 619 - 628 (1969)]
A. R. Hyland, E. E. Becklin, G. Neugebauer, G. Wallerstein.
Astrophys. Journ., Vol. 160, 381 (1970).

122.054 **Zirconium isotopic abundances in R Cygni.**
B. F. Peery, Jr., R. F. Beebe.
Astrophys. Journ., Vol. 160, 619 - 625 = Publ. Goethe Link

Obs., Indiana Univ., *Bloomington,* No. 102 (1970).

Relative isotopic abundances of zirconium have been found from the spectrum of the long-period-variable S star, R Cygni, by matching computed and observed profiles. These results are close to the abundances to be expected from production of Zr by the *s*-process after extensive neutron exposure. Improved molecular constants of $^{90}Zr^{16}O$ have been determined.

122.055 The variable star BD +63°141.
B. Ljunggren, T. Oja.
Astron. Astrophys., Vol. 5, 113 - 115 (1970).

The star BD +63°141 has been found to be a semi-regular variable of the type SRd with a period of 134 ± 4 days. The range of variation is in V about $0.^m3$, and in $B - V$ about $0.^m15$.

122.056 The zero-point of the period-luminosity-relation of Cepheids.   U. Geyer.
Astron. Astrophys., Vol. 5, 116 - 126 (1970).

The zero-point of the period-luminosity-relation has been redetermined by applying the method of secular parallaxes. The determination presented here is based on proper motions, radial velocities and photometric data of about a hundred Cepheids of population I. This approach was encouraged by the availability of improved proper motions and photometric data. The effects of galactic rotation, of incorrect precession, and of a motion of the equinox were taken into account. The radial velocities were discussed with respect to solar motion, differential galactic rotation and K-term. Adopting the slope of Kraft's period-luminosity-relation, the zero-point is found to be $1.^m88 ± 0.^m45$ (m.e.). With the slope of Fernie's period-luminosity-relation the zero-point is found to be $-2.^m05 ± 0.^m45$ (m.e.).

122.057 On the distance scale of the near-by classical Cepheids.   J. Jung.
Astron. Astrophys., Vol. 6, 130 - 137 (1970).

On the ground of new proper motions data, we have checked the distance scale of the Cepheids derived from the cluster Cepheid method, by applying the method of maximum likelihood (Jung, 1969), to a sample of 33 near-by Cepheids ($r \lesssim 1$ kpc). Special attention has been paid to the determination of interstellar absorption. The corrections of extrinsic unreddening have been discussed. The method of maximum likelihood gives a systematic correction to the period-luminosity relation as given by Sandage and Tammann (1968), of about 0.4 (± 0.4, standard deviation), after taking into account the bias caused by errors on the proper motions.

122.058 A simple model for Cepheid variability.
T. J. Rudd, R. M. Rosenberg.
Astron. Astrophys., Vol. 6, 193 - 205 (1970).

In this paper we propose to construct a simple one-zone model of Cepheid variables which is mathematically tractable, which is based on the known equations governing the pulsation process, and which reproduces qualitatively and quantitatively the pulsation phenomenon.

122.059 DI Car – A Cepheid?
W. Seggewiss.
Astron. Journ., Vol. 75, 345 - 346 (1970).

New photoelectric data of the variable star DI Car lead to the conclusion that it is a Cepheid, possibly of population II, instead of an RW Aur variable as previously supposed. The period is 29.210 days.

122.060 Catalogue des étoiles variables du type U Geminorum.   M. Petit.
Ciel et Terre, Vol. 86, 229 - 244 (1970).

122.061 Observations of faint variable stars near the North Galactic Pole.   L. Plaut.
Astron. Astrophys., Vol. 6, 486 - 487 (1970).

Magnitudes of 18 variable stars were determined from 46 pg and 10 pv plates taken by Borgman with the 122 cm (48″) Palomar Schmidt telescope. Earlier results by Kinman, Wirtanen and Janes (1966) were confirmed.

122.062 Continual photoelectric monitoring of flare stars. V. EV Lac and UV Cet (1969).
K. Ichimura, T. Noguchi, E. Watanabe.
Tokyo Astron. Bull., Second Series, No. 198, p. 2299 - 2305 (1970).

A continual photoelectric monitoring of the flare star EV Lac was done from 2 to 19 September 1969, and of UV Cet from 18 to 20 September and from 3 to 12 October 1969, with the 91 cm reflector of the Okayama Station.

122.063 Mira Ceti.   G. Comello.
Hemel en Dampkring, Vol. 68, 54 - 56 (1970).

122.064 De autocorrelatiemethode toegepast op waarnemingen van T Vul.   G. W. E. Beekman.
Hemel en Dampkring, Vol. 68, 68 - 70 (1970).

122.065 66 Ophiuchi (BD +4°3570).
F. M. Bateson.
Roy. Astron. Soc. New Zealand, Variable Star Section, Circ. No. 142, 2 pp. (1970).

The purpose of this note is to draw the attention of observers to 66 Ophiuchi which, on the basis of observations in 1969 August, is probably an early type flare star.

122.066 U Crucis.
F. M. Bateson, A. F. Jones, B. Menzies.
Roy. Astron. Soc. New Zealand, Variable Star Section, Circ. No. 141, 4 pp. (1970).

Observations of the Mira Ceti type variable, U Crucis, from 1961 July 24 to 1968 July 26 are now published. From eight maxima a period of 342.6 days is determined.

122.067 RY Piscis Austrini.
F. M. Bateson, A. F. Jones, B. Menzies.
Roy. Astron. Soc. New Zealand, Variable Star Section, Circ. No. 144, 5 pp. (1970).

Observations from 1955 December 20 to 1968 July 26 are tabulated in ten day means. A list of 21 observed maxima is given.

122.068 The semi-regular variable, T Reticuli.
F. M. Bateson, A. F. Jones, B. Menzies.
Roy. Astron. Soc. New Zealand, Variable Star Section, Circ. No. 145, 4 pp. (1970).

V magnitudes, determined photoelectrically, for a sequence for T Ret are listed.

122.069 66 Ophiuchi (B.D. +4° 3570).   F. M. Bateson.
Roy. Astron. Soc. New Zealand, Variable Star Section, Circ. No. 152, 3 pp. (1970).

122.070 A new RR Lyr variable with very high velocity.
J. B. Alexander, A. D. Thackeray.
Monthly Notices, Roy. Astron. Soc., Vol. 149, 59 - 64 (1970).

The star C.P.D. –74°214 (BV 1041) has been shown photometrically and spectroscopically to be an RR Lyr variable of type c with short period (0.287 days), very high radial velocity (+304 km s^{-1}) and Preston index Δs about 7. From the small proper motion an absolute magnitude brighter than +1.9 (M_v) is estimated.

122.071 UBV photometry of the β Cephei type variable

stars. I. γ Pegasi. M. Jerzykiewicz.
Acta Astron., Vol. 20, 93 - 97 (1970).

The light curves of γ Pegasi in the V, B, and U are presented. The $U - B$ colour index of the star is smallest around maximum light. In addition to the overall variability in the spectroscopic period of 3^h38^m, the B magnitude of γ Pegasi seems to undergo regular fluctuations with the amplitude of 0^m005 in a short period equal to about 44^m.

122.072 **Rayons des céphéides et la méthode d'Opolski.**
M. Takeuti.
Sci. Rep. Tôhoku Univ. Sendai, Japan, First Ser., Vol. 52, 149 - 153 (1969).

On vérifie la méthode d'Opolski pour déterminer des rayons des céphéides. La relation rayon-couleur des céphéides classiques est examinée aux données d'Opolski.

122.073 **An interpretation of δ Scuti stars.**
K.-C. Leung.
Astron. Journ., Vol. 75, 643 - 650 (1970).

Consistent absolute magnitudes are obtained for δ Scuti stars using Strömgren intermediate-band colors and Crawford's photoelectric $H\beta$ indices. A period-color-luminosity relation is found for these variables. From the locations of 34 known variables and 168 possible nonvariables, two distinct instability regions are found in the H-R diagram; the short-period group lies very close to the zero-age main sequence, while the long-period group lies about 2 mag above the main sequence.

122.074 **UBV observations of the magnetic variable HD 125248.** H. M. Maitzen, K. D. Rakosch.
Astron. Astrophys., Vol. 7, 10 - 16 (1970). In German.

Photoelectric observations in the UBV system of the Ap star HD 125248 were carried out with the 24-inch telescope of the Astronomical Institute of the University of Bochum on La Silla, Chile. The period of 9.295 days found earlier by Stibbs (1950) is confirmed within the limits of accuracy. The light curve in yellow has two maxima each in phase with a magnetic field maximum.

122.075 **Flare observations of UV Ceti stars.**
V. Oskanjan, A. Kubičela, J. Arsenijević.
Bull. Obs. Astron. Beograd, Vol. 27, No. 2, p. 87 - 90 (1969).

Observations of the flare stars UV Cet, Ross 882, Ross 154, EV Lac, and AD Leo are summarized.

122.076 **The light curve of BL Lac (= VRO 42.22.01).**
K. P. Tritton, R. A. Brett.
Observatory, Vol. 90, 110 - 113 (1970).

We here publish the photographic magnitudes of BL Lac for the period June – November 1969 determined as part of the Herstmonceux optical monitoring programme of quasars and galaxies.

122.077 **Polarimetric observations of the RV Tauri stars U Monocerotis and R Scuti.** K. Serkowski.
Astrophys. Journ., Vol. 160, 1107 - 1116 (1970).

The position angle of the intrinsic polarization for U Mon remains close to 0° around and before the deep light minima and close to 105° around and before the shallow light minima that occur halfway between the deep minima. This suggests nonspherical oscillations of the star. The changes in polarization of R Sct are less regular compared with those of U Mon. For both stars the largest amounts of intrinsic polarization were observed in the ultraviolet spectral region during the light minima.

122.078 **The spectrum of RW Aurigae, 3250 to 4900 Å.**
G. F. Gahm.
Astrophys. Journ., Vol. 160, 1117 - 1125 = Contr. Lick Obs.,

No. 311 (1970).

A catalog of emission and absorption lines and line intensities derived from three coudé spectrograms of RW Aurigae covering the wavelength region 3250 - 4900 Å is presented. Anomalous line intensities are discussed, and the presence of several fluorescent processes is suggested.

122.079 **The spectrum of VY Canis Majoris from 2.9 to 14 microns.**
F. C. Gillett, W. A. Stein, P. M. Solomon.
Astrophys. Journ. (*Letters*), Vol. 160, L173 - L176 (1970).

The bright infrared object VY CMa has been observed from $\lambda = 2.9\,\mu$ to $\lambda = 14\,\mu$ with a resolution $\Delta\lambda/\lambda = 0.01$ – 0.02. Three independent observations show an absorption feature extending from about $\lambda \approx 7.7\,\mu$ to $\lambda \approx 9\,\mu$. Preliminary analysis indicates that the band can be attributed to SiO molecules in the extended atmosphere of the star. It is difficult, however, to reconcile these results with previous models that have been suggested for the object.

122.080 **Synthesis of the parameters of the light curve of an irregular variable star.** L. M. Shulman.
Astrometriya i Astrofiz., *Kiev*, No. 8, (see 003.024), p. 13 - 26 (1969). In Russian.

The autocorrelation function and brightness distribution of an irregular variable star are obtained for the case of arbitrary distribution of single flares by amplitudes, increments and decrements. It is shown that all the stars, their irregularity being due to superposition of independent random flares, must be characterized by the non-negative asymmetry of the brightness distribution only.

122.081 **On a possible energy source of the flares of late-type stars.** I. A. Klimishin.
Astrometriya i Astrofiz., *Kiev*, No. 8, (see 003.024), p. 30 - 35 (1969). In Russian.

If the flares of late-type stars can be connected with the ejection of comparatively cold hydrogen clouds into the upper photospheric layers, then an essential part of the flare energy can release as a result of transition of hydrogen into molecular state.

122.082 **Two-colour observations of VX Cas.**
A. F. Pugach.
Astrometriya i Astrofiz., *Kiev*, No. 8 (see 003.024), p. 51 - 55 (1969). In Russian.

The diagram V–(B′ – V) for VX Cas is plotted using photographic observations. Brightness distributions for the two colours are obtained. The observations indicated a rapid brightness fading accompanied by an increase of colour temperature.

122.083 **Changes of period of chi Cygni.** F. Vaclík.
Říše hvězd, Vol. 51, 88 - 89 (1970). In Czech.

122.084 **Search for new flare stars.**
P. M. Corben, G. A. Harding, Y. Z. R. Thomas.
Monthly Notes Astron. Soc. Southern Africa, Vol. 29, 57 - 77 (1970).

An examination of Gliese's catalogue of stars within 20 pc (1969) and an extension of this catalogue to a distance of 25 pc being prepared at the Royal Greenwich Observatory showed that 6 southern M stars, which have been classified as having hydrogen emission lines, have not previously been observed as flare stars. It was therefore decided to observe Gliese 234 (Ross 614) for an extended period to see whether flares could be detected. On 1970 February 8 a UV Ceti type flare is likely to have occurred.

122.085 **Low intensity flares of AD Leo.**
A. H. Jarrett, J. P. Eksteen.

Monthly Notes Astron. Soc. Southern Africa, Vol. 29, 78 - 79 (1970).

Observations of AD Leo were carried out at Boyden Observatory over the period 1 - 11 March, 1970. Over the period mentioned, a total monitoring time of 41h 07m was obtained. Only three low intensity flares were recorded during this period.

122.086 The spectra of the symbiotic stars AG Peg and AG Dra in 1967. I. N. Chkhikvadze.
Astron. Tsirk., No. 546, p. 7 - 8 (1970). In Russian.

122.087 Intrinsic colours and absolute magnitudes of 32 halo population cepheids with periods from one to two days. O. E. Mandel.
Astron. Tsirk., No. 551, p. 1 - 2 (1970). In Russian.

122.088 Halo population cepheids with periods from one to two days and the Local System. O. E. Mandel.
Astron. Tsirk., No. 551, p. 3 - 4 (1970). In Russian.

122.089 On the variability of 30 Cygni. R. A. Botzula.
Astron. Tsirk., No. 551, p. 7 - 8 (1970).

122.090 Note on the unusual variable HR Aur.
E. A. Virtichenko.
Astron. Tsirk., No. 552, p. 7 - 8 (1970). In Russian.

122.091 Flare activity of UV Ceti.
A. H. Jarrett, J. P. Eksteen.
Inform. Bull. Variable Stars (I.A.U. Commission 27), Konkoly Obs., Budapest, No. 412 (1970).

Further observations of UV Ceti were made at Boyden Observatory. A total of 18 flares were recorded.

122.092 Elements of bright southern cepheids.
R. S. Stobie.
Inform. Bull. Variable Stars (I.A.U. Commission 27), Konkoly Obs., Budapest, No. 414 (1970).

122.093 High-frequency stellar oscillations. III. A brief report. B. M. Lasker, J. E. Hesser.
Inform. Bull. Variable Stars (I.A.U. Commission 27), Konkoly Obs., Budapest, No. 415 (1970).

122.094 Autocorrelation analysis of DQ Her and RW Tri.
A. F. Pugach.
Inform. Bull. Variable Stars (I.A.U. Commission 27), Konkoly Obs., Budapest, No. 418 (1970).

122.095 Photoelectric observations of the flare star YZ CMi.
S. Cristaldi, M. Rodono.
Inform. Bull. Variable Stars (I.A.U. Commission 27), Konkoly Obs., Budapest, No. 423 (1970).

122.096 The high velocity variable QT CrA.
M. W. Feast.
Inform. Bull. Variable Stars (I.A.U. Commission 27), Konkoly Obs., Budapest, No. 424 (1970).

122.097 Photographic observations of UV Ceti, (October, 3 - 18, 1969).
L. N. Mosidze, A. D. Chuadze.
Inform. Bull. Variable Stars (I.A.U. Commission 27), Konkoly Obs., Budapest, No. 425 (1970).

122.098 Flares of YZ CMi observed at Okayama, 31 January to 13 February, 1970.
K. Osawa, T. Noguchi, T. Okada, K. Ichimura, E. Watanabe, K. Okida.
Inform. Bull. Variable Stars (I.A.U. Commission 27), Konkoly

Obs., Budapest, No. 426 (1970).

122.099 List of probable Delta Scuti stars. M. S. Frolov.
Inform. Bull. Variable Stars (I.A.U. Commission 27), Konkoly Obs., Budapest, No. 427 (1970).

122.100 Fréquence des sursauts des étoiles du type UV Ceti. M. Petit.
Inform. Bull. Variable Stars (I.A.U. Commission 27), Konkoly Obs., Budapest, No. 430 (1970).

122.101 Etoiles naines rouges variables. M. Petit.
Inform. Bull. Variable Stars (I.A.U. Commission 27), Konkoly Obs., Budapest, No. 431 (1970).

122.102 AD Leo.
K. Osawa, T. Noguchi, T. Okada, K. Ichimura, E. Watanabe, K. Okida.
Inform. Bull. Variable Stars (I.A.U. Commission 27), Konkoly Obs., Budapest, No. 432 (1970).

122.103 Photometry of AD Leo.
A. H. Jarrett, J. P. Eksteen.
Inform. Bull. Variable Stars (I.A.U. Commission 27), Konkoly Obs., Budapest, No. 433 (1970).

122.104 Flare photometry of UV Ceti.
A. H. Jarrett, J. P. Eksteen.
Inform. Bull. Variable Stars (I.A.U. Commission 27), Konkoly Obs., Budapest, No. 434 (1970).

122.105 Photoelectric patrol of the flare star BD +13°2618.
S. Cristaldi, M. Rodono.
Inform. Bull. Variable Stars (I.A.U. Commission 27), Konkoly Obs., Budapest, No. 439 (1970).

122.106 Photoelectric observations of flare stars.
P. F. Chugainov, N. I. Shakhovskaya.
Inform. Bull. Variable Stars (I.A.U. Commission 27), Konkoly Obs., Budapest, No. 441 (1970).

122.107 66 Oph–a probable early type flare star.
A. A. Page, B. Page.
Proc. Astron. Soc. Australia, Vol. 1, 324 - 325 (1970).

122.108 Companions to RR Lyrae variables.
D. H. P. Jones.
Proc. Astron. Soc. Australia, Vol. 1, 329 - 330 (1970).

122.109 Photometric observations of FG Sagittae in 1967 - 1969. V. P. Arkhipova.
Astron. Tsirk., No. 553, p. 3 - 5 (1970). In Russian.

122.110 Spectral variations of CH Cyg in 1969.
G. N. Jimshelejshvili.
Astron. Tsirk., No. 557, p. 2 (1970). In Russian.

122.111 On the correlation between colour indices and amplitudes for RR Lyrae-type stars.
R. K. Kanishcheva.
Astron. Tsirk., No. 559, p. 1 - 3 (1970). In Russian.

122.112 New rapid variable SVS 1669 in Andromeda and improved elements of CSV 5879. G. A. Lange.
Astron. Tsirk., No. 559, p. 3 - 4 (1970). In Russian.

122.113 NP Her. H. Busch, K. Häussler.
Inform. Bull. Variable Stars (I.A.U. Commission 27), Konkoly Obs., Budapest, No. 417 (1970).

122.114 RR Lyrae stars. V. P. Tsesevich.

Zemlya i Vselennaya, No. 3, p. 46 - 49 (1970).
In Russian.

122.115 **Nine red variable stars in the Cygnus cloud, VV 242 - 250.** W. J. Miller.
Ric. Astron. Specola Vaticana, Vol. 8, (No. 3), 41 - 60 (1969).

Four old and five new red variable stars have been studied on a total of 12113 plates from the Vatican, Hamburg, Heidelberg, Harvard and Mount Wilson and Palomar Observatories. The variables range between the 14th and at least the 21st photographic magnitudes. One star is an irregular red variable, seven stars are long period variables, and the ninth star is probably an infrared LPV. The results are summarized in three tables and eight pages of fragmentary light curves.

122.116 **Photographische Reihenbeobachtungen von RR-Lyrae-Sternen und Bedeckungsveränderlichen.**
P. Ahnert.
MVS Sonneberg, Vol. 5, 112 - 115 (1970).

Observed epochs of maxima or minima are presented for the variable stars RU CVn, S Com, BV 138 CrB, TW Her, ST Leo, WZ And, XZ And, ZZ Cyg, VX Lac, RT Per, ST Per, SX Psc; revised elements are deduced for RU CVn, S Com, BV 138.

122.117 **Three-colour electrophotometric observations of NU Ori.** I. V. Shpychka.
Tsirk. L'vov. Astron. Obs., No. 44, p. 27 - 32 (1970).
In Russian.

122.118 **Variation in the amplitude and the period of RU Cam from December 1966 to July 1967.**
P. Broglia.
Rendiconti Ist. Lombardo, Ser. A, Vol. 103, No. 2, p. 235 - 246 (1969). In Italian.

A short description of a simple device for setting alternatively the variable and the comparison star is given. The B and V light curves of RU Cam measured from December 1966 to July 1967 are presented. The mean brightness and color of the variable are constant but the amplitude increases progressively.

122.119 **On the obscuration hypothesis of R Coronae Borealis.** A. Galatola.
Thesis, Univ. Pennsylvania, Philadelphia. [Available from Univ. Microfilms, Ann Arbor, Mi.], 96 pp. (1969).

The behavior of the light curve of R CrB is investigated on the basis of simple models in which it is assumed that a shell of gas is ejected from the star with some initial velocity. As the shell moves away from the star, carbon condenses within the shell, thereby obscuring the light from the star.

122.120 **Photometric studies of four Southern Hemisphere variable stars.** C. R. Chambliss.
Thesis, Univ. Pennsylvania, Philadelphia. [Available from Univ. Microfilms, Ann Arbor, Mi.], 193 pp. (1969).

122.121 **A preliminary search of stars of rapid variability.** E. Hardy, E. E. Mendoza V.
Asoc. Argentina Astron. Bol., No. 14 (see 012.020), p. 28 - 32 (1968).

122.122 **Changes in the activity of UV Ceti.** W. Kunkel.
Asoc. Argentina Astron. Bol., No. 14 (see 012.020), p. 36 (1968). In Spanish. – Abstract.

122.123 **R Coronae Australis.** E. Mendoza, M. Jaschek, C. Jaschek.
Asoc. Argentina Astron. Bol., No. 14 (see 012.020), p. 49 (1968). In Spanish.

122.124 **The light curve of l Carinae.** A. Feinstein, J. C. Muzzio.
Asoc. Argentina Astron. Bol., No. 14 (see 012.020), p. 53 - 54 (1968). In Spanish. – Abstract.

122.125 **AG Pegasi.** L. López, J. Sahade.
Asoc. Argentina Astron. Bol., No. 14 (see 012.020), p. 68 - 71 (1968).

122.126 **Radial velocities of WY Velorum.** C. Hernández, J. Sahade.
Asoc. Argentina Astron. Bol., No. 12, (see 012.019), p. 10 (1968). – Abstract.

Red giant mass loss and cepheid variables.
See Abstr. 065.078.

Adiabatic oscillations of stars.
See Abstr. 065.090.

Radial oscillations of stellar models: Density distributions which lead to three-term recurrence relations.
See Abstr. 065.091.

Proper motions of long-period variable stars. II.
See Abstr. 112.018.

Proper motions of two U Gem type variables and of stars of their surroundings. See Abstr. 112.019.

Photometry of classical cepheids η Aql, SU Cas and ζ Gem in the system UPXYZVS. See Abstr. 113.011.

***UBV* observations of Mira stars.**
See Abstr. 113.047.

Some remarks regarding photoelectric photometry.
See Abstr. 113.049.

Abundance analyses of cool carbon stars.
See Abstr. 114.046.

The rotational temperature of a TiO band in the spectrum of R Hydrae. See Abstr. 114.063.

Possibility of fluorescence phenomena in the ultraviolet spectrum of symbiotic stars and long period variables.
See Abstr. 114.098.

On the absolute magnitudes of U Gem type variables. See Abstr. 115.007.

Multi-colour photographic photometry of Orion flare stars. See Abstr. 132.032.

On the percentage of flare stars among the RW Aurigae type variables in the Orion association.
See Abstr. 152.009.

Flare stars in the Pleiades.
See Abstr. 153.010.

Flare stars in the Pleiades region. II.
See Abstr. 153.021.

New flare stars in the Pleiades. (A re-examination of the Tonantzintla photographic material: 1963 - 1970).
See Abstr. 153.022.

Photometrische Beobachtungen an RR Lyrae-Sternen. III. Die RR Lyrae-Veränderlichenlücke im Farben-

Helligkeitsdiagramm von Omega Centauri (NGC 5139).
See Abstr. 154.011.

On the incidence of Cepheids in globular clusters.
See Abstr. 154.013.

On the evolutionary phase of Cepheids in globular
clusters.   See Abstr. 154.012.

The galactic distribution of young cepheids.
See Abstr. 155.042.

123   Variable Stars: Lists of Observations, Individual Observations

123.001   Rho Persei im zweiten Halbjahr 1969.
C. Göttig.
VdS Nachrichtenblatt, 19. Jahrgang, p. 4 (1970).

123.002   **HD 4174.**   R. Ruhnow.
VdS Nachrichtenblatt, 19. Jahrgang, p. 31 - 32
(1970).

123.003   **Photographic observations of six variables in
Andromeda.**   I. A. Daube.
Peremennye Zvezdy, Vol. 17, 22 - 30 (1969).   In Russian.
   Photographic observations of the suspected variable
stars CSV 8876, CSV 8878, CSV 8882 and of the variables
WW And, AP And and AT And are given.

123.004   **Visual observations of variable stars.**
I. N. Latyshev.
Peremennye Zvezdy, Vol. 17, 64 - 75 (1969).   In Russian.
   Results are given of visual observations of SZ Tau,
SU Cyg, T Vul, U Aql, W Gem, S Sge, X Cyg, RR Lyr and
U Mon.

123.005   **SVS 1530.**   T. N. Kulikova.
Peremennye Zvezdy, Vol. 17, 90 - 93 (1969).
In Russian.

123.006   **FF Delphini.**   E. S. Raevskaya.
Peremennye Zvezdy, Vol. 17, 96 - 98 (1969).
In Russian.

123.007   **Photographic observations of two variable stars
in Cassiopeia.**   Muhamed Sinada, R. I. Chuprina.
Peremennye Zvezdy, Vol. 17, 98 - 102 (1969).   In Russian.

123.008   **Gamma Sagittae, étoile variable?**   A. Brun.
L'Astronomie, 84ᵉ année, p. 82 (1970).

123.009   **Beobachtungen des Halbregelmäßigen W Cygni.**
H. Schubert.
BAV Rundbrief, 19. Jahrgang, p. 7 - 10 (1970).

123.010   **Nebenprodukt von Nova-Del-Beobachtungen:
W Del.**   W. Braune.
BAV Rundbrief, 19. Jahrgang, p. 16 (1970).

123.011   **Nederlandse Vereniging voor Weer- en Sterrenkunde.
Observations of variable stars. Report No. 17.**
L. Plaut, H. Feijth.
Kapteyn Astron. Lab., Groningen–Netherlands. 8 pp. (1970).
3338 observations of 82 variable stars, July – December 1969.

123.012   **o Ceti, Mira, 1969.**   R. Germann.
Orion Schaffhausen, 28. Jahrgang, p. 53 (1970).

123.013   **Comparison stars for long period variables and RY
Sagittarii.**   A. W. J. Cousins, H. C. Lagerwey.

Monthly Notes Astron. Soc. Southern Africa, Vol. 29, 7 - 12
(1970).

123.014   **Observations of eclipsing binaries and RR-Lyrae-
type variables in 1965.**   O. Obůrka.
Contr. Public Obs. and Planetarium Brno 5, p. 1 - 20 (1968).
In Czech.

123.015   **Observations of eclipsing binaries.**
O. Obůrka.
Contr. Public Obs. and Planetarium Brno, 6, p. 1 - 46 (1968).
In Czech.

123.016   **377.1943 Sge - FG Sagittae.**
R. Ruhnow.
VdS Nachrichtenblatt, 19. Jahrgang, p. 65 (1970).

123.017   **Maxima von Mirasternen 1969.**
R. Lukas.
VdS Nachrichtenblatt, 19. Jahrgang, p. 67 - 68 (1970).

123.018   **TU Cas – ein interessanter Veränderlicher.**
R. Lukas.
SuW, Vol. 9, 156 (1970).

123.019   **Photoelectric observations.**
F. M. Bateson.
Roy. Astron. Soc. New Zealand, Variable Star Section,
Circ. No. 143, 1 pp. (1970).

123.020   **Observations of southern variable stars, 1967 July 1
to 1969 March 31.**   F. M. Bateson.
Roy. Astron. Soc. New Zealand, Variable Star Section,
Circ. No. 147, 23 pp. (1970).

123.021   **A new red variable star in Crucis.**   F. M. Bateson.
Roy. Astron. Soc. New Zealand, Variable Star
Section, Circ. No. 151, 4 pp. (1970).
   Observations of a new variable star in Crucis are presented
together with charts of the region and a provisional sequence.
The discovery was made by R. G. Welch of Auckland. Varia-
bility appears to be definitely established with an indication
that the period may be around 217 days.

123.022   **VY Canis Majoris.**   L. J. Robinson.
IAU Circ. No. 2257 (1970).

123.023   **A new variable in Cassiopeia.**   A. F. Pugach.
Astrometriya i Astrofiz., *Kiev,* No. 8, (see 003.024),
p. 56 - 58 (1969).   In Russian.
   A new variable star is found, the variations in the brightness of which occur mainly in the visual region of the spectrum. A two-colour diagram V–(B′–V) is given. The type of
variability is not determined.

123.024   **V 362 Her.**   H. Busch.

Inform. Bull. Variable Stars (I.A.U. Commission 27), Konkoly Obs., Budapest, No. 417 (1970).

123.025 KM Cassiopeiae. L. R. Wackerling.
Inform. Bull. Variable Stars (I.A.U. Commission 27), Konkoly Obs., Budapest, No. 436 (1970).

123.026 New observations and light curve of BL Lac.
N. E. Kurochkin.
Astron. Tsirk., No. 561, p. 4 - 6 (1970). In Russian.

123.027 BD +28°838. A. Brun.
Inform. Bull. Variable Stars (I. A. U. Commission 27), Konkoly Obs., Budapest, No. 443 (1970).

123.028 Neuentdeckte Veränderliche (S 10619 bis 10720).
G. A. Richter.
MVS Sonneberg, Vol. 5, 99 - 104 (1970).

123.029 Spektraltypen von Veränderlichen. Teil XVI.
W. Götz, W. Wenzel.
MVS Sonneberg, Vol. 5, 105 - 106 (1970).
In continuation of previous lists spectral types of 33 variable stars are given.

123.030 Bearbeitung von 44 Veränderlichen.
H. Geßner.
MVS Sonneberg, Vol. 5, 106 - 107 (1970).

123.031 XZ Apodis. H. Geßner.
MVS Sonneberg, Vol. 5, 108 - 109 (1970).

123.032 SU Lacertae. H. Huth, F. Splittgerber.
MVS Sonneberg, Vol. 5, 109 - 110 (1970).
In the years 1935 − 1945 a large variation of the period of SU Lac took place, the new value amounting to $319\overset{d}{.}14$ in contrast to $279\overset{d}{.}10$.

123.033 CSV 2229 Centauri. L. Meinunger.
MVS Sonneberg, Vol. 5, 111 (1970).

The star CSV 2229 = S 5003 Centauri, which was previously identified with a strong X-ray source, shows minima of brightness very similar to R Coronae Borealis; the minima obviously occur not very seldom.

123.034 VY Canis Maioris. L. Meinunger.
MVS Sonneberg, Vol. 5, 115 (1970).
Estimates on Sky Patrol plates (1957 to 1958) and photoelectric observations in nine nights 1970 show in V and B no brightness variations in excess of the observational errors.

123.035 Maxima der Mira-Sterne SX und Y Andromedae.
K. Kockel, G. Reimann.
MVS Sonneberg, Vol. 5, 116 (1970).

123.036 Identität Wr 99 ≡ S 5424. H. Huth.
MVS Sonneberg, Vol. 5, 116 (1970).

123.037 X and SV Vulpeculae.
O. S. Bojko.
Tsirk. L'vov. Astron. Obs., No. 44, p. 6 - 9 (1970).
In Russian.

123.038 CN Lyrae.
M. B. Girnyak.
Tsirk. L'vov. Astron. Obs., No. 44, p. 10 - 13 (1970).
In Russian.

123.039 On three little investigated variable stars.
V. V. Golovatyj.
Tsirk. L'vov. Astron. Obs., No. 44, p. 14 - 20 (1970).
In Russian.

123.040 Photographic observations of three variable stars.
A. T. Dul'tsev.
Tsirk. L'vov. Astron. Obs., No. 44, p. 21 - 26 (1970).
In Russian.

Be stars which recently showed large photometric changes. See Abstr. 113.003.

124 Novae

124.001 **Structure of envelopes ejected by Novae.**
E. R. Mustel, A. A. Boyarchuk.
Astrophys. Space Sci., Vol. 6, 183 - 204 (1970).
The paper contains an analysis of the structure of envelopes ejected during the outbursts of Novae. The data used for this purpose: (a) Direct photographs of envelopes and the photographs taken with the use of different colour filters: (b) Spectra of envelopes. The envelope of DQ Her is studied most carefully. The analysis of all available data for the envelopes around DQ Her and V 603 Aql permits us to outline a morphological model of these envelopes. The envelope ejected during the outburst of T Aur reveals the same properties, which are characteristic for the envelopes of DQ Her and V603 Aql. From this we conclude that the distribution of gases inside the envelopes of the majority of Novae is approximately of the same character. This speaks in favour of the presence of certain forces around many Novae, which guide the motion of ejected plasma along some quite definite directions inside rather small solid angles. It seems that the only conceivable forces of this type may be the forces of a magnetic nature. Comparing the velocity of expansion of the envelope of DQ Her and the rate of change of its angular size we computed that the distance to DQ Her is equal to 320 pc.

124.002 **Novae.** R. Głębocki.
Urania Kraków, Vol. 41, 98 - 103 (1970).
In Polish.

124.003 **Ultra-slow novae.** R. Głębocki.
Postępy Astron., Vol. 18, 59 - 80 (1970).
In Polish.
General description of photometric, spectral and kinematical characteristics of ultra-slow novae, called also RT Ser-type novae, is given. All suspected members of this group are described and suggestion is made that Nova Delphini 1967 is also of RT Ser type. It is pointed out that distribution of ultra-slow novae in the Galaxy shows population II characteristics.

124.004 **On the problem of thermal explosion in degenerate gas. II. Hydrogen combustion in a layer source.**
V. V. Porfiryev, Yu. N. Redkoborody.
Astrometriya i Astrofiz., Kiev, No. 8, (see 003.024), p. 3 - 5 (1969). In Russian.
The process of heating of degenerate gas at the presence of a layer source near the surface was analysed. The Mestel hypothesis is shown not to contradict the explosion of a nova.

124.005 **On the problem of thermal explosion in degenerate gas. III. The electrostatic screening of a Coulomb field of nuclei at thermonuclear reactions.**
V. V. Porfiryev, Yu. N. Redkoborody.
Astrometriya i Astrofiz., Kiev, No. 8, (see 003.024), p. 5 - 13 (1969). In Russian.
The effect of electron screening on thermonuclear fusion is considered. The energy field is shown to increase. The effective potential of electrostatic interaction is derived.

124.006 **About a similarity in the structure of envelopes ejected by novae and supernovae.** E. R. Mustel.
Astron. Tsirk., No. 544, p. 1 - 3 (1970). In Russian.

124.007 **New novae in the Andromeda nebula.**
A. S. Sharov, A. K. Alksnis.
Astron. Tsirk., No. 560, p. 1 - 3 (1970). In Russian.

The Dwarf Novae. See Abstr. 003.038.

A nova in the Large Magellanic Cloud.
See Abstr. 159.021.

124.100 **Nova Delphini 1967**

Spectrum of HR Del (Nova Delphini 1967) in the first two months following discovery. R. Głębocki.
Acta Astron., Vol. 20, 99 - 116 (1970).
In this paper an analysis of the spectrum of HR Del is given for the period extending from the end of July until the end of September 1967. Light curve as well as changes in the intensity of continuum for this period are shown in a figure. Identification and changes of intensities of absorption and emission lines are given. Radial velocity shifts of the absorption components of lines relatively to the emission peaks are discussed.

Photometric study of Nova Delphini 1967 and Nova Vulpeculae 1968 No. 1. A. Terzan.
Astron. Astrophys., Vol. 5, 167 - 170 (1970). In French.
The light curves (visual and photographic) of the novae Delphini 1967 and Vulpeculae 1968 No. 1 are drawn for periods going respectively from July 10, 1967 and April 15, 1968 to August 31, 1968.

Photoelectric observations of nova Del 1967 in 1967 - 1968. L. N. Kolesnik.
Astron. Tsirk., No. 562, p. 1 - 2 (1970). In Russian.

Raies nébulaires et coronales dans le spectre infrarouge de la Nova Delphini. Y. Andrillat, L. Houziaux.
Astrophys. Space Sci., Vol. 6, 36 - 44 (1970).
photographic infrared at the Haute-Provence Observatory from September 7 to November 12, 1968. Dispersions range from 230 Å/mm to 4 Å/mm. The slope of the continuous spectrum in the region $1.3 - 1.7\mu^{-1}$ has been measured. Lines of several elements have been indentified. Their intensities at various dates are given. Line profiles could be measured for a few lines.

Analyse des observations de Nova Delphini 1967.
J. Lecacheux.
Ciel et Terre, Vol. 86, 46 - 59 (1970).

Spectrographic observations of Nova Delphini 1967.
P. Galeotti, L. E. Pasinetti.
Mem. Soc. Astron. Italiana, Nuova Serie, Vol. 41, 47 - 56 (1970).
The results of spectrographic observations performed during 1967 - 1968, are reported. The identifications of the lines and the appearance of [OI] and HeI are discussed.

Observations of Nova (HR) Delphini 1967.
F. M. Bateson.
Roy. Astron. Soc. New Zealand, Variable Star Section, Circ. No. 153, 2 pp. (1970).

Observations of Nova Delphini. S. M. Azimov.
Stars, Nebulae, Galaxies, Symposium Byurakan 1968, (see 012.002), p. 139 - 143 (1969). In Russian.

124.101 Nova T Pyxidis

Photometric and spectroscopic observations of Nova T Pyxidis.   A. U. Landolt.
Publ. Astron. Soc. Pacific, Vol. 82, 86 - 92 = Contr. Louisiana State Univ. Obs. Baton Rouge, No. 31 = Contr. Cerro Tololo Inter-American Obs. No. 95 (1970).
    Photometric and spectroscopic observations obtained some two months after the December 1966 outburst of T Pyxidis are presented.

124.102 Nova Sagittarii 1969

Nova Sagittarii 1969.   F. M. Bateson.
Roy. Astron. Soc. New Zealand, Variable Star Section, Circ. No. 150, 2 pp. (1970).

Nova in Sagittarius.
Sky Telescope, Vol. 39, 102 (1970).

124.103 Nova Serpentis 1970

Identification of Nova Serpentis 1970.
M. S. Burkhead, M. A. Seeds.
Astrophys. Journ. *(Letters),* Vol. 160, L51 (1970).
    Nova Serpentis has been identified as a prenova star of $m_v = 16.1$ and $(B - V) = +0.8$.

Infrared observations of nova Serpentis 1970.
A. R. Hyland, G. Neugebauer.
Astrophys. Journ. (*Letters*), Vol. 160, L177 - L180 (1970).
    Preliminary infrared observations of nova Serpentis 1970 show that the total luminosity remained fairly constant while the visual luminosity decreased about 3.5 mag.

Nova Honda in Serpens.   N. M. Bronnikova.
Astron. Tsirk., No. 555, p. 1 - 2 (1970).   In Russian.

Nova Honda ($\alpha = 19^h 22^m 2$, $\delta = +4° 12'$ (1900)).
Astron. Tsirk., No. 558, p. 1 (1970).   In Russian.

Spectral observations of nova Ser 1970.
M. B. Babaev, M. S. Gadjiev, E. B. Gusev.
Astron. Tsirk., No. 560, p. 7 - 8 (1970).   In Russian.

Nova Serpens.
M. Honda.
British Astron. Ass. Circ., No. 518, 519 (1970).

Nova.
M. Honda.
IAU Circ. No. 2212 (1970).

Nova Serpentis 1970.
M. Honda, C. B. Ford.
IAU Circ. No. 2214 (1970).

Nova Serpentis 1970.
M. S. Burkhead, R. K. Honeycutt, V. J. Lee, W. S. Penhallow, M. A. Seeds, A. P. Cowley, K. Locher, C. B. Ford, J. Ashbrook, L. Jacchia.
IAU Circ. No. 2215 (1970).

Nova Serpentis 1970.
T. Seki, H. Kosai, M. Honda, E. Ichimura, M. Shimizu, E. Watanabe, K. Nariai, Y. Yamashita, Kanno, M. Sugano, L. C. Peltier, C. Hurless, R. Sweetsir, R. Hodson, E. Mayer,

K. Simmons, D. Rosebrugh, W. M. Lowder, F. Pilcher.
IAU Circ. No. 2216 (1970).

Nova Serpentis 1970.
P. Grosbøl, B. G. Jørgensen.
IAU Circ. No. 2218 (1970).

Nova Serpentis 1970.
M. S. Burkhead, M. A. Seeds, V. J. Lee, J. Grygar, J. B. Hutchings.
IAU Circ. No. 2220 (1970).

Nova Serpentis 1970.
S. Kříž, T. Seki, A. Mrkos, Petrovičová.
IAU Circ. No. 2225 (1970).

Nova Serpentis 1970.
F. Clayton, B. Yare, F. Pilcher.
IAU Circ. No. 2230 (1970).

Nova Serpentis 1970.
P. Tempesti, F. Ciatti, A. Mammano.
IAU Circ. No. 2245 (1970).

The spectrum of Nova Serpentis 1970.
U. K. Gehlich, J. Tremko, R. Wehmeyer.
Inform. Bull. Variable Stars (I.A.U. Commission 27), Konkoly Obs., Budapest, No. 428 (1970).

Nova Serpentis.   K. Osawa.
Inform. Bull. Variable Stars (I.A.U. Commission 27), Konkoly Obs., Budapest, No. 429 (1970).

Nova Serpentis 1970.   K. Locher.
Orion Schaffhausen, 28. Jahrgang, p. 52 - 53 (1970).

Nova Serpentis 1970.
F. M. Bateson.
Roy. Astron. Soc. New Zealand, Variable Star Section, Circ. No. 146, 1 pp. (1970).

Nova Serpentis 1970.   F. M. Bateson.
Roy. Astron. Soc. New Zealand, Variable Star Section, Circ. No. 149, 1 pp. (1970).

More about the new star.
Sky Telescope, Vol. 39, 334 (1970).

Nova Serpentis.   R. Lukas.
VdS Nachrichtenblatt, 19. Jahrgang, p. 40 (1970).

Nova Ser 70.
VdS Nachrichtenblatt, 19. Jahrgang, p. 54 (1970).

Nova Serpentis 1970.
W. Braune.
VdS Nachrichtenblatt, 19. Jahrgang, p. 66 - 67 (1970).

Nova Serpentis 1970.
Yamamoto Circ., Nos. 1715, 1720 (1970).   In Japanese.

124.104 Nova Vulpeculae 1968 No. 1

Observations of nova Vul 1968.   M. V. Babaev.
Astron. Tsirk., No. 555, p. 2 - 4 (1970).   In Russian.

Electrophotometry of N Vul 1968.
L. N. Kolesnik, A. F. Pugach.
Astron. Tsirk., No. 556, p. 3 - 5 (1970).   In Russian.

Nova Vulpeculae 1968.
J. Hübscher.
BAV Rundbrief, 19. Jahrgang, p. 12 - 13 (1970).

Nova Vulpeculae 1968 – Rückblick.
K. Locher.
Orion Schaffhausen, 28. Jahrgang, p. 18 - 19 (1970).

Observations of nova Vulpeculae 1968.
F. M. Bateson.
Roy. Astron. Soc. New Zealand, Variable Star Section, Circ.
No. 154, 1 pp. (1970).

**Photometric study of Nova Delphini 1967
and Nova Vulpeculae 1968 No. 1.** See Abstr. 124.100.

124.105 Nova WZ Sagittae

Origin of the ultra-short period binary WZ Sagittae.
See Abstr. 121.035.

124.106 Nova Aquilae 1970

On the identification of the nova V 368 Aql.
A. Sh. Khatisov.
Astron. Tsirk., No. 556, p. 5 - 7 (1970). In Russian.

Nova Aquilae 1970.
M. Honda, H. Kosai.
British Astron. Ass. Circ., No. 520 (1970).

Nova Aquilae 1970.
M. Honda, H. Kosai, L. Jacchia, J. Ashbrook.
IAU Circ. No. 2233 (1970).

Nova Aquilae 1970.
C. Y. Shao.
IAU Circ. No. 2235, 2242 (1970).

Nova Aquilae 1970.
A. P. Cowley, D. Lucas, B. G. Marsden, C. Y. Shao,
M. Mattei.
IAU Circ. No. 2237 (1970).

Nova Aquilae 1970.
F. M. Stienon, K. Locher, A. Fujii, H. Kato, K. Adachi,
M. Takeishi, R. A. Sweetsir, T. Seki, P. Grosbøl, B. G.
Jørgensen.
IAU Circ. No. 2239 (1970).

Nova Aquilae 1970.
M. Honda, H. Kato, O. Abe, H. Kosai, E. H. Mayer.
IAU Circ. No. 2241 (1970).

Nova Aquilae 1970.
H. Ohtani, K. Ichimura, S. Nishimura.
IAU Circ. No. 2243 (1970).

Nova Aquilae 1970. K. Locher.
Orion Schaffhausen, 28. Jahrgang, p. 95(1970).

Nova Aquilae 1970. F. M. Bateson.
Roy. Astron. Soc. New Zealand, Variable Star Section, Circ.
No. 155, 1 pp. (1970).

Nova Aquilae 1970.
Yamamoto Circ., Nos. 1717, 1718. 1719 (1970).
In Japanese.

124.107 Nova RS Ophiuchi

**RS Ophiuchi – Nachwirkungen des Ausbruchs von
1967.** K. Locher.
Orion Schaffhausen, 28. Jahrgang, p. 94 - 95 (1970).

124.108 Nova Cygni 1970

Nova Cygni 1970. F. M. Stienon.
IAU Circ. No. 2251 (1970).

Nova Cygni 1970. C. Bertaud.
IAU Circ. No. 2252 (1970).

Nova Cygni 1970.
L. Kohoutek, C. Y. Shao, M. Mattei.
IAU Circ. No. 2254 (1970).

Nova Cygni 1970.
F. Ciatti, A. Mammano, L. Rosino, D. Hoffleit, L. Lucignani.
IAU Circ. No. 2257 (1970).

Nova Cygni 1970.
Yamamoto Circ., No. 1721 (1970).
In Japanese.

124.109 Nova Herculis 1960

**Photoelectric observations of nova Herculis 1960
and some critical remarks on nova photometry.**
K. Gyldenkerne, V. Mejdahl, R. M. West.
Publ. mindre Medd. Københavns Obs., No. 201, 19 pp.
(1969).

Nova Her 1960 has been observed at Brorfelde by means
of standard B and V filters and six narrow interference filters.
A comparison of all available UBV observations made in the
period March—May 1960, revealed very large systematic
differences. They are due to the difficulty in transforming
UBV values for emission line objects from local instrumental
systems to the standard system. On the other hand it is
shown that suitably placed narrow-band filters may yield
valuable information about the emission features.

125 Supernovae, Supernova Remnants

125.001 Supernova remnants and hidden pulsars.
A. Cavaliere, F. Pacini.
Astrophys. Journ. (*Letters*), Vol. 159, L21 - L24 (1970).

If a supernova remnant draws its energy from a rotating neutron star, the total luminosity should decrease about 4 times faster than the rotation period of the central star. Parameters for the pulsar in Cas A (if there is one) are derived as a function of the luminosity and luminosity change. The same considerations could have given the parameters of the neutron star in the Crab before the discovery of the pulsar NP 0531. Some general remarks are made concerning the coupling of the stellar rotation with the nebular remnant.

125.002 Ejection of companion objects by supernovae.
S. A. Colgate.
Nature, Vol. 225, 247 - 248 (1970).

The possibility is recognized that supernovae may occur in conjunction with either a satellite planet or binary star. Previous calculations have considered only the change in binding energy of the combined system caused by the ejection of matter. In addition, it is shown that the shock heating of —and ablation from — the companion object is more decisive for survival. For a supernova that ejects $1/2$ $M_\odot$ at 2×10^9 cm sec^{-1} (leaving 1 $M_\odot$ behind as a neutron star), the critical survival radius for a main sequence star of $1/2$ $M_\odot$ is $\geqslant 10^{12}$ cm and for a planet like the earth $\geqslant 10^{13}$ cm.

125.003 An explanation for the broad supernova bands.
D. R. Huffman.
Nature, Vol. 225, 833 - 834 (1970).

Broad absorption features in the visible spectra of type I supernovae are assumed to be caused by dust clouds in the vicinity of the supernovae. Supernovae emission spectra are subtracted from an infinite temperature Planck curve to give an indication of absorption by the dust. This absorption curve shows seven bands and a rising absorption in the ultraviolet which correspond to laboratory spectra of iron ions in crystals such as garnets.

125.004 Radio observations of four supernova remnants.
J. L. Caswell.
Australian Journ. Phys., Vol. 23, 105 - 108 = Contr.
Dominion Obs., No. 287 (1970).

In this note, four of the sources, studied recently by Milne and Hill (1969), are investigated using observations from a survey at 178 MHz and other published data.

125.005 The neutral hydrogen near HB 21.
J. W. Erkes, K. C. Turner.
Bull. American Astron. Soc., Vol. 2, 191 (1970). — Abstr.
AAS.

125.006 Dust in supernova explosions.
F. Hoyle, N. C. Wickramasinghe.
Nature, Vol. 226, 62 - 63 (1970).

Solid particles of graphite, iron or silicates condensing during supernova explosions could produce significant effects on supernova light curves and also make a substantial contribution to the density of interstellar grains.

125.007 X-ray characteristics of three supernova remnants.
P. Gorenstein, E. M. Kellogg, H. Gursky.
Astrophys. Journ., Vol. 160, 199 - 208 (1970).

Spectral analyses were made of the X-ray data ($1 \leq E \leq 10$ keV) obtained during a rocket flight from two sources that have been identified with Cas A and SN 1572 (Tycho's supernova). The Crab nebula was also observed. The Crab is found to have the hardest spectrum of the three. Interstellar attenuation affects the spectrum of Cas A to an extent that is consistent with or perhaps less than presently accepted values. Comparison of X-ray and 21-cm absorption effects in the spectrum of the supernova remnants implies that the concentration of oxygen and neon relative to hydrogen along the line of sight to the sources is at most equal to presently accepted values of cosmic abundances.

125.008 Supernovae from ancient Korean observational records. S.-I. Chu.
Journ. Korean Astron. Soc., Vol. 1, No. 1, p. 29 - 36 (1968).

28 candidates of novae and supernovae were selected from ancient Korean observational records. 4 supernovae were confirmed. 1 guest star was suspected as the original explosion of Cas A. 9 positions were suggested for further study in the hope of finding additional supernovae-radio sources.

125.009 Detonation von ^{12} C **— ein möglicher Supernova-Mechanismus.** L. Kühn.
Sterne, 46. Jahrgang, p. 68 - 69 (1970).

125.010 Effects of central pulsars on supernova envelopes.
M. J. Rees.
Astrophys. Letters, Vol. 6, 55 - 58 (1970).

The luminosity of an expanding supernova envelope is estimated, taking into account the possible energy input from a newly formed central pulsar, and it is shown that the general shape of the type II supernova light curve can be reproduced. Other consequences of the hypothesis that pulsars form in type II supernovae are mentioned briefly.

125.011 The frequency of supernovae in our Galaxy, estimated from supernova remnants detected at 178 MHz. J. L. Caswell.
Astron. Astrophys., Vol. 7, 59 - 64 (1970).

A search for supernova remnants in our Galaxy, using a radio survey at 178 MHz, is described; from the results, the mean interval between supernova outbursts in our Galaxy is estimated to lie between 40 and 80 years. This is compared with previous radio determinations and with the rate derived for external galaxies using optical data.

125.012 Supernovae. C. Kowal, H. A. Dottori.
IAU Circ. No. 2226 (1970).

125.013 Nucleosynthesis in supernova models. I. The neutrino-transport model. W. D. Arnett, J. W. Truran.
Astrophys. Journ., Vol. 160, 959 - 970 (1970).

A critical analysis is presented of the nucleosynthesis conditions predicted by supernova models built upon the mechanism of the transport of energy by neutrinos. At the extreme densities demanded in these models to ensure that neutrinos leaving the core will interact with the surrounding medium, a substantial neutron enrichment occurs in the ejected material. The subsequent thermonuclear processing of this material is found to result in the production of very massive nuclei.

125.014 Dynamics of supernova explosions.
L. N. Ivanova, V. S. Imshennik, D. K. Nadezhin.
Nauchn. Informatsii, vyp. (No.) 13, p. 3 - 93 (1969).
In Russian.

The development of hydrodynamic instability after the exhaustion of the nuclear energy of a star (the gravitational collapse) is investigated. The hydrodynamic equations are solved numerically together with equations which describe the kinetics of oxygen burning. The neutrino emission as well as the photodesintegration of nuclei of the Fe-group are taken into account.

125.015 **Supernova remnants, pulsars and X-ray sources.**
D. K. Milne.
Proc. Astron. Soc. Australia, Vol. 1, 333 - 334 (1970).

125.016 **Nonthermal galactic radio sources.**
D. K. Milne.
Australian Journ. Phys., Vol. 23, 425 - 444 (1970).

A catalogue is presented of 97 supernova remnants and a distance scale, based on the radio surface brightness, is established for this type of radio source. The data are used to derive the galactic distribution and to test theories relating to the evolution of these objects.

125.017 **A search for gamma-ray emission from supernova remnants and its relation to the origin of cosmic rays.** T. C. May.
Thesis, Univ. Minnesota, Minneapolis. [Available from Univ. Microfilms, Ann Arbor, Mi.], 150 pp. (1969).

The physical significance of future observations of gamma rays from the supernova remnants, Cassiopeia A and the Crab nebula is discussed in relation to the origin of galactic cosmic rays. The experimental apparatus is described. The results of partially successful flight which produced a 20 minute observation of Cass A are presented.

Propagation of a strong shock wave into the extended envelope of supergiant stars and the light curves of supernovae. See Abstr. 064.054.

r-process nucleosynthesis of C^{12} seed nucleus.
See Abstr. 065.085.

On the effects of a supernova explosion in a close binary system. See Abstr. 117.011.

High-frequency stellar oscillations. III. A brief report. See Abstr. 122.093.

About a similarity in the structure of envelopes ejected by novae and supernovae.
See Abstr. 124.006.

Predicted parameters of supernova pulsars.
See Abstr. 141.008.

Pulsars and X-ray-emitting supernova remnants.
See Abstr. 141.036.

On the connexion between pulsars and supernovae.
See Abstr. 141.050.

3C 386: Radio galaxy or supernova remnant?
See Abstr. 141.061.

Percentage number of nonthermal sources in the northern and the southern halves of the Galaxy.
See Abstr. 141.200.

An X-ray survey of the Cassiopeia region and its implications concerning supernova remnants and the galactic source distribution. See Abstr. 142.043.

125.100 **Supernova in NGC 3389**

The supernova in NGC 3389.
A. D. Chuadze, T. I. Barblishvili.
Byull. Abastumansk. Astrofiz. Obs. No. 37, p. 9 - 12 (1969). In Russian.

The supernova in the galaxy NGC 3389 discovered by A. D. Chuadze in Abastumani on February 28-th, 1967, had on the same day a photovisual magnitude of 12.47 and a photovisual absolute one − 18.3. The magnitudes and light curve for the period February 28 − April 14 are given. The supernova belongs to type I.

125.101 **Supernova in NGC 1058**

Supernova in NGC 1058.
Astron. Tsirk., No. 542, p. 1 (1969).   In Russian.

125.102 **Supernova in IC 3476**

Supernova in IC 3476.
B. V. Kukarkin, D. Ja. Martynov, Grizunova.
IAU Circ. No. 2214 (1970).

Supernova in IC 3476.   K. Locher.
IAU Circ. No. 2229 (1970).

126 Low-luminosity Stars, Subdwarfs, White Dwarfs

126.001 Quasi-periodic outbursts in the white dwarf Haro-Luyten Taurus No. 76.
B. Warner, R. E. Nather.
Monthly Notices, Roy. Astron. Soc., Vol. 147, 21 - 26 (1970).

Photometric observations made with the 82-inch Struve reflector confirm Landolt's discovery that HL-Tau 76 is a variable star. The variations are quasi-periodic, with a time-scale of about 12.44 minutes and an amplitude ~0.3 magnitudes in white light. The light curve shows similarities to that of dwarf novae, but on a much shorter time-scale.

126.002 Rotation and the DC white dwarfs.
D. T. Wickramasinghe, P. A. Strittmatter.
Monthly Notices, Roy. Astron. Soc., Vol. 147, 123 - 131 (1970).

The effect of uniform rotation on spectra of white dwarfs is investigated to ascertain whether the apparently lineless DC stars may be accounted for in terms of extreme Doppler broadening but otherwise normal abundances and opacity sources. It is shown that, under these conditions, the hydrogen Balmer lines should be easily detectable even at maximum rotational velocity. The general absence of absorption lines of other elements in DA white dwarf spectra may, however, be understood in terms of even moderate rotation.

126.003 Hot white dwarfs.
W. B. Hubbard, R. L. Wagner.
Astrophys. Journ., Vol. 159, 93 - 100 (1970).

A sequence of models of white dwarfs including corrections for nonideality in the equation of state has been computed for a typical white-dwarf luminosity. The mass-radius relations found by Hamada and Salpeter are appreciably modified by thermal effects, particularly for white dwarfs of low mass. Our results for 40 Eri B, together with a recent determination of its radius, indicate a chemical composition heavier than helium.

126.004 Parent masses of cluster white-dwarfs.
E. M. Jones.
Astrophys. Journ., Vol. 159, 101 - 106 (1970).

An empirical limiting mass of 1.86 (–0.12, +0.48) $M_\odot$ is derived from the Hyades and from Procyon such that stars with an initial mass greater than the limit will become white dwarfs on the brighter of the two observed white-dwarf sequences.

126.005 The evolution of rotating white dwarfs.
R. Schwartz, S. Africk.
Astrophys. Letters, Vol. 5, 141 - 143 (1970).

The formation of a lattice by the ions in the interior of a differentially rotating white dwarf is important in changing the distribution of angular momentum. This effect can lead to the formation of an unstable, rigidly rotating core, in a time which may be less than 10^8 years. The lifetime of a differentially rotating white dwarf varies approximately as Z^{-6}, where Z is the charge on the ions in the interior.

126.006 Subluminous stars. VI. Photoelectric observations in the red and near-infrared. O. J. Eggen.
Astrophys. Journ., Vol. 159, 945 - 949 (1970).

The apparent bifurcation in the distribution of subluminous stars in the $(M_V, U - V)$-plane is shown, from observations in the (R, I)-system, to be a probable feature of the distribution in the $(M_{bol}, \log T_e)$-plane.

126.007 Nonequilibrium beta-processes as a source of the thermal energy of white dwarfs.
G. S. Bisnovaty-Kogan, Z. F. Seidov.
Astron. Zhurn. Akad. Nauk SSSR, Vol. 47, 139 - 144 (1970). In Russian. – English translation in Soviet Astron., AJ, Vol. 14, No. 1.

During the cooling and concentration of a white dwarf with a mass near limit mass at T = 0, two-stage reactions of beta-capture, for example, Fe → Mn → Cr begin in the centre of the star, in which the second stage proceeds outside equilibrium. This leads to emission of thermal energy, to slowing of contraction and to an increase of the evolution time of a star with mass in the interval $\Delta M/M \simeq 2 \times 10^{-4}$ near limit mass.

126.008 The stars of low luminosity.
W. J. Luyten.
Separate print Univ. Minnesota, Minneapolis, Minn., 48 pp. (1970).

A catalogue with a total number of 1055 low luminosity stars is presented.

126.009 On the theory of rapidly rotating white dwarfs.
D. M. Sedrakian, V. V. Papoian, E. V. Chubarian.
Monthly Notices, Roy. Astron. Soc., Vol. 149, 25 - 33 (1970).

The structure of rotating white dwarfs is investigated using a second-order perturbation analysis. It is shown that the second-order corrections are very small and that the first-order perturbation theory gives good results.

126.010 The atmospheres of helium-rich white dwarfs of spectral type DB. I. Bues.
Astron. Astrophys., Vol. 7, 91 - 108 (1970).

In order to explain the spectra of helium-rich white dwarfs, spectral type DB, flux constant model atmospheres are computed in the temperature range of 11000 - 21000° K with surface gravities log g = 7 and log g = 8. For 13 different helium-rich compositions with varied abundances of hydrogen as well as metals, pressure-temperature relations and absorption coefficients are calculated for a grid of 12 values in temperature, 24 values in electron pressure, and 57 wavelengths with a special program. Stratifications and fluxes of the non-blanketed models with flux constancy of $\leqq$ 1%, obtained by the method of Lucy are discussed for T_{eff} = 18000, 15500, 13200 and 12000° K.

126.011 The quadratic Zeeman effect and large magnetic fields in white dwarfs. G. W. Preston.
Astrophys. Journ. (*Letters*), Vol. 160, L143 - L145 (1970).

Comparison of the displacements of absorption lines in the spectra of white dwarfs with those produced by the quadratic Zeeman effect indicates that few if any white dwarfs with measurable absorption lines have surface magnetic fields as large as 5×10^5 gauss.

126.012 Magnetic observations of white dwarfs.
J. R. P. Angel, J. D. Landstreet.
Astrophys. Journ. (*Letters*), Vol. 160, L147 - L152 = Columbia Astrophys. Lab. Contr., No. 17 (1970).

A search has been made among the brighter DA-type dwarfs for magnetic fields by using a new, highly sensitive photoelectric polarimeter. No magnetic fields have been detected in the nine stars so far observed, and upper limits of about 10^5 gauss can be placed on fields that may be present.

Convection in envelopes of white dwarfs.
See Abstr. 064.035.

Nonradial oscillations of white dwarfs, hot white dwarfs, and 10 M$_\odot$ models. See Abstr. 065.032.

Interstellar Matter, Gaseous Nebulae, Planetary Nebulae

131 Interstellar Space, Interstellar Matter, Polarization of Starlight

131.001 **Retention of dust grains near galactic nuclei.**
N. C. Wickramasinghe.
Nature, Vol. 225, 145 - 147 (1970).
Grains responsible for the 2.2–22 μm infrared radiation
from galaxies are charged by the photoelectric effect and
effectively frozen into the ionized gas surrounding the galactic
nuclei.

131.002 **Interstellar dust and aromatic carbon.**
G. P. Vdovykin.
Nature, Vol. 225, 254 (1970).
Saslaw and Gaustad have suggested that the particles in
interstellar dust clouds are diamond grains. My investigations
show that the diamonds in meteorites are not present as mo-
nomineral grains but in the form of fine intergrowths of
diamond micromonocrystals with graphite.

131.003 **Ionization front interactions in interstellar gas.**
M. C. Marsh.
Monthly Notices, Roy. Astron. Soc., Vol. 147, 95 - 114
(1970).
An investigation is made into the interaction between
a rapidly expanding planar H II region, bounded by a weak
R-type ionization front, and a contact discontinuity which
separates two regions of unionized gas of differing densities
in the H I region. It is found that an interaction with high
density gas results in the evolution of a slow moving weak
D-type front with an isothermal shock reflected towards the
illuminating star.

131.004 **Formaldehyde absorption in the galaxy.**
J. B. Whiteoak, F. F. Gardner.
Astrophys. Letters, Vol. 5, 5 - 10 (1970).
Formaldehyde absorption was detected against 31 out
of 34 galactic sources or source components. The clouds with
the highest optical depths are found to be physically associated with HII regions. In addition to following the well-
known spiral arms, the galactic distribution of the absorbing
clouds delineates a spiral feature that is tangential to the
line of sight at $l^{\mathrm{II}} = 305$ °.

131.005 **New OH sources associated with H II regions.**
D. Downes.
Astrophys. Letters, Vol. 5, 55 - 58 (1970).
Eighty galactic sources have been examined for hydroxyl
lines. Two new sources have been observed in absorption and
eight in emission. The absorption-line sources are characterized
by moderately high opacities and are believed to be molecular
clouds only a few hundred pc distant. The emission-line sources lie close to galactic continuum sources, of which at least
five are compact H II regions.

131.006 **Stokes parameters for OH sources.**
W. A. Coles, V. H. Rumsey.
Astrophys. Journ., Vol. 159, 247 - 261 (1970).
Stokes parameters for certain of the 1665- and 1667-MHz
OH emission lines of the sources in W3, W22, W24B2, W33,
W43, W49, Ori A, Cas A, and NGC 6334; and for the 1612-
and 1720-MHz lines of W28, G5.9–04, W43, W44, and W51
were measured during 1968 February and May. At 1665 MHz

but not at 1667 MHz, W33 showed strong time variation from
February to May and during a 10-day period in May. There is
evidence of time variability in W3 at 1665 MHz.

131.007 **Collisions between H I clouds. I. One-dimensional
model.**   M. E. Stone.
Astrophys. Journ., Vol. 159, 277 - 292 (1970).
A numerical, hydrodynamical model has been developed
for head-on collisions between two identical interstellar clouds.
The clouds have been idealized by taking them to be slabs of
gas infinite in extent transverse to the direction of collision.
In all cases strong shocks form at impact and rapidly proceed
through the clouds. In the nonadiabatic cases the shocked gas
quickly cools and is compressed to very high density. When
the shocks reach the outer surface of the coalesced clouds,
they disappear, and the gas reexpands. At a point in the
reexpansion the surface satisfies the conditions for Rayleigh-
Taylor instability so that some gas must be ejected. In general,
the time for the expansion to halt and reverse is considerably
longer than the mean time between collisions. The gravitational
stability of the gas during the compression phase has been
examined analytically.

131.008 **Collisions between H I clouds. II. Two-dimensional
model.**   M. E. Stone.
Astrophys. Journ., Vol. 159, 293 - 307 (1970).
A two-dimensional numerical, hydrodynamical model
has been developed for head-on collisions between two
identical interstellar clouds. In the direction of collision
the motion is nearly identical with that of the one-dimensional
model discussed previously. In the transverse direction the
very outer layers expand after being shocked. The inner
layers contract for a while under the influence of gravity, but
the development of a transverse pressure gradient and the
reexpansion in the direction of collision prevent gravitational
instability. About 25 percent of the mass reaches escape
velocity in the expansion, probably breaking up into low-
mass cloudlets during the ejection.

131.009 **Microwave absorption of the $2_{12} \rightarrow 2_{11}$ rotational
transition in interstellar formaldehyde.**
N. J. Evans II, A. C. Cheung, R. M. Sloanaker.
Astrophys. Journ. (*Letters*), Vol. 159, L9 - L14 (1970).
The $2_{12} \rightarrow 2_{11}$ rotational transition of H_2CO at
14488.65 MHz has been observed in absorption in the direction of Sgr B2, Sgr A, and W51, and upper limits for its intensity set for various other directions. Implications of these
observations for radiation at a 2-cm wavelength and for the
occurrence of $H_2C^{12}O$ and $H_2C^{13}O$ are discussed.

131.010 **Radio search for interstellar H_2^+.**
K. B. Jefferts, A. A. Penzias, J. A. Ball, D. F.
Dickinson, A. E. Lilley.
Astrophys. Journ. (*Letters*), Vol. 159, L15 - L17 (1970).
A search was made for H_2^+ by using recently determined
frequencies. The results were negative, and the upper limits
are reported.

131.011 **Polarization of μ Cephei and graphite core-ice
mantle grains.**   J. Svatoš.

Bull. Astron. Inst. Czechoslovakia, Vol. 21, 54 - 55 (1970).

It is shown that the theoretical curve for graphite core-ice mantle grains having parameters R_0 = 0.05μ and R = 0.13μ gives a good fit with the observational points according to Coyne and Gehrels (1966).

131.012 The interstellar absorption in Selected Area 47.
J. J. Schreur.
Astron. Journ., Vol. 75, 38 - 41 (1970).

A study of the obscuration in the direction of Selected Area 47 (l^{II} = 159°, b^{II} = –21°) has been made. A total visual absorption of $2^m 5$ was found within a distance of 500 pc.

131.013 Wavelength dependence of polarization. XXI.
R Monocerotis.   B. Zellner.
Astron. Journ., Vol. 75, 182 - 185 (1970).

Strong, variable optical polarization is found for the diffuse object R Monocerotis, which is generally believed to be a protoplanetary system. The observations apparently rule out electron scattering or Rayleigh scattering as a source of the polarization, but may be compatible with scattering by larger grains. A survey of nine other objects located in cometary nebulae revealed no polarizations of comparable strength.

131.014 The motion of an interstellar grain.
J. G. Ireland.
Astrophys. Space Sci., Vol. 6, 107 - 117 (1970).

The motion of an interstellar grain is discussed, with account taken of the various resistive forces acting on it. Grains may survive and escape from an H II region surrounding a central star in periods of ~ 10^6 – 10^8 years, depending on the temperature of the exciting star. It is suggested that, under special circumstances, gas may diffuse out of the region faster than the grains.

131.015 The shape of the interstellar absorption band.
N. C. Wickramasinghe, K. Nandy.
Astrophys. Space Sci., Vol. 6, 154 - 156 (1970).

Profiles of the 4430 band are calculated for resonant absorbers distributed within graphite particles, silicate particles and solid H_2 grains. The sizes of grains adopted are those which give agreement with the interstellar extinction. Only in the latter two cases can satisfactory agreement be obtained with recent observational data.

131.016 Galactic longitude dependence of the intensity ratio of the diffuse interstellar absorption bands at λλ6180 and 4430.   M. Rudkjøbing.
Astrophys. Space Sci., Vol. 6, 157 - 160 (1970).

Equivalent widths, as published by Seddon, for the diffuse interstellar absorption band at λ6180 are compared with photoelectric λ4430 indices for ten stars in the northern Milky Way. The intensity ratio is found to depend on galactic longitude in a way similar to that found by Johnson for the ratio of total to selective interstellar absorption. The dependence found points to a direction for the galactic magnetic field that is in better agreement with the direction derived from rotation measures of extragalactic sources than with that based on the dust-hypothesis interpretation of interstellar optical polarization.

131.017 Effect of interstellar density fluctuations on signal dispersion measure.   I. Lerche.
Astrophys. Space Sci., Vol. 6, 287 - 292 (1970).

The effect of electron number density fluctuations in the interstellar medium on signals from pulsars is studied in terms of the frequency dependent signal dispersion. It is shown that if the density fluctuations are representative of long wavelength disturbances in the interstellar gas, then the observed signal dispersion is not a measure of the integral of the electron number density in the line of sight.

131.018 Intensity of the infrared radiation from interstellar grains in the solar neighborhood.
K. S. K. Swamy.
Astrophys. Space Sci., Vol. 6, 474 - 480 (1970).

The expected intensitiy distribution of the infrared radiation in the solar neighborhood from the grain models of dirty ice, graphite and graphite core-dirty ice mantle has been calculated. It is found that the expected intensity from grain models at 100 μ agrees reasonably well with the observations of Hoffmann and Frederick.

131.019 The effect of turbulent fluctuations in the interstellar gas and magnetic field on Faraday rotation.
I. Lerche.
Astrophys. Space Sci., Vol. 6, 481 - 491 (1970).

The effect of fluctuations in both the interstellar electron number density and galactic magnetic field on the propagation of high frequency radio waves is discussed in terms of the frequency dependent Faraday rotation. It is shown that when the fluctuations are representative of large scale disturbances (1–10^2 pc) in the interstellar medium, then the observed Faraday rotation is not a measure of the line of sight integral of the product of the magnetic field with the electron number density.

131.020 Atomic and molecular hydrogen in interstellar space.
G. R. Carruthers.
Space Sci. Rev., Vol. 10, 459 - 482 (1970).

This paper reviews the present state of knowledge of the abundances and physical state of interstellar atomic and molecular hydrogen. Much new data in this area have been obtained in recent rocket observations. There have also been new developments as a result of ground-based infrared and 21-cm observations, and theoretical research.

131.021 Interstellar masers.
D. F. Dickinson, M. M. Litvak, B. M. Zuckerman.
Sky Telescope, Vol. 39, 4 - 7 (1970).

131.022 The proper polarization of the radiation of a star, detected from its variability, and a new possibility for the discovery of circumstellar matter.
V. A. Dombrovskij.
Stars, Nebulae, Galaxies, Symposium Byurakan 1968, (see 012.002), p. 179 - 187 (1969).   In Russian.

131.023 On the luminescence of absorption regions of the Galaxy.   D. A. Rozhkovskij.
Stars, Nebulae, Galaxies, Symposium Byurakan 1968, (see 012.002), p. 189 - 199 (1969).   In Russian.

131.024 Composition de la poussière interstellaire.
J. E. Gaustad.
L'Astronomie, 84e année, p. 101 - 109 (1970). – Translated from Astron. Soc. Pacific, Leaflet No. 483 (1969).

131.025 Is magnetic alignment of interstellar dust really necessary?   M. Harwit.
Nature, Vol. 226, 61 - 62 (1970).

The evidence for magnetic field alignment of interstellar dust is weak and the main argument in its favour seems to have been that no reasonable alternative existed. The purpose of this letter is to show that there is a quite straightforward alignment mechanism, and that it produces a degree of polarization roughly consistent with observations.

131.026 Further evidence for the occurrence of a broad interstellar absorption band in the far ultraviolet.
K. Nandy, H. Seddon.
Nature, Vol. 226, 63 - 64 (1970).

The purpose of this letter is to present further data to

confirm the occurrence of an interstellar absorption band in the far ultraviolet.

131.027 Interstellar and atmospheric clouds.
G. R. Evans, M. V. Penston.
Nature, Vol. 225, 357 - 358 (1970).

Attention is drawn to the similarity between two apparently quite disparate phenomena, namely interstellar and atmospheric clouds. In both cases, one is concerned with a medium in a two-phase state, one a high-density state and the other a low density state. It is being suggested that the exchanges which are continuously taking place between the two phases must ultimately determine the behaviour of the system as a whole.

131.028 Interstellar solid hydrogen: How much and where?
V. C. Reddish.
Nature, Vol. 225, 367 (1970).

From his earlier suggestion that hydrogen will freeze onto interstellar grains and therefore control star formation and galactic evolution, the author estimates that 2×10^4 to 5×10^5 solar masses of solid hydrogen can be expected in the Galaxy. It probably occurs in the spiral arms in clouds of about 10^4 solar masses with apparent angular diameters averaging 2 arc minutes in a range up to 20 arc minutes. Suggestions for observational confirmation in the infrared and microwave regions are made.

131.029 Interstellar formaldehyde absorption near the galactic center.
F. F. Gardner, J. B. Whiteoak.
Astrophys. Letters, Vol. 5, 161 - 166 (1970).

Observations of formaldehyde absorption at 4829 MHz, with a resolution of 100 kHz, have been made at 24 positions in the galactic centre region. The results are compared with the corresponding OH and H I data. The radial velocities of the absorption maxima seem to fit continuous sequences in a velocity-longitude diagram and these sequences are interpreted in terms of a kinematic model containing expansion and contraction.

131.030 Electron temperatures of H II regions.
P. A. Shaver.
Astrophys. Letters, Vol. 5, 167 - 171 (1970).

Electron temperatures are determined from 85 MHz continuum observations for eleven nebulae. Evidence obtained from these and other radio frequency continuum observations concerning the electron temperatures of H II regions is summarized; generally the temperatures are considerably less than 10^4 °K, and range from 4000°K to 10000°K in different nebulae, in agreement with theoretical expectations and other methods of temperature determination.

131.031 Regioni galattiche compatte di idrogeno ionizzato.
M. Felli.
Mem. Soc. Astron. Italiana, Nuova Serie, Vol. 41, 95 - 117 (1970).

Optical and radioastronomical researches indicate directly and indirectly the existence of compact components of *HII* located inside galactic *HII* regions of lower electron density. Various theoretical models which explain their existence and evolution are discussed. Several physical phenomena which appear to be connected with these compact *HII* regions are also discussed.

131.032 Die Beobachtung der interstellaren Magnetfelder.
I. Appenzeller.
SuW, Vol. 9, 112 - 116 (1970).

131.033 Light pressure of stars on spherical dust particles of the interstellar space.

N. B. Divari, L. V. Reznova.
Astron. Zhurn. Akad. Nauk SSSR, Vol. 47, 166 - 171 (1970). In Russian. English translation in Soviet Astron. AJ, Vol. 14, No. 1.

Tables of values of the ratio of light pressure to the gravitational force of stars of various spectral classes for spherical particles of water, quartz and graphite with radii from 1×10^{-7} to 1×10^{-3} cm, are given. Calculations are carried out by precise formulae of Mie's theory for complex indices of refraction, depending on wavelength.

131.034 A proposal for an X-ray analysis of interstellar grains.
P. G. Martin, D. W. Sciama.
Astrophys. Letters, Vol. 5, 193 - 196 (1970).

A method is proposed for detecting the presence of light elements in interstellar grains. The method depends on being able to observe around an X-ray source a halo arising from small-angle scattering by the grains. Near the K edges of elements in the grains this scattering is anomalous. The resulting effect on the intensity of the halo is estimated and shown to be appreciable in favourable cases. Brief reference is also made to K-edge absorption phenomena which are specific to elements in grains.

131.035 Search for interstellar NO at radio frequencies.
B. E. Turner, C. E. Heiles, E. Scharlemann.
Astrophys. Letters, Vol. 5, 197 - 201 (1970).

A search has been conducted for interstellar NO radio lines, at approximately 431, 411, 226 and 206 MHz at the Arecibo Observatory and at the National Radio Astronomy Observatory. No definite detection has been made in 67 galactic sources covered. The observations were carried out at frequencies calculated on the basis of existing laboratory data. Measured frequencies which later became available were within the search band which was centered on the calculated frequencies. Conclusions are drawn on the negative results of the search and on the accuracy of the frequency calculation.

131.036 Impure graphite grains and the interstellar extinction curve.
K. S. K. Swamy.
Observatory, Vol. 90, 52 - 54 (1970).

131.037 Observations of the $^2\Pi_{1/2}$, $J = \frac{1}{2}$ state of interstellar OH.
B. Zuckerman, P. Palmer.
Astrophys. Journ. (*Letters*), Vol. 159, L197 - L201 (1970).

The 140-foot telescope of the National Radio Astronomy Observatory has been used to study 6-cm Λ-doublet radiation from the $^2\Pi_{1/2}$, $J = \frac{1}{2}$ state of interstellar OH. The presence of an $F = 1 \to 0$ transition in the source W49 has been confirmed, and additional features in the source W3 have been detected. The degree of polarization of these lines is low.

131.038 The calcium atom-graphite grain 4430 Å model.
H. A. J. McIntyre, D. A. Williams.
Monthly Notices, Roy. Astron. Soc., Vol. 148, 53 - 61 (1970).

The shift and the width of the calcium-atom resonant line produced by the proximity of a graphite surface are calculated as a function of atom-surface distance. The results indicate that the commonly suggested model of calcium atoms in association with interstellar graphite grains is unlikely to be the cause of the unidentified interstellar absorption line at 4430 Å.

131.039 Scattering of plane electromagnetic waves by infinite concentric circular cylinders at oblique incidence.
G. A. Shah.
Monthly Notices, Roy. Astron. Soc., Vol. 148, 93 - 102 (1970).

As a potential model of interstellar grains, the problem of scattering of plane electromagnetic waves by infinite concentric homogeneous cylinders at oblique incidence is given.

The appropriate electromagnetic field components for boundary conditions have been derived. The solution for the unknown expansion coefficients has been presented in a matrix form suitable for carrying out numerical work. Some preliminary results for extinction and polarization efficiencies by infinite concentric homogeneous cylinders have been presented for normal as well as oblique incidence.

131.040 Graphite-silicate grain mixtures and the diffuse galactic light. N. C. Wickramasinghe.
Publ. Astron. Soc. Japan, Vol. 22, 85 - 91 (1970).
The mean albedoes and phase parameters are computed for various combinations of graphite particles, silicate grains, and silicate grains covered with ice mantles. Graphite-silicate grain mixtures which yield good agreement to the interstellar extinction curve and to the wavelength dependence of interstellar polarization possess albedoes and phase functions which are consistent with available data on the diffuse galactic light.

131.041 Anomalous interstellar OH lines. A. Żytkow.
Postępy Astron., Vol. 18, 226 - 234 (1970).
In Polish. – Short report.

131.042 Observations of interstellar Lyman-Alpha absorption with the Orbiting Astronomical Observatory. R. C. Bless, A. D. Code, T. E. Houck, C. F. Lillie, B. D. Savage.
Bull. American Astron. Soc., Vol. 2, 183 (1970). – Abstr. AAS.

131.043 Absorption of X rays in the interstellar gas. R. L. Brown, R. J. Gould.
Bull. American Astron. Soc., Vol. 2, 184 - 185 (1970). Abstr. AAS.

131.044 Interstellar formaldehyde abundances. D. Buhl, L. E. Snyder, B. Zuckerman, P. Palmer.
Bull. American Astron. Soc., Vol. 2, 185 (1970). – Abstr. AAS.

131.045 High-velocity gas collisions. T. L. Chow, M. P. Savedoff.
Bull. American Astron. Soc., Vol. 2, 187 (1970). – Abstr. AAS.

131.046 Observations of radio recombination lines of helium and "carbon" in galactic H II regions. E. Churchwell, P. G. Mezger.
Bull. American Astron. Soc., Vol. 2, 188 (1970). – Abstr. AAS.

131.047 Polarization and extinction shapes of diffuse interstellar bands. J. M. Greenberg, R. Stoeckly.
Bull. American Astron. Soc., Vol. 2, 194 (1970). – Abstr. AAS.

131.048 Far-ultraviolet wide-angle electronographic photography of the Orion region. R. C. Henry, G. R. Carruthers.
Bull. American Astron. Soc., Vol. 2, 198 (1970). – Abstr. AAS.

131.049 Observations of H II regions at 0.96, 1.65, and 2.73 cm. R. W. Hobbs, K. J. Johnston.
Bull. American Astron. Soc., Vol. 2, 199 - 200 (1970). – Abstr. AAS.

131.050 Laboratory studies of interstellar grains: The low-temperature absorption spectra of silicates and of "ices". C. Hunter, B. Donn.
Bull. American Astron. Soc., Vol. 2, 201 (1970). – Abstr. AAS.

131.051 Infrared pumping of microwave lines of OH, H_2O, and H_2CO near IR stars and shockwaves. M. M. Litvak.
Bull. American Astron. Soc., Vol. 2, 206 (1970). – Abstr. AAS.

131.052 Observations of 18-cm OH absorption. R. N. Manchester, B. J. Robinson, W. M. Goss.
Bull. American Astron. Soc., Vol. 2, 207 (1970). – Abstr. AAS.

131.053 Long-baseline interferometric measurements of the OH source NML Cygnus. J. M. Moran, A. H. Barrett, W. J. Wilson.
Bull. American Astron. Soc., Vol. 2, 209 - 210 (1970). Abstr. AAS.

131.054 Helium abundances from 3-cm recombination lines. P. Palmer, B. Zuckerman, R. H. Rubin, J. A. Ball.
Bull. American Astron. Soc., Vol. 2, 212 - 213 (1970). Abstr. AAS.

131.055 Search for new microwave spectral lines from interstellar molecules and atoms. J. M. Pasachoff, C. A. Gottlieb, L. E. Snyder, D. Buhl, P. Palmer, B. Zuckerman, D. F. Dickinson.
Bull. American Astron. Soc., Vol. 2, 213 (1970). – Abstr. AAS.

131.056 Interstellar formaldehyde absorption in dark nebulae. L. E. Snyder, P. Palmer, D. Buhl, B. Zuckerman.
Bull. American Astron. Soc., Vol. 2, 218 (1970). – Abstr. AAS.

131.057 Excitation of interstellar formaldehyde. P. M. Solomom, P. Thaddeus.
Bull. American Astron. Soc., Vol. 2, 218 - 219 (1970). Abstr. AAS.

131.058 Laboratory studies of interstellar molecules: Formaldehyde. L. J. Stief, S. Glicker, B. Donn, E. P. Gentieu, J. E. Mentall.
Bull. American Astron. Soc., Vol. 2, 219 - 220 (1970). Abstr. AAS.

131.059 Forty-seven new galactic OH sources. – Distribution and preliminary properties. B. E. Turner.
Bull. American Astron. Soc., Vol. 2, 222 (1970). – Abstr. AAS.

131.060 Observations of the anomalous microwave recombination line at 21-cm wavelength. B. Zuckerman, J. A. Ball.
Bull. American Astron. Soc., Vol. 2, 227 (1970). – Abstr. AAS.

131.061 Diamonds and the interstellar extinction curve. R. Landau.
Nature, Vol. 226, 924 (1970).
It is concluded that a simple size distribution of diamond particles can be found which gives a calculated extinction curve satisfying the observational data as satisfactorily as any other model of interstellar grains so far proposed. It should be noted, however, that the existence of such a distribution does not constitute a strict test for the position of the dust grains.

131.062 Das dynamische Verhalten interstellarer Staub-teilchen beim Wolkenstoß. Teil I. Ohne Berück-sichtigung von Masseänderungen. H. Zimmermann.
Astron. Nachr., Vol. 292, 17 - 29 = Mitt. Univ.-Sternw. Jena, No. 91 (1970).

During the collision of interstellar clouds a partial separation between gas and dust occurs. It can be expected that also a separation between heavier and lighter dust particles takes place. To determine the ratio of this dynamical effect the way of dust particles with different values of the product $\alpha \times \rho_p$ (α radius; ρ_p density of the particles) during the three successive cooling periods is numerically calculated. It is shown that the heavier particles ($\alpha \times \rho_p \gtrsim 5 \times 10^{-5}$ g/cm^2), at the end of the collision and the expansion period are gathered in a thin sheet in the inner parts of the new-built cloud whereas the lighter ones ($\alpha \times \rho_p \lesssim 1 \times 10^{-5}$ g/cm^2) are distributed more or less uniformly among the gas of the cloud.

131.063 A study of interstellar matter in the Cassiopeia-Perseus region. II. Structure of interstellar neutral hydrogen gas in the Perseus arm. F. Sato.
Annals Tokyo Astron. Obs., Second Ser., Vol. 12, 1 - 33 (1970).

The structure of neutral hydrogen gas in the Perseus arm region is investigated in the present paper. The study is based upon the first edition of the *Maryland–Green Bank Galactic 21-cm Line Survey* and the equal-velocity contour diagrams constructed from it. Equal-velocity contour diagrams in the vicinity of five galactic clusters and/or H II regions located in the studied region are examined to find any neutral hydrogen gas cloud associated with them.

131.064 A study of interstellar matter in the Cassiopeia-Perseus region. III. A search for early-type stars and a study of interstellar absorption in the region of emission nebula IC 1795. F. Sato.
Annals Tokyo Astron. Obs., Second Ser., Vol. 12, 34 - 50 (1970).

The present paper deals with the photometric and spectrographic observations of the stars in the emission nebula IC 1795, which was identified with the thermal radio source W3. The observations were made in order to search for the exciting stars of the nebula and to investigate the interstellar K-line of Ca II in its direction.

131.065 Observations of interstellar water vapor. B. E. Turner, D. Buhl, E. B. Churchwell, P. G. Mezger, L. E. Snyder.
Astron. Astrophys., Vol. 4, 165 - 172 (1970).

A search has been made for interstellar H_2O emission in 53 galactic sources and for NH_3 in 16 galactic sources using the NRAO 140-foot telescope at 1.35 cm wavelength. Two new H_2O emission sources have been found, one associated with an IR star and the other with the Cygnus-1 OH emission source. Observations at high frequency resolution have been made of three known H_2O emission sources. Analysis of these spectra shows that the known hyperfine components are in no case present with LTE intensity distribution.

131.066 Observations of the anomalous microwave recom-bination line at 11 cm wavelength. B. Zuckerman, P. Palmer.
Astron. Astrophys., Vol. 4, 244 - 247 (1970).

The anomalous microwave recombination line at 11 cm wavelength has been detected in NGC 2024, Orion A, and IC 1795 and not detected in M 17, W 51, and W 49. In NGC 2024 the peak antenna temperature of the anomalous line increases between 6 and 11 cm wavelengths. This increase is not expected if the line is emitted under the conditions of local thermodynamic equilibrium.

131.067 Interaction between interstellar hydrogen and the solar wind. P. W. Blum, H. J. Fahr.
Astron. Astrophys., Vol. 4, 280 - 290 (1970).

A consistent calculation concerning the interplanetary density distribution of cold hydrogen of interstellar origins is given. The calculations take account of the macroscopic velocity v_0 of the interstellar gas relative to the solar system, of the focusing effect of the gravitational field of the sun and of loss processes by charge exchange with solar wind protons and EUV-ionization. The results show a density decrease from outer to inner regions of the system that depends on the solar EUV-flux. During the solar cycle the depth of penetration of interstellar hydrogen into the solar system varies between 6 and 1 AU. The production of secondary fast hydrogen by charge exchange processes of the cold hydrogen with solar wind protons has been investigated. The results concerning density and flux of secondary neutrals give some indication that the magnetic shock front should be located near 80 AU.

131.068 Equivalent widths and optimum regions of space for the detection of interstellar H_2. D. C. Cartwright, S. Drapatz.
Astron. Astrophys., Vol. 4, 443 - 451 (1970).

The equivalent widths for the rotational lines from $J'' = 0.1$ of the Lyman and Werner bands of H_2 originating from $v'' = 0$ have been calculated for kinetic temperatures of 10 °K and 100 °K, and H_2 column densities from 10^{+11} to 10^{+18} cm^{-2}. From examination of data on the character and location of dust clouds and UV stars, a number of dust clouds with bright background stars have been determined which are promising candidates for the detection of absorption by H_2. Those H_2 absorption lines which are least likely to be obscured by other spectral lines are given as determined by comparison with recent rocket UV spectra from typical background stars.

131.069 Motions and physical conditions inside dust clouds. C. Heiles.
Astrophys. Journ., Vol. 160, 51 - 58 (1970).

OH profiles have been obtained at various positions within four dust clouds. No variation of velocity within any cloud is visible, an observation which indicates little rotation. All clouds show profiles which are probably double or multiple; the lack of velocity variation with position permits the conclusion that the components arise from independent entities.

131.070 Observations of the spatial structure of interstellar hydrogen. II. Optical determination of distances in a small region. S. Ames, C. Heiles.
Astrophys. Journ., Vol. 160, 59 - 64 (1970).

A configuration of interstellar neutral hydrogen, suggested earlier to be in the form of sheets, is examined in more detail, by using both the 21-cm radio data and optical spectra of stars in that region of the sky showing interstellar absorption lines. We conclude that the sheets are fairly thin, with a thickness of the order of less than 10 pc, and that they are approaching each other and are in fact colliding supersonically in one region.

131.071 Internal kinematics of two bright H II regions. M. G. Smith, D. W. Weedman.
Astrophys. Journ., Vol. 160, 65 - 74 = Contr. Kitt Peak National Obs., No. 507 (1970).

A pressure-scanned, Fabry-Pérot interferometer has been used to determine profiles of the [O III] $\lambda 5007$ emission line in the Orion nebula and M8 with a spectral resolution of 4 km sec^{-1} and angular resolutions varying from 7$''$ to 21$''$. Most of the turbulent motions in the nebulae are found to be subsonic, with an average most probable speed of 7.6 km sec^{-1} in the Orion nebula. Line profiles from two condensations (the Hourglass) in the center of M8 indicate that these condensations are apparently expanding, and slitless spectra are used to derive the density and structure of the Hourglass.

131.072 The ratio of atomic hydrogen to dust in the direction of the Omega nebula.
R. J. Quiroga, C. M. Varsavsky.
Astrophys. Journ., Vol. 160, 83 - 88 = Contr. Argentine – Carnegie Radio Astron. Station, Inst. Argentino Radioastron., Dep. Terr. Magnetism, No. 5 (1970).

The 21-cm line of atomic hydrogen was observed in absorption against the continuum background of the thermal source known as the Omega nebula. The absorption profile was observed in directions where there is relatively little dust and also in the direction of a dense dust cloud. A decrease in the optical depth of the 21-cm line in the latter direction indicated that in the presence of dust part of the hydrogen may become molecular. Because of the size of the beam it is not possible to make a precise quantitative determination of the decrease in the number of hydrogen atoms.

131.073 Search for ^{18}OH in emission.
J. A. Ball, H. Penfield.
Astrophys. Journ., Vol. 160, 349 - 351 (1970).

A search was made for radio emission from ^{18}OH at 1584.33 MHz in the direction of NML Cygnus, VY Canis Majoris, and W3-OH at velocities that correspond to the 1612 MHz emission from ^{16}OH in these sources. No ^{18}OH emission was found. For NML Cygnus, the upper limit is less than the terrestrial isotopic abundance ratio times the strength of the 1612-MHz emission.

131.074 On the question of interstellar silicate absorption.
B. Donn, K. S. K. Swamy, C. Hunter.
Astrophys. Journ., Vol. 160, 353 - 355 (1970).

The observations of the absorption feature in the infrared spectrum by Knacke, Gaustad, Gillett, and Stein are discussed; and doubts concerning their identifying silicate minerals in interstellar grains are raised.

131.075 Studies of H_2O sources by means of a very-long-baseline interferometer.
B. F. Burke, D. C. Papa, G. D. Papadopoulos, P. R. Schwartz, S. H. Knowles, W. T. Sullivan, M. L. Meeks, J. M. Moran.
Astrophys. Journ. (Letters), Vol. 160, L63 - L68 (1970).

Observations of the H_2O sources associated with the objects W3 (OH position), Orion nebula, W49, and VY Canis Majoris with coherent interferometers of baseline lengths 1.7×10^7 and 5.0×10^7 wavelengths have been made. None of the sources have been resolved at the limit of our resolution of about $0.''003$ of arc. The angular separation of two features in W49 is found to be $0.''7$.

131.076 On the initial excitation temperature of interstellar H_2CO molecules.
T. Oka.
Astrophys. Journ. (Letters), Vol. 160, L69 - L71 (1970).

The excitation temperature determined from the lowest K-type doubling line of the interstellar H_2CO molecules is discussed. It is shown that, because of the rigorous dipole selection rules, the molecules are "adiabatically cooled" as they make transitions to lower levels by spontaneous emission. If a Boltzmannian rotational distribution is assumed at the time of formation, the initial excitation temperature is given by $kT_e = 2hB$, regardless of the temperature of formation.

131.077 Absorption interstellaire dans la direction du Grand Nuage de Magellan.
C. Fehrenbach, J.-P. Brunet, E. Maurice, L. Prévot.
Comptes Rendus, Acad. Sci. Paris, Sér. B, Vol. 270, 1504 - 1506 (1970).

Détermination de l'absorption interstellaire $\overline{A}_v = 0.30 \pm$ 0.02 dans la direction du Grand Nuage de Magellan à partir de la classification spectrale et des mesures UBV de 60 étoiles membres du Grand Nuage.

131.078 Observations of interstellar formaldehyde.
B. Zuckerman, D. Buhl, P. Palmer, L. E. Snyder.
Astrophys. Journ., Vol. 160, 485 - 506 (1970).

In this paper we present a detailed report of observations of $H_2^{12}C^{16}O$ undertaken with the 140-foot telescope at the National Radio Astronomy Observatory in Greenbank, West Virginia, during 1969 March and June. The $1_{11} \rightarrow 1_{10}$ transition has been observed in absorption in the direction of numerous background sources of various types. H_2CO radial velocities have been compared with those of other atomic and molecular constituents of the interstellar medium.

131.079 The interstellar extinction of stars in H II regions.
C. M. Anderson.
Astrophys. Journ., Vol. 160, 507 - 517 (1970).

Observations are presented which show that several stars involved in H II regions exhibit a color anomaly similar to that found for the Trapezium. Alternative explanations of the phenomenon are reviewed.

131.080 An experimental study of the dust of iron, carbon, silicon carbide and silica.
J. Lefèvre.
Astron. Astrophys., Vol. 5, 37 - 44 (1970).

Grains of iron, carbon, silicon carbide and silica have been produced by striking an electric arc in argon. Their shapes, sizes and grouping have been studied with an electron microscope. The size of the grains varies from 100 Å to 2000 Å. They are spherical only for silica and in every case are associated in chain like structures which can be explained by van der Waals forces for the non-metallic particles. The optical absorption of the clouds of dust has been measured at wavelengths between 3600 Å and 7000 Å and the results compared with Mie's theory for spheres of infinite cylinders of the same size. A good agreement is obtained only for silica.

131.081 Properties of H II regions obtained from non-LTE analysis of radio recombination line data.
R. M. Hjellming, R. D. Davies.
Astron. Astrophys., Vol. 5, 53 - 67 (1970).

A non-LTE analysis of all available α-, β-, γ-, δ-, and ϵ-radio recombination line data determines electron temperatures, electron concentrations, and emission measures for the dominant emitting gas in seven H II regions. For the Orion nebula, M17, W51, and W3, the radio lines are emitted from extremely small dense regions having temperatures of the order of 10^4 °K.

131.082 A recombination line of mapping the H II region W3 (IC 1795).
R. H. Rubin, P. G. Mezger.
Astron. Astrophys., Vol. 5, 407 - 412 (1970).

The H109α-recombination-line emission from the H II region W3 (IC 1795) was observed in a grid of 43 points which are separated by 3' (approximately one-half the half-power beamwidth of the telescope). The observed quantities are displayed as contour maps.

131.083 Search for an effect of the sun on the frequency of 18-centimeter radiation.
J. A. Ball, D. F. Dickinson, A. E. Lilley, H. Penfield, I. I. Shapiro.
Science, Vol. 167, 1755 - 1757 (1970).

The frequency of an OH emission source near W28 was monitored during a close approach by the sun. The anomalous frequency shift previously reported by Sadeh et al. was not seen although the present experiment was somewhat more precise.

131.084 Observations of radio recombination lines at λ = 18 cm. E. Churchwell, J. Edrich.

Astron. Astrophys., Vol. 6, 261 - 267 (1970).

Measurements of radio recombination lines at λ = 18 cm are reported in six galactic H II regions. A test for systematic line broadening over that expected from Doppler broadening is made and found to be negligible in all H II regions considered except in Orion A, for which a comparison is made with Griem's (1967) theory of Stark broadening. It is shown that the "carbon" recombination line intensity is definitely frequency dependent in Orion A and IC 1795, but the helium-to-hydrogen number density ratio is not frequency dependent within the measuring accuracy.

131.085 Ultraviolet interstellar extinction from a comparison of ϵ Persei and ζ Persei. T. P. Stecher.

IAU Symposium No. 36, (see 012.014), p. 24 - 27 (1970).

Hall's and Stebbins and Whitford's extinction pair has been used to determine interstellar extinction in the ultraviolet.

131.086 Observations of interstellar extinction in the ultraviolet with the OAO satellite.

R. C. Bless, B. D. Savage.

IAU Symposium No. 36, (see 012.014), p. 28 - 35 (1970).

We present a preliminary analysis of a number of spectrophotometric scans of reddened and unreddened early-type stars obtained with the OAO-A2 satellite. A brief comparison of our observed extinction curves is made with theoretical particle models.

131.087 On dielectric models of interstellar grains.

J. M. Greenberg, R. Stoeckly.

IAU Symposium No. 36, (see 012.014), p. 36 - 41 (1970).

Silicate, ice, and silicate core-ice mantle particles are considered with a view to describing or predicting ultraviolet features in both extinction and polarization which depend on the dielectric nature of the particles.

131.088 Extinction curves for graphite-silicate grain mixtures.

N. C. Wickramasinghe.

IAU Symposium No. 36, (see 012.014), p. 42 - 49 (1970).

Interstellar grains may be regarded as a mixture of at least two distinct, highly refractory components – graphite and silicates. We present here a discussion of the optical properties of such mixtures with particular reference to the interstellar extinction curve in the rocket ultraviolet.

131.089 Measurement of interstellar extinction in emission line stars. R. Viotti.

IAU Symposium No. 36, (see 012.014), p. 50 - 51 (1970).

131.090 The interstellar extinction curve from 4000 Å to 6500 Å.

G. A. H. Walker, J. B. Hutchings, P. F. Younger.

IAU Symposium No. 36, (see 012.014), p. 52 - 56 (1970).

Interstellar extinction curves of 20 Å resolution have been obtained at the DAO from photoelectric scanner observations in the range 4000 Å to 5000 Å for five stars, and of 50 Å resolution for four stars in the range 4000 Å to 6500 Å from Willstrop's photoelectric data.

131.091. Interstellar molecular hydrogen. P. M. Solomon.

IAU Symposium No. 36, (see 012.014), p. 320 (1970). – Abstract.

131.092 On the physical conditions in interstellar H I gas.

M. Grewing, U. Mebold, K. Rohlfs.

IAU Symposium No. 38, (see 012.013), p. 182 - 185 (1970).

Weighted average values of the ionization ratio n_e/n_H for the interstellar gas can be obtained from a comparison of pulsar data and 21-cm emission measurements. For high latitude pulsars the resulting ionization ratios are high (n_e/n_H = 0.39), temperatures for the 'neutral' gas above 10^4 K are obtained. For low latitude pulsars the values found are n_e/n_H = 0.07 and T_e = 2000 K.

131.093 Polarization of southern OB-stars.

G. Klare, T. Neckel.

IAU Symposium No. 38, (see 012.013), p. 449 - 451 (1970).

The polarization data of 1421 southern OB-stars of the Heidelberg catalogue have been measured and plotted in a galactic l^{II}, b^{II}-diagram. For some longitude intervals the relationship of the standard deviation of the electric vector alignment and the galactic longitude was computed.

131.094 Radial velocities and distances of galactic H II regions. Y. P. Georgelin, Y. M. Georgelin.

Astron. Astrophys., Vol. 6, 349 - 363 (1970).   In French.

174 radial velocities and distances of optical H II regions are given here.

131.095 A survey of H 109α recombination line emission in galactic H II regions of the southern sky.

T. L. Wilson, P. G. Mezger, F. F. Gardner, D. K. Milne.

Astron. Astrophys., Vol. 6, 364 - 384 (1970).

A survey of hydrogen 109α-recombination lines was carried out during 1968 using the 210-foot radio telescope of the CSIRO Division of Radiophysics, together with the NRAO 6-cm cooled parametric amplifier. One hundred and forty-nine galactic sources and one extragalactic source were observed. Line emission was detected in a total of 130 of the sources in the survey. Kinematic distances, using the Schmidt model, have been calculated for those sources exhibiting the H 109α line. In addition, intrinsic source parameters, such as the mass of ionized hydrogen and the average electron density, have been calculated, assuming the source is a uniform, spherical region whose distance is given from kinematics. The percentage number of non-thermal sources is compared with that of the NRAO-MIT survey.

131.096 On the electron temperature in H II regions.

R. Louise.

Astron. Astrophys., Vol. 6, 460 - 463 (1970).

In this paper new results obtained by the method using widths at half-intensity of Hα and [N II] lines are presented. The observations are done with a high resolution Fabry-Pérot interferometer at the 80 cm and 120 cm telescopes of the Haute Provence Observatory. From these observations the electron temperature and the turbulence in five H II regions are found. Using the spectral type given by Georgelin (1969) for the exciting stars, we have compared our results with Hjellming's theoretical predictions.

131.097 Der kosmische Staub in der Galaxis.

G. N. Sjagajlo.

Astron. in der Schule, 7. Jahrgang, p. 60 - 65 (1970).

131.098 Determination of the coordinate (R. A.) of some galactic sources of anomalously excited hydroxyl.

N. V. Bystrova, I. V. Gosachinskii, T. M. Egorova, N. V. Karlov, B. B. Krynetskii, N. F. Ryzhkov.

Dokl. Akad. Nauk SSSR, Ser. Mat. Fiz., Vol. 191, 791 - 794 (1970).   In Russian.

131.099 Ultraviolet absorption lines in H I regions heated by cosmic rays. J. Silk.

Astrophys. Letters, Vol. 5, 283 - 285 (1970).

A small but significant fraction of the more abundant atoms in H I regions will be ionized by low energy cosmic rays, if such particles are indeed responsible for heating and ionizing the neutral hydrogen. Some of the higher ion states in H I regions should be detectable by rocket or satellite spectroscopic observations of the ultimate absorption lines.

131.100 Erratum: Recombination lines in thermal and non-thermal galactic sources. [Astrophys. Letters, Vol. 4, 121 - 127 (1969)].   D. K. Milne, T. L. Wilson, F. F. Gardner, P. G. Mezger.
Astrophys. Letters, Vol. 5, 287 (1970).

131.101 Staubteilchen im interstellaren Raum. K.-H. Schmidt.
Sterne, 46. Jahrgang, p. 49 - 53 (1970).

131.102 Observations of high-velocity hydrogen clouds at 21 cm.   S. Y. Meng, J. D. Kraus.
Astron. Journ., Vol. 75, 535 - 562 (1970).

A 21-cm sky survey of high-velocity neutral hydrogen clouds has been conducted with the Ohio State University 260 × 70 ft radio telescope using a 16-channel receiver. Hydrogen clouds were found as follows: four principal groups of clouds, 16 individual clouds and a large, continuous group of clouds of neutral hydrogen with high negative velocities, and seven groups and a few isolated clouds with medium negative velocities. The results indicate that the high- and medium-velocity clouds seem to be associated. Cloud characteristics are tabulated and maps of the clouds are presented.

131.103 The interstellar reddening law in the direction of Ara OB1.   R. K. Honeycutt, R. S. Chaldu.
Astron. Journ., Vol. 75, 600 - 601 = Publ. Goethe Link Obs., No. 107 (1970).

The wavelength dependence of interstellar absorption has been studied for a region of the southern Milky Way in the constellation Ara using photographic spectrophotometry of objective-prism spectra. The reddening law was found to be similar to the law in Cygnus.

131.104 Is the diffuse interstellar absorption band at 4430 Å caused by trivalent iron?   P. G. Manning.
Nature, Vol. 226, 829 - 830 (1970).

The wavelength, half-width and asymmetry of the diffuse interstellar absorption at 4430 Å are reminiscent of $^6A_1(S) \rightarrow$ $^4A_1\,^4E(G)$ crystal-field transition in trivalent iron ions in terrestrial silicates. A silicate matrix is suggested from the position of the ultraviolet absorption edge. Because the above electronic transition is independent of the crystal field, spectra of Fe^{3+} are similar in tetrahedral and octahedral fields, and it is not possible therefore to identify the silicate positively. It is also shown that absorption bands at other wavelengths in the interstellar spectra are not incompatible with known features of Fe^{3+} spectra.

131.105 An argument against cooling saturation during the collapse of interstellar gas clouds.   A. E. Wright.
Observatory, Vol. 90, 99 - 101 (1970).

It is the purpose of this paper to present a simple argument why this should be so. It is shown that a general cooling mechanism does not "saturate", i.e. that the rate of cooling does not increase more slowly than the rate of compressional heating.

131.106 Interstellar carbon monoxide.   K. B. Jefferts, N. J. Penzias, A. A. Penzias, R. W. Wilson.
IAU Circ. No. 2231, 2242 (1970).

131.107 Interstellar molecular hydrogen. G. R. Carruthers.
IAU Circ. No. 2250 (1970).

131.108 Radio emission from HCN.

W. E. Howard, D. Buhl, L. E. Snyder.
IAU Circ. No. 2251 (1970).

131.109 On the interaction of cosmic X-rays with interstellar grains.   P. G. Martin.
Monthly Notices, Roy. Astron. Soc., Vol. 149, 221 - 235 (1970).

An investigation of the interaction of cosmic X-rays with interstellar grains has been undertaken to determine what might be learned about the grains by observations at X-ray energies. To begin, a review of the extinction of X-rays by grains is given. Then the phenomenon of the X-ray halo is discussed. The many factors which can affect the intensity distribution of the halo are stressed so that in interpreting the observations a good estimate of the grain size and shape can be made.

131.110 Intrinsic polarization of early-type stars with extended atmospheres.   K. Serkowski.
Astrophys. Journ., Vol. 160, 1083 - 1105 (1970).

Polarimetric observations were made for forty Be and shell stars. About half of these stars show evidence of intrinsic polarization, in most cases changing with time. There is no evidence of intrinsic polarization in the early-type stars which are not surrounded by extended envelopes of shells. Polarimetric observations were made for the stars for which diameters were measured with an intensity interferometer. An attempt has been made to separate intrinsic and interstellar polarization under the assumption that the color-dependence law of intrinsic as well as interstellar polarization is already sufficiently known.

131.111 Superbright radio "knots" in the H II region W51. G. K. Miley, B. E. Turner, B. Balick, C. Heiles.
Astrophys. Journ. (*Letters*), Vol. 160, L119 - L123 (1970).

Fringes have been detected from the H II region W51 at a wavelength of 11 cm with an interferometer of lobe separation $\sim 0\overset{''}{.}6$. These observations indicate that the brightness temperature may be in excess of $10^{6\,\circ}K$. Some possible implications of this result are discussed.

131.112 Detection of the $^2\Pi_{3/2}$, $J = ^7/_2$ state of interstellar OH at a wavelength of 2.2 centimeters.
B. E. Turner, P. Palmer, B. Zuckerman.
Astrophys. Journ. (*Letters*), Vol. 160, L125 - L129 (1970).

The $F = 4 \rightarrow 4$ Λ-doublet transition in the $^2\Pi_{3/2}$, $J = ^7/_2$ state of interstellar OH has been detected in the source W3. The other three transitions ($F = 4 \rightarrow 3$, $3 \rightarrow 3$, and $3 \rightarrow 4$) were not detected in W3, and none of the four transitions were observed in the eight other sources examined.

131.113 Anomalous absorption of microwaves by interstellar H_2CO.   M. M. Litvak.
Astrophys. Journ. (*Letters*), Vol. 160, L133 - L138 (1970).

It is suggested that gravitationally unstable dust clouds may contain a shock-heated layer where H_2CO molecules are formed. Their infrared emission lines pump the downstream H_2CO to absorb the 2.7°K cosmic background at 6 cm, as observed.

131.114 The structure of the stellar field in the direction of the cluster NGC 6913. I. Interstellar absorption.
N. B. Kalandadze, L. N. Kolesnik.
Astrometriya i Astrofiz., *Kiev*, No. 8, (see 003.024), p. 58 - 62 (1969).   In Russian.

An investigation of interstellar absorption was undertaken in a Milky Way field around NGC 6913. By means of BV-photometry and spectral classification the spatial distribution of absorbing matter was derived. The distance modulus of the open cluster NGC 6913 based on BV-photometry and spectra is found to be $11\overset{m}{.}4$.

131.115 Interstellar absorption in the direction of the star cluster NGC 6823. G. L. Fedorchenko.
Astrometriya i Astrofiz., *Kiev*, No. 8, (see 003.024), p. 62 - 64 (1969). In Russian.

Interstellar absorption is investigated by the colour-excess method in a region around the open cluster NGC 6823.

131.116 Interstellar extinction by graphite, iron and silicate grains. N. C. Wickramasinghe, K. Nandy.
Nature, Vol. 227, 51 - 53 (1970).

Recent data on the interstellar extinction curve and on the diffuse galactic light may be accounted for by a mixture of graphite, iron and silicate grains.

131.117 Polarization-wavelength profile of the interstellar 4430 Å absorption band. K. Nandy, H. Seddon.
Nature, Vol. 227, 264 - 265 (1970).

By the use of a Wollaston prism at the Cassegrain focus of the 36-inch telescope of the Royal Observatory, Edinburgh, we have been able to derive a polarization profile across the 4430 band. This report describes the method and presents the preliminary results.

131.118 Self-polarization of the light of late type stars. R. A. Vardanian.
Astron. Tsirk., No. 550, p. 4 - 7 (1970). In Russian.

131.119 Interstellar matter and planetary nebulae. D. E. Osterbrock.
Trans. IAU, Vol. 14A, (see 003.028), 387 - 404 (1970). Report of Commission 34.

131.120 Light scattering in reflection nebulae. I. R. Shortt.
Proc. Astron. Soc. Australia, Vol. 1, 330 - 332 (1970).

131.121 Measurements of OH in Sagittarius A and B2. R. X. McGee, F. F. Gardner, M. W. Sinclair.
Proc. Astron. Soc. Australia, Vol. 1, 334 - 336 (1970).

131.122 On the polarization of the 18 cm OH emission. A. G. Bromley.
Proc. Astron. Soc. Australia, Vol. 1, 347 - 348 (1970).

131.123 18 cm observations of galactic OH from longitudes 350° to 50°.
B. J. Robinson, W. M. Goss, R. N. Manchester.
Australian Journ. Phys., Vol. 23, 363 - 404 (1970).

Observations have been made of OH emission and absorption for 15 sources between galactic longitudes 350° and 50°. The four Stokes parameters of the OH emission have been measured on all four 18 cm transitions. Positions have been measured for the 12 OH sources which have not had their positions determined by interferometry or lunar occultations. The relationship of the OH to the background continuum has been investigated from the positional information and from a comparison of OH radial velocities with recombination line velocities of the H II regions. Secular variations have been observed in five cases.

131.124 OH absorption at 1667 MHz near the galactic centre. B. J. Robinson, R. X. McGee.
Australian Journ. Phys., Vol. 23, 405 - 423 (1970).

Absorption by OH at 1667 MHz has been measured for galactic longitudes between 357°30′ and 3°20′ and latitudes between +0°20′ and −0°40′. The results for $0°10′ \geqslant b^{II} \geqslant -0°20′$ are presented as contours of absorption temperature in the longitude–radial velocity plane. A comparison with 1420 MHz observations of neutral hydrogen reveals many differences between the distributions and opacities of H and OH. The motions of the OH near the galactic nucleus differ markedly from those deduced from the H line measurements.

131.125 Scattering of cosmic X-rays by interstellar dust grains. S. Hayakawa.
Progr. Theor. Phys. Japan, Vol. 43, 1224 - 1230 (1970).

The present paper deals with the scattering of soft X-rays for a number of representative models of interstellar grains.

131.126 Optical observation of galactic H II regions in Hα emission, (II). K. Ishida, M. Ohashi.
Tokyo Astron. Bull., Second Series, No. 200, p. 2317 - 2336 (1970).

Hα photometry has been made for the six galactic H II regions, IC 1848, IC 434 region, NGC 2174–5, NGC 2244 region, NGC 7000 region, and NGC 7822.

131.127 Hydrogenic spectral lines in radio astronomy. P. G. Mezger.
National Radio Astron. Obs., *Green Bank*, Repr. Ser. B, No. 160 [Reprinted from *"Physics of the One- and Two-Electron Atoms"*, North-Holland, 1969, p. 801 - 823], 23 pp. (1969). – Review article.

131.128 Observations of radio emission from galactic H_2O. A. H. Barrett, P. Rosenkranz, P. R. Schwartz, J. W. Waters, W. J. Wilson, C. A. Zapata.
Mass. Inst. Technol. Res. Lab. Electronics Quarterly Progr. Rep., No. 93, p. 29 - 32 (1969).

Observations of 1.35 cm line from the $6_{16} \rightarrow 5_{23}$ rotational transition of water vapour. Eight of the old regular northern sources came good with spectra. For W49 the Haystack antenna temperature was greater than 1500 deg K. – *RXM*

131.129 Further observations of galactic water-vapor emission. A. H. Barrett, P. R. Schwartz.
Mass. Inst. Technol. Res. Lab. Electronics Quarterly Progr. Rep., No. 94, p. 48 - 54 (1969).

New results concerning time variations, linear polarization and high resolution spectroscopy in water vapour emission sources from Haystack at 22 GHz. Variable features are not polarized while linearly polarized ones are steady. – *GD*

131.130 Observations of OH emission associated with infrared stars. W. J. Wilson, A. H. Barrett.
Mass. Inst. Technol. Res. Lab. Electronics Quarterly Progr. Rep., No. 93, p. 26 - 28 (1969).

Report on search for OH emission in 66 infrared stars. Eight successful detections were made. The satellite 1612 MHz line is the strongest, the 1720 MHz apparently not present. – *RXM*

131.131 Kinematics of the interstellar gas and the brightness temperature of the 21-cm line emission. K. Rohlfs.
Bonn, Max-Planck-Institut für Radioastronomie, 35 pp. (1970).

Two models are proposed: A one component model with uniform spin temperature and a two component model with (a) cold clouds (b) hot intercloud medium. – *RXM*

131.132 Anomalous recombination line 166α. A. Pedlar.
Nature, Vol. 226, 830 - 831 (1970).

A spectral line corresponding to the "anomalous" recombination line at 5.011 MHz (109α) described by Palmer *et al.* has been detected at 1.425 MHz (166α) in the HII regions NGC 2024, W3 and possibly W43. The intensity of the anomalous line compared with that of hydrogen is much greater for the 166α line than for the 109α line.

131.133 Discovery of interstellar water vapour.
A. H. Barrett.
Comments Atomic Molecular Phys., Vol. 1, No. 5, p. 93 -
96 (1969). − See Phys. Abstr., Vol. 73, No. 13025 (1970).

131.134 Microwave spectroscopy of interstellar ammonia.
A. H. Barrett.
Comments Atomic Molecular Phys., Vol. 1, No. 2, p. 60 -
63 (1969). − See Phys. Abstr., Vol. 73, No. 13026 (1970).

131.135 An empirical determination of the interstellar field.
C. F. Lillie.
Thesis, Univ. Wisconsin, Milwaukee. [Available from Univ.
Microfilms, Ann Arbor, Mi.], 110 pp. (1968).
  The interstellar radiation field in the λ 2000 − 5560 Å
region has been determined from photoelectric observations
of the night sky brightness obtained with an Aerobee rocket
in the ultra-violet and with a ground based telescope in the
visual region. The interstellar radiation density was determined at each wavelength and has a mean value of
5×10^{-17} ergs/cm^{-3}-Å.

131.136 The interstellar extinction of stars in H II regions.
C. M. Anderson.
Thesis, California Inst. Technol., Pasadena. [Available from
Univ. Microfilms, Ann Arbor, Mi.], 139 pp. (1969).

131.137 Polarization of interstellar OH emission.
W. A. Coles.
Thesis, Univ. California, San Diego. [Available from Univ.
Microfilms, Ann Arbor, Mi.], 89 pp. (1969).
  A polarimeter feed of novel design has been constructed
for use with the 85-foot Hat Creek radio-telescope of the
University of California. This instrument was used, during
February and May of 1968, to measure the Stokes parameters
of a group of 18 cm OH-line sources.

131.138 Interstellar OH molecules.
Y. Terzian, E. Scharlemann.
Earth Extraterrestr. Sci., Vol. 1, No. 4, p. 103 - 122 (1970).
  This paper summarizes research on the interstellar OH
molecule. The molecular lines so far observed are tabulated
along with other pertinent data.

131.139 Observations of interstellar molecular hydrogen.
M. W. Werner.
Thesis, Cornell Univ., Ithaca, N. Y. [Available from Univ.
Microfilms, Ann Arbor, Mi.], 203 pp. (1969).
  An observational program for the detection of interstellar
molecular hydrogen was conceived and carried out. The obser-
vational technique involves a search for vibration-rotation
emission lines from the ground electronic state of H$_2$.

131.140 Considérations sur l'origine des particules de Platt.
S. Codina.
Dep. Astron. Fis., Fac. Humanidades Ciencias, Univ. Montevi-
deo, Publ. No. 34, 21pp. (1969).

131.141 The interstellar reddening law in the direction of
Cep OB3. R. K. Honeycutt.
Thesis, Case Western Reserve Univ. [Available from Univ.
Microfilms, Ann Arbor, Mi.], 114pp. (1969).
  The interstellar reddening law between 3360 Å and
8400 A has been derived for the members of the association
Cep OB3 and for an adjacent comparison region.

131.142 Contribution à l'étude de l'hydrogène ionisé
interstellaire.   D. Hoang-Binh.
Thesis, Doct. Sci. Phys., Paris [Centre Document. C.N.R.S.
No. 3769], 66 pp. (1969). − See Bull. Signalét., Vol. 31,
Section 120, No. 7361 (1970).

The Interstellar Medium.   See Abstr. 003.039.

Microwave detection of H$_2$ ^{18}O.
See Abstr. 022.012.

Probabilities for radiation and predissociation.
II. The excited states of CH, CD, and CH$^+$, and some astrophysical implications.   See Abstr. 022.022.

Non-similarity linear wave cloud collapse solutions.
See Abstr. 065.064.

Time delays of electromagnetic pulses due to molecular resonances in the atmosphere and the interstellar medium.
See Abstr. 082.004.

Exploration of the solar system and of interstellar
space.   See Abstr. 091.048.

Interstellar hydrogen densities in the surroundings
of the solar system.   See Abstr. 106.020.

Delta Scorpii, an infrared deficient star and the
value of R for the Scorpius region.
See Abstr. 113.014.

The intrinsic colours of stars and two-colour reddening lines.   See Abstr. 113.018.

Groupes physiques très jeunes. III. L'influence du
« coude » dans les lois d'extinction.   See Abstr. 113.044.

Effects of reddening on colour transformations.
See Abstr. 113.053.

Observations of interstellar sodium lines in stars in
the direction of the galactic center.
See Abstr. 114.018.

OH radio emission associated with infrared stars.
See Abstr. 114.073.

Observations of interstellar Lyman-α absorption.
See Abstr. 114.100.

Observations of interstellar Lyman-α with the
Orbiting Astronomical Observatory.
See Abstr. 114.101.

Interstellar lines other than hydrogen.
See Abstr. 114.102.

The nebulosity near the infra-red, OH and H_2O
source, VY CMa.   See Abstr. 122.024.

Interstellar reddening of RR Lyrae.
See Abstr. 122.049.

Infrared line emission from H II regions.
See Abstr. 132.009.

Properties of H II regions obtained from non-LTE
analysis of radio recombination lines.
See Abstr. 132.015.

Detection of He 109α and C 109α recombination
lines in southern galactic sources.   See Abstr. 132.031.

Size and motion of the interstellar scintillation
pattern from observations of CP 1133.
See Abstr. 141.053.

High resolution measurements of 21-cm absorption in the direction of W49.   See Abstr. 141.115.

Pulsar distances, spiral structure and the interstellar medium.   See Abstr. 141.144.

Search for interstellar formic acid at 1638.806 MHz in radio sources.   See Abstr. 141.174.

X-ray scattering by grains in the direction of the Crab pulsar and Sco XR-1.   See Abstr. 142.036.

Galactic component of the diffuse X-ray background.   See Abstr. 142.048.

Low-temperature regions of interstellar gas and formation of stellar associations.   See Abstr. 152.007.

Investigation of absorption in the region of the open cluster NGC 7086 and determination of the cluster distance.   See Abstr. 153.028.

Fluctuations in brightness of the Milky Way and interstellar clouds.   See Abstr. 155.016.

Space distribution of interstellar dust in connection with the galactic spiral structure.   See Abstr. 155.040.

A survey of H 109 α recombination line emission in galactic H II regions of the northern sky. See Abstr. 157.011.

H II–zones in central regions of nine normal galaxies.   See Abstr. 158.016.

Formation of gas clouds in galactic nuclei. See Abstr. 158.047.

The extinction law in the Magellanic Clouds. See Abstr. 159.003.

132 Emission Nebulae, Reflection Nebulae

132.001 Internal dust in gaseous nebulae.
J. S. Mathis.
Astrophys. Journ., Vol. 159, 263 - 275 (1970).

The emergent flux from spherical nebulae is presented as a function of optical depth and albedo of internal isotropically scattering particles. Several different structures of the emission and absorption, including emission-free shells, are considered. In color-color plots, the tracks of increasing optical depth are remarkably insensitive to the geometry of the nebula. For some proposed interstellar particles, such as graphite with and without coatings of ice and hydrogen, the paths in a color-color plot are about the same for both internal and external dust, although the optical depth associated with a given point is about twice as great for internal as for external extinction. In the Orion Nebula, the intensity and color of the scattered light are consistent with about 1.8 mag of internal extinction and 0.35 mag of external extinction.

132.002 Internal radial velocities of some selected small diameter H II regions. R. A. Williamson.
Astrophys. Space Sci., Vol. 6, 45 - 59 (1970).

A photographic Fabry-Pérot interferometer was built for the University of Maryland's 20-in. telescope for the study of the radial velocities of diffuse emission nebulae. Observations of S101, S112, IC5146, NGC7380, NGC281, and IC1795 were made and radial velocities at many points within each nebula were determined with an estimated error of ±3 km/sec. Average heliocentric radial velocities were obtained for each nebula and they agree well with those found by other observers.

132.003 Fast motions in nebulae, and the stellar wind.
S. B. Pikel'ner.
Stars, Nebulae, Galaxies, Symposium Byurakan 1968, (see 012.002), p. 159 - 163 (1969).   In Russian.

132.004 Investigation of instationary objects in comet-shaped nebulae. Eh. A. Dibaj.
Stars, Nebulae, Galaxies, Symposium Byurakan 1968, (see 012.002), p. 165 - 172 (1969).   In Russian.

132.005 Peculiarities of the continuous spectrum and the structure of the nebula (IC 349) surrounding
Merope. Yu. I. Glushkov.
Stars, Nebulae, Galaxies, Symposium Byurakan 1968, (see 012.002), p. 201 - 206 (1969).   In Russian.

132.006 Level populations of excited states of hydrogen in gaseous nebulae. M. Brocklehurst.
Nature, Vol. 225, 618 - 619 (1970).

The level populations of excited states of hydrogen are calculated for temperatures and densities typical of gaseous nebulae, and for $20 \leqslant n \leqslant 300$. Collisional transitions between any two principal quantum levels are allowed for. The infinite set of simultaneous equations is reduced to a convenient size by a matrix condensation process. It is shown that the results differ considerably from those obtained allowing for $n \rightarrow n \pm 1$ collisions only.

132.007 Detection of the H65α radio recombination line in
Orion A. E. Churchwell, P. Mezger, E. Reifenstein III, R. Rubin, B. Turner.
Astrophys. Letters, Vol. 5, 157 - 160 (1970).

We report on observations of the H65α-recombination line at 23.404 GHz in the center of Orion A. The observed set of parameters is consistent with α-line parameters observed at other frequencies. The LTE-electron temperature is

7200^{+3300}_{-2000} °K. The implications of the H65α line results on the non-LTE theory of radio recombination lines are discussed.

132.008 Infrared studies of galactic nebulae. I. NGC 6523,
NGC 6572, and BD 30°3639.
F. C. Gillett, W. A. Stein.
Astrophys. Journ., Vol. 159, 817 - 822 = Contr. Kitt Peak National Obs. No. 505 (1970).

Observations from $\lambda = 3\mu$ to $\lambda = 14\mu$ have been made of the galactic diffuse nebula NGC 6523 (M8) and two planetary nebulae BD 30°3639 and NGC 6572. The results are compared with earlier observations of the Orion nebula and the planetary nebula NGC 7027. The possible sources of the infrared energy observed from the "hourglass" region of M8 are discussed. The observations show differences in the shape of the spectra of the planetaries which indicate either variations in temperature or differences in chemical composition.

132.009 Infrared line emission from H II regions.
V. Petrosian.
Astrophys. Journ., Vol. 159, 833 - 846 (1970).

Formulae are given relating the infrared line intensities to radio and optical fluxes of H II regions. The line intensities expected from the Orion nebula are calculated. With present techniques a number of fine-structure lines and only a few infrared recombination lines of hydrogen could be detected The ratios of infrared to optical recombination-line intensities could be used to determine the interstellar absorption.

132.010 Interferometric observations of the thin-filamentary
nebula NGC 6888. T. A. Lozinskaya.
Astron. Zhurn. Akad. Nauk SSSR, Vol. 47, 122 - 128 (1970). In Russian. English translation in Soviet Astron. AJ, Vol. 14, No. 1.

A large series of NGC 6888 observations with a high contrast Fabry-Pérot étalon and image converter in Hα, Hβ, [NII] lines was carried out. A splitting of a spectral line profile into several components with a half-width of 30—40 km/sec was detected, corresponding to galactic background emission, approaching and receding parts of the expanding nebula. The velocity of expansion was determined to be in the range from 55 to 110 km/sec. From observed half-widths of Hα and [NII] ($\lambda = 6583$ Å) the temperature $T_e = 19000 \pm 4000$ °K was determined.

132.011 On the determination of helium abundance from
radio recombination lines.
E. Churchwell, P. G. Mezger.
Astrophys. Letters, Vol. 5, 227 - 231 (1970).

New observational evidence is presented that (1) the relative abundance of singly ionized helium, as determined from radio recombination lines, is not affected by possible non-LTE effects; and that (2) the ionization state of helium changes – if at all – only very slowly with position in M17 and M42. The agreement of radio determinations with recent optical determinations of helium abundances in M8, M17, and M42 is shown to be very good.

132.012 Aperture synthesis observations of Orion A at
2.695 GHz. W. J. Webster, Jr.,W. J. Altenhoff.
Astrophys. Letters, Vol. 5, 233 - 238 (1970).

The radio source Orion A has been mapped with the 2.7-km interferometer of the National Radio Astronomy Observatory. With the synthesized half-power beamwidth of 7 by 35 arc sec, a considerable amount of structure has been detected. In addition to a broad component apparently associated with the structure detected in single antenna observations, smaller components apparently associated with each of the O stars in the θ Orionis complex have been detected. The small structures can be interpreted as the remnants of the protostar cloud complex from which the stars of the Orion nebula formed.

132.013   Evidence of expansion in NGC 6888.
          Y. P. Georgelin, G. Monnet.
Astrophys. Letters, Vol. 5, 239 - 243 (1970).
      Interferometric Hα radial velocities in the shell nebula NGC 6888 show an expansion of 50 km sec⁻¹ around the exciting Wolf-Rayet star. NGC 6888 is thus interpreted as coming from ejection of material from the central exciting star, and physical data on this process are given.

132.014   The study of the H_α half-width and the radial
          velocities in the Cygnus loop.
V. T. Doroshenko.
Astron. Zhurn. Akad. Nauk SSSR, Vol. 47, 292 - 296 (1970). In Russian. – English translation in Soviet Astron., AJ, Vol. 14, No. 2.
      On the basis of the analysis of Hα half-widths it is shown that the mean half-width of the Hα line is equal to 40 km/sec in filaments, and 63 km/sec in diffuse patterns.

132.015   Properties of H II regions obtained from non-LTE
          analysis of radio recombination lines.
R. M. Hjellming, R. D. Davies.
Bull. American Astron. Soc., Vol. 2, 199 (1970). – Abstr. AAS.

132.016   Radio interferometer observations of galactic
          H II regions.   W. J. Webster, Jr., G. K. Miley,
J. E. Wink, R. M. Hjellming, P. G. Mezger, W. J. Altenhoff.
Bull. American Astron. Soc., Vol. 2, 224 (1970). – Abstr. AAS.

132.017   A comparison of a 15.4-GHz map of M16 with the
          OB stars in the cluster NGC 6611.
M. Felli, E. Churchwell.
Astrophys. Journ., Vol. 160, 43 - 49 (1970).
      The 15.4-GHz map of M 16 indicates the existence of three separate components. The flux density of each component was used to calculate the Lyman-continuum flux required to give the measured radio-flux density. A comparison with the Lyman-continuum flux emitted by the OB stars in the nebula indicates that not all the sources of ionization have been identified.

132.018   On the value of R for NGC 2244.
          R. J. Dufour, P. Lee.
Astrophys. Journ., Vol. 160, 357 - 361 = Contr. Louisiana State Univ. Obs., *Baton Rouge,* No. 33 = Contr. Kitt Peak National Obs., No. 541 (1970).
      Comparison of the surface brightness of the Rosette nebula at 10 cm, Hβ, and Hγ indicates an average value of R to be 3.96.

132.019   On the motions of gas and dust in NGC 2068.
          A. Stockton, G. Chapman.
Publ. Astron. Soc. Pacific, Vol. 82, 306 - 312 = Contr. Kitt Peak National Obs. No. 517 (1970).
      The spectrum of the diffuse nebula NGC 2068 is composite, showing nebular emission lines as well as the scattered spectrum of the illuminating stars. The difference in radial velocity between the emission and absorption spectra is ~ 100 km/sec; this difference can be interpreted as resulting from the expansion of the dust cloud about the central star.

132.020   The helium spectrum and dust in the Orion nebula.
          R. R. Robbins.

Astrophys. Journ., Vol. 160, 519 - 530 (1970).
      Capture-cascade intensities appropriate to diffuse nebulae are calculated and then modified for self-absorption in the helium triplets by the construction and solution of an integral equation of transfer. The nebula is considered to be a spherically symmetric gas cloud, either static or expanding with a constant velocity gradient. The application of such calculations to the Orion nebula make it possible to infer the total abundance of helium and the population of all levels (n, l). Since motions within the nebula encourage photon escape through Doppler shifting, some insight into the state of motion may be obtained. Finally, it will be shown below that conclusions may be drawn concerning the dust opacity and its effects in the nebula.

132.021   A study of two bright rims in the North America
          nebula.   R. Louise.
Astron. Astrophys., Vol. 5, 35 - 36 (1970).
      A plate of NGC 7000 has been taken with a narrow interference filter centered on the Hα line. From this, both the electron density and the brightness profile of the bright rim are derived. We found that the bright rim profile presented a double peak.

132.022   Peculiar motions of neutral hydrogen in the vicinity
          of the Orion nebula.   T. K. Menon.
Astron. Astrophys., Vol. 5, 240 - 243 (1970).
      A detailed study of the distribution of 21-cm emission in the vicinity of the Orion nebula shows the presence of a neutral hydrogen cloud close to the nebula which is approaching the nebula with a velocity of about 14.2 km/s. It is suggested that the conical absorbing band seen in projection against the nebula is part of the above cloud. The minimum density of neutral hydrogen atoms in the cloud is about 100 cm^{-3}.

132.023   Submillimeter radiation from the Orion nebula.
          W. M. Fark, D. G. Vickers, P. E. Clegg.
Astron. Astrophys., Vol. 5, 325 - 327 (1970).
      The flux from the Orion nebula at 900 μm has been measured to be $(4.5 \pm 1.6) \times 10^{-22} \text{ W/m}^2 / \text{Hz}$. This result is discussed in the context of previously reported results.

132.024   Level populations of hydrogen in gaseous nebulae.
          M. Brocklehurst.
Monthly Notices, Roy. Astron. Soc., Vol. 148, 417 - 434 (1970).
      The departures of the populations of excited levels of hydrogen, from those under conditions of thermodynamic equilibrium, are calculated. Corresponding tables are given for $n \leqslant 300$ and for a wide range of electron temperatures and densities.

132.025   Lyman-α radiation from nebular objects.
          N. Panagia, M. Fulchignoni.
IAU Symposium No. 36, (see 012.014), p. 349 - 354 (1970).
      For a simplified model of a gaseous nebula, that is spherical, isothermal, homogeneous and composed of pure hydrogen, the amount of Lyman-α radiation emitted from the surface of the nebula is evaluated for various physical conditions. The problem of observability of the Lyman-α emission line is examined, taking into account absorption by dust and by interstellar neutral hydrogen.

132.026   Hα photography of the Orion nebula with a
          half-angstrom filter.
R. R. Fisher, R. A. Williamson.
Astron. Journ., Vol. 75, 347 - 350, 517 - 520 (1970).
      The Orion nebula has been photographed at a number of wavelengths near the Hα line with a filter of half-width at half-maximum transmission of 0.47 Å. Structural differences

in the nebula are revealed on examination of filtergrams obtained at different wavelength settings of the filter. Bright knots, condensations, and sharp intensity gradients are visible at different positions in some of the filtergrams. Many of these features do not appear in a wider-bandpass filtergram (10 Å) of the same region. A Doppler-velocity image, constructed from the photographic subtraction of pairs of filtergrams, is also presented.

132.027 Further observations of the filamentary nebulosity around the Cetus Arc (Loop 2).
K. H. Elliott, J. Meaburn.
Astrophys. Space Sci., Vol. 7, 252 - 260 (1970).

Some further observations have now been made of this filamentary nebulosity. More photographs have been added to the original mosaic. No proper motion greater than 0."5/year was detected. Some further ridges of nebulosity were discovered. A shell model for Loop 2 based on this upper limit to the proper motion of the nebulosity and its similarity to the Cygnus Loop is presented. Some comments on the excitation mechanism of the nebulosity are also made.

132.028 Observations of the Orion nebula at low radio frequencies.
Y. Terzian, A. Parrish.
Astrophys. Letters, Vol. 5, 261 - 265 (1970).

New, low-frequency flux density measurements are reported at 73.8, 111.5 and 196.5 MHz for the Orion nebula. These observations were made with the 1000-ft radio telescope at the Arecibo Observatory. A short discussion on the spectrum and electron temperature of the Orion nebula is given.

132.029 11 cm observations of small nebulae associated with Wolf-Rayet stars.
L. F. Smith, R. A. Batchelor.
Australian Journ. Phys., Vol. 23, 203 - 216 (1970).

Total masses of three small H II regions are derived from (1) total flux densities obtained from 11 cm observations with the 210 ft radio telescope of the Australian National Radio Astronomy Observatory, (2) electron densities derived from spectra of the λ 3727 doublet of [O II], and (3) photometric distances for the stars associated with the nebulae. The nebulae are of particular interest because they are among seven now known which consists of an arc or arcs of nebulosity centred on a Wolf-Rayet star. Mass estimates made by Johnson and Hogg (1965) from radio observations at 21 cm (continuum) of two other of the seven nebulae have been revised in accord with new photometric distances for the stars concerned. The derived masses range from 6 to 1000 solar masses.

132.030 Recombination lines at λ3 cm from the core of the Orion nebula.
M. A. Gordon.
Astrophys. Letters, Vol. 6, 27 - 34 (1970).

Observations of the 85α and 106β recombination lines show helium and hydrogen to have similar states of thermodynamic equilibrium and imply the ratio of helium to hydrogen number densities to be 0.073 ± 0.008. The 85α, 106β, 121γ, and 133δ hydrogen line intensities are consistent with a mean electron temperature of 10^4 °K over the core of the nebula but imply variations in temperature and hence variations in physical conditions over intervals of 1.5 arc min.

132.031 Detection of He 109α and C 109α recombination lines in southern galactic sources.
P. G. Mezger, T. L. Wilson, F. F. Gardner, D. K. Milne.
Astrophys. Letters, Vol. 6, 35 - 37 (1970).

Seventeen sources included in the NRAO-CSIRO-MIT survey have been surveyed for the He 109α emission. The line was observed in all sources investigated except the two components G0.5-0.0 and G0.7-0.0 in the Sagittarius region.

The mean value of the helium-to-hydrogen ratio in all H II regions which show emission of the He 109α line is 0.088. The C 109α line was detected in three H II regions.

132.032 Multi-colour photographic photometry of Orion flare stars.
A. D. Andrews.
Bol. Obs. Tonantzintla y Tacubaya, Vol. 5 (No. 34), 195 - 207 (1970).

Photographic photometry of flare stars in the Orion aggregate, based on two $UBVR$ photoelectric sequences, is presented. The photometric accuracy outside the main nebulosity is sufficient to discuss the ultraviolet and blue excesses found by Haro, and to examine their effect on the colour-magnitude diagram. From the evidence of a restricted band in the $V - R/V$ diagram, due probably to the fact that colour anomalies in the visual and red are minimal, it is concluded that by far the majority of flare stars are members of the Orion aggregate.

132.033 The extinction law in the Orion nebula.
R. Costero, M. Peimbert.
Bol. Obs. Tonantzintla y Tacubaya, Vol. 5 (No. 34), 229 - 236 (1970).

An average extinction law for the Orion nebula is obtained by comparing emission line optical observations and continuum radio observations with theoretical computations for four regions in the nebula. The derived law is abnormal and yields an $R = 5.5 ± 0.7$.

132.034 Some forbidden-line intensity ratios in gaseous nebulae.
T. K. Krueger, L. H. Aller, S. J. Czyzak.
Astrophys. Journ., Vol. 160, 921 - 927 (1970).

The equations of statistical equilibrium for the ground $3p^3$ configuration of the isoelectronic sequence S II, Cl III, Ar IV, are solved as a function of T_e and N_e over the range of values normally encountered in gaseous nebulae. Applications to observed line ratios, e.g., $I(\lambda 6716)/I(\lambda 6730)$ and $I(\lambda 4068) / [I(\lambda 6730) + I(\lambda 6716)]$ of [S II] are made for typical nebulae.

132.035 Spectrophotometric studies of gaseous nebulae. XVI. The moderately high-excitation planetaries IC 2003, NGC 2022, and CD –23° 12238.
L. H. Aller, S. J. Czyzak.
Astrophys. Journ., Vol. 160, 929 - 938 (1970).

Line-intensity data and corresponding electron densities, temperatures, and ionic concentrations are presented for a number of nebulae which seem to have relatively simple spatial structures.

132.036 Theoretical continuous spectra of gaseous nebulae.
R. L. Brown, W. G. Mathews.
Astrophys. Journ., Vol. 160, 939 - 946 (1970).

Tables are given from which the continuous spectrum of emission nebulae can be computed from a knowledge of the electron temperature, the electron density, and the abundance of helium ions. The recombination continua of He I are determined from photoionization cross-sections calculated by the quantum-defect method as described by Peach. Total recombination coefficients to low levels of He⁰ are also computed.

132.037 Filamentary nebulosity in the vicinity of Loop III.
K. H. Elliott.
Nature, Vol. 226, 1236 (1970).

An extensive photographic survey of radio loops has recently been made at the high altitude observatory at the Pic du Midi. This was carried out with a widened field filter camera. Two arcs of filamentary nebulosity in the vicinity of Loop III were discovered. The filaments are at +70° and +60° respectively, which are the greatest galactic latitudes at which any nebulosity has been discovered so far.

132.038 The excitation of M 8. A. D. Thackeray.
Observatory, Vol. 90, 30 = Radcliffe Obs. Repr., No. 78 (1970). — Letter.

Far-ultraviolet intensities of Orion stars.
See Abstr. 113.032.

Some remarks regarding photoelectric photometry.
See Abstr. 113.049.

The spectrum of R Coronae Australis.
See Abstr. 114.042.

Cosmic sources of infrared radiation.
See Abstr. 114.074.

On some theoretical questions on the origin of magnetic fields of stars and nebulae. See Abstr. 116.006.

Flare stars in the Orion nebula region.
See Abstr. 122.028.

On the photoelectric photometry of some Orion flare stars. See Abstr. 122.029.

On the variability of O, B stars within nebulae.
See Abstr. 122.031.

Wavelength dependence of polarization. XXI. R Monocerotis. See Abstr. 131.013.

Interstellar masers. See Abstr. 131.021.

Electron temperatures of H II regions.
See Abstr. 131.030.

Regioni galattiche compatte di idrogeno ionizzato.
See Abstr. 131.031.

Helium abundances from 3-cm recombination lines. See Abstr. 131.054.

Interstellar formaldehyde absorption in dark nebulae. See Abstr. 131.056.

Observations of the anomalous microwave recombination line at 21-cm wavelength.

See Abstr. 131.060.

A study of interstellar matter in the Cassiopeia-Perseus region. III. A search for early-type stars and a study of interstellar absorption in the region of emission nebula IC 1795. See Abstr. 131.064.

Observations of the anomalous microwave recombination line at 11 cm wavelength. See Abstr. 131.066.

Search for ^{18}OH in emission.
See Abstr. 131.073.

Properties of H II regions obtained from non-LTE analysis of radio recombination line data.
See Abstr. 131.081.

Observations of radio recombination lines at λ = 18 cm. See Abstr. 131.084.

Light scattering in reflection nebulae.
See Abstr. 131.120.

Hydrogenic spectral lines in radio astronomy.
See Abstr. 131.127.

Anomalous recombination line 166α.
See Abstr. 131.132.

Excitation of C III] λ1909 and other semiforbidden emission lines in QSOs and nebulae. See Abstr. 141.120.

Excitation of semi-forbidden 2s^2 ^{1}S–2s2p ^{3}P lines observed in quasars and nebulae.
See Abstr. 141.223.

The absorption in the continuous spectrum of the radio source 18 SIA associated with the gaseous nebula NGC 6618. See Abstr. 141.232.

Reflection nebulae and spiral structure.
See Abstr. 155.037.

A survey of H 109 α recombination line emission in galactic H II regions of the northern sky.
See Abstr. 157.011.

6-cm recombination line observations of the 30 Doradus nebula. See Abstr. 159.002.

133 Planetary Nebulae

133.001 **Ejection of mass by radiation pressure and formation of planetary nebulae.** A. Finzi, R. A. Wolf. Astrophys. Letters, Vol. 5, 63 - 66 (1970).
We have constructed time independent models representing the light outflowing envelope that surrounds the core of the star. In our models, the rate of mass ejection is in the approximate range from 10^{21} g sec^{-1} to 10^{23} g sec^{-1}. The velocity at infinity of the ejected material is a few tens of km sec^{-1}, comparable to the observed velocities of expansion of planetary nebulae.

133.002 **The planetary nebulae – IX.** L. H. Aller. Sky Telescope, Vol. 39, 15 - 18 (1970).

133.003 **The planetary nebulae – X.** L. H. Aller. Sky Telescope, Vol. 39, 163 - 166 (1970).

133.004 **The planetary nebulae – XI.** L. H. Aller. Sky Telescope, Vol. 39, 220 - 223 (1970).

133.005 **Recombination lines in NGC 7027.** L. Goldberg. Astrophys. Letters, Vol. 5, 151 - 152 (1970).
The absence of the hydrogen 109α recombination line in the planetary nebula NGC 7027 (Terzian and Balick 1969) suggests that the populations of the levels $n = 109$ and 110 are very close or equal to their LTE values and that the nebula is optically thick in continuum radiation at $\nu = 5$ GHz. The near LTE populations imply a lower limit to the electron density of 10^6 cm^{-3}. If the suggested explanation for the absence of the 109α line is correct, recombination lines should be observed at frequencies of 10–12 GHz or greater.

133.006 **A microwave survey of planetary nebulae.** L. A. Higgs. Bull. Radio Electr. Engineering Div., National Res. Council, Canada, Vol. 19, No. 2, p. 19 - 26 (1969).
During the past three years, an extensive program of radio observations of planetary nebulae has been in progress at the Algonquin Observatory. This account covers the work done so far.

133.007 **Emission-line profiles in the planetary nebula IC 418.** D. E. Osterbrock. Astrophys. Journ., Vol. 159, 823 - 827 (1970).
Measurements are presented of H I [N II], and [O III] emission-line profiles at the center of IC 418. The observed profiles are in qualitative agreement with published theoretical models of expanding planetary nebulae. However, detailed comparison shows discrepancies, particularly in the overlap of the [O III] and [N II] distribution functions of the emission coefficient, which indicate important deviations from spherical symmetry.

133.008 **[O I] λ6300 emission in planetary nebulae.** R. E. Williams. Astrophys. Journ., Vol. 159, 829 - 832 (1970).
The presence of [O I] λ6300 radiation in planetary nebulae is explained in terms of high-density condensations. It is found from ionization calculations of model nebulae that inhomogeneities can contribute appreciably to the emission of [O I], [O II], and [N II] lines, particularly in the more evolved, density-bounded nebulae.

133.009 **Les nébuleuses planétaires.** V. P. Arkhipova. L'Astronomie, 84e année, p. 141 - 152 (1970). – Translated from Russian by J. Kovalevsky.

133.010 **On the spatial orientation of planetary nebulae.** G. V. Akhundova, Z. F. Seidov. Astron. Zhurn. Akad. Nauk SSSR, Vol. 47, 129 - 131 (1970). In Russian. English translation in Soviet Astron. AJ, Vol. 14, NO. 1.
The orientation of the axes of the planetary nebulae in space is shown to be random.

133.011 **On the infra-red line spectra of planetary nebulae.** D. R. Flower. Monthly Notices, Roy. Astron. Soc., Vol. 147, 245 - 252 (1970).
Results of detailed calculations of the infra-red line spectra of NGC 7662, IC 418 and NGC 2392 are presented. The calculated intensity of the [Ne II] 12.79μ line in IC 418 is found to be in good agreement with observations and estimates of the intensity of this line are given for four other planetary nebulae in which it should be relatively intense.

133.012 **Spectra of O- and Of-type central stars of planetary nebulae.** S. R. Heap. Bull. American Astron. Soc., Vol. 2, 197 (1970). – Abstr. AAS.

133.013 **Microwave observations of the "planetary" nebula, K3 – 50.** L. A. Higgs. Bull. American Astron. Soc., Vol. 2, 199 (1970). – Abstr. AAS.

133.014 **Oxygen and helium abundances in planetary nebulae.** J. B. Kaler. Bull. American Astron. Soc., Vol. 2, 202 (1970). – Abstr. AAS.

133.015 **Extinctions and electron temperatures of planetary nebulae.** J. B. Kaler, J. H. Cahn. Bull. American Astron. Soc., Vol. 2, 202 (1970). – Abstr. AAS.

133.016 **Multicolor observations of the nuclei of planetary nebulae.** W. Liller, C.-Y. Shao. Bull. American Astron. Soc., Vol. 2, 205 (1970). – Abstr. AAS.

133.017 **A mechanism for producing "condensations" in planetary nebulae.** R. E. Wilson, S. Sofia. Bull. American Astron. Soc., Vol. 2, 225 (1970). – Abstr. AAS.

133.018 **The planetary nebulae – XII.** L. H. Aller. Sky Telescope, Vol. 39, 299 - 303 (1970).

133.019 **An attempt to detect neutral hydrogen in planetary nebulae.** A. R. Thompson, R. S. Colvin. Astrophys. Journ., Vol. 160, 363 - 368 (1970).
Six planetary nebulae were observed at the frequency of the 21-cm hydrogen line in an attempt to detect absorption of the thermal radio emission in neutral hydrogen in the outer parts of the nebulae. No evidence of such neutral gas was detected, and in two cases an upper limit of 0.12 $M_\odot$ can be placed on the H I mass if the observations are interpreted in terms of a spherically symmetrical model and the temperature of the neutral gas is $\leq 100°$K.

133.020   Electron densities in planetary nebulae.
H. E. Saraph, M. J. Seaton.
Monthly Notices, Roy. Astron. Soc., Vol. 148, 367 - 381 (1970).

Emissivities are tabulated for lines of [O II] (configuration $2p^3$) and [S II], [Cl III], [Ar IV] and [K V] (configuration $3p^3$), using the most accurate atomic data available. Densities are obtained from observed ratios in eleven planetary nebulae. The results obtained from [O II], [S II] and [Cl III] are in good agreement. Larger densities are obtained from [Ar IV] and [K V] in high excitation planetaries; this may be evidence for large-scale density variations.

133.021   The planetary nebulae – XIII.
L. H. Aller.
Sky Telescope, Vol. 39, 368 - 370 (1970).

133.022   HBV 475: Evolution stage of a planetary nebula?
D. Crampton, J. Grygar, L. Kohoutek, R. Viotti.
Astrophys. Letters, Vol. 6, 5 - 9 (1970).

Preliminary analysis of recent spectroscopic observations of HBV 475 suggests that a Wolf-Rayet star may be exciting a small nebula of rather high density ($N_e > 10^6$ cm^{-3}). It is also suggested that the forbidden lines are formed in several condensations of matter moving away from the central star with velocities of up to 250 km sec^{-1}. A new planetary nebula could be the result.

133.023   Microwave observations of the 'planetary' nebula K3-50.   L. A. Higgs.
Astrophys. Letters, Vol. 6, 11 - 15 (1970).

It has recently been discovered that the radio source in the direction of the planetary nebula K3-50 consists of several components. The main component, coincident with K3-50, has properties similar to DR21 and other 'compact' H II regions. At the Algonquin Radio Observatory, the K3-50 radio source has been observed at 9.3, 4.5 and 2.8 cm, and a detailed map of the source has been made at the latter wavelength. These results, combined with those obtained by other observers, allow the determination of the radio spectrum of the composite source from 0.4 to 15 GHz.

133.024   A small radio component in the planetary nebula NGC 7027.   G. K. Miley, W. J. Webster, Jr., J. W. Fullmer.
Astrophys. Letters, Vol. 6, 17 - 19 (1970).

Observations have been made of NGC 7027 at 2695 MHz with an interferometer of lobe spacing 0.6 arc sec. Fringes were detected over a wide range of hour angle, implying that a small component is contained within the main body of the nebula, located near the radio center. The observed

fringe amplitude suggests that the brightness temperature of the small component is well in excess of the 13900 °K electron temperature for NGC 7027 estimated from the absence of recombination lines.

133.025   The planetary nebula in M15.
W. A. Feibelman.
Journ. Royal Astron. Soc. Canada, Vol. 64, 193 - 194 (1970).

133.026   Chemical abundances and the parameters of planetary nebulae.   J. B. Kaler.
Astrophys. Journ., Vol. 160, 887 - 913 (1970).

All known existing observations of spectral-line intensities of planetary nebulae have been compiled into a catalog which consists of 250 nebulae, including four well-observed H II regions. The principal conclusions reached are as follows: 1. The oxygen-to-hydrogen ratio should be based upon an ionization equilibrium established by the He^{2+}/He$^+$ ratio. 2. The electron temperatures near the central stars are quite high for high-excitation nebulae. 3. The oxygen-to-hydrogen ratio is dependent upon population type, decreasing toward population II. 4. The helium-to-hydrogen and oxygen-to-hydrogen ratios are negatively correlated with one another.

133.027   The planetary nebula BD +30°3639.
C. R. O'Dell, Y. Terzian.
Astrophys. Journ., Vol. 160, 915 - 920 (1970).

We report on a new measurement of the Hβ emission-line flux from BD +30°3639 and combine this with the existing radio data to discuss the physical nature of this unusually low-excitation planetary nebula.

133.028   The planetary nebulae – XIV.   L. H. Aller.
Sky Telescope, Vol. 40, 25 - 27 (1970).

Total transition probabilities for the Bowen levels of O III.   See Abstr. 022.004.

Model atmospheres for the central stars of planetary nebulae.   See Abstr. 064.023.

Interstellar matter and planetary nebulae.
See Abstr. 131.119.

Infrared studies of galactic nebulae. I. NGC 6523, NGC 6572, and BD 30°3639.
See Abstr. 132.008.

Spectrophotometric studies of gaseous nebulae. XVI. The moderately high-excitation planetaries IC 2003, NGC 2022, and CD –23°12238.   See Abstr. 132.035.

134 Crab Nebula

134.001 The expansion energy of the Crab nebula.
V. Trimble, M. Rees.
Astrophys. Letters, Vol. 5, 93 - 97 (1970).

Calculations show that the kinetic energy input required to produce the observed acceleration of the expansion of the Crab nebula is between 2 and 4 X 10^{38} ergs/sec. This value is consistent with a model for pulsars which, at present, converts virtually all of the rotation energy of a neutron star into relativistic particle energy. Rather stringent limits can, however, be placed upon the relativistic particle energy accumulated by the nebula over its lifetime and, thus, upon input rates in the past.

134.002 A search for a consistent model for the electro-magnetic spectrum of the Crab nebula.
R. Cowsik, Y. Pal, T. N. Rengarajan.
Astrophys. Space Sci., Vol. 6, 390 - 395 (1970).

An attempt is made to search for a consistent model to explain the electromagnetic spectrum of the Crab nebula (Tau A). It is assumed that there is a continuous injection of electrons at the centre of the nebula with an energy spectrum $E^{-1.54}$ as evidenced by radio data. Two types of models are considered: Class I, in which the whole nebula is characterised by a uniform magnetic field, and Class II, in which besides the general field H_0, small filamentary regions of strong field H_s are postulated. Experiments to distinguish between the various models are indicated.

134.003 What is new in the Crab nebula?
A. Spodenkiewicz.
Urania Kraków, Vol. 41, 74 - 80 (1970).   In Polish.

134.004 Coherent synchrotron emission in the Crab nebula.
L. J. Caroff, J. D. Scargle.
Nature, Vol. 225, 168 (1970).

A mechanism is proposed for generating coherent synchrotron radiation and applied to the low-frequency radio source in the Crab nebula. Coherence is obtained by injecting a streaming, inhomogeneous plasma into a background magnetic field. Injection must occur in a time small compared to the Larmor period; and the degree of coherence is proportional to the amplitude of density fluctuations with wavelengths $\approx$ the radiation wavelength.

134.005 Activity in the Crab nebula following the pulsar spin-up of 1969 September.
J. D. Scargle, E. A. Harlan.
Astrophys. Journ. (Letters), Vol. 159, L143 - L146 = Contr. Lick Obs. No. 313 (1970).

The wisps near the center of the Crab have undergone definite changes which appear to have originated near the pulsar at the time of its recent spin-up. The amount of energy liberated in this activity (between 10^{41} and 10^{45} ergs) suggests that the event was more than a simple internal change of the neutron star.

134.006 A balloon observation of pulsed X rays from the Crab.   I. S. Glass, K. W. Schnopper.
Bull. American Astron. Soc., Vol. 2, 194 (1970). – Abstr. AAS.

134.007 High-precision radio observations of the period of pulsar NP 0532.   D. W. Richards, J. M. Rankin, C. C. Counselman III.
Bull. American Astron. Soc., Vol. 2, 215 (1970). – Abstr. AAS.

134.008 Search for polarization in the X-ray emission of the Crab nebula.   R. S. Wolff, J. R. P. Angel, R. Novick, P. Vanden Bout.
Astrophys. Journ. (Letters), Vol. 160, L21 - L25 (1970).

A rocket-borne X-ray polarimeter was flown in search of polarization in Taurus X-1. Although a result consistent with zero polarization was obtained, the statistics were such that X-ray polarization comparable in magnitude to that of radio or optical emission cannot be excluded.

134.009 A jetlike structure associated with the Crab nebula.
S. van den Bergh.
Astrophys. Journ. (Letters), Vol. 160, L27 (1970).

A faint jetlike structure has been found outside the main body of the Crab nebula.

134.010 Low-frequency spectrum of the Crab nebula.
A. H. Bridle.
Nature, Vol. 225, 1035 - 1037 (1970).

The spectrum of the radio emission of the Crab nebula between 10 MHz and 15.5 GHz is derived by correcting the observed spectrum for interstellar free-free absorption. Enhancement of the total intensity at frequencies below 30 MHz indicates that the spectrum of the compact source in the nebula can be fitted by a power law of index $\alpha = 1.76 \pm 0.15$ between 10 and 1407 MHz. Implications for models of the source are considered.

134.011 Observed shapes of Crab nebula radio pulses.
D. H. Staelin, J. M. Sutton.
Nature, Vol. 226, 69 - 70 (1970).

Pulse shapes of isolated strong radio pulses from the Crab nebula have been measured using a 50-channel spectrometer and a swept-frequency local oscillator to compensate for dispersion. The resulting time resolutions of 10 ms and 3.6 ms at 115 MHz and 157.5 MHz yielded pulse shapes characterized by an unresolved rise followed by exponential decays averaging 13 ms and 3.8 ms, respectively. The $\lambda^{4 \pm 1}$ dependence of the decay time is consistent with multipath effects in the intervening plasma.

134.012 Sur la nébuleuse du Crabe.
L'Astronomie, 84e année, p. 220 - 222 (1970).

134.013 Observation of primary X-rays spectra from the Crab nebula and diffuse background.
G. Ducros, R. Ducros, R. Rocchia, A. Tarrius.
Astron. Astrophys., Vol. 7, 162 - 166 (1970).   In French.

We present the results of an astronomical experiment in the X-rays energy-range from 2.5 to 30 keV. It has been performed in December 1968 from a rocket with an attitude control system.

134.014 Possible pulsed gamma ray emission from the Crab nebula pulsar.   W. N. Charman, G. M. White.
Nature, Vol. 226, 1233 (1970).

Some comments to the possible detection of pulsed gamma emission from the Crab nebula by J. Vasseur et al. (Nature, Vol. 226, 534 - 535, 1970) are given.

134.015 Pulsed gamma rays from the Crab nebula.
J. P. Delvaille, B. McBreen.
Nature, Vol. 226, 1233 - 1234 (1970).

The purpose of this letter is to point out that the method

of analysis, used by Vasseur *et al.* (Nature, Vol. 226, 534 - 535, 1970), leads to a much higher probability for these pulsed effects to arrise from random fluctuations as stated by these authors.

134.016 **80 MHz observations of the coronal broadening of the Crab nebula.**
J. R. Harries, R. G. Blesing, P. A. Dennison.
Proc. Astron. Soc. Australia, Vol. 1, 319 - 320 (1970).

134.017 **On the orientation of magnetic fields in the Crab nebula relative to the rotation axis of the pulsar NP 0532.** E. R. Mustel.
Astron. Tsirk., No. 556, p. 1 - 3 (1970). In Russian.

134.018 **Optical studies of the Crab nebula. – Line emission component.** V. Trimble.
Publ. Astron. Soc. Pacific, Vol. 82, 375 - 387 (1970).

134.019 **Optical continuum studies.**
J. D. Scargle.
Publ. Astron. Soc. Pacific, Vol. 82, 388 - 394 (1970).

134.020 **Radio observations of the Crab nebula.**
F. Drake.
Publ. Astron. Soc. Pacific, Vol. 82, 395 - 411 (1970).

134.021 **X-ray observations of the Crab nebula.**
L. E. Peterson, A. S. Jacobson.
Publ. Astron. Soc. Pacific, Vol. 82, 412 - 430 (1970).
The results of X-ray observations on the Crab nebula since 1964 are briefly reviewed. Recently, pulses emission at the same period and phase as that of the radio and optical pulsar has been discovered.

134.022 **The nebula: X-ray data.** P. Morrison.
Publ. Astron. Soc. Pacific, Vol. 82, 431 - 437 (1970).

134.023 **Radio observation of the pulsar NP 0532.**
D. W. Richards.
Publ. Astron. Soc. Pacific, Vol. 82, 438 - 455 (1970).

134.024 **The central star: Photometric optical observations.**
J. Kristian.
Publ. Astron. Soc. Pacific, Vol. 82, 456 - 469 (1970).

134.025 **Spectroscopic optical observations of the central star.** R. Minkowski.
Publ. Astron. Soc. Pacific, Vol. 82, 470 - 478 (1970).

134.026 **Comments on the velocity and distance of the Crab nebula.** L. Woltjer.
Publ. Astron. Soc. Pacific, Vol. 82, 479 - 482 (1970).

134.027 **Remarks at discussion section of Crab nebula symposium.** S. P. Maran.
Publ. Astron. Soc. Pacific, Vol. 82, 483 - 486 (1970).

134.028 **Activity in the Crab nebula.**
J. D. Scargle.
Thesis, California Inst. Technol., Pasadena. [Available from Univ. Microfilms, Ann Arbor, Mi.], 174 pp. (1969).
Changes in the structure shown by the continuous emission from the Crab nebula are investigated observationally and theoretically. Near the center of the nebula is a striking series of strongly polarized, elongated features called 'wips', the motions of which are described.

134.029 **X-ray pulsar in the Crab nebula.**
G. Fritz, R. C. Henry, J. F. Meekins, T. A. Chubb, H. Friedman.

Naval Res. Lab., Prog. Rep., 1969, June, p. 1 - 5 (1969).
See Phys. Ber. Vol. 49, No. 2 - 3648 (1970).

134.030 **Motions and structure of the filamentary envelope of the Crab nebula.** V. L. Trimble.
Thesis, California Inst. Technol., Pasadena. [Available from Univ. Microfilms, Ann Arbor, Mi.], 138 pp. (1969).
Proper motions have been measured for 132 line-emitting filaments in the Crab nebula. These motions, if assumed constant and extrapolated backwards in time, converge toward a point about 12″ southeast of the double star near the center of the nebula.

X-ray characteristics of three supernova remnants.
See Abstr. 125.007.

Pulsar NP 0532 and the injection of relativistic particles into the Crab nebula. See Abstr. 141.023.

Pulsar NP 0532: Average polarization and daily variability at 430 MHz. See Abstr. 141.052.

Scattering of X-rays from pulsar NP 0532 by interstellar grains. See Abstr. 141.056.

Pulsar NP 0532: Properties and systematic polarization of individual strong pulses at 430 MHz.
See Abstr. 141.065.

Possible pulsed gamma ray emission above 50 MeV from the Crab pulsar.
See Abstr. 141.068.

Scattering of the decameter wavelength radiation of the Crab nebula by the solar corona.
See Abstr. 141.075.

The polarization of radio sources. II. Observations of extended radio sources. See Abstr. 141.081.

Evidence for identification of the compact low-frequency source in the Crab nebula with pulsar NP 0532. See Abstr. 141.092.

Runaway stars and the pulsar near the Crab nebula. See Abstr. 141.130.

Shape of the Crab pulsar and its period fluctuations. See Abstr. 141.137.

Optical and X-ray synchrotron radiation from the Crab pulsar. See Abstr. 141.151.

On the relation between the pulsar NP 0532 and the small-angular-diameter source in the Crab nebula. See Abstr. 141.155.

Interstellar scattering of pulsar radiation and its effect on the spectrum of NP 0532. See Abstr. 141.161.

Possibilité d'une émission pulsée de rayonnement gamma d'énergie supérieure à 50 MeV dans la direction de la nébuleuse du Crabe. See Abstr. 141.165.

Pulsar wobble and neutron starquakes. See Abstr. 141.217.

Polarization of the radio pulses from the Crab nebula pulsar. See Abstr. 141.218.

The soft X-ray spectra of three cosmic sources and simultaneous optical observations of Sco XR-1. See Abstr. 142.003.

Angular size of the high energy X-ray source in the Crab nebula.   See Abstr. 142.014.

A search for cosmic sources of gamma rays with energies between 10^{11} and 10^{14} ev. See Abstr. 142.020.

X-ray scattering by grains in the direction of the Crab pulsar and Sco XR-1.   See Abstr. 142.036.

X-ray pulsar in the Crab nebula. See Abstr. 142.040.

Scattering of X-rays from the Crab nebula. See Abstr. 142.046.

Pulsations in the X-radiation of the Crab nebula. See Abstr. 142.051.

Radio Sources, Quasars, Pulsars, X Ray-, Gamma Ray-Sources, Cosmic Radiation

141 Radio Sources, Quasars, Pulsars

141.001 Bouncing-core theory of pulsars.
E. R. Harrison.
Nature, Vol. 225, 44 - 46 (1970).
The catastrophic collapse of a star probably results in a "bouncing-core" configuration rather than a neutron star. Bouncing-cores explain many of the properties of pulsars in a more natural way than rotating neutron stars.

141.002 Five new pulsars.
A. E. Vaughan, M. I. Large.
Nature, Vol. 225, 167 - 168 (1970).
Five new pulsars have been found in the course of the pulsar search programme at the Molonglo Radio Observatory. The pulsar search at Molonglo has now covered about 85 per cent of the sky south of $\delta = + 20°$ at least once.

141.003 Low mode coherent synchrotron radiation and pulsar phenomena.
B. J. Eastlund.
Nature, Vol. 225, 430 - 434 (1970).
Information on the properties and distributions of the charges responsible for pulsar emission can be deduced from an appraisal of the coherent synchrotron radiation from an oblique rotator model.

141.004 Long baseline interferometer measurements of CP 0329.
J. A. Galt, N. W. Broten, T. H. Legg, J. L. Locke, J. L. Yen.
Nature, Vol. 225, 530 - 531 (1970).
The pulsar CP 0329 has been observed with an interferometer of $4.2 \times 10^6 \lambda$ baseline at a frequency of 408 MHz.

141.005 Possible mechanism for the pulsar radio emission.
M. M. Komesaroff.
Nature, Vol. 225, 612 - 614 (1970).
An attempt to account quantitatively for the pulse width, polarization and spectrum of the pulses.

141.006 The evolution of radio sources at large redshifts.
A. G. Doroshkevich, M. S. Longair, Ya. B. Zeldovich.
Monthly Notices, Roy. Astron. Soc., Vol. 147, 139 - 148 (1970).
Evolutionary cosmological models in which the radio source population can extend to large redshifts are discussed. Models incorporating exponential and power-law forms of evolution of the average properties of radio sources with cosmological epoch have been computed for world models with $\Lambda = 0$; $\Omega = 1$ and $\Omega = 0$. These are compared with the latest observational data and previous analyses of the source counts. It is shown that exponential models are compatible with the observations and do not require a sharp cut-off in the radio source distribution at large redshifts.

141.007 Measurements of the integrated linear polarization of discrete radio sources at 610 MHz.
P. P. Kronberg, R. G. Conway.
Monthly Notices, Roy. Astron. Soc., Vol. 147, 149 - 160 (1970).
The results show an overall trend to lower polarizations than at wavelengths of 21 cm and 10 cm. For the majority of sources measured, the degree of polarization at 49.1 cm is not affected by depolarization in the Galaxy. Correlations with optical identification and radio spectral index are discussed.

141.008 Predicted parameters of supernova pulsars.
M. I. Large.
Astrophys. Letters, Vol. 5, 11 - 15 (1970).
Thirty-five known or suspected supernova remnants are selected as promising candidates for the detection of an associated pulsar, and tentative predictions of the pulsar parameters are given.

141.009 Short term pulsar energy fluctuations.
B. Jones, R. Wielebinski.
Astrophys. Letters, Vol. 5, 17 - 20 (1970).
Observations of a number of pulsars have been analyzed so as to compare the statistics of their short term pulse energy fluctuations.

141.010 Alignment of oblique rotators.
F. C. Michel, H. C. Goldwire, Jr.
Astrophys. Letters, Vol. 5, 21 - 24 (1970).
Pulsars may be slowed down by radiation field torques. We have examined the torques on a rigid uniformly-magnetized sphere rotating in a vacuum and find an accompanying torque that aligns the magnetic moment with the spin axis.

141.011 Compton-synchrotron radiation from pulsars.
K. M. V. Apparao, J. Hoffman.
Astrophys. Letters, Vol. 5, 25 - 28 (1970).
Compton scattering of synchrotron radiation from its parent electrons in pulsars is discussed, and Compton-synchrotron radiation from the pulsar NP 0532 is calculated. Comparison with observations leads to a lower limit of a few times 10^3 Gauss for the magnetic field near the radiation-emitting region of the pulsar.

141.012 Improved parameters for seven pulsars.
R. N. Manchester.
Nature, Vol. 225, 1124 - 1125 (1970).
Observations at 410 MHz using the 300 foot transit telescope of the National Radio Astronomy Observatory have been undertaken to determine more accurate declinations and periods for several pulsars. Mean pulse energies at 410 MHz have also been measured.

141.013 Linear polarization and pulse shape measurements on nine pulsars.
M. M. Komesaroff, D. Morris, D. J. Cooke.
Astrophys. Letters, Vol. 5, 37 - 41 (1970).
Average pulse shapes are given for nine pulsars while for six the average linear polarization has been measured. In the 'double pulses' a very rapid change in the direction of polarization is found near the time of bifurcation of the pulse. Its relation to the beaming pattern of the radiation is discussed.

141.014 O^{18}H in Sagittarius.
F. F. Gardner, R. X. McGee, M. W. Sinclair.

Astrophys. Letters, Vol. 5, 67 - 71 (1970).

Absorptions of the two principal $O^{18}H$ lines have been measured for the sources Sgr A and Sgr B2 with an antenna beam of 12.5 arc min and a frequency resolution of 100 kHz. The abundance ratios $O^{18}H/O^{16}H$ appear to be higher than terrestrial for the +40 km/sec OH cloud in SgrA and the +60 km/sec cloud in Sgr B2, and lower for the negative velocity features –130 km/sec and –90 km/sec in Sgr A and Sgr B2, respectively.

141.015 On the motion of current sheets, and the radio, optical, and X-ray emission from pulsars.
I. Lerche.
Astrophys. Journ., Vol. 159, 229 - 237 (1970).

We suggest that an oscillating current sheet is the source of pulsar radiation produced over a broad band of high frequencies. The radiation extends through the radio, optical, and X-ray regions. The efficiency of the radiation mechanism is so high that the oscillation energy goes mainly into radiation in each cycle of the oscillation.

141.016 Absolute spectral energy distribution of quasi-stellar objects from 0.3 to 2.2 microns.
J. B. Oke, G. Neugebauer, E. E. Becklin.
Astrophys. Journ., Vol. 159, 341 - 355 (1970).

The absolute spectral energy distribution from 0.32 to 2.2μ has been obtained for twenty-eight quasi-stellar sources. Photometry at 2.2μ has been obtained for an additional fifteen objects. The data show that the continua over this wavelength range generally can be described with a power-law spectrum; the index varies from –0.2 to –1.6, with the entire range being populated. For quasi-stellar objects which are known to have large-amplitude variations in visual magnitude, the energy distributions remain sensibly unchanged during the variations. No characteristic of the energy distribution in the observed range of wavelengths which distinguishes between radio-quiet and radio-active quasi-stellar objects is found. The near constancy of the observed ratio of the line intensities of $L\alpha$ to the C IV line can be interpreted as implying an electron temperature of $20000°K$.

141.017 The fine-scale structure of Virgo A.
G. K. Miley, D. E. Hogg, J. Basart.
Astrophys. Journ. (Letters), Vol. 159, L19 - L20 (1970).

The nucleus of M87 is shown to coincide with a small radio component of strength 3.9 ± 0.4 flux units. A second component associated with the tip of the optical jet must have an angular size of between $0.''5$ and $3''$.

141.018 Pulsar planetary systems. F. C. Michel.
Astrophys. Journ. (Letters), Vol. 159, L25 - L28 (1970).

We show that any planets surrounding a pulsar should be in eccentric orbits and that one effect of such planets would be to give apparently discontinuous changes in the frequency of the pulsar, as have been observed for NP 0532 and PSR 0833.

141.019 The possible effects of scattering of pulsar radiation. A. D. Code.
Astrophys. Journ. (Letters), Vol. 159, L29 - L34 = Contr. Kitt Peak National Obs. No. 511 (1970).

The modification of the light curve of a pulsar by a scattering atmosphere is investigated from the point of view of the time-dependent equation of transfer. The pulses are asymmetrically broadened as the contribution from diffusely scattered radiation is increased. It is suggested that a pulsar may possess an envelope of sufficient density that the free-free opacity is important for frequencies less than about 100 MHz. In this case the pulse widths would increase with decreasing frequency, and at lower frequencies the pulsar would not

longer exhibit the pulse characteristic.

141.020 Time-of-arrival observations of eleven pulsars.
P. E. Reichley, G. S. Downs, G. A. Morris.
Astrophys. Journ. (Letters), Vol. 159, L35 - L40 (1970).

The analysis of high-precision time-of-arrival measurements has yielded the position, period, and first derivative of the period for ten pulsars and an improved period for one other. An upper bound was obtained for the second derivative of the period for six pulsars.

141.021 Spectral data on some Ohio radio sources.
J. D. Kraus, B. H. Andrew.
Astrophys. Journ. (Letters), Vol. 159, L41 - L43 (1970).

Flux densities are given for seventy-one sources in the frequency range 0.6 - 10.6 GHz. Most of the sources have been discovered in the Ohio survey, and the majority have low spectral indices or peaked spectra.

141.022 Radio sources with flat spectra.
B. H. Andrew, J. D. Kraus.
Astrophys. Journ. (Letters), Vol. 159, L45 - L49 (1970).

The spectra of sixty-three radio sources discovered in surveys conducted at 1.4 GHz are studied in the frequency range 0.6 - 10.6 GHz. Many of the spectra are flat at high frequencies, and very few show evidence of having more than one spectral component. It is argued that a large fraction of the population of sources with flat high-frequency spectra are optically thin sources with low values of the electron-energy distribution index γ.

141.023 Pulsar NP 0532 and the injection of relativistic particles into the Crab nebula.
I. S. Shklovsky.
Astrophys. Journ. (Letters), Vol. 159, L77 - L80 (1970).

The purpose of the present letter is to improve the calculation and to connect the problem of synchrotron emission from NP 0532 with the general problem of the injection of relativistic particles into the Crab nebula.

141.024 Magnetic-dipole alignment in pulsars.
L. Davis, M. Goldstein.
Astrophys. Journ. (Letters), Vol. 159, L81 - L85 (1970).

If the explanation of pulsars involves classical magnetic-dipole radiation of a rotating dipole that makes the angle ψ with the rotation axis, the radiation reaction causes ψ to decrease with time and the frequency to approach $\cos \psi$ times the current frequency.

141.025 The period distribution of pulsars.
G. Setti, L. Woltjer.
Astrophys. Journ. (Letters), Vol. 159, L87 - L88 (1970).

The assertion that decay of a pulsar magnetic field is needed to provide an explanation for the period distribution is unfounded.

141.026 Three pulsars with marching subpulses.
J. M. Sutton, D. H. Staelin, R. M. Price, R. Weimer.
Astrophys. Journ. (Letters), Vol. 159, L89 - L93 (1970).

High time-resolution studies of individual pulses from the pulsars MP 0031, CP 0808, and CP 0834 reveal subpulse structures which systematically march across the average pulse window from one pulse to the next. It is shown that periodic fluctuations in pulse intensities are manifestations of the same phenomenon.

141.027 On the masses of quasi-stellar sources.
J. Barnothy.
Astrophys. Journ. (Letters), Vol. 159, L133 (1970).

The method proposed by Bahcall and Salpeter to compute the mass of a QSO belonging to a cluster of galaxies is self-contradictory. The authors did not consider the gravitational redshift of heavy objects of small size.

141.028 Upper limits on the masses of quasi-stellar sources.
J. N. Bahcall, E. E. Salpeter.
Astrophys. Journ. *(Letters)*, Vol. 159, L135 - L136 (1970).

There are a number of independent methods for estimating upper limits to the masses of quasi-stellar sources. Some of these methods are listed below.

141.029 The Sturrock-Feldman model of quasi-stellar sources.
J. M. Barnothy.
Astrophys. Journ. *(Letters)*, Vol. 159, L137 - L138 (1970).

The influx of matter from the intergalactic space, accelerated by the gravitational field of a quasi-stellar object with a mass of $10^{43.5}$ g, is at least four to five orders of magnitude lower than needed to explain the intrinsic luminosity of 3C 273.

141.030 On the possibility of accretion by quasi-stellar objects. P. A. Sturrock.
Astrophys. Journ. *(Letters)*, Vol. 159, L139 (1970).

An estimate of the rate of accretion by a QSO depends upon the hypotheses on which the calculation is based.

141.031 Distribution of quasistellar radio sources on the sky. H. Arp.
Astron. Journ., Vol. 75, 1 - 12 (1970).

The paper analyzes the distribution of QSR's on the sky as a function of apparent magnitude. The inter-associations between QSR's themselves and QSR's and bright galaxies which result indicate that this approach is successful and enables QSR's in general to be identified with their galaxies of origin.

141.032 Flux measurements of radio sources at 2695 MHz.
S. Ames.
Astron. Journ., Vol. 75, 71 - 75 (1970).

The flux has been measured at 2695 MHz of the radio sources in a 5000-MHz survey. The spectral indices between these frequencies are found, and their distribution shows an excess of flat spectra. Approximately one-third of the sources appear to have convex spectra at these higher frequencies.

141.033 High-frequency-resolution studies of H I, OH, and CH_2O absorption spectra in Sagittarius A.
A. Sandqvist.
Astron. Journ., Vol. 75, 135 - 140 (1970).

Absorption profiles towards Sgr A have been observed in the 1420-, 1667-, and 4830-MHz lines of H I, OH, and CH_2O, with velocity resolutions of 1.67, 1.42, and 0.49 km/sec, respectively. The 0-, –50-, and +45-km/sec H I absorption features and the +40- and –130-km/sec OH absorption features were also observed with higher velocity resolutions ranging from 0.42 to 0.10 km/sec. Some astrophysical implications of these profiles are discussed.

141.034 Survey of the south celestial polar region at 408 MHz. R. M. Price.
Astron. Journ., Vol. 75, 144 - 147 (1970).

A survey of the region within 15° of the south celestial pole at a frequency of 408 MHz is compared with a source survey at 1410 MHz to show the contribution of discrete sources to the over-all brightness distribution in such a medium-latitude region. Smoothing the 408-MHz survey and comparing it with the absolute brightness temperature at the south celestial pole as measured with a large horn antenna at the same frequency yields the absolute base level for the region.

141.035 A survey of the sky at 610.5 MHz. III. The region

between Dec. +22° and +27°.
H. J. Wendker, J. R. Dickel, K. S. Yang, and Staff.
Astron. Journ., Vol. 75, 148 - 157 (1970).

The third section of the catalogue of radio sources being made with the 400-ft radio telescope at the Vermilion River Observatory of the University of Illinois has been completed. The declination range is +22 to +27°; approximately 7.3% of this region has been omitted to avoid sidelobe confusion. All sources are tabulated whose 610.5-MHz flux densities are greater than 0.8 f.u. Spectral indices are also given for the sources.

141.036 Pulsars and X-ray-emitting supernova remnants.
S. S. Holt, R. Ramaty.
Astrophys. Letters, Vol. 5, 89 - 91 (1970).

X-ray emission from a supernova remnant may indicate the existence of a rotating neutron star within the nebula. The properties of candidate pulsars in Cassiopeia A and Tycho's supernova are investigated.

141.037 H109α recombination line measurements of the W51 region. T. L. Wilson, P. G. Mezger,
F. F. Gardner, D. K. Milne.
Astrophys. Letters, Vol. 5, 99 - 103 (1970).

A search for H109α emission at nine positions within the W51 complex with the 210-ft radio telescope at Parkes, beamwidth 4 arc min at 5 GHz, indicates (i) the thermal component G 48.6 + 0.0 has a very low radial velocity (+21 km/sec) and is not a member of the W51 complex; (ii) six thermal components have radial velocities ranging from 53 to 72 km/sec and appear to be members of two associations of giant H II regions located close to the tangential point of the Sagittarius spiral arm; (iii) since no H109α emission was found at two positions on the extended component, it is assumed to be non-thermal; and we consider it to be unrelated to the W51 complex.

141.038 A model of extragalactic radio sources.
D. M. Mills, P. A. Sturrock.
Astrophys. Letters, Vol. 5, 105 - 110 (1970).

We propose a model for the compact bright radio sources ejected from radio galaxies and quasars, and show that this model gives good agreement with the characteristic brightness distribution and polarization behavior of these sources. Specific application is made to the radio source Cygnus A.

141.039 A search for circular polarization in extragalactic compact radio sources at 3240 MHz.
E. R. Seaquist.
Astrophys. Letters, Vol. 5, 111 - 115 (1970).

A search has been carried out for circular polarization in twenty-nine compact radio sources at 3240 MHz. Although most sources indicate no circular polarization exceeding 1 per cent, there is some evidence for left-hand circular polarization of this order in the sources DA 344 and NRAO 140.

141.040 Flux densities of some radio sources in the frequency range 12–25 MHz. S. Ya. Braude,
O. M. Lebedeva, A. V. Megn, B. P. Ryabov, I. N. Zhouck.
Astrophys. Letters, Vol. 5, 129 - 132 (1970).

This paper presents measurements of the flux densities of 33 discrete radio sources, made with the radio-telescope UTR-1 in Grakovo, at frequencies ranging from 12.6 to 25 MHz.

141.041 Are pulsars single stars?
V. A. Krat.
Astrophys. Space Sci., Vol. 6, 420 - 421 (1970). – Research note.

141.042 Nochmals Pulsare. W. Kranzer.

Phys. Blätter, 26. Jahrgang, p. 32 - 34 (1970).

141.043 Pulsare. G. M. Richter.
Sterne, 45. Jahrgang, p. 217 - 226 (1969).

141.044 On the nature of evolutionary effects associated with quasi-stellar radio sources. II. M. A. Arakelian.
Astrofizika, Vol. 5, 603 - 614 (1969). In Russian. — English translation in Astrophysics, Vol. 5, No. 4.

The results of the investigation of evolutionary effects associated with quasi-stellar radio sources based on the data of the Parkes Survey are presented. The conclusion about the uniform distribution of quasi-stellar radio sources in the co-moving volume, their constant optical luminosity and evolving radio luminosity made formerly is confirmed.

141.045 Pulsars. L. Zaleski.
Urania Kraków, Vol. 41, 80 - 85 (1970).
In Polish.

141.046 Excess surface density of extragalactic radio sources at 8 GHz. G. W. Brandie.
Nature, Vol. 225, 352 - 353 (1970).

Over 97% of the area between declinations +5° and -4° and with $|b^{II}| > 15°$ has been surveyed at a frequency of 8 GHz (λ 3.8 cm). Analysis of the source counts shows that the number of sources per steradian with flux density greater than or equal to 1 flux unit is $61.5 \, {}^{+13.7}_{-10.0}$. This surface density of sources is twice the number predicted by extrapolating lower-frequency data, and is interpreted as indicating that a majority of the sources found at 8 GHz have flat or inverted radio spectra.

141.047 Can pulsar masses be determined?
P. Sutherland, G. Baym, C. Pethick, D. Pines.
Nature, Vol. 225, 353 - 354 (1970).

In the context of a simple starquake model proposed by Baym, Pethick, Pines and Ruderman we calculate pulsar masses assuming (1) a cold matter equation of state, (2) a starquake which takes the pulsar from one essentially quadrupolar distorted form to another, less distorted, form (to relieve stresses built up because of slowdown of the pulsar's rotation rate). The mass is determined in terms of a parameter obtained through observation of the changes in the rotation rate and its time derivatives immediately after a starquake.

141.048 Decametric variability of Cas A and 3C84.
A. H. Bridle, J. L. Caswell.
Nature, Vol. 225, 356 - 357 (1970).

It is suggested that ionospheric phenomena are responsible for apparent variability of the decametric radio emission from Cas A and 3C84 reported by Braude et al. The effects of scintillation, multiple scattering and refraction are discussed, and it is suggested that Braude et al. may have made inadequate allowance for these phenomena in attributing variability to the sources themselves. 10.03 MHz observations of 3C84 made at times when good ionospheric conditions prevail show no evidence for intrinsic variability of this source greater than 20 per cent of its intensity at this frequency.

141.049 Pulsar periods and rapid changes in the terrestrial rotation rate: a reply. J. Pfleiderer.
Nature, Vol. 225, 437 - 438 (1970).

The relation between accuracies in determining mean pulse arrival times, periods, rates of period changes, and changes of period per period are discussed. The influence on these quantities of any short-scale change in the terrestrial rotation rate is far below the present detection level.

141.050 On the connexion between pulsars and supernovae.
A. J. R. Prentice.

Nature, Vol. 225, 438 - 439 (1970).

Evidence is presented that most pulsars may have been formed in Type II supernova explosions and initially possessed extremely high velocities, or order 1000 km/sec. On this basis, four new pulsar-supernovae identifications are suggested. An explanation is offered for certain features in the observed distribution of pulsar galactic latitudes.

141.051 Variations of linear polarization in extragalactic radio sources. H. D. Aller.
Nature, Vol. 225, 440 (1970).

The characteristics of the observed variations in polarization of variable radio sources are briefly reviewed and it is pointed out that these characteristics can be accounted for by the so called expanding source model. The epochs of observed polarization variations suggest that the variations are caused primarily by the decrease with time of absorption depths within the sources. Two processes which produce variations in linear polarization as a result of a change in absorption depth are discussed.

141.052 Pulsar NP 0532: Average polarization and daily variability at 430 MHz.
D. B. Campbell, C. Heiles, J. M. Rankin.
Nature, Vol. 225, 527 - 528 (1970).

The four Stokes parameters of NP 0532 have been simultaneously measured at 430 MHz. Two of the three components of this pulsar are linearly polarized. The precursor, which is essentially 100% linearly polarized, shows no rotation of position angle, in contrast to the optical behavior of NP 0532 and the radio behavior of other pulsars. The total intensities of the three pulse components vary from day-to-day in an uncorrelated manner. This cannot result from scintillation, and therefore must be basic to the pulsar itself.

141.053 Size and motion of the interstellar scintillation pattern from observations of CP 1133.
K. R. Lang, B. J. Rickett.
Nature, Vol. 225, 528 - 530 (1970).

The 406 MHz radiation intensity from the pulsar CP 1133 has been observed simultaneously at the Arecibo and Jodrell Bank observatories. The intensity variations seen at the two observatories were highly correlated but displaced in time by 50 ± 10 pulse periods. This is interpreted to be due to a stable interstellar scintillation pattern moving past the two sites. The pattern scale size must lie in the range (0.5 – 3) × 10⁹ cm. The pulsar distance deduced from these scale sizes and the decorrelation frequency agrees with that based on dispersion measurements. The relative velocity of the interstellar medium with respect to the pulsar and/or the earth must lie in the range 22 – 92 km s⁻¹.

141.054 Apparent periodicities in the redshifts of quasi-stellar objects. T. J. Deeming.
Nature, Vol. 225, 620 - 621 (1970).

Possible periodicities in the distribution of the redshifts of quasi-stellar objects, found by the application of power spectrum analysis, are shown not to be statistically significant. The observed distribution of peaks in the power spectrum agrees well with what would be expected from white noise.

141.055 Long term variations of pulsar intensities.
T. W. Cole, H. K. Hesse, C. G. Page.
Nature, Vol. 225, 712 - 713 (1970).

Pulsars CP 0808, CP 0834, CP 0950, CP 1133 and CP 1919 were observed daily at 81.5 MHz during 1968. Weekly averages of intensity revealed variations in intensity on time scales of weeks and months. It is shown that interstellar scintillation cannot account for the variations, which must be intrinsic to the source.

141.056 Scattering of X-rays from pulsar NP 0532 by interstellar grains. S. Naranan, G. A. Shah.
Nature, Vol. 225, 834 - 836 (1970).

The present investigation has been carried out to see if the diffuse component of X-rays from the Crab nebula could originate from the scattering, by interstellar grains, of pulsed X-rays from the pulsar NP 0532 as suggested by Slysh (Nature, Vol. 224, 159 (1969)). The extinction efficiencies have been calculated with due regard to the wavelength dependent indices of refraction in X-ray region, and the relevant approximations to the theories of scattering by single and concentric spheres consisting of dirty ice dielectric and graphite (core). The column density of the grains has been obtained from the theoretical extinction cross section and the observed visual extinction towards the Crab nebula. It has been found that the optical depth for scattering in the 1–10 keV region is small and quite insensitive to particular grain model. The results imply that the spectrum of the pulsed X-rays from NP 0532 (in 1–100 keV region) is affected only very slightly due to scattering by interstellar grains. The diffuse component in the nebula owes very little or probably nothing to the pulsar source.

141.057 Pulsars and the universal X-ray background.
M. V. K. Apparao.
Nature, Vol. 225, 836 - 837 (1970).

It is shown that the universal X-ray background arises as a superposition of X-radiation from pulsars in various galaxies in the universe.

141.058 Secondary periodicities in pulsars.
K. H. Hesse.
Nature, Vol. 225, 837 - 838 (1970).

Pulsars CP 0808, CP 0834, CP 0950, CP 1133 and CP 1919 have been studied for periodic fluctuations in pulse intensities. The paper reports such an effect for the above pulsars with the exception of CP 0950, reveals a very pronounced periodicity for CP 0834, and sets an upper limit to the stability of this periodicity of CP 1919.

141.059 Search for high energy gamma rays from four pulsars. B. K. Chatterjee, G. T. Murthy,
P. V. Ramana Murthy, B. V. Sreekantan, S. C. Tonwar.
Nature, Vol. 225, 839 - 840 (1970).

Results of an experiment to detect the possible emission of high energy pulsed gamma rays ($\geqslant 10^{13}$ eV) from four pulsars, viz. CP 0950, CP 1133, AP 1541 and PSR 1642 are reported. It is concluded that of the four pulsars, CP 0950 appears to be a good candidate for further observations on pulsed gamma ray emission.

141.060 Measurements of neutral-hydrogen absorption in the spectra of four pulsars.
K. J. Gordon, C. P. Gordon.
Astrophys. Letters, Vol. 5, 153 - 156 (1970).

Upper limits are placed on the strength of the 21-cm line absorption of the signals from four pulsars at low galactic latitudes. On a 2σ basis, the limits are: NP 0527, $\tau \lesssim 0.3$; PSR 0833-45, $\tau \lesssim 0.08$; PSR 1929 + 10, $\tau \lesssim 0.12$; and JP 1933, $\tau \lesssim 0.8$. The apparent lack of absorption in these observations suggests that many of the absorption lines found in the spectra of other radio sources are produced by dense, cool clouds of neutral hydrogen smaller than the angular size of the radio sources.

141.061 3C 386: Radio galaxy or supernova remnant?
C. D. Mackay.
Astrophys. Letters, Vol. 5, 173 - 176 (1970).

Recent radio observations of 3C 386 support the suggestion of Pskovskii that it is the remnant of a Crab-like supernova and not an external galaxy as it was originally identified. It is shown, however, that it is still not possible to decide which interpretation is correct.

141.062 Measurements of the linear polarization of seven pulsars at 11-cm wavelength.
D. Morris, U. J. Schwarz, D. J. Cooke.
Astrophys. Letters, Vol. 5, 181 - 186 (1970).

Simultaneous measurements have been made of the first three Stokes parameters specifying the linear polarization of the 11-cm radiation from seven pulsars. The results are averages over at least 500 pulses. The variation of position angle of polarization during the pulse is compared with the predictions of Komesaroff (1970). The data support the view that the radiation from some pulsars, PSR 2045 – 16 in particular, is beamed into a hollow cone centered on a magnetic pole.

141.063 Improved positions and periods for five southern pulsars. D. Morris, U. J. Schwarz, O. B. Slee.
Astrophys. Letters, Vol. 5, 187 - 188 (1970).

More accurate declinations and periods have been found for five Molonglo pulsars, using a multi-period integration technique.

141.064 Pulsars – is a nod as good as a wink?
W. I. Axford, H. E. Johnson, D. A. Mendis, T. Yeh.
Comments Astrophys. Space Phys., Vol. 2, 53 - 58 (1970).

The suggestion that pulsars might be rotating and precessing neutron stars has recently been published by many authors (Vila, Pacini, Urey and ourselves). We describe some of the objections which can be raised against this suggestion, and which leave us with the opinion that it is probably untenable despite some apparently attractive features.

141.065 Pulsar NP 0532: Properties and systematic polarization of individual strong pulses at 430 MHz.
C. Heiles, D. B. Campbell, J. M. Rankin.
Nature, Vol. 226, 529 - 531 (1970).

This paper reports further detailed measurements of strong radio pulses from the pulsar NP 0532 in the Crab nebula, recorded at Arecibo.

141.066 The unusually large redshift of 4C 05.34.
R. Lynds, D. Wills.
Nature, Vol. 226, 532 (1970).

The quasi-stellar object identified with the radio source 4C 05.34 has been recently observed spectroscopically by us at Kitt Peak National Observatory and has been found to have a redshift of $z = 2.877$, a value considerably higher than that of any quasi-stellar object reported so far.

141.067 Upper limits for CP 1919 X-ray emission.
D. Gruber, J. L. Matteson, M. R. Pelling,
L. E. Peterson.
Nature, Vol. 226, 532 - 534 (1970).

We report here a search in the direction of the radio pulsar CP 1919 which has resulted in upper limits on possible X-ray emission from this object at 3.9×10^{-33} J m^{-1} Hz^{-1} per pulse over the 30 – 100 keV range.

141.068 Possible pulsed gamma ray emission above 50 MeV from the Crab pulsar.
J. Vasseur, J. Paul, B. Parlier, J. P. Leray, M. Forichon,
B. Agrinier, G. Boella, L. Maraschi, A. Treves, R. Buccheri,
L. Scarsi.
Nature, Vol. 226, 534 - 535 (1970).

A search for pulsed gamma ray emission from NP 0532 has been carried out with a balloon-borne experiment. About 30 per cent of the recorded events show the characteristic electron pairs from gamma rays, while another 30 per cent show single tracks originating in the chamber plates. The remaining fraction corresponds to spurious or unidentified events.

141.069 Observations of PP 0943.
        G. A. Zeissig.
Nature, Vol. 226, 536 (1970).
    The purpose of this letter is to provide more accurate values of the basic parameters (position, period, dispersion) of this pulsar.

141.070 Pulsare: Neutronensterne?
        L. Biermann.
Umschau, 70. Jahrgang, p. 231 - 238 (1970). − Review article.

141.071 Los pulsares.   C. Jaschek.
    Revista Astron., Vol. 41, No. 170, p. 35 - 37 (1969).

141.072 On a possible anisotropy of the radio emission of some quasars.   V. N. Kurilchik.
Astron. Zhurn. Akad. Nauk SSSR, Vol. 47, 27 - 31 (1970).
In Russian. English translation in Soviet Astron. AJ, Vol. 14, No. 1.
    On the basis of an analysis of spectra of the radio emission of quasars from the Catalogue 3CR, the origin of anomalously flat spectra of radio emission is considered.

141.073 A possible explanation of the observed features of the distribution of quasars in relation to redshifts.
E. A. Karitskaya, B. V. Komberg.
Astron. Zhurn. Akad. Nauk SSSR, Vol. 47, 43 - 45 (1970).
In Russian. English translation in Soviet Astron. AJ, Vol. 14, No. 1.
    It is shown that the large-scale distribution of quasars in relation to redshifts could be explained by the influence of strong emission lines on the conditions of identification of these objects.

141.074 Scintillations of radio sources with finite angular dimensions.   V. I. Shishov.
Astron. Zhurn. Akad. Nauk SSSR, Vol. 47, 182 - 187 (1970).
In Russian. English translation in Soviet Astron. AJ, Vol. 14, No. 1.
    Expressions determining spectra of scintillations of sources with finite angular dimensions are obtained. It is shown that spectra of scintillations of sources with large angular dimensions are determined by a medium near the earth. Theoretical results are compared with the observed data obtained from observations of the radio source 3C 273. It is shown that at the frequency of 195 MHz the source 3C 273 has a component with a dimension of 5″ and one with 0.″02.

141.075 Scattering of the decameter wavelength radiation of the Crab nebula by the solar corona.
L. L. Bazelyan, S. Ya. Braude, A. V. Men'.
Astron. Zhurn. Akad. Nauk SSSR, Vol. 47, 188 - 200 (1970).
In Russian. English translation in Soviet Astron. AJ, Vol. 14, No. 1.
    Presented are results of investigations of the Tau A occultation by the solar corona in 1965 and 1966. The measurements were made at 20, 25, 31, and 38 MHz with the aid of two interferometers whose bases were 88 and 470 m. At angular separations from the sun of 5 - 20 $R_\odot$, the decrease of the total flux was determined. The dependence of the increase of the angular diameters on wavelength is different from that observed for meter waves.

141.076 Frequency of pulsar starquakes.
        R. Smoluchowski.
Phys. Rev. Letters, Vol. 24, 923 - 925 (1970).
    A relationship between the frequency of the occurrence of starquakes on pulsars and the shear strength of the solid crust of these objects is derived assuming various mechanisms of damping of their rates of rotation. Tentative conclusions concerning the shear strength, the age, and the composition of the Crab and the Vela pulsars are made.

141.077 The nature of pulsar radiation.
        F. Pacini, M. J. Rees.
Nature, Vol. 226, 622 - 624 (1970).
    Pulsar radiation may be due to particles which gyrate while moving in bunches along curved orbits. If emitted by electrons, the pulses should come from near the speed of light circle; if emitted by protons, they should arise within a few stellar radii of the underlying star.

141.078 Pulsare, magnetische Verstärkung und Quasare.
        J. H. Piddington.
Umschau, 70. Jahrgang, p. 316 (1970).

141.079 Pulsare. , L. Biermann.
    Plenarvorträge, 34. Physikertagung 1969 Salzburg (B. G. Teubner, Stuttgart), p. 26 - 51 (1969). − Conference paper.

141.080 Plasma turbulence as a mechanism of pulsar radiation.   S. Ichimaru.
Nature, Vol. 226, 731 - 733 (1970).
    This report is about a pulsar model in which anisotrpic plasma turbulence, produced by the onset of a two-stream instability near the light cylinder of a rotating magnetic star, is the chief radiation mechanism.

141.081 The polarization of radio sources. II. Observations of extended radio sources.
R. D. Davies, F. F. Gardner.
Australian Journ. Phys., Vol. 23, 59 - 78 = Separate print Division Radiophys. C.S.I.R.O. Sydney (1970).
    A polarization survey at 11.3 cm wavelength with beamwidth 7.′5 × 7.′2 arc has been made of 13 extended extragalactic sources and of 7 extended galactic objects presumed to be supernova remnants. The observations have been supplemented in some cases by measurements at 21.3 cm wavelength. After correction for these effects the observed polarization distributions for all the extragalactic sources were satisfactorily fitted to double or triple component models; in half the cases magnetic field configurations could be derived. Measurements of Taurus A made at both 6 and 11.3 cm showed that the polarized region was about 2′ arc in extent and was displaced from the centre of the object.

141.082 De-convolution of barely resolved radio sources mapped with aerial beams of elliptical cross section.   J. P. Wild.
Australian Journ. Phys., Vol. 23, 113 - 115 = Separate print Division C.S.I.R.O. Sydney (1970). − Short communication.

141.083 A high-resolution map of Cassiopeia A at 2.7 GHz.
        I. Rosenberg.
Monthly Notices, Roy. Astron. Soc., Vol. 147, 215 - 230 (1970).
    Observations of Cas A (3C 461) have been made using the Cambridge One-mile telescope at a frequency of 2695 MHz with a beamwidth of 12″× 14″ arc. By comparing the new observations with earlier ones made at 1407 MHz variations of structure with frequency across the source have been found, with spectral indices higher on average on the west half of the source than on the east. Comparisons with optical photographs show a remarkable correlation between the gaps in the radio shell and some of the high velocity optical filaments. Estimates of some physical parameters of the main source and of the compact components are derived in the last

part of the paper.

141.084   On absorption lines in quasi-stellar objects.
   E. M. Burbidge, G. R. Burbidge.
Astrophys. Journ. (*Letters*), Vol. 159, L185 - L186 (1970).
   Four of the QSOs listed by Bahcall as being in the fields of comparatively nearby clusters of galaxies and therefore as being worth searching for intergalactic absorption lines in their spectra, have already been noted as having absorptions, but in all cases absorption redshifts are much larger than the cluster redshifts and are likely to arise in the objects. New observations of two of the QSOs are described.

141.085   On some implications of the m(z) and M(B—V)
   relations for quasars.   M. Abramowicz.
Postępy Astron., Vol. 18, 201 - 207 (1970).   In Polish.
Abstr. Polish Astron. Soc.

141.086   Spin-up in the Vela pulsar:Microscopic descrip-
   tion.   G. Baym, D. Pines, C. Pethick,
M. Ruderman.
Bull. American Astron. Soc., Vol. 2, 182 (1970). – Abstr.
AAS.

141.087   Precision measurement of the frequency decay of
   the Crab nebula pulsar, NP 0532.
P. E. Boynton, E. J. Groth III, R. B. Partridge, D. T. Wilkinson.
Bull. American Astron. Soc., Vol. 2, 183 (1970). – Abstr.
AAS.

141.088   An excess surface density of radio sources at
   8 GHz.   G. W. Brandie.
Bull. American Astron. Soc., Vol. 2, 183 - 184 (1970).
Abstr. AAS.

141.089   Theory of pulsars. II. Beamed electromagnetic
   radiation in a magnetized plasma.
V. Canuto.
Bull. American Astron. Soc., Vol. 2, 186 (1970). – Abstr.
AAS.

141.090   Theory of pulsars. III. Laser process and pulsar
   radiation.   H. Y. Chiu, A. Muriel.
Bull. American Astron. Soc., Vol. 2, 187 (1970). – Abstr.
AAS.

141.091   The structure of 3C9.
   B. G. Clark, G. K. Miley.
Bull. American Astron. Soc., Vol. 2, 188 (1970). – Abstr.
AAS.

141.092   Evidence for identification of the compact low-
   frequency source in the Crab nebula with pulsar
NP 0532.   W. M. Cronyn.
Bull. American Astron. Soc., Vol. 2, 189 - 190 (1970).
Abstr. AAS.

141.093   21-cm absorption of pulsar signals.
   K. J. Gordon, C. P. Gordon.
Bull. American Astron. Soc., Vol. 2, 194 (1970). – Abstr.
AAS.

141.094   Concerning the minimum angular diameter of
   radio sources.   D. E. Harris, G. Zissig,
R. Lovelace.
Bull. American Astron. Soc., Vol. 2, 197 (1970). – Abstr.
AAS.

141.095   Results of a search for visible pulsars.
   P. Horowitz, C. D. Papaliolios, N. P. Carleton.

Bull. American Astron. Soc., Vol. 2, 200 (1970). – Abstr.
AAS.

141.096   Variations in pulsar radiation intensity.
   K. R. Lang.
Bull. American Astron. Soc., Vol. 2, 204 - 205 (1970).
Abstr. AAS.

141.097   Alignment of oblique rotators.
   F. C. Michel, H. C. Goldwire, Jr.
Bull. American Astron. Soc., Vol. 2, 209 (1970). – Abstr.
AAS.

141.098   A model for extragalactic radio sources.
   D. M. Mills, P. A. Sturrock.
Bull. American Astron. Soc., Vol. 2, 209 (1970). – Abstr.
AAS.

141.099   Theory of pulsars. I. General relativistic electro-
   dynamics.   F. Occhionero.
Bull. American Astron. Soc., Vol. 2, 211 (1970). – Abstr.
AAS.

141.100   Rapid variations in the radio polarization of BL
   Lac.   E. T. Olsen.
Bull. American Astron. Soc., Vol. 2, 212 (1970). – Abstr.
AAS.

141.101   Spin-up in the Vela pulsar: A two-component
   model.   D. Pines, G. Baym, C. Pethick,
M. Ruderman.
Bull. American Astron. Soc., Vol. 2, 214 - 215 (1970). – Abstr
AAS.

141.102   On multiple absorption redshifts in quasar spectra.
   M. J. Rees.
Bull. American Astron. Soc., Vol. 2, 215 (1970). – Abstr.
AAS.

141.103   An interpretation of pulsar dispersion-measure
   statistics.   M. P. Savedoff.
Bull. American Astron. Soc., Vol. 2, 216 - 217 (1970).
Abstr. AAS.

141.104   Particle acceleration during the 1966-67 radio
   burst of 3C273.   M. Simon.
Bull. American Astron. Soc., Vol. 2, 218 (1970). – Abstr.
AAS.

141.105   Model for pulsar radio emission.
   D. H. Staelin.
Bull. American Astron. Soc., Vol. 2, 219 (1970). – Abstr.
AAS.

141.106   Possible radiation mechanisms for pulsars.
   P. A. Sturrock.
Bull. American Astron. Soc., Vol. 2, 221 (1970). – Abstr.
AAS.

141.107   Pulsar observations using a swept-frequency local
   oscillator.   J. M. Sutton, D. H. Staelin,
R. M. Price.
Bull. American Astron. Soc., Vol. 2, 221 (1970). – Abstr.
AAS.

141.108   Synthesis of the polarization properties of 3C10
   and 3C58 at 1420 and 2880 MHz.
K. W. Weiler, G. A. Seielstad.
Bull. American Astron. Soc., Vol. 2, 224 - 225 (1970).
Abstr. AAS.

141.109 Inverse Compton process in quasars and Seyfert-type nuclei. K. Takarada.
Progress Theor. Phys. Japan, Vol. 43. 303 - 318 (1970).

In the present paper, the inverse Compton spectral powers are calculated for infrared sources with a large optical depth due to the synchrotron self-absorption, and we interpret the optical continuum radiations of quasars and also obtain physical parameters of infrared sources based on our interpretation.

141.110 Pulsars may be neutron stars.
S. Tilson.
IEEE Spectrum, Vol. 7, No. 2, p. 42 - 55 (1970).

Radio astronomers have detected several-score pulsars — amazingly periodic radio sources in the Milky Way — since the first was discovered, entirely by accident, some two and a half years ago. They may be tiny, superdense neutron stars whose rapid spin could also explain galactic cosmic rays and the radiation from supernova remnants.

141.111 The radio astronomer's universe.
E. G. Bowen.
Journ. Astron. Soc. Victoria, Vol. 23, 35 - 40 (1970).

141.112 Quasars as protoclusters of galaxies.
B. A. Vorontsov-Velyaminov.
Astrofizika, Vol. 6, 101 - 107 (1970). In Russian. — English translation in Astrophysics, Vol. 6, No. 1.

The hypothesis is suggested that the dense objects QSG, from time to time transforming into QSS, disrupt into components. Their subsequent cascading fragmentation leads to the formation of groups and of clusters of galaxies. The hypothesis explains the absence of QSS in the clusters of galaxies, their complicated radio structure, details of the structure of clusters and of their positions in the radio galaxies.

141.113 The angular dimensions of the quasar 3C 298 and of the scintillating component 3C 273 at 6 MHz.
V. G. Panajian.
Astrofizika, Vol. 6, 165 - 167 = Byurakan Astrophys. Obs. No. 55 (1970). In Russian. — English translation in Astrophysics, Vol. 6, No. 1.

The angular dimensions of the quasar 3C 298 and of the scintillating component 3C 273 at 60 MHz are estimated by shifts of histograms of mean quasi-periods of scintillations. In the case of a symmetrical source with a Gaussian distribution of the radio luminosities magnitude the angular dimensions of the quasar 3C 298 and of the scintillating component of 3C 273 (the core of component) are equal to $0.^m 7$ and $0.^m 5$.

141.114 On a possibility of testing Arp's hypothesis.
M. K. Babadzhanyianz.
Astrofizika, Vol. 6, 170 - 173 (1970). In Russian. — English translation in Astrophysics, Vol. 6, No. 1.

A test of Arp's hypothesis on the ejection of radio sources from the peculiar galaxies is proposed. The application of the test leads to the conclusion that the hypothesis seems to be erroneous.

141.115 High resolution measurements of 21-cm absorption in the direction of W49.
I. Kazès, Nguyen-Quang-Rieu.
Astron. Astrophys., Vol. 4, 111 - 114 (1970).

New measurements of the hydrogen line absorption of the double radiosource W49 have been made with the Nançay radiotelescope. They reveal two H I clouds, one with radial velocity of 12 km/s already pointed out by different authors, and a second one, at about 63 km/s, both responsible for an enhanced absorption in the direction of the thermal component. The detection of the cloud supports the non-physically related double source hypothesis.

141.116 Curve of growth analysis of the absorption line spectrum of PHL 938.
M. H. Demoulin, N. Doras.
Astron. Astrophys., Vol. 4, 339 - 340 (1970).

The equivalent widths of 9 absorption lines of the quasi-stellar object PHL 938 have been measured on two spectra taken by E. M. Burbidge at the Lick 120-inch telescope. The lines are saturated. The curve of growth analysis applied to 5 lines of Fe II gives a lower limit of the number of absorbing ions in the line of sight of $3 \times 10^{13} \mathrm{cm}^{-2}$ and an upper limit of the more probable velocity of the absorbing ions of $45 \mathrm{~km~s}^{-1}$.

141.117 On the question of the distance to the galactic radio source W 31. T. L. Wilson.
Astron. Astrophys., Vol. 4, 487 - 488 (1970).

On the basis of OH, H I and formaldehyde absorption spectra, we conclude the galactic radio source W 31 is more than 2.4 kpc from the sun, the near kinematic distance estimated from the H 109α recombination line radial velocity.

141.118 A catalogue of 3235 radio sources at 408 MHz.
G. Colla, C. Fanti, R. Fanti, A. Ficarra, L. Formiggini, E. Gandolfi, G. Grueff, C. Lari, L. Padrielli, G. Roffi, P. Tomasi, M. Vigotti.
Astron. Astrophys., Suppl. Series, Vol. 1, 281 - 317 (1970).

The catalogue lists 3235 radio sources observed at 408 MHz with the Bologna Northern Cross telescope. It covers an area of 0.44 sterad between declinations +34°02' and +29°18', and is complete down to 0.2 flux units.

141.119 Additional occultation studies of weak radio sources at Arecibo Observatory: List 4.
K. R. Lang, J. Sutton, C. Hazard, S. Gulkis.
Astrophys. Journ., Vol. 160, 17 - 23 (1970).

Positions, structures, and optical identifications are derived from the lunar-occultation observations of ten weak radio sources.

141.120 Excitation of C III] λ1909 and other semiforbidden emission lines in QSOs and nebulae.
D. E. Osterbrock.
Astrophys. Journ., Vol. 160, 25 - 30 (1970).

Newly calculated collision strengths of C^{+2} and other isoelectronic ions are integrated over a Maxwellian distribution to give effective collision strengths. The fraction of excitations leading to emission of a line photon is calculated in various density ranges. The possibility of determining electron densities in the range $10^5 - 10^6 \mathrm{cm}^{-3}$ from the relative strengths of the two components $^1S_0-^3P_{1\,2}$ is pointed out.

141.121 High-resolution radio interferometry at 610 MHz.
D. L. Jauncey, C. C. Bare, B. G. Clark, K. I. Kellermann, M. H. Cohen.
Astrophys. Journ., Vol. 160, 337 - 339 (1970).

High-resolution interferometer observations, with a baseline of 5.2×10^6 wavelengths at 610 MHz, are reported. Fringes were detected at a level of greater than 1 flux unit on twelve radio sources.

141.122 Quasi-sinusoidal oscillation in arrival times of pulses from NP 0532.
D. W. Richards, G. H. Pettengill, C. C. Counselman III, J. M. Rankin.
Astrophys. Journ. (Letters), Vol. 160, L1 - L6 (1970).

In 1969 from May 10 until September 16, times of arrival of radio pulses from the pulsar NP 0532 in the Crab nebula had a quasi-sinusoidal oscillation of 760 μsec amplitude from peak to peak with an 11-week period, in addition to previously reported secular deceleration.

141.123 Slow periodic variations in pulsars.
D. H. Staelin, M. S. Ewing, R. M. Price, J. M. Sutton. Astrophys. Journ. *(Letters)*, Vol. 160, L7 - L10 (1970).

The pulsars MP 0031 and CP 1919 exhibit periodicities in pulse intensity of approximately 2 and 6 min, respectively. In each pulsar strong bursts of pulses are often follwed by one or more similar bursts spaced at the appropriate 2- or 6-min period. Sometimes there is more than one burst per 2- or 6-min period, and this combination of bursts then repeats as a unit until the component bursts independently diminish. These long "circulation periods" are approximately equal to the time required for the marching subpulses observed in each pulsar to march one fundamental pulsar period. Radio-emission models involving plasma waves circling around pulsars are favored by these results.

141.124 On multiple absorption redshifts in quasi-stellar objects. M. J. Rees.
Astrophys. Journ. *(Letters)*, Vol. 160, L29 - L32 (1970).

The multiple absorption redshifts observed in PKS 0237-23, Ton 1530, and other objects may arise from gaseous filaments trapped within rapidly expanding remnants of radio clouds ejected from QSOs.

141.125 Further spectroscopic observations of quasi-stellar objects and radio galaxies. E. M. Burbidge.
Astrophys. Journ. *(Letters)*, Vol. 160, L33 - L39 (1970).

Spectroscopic observations of nineteen quasi-stellar objects and five radio galaxies have been made with the Lick 120-inch telescope. Redshifts were determined for fourteen QSOs and all of the radio galaxies; in five QSOs only one emission was measured. Four and possibly five of the QSOs of largest redshift have absorption lines. In the three cases analyzed, the absorptions are at redshifts close to the emission-line redshifts, and presumably arise in gas associated with the QSO. The QSO 5C 2.56, with probably the largest observed redshift, has become extremely faint and should be continuously monitored. Some misidentified radio sources are listed.

141.126 On radio emission from quasi-stellar objects.
G. Grueff.
Astrophys. Journ. *(Letters)*, Vol. 160, L41 - L42 (1970).

Some recent identifications of BSOs *(Blue Stellar Objects)* with faint radio sources are questioned.

141.127 Polarized-brightness distribution for 3C 20.
E. B. Fomalont.
Astrophys. Journ. *(Letters)*, Vol. 160, L73 - L77 (1970).

High-resolution maps at 2695 MHz using the NRAO three-element interferometer have been obtained for the linearly polarized and total radiation of 3C 20. The synthesized beam width for the maps is about 6" in diameter. The total radiation shows a common structure composed of two unresolved components with larger-scale radiation between them. The polarized distribution shows good correspondence with the total distribution. The intense components are only slightly polarized whereas the central regions have polarizations as large as 20 percent.

141.128 Spectrographic observations of six radio sources.
M.-H. Demoulin.
Astrophys. Journ. *(Letters)*, Vol. 160, L79 - L81 (1970).

From spectrographic observations, we have obtained (1) the central density and the mass within 1 kpc of the elliptical galaxy NGC 3998 (10^{-21} g cm^{-3} and 7.4×10^9 $M_\odot$, respectively) and (2) the redshifts of four radio galaxies.

141.129 A search for redshifted 21-cm absorption in quasi-stellar objects. C. Heiles, G. K. Miley.
Astrophys. Journ. *(Letters)*, Vol. 160, L83 – L86 (1970).

The QSOs 3C 191 and PKS 1116 + 12 were investigated for redshifted neutral-hydrogen absorption. In no case was any absorption line detected, a fact providing further evidence that the optical absorption does not arise from interstellar matter in intervening galaxies.

141.130 Runaway stars and the pulsars near the Crab nebula. J. R. Gott III, J. E. Gunn, J. P. Ostriker.
Astrophys. Journ. *(Letters)*, Vol. 160, L91 - L95 (1970).

The "runaway stars" formed when one member of a binary system explodes will be followed by the runaway remnants of the exploded stars; thus many pulsars should be high-velocity (~ 100 km sec^{-1}) objects. Treating the pulsars 0527 and 0531 as the remnants of a binary system which was originally in the association I Gem, we predict the proper motion and rate of period change for 0527. Proper-motion observations will enable a check to be made of the runaway-remnant hypothesis; observations of P, $\dot{}$ will check this hypothesis and provide a constraint on pulsar theories.

141.131 Erratum: The period and hard-X-ray spectrum of NP 0532 in 1967. [Astrophys. Journ. *(Letters)*, Vol. 158, L61 - L64 (1969)]. G. J. Fishman, F. R. Harnden, Jr., W. N. Johnson III, R. C. Haymes.
Astrophys. Journ. *(Letters)*, Vol. 160, L117 (1970).

141.132 Evidence of radio luminosity evolution of quasi-stellar radio sources. M. A. Arakelian.
Nature, Vol. 225, 358 - 359 (1970).

It is shown that the mean radio luminosity of any number of quasars in the shell with $0.735 < z < 1.12$ is systematically greater than that of the same number of objects in the sphere of the same volume with $z > 0.735$. The optical luminosity does not show such a feature. The difference between logarithms of radio and optical luminosities is of the form $\log F_{rad} - \log F_{opt} = 3.4 \log (1 + z) + 3.4$.

141.133 Bremsstrahlung radiation in an intense magnetic field and emission from pulsars.
H.-Y. Chiu, V. Canuto.
Nature, Vol. 225, 1230 - 1231 (1970).

It is shown that the criticism of Strange and Simon is not valid since their argument is based on an incorrect expression for the absorption coefficient. The correct criterion for the occurrence of an instability is given and the magnetic bremsstrahlung is discussed thereof.

141.134 Redshift and the size of double radio sources.
T. H. Legg.
Nature, Vol. 226, 65 - 67 (1970).

Published data have been collected on 32 radio galaxies and 25 quasi-stellar sources that have a measured redshift and predominantly double structure. The quasi-stellar sources are found to fit smoothly into a redshift−size relationship with the radio galaxies, supporting the cosmological interpretation of the quasi-stellar redshifts. The distribution of linear separation of components is found to be approximately uniform from about 80 to 320 kpc for the radio galaxies and from about 140 to 230 kpc for the quasi-stellar sources.

141.135 Linear polarization of distant quasars.
R. G. Conway, J. A. Gilbert.
Nature, Vol. 226, 332 - 333 (1970).

Measurements at $\lambda 49$ cm of the linear polarization of quasars show that the degree of polarization is systematically lower for distant quasars ($z > 1.25$) than would be expected from the polarization of nearby quasars, after allowing for the change of wavelength. This effect is most marked for quasars showing red-shifted absorption lines, suggesting that the regions responsible for the absorption also cause Faraday rotation, from which the product (B. Ne. L) may be deduced.

141.136 Quasars behind clusters of galaxies.
J. M. Barnothy, M. F. Barnothy.
Nature, Vol. 226, 335 - 337 (1970).

The chance to find a gravitational lens image, having the characteristic of a quasi-stellar object behind a cluster of galaxies is about 50%, while the chance to find behind a given cluster of galaxies a quasi-stellar object which is a gravitational lens image, is about 1 in 400, or 1 in 10, should the cluster contain dead galaxies. The brightness variations of 3C273 suggest that the gravitational lens of this QSO is a compact galaxy of the type Zw 0930–5527, or Zw 1117–5141, located in the Virgo cluster.

141.137 Shape of the Crab pulsar and its period fluctuations.
C. Chiuderi, F. Occhionero.
Nature, Vol. 226, 337 - 338 (1970).

The recently discovered sinusoidal variation in the arrival times of the Crab pulsar's pulses may be interpreted as an intrinsic free nutation of the pulsar. For this to occur, it is necessary to postulate slight deviations from spherical symmetry of the neutron star solid crust. Numerical estimates for the required deformation show that no substantial loss of gravitational radiation is implied.

141.138 Spectral index–flux density relation for quasars.
R. D. Dagkesamanskii.
Nature, Vol. 226, 432 (1970).

It was found that the mean spectral index (α) of quasars from 3CR catalogue tends to increase with the decrease in 178 MHz flux density (S_{178}). The corresponding relationship for radio galaxies and unidentified radio sources is either much weaker or lacking completely. It seems that the relationship $\alpha - S_{178}$ for quasars becomes somewhat weaker as we proceed to the spectral indices measured at higher frequencies.

141.139 Planetary companions of pulsars.
J. G. Hills.
Nature, Vol. 226, 730 - 731 (1970).

The change in the eccentricity and the semi-major axis of the orbit of a planet resulting from a supernova explosion is calculated. The new orbit is highly eccentric, and it is unbound if more than half of the mass of the star is lost in the explosion. The effect of the orbital motion of the planet on the apparent rotational period of the pulsar is calculated. The sudden decrease in the period of the Crab nebula pulsar can be explained by the orbital motion of its recently discovered planetary companion.

141.140 On the optical identifications of a complete sample of faint radiosources.
A. Braccesi, A. Ficarra, L. Formiggini, E. Gandolfi, C. Lari, L. Padrielli, P. Tomasi, C. Fanti, R. Fanti.
Astron. Astrophys., Vol. 6, 268 - 274 (1970).

The radio and optical properties of 214 extra-galactic radiosources are discussed. It is shown that the bulk of the unidentified sources is likely to be due to powerful radiogalaxies. Their $\log S - \log N$ slope implies a strong evolution, similar to the one found for the 3CR QSS's. On the other hand the small slope for the QSS's counts at our flux level seems to suggest, for them, a cut-off in the red-shift, around 2.5.

141.141 The Ohio survey between declinations of 0° and 36° south.
J. R. Ehman, R. S. Dixon, J. D. Kraus.
Astron. Journ., Vol. 75, 351 - 506 (1970).

A 1415-MHz continuum survey with the OSU 260- X

70-ft radio telescope has been made between declinations of 0° and 36° south covering 7765 deg^2 of sky. Results are presented by maps of the region surveyed and by a list of 4550 sources above 0.16 f.u. Of these sources, 3354 are previously uncatalogued.

141.142 Galactic continuum source counts. T. L. Wilson.
IAU Symposium No. 38, (see 012.013), p. 140 - 141 (1970).

141.143 The distribution in galactic longitude of observable pulsars. G. Lyngå.
IAU Symposium No. 38, (see 012.013), p. 177 (1970).

141.144 Pulsar distances, spiral structure and the interstellar medium. B. Y. Mills.
IAU Symposium No. 38, (see 012.013), p. 178 - 181 (1970).

The distances of all pulsars are calculated on the assumption that they are immersed in a uniform medium of average electron density 0.06 cm^{-3}. It then appears that the pulsars are concentrated towards the local and Sagittarius spiral features and that their mean height above the plane is consistent with that of known supernova remnants. The mean distances appear to be approximately correct, but individual distances are uncertain by about a factor of two. Evidence from radio continuum results supports this model of the ionized interstellar medium.

141.145 Short-period pulsations of the CP 0808 pulsar and the main characteristics of its radiowave emission in metre-wave range. V. V. Vitkevich, Iu. P. Shitov.
Dokl. Akad. Nauk SSSR, Ser. Mat. Fiz., Vol. 191, 553 - 556 (1970). In Russian.

Results of observations in the 60 - 110 MHz range of the CP 0808 pulsar in Puschchino are analysed. A periodic process of class 2, according to the classification of Drake and Craft, Jr., with $P_2 = 0\overset{s}{.}053642$ has been discovered. In a group of about 11 pulses subpulses repeat with a mean period of 24 P_2 which is slightly shorter then the main period $P_1 = 1\overset{s}{.}292241$. Then subpulses of this group decay and a new group shifted by P_2 appears. Both short- and long-period variations in time of subpulses with respect to multiples of P_2 have been detected. The width and the form of the radio emission diagrams of subpulses are studied in connection with a pulsating-rotating star model. The fine structure, form, and time delay of subpulses are described.
B. Onderlička

141.146 Pulsars as sources of cosmic rays.
J. R. Wayland.
Astrophys. Space Sci., Vol. 7, 201 - 210 (1970).

The observed characteristics of pulsars have been considered with respect to how they would affect cosmic rays if pulsars are the sources of cosmic radiation. Particular attention was given to the various time properties of suggested acceleration mechanisms. They are divided into three general cases: (a) constant with time, (b) sudden injections at discrete times, (c) slowly varying with time. In each case the possibility of either a monoenergetic or a power law injection spectrum with respect to energy was taken into account.

141.147 Linear polarization of MP 0628 and its emission at metre wavelengths.
V. V. Vitkevich, Y. P. Shitov.
Nature, Vol. 226, 1235 - 1236 (1970).

The detection of pulsar MP 0628 and some characteristics of its radio emission at 408 MHz is reported in an earlier paper. We have studied this pulsar at metre wavelengths using the east-west arm of the broadband cross-type radio telescope at Pushchino, and linear polarization was detected.

141.148   Les « pulsars ».
Gaz. astron., No. 2, p. 18 - 22 (1969).

141.149   Observations des pulsars à 1420 MHz. Étude de la structure fine des impulsions.
P. Encrenaz, F. Foy, M. Guélin.
Comptes Rendus Acad. Sci. Paris, Sér. B, Vol. 270, 1585 - 1588 (1970).

Nous avons entrepris depuis décembre 1969, une étude systématique de ces objets, bénéficiant d'un rapport signal sur bruit amélioré par rapport à 1968 du fait de la diminution de la température du système et de la constante de temps des corrélateurs et de l'utilisation d'une procédure d'intégration plus élaborée. Nous en publions ici les premiers résultats.

141.150   Polarization of pulsating radio sources.
R. D. Ekers, A. T. Moffet.
Observations Owens Valley Radio Obs., 1969 No. 8, 13 pp. (1969).

A somewhat qualitative discussion of the polarization characteristics of pulsars CP 0328, AP 0823, PSR 0833-45, CP 1133 at 13 cm. A variation of direction of polarization across a pulse is common. – JBW

141.151   Optical and X-ray synchrotron radiation from the Crab pulsar.   J. Trümper.
Astrophys. Letters, Vol. 5, 271 - 274 (1970).

It is assumed that the Crab pulsar is a rotating magnetic neutron star whose magnetic moment is approximately at right angles to the rotation axis. The synchrotron radiation is calculated from the electrons which are accelerated in the wave zone. It is shown that the synchrotron spectrum extends up to the hard X-ray range if magnetic fields in the wave zone are assumed to be of the order of 10^6 Gauss. The observed ratio of pulsar energy flux to that of the nebula can be explained in this model.

141.152   Quasare.
F. Baier.
Sterne, 46. Jahrgang, p. 53 - 65 (1970).

141.153   Der entfernteste Quasar.
H. Oleak.
Sterne, 46. Jahrgang, p. 66 - 67 (1970).

141.154   Pulsars and non-thermal radio sources.
N. S. Kardashev.
Astron. Zhurn. Akad. Nauk SSSR, Vol. 47, 465 - 478 (1970). In Russian.   English translation in Soviet Astron. AJ, Vol. 14, No. 3.

Physical processes in galactic and extra-galactic sources of non-thermal radio emission are considered assuming that pulsars are principal sources of the magnetic field and cosmic rays.

141.155   On the relation between the pulsar NP 0532 and the small-angular-diameter source in the Crab nebula.   L. I. Matveyenko, N. A. Lotova.
Astron. Zhurn. Akad. Nauk SSSR, Vol. 47, 483 - 486 (1970). In Russian.   English translation in Soviet Astron. AJ, Vol. 14, No. 3.

It is shown, on the basis of the spectra and the pulse duration, that the source of small angular diameter in the central part of the Crab nebula may be associated with the pulsar and its magnetosphere. The size of the pulsar in the supernova remnant Vela X is also estimated.

141.156   The beaming of radio waves from pulsars.
F. G. Smith.
Monthly Notices, Roy. Astron. Soc., Vol. 149, 1 - 15 (1970).

The radio pulses from pulsars are formed by a beam of directed emission which sweeps across the observer. The structure of individual pulses, rather than the average pulse shape, is considered to be important. The beam could be formed by an anisotropic emission process, such as synchrotron radiation, or by anisotropic propagation, or by the relativistic effect of the rapid orbital motion of the source. Pulse shapes to be expected from synchrotron radiation and from the relativistic effect are calculated. Comparison with observed widths and polarization of radio pulses suggests that they are formed by the relativistic effect, and that the radiation mechanism is through some process akin to cyclotron radiation.

141.157   Observations of radio sources near 2 f.u. at 408 MHz.   R. W. Hunstead, D. L. Jauncey.
Monthly Notices, Roy. Astron. Soc., Vol. 149, 91 - 100 (1970).

Comprehensive data, namely, accurate radio positions, optical identifications, angular size estimates and flux densities, are presented for a sample of 35 sources, 33 of which are near 2.0 f.u. at 408 MHz.

141.158   Identification of southern extragalactic radio sources.   R. D. Ekers.
Australian Journ. Phys., Vol. 23, 217 - 221 (1970).

Field classifications are given for 46 radio sources south of declination -44° for which suitable optical observations were available in September 1967. Identifications are suggested for 21 objects and finding charts are given for 7 objects not previously published.

141.159   Modulation effects in the pulsars PSR 1451-68, 1749-28 and 2045-16.   T. W. Cole, K. H. Hesse.
Astrophys. Letters, Vol. 6, 59 - 60 (1970).

Examination of the modulation effects associated with the 'second periodicity' in a pulsar with an interpulse suggests that there may be interpulses in PSR 1451-68, 1749-28 and 2045-16. The true periods of these objects may therefore be twice the previously accepted values.

141.160   The period-latitude correlation of pulsars and its implications.   P. Notni, H. Oleak, K.-H. Schmidt.
Astrophys. Letters, Vol. 6, 61 - 63 (1970).

The correlation between the periods and the concentration of pulsars to the galactic equator is investigated. Possible explanations are discussed; (1) the existence of a period-radio luminosity relation or (2) a division into groups of various spatial distribution.

141.161   Interstellar scattering of pulsar radiation and its effect on the spectrum of NP 0532.
W. M. Cronyn.
Science, Vol. 168, 1453 - 1455 (1970).

Angular scattering in the interstellar medium results in multipath dispersion which can amount to more than one pulse period for pulsars of short period and high dispersion measure. The dispersion, if operative, imposes on the pulsation flux a cutoff inversely proportional to the fourth power of the observing wavelength. The low-frequency pulse shape of pulsar NP 0532 suggests that this pulsar is subject to such scattering and that the observed low-frequency cutoff in the apparent spectrum is not an intrinsic property of the pulsar.

141.162   Declination and flux density measurements of selected 4 C sources at 430 MHz using the Arecibo 1000-ft reflector. II.   D. C. Backer, C. Hazard, D. L. Jauncey, J. Sutton.
Astron. Journ., Vol. 75, 529 - 534 (1970).

Declination and flux density measurements at 430 MHz are presented for 412 sources from the 4C Catalogue. Examination of the error distribution indicates a standard deviation of 25 arc sec for the Arecibo declination measurements.

141.163 Discovery of the pulsar PP 0943 by means of the difference method. V. E. Zhuravlev.
Priroda, No. 6.70, p. 66 - 68 (1970). In Russian.

141.164 Maps of six extra-galactic radio sources at 5 GHz. S. Mitton.
Monthly Notices, Roy. Astron. Soc., Vol. 149, 101 - 109 (1970).
3C 123, 196, 247, 275.1, 295 and 336 have been mapped with the Cambridge one-mile telescope at a frequency of 5 GHz. The maps provide new positional information and physical data for the six sources. A comparison of the structural information derived from the 5 GHz work with previous observations using long baseline interferometers, and with observations of interplanetary scintillation, is made.

141.165 Possibilité d'une émission pulsée de rayonnement gamma d'énergie supérieure à 50 MeV dans la direction de la nébuleuse du Crabe.
J. Vasseur, J. Paul, B. Parlier, J.-P. Leray, M. Forichon, B. Agrinier, G. Boella, L. Maraschi, A. Treves, L. Buccheri, L. Scarsi.
Comptes Rendus, Acad. Sci. Paris, Sér. B, Vol. 271, 61 - 64 (1970).
Le traitement des données d'une expérience faite en ballon stratosphérique le 11 juillet 1969 nous a permis d'observer une pulsation dont les caractéristiques sont très proches de celles du «pulsar» NP 0532. Une étude statistique portant sur la fréquence d'apparition de pulsations, d'amplitudes diverses, nous a permis d'apprécier dans quelle mesure la pulsation observée pourrait n'être due qu'à une fluctuation statistique.

141.166 On mechanisms of formation of ejections in quasars. I. Nedyalkov.
Izv. Fiz. in-t s ANEB, Vol. 18, 105 - 113 (1969). In Bulgarian.
Abstr. in Referativ. Zhurn. 51. Astron., 4.51.801 (1970).

141.167 Statistics of quasistellar objects. O. V. Dobrovol'skij.
Dokl. AN Tadzh. SSR, Vol. 12, No. 9, p. 16 - 20 (1969). In Russian. – Abstr. in Referativ. Zhurn. 51. Astron., 4.51.806 (1970).

141.168 On discrete distances of extragalactic objects. M. Kalinkov, N. Cholakova.
Izv. Fiz. in-t s ANEB, Vol. 18, 119 - 149 (1969). In Bulgarian. – Abstr. in Referativ. Zhurn. 51. Astron., 4.51.810 (1970).

141.169 Counts of radio sources. M. S. Longair.
Uspekhi fiz. nauk, Vol. 99, 229 - 248 (1969). In Russian. – Abstr. in Referativ. Zhurn. 51. Astron., 4.51.839 (1970).

141.170 Periods of four southern pulsars. M. M. Komesaroff, J. G. Ables, D. Morris, D. J. Cooke, U. J. Schwarz, P. A. Hamilton.
IAU Circ. No. 2201 (1970).

141.171 On the nature of pulsars. III. Analysis of observations. J. E. Gunn, J. P. Ostriker.
Astrophys. Journ., Vol. 160, 979 - 1002 (1970).
In this paper we formulate statistical methods for the analysis of pulsar observations based upon the magnetic-dipole model presented in the first paper in this series. Applying this formalism to the extant observations, we reach some tentative conclusions. With these results we consider the evolution of the nuclear region of the Galaxy and conclude that during its early history it might have been a small nonthermal source with a total luminosity between that of Seyfert nuclei and

QSOs, fluctuating on a time scale of approximately 1 year.

141.172 On the motion of current sheets, and the radio, optical, and X-ray emission from pulsars. II. Pulse structure, polarization, time-varying features, and tight-beam emission. I. Lerche.
Astrophys. Journ., Vol. 160, 1003 - 1017 (1970).
The physical mechanism of an oscillating interface carrying a steady current produces high-frequency radiation, and was proposed in an earlier paper as the generator of pulsar emission. The mechanism is investigated in greater detail in this paper. We use simple models of finite-thickness interfaces between the vacuum radiating dipole field and the surrounding ultrarelativistic plasma to elucidate various facets of real pulsars.

141.173 Flux of Cassiopeia A at 1415 MHz. P. J. Encrenaz, A. A. Penzias, R. W. Wilson.
Astrophys. Journ., Vol. 160, 1185 - 1186 (1970).
We have measured the flux of Cas A at 1415 MHz at an average date of 1969.5, and find it to be 2369 flux units with a probable error of 60 flux units.

141.174 Search for interstellar formic acid at 1638.806 MHz in radio sources.
M. Cato, T. Cato, P. Landgren, A. Sume.
Astrophys. Journ. (Letters), Vol. 160, L131 - L132 (1970).
A search for interstellar formic acid in the frequency range of 1.2 MHz has led to negative results.

141.175 Search for optical counterparts of pulsars CP 0328 and CP 0950. W. J. Cocke, M. J. Disney.
Astrophys. Journ. (Letters), Vol. 160, L139 - L142 (1970).
Photoelectric multichannel signal averaging was used to search for pulsed optical components of CP 0328 and CP 0950. We obtain limiting visual magnitudes of 25.5 for CP 0328 and 24.5 for CP 0950.

141.176 The optical position of 3C345. C. Barbieri, M. Capaccioli, G. Pinto.
Mem. Soc. Astron. Italiana, Nuova Serie, Vol. 41, 247 - 249 (1970).

141.177 At the frontier of universe. J. Olmr.
Říše hvězd, Vol. 51, 29 - 31 (1970). In Czech.

141.178 Interpreting the curvature in the extragalactic radio source spectrum by means of a high cut-off energy in the electron spectrum. D. Ristow.
Beiträge Radioastronomie, Max-Planck-Inst. Radioastronomie, Bonn, Vol. 1, (No. 6), p. 149 - 160 (1970).
An analysis of the 3C-sources with convex spectra is given. An attempt is made to describe the spectrum by a power law above 178 MHz and examine the distribution of the "critical" frequency f_c. The distribution function increases with f_c. There is no significant difference between quasars and radiogalaxies. Also brightness temperature, source diameter, and redshift do not show the correlation with the "critical" frequency.

141.179 Messungen von Radioquellen bei 10.69 GHz. P. Zimmermann.
Beiträge Radioastronomie, Max-Planck-Inst. Radioastronomie, Bonn, Vol. 1, (No. 6), 161 - 168 (1970).
The flux density of 74 radio sources was measured at 10.69 GHz. The measurements were made with the 20-m-telescope of the Institut für Weltraumforschung in Bochum. The telescope is protected by a radom. A crygenically cooled parametric amplifier was used as receiver. By applicating the dual beam method, the influence of the varying noise contribution of the atmosphere could be largely reduced. All the sources of

the NRAO catalogue with an expected flux density of $\geq 10^{-26}$ W m^{-2} Hz^{-1} were measured.

141.180 New analysis of the radio source counts.
W. Davidson.
Nature, Vol. 227, 357 - 359 (1970).
Luminosity evolution and not density evolution emerges as the dominating influence explaining the radio source counts.

141.181 Spectral flux densities of radio emission from discrete sources at 3.5 cm wavelength. II.
A. G. Gorshkov, E. E. Spangenberg, I. E. Valtz, L. M. Gindilis, V. K. Konnikova, V. I. Koptjaev, I. G. Moiseev, V. V. Nikitin, V. A. Soglasnov, V. A. Soglasnova.
Astron. Tsirk., No. 545, p. 1 - 3 (1970). In Russian.

141.182 On the clusters of quasars. N. F. Sleptsova.
Astron. Tsirk., No. 545, p. 5 (1970). In Russian.

141.183 Frequency correlation of scintillations of pulsars.
V. I. Shishov.
Astron. Tsirk., No. 547, p. 1 - 3 (1970). In Russian.

141.184 On double radio structures.
V. N. Kurilchik.
Astron. Tsirk., No. 550, p. 1 - 4 (1970). In Russian.

141.185 Evolution of the luminosity of quasars.
M. A. Arakelian.
Astron. Tsirk., No. 552, p. 1 - 3 (1970). In Russian.

141.186 Once more about quasars and nuclei of galaxies.
N. E. Kurochkin.
Astron. Tsirk., No. 552, p. 3 - 5 (1970). In Russian.

141.187 On the radio source in M82. V. N. Kurilchik.
Astron. Tsirk., No. 552, p. 5 - 7 (1970).
In Russian.

141.188 Secondary periodicities in four strong pulsars at 80 MHz. O. B. Slee, P. S. Mulhall.
Proc. Astron. Soc. Australia, Vol. 1, 322 - 324 (1970).

141.189 The use of HI absorption to determine distance for 10 galactic radio sources. W. M. Goss,
J. D. Murray, V. Radhakrishnan.
Proc. Astron. Soc. Australia, Vol. 1, 332 - 333 (1970).

141.190 Extra-galactic radio sources with extended structure.
R. T. Schilizzi, W. B. McAdam.
Proc. Astron. Soc. Australia, Vol. 1, 337 - 340 (1970).

141.191 The Molonglo catalogue survey. I. M. Davies.
Proc. Astron. Soc. Australia, Vol. 1, 340 - 341 (1970).

141.192 Frequency dependence of 4C source positions.
R. E. B. Munro, D. G. Hoskins.
Proc. Astron. Soc. Australia, Vol. 1, 341 - 344 (1970).

141.193 Observations of Centaurus A at 80 MHz.
I. A. Lockhart, K. V. Sheridan.
Proc. Astron. Soc. Australia, Vol. 1, 344 - 345 (1970).

141.194 An analysis of the redshift-apparent magnitude relation for 147 quasi-stellar objects.
N. V. Zotov, W. Davidson.
Australian Journ. Phys., Vol. 23, 351 - 362 (1970).
Optical data for 147 quasi-stellar objects are analysed to obtain an estimate of q_0, the present value of the deceleration parameter of the universe.

141.195 Autocorrelation analysis of the brightness of the quasar 3C 345. F. I. Lukatskaja.
Astron. Tsirk., No. 553, p. 2 - 3 (1970). In Russian.

141.196 A review of theories of pulsars. H.-Y. Chiu.
Publ. Astron. Soc. Pacific, Vol. 82, 487 - 533 (1970).
It is the purpose of this paper to discuss, on the basis of observed data and current knowledge of physics, the present theoretical status of pulsars.

141.197 A Master List of radio sources.
R. S. Dixon.
Astrophys. Journ., Suppl. Series, Vol. 20, (No. 180), 503 pp. (1970).
A Master List of radio sources has been prepared by combining about thirty catalogs in a common format. Approximately 25000 listings are included for some 12000 separate sources. The list is in order of increasing right ascension and is arranged for maximum utility and minimum complexity. Each entry gives the source name, 1950.0 position, flux density, and the frequency at which it was observed. This list will be maintained and updated indefinitely at Ohio State University Radio Observatory, and magnetic-tape copies of the latest versions will be available at any time.

141.198 Statistical investigation of the distribution of pulsars in space. T. Gold, H. M. Newman.
Nature, Vol. 227, 151 - 152 (1970).
The distribution of angular coordinates in the sky, in relation to the galactic plane, can be used to infer information about the distribution in distance. We have therefore divided the fifty pulsars into two groups with periods longer and shorter than one second. Although the statistical weight that can be attributed to this limited set of data is not very great, it still makes absolutely clear that long period pulsars as a group are intrinsically very much fainter than short period ones, and the comparison with the calculations gives distance estimates which are not likely to be in error by a large factor.

141.199 Experimental method for distinguishing pulsar models. P. E. Roe.
Nature, Vol. 227, 154 - 156 (1970).
In this letter I suggest that the fact that Huygen's principle is not valid in large gravitional fields gives rise to detectable effects.

141.200 Percentage number of nonthermal sources in the northern and southern halves of the Galaxy.
W. J. Altenhoff, T. L. Wilson.
Nature, Vol. 225, 245 - 246 (1970).
The apparent asymmetry between the northern and the southern parts of the Galaxy is discussed. It is suggested that any difference could be exaggerated by the use of different radio telescopes for the survey.

141.201 Short duration pulsations of CP 0808 and main features of its radio emission in the metre band.
V. V. Vitkevich, Yu. P. Shitov.
Nature, Vol. 225, 248 - 251 (1970).
Details of short duration sub-pulses in CP 0808 are discussed and suggestions for their probable explanation are given.

141.202 Reply to Kinman concerning 3C 345.
J. H. Hunter, Jr., P. K. Lü.
Nature, Vol. 225, 366 - 367 (1970).
This is a reply to Kinman's criticisms of a light curve for 3C 345 published earlier by the authors (Nature, Vol. 223, 1045 - 1046, 1969).

141.203 Cassiopeia A absorption spectrum and the interarm medium. G. de Jager.
Nature, Vol. 225, 622 - 623 (1970).

A neutral hydrogen spin temperature of 540 ± 230° K is derived for the interarm region along the line of sight to Cas A.

141.204 Possible pulsar formation mechanism.
J. C. Wheeler.
Nature, Vol. 226, 1043 (1970).

Evidence is cited which links the degenerate cores of all red giants in the range $3.5 \lesssim M/M_\odot \lesssim 8$ to pulsars. The implied connection may be understood if carbon burning does not result in violent detonation. If the star survives carbon burning the core will collapse due to electron capture on ^{24}Mg. During collapse, ^{16}O will detonate, sending an energetic shock wave into the loosely bound hydrogen envelope. The core will collapse toward a neutron star state due to rapid beta processes in the detonated material.

141.205 The structure and evolution of extragalactic sources. K. I. Kellermann.
National Radio Astron. Obs. , *Green Bank,* Repr., Ser. B, No. 162 [Reprinted from *"Plasma Instabilities in Astrophysics"*, Gordon and Breach, 1969, p. 283 - 296], 14 pp.(1969).

141.206 The radio variability of VRO 42.22.01 at wavelengths of 2.8 and 4.5 cm.
B. H. Andrew, J. M. Macleod, J. L. Locke, W. J. Medd, C. R. Purton.
Bull. Radio Electr. Engineering Div.,National Res. Council, Canada, Vol. 19, (No. 1), p. 33 - 35 (1969).

Presents an extensive set of measurements of the radio variability of the source VRO 42.22.01 at 4.5 and 2.8 cm. – *MWS.*

141.207 Variation of six radio sources at four and two centimeters. P. R. Schwartz, A. H. Barrett.
Mass. Inst. Technol. Res. Lab. Electronics Quarterly Progr. Rep., No. 94, p. 43 - 48 (1969).

An investigation of the variability of 24 continuum radio sources at 4 and 2 cm. – *GD.*

141.208 Pulsar optical and radio emissions observed simultaneously. C. N. Taubman.
Hewlett-Packard Journ., Vol. 20, (No. 10), p. 17 - 20 (1969).

141.209 Dead quasar found in our Galaxy?
Science Journ., Vol. 5A, No. 4, p. 17 - 18 (1969).
A discussion on the Lynden-Bell suggestion that our galactic nucleus is a dead quasar. – *JBW*

141.210 Planetary satellite orbits around pulsar.
New Scientist, Vol. 44, (No. 673), 222 (1969).
A report on the suggestion that the periodicity in the variation of arrival times of individual pulses from NP 0532 is due to a planet orbiting the pulsar. – *JBW*

141.211 Pulsar observations.
M. S. Ewing, B. F. Burke, R. M. Price, D. H. Staelin, J. Sutton.
Mass. Inst. Technol. Res. Lab. Electronics Quarterly Progr. Rep., No. 95, p. 8 - 10 (1969).

Brief report of some measurements of pulsar spectra using the NRAO 300-ft antenna in the 110 to 170 MHz frequency range. – *BFC*

141.212 Pulsar observations with a swept-frequency local oscillator. D. H. Staelin, J. Sutton, M. S. Ewing.
Mass. Inst. Technol. Res. Lab. Electronics Quarterly Progr. Rep., No. 95, p. 3 - 6 (1969).

Observations show that the trailing edge of the intrinsic pulse emitted by NP 0532 decays much more slowly at 115 MHz than at 157.5 MHz. – *BFC*

141.213 Compact radio source in the nucleus of M 87.
M. H. Cohen, A. T. Moffet, D. Shaffer, B. G. Clark, K. I. Kellermann, D. L. Jauncey, S. Gulkis.
Observations Owens Valley Radio Obs.,1969, No. 10, 6 pp.
VLBI, Goldstone-Canberra, shows a small component source in the nucleus of M 87. Its diameter is either 0.0016 sec arc or 0.001 sec arc depending on the model adopted. – *DKM*

141.214 Radio astronomers measure distance and power of pulsars. E. Ashpole.
Electronics Weekly, No. 457, p. 7 - 8 (1969).
Popular description of measurements of distance and power of pulsar CP 0328 undertaken at Jodrell Bank. – *DNC*

141.215 Circular polarization measurements of compact radio sources at 1420 MHz.
G. A. Seielstad.
Observations Owens Valley Radio Obs., 1969, No. 6, 8 pp.
Circular polarization of seventeen small-diameter sources measured at 1420 MHz. None had detectable polarization. – *GD*

141.216 Observations of compact radio-sources using radio-interferometer with Green-Bank-Crimea's base line.
J. J. Broderick, B. G. Clark, M. H. Cohen, V. A. Efanov, B. Hansen, J. L. Jauncey, K. I. Kellermann, L. R. Kogan, V. I. Kostenko, L. I. Matveenko, I. G. Moiseev, J. Payne, V. V. Vitkevich.
Moscow, Akad. Nauk SSSR Inst. Kosmich. Issled., 20 pp. (1970). In Russian and English.

A short report of the results of very long base line interferometry at wavelengths of 6 cm and 2,8 cm between Green Bank and the Crimea in September–October 1969. The fringe invisibilities of 12 sources were measured at 6 cm and 2 sources were detected at 2,8 cm. – *OBS*

141.217 Pulsar wobble and neutron starquakes.
M. Ruderman.
Nature, Vol. 225, 838 - 839 (1970).

Two types of non-secular changes in pulsar frequency have been reported. One is a sudden small frequency jump in NP 0532 and in PSR 0833. The other is a slow small amplitude "wobble" of frequency in the Crab pulsar. A number of explanations for these jumps have been proposed based on the model of a pulsar as a rotating neutron star. The author discusses some consequences of the assumption of free precession and starquake frequency jumps.

141.218 Polarization of the radio pulses from the Crab nebula pulsar. D. A. Graham, A. G. Lyne, F. G. Smith.
Nature, Vol. 225, 526 (1970).

The results for the Crab nebula pulsar, NP 0532, are of special interest, because they may be compared with the optical polarization recently measured by Wampler *et al.*. We therefore present a preliminary analysis for the main radio pulse, both in the form of an integration over many pulses and for individual pulses recorded photographically.

141.219 Nature of emission from pulsar NP 0532.
I. S. Shklovsky.
Nature, Vol. 225, 251 - 252 (1970).

Several authors have proposed that the optical and X-radiation of NP 0532 can be explained by a synchrotron mechanism in the dipole magnetic field of a rotating neutron star. It seems to me that there is important recent support of this mechanism which makes possible a qualitative development of the theory of nonthermal radiation from NP 0532.

141.220   Age of pulsar PSR 0833–45.
        I. S. Shklovsky.
Nature, Vol. 225, 252 - 253 (1970).
    PSR 0833–45 has been identified with the radio nebula
Vela X. This nebula is a typical remnant of a type II supernova, like the Cygnus Loop and IC 443, for instance. The
radio and optical properties of this nebula were described by
Milne. From the observed characteristics of Vela X we can
calculate the age of the nebula.

**141.221   How significant are the sharp peaks in the frequency
        distribution of QSO redshifts?**
A. J. Wesselink.
Nature, Vol. 225, 927 - 928 (1970).
    Using the observational material, which was taken from
a list by Burbidge and Burbidge (Nature, Vol. 224, 21 - 24
(1969)), I find that the well known peak at $z = 0.060$ cannot
be due to chance, although there is some uncertainty of the
significance level. On the other hand, the conclusion of Burbidge that the peak at $z = 1.95$ is highly significant is not confirmed. It can be shown that the difference between our result
and that of Burbidge is due to the interpretation of the statistics.

141.222   Quarks in quasars and a false analogy.
        B. Kuchowicz.
Acta Phys. Polon., Vol. 36, 1105 - 1107 (1969).

**141.223   Excitation of semi-forbidden $2s^2$ $^1S - 2s2p$ ^3P
        lines observed in quasars and nebulae.**
D. E. Osterbrock.
Journ. Phys. (Proc. Phys. Soc. B), Vol. 3, (No. 2), 149 -
160 (1970). – See Phys. Abstr., Vol. 73, No. 35141.

141.224   Pulsar periods and Kepler's third law.
        H. L. Poss.
American Journ. Phys., Vol. 38, (No. 1), 109 - 110 (1970).
    The periods of currently known pulsars range from 3.3 ×
10^{-2} sec–3.5 sec. If one assumes that the nature of these
objects is such that their periods are direct measures of their
rotational frequencies, then it is possible to demonstrate that
pulsars are probably superdense bodies, surpassing white
dwarfs in this regard. Kepler's third law is derived for the
special case of a test mass in circular orbit close to the surface of a sphere.

**141.225   Polarization measurements of extragalactic radio
        sources at 3.12 cm wavelength.**
G. L. Berge, G. A. Seielstad.
Report 1969-2, California Inst. Technol., Pasadena.
[Available from CFSTI, Springfield], 23 pp. (1969).
    The two-element interferometer at the Owens Valley
Radio Observatory was used in late 1967 to measure the
integrated polarization, both linear and circular, of 32
extragalactic radio sources at a wavelength of 3.12 cm.
In addition, flux densities of 103 radio sources were also
measured.

**141.226   A general discussion of the brightness distribution
        of extragalactic radio sources. III.**
E. B. Fomalont.
Report 1969-3, California Inst. Technol., Pasadena. [Available
from CFSTI, Springfield, Va.], 52 pp. (1969).
    The paper contains a description and discussion of the
brightness distribution of extra-galactic radio sources based
mainly on recent observations at the California Institute of
Technology.

141.227   High resolution observations of radio sources. IV.

M. H. Cohen.
Report 1969-4, California Inst. Technol., Pasadena. [Available
from CFSTI, Springfield, Va.], 106 pp. (1969).
    Three principal methods are used to obtain high angular
resolution at radio wavelengths: interferometry, lunar occulta -
tions, and interplanetary scintillations. These methods are
reviewed and compared.

**141.228   An interferometric study of the polarization of
        two radio sources at a wavelength of 9.8 cm.**
G. S. Downs.
Thesis, Stanford Univ., Stanford, Calif. [Available from Univ.
Microfilms, Ann Arbor, Mi.], 200 pp. (1969).
    A two-element radio interferometer has been used to
map distribution of linear polarization over the radio sources
Taurus A and Cassiopeia A, both supernova remnants.

141.229   Quasar red shifts.   J. Yates.
        International Journ. Theor. Phys., Vol. 2, 293 - 296
(1969).
    The results of techniques developed in earlier papers are
used in a discussion of quasar red shifts.

141.230   The pulsars.   F. G. Smith.
        Proc. Roy. Instn., Vol. 42, No. 197, Part 4, p.288 -
294 (1969).
    The author discusses the discovery and observation of
pulsars. Recent search for more pulsars is reported.

141.231   On the red-shift of quasars.   R. F. Sisteró.
        Asoc. Argentina Astron. Bol., No. 12, (see 012.019),
p. 20 - 25 (1968). – Note.

**141.232   The absorption in the continuous spectrum of the
        radio source 18 SIA associated with the gaseous
nebula NGC 6618.   R. Quiroga.**
Asoc. Argentina Astron. Bol., No. 14 (see 012.020), p. 48 -
49 (1968). In Spanish. – Abstract.

**141.233   Measurements of position, size, and polarization
        of radio sources.   R. F. Colomb.**
Asoc. Argentina Astron. Bol., No. 14 (see 012.020), p. 52 -
53 (1968). In Spanish. – Abstract.

141.234   Optical identification of radio sources.
        E. Bajaja.
Asoc. Argentina Astron. Bol., No. 14 (see 012.020), p. 67 -
68 (1968). In Spanish. – Abstract.

141.235   Quasars.   F. D. Kahn.
        Phys. Bull., G. B., Vol. 20, 261 - 264 (1969). – See
Bull. Signalét., Vol. 31, Section 120, No. 3814 (1970).

**Radio Astrophysics. Nonthermal Processes in
Galactic and Extragalactic Sources.   See Abstr. 003.155.**

    Synchrotron emission at strong radiative damping.
See Abstr. 022.011.

    Optical search techniques for pulsars.
See Abstr. 034.026.

    **Astrophysical implications of the continuity
equation plasma oscillation.   See Abstr. 062.016.**

    Relativistic stellar winds and the Crab pulsar.
See Abstr. 064.020.

    **Neutron star crusts and alignment of magnetic
axes in pulsars.   See Abstr. 065.047.**

Long period oscillations in rotating neutron stars.
See Abstr. 065.094.

The 1969 occultation of Taurus A by solar corona.
See Abstr. 074.061.

Radioastronomie der Flaresterne.
See Abstr. 122.006.

The light curve of BL Lac (= VRO 42.22.01).
See Abstr. 122.076.

High-frequency stellar oscillations. III. A brief
report.   See Abstr. 122.093.

Supernova remnants and hidden pulsars.
See Abstr. 125.001.

Effects of central pulsars on supernova envelopes.
See Abstr. 125.010.

Supernova remnants, pulsars and X-ray sources.
See Abstr. 125.015.

Observations of H II regions at 0.96, 1.65, and
2.73 cm.   See Abstr. 131.049.

Observations of the anomalous microwave recombination line at 21-cm wavelength.
See Abstr. 131.060.

A recombination line of mapping the H II region
W3 (IC 1795).   See Abstr. 131.082.

Superbright radio "knots" in the H II region W51.
See Abstr. 131.111.

Anomalous recombination line 166α.
See Abstr. 131.132.

Microwave observations of the "planetary" nebula,
K3 – 50.   See Abstr. 133.013.

Microwave observations of the 'planetary' nebula
K3-50.   See Abstr. 133.023.

Observed shapes of Crab nebula radio pulses.
See Abstr. 134.011.

Radio observation of the pulsar NP 0532.
See Abstr. 134.023

The central star: Photometric optical observations.
See Abstr. 134.024.

An X-ray survey of the Cassiopeia region and its

implications concerning supernova remnants and the galactic
source distribution.   See Abstr. 142.043.

Nonthermal radiation emission from X stars and
pulsars.   See Abstr. 142.023.

The Cygnus X region. VI. A new 2695 MHz
continuum survey.   See Abstr. 142.031.

A comparison of radio observations of NGC 7822
(W 1) with the OB stars in the Ceph IV association.
See Abstr. 152.010.

A 408 MHz survey of the galactic anticentre region.
See Abstr. 157.006.

Detection of 166α recombination-line radiation in
the direction of the galactic center.   See Abstr. 157.007.

Surveys of the galactic plane at 1.414, 2.695 and
5.000 GHz.   See Abstr. 157.012.

A low latitude galactic survey from $l^{\text{II}} = 6°$ to
26° at 2.7 GHz.   See Abstr. 157.013.

A low latitude galactic survey from $l^{\text{II}} = 37°$ to 47°
at 2.7 GHz.   See Abstr. 157.014.

Radio astronomy.   See Abstr. 157.027.

Spectrum and redshift of the optically variable,
compact radio galaxy 3C371.   See Abstr. 158.002.

A large, asymmetrical light pulse from the N galaxy
PKS 0521–36.   See Abstr. 158.005.

On the origin of gas with high radial velocity in the
radio galaxy NGC 1275.   See Abstr. 158.015.

Galaxies and quasars: Puzzling observations and
bizarre theories.   See Abstr. 158.023.

Yearly variations of 3C 120.
See Abstr. 158.024.

A high-resolution 5-GHz map of 3C 33.
See Abstr. 158.040.

An electrodynamic model of radio galaxies and
quasars.   See Abstr. 158.057.

Radio observations of E and S0 galaxies.
See Abstr. 158.096.

The log S–log z diagram for radio galaxies and its
relation to cosmology.   See Abstr. 158.102.

142 X Ray-, Gamma Ray-Sources

142.001 On the two distance estimates to Sco X-1.
O. P. Manley.
Astrophys. Letters, Vol. 5, 43 - 44 (1970).
Two disparate distance estimates to Sco X-1 are discussed. It is suggested that very likely the larger of the two is correct.

142.002 Magnetic funnelling for accretion on neutron stars.
S. Sofia.
Astrophys. Letters, Vol. 5, 45 - 46 (1970).
A funnelling mechanism is proposed which increases the efficiency of the accretion of matter onto a neutron star to such a degree that the X-ray luminosity of Scorpius X-1, for example, can be produced by a neutron star moving through an interstellar cloud, if the velocity of the star is $<10^6$ cm sec^{-1}, and the cloud density high. By not requiring the presence of a large companion star to produce the accreting matter, this model overcomes the difficulty introduced by the low optical luminosity recently determined for the source.

142.003 The soft X-ray spectra of three cosmic sources and simultaneous optical observations of Sco XR-1.
R. J. Grader, R. W. Hill, F. D. Seward, W. A. Hiltner.
Astrophys. Journ., Vol. 159, 201 - 214 = Contr. Cerro Tololo Inter-American Obs. No. 66 (1970).
New X-ray spectra have been obtained for Sco XR-1, the Crab Nebula, and a low-energy source currently designated Vela XR-2. The range of photon energies covered was ~ 150 eV–10 keV. The Sco XR-1 data are interpreted as thin-source, thermal bremsstrahlung with absorption in cool material. The new data are compared with previous low-energy spectra of Sco XR-1, and this source is shown to be probably variable in the soft X-ray region. The spectrum of the Crab Nebula shows absorption in cool material at low energies. Vela XR-2 is the most intense source of the three at photon energies $\lesssim 0.5$ keV.

142.004 The diffuse cosmic X-ray background from 20 to 220 keV. J. A. M. Bleeker, A. J. M. Deerenberg.
Astrophys. Journ., Vol. 159, 215 - 228 (1970).
Balloon experiments with identical X-ray detectors, carried out from three different geomagnetic latitudes resulted in an accurate determination of the energy spectrum of the diffuse cosmic X-ray background between 20 and 220 keV. The spectrum closely fits a single power law. Thus we conclude that the flattening of the spectrum observed at lower energies occurs below 20 keV, with a change in spectral index of one power.

142.005 Simultaneous X-ray and optical observations of Sco X-1 flares.
H. S. Hudson, L. E. Peterson, D. A. Schwartz.
Astrophys. Journ. (Letters), Vol. 159, L51 - L55 (1970).
The X-ray telescope on the Orbiting Solar Observatory III has made extensive observations of Sco X-1 during 1967 May and June. Two occasions of X-ray flaring, happened to coincide with optical observations of the same phenomena, thus proving the connection between optical and X-ray flares on Sco X-1. The X-ray enhancements amounted to about a factor of 2 over the quiescent emission in the energy range 7.7 – 12.5 keV.

142.006 Observations of the development and disappearance of the X-ray source Centaurus XR-4.
W. D. Evans, R. D. Belian, J. P. Conner.
Astrophys. Journ. (Letters), Vol. 159, L57 - L60 (1970).
The intensity of 3 - 12-keV photons from the cosmic X-ray source that appeared in 1969, herein called Cen XR-4, has decreased to less than 0.5 percent of its maximum value; and since September 24 the source has not been detectable by the Vela satellite instruments that first observed it. Initially the decay of this source was similar to that observed for Cen XR-2.

142.007 Upper limits on the angular sizes of three X-ray sources in the Sagittarius region.
G. Polucci, H. V. Bradt, W. Mayer, S. Rappaport.
Astrophys. Journ. (Letters), Vol. 159, L109 - L113 (1970).
A large-area modulation collimator aboard a sounding rocket was used to study the angular sizes of three galactic X-ray sources (GX 349+2, GX 9+1, GX 17+2) in the Sagittarius region. In each case an upper limit of $120''$ was obtained for the full-width angular size.

142.008 Positions of X-ray sources in the Sagittarius region.
W. Mayer, H. V. Bradt, S. Rappaport.
Astrophys. Journ. (Letters), Vol. 159, L115 - L120 (1970).
A rocket experiment yielded data concerning the celestial positions and intensities (1.5–8 keV) of twelve X-ray sources in the region of Sagittarius. Comparison with data from other experiments yields no clear evidence for gross time variations in the intensity of these sources.

142.009 Power spectra of the optical fluctuations of Scorpius X-1. M. A. Seeds.
Astrophys. Journ. (Letters), Vol. 159, L121 - L122 = Publ. Goethe Link Obs., Indiana Univ., Bloomington, No. 104 (1970).
Power spectra of the optical fluctuations of Sco X-1 are presented for five nights of observations obtained by Sandage, Westphal, and Kristian in 1967. No harmonic features are found in the frequency range from 10 to 360 cycles per hour.

142. 010 Intensity of the soft X-ray background flux: Reply to Bowyer and Field.
R. C. Henry, G. Fritz, J. F. Meekins, H. Friedman, E. T. Byram, with a reply by C. S. Bowyer, G. B. Field.
Nature, Vol. 225, 362 - 363 (1970).
Criticisms of our experiment by Bowyer and Field are shown not to be valid. Their detector field-of-view is shown to be incorrectly calculated, and the quoted value of their detector entrance-window transmission is implausibly high. Lack of detailed publication of their data on higher-energy X-rays raises additional doubts concerning their published value for the intensity of the very soft X-rays.

142.011 Galactic Gamma-ray observations and scattering of electrons on the millimeter radiation.
L. Maraschi, A. Treves.
Astrophys. Letters, Vol. 5, 177 - 180 (1970).
The galactic γ-ray flux which can arise by Compton scattering of electrons on the diffuse infrared radiation observed by Shivanandan et al. (1968) is compared with experimental results of Clark et al. (1968), within the range of values indicated by the experimental errors quoted in both observations. Indications on the thickness of the γ-ray production region are derived from the angular distribution of the anisotropic γ-ray flux, and the possibility is discussed of the existence of a central source and of an extragalactic isotropic component.

142.012 The diffuse cosmic X-ray background radiation.
K. Brecher, G. Burbidge.
Comments Astrophys. Space Phys., Vol. 2, 75 - 83 (1970).

The purpose of this comment is to examine the present state of the observations of the isotropic component of the X- and γ-ray background radiation, the numerous ingenious schemes which have been proposed to explain it, and to discuss what, if any, conclusions can be drawn about the nature of the universe from the data so far available.

142.013   A catalogue of discrete celestial X-ray sources.
J. F. Dolan.
Astron. Journ., Vol. 75, 223 - 230 (1970).

A comprehensive list is given of all sources of discrete celestial X radiation observed in the energy region above 2 keV whose positions were published before 1 December 1969. Identities have been pointed out between the sources observed by different observers. Fluxes are given for each source in the 2-10- and 20-35-keV regions.

142.014   Angular size of the high energy X-ray source in the Crab nebula.   F. W. Floyd, Jr.
Nature, Vol. 226, 733 - 734 (1970).

This article reports new data on the spatial structure of the source of 25 to 100 keV X-rays in the Crab nebula, obtained with a balloon-borne X-ray modulation collimator.

142.015   Low-energy gamma radiation from Cygnus.
R. C. Haymes, F. R. Harnden, Jr.
Astrophys. Journ., Vol. 159, 1111 - 1114 (1970).

The low-energy γ-ray spectrum of Cyg XR-1 and XR-3 was remeasured in a balloon flight launched 1969 June 4; the measurement employed the same instrumentation as that used 21 months earlier. No change in the spectrum at energies between 34 and 567 keV was detected. The power-law energy spectrum steepens by approximately one power between 100 and 150 keV.

142.016   Production of the diffuse background X-ray flux at 44 Å by suprathermal proton bremsstrahlung.
R. L. Brown.
Astrophys. Journ. (*Letters*), Vol. 159, L187 - L192 (1970).

It is shown that the excess of soft X-rays observed at 44 Å can be interpreted as bremsstrahlung emission from collisions of suprathermal protons with an ambient distribution of electrons. Three models are constructed in which the demodulated spectrum of cosmic-ray protons is assumed confined to the Metagalaxy, to the Local Group, and to the galactic halo; X-ray production by the bremsstrahlung interaction of this proton flux with the electrons residing in the respective media is discussed. It is concluded that present observations cannot distinguish between these models and ones presented previously based on free-free emission in a hot, dense intergalactic plasma.

142.017   Decrease in the high-energy X-ray flux from Centaurus XR-2.
W.·H. G. Lewin, J. E. McClintock, W. B. Smith.
Astrophys. Journ. (*Letters*), Vol. 159, L193 - L196 (1970).

On 1969 March 20 we carried out balloon X-ray observations (>20 keV) from Mildura, Australia. We found no evidence for X-ray emission from Cen XR-2. This means that the energy flux from Cen XR-2 in the range 20 - 100 keV has decreased to a level at least 7 times below the one observed by Lewin, Clark, and Smith on 1967 October 15.

142.018   Interpretation of cosmic X-ray and gamma ray spectra.   J. I. Trombka.
Nature, Vol. 226, 827 - 828 (1970).

Changes in the slope of a measured distribution may be explained partially by the effect of the measurement process itself.

142.019   The variable X-ray spectrum of Cyg XR-1.
J. F. Dolan.
Space Sci. Rev., Vol. 10, 830 - 868 (1970).

The variability of the X-ray spectrum of the discrete source Cyg XR-1 is reviewed. The variations observed in the energy region accessible to balloon borne detectors can be explained by assuming them to be caused by the eclipsing properties of a binary system. It is suggested that the system is composed of a source of small angular extent having a spectrum similar to that of a black body at approximately 1.5×10^8 K and a non X-radiating companion which eclipses it at intervals of 2.9850 days. The system would be surrounded by an X-radiating plasma whose photon flux between 1 and 100 keV can be approximated by a power law spectrum whose exponent is –1.7.

142.020   A search for cosmic sources of gamma rays with energies between 10^{11} and 10^{14} ev.
G. H. Rieke.
SAO, *Cambridge, Mass.*, Special Rep. No. 301, 11 + 86 pp. (1969). – Thesis Dep. Phys. Harvard Univ.

Theoretical estimates of gamma-ray fluxes from a number of cosmic objects are presented; conditions under which detectable fluxes would be generated by M 87, the Crab nebula, M 82, and Cas A are discussed in detail. No statistically significant effects have been recorded. In particular, reports of possible fluxes from the Crab nebula and CP 1133 have not been verified. It is concluded from the upper limit established for the Crab nebula that the effective magnetic-field strength in this object is greater than 1.5×10^{-4} gauss.

142.021   On the radiative transfer of isotropic X rays and gamma rays.   J. Arons.
Bull. American Astron. Soc., Vol. 2, 180 (1970). – Abstr. AAS.

142.022   The birth and death of a cosmic X-ray source.
J. P. Conner, W. D. Evans, R. D. Belian.
Bull. American Astron. Soc., Vol. 2, 188 (1970). – Abstr. AAS.

142.023   Nonthermal radiation emission from X stars and pulsars.   A. Ferrari, B. Coppi.
Bull. American Astron. Soc., Vol. 2, 192 (1970). – Abstr. AAS.

142.024   The optical power spectra of Sco X-1 and WX Cen.
J. E. Hesser, B. M. Lasker.
Bull. American Astron. Soc., Vol. 2, 198 - 199 (1970). Abstr. AAS.

142.025   The time variability of X-ray emission from Sco X-1.   H. S. Hudson, R. M. Pelling,
L. E. Peterson, D. A. Schwartz.
Bull. American Astron. Soc., Vol. 2, 200 (1970). – Abstr. AAS.

142.026   Precise location of Sagittarius X-ray sources with a rocket-borne rotating modulation collimator.
H. W. Schnopper, H. V. Bradt, S. Rappaport.
Bull. American Astron. Soc., Vol. 2, 217 (1970). – Abstr. AAS.

142.027   The cosmic gamma-ray spectrum from secondary particle production in cosmic-ray interactions.
F. W. Stecker.
Bull. American Astron. Soc., Vol. 2, 219 (1970). – Abstr. AAS.

142.028   Observation of galactic X-ray sources on November 3, 1968.   I. Turiel, G. A. MacGregor,

F. D. Seward.
Bull. American Astron. Soc., Vol. 2, 222 (1970). — Abstr. AAS.

142.029 On the gamma and radio radiation from neutron stars in the state of accretion.
V. F. Schwarzman.
Astrofizika, Vol. 6, 123 - 134 (1970). In Russian. — English translation in Astrophysics, Vol. 6, No. 1.

The accretion of gas on a neutron star must be accompanied by intensive gamma radiation because of production of π°-, π^+-, π^--mesons as well as by nuclear and thermonuclear reactions in the stellar atmosphere. The recording of the γ-quanta spectrum will make it possible to distinguish such objects from other X-ray sources and to fix the value of the gravitational potential on the surface of the neutron stars. A possibility of an interpretation for the radio luminosity of Sco X-1 is noted.

142.030 UBV photometry of Sco-X1 during the time of a balloon observation on April 18, 1969.
L. O. Lodén.
Astron. Astrophys., Vol. 4, 337 - 338 (1970). — Research note.

142.031 The Cygnus X region. VI. A new 2695 MHz continuum survey. H. J. Wendker.
Astron. Astrophys., Vol. 4, 378 - 386 (1970).

A 2695 MHz continuum survey of the Cygnus X region was undertaken with the NRAO 140-ft telescope with an angular resolution of 11'. A reference frame was established relative to which all intensities were measured. The improved positional accuracy helped to identify the resolved sources more accurately with optical features than had been possible before. From a comparison between radio and H_α emission the visual absorption is estimated in a few cases. For IC 1318a a pair of possibly exciting stars is suggested.

142.032 The sky near the brightest X-ray source in Scorpius. II. H. M. Johnson.
Astrophys. Journ., Vol. 160, 193 - 197 = Contr. Kitt Peak National Obs., No. 509 (1970).

The luminous nebulae within 18° of Sco X-1 have been observed with an $H\beta$ photometer. Twenty out of fifty-one of them definitely show $H\beta$ in emission. Nearly all appear to be photoionized by hot stars in the Sco-Cen association rather than by Sco X-1, although Sco X-1 is also a probable member of the association and is a source of high-energy photons. Specific stars are proposed for association with the nebulae.

142.033 A measurement of the optical and X-ray emission from Scorpius X-1 and the X-ray diffuse background.
A. Toor, F. D. Seward, L. R. Cathey, W. E. Kunkel.
Astrophys. Journ., Vol. 160, 209 - 213 = Contr. Cerro Tololo Inter-American Obs., No. 66 (1970).

Another simultaneous measurement of the optical and X-ray emission from Sco X-1 has been made. The optical intensity at the time of the X-ray measurements was $B = 12.97$ mag. The X-ray spectrum looks like thin-source thermal bremsstrahlung with an apparent plasma temperature of 7 ± 1 keV. These data are compared with previous measurements of X-ray and optical emission. It is possible that total energy emission from Sco X-1 follows the B-magnitude in a nearly linear fashion. The diffuse X-ray background was measured in the energy range 4 - 70 keV.

142.034 The X ray source in Scorpius.
A. Spodenkiewicz.
Urania Kraków, Vol. 41, 162 - 168 (1970). In Polish.

142.035 Short-term periodic variations following an optical flare of Sco X-1.
J. R. Gribbin, P. A. Feldman, S. H. Plagemann.
Nature, Vol. 225, 1123 - 1124 (1970).

Power spectrum analysis of the optical variations of Sco X-1 reveals the presence of short-term periodic components in the optical 'flickering' of this source. In particular, a component with frequency 0.006 Hz is found in the variations immediately following a large optical flare on 3 April 1967. This component appears to decay over some 80 minutes until it is no longer detectable.

142.036 X-ray scattering by grains in the direction of the Crab pulsar and Sco XR-1.
C. S. Bowyer, J. Mack, M. Lampton.
Nature, Vol. 225, 1125 - 1127 (1970).

Small angle scattering of X-rays by interstellar grains is studied. Upper limits to this effect in the direction of Sco XR-1 are established from data in the literature, and limits on grain parameters are derived from these limits. It is concluded that the finite angular extent of the X-ray source in the Crab nebula can not be due to this effect.

142.037 Predicted high energy break in the isotropic gamma ray spectrum: A test of cosmological origin.
G. G. Fazio, F. W. Stecker.
Nature, Vol. 226, 135 - 136 (1970).

Recently Stecker (Nature, Vol. 224, 870 - 872 (1969)) proposed the hypothesis that the isotropic component of cosmic gamma-rays above 1 MeV energy is of cosmological significance and may have originated at a redshift of approximately z = 100. Such a model yields a theoretical gamma-ray spectrum which is in agreement with recent measurements above 1 MeV. A test is proposed for this model which lies in the detection of a break in the spectrum of isotropic gamma-radiation at approximately 7 GeV. Gamma-rays above a present energy of 7 GeV will be attenuated by interactions with the 2.7°K universal radiation field.

142.038 Upper limit on the intensity of Sgr γ-1 in soft X-rays. S. Naranan.
Nature, Vol. 226, 333 (1970).

Assuming that the discrete γ-ray source, Sgr γ-1 reported by Frye et al (Nature, Vol. 223, 1320 - 1321 (1969)) is real, it is possible to set an upper limit on its intensity in soft X-rays (1.5 - 6 keV) from a rocket experiment (Bradt et al, Astrophys. Journ., Vol. 152, 1005 - 1013 (1968)). The intensity of Sgr γ-1, is less than one-fifth of Crab nebula in soft X-rays; below 50 MeV, it has a spectrum flatter than that of Crab nebula.

142.039 Characteristic variation of Sco X-1.
R. E. Wilson, L. W. Twigg.
Nature, Vol. 226, 734 - 735 (1970).

The published photometry of Scorpio X-1 is examined for repetitions. Two figures show the brightness variation for three nights on which a consistent pattern, on a time scale of about six hours, occurred. A broad eclipse-like depression is followed by a gradual rise, then a small hill of about 0^m1, and finally a second depression on each of the three nights. Shorter runs showing parts of the characteristic variation were found but are not illustrated. The phenomenon appears to be non-periodic.

142.040 X-ray pulsar in the Crab nebula.
G. Fritz, R. C. Henry, J. F. Meekins, T. A. Chubb, H. Friedman.
Rep. Naval Res. Lab. Progr., June 1969, p. 1 - 4 (1969).

Reports detection of X-ray pulsations similar to the optical pulsations. Five percent of total X-ray power appears in the pulsed component. The pulsed X-ray power is 200 times the optical power and 20.000 times the radio power.

142.041 On the nature of the underlying star associated with Sco X-1. P. A. Feldman, J. R. Gribbin, S. H. Plagemann.
Nature, Vol. 226, 432 - 434 (1970).

The short-term periodic variations in the optical "flickering" of Sco X-1 following a large optical flare (Gribbin, Feldman, and Plagemann: Nature, Vol. 225, 1123 - 1124 (1970)) are interpreted as acoustic oscillations of a hot, bounded, bremsstrahlung-emitting plasma atmosphere in the gravitational field of an underlying star. The effective gravitational acceleration is estimated to be $\approx 5 \times 10^6$ cm sec^{-2}, which implies that the underlying star is a condensed object – either a white dwarf or a neutron star.

142.042 A possible mechanism of formation of the spectra of cosmic X-ray sources.
Yu. N. Gnedin, A. Z. Dolginov, A. I. Tsygan.
Pis'ma v ZhEhTF, Vol. 10, 441 - 444 (1969). In Russian.
Abstr. in Referativ. Zhurn. 51. Astron., 4.51.648 (1970).

142.043 An X-ray survey of the Cassiopeia region and its implications concerning supernova remnants and the galactic source distribution.
P. Gorenstein, H. Gursky, E. M. Kellogg, R. Giacconi.
Astrophys. Journ., Vol. 160, 947 - 957 (1970).

An X-ray survey (1 - 10 keV) was made of the Cassiopeia region, from a sounding rocket using detectors of high angular resolution. Two sources were observed, and their positions were determined to a precision of about one-tenth of a square degree. The source positions are consistent with those of the supernova remnants Cas A and SN 1572. Observations from this flight and two preceding flights suggest that observable X-ray emission comes only from the supernova remnants whose intrinsic diameter at radiofrequencies is relatively small.

142.044 A new component of cosmic gamma rays near 1 MeV observed by the ERS-18.
J. I. Vette, D. Gruber, J. L. Matteson, L. E. Peterson.
Astrophys. Journ. (*Letters*), Vol. 160, L161 - L170 (1970).

We wish to confirm and discuss in detail new data on the total omni-directional flux of cosmic γ-rays in the range of 0.25–6 MeV obtained by the Environmental Research Satellite 18 (ERS-18) which were presented in a preliminary form earlier.

142.045 Radial velocities of the X-ray candidate star S5003 Centauri. M. W. Feast.
Astrophys. Journ. (*Letters*), Vol. 160, L171 - L172 (1970).

No large variations in radial velocity were detected for Hα emission in S5003 Cen for the period 1970 February 27 – April 1. However, there is a large change in velocity compared with observations by Eggen and Rodgers in 1969 September.

142.046 Scattering of X-rays from the Crab nebula.
C. Ryter.
Nature, Vol. 226, 1040 - 1041 (1970).

Scattering of X-rays by interstellar dust grains affects both the travel time of the photons and the apparent direction of the source. If the amount of dust in the line of sight is sufficent, an intrinsically pulsating star-like source may appear continuous and diffuse. This is what is observed in the Crab system, where an X-ray pulsar and an extended source appear. However, various reasons are given to show that the diffuse source is not likely to be accounted for by interstellar scattering, and that the extension of the source should be real.

142.047 Origin of galactic gamma rays.
T. P. Stecher, F. W. Stecker.
Nature, Vol. 226, 1234 - 1235 (1970).

Two important conclusions about the origin of galactic

gamma rays are given. (1) Gamma rays originating in the galactic disk most likely result from the decay of neutral pions produced in interstellar cosmic ray interactions. The excess originating in the region of the galactic centre can be produced by a combination of the pion decay process and Compton interactions between cosmic ray electrons and infrared radiation. (2) The explanations offered here obviate the need to invoke strong gamma ray point sources, large gradients in the galactic cosmic ray flux, or 8 K greybody radiation fields of galactic extent in order to understand the revised gamma ray observations.

142.048 Galactic component of the diffuse X-ray background. N. C. Wickramasinghe.
Nature, Vol. 227, 265 - 266 (1970).

Cooke, Griffiths and Pounds have recently detected an increase of intensity of the cosmic X-ray background in directions close to the galactic plane, indicating the possible existence of a diffuse galactic X-ray component. The similarity of this galactic component, in both intensity and spectrum, to the general isotropic background would tend to suggest a possible causal connexion. This report argues that interstellar dust occurring near the galactic plane would scatter the isotropic background so as to produce the observed phenomenon.

142.049 On the optical emission of Sco X-1.
F. I. Lukatskaja.
Astron. Tsirk., No. 547, p. 3 - 4 (1970), with a correction in No. 557, p. 8 (1970). In Russian.

142.050 Balloon observations of a new-born X-ray source. R. M. Thomas, G. Buselli, M. C. Clancy, P. J. N. Davison.
Proc. Astron. Soc. Australia, Vol. 1, 345 - 347 (1970).

The results of an attempt to detect the source during a balloon flight made from Mildura, Australia, on 1969 August 25 are reported.

142.051 Pulsations in the X-radiation of the Crab nebula. G. Ducros, R. Ducros, R. Rocchia, A. Tarrius.
Nature, Vol. 227, 152 - 154 (1970).

This report presents the final results of observations of pulsations in the X-ray flux in the 2·5 to 30 keV range coming from the direction of the Crab nebula.

142.052 Hard X-rays from Sco X-1.
G. R. Riegler, E. Boldt, P. Serlemitsos.
Nature, Vol. 226, 1041 - 1043 (1970).

During a balloon flight launched from Mildura, Australia, on December 15, 1966 hard X-rays from Sco X-1 have been observed. The results of these and more recent observations show that the hard X-ray component is definitely not compatible with a simple extension of the thermal spectrum at lower energies.

142.053 High-energy emission from X-stars and pulsars. B. Coppi, A, Ferrari.
Nuovo Cimento Lettere, Vol. 3, 93 - 97 (1970).

The authors present nonthermal mechanism for high-energy radiation and particle emission which, in fact, interprets more generally X-ray sources as collapsed stars transforming rotational energy into electromagnetic radiation from the surrounding plasma.

142.054 Equipment for X-ray emission of the galactic center region by lunar occultation.
C. S. Bowyer.
Report Series 10, Issue 2, California Univ., Berkeley.
[Available from CFSTI, Springfield, Va.], 13 pp. (1969).

An experimental program with the following objectives is described: To monitor short term variations in the flux of Sco XR-1 and to correlate these variations with variations in the light flux from these sources; to obtain the precise locations and the high energy spectra for known X-ray sources; to scan unusual astronomical objects for evidence of X-ray emission; and to obtain information on the gamma-ray background flux.

Use of highly reflecting crystals for spectroscopy and polarimetry in X-ray astronomy.   See Abstr. 034.019.

Solar X-ray and diffuse cosmic X-ray spectra measured with a satellite-borne instrument.
See Abstr. 076.028.

Supernova remnants, pulsars and X-ray sources.
See Abstr. 125.015.

On the interaction of cosmic X-rays with interstellar grains.   See Abstr. 131.109.

Scattering of cosmic X-rays by interstellar dust grains.   See Abstr. 131.125.

A balloon observation of pulsed X rays from the Crab.   See Abstr. 134.006.

Search for polarization in the X-ray emission of the Crab nebula.   See Abstr. 134.008.

Observation of primary X-ray spectra from the Crab nebula and diffuse background.
See Abstr. 134.013.

Possible pulsed gamma ray emission from the Crab nebula pulsar.   See Abstr. 134.014.

Pulsed gamma rays from the Crab nebula?
See Abstr. 134.015.

X-ray observations of the Crab nebula.
See Abstr. 134.021.

The nebula: X-ray data.
See Abstr. 134.022.

Search for high energy gamma rays from four pulsars.   See Abstr. 141.059.

Upper limits for CP 1919 X-ray emission.
See Abstr. 141.067.

Possible pulsed gamma ray emission above 50 MeV from the Crab pulsar.
See Abstr. 141.068.

Optical and X-ray synchrotron radiation from the Crab pulsar.   See Abstr. 141.151.

On the motion of current sheets, and the radio, optical, and X-ray emission from pulsars. II. Pulse structure, polarization, time-varying features, and tight-beam emission.
See Abstr. 141.172.

The cosmic γ-ray spectrum from secondary particle production in cosmic-ray interactions.
See Abstr. 143.006.

Cosmic rays and cosmic X-rays. I, II, III.
See Abstr. 143.029.

The sub-millimeter nightsky radiation background and its observable consequences on cosmic ray phenomena.
See Abstr. 143.042.

Compact nonthermal sources in M87.
See Abstr. 158.006.

143 Cosmic Radiation

143.001 **The diffusion of cosmic rays.**
J. Skilling.
Monthly Notices, Roy. Astron. Soc., Vol. 147, 1 - 12 (1970).
Micro-instabilities are shown to play a dominant role in the diffusion of cosmic rays through the interstellar plasma. Some implications of this are considered, including possible explanations for the 'break' in the cosmic ray energy spectrum at around 10^{15} eV, and for the 'two-component' model.

143.002 **Study of solar modulation of low-energy cosmic rays using differential spectra of protons, ^{3}He, and ^{4}He at $E \lesssim 100$ MeV per nucleon during the quiet time in 1965 and 1967.** K. C. Hsieh.
Astrophys. Journ., Vol. 159, 61 - 76 (1970).
The differential spectra of galactic cosmic-ray protons, ^{3}He, and ^{4}He at energies between ~ 20 and 100 MeV per nucleon have been measured during quiet periods between 1967 July and October. The differential spectra of ^{3}He and ^{4}He between 30 and 100 MeV per nucleon measured during solar minimum, 1965 May−September, have been revised and compared with the results of higher-energy measurements made by balloon experiments. These satellite measurements, shown to be free of detectable solar contamination, are used for a study of solar modulation of low-energy cosmic rays.

143.003 **Low-energy cosmic ray positrons and 0.51-Mev Gamma rays from the galaxy.**
R. Ramaty, F. W. Stecker, D. Misra.
Journ. Geophys. Res., Vol. 75, 1141 - 1149 (1970).
Large fluxes of low-energy cosmic rays in interstellar space may produce, via unstable CNO beta emitters, large fluxes of low-energy positrons and a detectable intensity of 0.51-Mev γ rays. Although on a galactic scale these cosmic-ray intensities are probably untenable because they conflict with the dynamics of the interstellar medium, they may exist on a smaller scale of the order of the stopping distances of both low-energy cosmic rays and positrons. We compare the results of a detailed calculation with direct positron measurements at the earth, and we discuss the observability of the annihilation radiation above the isotropic X-ray background.

143.004 **The variation with a period of two solar cycles in the cosmic ray diurnal anisotropy for the nucleonic component.** S. P. Duggal, S. E. Forbush,
M. A. Pomerantz.
Journ. Geophys. Res., Vol. 75, 1150 - 1156 (1970).
It has been shown previously that annual means of the diurnal anisotropy determined from 30 years of ionization chamber data result from the superposition of two distinct independent components W and V. The resulting implication that V and W were each about 1.35 times greater in the neutron monitor data is explicitly established. The objectives of the present work are to consider this implication in detail and to examine its consequences, in the light of results based on the analysis of data from a geographical distribution of neutron monitors.

143.005 **Latitude dependence of charged particle albedo estimated from balloon-borne Geiger telescopes.**
P. J. König, J. P. Maree, P. H. Stoker, A. J. van der Walt.
Journ. Geophys. Res., Vol. 75, 1172 - 1177 (1970).
From the atmospheric depth-intensity distributions for a series of balloon flights with standard Geiger telescopes during the IQSY, the total charged particle intensity at the top of the atmosphere was obtained. The resulting charged albedo intensity, as function of cutoff rigidity, compares well

with the estimates by Webber (1967).

143.006 **The cosmic γ-ray spectrum from secondary particle production in cosmic-ray interactions.**
F. W. Stecker.
Astrophys. Space Sci., Vol. 6, 377 - 389 (1970).
The cosmic γ-ray spectrum below 1 GeV arising from cosmic ray p-p interactions is calculated. A model is chosen for numerical calculations in which the two dominant modes of neutral pion production at accelerator energies are the production of the Δ (1.238) isobar and one fireball. The effect of α-p and p-α interactions on the cosmic γ-ray spectrum is also calculated. The final results are given in terms of both differential and integral γ-ray energy spectra.

143.007 **Homogeneous metagalactic origin of cosmic rays.**
S. Lal, K. Brunstein.
Astrophys. Space Sci., Vol. 6, 415 - 419 (1970).
Starting with the hypothesis that cosmic rays are evenly distributed in the metagalaxy, it is shown that the flux of the electron-positron component, which is produced through $\pi - \mu - e$ decays, following the nuclear collisions of the cosmic ray beam with the intergalactic medium, takes $\leqslant 4 \times 10^{16}$ sec to reach steady state. The corresponding value of the flux of the positron component and its implications regarding the homogeneous model of the metagalactic origin of cosmic rays are discussed.

143.008 **The Compton-Getting effect for cosmic-ray particles and photons and the Lorentz-invariance of distribution functions.** M. A. Forman.
Planet. Space Sci., Vol. 18, 25 - 31 (1970).
The anisotropy introduced by the motion of an observer with velocity V through a particle or photon flux is derived from the Lorentz-invariance of distribution functions in phase space. For massive particles whose speed, v, is much greater than V, the Compton-Getting anisotropy is added to the angular part of the distribution function, and the isotropic part is unchanged, to order V/v. The cosmic-ray proton spectrum at solar minimum is observed to have a region where the Compton-Getting anisotropy is zero.

143.009 **A study of the long-term modulation of galactic cosmic ray intensity.**
P. N. Pathak, V. Sarabhai.
Planet. Space Sci., Vol. 18, 81 - 94 (1970).
The present investigation attempts to relate the conditions in the solar corona with the solar wind parameters and the electromagnetic state of the interplanetary space relevant to the 11-yr modulation of the intensity of the galactic cosmic rays observed at the earth.

143.010 **Cosmic electrons and related astrophysics.**
R. R. Daniel, S. A. Stephens.
Space Sci. Rev., Vol. 10, 599 - 671 (1970).
The review article contains the following sections: Introduction; Experimental methods; Cosmic electrons in the solar system, observational data; Cosmic electrons in interstellar space; Cosmic electrons and the galactic continuum radio emission; Propagation and confinement of cosmic electrons; Electrons in the metagalaxy; I: Atmospheric electrons; II: Interaction of electrons with matter, magnetic fields and radiation fields existing in cosmic space.

143.011 **Identification of a highly variable component in low-energy cosmic rays at 1 A. U.**
J. H. Kinsey.

Phys. Rev. Letters, Vol. 24, 246 - 249 (1970).

Evidence is given for the existence of two distinct populations of cosmic-ray protons and alpha particles in the energy range of 4 to 80 MeV/nucleon: (1) a highly variable component with an inverse energy dependence and (2) a quasisteady "residual" component which shows a positive dependency on energy. It is suggested that the highly variable component is probably entirely of solar origin. The "residual" component is most likely of galactic origin although below 10 MeV it may also contain some solar particles.

143.012 Measurements of streams of γ-quanta with energies above 50 MeV in the primary cosmic radiation with the artificial earth satellite Cosmos 208.
L. S. Bratolyubova-Tsulukidze, N. L. Grigorov, L. F. Kalinkin, A. S. Melioranskij, E. A. Pryakhin, I. A. Savenko, V. Ya. Yu-farkin.
Kosmich. Issled., Vol. 8, 136 - 139 (1970). In Russian.

143.013 The abundances of cosmic ray particles with $Z \geqslant 3$. S. A. Fody, M. W. Friedlander, H. Hasegawa, J. Klarmann, W. C. Wells.
Planet. Space Sci., Vol. 18, 265 - 270 (1970).

Two large stacks of nuclear photographic emulsions were exposed to the primary cosmic radiation on a high altitude balloon flight over Palestine, Texas on April 24, 1965, during the period of minimum solar activity. The intensities of the fluxes of the groups of heavy particles having rigidities above about 4.3 GV were found. From these, the L/M ratio is derived as 0.32 ± 0.07. The relative abundances of the elements within the L group have also been estimated.

143.014 Age of cosmic rays and abundance of antimatter in the Galaxy. L. M. Libby.
Nature, Vol. 225, 166 - 167 (1970).

Acceptance of a disc-halo structure for the Galaxy leads to the conclusion that cosmic rays are very old, 100 million years or more, and to the conclusion that the abundance of antimatter in the entire Galaxy is less than 3×10^{-4}.

143.015 Origin of cosmic rays. I. Introduction. Metagalactic models. V. L. Ginzburg.
Comments Astrophys. Space Phys., Vol. 2, 1 - 6 (1970).

To solve the problem of the origin of cosmic rays near the earth four models are discussed: (1) Universal (or uniform) metagalactic model; (2) Local metagalactic model; (3) Galactic model with halo; (4) Galactic model of disc type.

143.016 Origin of cosmic rays. II. The halo problem. Galactic models. V. L. Ginzburg.
Comments Astrophys. Space Phys., Vol. 2, 43 - 52 (1970).

In order to consider the galactic models of the origin of cosmic rays a model with halo is discussed. The existence of an extended region (halo) around the Galaxy is reviewed.

143.017 Charge spectrum and solar minimum intensity of heavy ($Z \geqslant 3$) primary cosmic-ray nuclei at 41° N mag. S. A. Fody.
Thesis, Washington Univ., St. Louis, Mo. 142 pp. (1969).

It is during times of minimum solar activity that the galactic cosmic radiation reaching the earth is least altered by the interplanetary medium. The main purpose of this work is to study the fluxes and composition of the $Z \geqslant 3$ primary cosmic-ray nuclei observed during the 1965 solar minimum.

143.018 Emission of cosmic rays by a dipole magnetosphere. G. Pizzella.
Nature, Vol. 226, 434 - 435 (1970).

This report shows that the energy spectrum of the particles which abandon the magnetosphere is very similar to the observed energy spectrum of the primary cosmic radiation, suggesting that cosmic rays could be generated in some earth-like magnetosphere, for instance, in magnetospheres associated with neutron stars.

143.019 The influence of ionization losses on the spectrum of cosmic rays under a statistical acceleration mechanism. P. Velinov.
Dokl. Bolg. AN, Vol. 22, 847 - 850 (1969). In Russian. Abstr. in Referativ. Zhurn. 51. Astron., 3.51.607 (1970).

143.020 On cosmic electrons in the galaxy. V. A. Razin.
Astron. Zhurn. Akad. Nauk SSSR, Vol. 47, 56 - 59 (1970). In Russian. English translation in Soviet Astron. AJ, Vol. 14, No. 1.

The frequency spectrum of the intensity of cosmic radio emission calculated for the case, when relativistic electrons have the energy spectrum given by Razin (1960), conforms rather satisfactorily to the data of observations. It testifies the similarity of energy spectra of cosmic electrons near the earth and in interstellar space, as well as, probably, the fact that cosmic particles are situated mainly in the galactic disc or in galactic arms.

143.021 Statistical discrete-source model of local cosmic rays. R. Ramaty, D. V. Reames, R. E. Lingenfelter.
Phys. Rev. Letters, Vol. 24, 913 - 916 (1970).

The anisotropy, lifetime, and fluctuations of the cosmic rays are considered for a model in which the cosmic-ray sources are random discrete events in space-time.

143.022 On the diffusion-loss model of cosmic ray electron propagation in the Galaxy. A. S. Webster.
Astrophys. Letters, Vol. 5, 189 - 192 (1970).

It is shown to be unlikely that the observed break in the cosmic ray electron spectrum at about 3 GeV is caused by a diffusion-loss process.

143.023 Trapping regions for cosmic rays of the highest energies. M. Johnson.
Observatory, Vol. 90, 31 - 33 (1970). – Letter.

143.024 On the energy spectrum of cosmic rays in the interstellar space.
A. N. Charakhchyan, T. N. Charakhchyan.
Geomagn. Aeronom., Vol. 10, 240 - 243 (1970). In Russian.

143.025 Semi-diurnal anisotropy of cosmic radiation in the energy range 1 to 200 Gev.
U. R. Rao, S. P. Agrawal.
Journ. Geophys. Res., Vol. 75, 2391 - 2401 (1970).

In this paper we have utilized the data from all the super neutron monitors (including the super neutron monitor at Ahmedabad, India), which are located near the geomagnetic equator and scan the equatorial region of the sky, to make a detailed study of the general properties of the semi-diurnal anisotropy of cosmic radiation.

143.026 Periodic variations of the cosmic radiation – I. W. Messerschmidt.
Planet. Space. Sci., Vol. 18, 509 - 524 (1970).

Since 1957, measurements of the cosmic radiation have been carried out by different instruments at Halle and Lindau. The measurements include an energy spectrum from 5 to 75 GeV. The data were corrected by the pressure coefficient. The periodic variations of the cosmic radiation during the last period of solar activity are discussed.

143.027 **Radial gradients and anisotropies of cosmic rays in the interplanetary medium.**
L. A. Fisk, W. I. Axford.
Solar Physics, Vol. 12, 304 - 316 (1970).

Approximate equations which describe the behavior of cosmic rays in the interplanetary medium under suitable conditions are used to make comparisons between observations and theoretical predictions of radial gradients and radial anisotropies. In the high energy region there appear to be no inconsistencies between theory and observations. In the low energy region it is shown that theoretical predictions of the radial anisotropy expected from large radial gradients of the intensity are not inconsistent with observed radial anisotropies.

143.028 **Dynamics of the eleven-year modulation of galactic cosmic rays.** J. R. Wang.
Astrophys. Journ., Vol. 160, 261 - 281 (1970).

In this paper we study the buildup and relaxation of the effective cosmic-ray modulating region and the time variations in the intensity of the primary cosmic radiation by the simultaneous study of the intensity of the nucleonic components and the intensity of the green coronal emission line at 5303 Å which we argue is related to the coronal temperature and the solar-wind velocity.

143.029 **Cosmic rays and cosmic X-rays. I, II, III.**
J. E. Felten.
Journ. Roy. Astron. Soc. Canada, Vol. 64, 33 - 49, 73 - 89 = Commun. David Dunlap Obs., Univ. Toronto, No. 162 (1970).

The intention of this paper is to explore some aspects of cosmic-ray physics and the physics of particles and fields as they bear upon astronomy, particularly the newer disciplines of radio and X-ray astronomy.

143.030 **Cosmic ray muons from low galactic latitude.**
W. T. Chu, Y. S. Kim, G. Reber.
Publ. Astron. Soc. Pacific, Vol. 82, 339 - 344 (1970).

A sample of 1855 cosmic ray muons with energy greater than 10 GeV is studied in the pictures of the Michigan-Argonne 40-inch heavy-liquid bubble chamber. Above 100 GeV, we find an excess of muons arriving from low galactic latitude.

143.031 **On the convection, diffusion, and adiabatic deceleration of cosmic rays in the solar wind.**
J. R. Jokipii, E. N. Parker.
Astrophys. Journ., Vol. 160, 735 - 744 (1970).

We present a generalization and clarification of work on the transport of cosmic rays in the solar system. The general equations for the cosmic-ray density and flux, in the anisotropic-diffusion approximation, are derived in the frame of reference moving with the solar wind, and the transformation to the fixed frame is discussed.

143.032 **Short-period fluctuations of cosmic ray intensity at the geomagnetic equator and their solar and terrestrial relationship.** M. S. Dhanju, V. Sarabhai.
Journ. Geophys. Res., Vol. 75, 1795 - 1801 (1970).

By using power spectrum analysis on the mu meson intensity records of a high counting rate instrument (10^6 counts per minute) operated at Chacaltaya, Bolivia, during 1965–66, it has been possible to identify the presence of cosmic ray fluctuations of the order of few minutes. The integrated power of cosmic ray oscillations in the frequency range of 6 to 30 cycles per hour has been studied at various periods and its solar-terrestrial relationship is examined. The association of observed frequencies in cosmic rays with frequencies in the solar photosphere and the interplanetary magnetic field as well as the resonance frequencies of the magnetosphere are discussed.

143.033 **The influence of Forbush effects on the state of the cosmic ray layer in the lower ionosphere.**
P. Velinov, L. I. Dorman, G. Nesterov.
Dokl. Akad. Nauk SSSR, Ser. Mat. Fiz., Vol. 190, 1063 - 1065 (1970). In Russian.

The state of the galactic cosmic ray layer in the lower ionosphere at heights 50 - 80 km has been investigated by means of radio waves at a frequency 155 kHz. The observations have been carried out at night times during the period April 25 - May 15, 1960. Surprisingly the ionization of the layer increased during periods of Forbush effects when cosmic ray intensity decreased. An analysis of the formation of electrons in the cosmic ray layer shows that its rate may increase owing to a lower cutting off cosmic rays during strong geomagnetic storms even if the cosmic ray flux decreases. The derived equation for the cutting off cosmic rays may be used for an investigation of current systems outside the ionosphere. *B. Onderlička*

143.034 **The relative abundance of the carbon isotopes ^{12}C and ^{13}C in primary cosmic radiation.**
G. Jönsson, K. Kristiansson, L. Malmqvist.
Astrophys. Space Sci., Vol. 7, 231 - 251 (1970).

This paper reports the results of new measurements of the isotopic composition of carbon with a higher statistical accuracy and with methods which were not used in the earlier studies. Our investigation is made on carbon nuclei stopping in a nuclear emulsion stack which had been exposed to the primary radiation in a high altitude balloon flight from Fort Churchill, Manitoba, Canada, in July 1963.

143.035 **Some comments on the existence of a density gradient of the cosmic radiation perpendicular to the plane of the ecliptic.** P. L. Marsden.
Ann. IQSY, (see 012.016), Vol. 4, 131 - 137 (1969).

The implications of a density gradient perpendicular to the plane of the ecliptic are discussed particularly in relation to seasonal and diurnal variations in cosmic-ray intensity. Experimental observations on these features are examined and it is concluded that there is reasonable evidence for the existence of the perpendicular gradient.

143.036 **Ground-based synoptic observations of cosmic rays.** H. Carmichael.
Ann. IQSY, (see 012.016), Vol. 4, 141 - 174 (1969).

143.037 **Some observations on the 11-year modulation of galactic cosmic rays.** W. R. Webber.
Ann. IQSY, (see 012.016), Vol. 4, 181 - 186 (1969).

We have examined and compared the recent extensive balloon and satellite measurements on the 11-year solar modulation of low-energy protons and helium nuclei. None of the presently accepted theoretical models for this modulation can satisfactorily explain all of the data. It is suggested that other processes must play an important role in the 11-year modulation.

143.038 **IQSY observations of low-energy galactic and solar cosmic rays.** F. B. McDonald.
Ann. IQSY, (see 012.016), Vol. 4, 187 - 216 (1969).

Space probe missions have provided detailed, synoptic observations of both low-energy galactic and solar cosmic rays. In this paper some results are summarized.

143.039 **Cosmic ray variations: Theory.**
L. I. Dorman.
Ann. IQSY, (see 012.016), Vol. 4, 217 - 268 (1969).

In this paper are presented data on the coupling of solar activity with large-scale characteristics of the solar wind during the 11-year cycle.

143.040 **Some problems of detecting galactic anisotropies.**
A. J. Somogyi.
Ann. IQSY, (see 012.016), Vol. 4, 269 - 273 (1969).

Attention is drawn to the fact that there are no observations concerning galactic anisotropies within the energy region $10^{11} - 10^{14}$ eV. At the same time, the importance of this energy region is emphasized. The second part of the paper deals with some general problems connected with the statistical methods of detecting periodicities.

143.041 **Modulation of cosmic-ray electrons.**
S. Hayakawa.
Ann. IQSY, (see 012.016), Vol. 4, 274 - 278 (1969).

It is emphasized that the mechanism of the solar modulation of galactic rays will be best studied through the observation of cosmic-ray electrons. If the knee of the electron spectrum at about 3 GeV is interpreted as due to the solar modulation, the spectra of electrons and nuclear particles are accounted for with a common modulation factor. However, the conclusion has to be reserved on account of the absence of long-term variations of the electron intensity.

143.042 **The sub-millimeter nightsky radiation background and its observable consequences on cosmic ray phenomena.** R. Cowsik.
Astrophys. Letters, Vol. 6, 39 - 43 (1970).

In an attempt to find confirmatory evidence for the existence of the strong radiation background at sub-millimeter wavelengths, its observable consequences on various cosmic-ray phenomena are examined. The existence of such a radiation background implies an enhanced hard X-ray and gamma-ray emission, from supernova remnants and from the galactic disc, through Compton scattering by relativistic electrons; and also a decrease in the mean residence time of cosmic rays with increasing energy.

143.043 **Investigation of the cosmic radiation and the radiation belts on board a vertical cosmic probe.**
S. N. Vernov, P. V. Vakulov, S. N. Kuznetsov, Yu. I. Logachev, G. B. Lopatina, Yu. A. Rozental', V. G. Stolpovskij.
Kosmich. Issled., Vol. 8, 408 - 417 (1970).   In Russian.

143.044 **Some remarks on the possible origin of superheavy nuclei in primary cosmic rays.**
P. Bandyopadhyay, P. R. Chaudhuri.
Journ. Phys. A, General Phys., Ser. 2, Vol. 3, L33 - L36 (1970).

Neutron stars (pulsars) are here proposed as the sources of superheavy (SH) nuclei $Z \geq 110$ in primary cosmic rays. Taking into account the 1969 result of Berlovich and Novikov that these SH nuclei can be formed by the r process when the temperature is greater than 1.8×10^9 K and at sufficiently high neutron number density, it is here pointed out that this temperature condition can prevail in a neutron star for approximately 10^3 years when the cooling behaviour is governed by the synchrotron radiation of neutrinos according to the photon-neutrino weak coupling theory.

143.045 **Variations of cosmic rays and the interplanetary medium.** G. F. Krymskij.
Izv. AN SSSR. Ser. fiz., Vol. 33, 1858 - 1869 (1969).
In Russian. – Abstr. in Referativ. Zhurn. 51. Astron., 4.51.421 (1970).

143.046 **On a reverse effect of cosmic rays on the solar wind.**
I. V. Dorman, L. I. Dorman.
Izv. AN SSSR. Ser. fiz., Vol. 33, 1908 - 1917 (1969).
In Russian. – Abstr. in Referativ. Zhurn. 51. Astron., 4.51.422 (1970).

143.047 **Cosmic-ray variations.** L. I. Dorman.
Izv. AN SSSR. Ser. fiz., Vol. 33, 1832 - 1857

(1969). In Russian. – Abstr. in Referativ. Zhurn. 51. Astron., 4.51.455 (1970).

143.048 **On the anisotropy of cosmic rays during the periods following proton flares.**
L. I. Dorman, O. I. Inozemtseva.
Izv. AN SSSR, Ser. fiz., Vol. 33, 1898 - 1907 (1969).
In Russian. – Abstr. in Referativ. Zhurn. 51. Astron., 4.51.467 (1970).

143.049 **Processes causing deviations of the spectra of discrete sources of cosmic rays from the power law. (Review).**
S. Ya. Braude, I. N. Zhuk, O. M. Lebedeva, A. V. Men', B. P. Ryabov.
Ukr. fiz. zhurn., Vol. 14, 1761 - 1785 (1969). In Russian. Abstr. in Referativ. Zhurn. 51. Astron., 4.51.689 (1970).

143.050 **Some problems of the origin of cosmic rays.**
V. L. Ginzburg, S. I. Syrovatskij.
Izv. AN SSSR. Ser. fiz., Vol. 33, 1770 - 1775 (1969).
In Russian. – Abstr. in Referativ. Zhurn. 51. Astron., 4.51.692 (1970).

143.051 **Some new data on temporal and spatial variations of cosmic rays in the solar system (according to meteoritic data).** A. K. Lavrukhina.
Izv. AN SSSR. Ser. fiz., Vol. 33, 1870 - 1876 (1969).
In Russian. – Abstr. in Referativ. Zhurn. 51. Astron. 4.51.693 (1970).

143.052 **Cosmic rays and plasma turbulence.**
V. N. Tsytovich.
Izv. AN SSSR. Ser. fiz., Vol. 33, 1800 - 1816 (1969).
In Russian. – Abstr. in Referativ. Zhurn. 51. Astron., 4.51.694 (1970).

143.053 **27-day and seasonal changes of the anisotropy of cosmic rays.**
E. V. Kolomeets, G. A. Sergeeva, V. V. Fedoseenko, R. A. Chumbalova.
Cosmic Rays No. 11, Moscow, p. 23 - 25 (1969). In Russian.

143.054 **Increase of cosmic-ray intensity not connected with chromospheric flares.**
V. K. Budilov, E. V. Kolomeets, V. A. Likhoded.
Cosmic Rays No. 11, Moscow, p. 26 - 28 (1969). In Russian.

According to the data of cosmic ray registrations in the stratosphere on the Soviet stations for the period 1957 - 1965 146 cases of the increase of cosmic-ray intensity were revealed and analysed. The paper gives a description and a tentative interpretation of the observed effects.

143.055 **Changes of intensity of primary cosmic rays during the half-cycle of solar activity.**
Yu. M. Nikolaev.
Cosmic Rays No. 11, Moscow, p. 29 - 32 (1969). In Russian.

The paper suggests a model of the quasi-stationary diffusion of primary cosmic rays on the supposition of a stationary solar wind with spherical symmetry.

143.056 **27-day variations of cosmic rays in the minimum of solar activity.** M. V. Alaniya, L. Kh. Shatashvili.
Cosmic Rays No. 11, Moscow, p. 33 - 35 (1969). In Russian.

The present paper makes an attempt to distinguish long-life active formations on the sun which cause 27-day cosmic-ray variations in the wide energy range of primary particles. The analysis of obtained results is given.

143.057 **Local decreases in the cosmic-ray intensity and the Forbush effects in different periods of solar activity.**

K. Imazhanova, E. V. Kolomeets, V. T. Pivneva.
Cosmic Rays No. 11, Moscow, p. 36 - 39 (1969). In Russian.
The paper discusses a local decrease of cosmic-ray intensity in the stratosphere and decrease of intensity observed only in high latitudes. A possible interpretation of the observed effects is given.

143.058 **Distortion in interplanetary space of the anisotropy of cosmic rays of galactic origin.**
L. I. Dorman, O. I. Inozemtseva, S. F. Ilgach, E. A. Mazariuk.
Cosmic Rays No. 11, Moscow, p. 40 - 44 (1969). In Russian.

143.059 **On the interaction of cosmic rays with photons.**
G. T. Zatsepin, V. A. Kuzmin.
Cosmic Rays No. 11, Moscow, p. 45 - 47 (1969). In Russian.
The paper analyses the interaction of cosmic rays of ultra-high energies with isotropic thermal radiation of the universe.

143.060 **Determination of fragmentation parameters of the multi-charged component of primary cosmic rays.**
Yu. F. Gagarin, N. S. Ivanov.
Cosmic Rays No. 11, Moscow, p. 48 - 50 (1969). In Russian.
The paper presents parameters of fragmentation which characterize the transition of the nuclei from the H group to the groups H, M, h and d and from the M group to the groups M, h, d.

143.061 **Modulation of the solar-diurnal and half-diurnal variations of cosmic-ray intensity caused by the discrepancy between the rotation axis of the earth and the perpendicular to the plane of the ecliptic.**
L. I. Dorman, S. Fisher.
Cosmic Rays No. 11, Moscow, p. 64 - 73 (1969). In Russian.
The paper analyses modulations of the diurnal variation due to discrepancy between the rotation axis of the earth and the perpendicular to the ecliptic plane. Four geometric forms of sources of anisotropy are studied.

143.062 **Calculations of the rate of formation of electrons by cosmic rays in the earth's atmosphere with allowance for the changes of energy of the falling primary particles.** L. I. Dorman, T. M. Krupitskaya.
Cosmic Rays No. 11, Moscow, p. 91 - 94 (1969). In Russian.

143.063 **Influence of meteor streams on the level of cosmic radiation on the earth.** S. A. Belskiy.
Cosmic Rays No. 11, Moscow, p. 95 - 99 (1969). In Russian.
The influence of meteor streams on the intensity of cosmic rays on the earth is studied by the method of superposition of epochs. It has been found that the meteor variation of cosmic-ray intensity has the largest amplitude in the years of the maximum solar activity and is caused mainly by particles with the energy up to 14 BeV.

143.064 **Radioactivity caused by cosmic rays in bodies of asteroid size.**
A. K. Lavrukhina, G. K. Ustinova, T. A. Ibraev.
Cosmic Rays No. 11, Moscow, p. 100 - 104 (1969). In Russian.
The paper suggests formulas for deep distribution of separate components of nuclei-active particles generated by cosmic rays in the bodies of the solar system with different chemical composition and size. The paper presents results of calculation of streams of primary and secondary particles in the surface layers of cosmic bodies, greater than that of

asteroids in size, and for four chemical compositions: chondrite, granites, basalts and tectites.

143.065 **Optimal methods of calculation of the primary streams of cosmic rays in interaction with the earth's magnetic field.** V. S. Smirnov.
Cosmic Rays No. 11, Moscow, p. 109 - 111 (1969). In Russian.
The paper presents the study of the equations of motion of high-energy charged particles in a quiet interplanetary magnetic field. The results of this research are used to obtain the function of cosmic-ray distribution, which characterizes the distribution of densities and currents of high-energy charges particles in the interplanetary medium.

143.066 **Cosmic radiation and evolution: A lunar experimental test.** R. Gold.
Radiation Effects, Vol. 1, (No. 3), 178 - 186 (1969).
It is proposed that mineral samples from the lunar surface be examined for possible measurement of the long-term lunar cosmic-ray environment. Observation of radiation-induced tracks in suitable lunar specimens could reveal, in principle, the primary cosmic-ray influence for an epoch comparable with the lunar age.

143.067 **Elementary composition and origin of primary cosmic rays.** L. Koch.
Nucleus, Vol. 10, (No. 5), 324 - 338 (1969).
Discusses the origin of cosmic rays and describes their detection on the earth. Cosmic rays in the galactic magnetic field and the action of elementary particles and nuclei is also discussed.

Cosmic Ray Physics: Nuclear and Astrophysical Aspects. See Abstr. 003.053.

On the spectrum of relativistic particles accelerated by plasma turbulence. See Abstr. 062.017.

Galactic cosmic ray origin of Li, Be and B in stars. See Abstr. 064.019.

The Forbush decrease associated with the proton event of 7 July 1966 as recorded by neutron monitors. See Abstr. 073.108.

Underground cosmic-ray measurements. See Abstr. 078.010.

The altitude dependence of the quiettime cosmic ray ionization over the polar regions at solar minimum. See Abstr. 083.061.

Observations of the cosmic ray knee with a polar orbiting ionization chamber. See Abstr. 083.062.

Argon 37 and argon 39 in recently fallen meteorites and cosmic-ray variations. See Abstr. 105.098.

Pulsars as sources of cosmic rays. See Abstr. 141.146.

Evolution of galaxies and secular variation of cosmic rays, magnetic fields and turbulence. See Abstr. 162.050.

Stellar Systems

151 Kinematics and Dynamics of Stellar Systems

151.001 **A necessary and sufficient condition for the collisionless stability of a stellar system in the post-Newtonian approximation.** A. W. Sudbury.
Monthly Notices, Roy. Astron. Soc., Vol. 147, 187 - 199 (1970).

Using an extension of a method of Antonov, a necessary and sufficient condition for the collisionless stability of stellar systems whose distribution function depends only on the energy of each star is given for the post-Newtonian approximation. It is shown that in this approximation, systems that are spherically symmetric, with $F_E < 0$ in the steady state, are stable against non-radial perturbations.

151.002 **Gravitational polarization in spherical stellar systems.** I. H. Gilbert.
Astrophys. Journ., Vol. 159, 239 - 246 (1970).

The distortion or "polarization" of a spherical and isotropic stellar system caused by the presence of a test star at its center is analyzed. An integral equation for the "polarization" is developed, and a specific solution is found in the case of the isochrone cluster of Henon. The relation between this phenomenon and the theory of collisional relaxation in stellar systems is discussed. It is concluded that collective effects might decrease relaxation times by a factor of 2 or 3.

151.003 **The motions of high-latitude hydrogen clouds.** J. W. Mast, S. J. Goldstein, Jr.
Astrophys. Journ., Vol. 159, 319 - 324 (1970).

We have obtained the solar motion with respect to 268 well-resolved (in velocity) 21-cm hydrogen clouds with galactic latitudes numerically greater than $20°$ and declinations greater than $-40°$. The result is a velocity of 21.2 km sec^{-1} toward $l = 52°.5$, $b = +21°.3$. The mean distance times the galactic-rotation constant is 6.8 km sec^{-1}, and the apparent longitude of the center of rotation is $346°.5$.

151.004 **Orbits in highly perturbed dynamical systems. I. Periodic orbits.** G. Contopoulos.
Astron. Journ., Vol. 75, 96 - 107 (1970).

We find the characteristic curves of many families of periodic orbits in a "galactic-type" potential with increasing perturbation. There are two types of families: (a) regular families which are generated from the unperturbed system, and (b) irregular ones, which are independent of the above. The appearance of irregular families of periodic orbits seems to be connected with the "dissolution" of the invariant curves of nearby nonperiodic orbits. The number of families crossing the $\bar{x}$ axis a given number of times increases considerably as the perturbation increases. All families seem to continue to exist as the perturbation increases and tends to infinity.

151.005 **Orbits in highly perturbed dynamical systems. II. Stability of periodic orbits.** G. Contopoulos.
Astron. Journ., Vol. 75, 108 - 130 (1970).

We study the characteristics of orbits near periodic orbits by using the variational equations. A number of relations between the solutions of the variational equations are found. The analysis is used in finding the exchange of stability near critical points. We found examples of three types of

nonresonant and four types of resonant critical points. Any given family becomes extremely unstable when the perturbation ϵ becomes large. For large ϵ there are stable orbits but their regions of stability are extremely small.

151.006 **Potential energy of continuous and discrete distributions of matter from the point of view of the application of the virial theorem.** D. Sher.
Astrophys. Space Sci., Vol. 6, 275 - 286 = Contr. Louisiana State Univ. Obs., Baton Rouge, No. 30 (1970).

The potential energy of clusters of stars in which the distribution of matter is taken to be continuous is compared with that of static model clusters in which the distribution of matter is discrete, the comparison being made from the point of view of applying the virial theorem to estimate the masses of the clusters.

151.007 **Stability of the system of stars, gas and magnetic fields.** H. Niimi.
Astrophys. Space Sci., Vol. 6, 297 - 314 (1970).

Linear stability of a system of stars, gas and magnetic fields under the existence of a relative motion between the stars and the gas is investigated by the use of the magneto-hydrodynamic and the polytropic equations for the gas and the collisionless Boltzmann equation for the stars together with the Poisson equation. The star system is supposed to have the anisotropic Schwarzschild distribution. The critical wavenumber is calculated and it is found that the system becomes universally unstable under some conditions.

151.008 **Structure and dynamics of stellar systems.** G. Kuzmin.
Astron. Geod. Eston. SSR, Tartu, (see 003.004), p. 52 - 76 (1969). In Russian.

151.009 **Rotating systems of gravitating bodies in quasi-stationary state.**
T. A. Agekian, I. M. Micheile.
Astrofizika, Vol. 5, 623 - 636 (1969). In Russian. – English translation in Astrophysics, Vol. 5, No. 4.

The system of equations describing the state and the evolution of a rotating quasi-stationary system of gravitating bodies is given. Unlike previous results the theory is free from the assumption of equality of the components of the peculiar velocity dispersion. It is shown that there exists a rotating quasi-stationary system the evolution of which is homological. The surplus angular momentum carried out by dissipating stars in such a system is equal on an average to 3/2 of the mean angular momentum of the stars.

151.010 **Barred spiral galaxies.**
E. M. Burbidge.
Comments Astrophys. Space Phys., Vol. 2, 25 - 34 (1970).

A discussion on theoretical papers on the density-wave theory of spiral structure is presented. The formation of barred spiral galaxies using this theory is discussed and a review is given of what is known observationally about them.

151.011 **Forced spiral structure in a uniformly rotating disk of gas.** S. M. Simkin.

Astrophys. Journ., Vol. 159, 463 - 471 (1970).

It is demonstrated in a linear analysis that a uniformly rotating infinite disk of gas in equilibrium, when disturbed by a gravitating particle moving supersonically about its center, will develop two spiral density waves which are coherent on a large scale.

151.012 Numerical calculation of the evolution of a rotating globular galaxy.
B. M. Dzjuba, V. B. Yakubov.
Astron. Zhurn. Akad. Nauk SSSR, Vol. 47, 3 - 9 (1970). In Russian. English translation in Soviet Astron. AJ, Vol. 14, No. 1.

A numerical calculation of the equations of the motion of 1000 material points approximately modeling the mass distribution in the galaxy has been carried out. As a primary state, a rotating globular galaxy with accidental, but on an average, even distribution of particles was chosen.

151.013 The evolution of a protostellar cluster within the limits of a standard problem of 25 bodies.
V. I. Aleshin, S. A. Kaplan.
Astron. Zhurn. Akad. Nauk SSSR, Vol. 47, 10 - 15 (1970). In Russian. English translation in Soviet Astron. AJ, Vol. 14, No. 1.

Results of the numerical solution of the system of equations describing the motion in a cluster of 25 protostars (matter condensations) are given; both the Newtonian interaction and the influence of gaseous and magnetic pressures on the motion of these protostars have been taken into account.

151.014 On thermodynamic irreversibility and relaxation in galaxies. L. S. Marochnik.
Astron. Zhurn. Akad. Nauk SSSR, Vol. 47, 46 - 55 (1970). In Russian. English translation in Soviet Astron AJ, Vol. 14, No. 1.

The interaction of stars with unstable collective motions of stars and interstellar gas is one of the reasons of the origin of the thermodynamic irreversibility and relaxation in stellar systems without encounters. The instability can be of arbitrary origin. The dispersion of the peculiar velocities of stars and temperature of interstellar gas can be increased due to the decrease of gravitational energy in the instability process. The increase of dispersions of stellar velocities of the population I in the galactic plane of our and other spiral galaxies with age can take place due to "collisions" with unstable density waves, which, probably, have to be identified with spiral arms.

151.015 Le problème des N corps en astronomie et en physique des plasmas. M. Hénon.
Journ. Physique, Vol. 30 (Colloque C3, Suppl. No. 11 - 12), C3-27 - C3-41 (1969).

A formal similarity exists between a stellar system and a fully ionized plasma: in both cases, the force between the «particles» is proportional to $1/r^2$. A detailed correspondence can be set up between the concepts, quantities and equations used in both fields. This had led during recent years to a fruitful exchange of theoretical results and computational schemes. This last point is illustrated by some numerical experiments for «two-dimensional» systems.

151.016 A method for computing the evolution of star clusters. R. B. Larson.
Monthly Notices, Roy. Astron. Soc., Vol. 147, 323 - 337 (1970).

We describe a method for computing the evolution of a spherical stellar system, using a fluid-dynamical approach based on the numerical solution of moment equations derived from the Boltzmann equation. Moments of the velo-city distribution up to the fourth order are included in this treatment. In order to represent relaxation effects, 'collision terms' are included and are evaluated using the Fokker-Planck equation.

151.017 Dynamics of stellar systems. Part II. K. Rudnicki.
Postępy Astron., Vol. 18, 81 - 100 (1970). In Polish.

An introduction to the contemporary achievements of the dynamics of stellar systems is presented.

151.018 On the stability of stellar systems. J. Stodółkiewicz.
Postępy Astron., Vol. 18, 115 - 116 (1970). In Polish. Short note.

151.019 Spiral structures of disc galaxies as non-axisymmetric perturbations – II. S. Yabushita.
Monthly Notices, Roy. Astron. Soc., Vol. 148, 87 - 91 (1970).

A stability analysis is made of a differentially rotating disc whose unperturbed tangential velocity is a constant by making use of a method adopted in paper I. It is shown that the non-axisymmetric perturbations in density of the disc show spiral patterns that are leading. Comparing the present result with that obtained in the previous paper it appears almost certain that growing perturbations in density are leading spiral patterns when they are obtained using pressure-free hydrodynamical equations.

151.020 The possibility of the formation of bars and tails as a result of tidal interaction of galaxies. II.
N. Tashpulatov.
Astron. Zhurn. Akad. Nauk SSSR, Vol. 47, 277 - 291 (1970). In Russian. – English translation in Soviet Astron., AJ, Vol. 14, No. 2.

The hydrodynamic outflow of matter from a vertex of a uniform prolate ellipsoidal galaxy after a close passage by another (point-like) galaxy is considered. The relative orbit of the point-like galaxy is assumed to be parabolic. It is assumed also that the ellipsoidal galaxy rotates with constant and appropriately chosen uniform angular velocity. A series of numerical calculations has been carried out for two values of the mass times three values of the pericentric distance of the parabolic orbit of the point-like galaxy. Differential equations of motions of the hydrodynamic mass elements were integrated by the Runge–Kutta method. These calculations led in all cases to the formation of bars and tails by interacting galaxies.

151.021 On a study by Podurets of the stability of relativistic star clusters. J. R. Ipser.
Astron. Zhurn. Akad. Nauk SSSR, Vol. 47, 452 - 453 (1970). In Russian. – English translation in Soviet Astron., AJ, Vol. 14, No. 2.

There is a contradiction between the results of investigations by Podurets and by Ipser of the stability against radial perturbations of a class of relativistic isothermal clusters with truncated Maxwellian distribution. Here it is pointed out that Podurets' criterion for stability is correct, but that there exists an important error in his application of that criterion to the isothermal clusters. When Podurets' error is corrected, his criterion is found to yield no information about the stability of the isothermal clusters. The connection between Podurets' criterion for stability and the criterion due to Ipser is elucidated.

151.022 Recent evidence for the magnetogravitational theory of galaxies. H. D. Greyber.
Bull. American Astron. Soc., Vol. 2, 194 - 195 (1970). Abstr. AAS.

151.023 **Coalescence vs disruption of colliding stars in dense stellar systems.** R. H. Sanders.
Bull. American Astron. Soc., Vol. 2, 216 (1970). − Abstr. AAS.

151.024 **The haloing effect of the third integral.**
A. Deprit, J. Henrard.
Astrophys. Space Sci., Vol. 7, 54 - 70 (1970).

Whether Contopoulos's galactic system is separable (unlikely) or not (likely), the fact is that there exists a vicinity of the equilibrium in which numerical integration of high accuracy cannot separate the system from its image through Birkhoff's normalization of high order. To all practical purposes, stellar dynamics is then justified in pretending that the model is, in that region, structured by a so-called third integral.

151.025 **The existence of the third integral for a family of high-velocity stars.**
K. A. Innanen, F. C. House.
Astrophys. Space Sci., Vol. 7, 139 - 150 (1970).

A family of high-velocity stars (energy − 19000 km^2/sec^2, angular momentum 1150 kpc km/sec) were computed using Innanen's (1966) mass model of the Galaxy. The stars were set initially on the curve of zero velocity. The inclination diagram indicates a quasi-isolating third integral.

151.026 **The velocity variation of a star as a purely discontinuous random process. I. Zero mass stars.**
V. S. Kaliberda, I. V. Petrovskaya.
Astrofizika, Vol. 6, 135 - 147 (1970). In Russian.
English translation in Astrophysics, Vol. 6, No. 1.

The variation of the velocity module of a star in a system is considered as a purely discontinuous random process. Using the second Kolmogorov-Feller equation the evolution of the velocity distribution function of zero mass stars in an open cluster is investigated without taking into account the regular potential. The escape rate of stars and the amount of energy, taken away by the dissipated stars in different moments of time are also found.

151.027 **On the density-wave theory of galactic spirals. I. Spiral structure as a normal mode of oscillation.**
F. H. Shu.
Astrophys. Journ., Vol. 160, 89 - 97 (1970).

An exact formulation of the linearized problem, including appropriate boundary conditions, is developed to explore whether extensive galactic density waves of spiral form are permissible normal modes of oscillation for a stellar disk. An "anti-spiral theorem," of the type reported previously by Lynden-Bell and Ostriker for neutral modes in a gaseous disk, holds here with limited validity − namely, whenever the effects of stellar resonances can be ignored.

151.028 **On the density-wave theory of galactic spirals. II. The propagation of the density of wave action.**
F. H. Shu.
Astrophys. Journ., Vol. 160, 99 - 112 (1970).

The properties of galactic density waves are studied in the WKBJ approximation. In the lowest order of approximation, we reproduce the dispersion relation reported by Lin and Shu in an earlier communication. In the next order, we demonstrate explicitly that the density of "wave action" is transported with the group velocity derived by Toomre. Some general implications are drawn for mechanisms proposed for the origin of spiral structure.

151.029 **Resonance effects in spiral galaxies.**
G. Contopoulos.
Astrophys. Journ., Vol. 160, 113 - 133 (1970).

We consider various resonances expected to occur in our Galaxy. We study in some detail the forms of the orbits and invariant curves near the inner Lindblad resonance; we find two stable resonant periodic orbits and tube orbits around them. A comparison between the theoretical results and the numerical calculations is made. Finally, we see how the stars respond to a growing spiral wave. As the wave is established, the orbits near resonance change from their original epicyclic form to a resonant form near one or the other of the two resonant periodic orbits. The resulting density distribution has roughly a quadruple symmetry.

151.030 **Collective instabilities and waves for inhomogeneous stellar systems. I. The necessary and sufficient energy principle.** R. M. Kulsrud, J. W-K. Mark.
Astrophys. Journ., Vol. 160, 471 - 483 (1970).

In this paper an energy principle is derived that gives a necessary and sufficient condition for the linear stability of inhomogeneous collisionless nonrelativistic stellar systems whose distribution function is a monotonic decreasing function of the energy. The energy principle is applied to the case of an equilibrium with plane-parallel symmetry.

151.031 **Dynamical mixing of orbits in the spherical waterbag model.** P. Bouvier, G. Janin.
Astron. Astrophys., Vol. 5, 127 - 134 (1970).

This paper deals with the evolution of a self-gravitating spherical system during its preliminary stage of orbital mixing. Initial conditions are chosen according to the so-called waterbag model; numerical integration of the equations of motion discloses the detailed dynamical evolution of the system.

151.032 **A numerical study of two families of globular cluster orbits.**
K. A. Innanen, R. V. Hodder, C. E. Smith.
Astron. Astrophys., Vol. 5, 206 - 208 (1970).

A description is given of two numerically integrated groups of galactic orbits related to previous calculations (Innanen, 1966). It is found that periodic or nearly periodic orbits in the ($\tilde{\omega}$ − z) plane are a standard feature of these orbits for velocity vectors initially highly inclined to the galactic plane whereas tube and perhaps box orbits are confined to velocity vectors of low initial inclination to the galactic plane. In the first group, where inclination and angular momentum are held constant, but where energy increases, periodic or nearly periodic orbits and tube-like orbits appear alternately.

151.033 **A model of a stellar system with constant density in phase space. The stationary state.**
J. P. Doremus, M. R. Feix, G. Baumann.
Astron. Astrophys., Vol. 5, 280 - 285 (1970). In French.

We study the steady state of a collisionless stellar system with spherical symmetry. The distribution function is a function of the two integrals of the motion.

151.034 **A model for the early evolution of galaxies.**
P. Brosche.
Astron. Astrophys., Vol. 6, 240 - 253 (1970).

The aim of this paper is to show how the evolution from protogalaxies with equal starting values but different angular momenta can lead to galaxies with different time scales of stellar formation.

151.035 **Gravitational theories of spiral structure.**
G. Contopoulos.
IAU Symposium No. 38, (see 012.013), p. 303 - 316 (1970).

The basic ideas and some of the most important recent developments of the gravitational theories of spiral structure are described. The linear self consistent problem consists of the problem of modes and of the initial value problem, which is discussed here in some detail. More emphasis is put on the non linear problem near resonances and in particular the inner Lindblad resonance.

151.036 The theory of spiral structure of galaxies.
L. S. Marochnik.
IAU Symposium No. 38, (see 012.013), p. 317 (1970). Abstract.

151.037 Small amplitude density waves on a flat galaxy.
A. J. Kalnajs.
IAU Symposium No. 38, (see 012.013), p. 318 - 322 (1970).

By numerical methods we have found an unstable two-armed density wave on a flat galactic model. We present the results in a form of four plots, and briefly discuss the observational implications as well as the uncertainties involved in the models and the calculations.

151.038 The propagation and absorption of spiral density waves.
F. H. Shu.
IAU Symposium No. 38, (see 012.013), p. 323 - 325 (1970).

An 'anti-spiral theorem' holds with limited validity for the neutral modes of oscillation in a stellar disk – namely, whenever the effects of stellar resonances can be ignored. In the regions between Lindblad resonances, a group of spiral waves will propagate in the radial direction with the group velocity found by Toomre.

151.039 Large scale oscillations of galaxies.
C. Hunter.
IAU Symposium No. 38, (see 012.013), p. 326 - 330 (1970).

The observational evidence that may indicate the presence of large scale modes of oscillation in galaxies is reviewed, and some results of theoretical modal calculations are described.

151.040 Generating mechanisms for spiral waves.
D. Lynden-Bell.
IAU Symposium No. 38, (see 012.013), p. 331 - 333 (1970).

151.041 Spiral waves caused by a passage of the LMC?
A. Toomre.
IAU Symposium No. 38, (see 012.013), p. 334 - 335 (1970).

151.042 Non-linear density waves in pressureless disks.
C. L. Berry, P. O. Vandervoort.
IAU Symposium No. 38, (see 012.013), p. 336 - 340 (1970).

This is a report on our first results in a non-linear theory of density waves. We have considered the hydrodynamics of a pressureless, self-gravitating disk of infinitesimal thickness in a given external gravitational field which is time-independent and axisymmetric; and we have constructed spiral waves of finite amplitude in that disk.

151.043 Density waves in galaxies of finite thickness.
P. O. Vandervoort.
IAU Symposium No. 38, (see 012.013), p. 341 - 342 (1970).

The effect of the finite thickness of a galaxy on the propagation of density waves of the type described by Lin and his collaborators has been calculated. The calculated effect does not differ appreciably from what has been estimated previously on the basis of heuristic arguments.

151.044 Particle resonance in a spiral field.
B. Barbanis.
IAU Symposium No. 38, (see 012.013), p. 343 - 347 (1970).

Plane galactic orbits near the particle resonance of a spiral field have been calculated numerically. Besides the ring type orbits, i.e. orbits filling a ring around the galactic center, there are orbits librating near the potential maxima at the co-rotation distance (Lagrangian points).

151.045 Numerical experiments in spiral structure.
R. H. Miller, K. H. Prendergast, W. J. Quirk.
IAU Symposium No. 38, (see 012.013), p. 365 - 367 (1970).

Results of an n-body calculation, containing about 120000 particles, were shown as a motion picture. Some of the particles are treated as 'gas', obeying a special dissipative dynamics, the rest as 'stars'. The system was started as pure 'gas', and 'stars' were made out of the 'gas' in a manner closely mimicking real galaxies.

151.046 Computer models of spiral structure.
F. Hohl.
IAU Symposium No. 38, (see 012.013), p. 368 - 372 (1970).

A computer model for isolated disks of stars is used to study the self-consistent motion of large numbers of point masses as they move in the plane of the galactic disk. The results are presented in the form of a motion picture.

151.047 Interpretation of large-scale spiral structure.
C. C. Lin.
IAU Symposium No. 38, (see 012.013), p. 377 - 390 (1970).

The present paper consists of three parts: (1) A general explanation, from a semi-empirical point of view, of the density wave theory and its ramifications, with only a few remarks on those detailed features which have not been stressed before; (2) a statement of our conclusions about the Milky Way System; and (3) a discussion of the problem of the origin of density waves of spiral form.

151.048 Large-scale galactic shock phenomena and the implications on star formation.
W. W. Roberts, Jr.
IAU Symposium No. 38, (see 012.013), p. 415 - 422 (1970).

The possible existence of a stationary two-armed spiral shock pattern for a disk-shaped galaxy, such as our own Milky Way System, is demonstrated. It is therefore suggested that large-scale galactic shock phenomena may very well form the large-scale triggering mechanism for the gravitational collapse of gas clouds, leading to star formation along narrow spiral arcs within a two-armed grand design of spiral structure.

151.049 Magnetohydrodynamical models of helical magnetic fields in spiral arms.
M. Fujimoto, M. Miyamoto.
IAU Symposium No. 38, (see 012.013), p. 444 - 447 (1970).

A circular arm with elliptical cross-section is used as a model of the spiral arm. It is demonstrated magnetohydrodynamically that interstellar gas may flow in a helical path along the axis of the arm and interstellar helical magnetic lines of force can spiral around it. Some observational supports to the present model are given.

151.050 Exchange of gravitational energy during a collision of galaxies.
K. S. Sastry, S. M. Alladin.
Astrophys. Space Sci., Vol. 7, 261 - 271 (1970).

The problem of the change in internal energy of a colliding galaxy due to tidal effects is considered, assuming that the galaxies may be regarded as spherical stellar systems whose over-all structure remains unchanged during the collision and that the stars move in circular orbits. The numerical estimates thus made for the energy gained by the stars during the collision are compared with those derived on the basis of the assumption that the motions of the stars may be neglected during the encounter (the 'impulsive approximation') to test the adequacy of the latter approximation.

151.051 On the role of the magnetic field in a wave theory of the spiral structure of galaxies.
L. S. Marochnik, N. G. Ptitsina.

Astrophys. Space Sci., Vol. 7, 437 - 445 (1970).

The spiral waves in a model galaxy consisting of the differentially rotating interstellar gas and population I are considered. The instability of spiral waves in the presence of differential rotation and magnetic field is found.

151.052　**Système homogène en phase à symétrie cylindrique.**
P. Bouvier.
C.R. Séances, SPHN Genève, NS, Vol. 4, Fasc. 2, p. 142 - 146 (1969) = Publ. Obs. Genève, Sér. A, Fasc. 77/III (1970).

Examinons le problème à deux dimensions, comme nous l'avons déjà partiellement fait dans un article récent (Publ. Obs. Genève, Sér. A, Fasc. 75 (1968)) qui contient toutefois une erreur de calcul qu'il convient de rectifier ici.

151.053　**Modèles numériques de systèmes autogravitants.**
I. Couches sphériques concentriques.
G. Janin.
C.R. Séances, SPHN Genève, NS, Vol. 4, 146 - 149 (1969) = Publ. Obs. Genève, Sér. A, Fasc. 77/IV (1970).

151.054　**Modèles numériques de systèmes autogravitants.**
II. Masses ponctuelles.　G. Janin.
Arch. Sci. Genève, Vol. 23, Fasc. 1, p. 27 - 32 (1970) =Publ. Obs. Genève, Sér. A, Fasc. 77/V (1970).

151.055　**Modèles numériques de systèmes autogravitants.**
III. Couches planes parallèles.　G. Janin.
Arch. Sci. Genève, Vol. 23, Fasc. 1, p. 33 - 40 (1970) = Publ. Obs. Genève, Sér. A, Fasc. 77/VI (1970).

151.056　**On the instability of galactic spiral density waves**
in the presence of a magnetic field and differential
rotation.　N. G. Ptitsina.
Astrophys. Letters, Vol. 5, 279 - 281 (1970).

Spiral (and ring) density waves in an infinitesimally thin disc (and a cylinder) are unstable in the presence of a magnetic field and differential rotation. The instability is of a non-gravitational nature. It may possibly lead to the increase of amplitude of spiral waves in galaxies up to observable magnitudes.

151.057　**Nonlinear waves of stellar density and the spiral**
structure of the Galaxy.　N. G. Ptitsina.
Astron. Zhurn. Akad. Nauk SSSR, Vol. 47, 499 - 502 (1970). In Russian. English translation in Soviet Astron. AJ, Vol. 14, No. 3.

It is possible that density waves of finite amplitude, the shape of which is changed in the presence of differential rotation, may exist in a self-gravitating one-component cylinder. The shape of front of these waves is a logarithmical spiral or straight line. It is shown that waves of small amplitude with the same shape of front can exist in such a model.

151.058　**On the character of motions in pairs of galaxies.**
I. D. Karachentsev.
Astron. Zhurn. Akad. Nauk SSSR, Vol. 47, 509 - 515 (1970). In Russian. English translation in Soviet Astron. AJ, Vol. 14, No. 3.

Three types of motions of the components of double galaxies are examined: 1) the pairs have a positive total energy and disrupt, 2) the members of the pairs rotate on circular orbits, 3) the members of double galaxies move on radial directions with negative total energy (oscillations). The assumption of the disruption of pairs gives the best agreement with observational data.

151.059　**On the character of the distribution of peculiar**
velocities in differentially rotating stellar systems.
M. N. Maksumov.
Astron. Zhurn. Akad. Nauk SSSR, Vol. 47, 668 - 671 (1970).

In Russian. English translation in Soviet Astron. AJ, Vol. 14, No. 3.

On a simple model of a stationary differentially rotating stellar system it is shown that the distribution of peculiar velocities must be nonisotropic.

151.060　**Numerical experiments on the escape from non**
isolated clusters.　A. Hayli.
Astron. Astrophys., Vol. 7, 17 - 23 (1970).

Isolated and non isolated clusters with a mass distribution have been studied by numerical techniques. The rates of escape of stars and of kinetic energy are compared with Hénon's theoretical expressions. Multiple encounters play a very important role in the excape phenomenon, at least for clusters with a small number of stars. This leads to a theoretical underestimate of the rates of escape when the stars have equal masses and to an overestimate when masses are unequal.

151.061　**Sur l'instabilité gravitationnelle.**　F. Nahon.
Comptes Rendus Acad. Sci. Paris, Sér. A, Vol. 271, 203 - 205 (1970).

On sait que les systèmes stellaires sont instables pour les petites perturbations de longueur d'onde supérieure à une distance critique appelée distance de Jeans. Nous donnons une définition de la distance de deux solutions d'une équation aux dérivées partielles, et nous pouvons ainsi préciser le critère de Jeans.

151.062　**On the disks of spiral and S0 galaxies.**
K. C. Freeman.
Astrophys. Journ., Vol. 160, 811 - 830 (1970).

The main problems pointed out in this paper are: (1) Why are the disks of spiral and S0 systems exponential? (2) Why is the surface-brightness scale for these disks approximately the same for about three-quarters of the sample, despite the great range in absolute magnitude?

151.063　**The intrinsic flattening of E, S0, and spiral galaxies**
as related to galaxy formation and evolution.
A. Sandage, K. C. Freeman, N. R. Stokes.
Astrophys. Journ., Vol. 160, 831 - 844 (1970).

Evolutionary questions concern (a) why S0 → Im galaxies have flattened to a disk, while E galaxies have not; (b) why the spheroidal components of all galaxies contain only old stars, probably formed at a single epoch; and (c) why S0 and early Sa galaxies have lost their young spiral-arm population, while Sb → Im galaxies have not. Analysis of the four properties, (1) the disk blue-light surface brightness at $r = 0$, (2) the absolute length scale of the disk, (3) the size ratio of the spheroidal component to the total radius of the galaxy and (4) the mean total mass density, shows that galaxy type does not depend uniquely on systematic variations of these parameters along the Hubble sequence.

151.064　**Rotational alignment of disturbances in the nuclei**
of variable galaxies.　P. A. A. Clark.
Astrophys. Journ., Vol. 160, 845 - 857 (1970).

A model is presented here in which the action of the galactic rotation focuses an initially spherical disturbance into a pair of gas clouds (eventually jets) aligned symmetrically along the rotation axis. The nature of the rotationally induced focusing mechanism is described. A model is developed and solved in order to illustrate the mechanism. The model is discussed and a comparison of the theoretical predictions with observations of NGC 1068 is given.

151.065　**Relativistic, spherically symmetric star clusters.**
IV. A sufficient condition for instability of isotropic
clusters against radial perturbations.　E. D. Fackerell.
Astrophys. Journ., Vol. 160, 859 - 874 (1970).

In this paper a one-dimensional eigenvalue equation of

the Sturm-Liouville type is obtained which optimizes the trial functions used by Ipser in paper III of this series to analyze the stability against radial perturbations of isotropic, spherically symmetric, relativistic star clusters. This eigenequation provides a very powerful, sufficient criterion for the instability of such a star cluster. Using this criterion, we show that a polytropic star cluster of indices $n = 4$ and $\Gamma_4 = {}^5/_4$ is unstable if the redshift of a photon emitted from its center and received at infinity is greater than 0.7302.

151.066 On the galactic orbit of the cluster M67.
V. V. Syrovoj.
Uch. zap. Ural'skogo un-ta, 1969, No. 70, p. 202 - 205. In Russian. – Abstr. in Referativ. Zhurn. 51. Astron., 6.51.716 (1970).

151.067 On the theory of interaction between stars and cooperative motions of stellar "gas".
L. S. Marochnik.
Byull. Inst. Astrofiz., *Dushanbe*, No. 50, p. 3 - 15 (1968). In Russian.

Equations are obtained for the interaction of stars and cooperative motions of stellar "gas". A comparative analysis of the situations for plasma and gravitation is given.

151.068 Influence of the star-star collisions on the cooperative phenomena in rotating stellar systems.
L. S. Marochnik, V. K. Babkov.
Byull. Inst. Astrofiz., *Dushanbe*, No. 50, p. 16 - 19 (1968). In Russian.

It is shown that the stability of stellar systems for small perturbations of density is increased by star-star collisions.

151.069 The stability problem of oscillations along the axis of symmetry in a galaxy. II. The first order perturbations in a general resonance case.   P. Andrle.
Bull. Astron. Inst. Czechoslovakia, Vol. 21, 132 - 139 (1970).

The present papers are concerned with the properties of galactic orbits of a testing particle (a star) being in a close vicinity of the axis of symmetry. In an unperturbed case the considered movement is supposed to be created by two oscillations perpendicular to each other. In this paper the ratio of the frequencies of booth oscillations is equal to r/s, where r and s are natural numbers. Two special examples are solved.

151.070 Non-linear density waves in the galaxy-gasdynamical approximation.   H. Niimi.
Journ. Phys. Soc. Japan, Vol. 28, (No. 1), 232 - 237 (1970).

Non-linear density waves in a rotating stellar system are investigated within the framework of the gasdynamic approximation endowing the system with an isotropic pressure.

151.071 Dynamical relaxation in galaxy formation.
B. F. Smith.
Thesis, Univ. Maryland, Cantonsville. [Available from Univ. Microfilms, Ann Arbor, Mi.], 65 pp. (1969).

The initial evolution of a galaxy condensing out of the universe is studied by numerical methods to investigate a possible relaxation effect in the early stages of evolution.

151.072 Treatment of close approaches in stellar dynamics.
C. F. Peters.
Thesis, Yale Univ., New Haven, Conn. [Available from Univ. Microfilms, Ann Arbor, Mi.], 97 pp. (1969).

The goal of the present work is the development of a rigorous method of treating binary encounters within a gravitational system of n bodies.

151.073 Problems of stellar dynamics.   G. Contopoulos.
Lectures Applied Math., Vol. 5, Space Math., Part I, p. 169 - 258 = Contr. Astron. Dep. Univ. Thessaloniki, No. 28 (1966).

Escape from a gravitational system of positive energy.   See Abstr. 042.012.

Numerical study of dynamical systems with three degrees of freedom. I. Graphical displays of four-dimensional sections.   See Abstr. 042.036.

On a case of generalization of the problem of two fixed centres.   See Abstr. 042.051.

The velocity dispersion of O- and B-stars within a few kpc.   See Abstr. 112.012.

Collisions between H I clouds. I. One-dimensional model.   See Abstr. 131.007.

Collisions between H I clouds. II. Two-dimensional model.   See Abstr. 131.008.

Mechanisms in galactic evolution and spiral structure.   See Abstr. 155.013.

Possible influence of the spiral galactic structure on the local distributions of residual stellar velocities. See Abstr. 155.021.

Neutral hydrogen in M31 – II. The dynamics and kinematics of M31.   See Abstr. 158.099.

152 Stellar Associations

152.001 **Nouvelles étoiles lointaines dans la région de l'association I Scorpion.** A. Laval.
Comptes Rendus Acad. Sci. Paris, Sér. B, Vol. 270, 769 - 772 (1970).

L'étude des vitesses radiales dans l'association I Sco (l_{II} = 343°, b_{II} = 1°,7) a permis de mettre en évidence la présence, parmi les étoiles de l'association, d'étoiles supergéantes lointaines. Ces étoiles se placent à une distance moyenne de 4500 pc, dans le prolongement du bras spiral Norma–Centaure.

152.002 **Chemical homogeneity and absolute magnitudes of members of the Hyades and Wolf 630 groups.** O. J. Eggen.
Publ. Astron. Soc. Pacific, Vol. 82, 99 - 121 (1970).

Narrow-band $uvby\beta$ photometry is used to confirm group membership of some 70 to 80 percent of the A- to F-type stars that were assigned to the Hyades and Wolf 630 groups using other criteria. A lower percentage of confirmation as group members among the early type stars than previously found for the later types is not unexpected, in view of the greater reliability (for late-type stars) of spectroscopic and trigonometric parallaxes used in originally choosing group members.

152.003 **Catalogue of spectral and luminosity classes of stars in the O-associations Cas III, Cas IV, Cas VII.** R. A. Bartaya, E. K. Kharadze.
Byull. Abastumansk. Astrofiz. Obs. No. 38, p. 127 - 143 (1969). In Russian.

152.004 **Spectral classes and luminosities of stars in the O-association Cam I.** R. A. Bartaya, E. K. Kharadze.
Byull. Abastumansk. Astrofiz. Obs. No. 38, p. 145 - 150 (1969). In Russian.

152.005 **Low dispersion spectral classification of stars in T-associations Oph TI and And TI.** M. A. Shiukashvili.
Byull. Abastumansk. Astrofiz. Obs. No. 38, p. 151 - 169 (1969). In Russian.

152.006 **Busqueda de anillos estelares.** R. H. Méndez.
Revista Astron., Vol. 41, No. 171, p. 34 (1969).

152.007 **Low-temperature regions of interstellar gas and formation of stellar associations.** S. B. Pikelner.
Astron. Zhurn. Akad. Nauk SSSR, Vol. 47, 254 - 264 (1970). In Russian. – English translation in Soviet Astron., AJ, Vol. 14, No. 2.

Accepted ideas on gravitational condensation and fragmentation of dense clouds meet with some difficulties in explaining the formation of extended associations of low mean density. Therefore another possible mechanism is discussed. Interstellar gas is heated and ionized by low-energy cosmic rays or more probably by low-energy X-rays.

152.008 **Development of stellar associations: A simple model.** T. Arny.
Bull. American Astron. Soc., Vol. 2, 180 (1970). – Abstr. AAS.

152.009 **On the percentage of flare stars among the RW Aurigae type variables in the Orion association.** V. A. Ambartsumian.
Astrofizika, Vol. 6, 31 - 38 = Byurakan Astrophys. Obs. No. 52 (1970). In Russian. – English translation in Astrophysics, Vol. 6, No. 1.

A statistical study of a certain sample of RW Aurigae variables belonging to the Orion association which have amplitudes of variations $>1^m$ has shown that only a quarter of them are flare stars, with flares observable by means of the usual photographic method. There are indications that such flares occur only in the late phase of the variability.

152.010 **A comparison of radio observations of NGC 7822 (W 1) with the OB stars in the Ceph IV association.** E. Churchwell, M. Felli.
Astron. Astrophys., Vol. 4, 309 - 314 (1970).

Radio maps made at 5.0 and 1.4 GHz of W 1 resolve two main components. A non-thermal contribution to the radio emission is indicated when the data at both frequencies are compared. The Lyman continuum photon fluxes of the early-type stars in the Ceph IV association are given, but the question of whether the visible O stars can fully account for the thermal emission cannot be answered until the contribution of the thermal emission is separated from that of the non-thermal component.

152.011 **7 Sextantis, a possible run-away star from upper Centaurus Lupus.** L. Martinet.
Astron. Astrophys., Vol. 4, 331 - 334 (1970).

It is suggested that the high-velocity A star 7 Sextantis could have been formed in the upper Centaurus Lupus subgroup of the Scorpio-Centaurus association and ejected from it as "run-away" Ap star, about 1.3×10^6 years.

152.012 **Motion of neutral hydrogen in connection with the association II (ζ) Per.** R. Sancisi.
Astron. Astrophys., Vol. 4, 387 - 390 (1970).

Observations of the 21-cm hydrogen emission line in the Perseus region have been made with the 25-metre radiotelescope at Dwingeloo. In the direction of the association of O and B stars II Per the radial velocity of the neutral hydrogen referred to the local standard of rest reaches a maximum value of +8 km/s, which is a few km/s higher than the velocity in the surroundings. The average radial velocity of the stars of the association is about +15 km/s, referred to the local standard of rest.

152.013 **An elongated neutral-hydrogen emission feature in Scorpius and Ophiuchus.** R. Sancisi, H. van Woerden.
Astron. Astrophys., Vol. 5, 135 - 139 (1970).

Hydrogen 21-cm line profiles in the Sco-Oph region show a narrow emission feature having a radial velocity of -12 km/s with respect to the local standard of rest. The feature extends over a large area of sky; it has a filamentary shape and considerable internal structure. Comparisons with interstellar sodium lines place an upper limit of 170 pc on its distance from the sun.

152.014 **Photometric investigation of the association Car OB 2.** W. Seggewiss.
IAU Symposium No. 38, (see 012.013), p. 265 - 269 (1970).

A photometric investigation of the association Car OB 2 was carried out with a photoelectric standard sequence observed with the ESO photometric telescope and plates taken with the ADH telescope of Boyden Observatory. The luminosity function was determined and compared with the luminosity functions of very young open clusters observed by Walker (1957). The relation of Car OB 2 and group c to the structure of the Carina arm is regarded.

152.015 **Motions of groups of early stars and Orion variables respective to the galactic plane.**
N. M. Artiukhina.
Astron. Zhurn. Akad. Nauk SSSR, Vol. 47, 667 - 668 (1970). In Russian. English translation in Soviet Astron. AJ, Vol. 14, No. 3.

It is shown that two groups of Orion variables in Taurus and the groups of O - B stars of the Cep II association approach to the galactic plane. V_z values for groups and stars in Orion and in the ζ Per association are nearly zero.

152.016 **Supergiants in stellar associations and galactic clusters.** R. M. Humphreys.
Astrophys. Letters, Vol. 6, 1 - 3 (1970).

Observational data are presented which support recent suggestions that the photoneutrino process is occurring in some of the M supergiants.

152.017 **Scorpii I and NGC 6231.** L. A. Milone.
Asoc. Argentina Astron. Bol., No. 14 (see 012.020), p. 14 - 17 (1968). In Spanish.

On the evolution speed of young stars in associations. See Abstr. 065.009.

Formation of stellar associations.
See Abstr. 065.074.

On the distribution of space velocities of OB stars.
See Abstr. 112.013.

Application of the dispersion analysis for the search of variable stars. See Abstr. 122.013.

The interstellar reddening law in the direction of Ara OB1. See Abstr. 131.103.

The sky near the brightest X-ray source in Scorpius. II. See Abstr. 142.032.

Star clusters and associations.
See Abstr. 153.029.

Spiral structure and distribution of stellar associations in NGC 6946. See Abstr. 158.082.

A catalogue of stellar associations in the Large Magellanic Cloud. See Abstr. 159.001.

153 Galactic Clusters

153.001 Observations of highly eccentric orbits in the Pleiades cluster. B. F. Jones.
Astrophys. Journ. *(Letters)*, Vol. 159, L129 - L131 (1970).

New relative proper motions have been determined for fifty-two Pleiades stars on plates taken with the 40-inch refracting telescope at Yerkes Observatory, which have a time baseline of up to 66 years. These motions show that stars at distances greater than 20 minutes of arc from the center of the cluster have internal motions that are almost completely in a radial direction, an observation which indicates that these stars are in highly eccentric orbits.

153.002 Dust and neutral hydrogen in open clusters. S. D'Odorico, M. Felli.
Mem. Soc. Astron. Italiana, Nuova Serie, Vol. 41, 89 - 93 (1970).

The correlation between the mass of dust in open clusters in the form of interstellar grains and the mass of neutral hydrogen present in the same region, and possibly associated with the clusters, is investigated. A value of 35 for the ratio is found.

153.003 The frequency of spectroscopic binaries in NGC 6475.
H. A. Abt, S. G. Levy, L. A. Baylor, R. R. Hayward, C. P. Jewsbury, C. M. Snell.
Astrophys. Journ., Vol. 159, 919 - 926 = Contr. Cerro Tololo Inter-American Obs. No. 94 (1970).

This cluster is known to have unusually low projected rotational velocities among its B stars, in contrast to the Pleiades, a cluster of the same age, which has unusually high rotational velocities. In NGC 6475 eight of the nineteen brightest main-sequence stars are found to be short-period spectroscopic binaries; orbital elements are given for these. This frequency is high relative to that for field stars and especially high relative to that for the Pleiades, among whose B stars no short-period binaries were found. It is concluded that low rotational velocities are probably caused primarily by tidal interactions in binary systems and that open clusters differ significantly in their binary frequencies.

153.004 Structure of the cluster M 39. N. M. Artyukhina.
Astron. Zhurn. Akad. Nauk SSSR, Vol. 47, 162 - 165 (1970). In Russian. English translation in Soviet Astron.,AJ, Vol. 14, No. 1.

On the basis of proper motions obtained earlier, curves of the radial distribution of the apparent density of the possible cluster members brighter than $11.^{m}2$ pg, picked out by their proper motions, are constructed. It is shown that the cluster consists of a nucleus (radius about $22.'5$) and a corona (radius about $1.°2$). About 50% of members of the cluster brighter than $11.^{m}2$ are in the corona.

153.005 Photometry in UBVRIJHKL of the early main sequence in the Pleiades down to G0V.
B. Iriarte Erro.
Bol. Obs. Tonantzintla y Tacubaya, Vol. 5 (No. 32), 89 - 100 (1969).

Photoelectric photometry in UBVRIJHKL of thirty-nine members of the Pleiades cluster was obtained in an attempt to deterime the dependence of interstellar extinction on wavelength throughout the cluster. In the process of the investigation infrared excess was found in some of the members not totally due to interstellar reddening. For some B stars, the infrared excesses were interpreted as infrared emission from circumstellar shells. It was found that not only excesses are present, but deficiencies as well, and finally, infrared excess was found for a flare star in the Pleiades.

153.006 Accurate positions of 502 stars in the region of the Pleiades.
H. Eichhorn, W. D. Googe, C. F. Lukac, J. K. Murphy.
Mem. Roy. Astron. Soc., Vol. 73, (Part 2), 125 - 151 (1970).

This paper contains the right ascensions and declinations of 502 stars in a region of about 1.5 degrees square in the Pleiades cluster and describes the unconventional procedures used to obtain their high relative accuracy which is on the order of $0.''01$. Relative proper motions for the reference stars were computed by comparison of the positions given in this paper with those previously determined by König. Means for the reduction of the positions and proper motions to the FK4 system are provided. Other results of significance for astrometric technique are discussed.

153.007 Occultation observations of the Pleiades. R. A. Berg.
Bull. American Astron. Soc., Vol. 2, 182 (1970). – Abstr. AAS.

153.008 Short-period light variations in stars in h and χ Persei. J. R. Percy.
Bull. American Astron. Soc., Vol. 2, 214 (1970). – Abstr. AAS.

153.009 The distances of the galactic clusters NGC 1664 and NGC 1605. C. Fang.
Astron. Astrophys., Vol. 4, 75 - 77 (1970).

The distances of the two galactic clusters NGC 1664 and NGC 1605 are determined with the method of *RGU* three-color photometry. The distances are found to be 1240 and 2650 pc, respectively.

153.010 Flare stars in the Pleiades.
V. A. Ambartsumian, L. V. Mirzoyan, E. S. Parsamian, O. S. Chavushian, L. K. Erastova.
Astrofizika, Vol. 6, 7 - 30 = Byurakan Astrophys. Obs. No. 51 (1970). In Russian. – English translation in Astrophysics, Vol. 6, No. 1.

A statistical study of 145 flare stars in the Pleiades is given. It is shown that the total number of flare stars in the Pleiades should be greater than 600. The distribution of flare stars according to the number of observed flares is well represented by the sum of two Poisson distributions with different mean frequencies. All, or almost all of the members of the Pleiades with $V \geq 13.3$ are flare stars. The mean frequency of large flares ($A > 0.^{m}6$) for the majority of stars is of the order of $4 \times 10^{-4} h^{-1}$. The total mass of the Pleiades is larger than the value determined by usual dynamical methods (400 $M_\odot$).

153.011 Infra-red photometry of a heavily reddened cluster in Ara.
J. Borgman, J. Koornneef, J. Slingerland.
Astron. Astrophys., Vol. 4, 248 - 252 (1970).

Westerlund's (1968) identification of an extended infra-red object found by Price (1968) is confirmed. At least 12 objects brighter than $K = 5.^{m}0$ are found in the cluster area, 9 of which can be positively identified as supergiants in an extremely reddened cluster; the visual absorption is at least $12.^{m}8$.

153.012 Radial velocities of stars in the region of the open

cluster NGC 752.   E. Rebeirot.
Astron. Astrophys., Vol. 4, 404 - 414 (1970).   In French.
Radial velocities and spectral types are determined for 236 stars in the region of NGC 752. From this results we have deduced for the cluster a mean radial velocity of +9 km s^{-1} and a true distance modulus of 7.^{m}75. Ten stars are indicated as new probable members. The evolved F type stars and K type giants give a mean radial velocity faintly smaller than the mean value found for unevolved stars of the main sequence.

153.013   A study of the southern open clusters: Tr 27, Tr 28, NGC 6416 and NGC 6425.
P. S. Thé, N. Stokes.
Astron. Astrophys., Vol. 5, 298 - 311 (1970).
A photometric study on the UBV-system of these southern open clusters, based on a photoelectric sequence in the open cluster NGC 6405, was undertaken photographically using plates taken with the Uppsala and Lembang Schmidt telescopes. For zeropoint corrections of the photographic photometry at least 7 stars in each cluster were measured photoelectrically. Fundamental data of the open clusters are obtained using Becker's method. They indicate that Tr 27 and Tr 28 are located at the Sagittarius spiral arm as delineated by optical spiral tracers, while NGC 6416 and NGC 6425 are interarm objects. The cluster membership of several interesting stars located in the cluster area is discussed.

153.014   Photoelectric UBV photometry of the open cluster NGC 2516.   J. Dachs.
Astron. Astrophys.,Vol. 5, 312 - 317 (1970).
70 stars in the region of the open cluster NGC 2516 have been investigated by photoelectric UBV photometry down to magnitude $V = 12^{m}$.

153.015   Hydrogen in the regions of four stellar clusters.
H. M. Tovmassian.
IAU Symposium No. 38, (see 012.013), p. 173 - 176 (1970).
The present report gives preliminary results of observations of the hydrogen line in the regions of four clusters, aiming to reveal the gas clouds associated with these clusters.

153.016   Galactic clusters and H II regions.
W. Becker, R. Fenkart.
IAU Symposium No. 38, (see 012.013), p. 205 - 208 (1970).

153.017   Photometric and spectroscopic data for some distant O and B stars.
G. Lyngå.
IAU Symposium No. 38, (see 012.013), p. 270 - 275 (1970).
The open cluster Stock 16 and some distant OB stars in Crux have been studied. Lack of absorption between spiral arms –I and –II is apparent. Negative residual radial velocities are found at distances of about 6 kpc at l^{II} = 298°.

153.018   Stellar rings and galactic structure.
J. Isserstedt.
IAU Symposium No. 38, (see 012.013), p. 287 - 289 (1970).
Stellar rings are shell type prolate ellipsoidal stellar aggregates. The most important property is the constancy of the minor diameters which allows precise geometric distance determinations up to great distances. New photoelectric and photographic UBV-observations of two stellar rings confirm the reality and the diameter constancy of these objects. All A- and F-stars of a nearby ring in Aquila exhibit an UV-excess, similar to that observed in the very young open cluster NGC 2264.

153.019   A method of the determination of selective absorption and distance modulus for open clusters, based on colour-magnitude diagram only, with evolutionary deviations taken into account.   T. A. Uranova.
Astron. Zhurn. Akad. Nauk SSSR, Vol. 47, 560 - 565 (1970). In Russian. English translation in Soviet Astron.,AJ, Vol. 14, No. 3.
For the cases without data of three-colour photometry for open clusters, a method for the determination of E_{B-V} and ρ_0, based on a V- (B – V) diagram only, is proposed. Taking into account the evolutionary deviations of cluster stars from the zero-age main sequence allows to avoid systematical errors.

153.020   Internal motions in the Pleiades.
B. F. Jones.
Astron. Journ., Vol. 75, 563 - 574 (1970).
An observational determination of the velocity dispersion in the Pleiades has been made. This has been combined with other observations to give an observational model of the cluster.

153.021   Flare stars in the Pleiades region. II.
G. Haro, E. Chavira.
Bol. Obs. Tonantzintla y Tacubaya, Vol. 5 (No. 34), 181 - 190 (1970).
The multiple exposure ultraviolet photographic material obtained on the Pleiades region during the period October 7, 1969 - March 5, 1970 comprises 104 plates with 600 different exposures taken in 150 hours of observations. The results of 19 new flare stars are summarized in a table. With the new and old data on hand the authors try to clarify and refine some aspects of the flare star problem in the Pleiades.

153.022   New flare stars in the Pleiades. (A re-examination of the Tonantzintla photographic material: 1963 - 1970).   G. Haro, G. González.
Bol. Obs. Tonantzintla y Tacubaya, Vol. 5 (No. 34), 191 - 194 (1970).
In the present work our attention has been focused mainly in the stars for which either Kraft and Greenstein or McCarthy have obtained slit spectrograms, and in a rather small sample of known very faint flare objects – namely flare stars Nos. 2, 62, 63, 101 and 102 in the Tonantzintla lists – looking for possible outbursts or flare-up repetitions. The results to be described here show, among other things, that rather conspicuous outbursts in some of the brightest Pleiades flare stars escaped detection in our previous works.

153.023   Erratum: Photometry in UBVRIJHKL of the early main sequence in the Pleiades down to G0V.
[Bol. Obs. Tonantzintla y Tacubaya, Vol. 5 (No. 32), 89 - 100 (1969), see 03.153.005].   B. Iriarte Erro.
Bol. Obs. Tonantzintla y Tacubaya, Vol. 5 (No. 34), 218 (1970).

153.024   Studies on star clusters. A new small cluster near NGC 2175 and some remarks on the latter.
P. Pişmiş.
Bol. Obs. Tonantzintla y Tacubaya, Vol. 5 (No. 34), 219 - 227 (1970).
A small cluster was discovered in the vicinity of the nebulous cluster NGC 2175. A bright patch of nebulosity encircles the small cluster which we designate as NGC 2175s. On long exposure plates the H II region of NGC 2175 has such extension that it overruns the small cluster. The distances found for NGC 2175s and NGC 2175 are 3500 and 1950 pcs respectively.

153.025   On the application of two methods of star counts to galactic star clusters.   V. V. Syrovoj.
Uch. zap. Ural'skogo un-ta, 1969, No. 70, p. 198 - 201. In Russian. – Abstr. in Referativ. Zhurn. 51. Astron. 6.51.696 (1970).

153.026 Investigation of the system of the open star clusters NGC 1513, NGC 1528, NGC 1545.
K. A. Barkhatova, V. A. Kuz'mina, L. P. Shashkina.
Uch. zap. Ural'skogo un-ta, 1969, No. 70, p. 7 - 158.
In Russian. − Abstr. in Referativ. Zhurn. 51. Astron.,
6.51.701 (1970).

153.027 On the structure of the galactic star cluster NGC 6819.   A. E. Vasilevskij.
Uch. zap. Ural'skogo un-ta, 1969, No. 70, p. 159 - 162.
In Russian. − Abstr. in Referativ. Zhurn. 51. Astron.,
6.51.706 (1970).

153.028 Investigation of absorption in the region of the open cluster NGC 7086 and determination of the cluster distance.   E. P. Polishchuk.
Astrometriya i Astrofiz., *Kiev*, No. 8, (see 003.024), p. 65 - 69 (1969).   In Russian.

153.029 Star clusters and associations.   M. Golay.
Trans. IAU, Vol. 14A, (see 003.028), 437 - 450 (1970). − Report of Commission 37.

A catalog of positions for 502 stars in the region of the Pleiades.   See Abstr. 041.003.

Studies in stellar evolution. IX. Theoretical iso-chrones for early-type clusters.   See Abstr. 065.017.

Photoelectric observations of early A stars.
See Abstr. 113.024.

Dependence of chromospheric emission upon bolometric luminosity for the Hyades.
See Abstr. 114.067.

A high-dispersion velocity curve of U Sagittarii.
See Abstr. 122.018.

The structure of the stellar field in the direction of the cluster NGC 6913. 1. Interstellar absorption.
See Abstr. 131.114.

Interstellar absorption in the direction of the star cluster NGC 6823.   See Abstr. 131.115.

A comparison of a 15.4-GHz map of M16 with the OB stars in the cluster NGC 6611.   See Abstr. 132.017.

On the value of R for NGC 2244.
See Abstr. 132.018.

On the galactic orbit of the cluster M67.
See Abstr. 151.066.

Chemical homogeneity and absolute magnitudes of members of the Hyades and Wolf 630 groups.
See Abstr. 152.002.

Supergiants in stellar associations and galactic clusters.   See Abstr. 152.016.

154   Globular Clusters

154.001   An analysis of the bright O star in the globular cluster M3.   S. E. Strom, K. M. Strom.
Astrophys. Journ., Vol. 159, 195 - 200 (1970).

Von Zeipel 1128 is shown to be an unusual late O-type member of the globular cluster M3. From an analysis of its spectrum it is concluded that $M > 0.6\,M_\odot$ and that the He/H ratio in Von Zeipel 1128 is close to that of O and B stars of population I. Despite the fact that other stars in M3 have metal deficiencies of at least a factor of 10, the measured strengths of N III and O II lines suggest much higher metal abundances in Von Zeipel 1128. Various speculations on the evolutionary history of this star are briefly presented.

154.002   The globular clusters NGC 6752 and NGC 6362.   G. Alcaino.
Astrophys. Journ., Vol. 159, 325 - 331 (1970).

A photoelectric investigation in *UBV* on the southern globular clusters NGC 6752 and NGC 6362 was carried out at the Cerro Tololo Inter-American Observatory with the 36-inch and 60-inch reflectors. Seventy-one stars were observed an average of 1.7 times in NGC 6752 to apparent magnitude $V = 15.50$, and sixty-one stars were observed an average of 1.9 times in NGC 6362 to magnitude $V = 15.73$. The distance of NGC 6752 is 5.01 kpc, and it is 7.08 kpc for NGC 6362.

154.003   Erratum: UBV photometry of globular clusters.   [Astron. Journ., Vol. 72, 70 - 81 (1967)].
S. van den Bergh.
Astron. Journ., Vol. 75, 131 (1970).

154.004   Secondary *UBV* comparison stars near some southern globular clusters. I. Stars brighter than 16th magnitude.   R. E. White.
Astron. Journ., Vol. 75, 167 - 168, 219 - 221 = Contr. Cerro Tololo Inter-American Obs. No. 101 (1970).

The results concerning photoelectrically observed *UBV* magnitudes and colors of bright to intermediately faint stars near three southern clusters (NGC 2808, 5927, and 6101) are assembled in tables. Finder charts for the stars observed in the vicinity of each cluster are also included.

154.005   Stellar evolution in globular clusters.   M. Schwarzschild.
Quarterly Journ. Roy. Astron. Soc., Vol. 11, 12 - 22 (1970). George Darwin lecture delivered in Burlington House on 1969 October 10.

154.006   The blue horizontal-branch stars of M15, M92, M13, and M3.   E. B. Newell.
Astrophys. Journ., Vol. 159, 443 - 457 (1970).

The physical characteristics of blue horizontal-branch stars in the northern globular clusters M15, M92, M13, and M3 are investigated on the basis of Sandage's (1969) *UBV* observations. These data are interpreted by means of the empirical correlations between broad-band colors and atmospheric parameters, found for the blue horizontal-branch stars of NGC 6397 and ω Cen by Newell, Rodgers, and Searle. Use is made also of Sargent's low-dispersion spectra of several stars in M15 and M92. A discussion of the results for the four northern clusters, together with the results previously obtained for NGC 6397 and ω Cen is presented.

154.007   The ellipticity of Omega Centauri determined by equidensitometry.
R. F. Sisteró, C. R. Fourcade.
Astron. Journ., Vol. 75, 34, 133 (1970).

Equidensitometry curves based on the Sabattier effect have been obtained for the globular cluster Omega Centauri. They show elliptical flattening. Measurements of these curves give excentricities from $e = 0.42$ to 0.58 at distances from the center of the cluster (semimajor axis) in the $3' < a < 8'$ range. The mean position angle of the semimajor axes is P.A. = $103 \pm 6°$.

154.008   Two Mira variables in the globular cluster NGC 6637 (M69).
R. M. Catchpole, M. W. Feast, J. W. Menzies.
Observatory, Vol. 90, 63 - 65 (1970). − Note.

154.009   The globular clusters M3, M13, M15 and M92.   K. Stępień.
Postępy Astron., Vol. 18, 222 - 226 (1970).   In Polish. Short report.

154.010   Derivation of stellar properties by comparison between theoretical models and globular cluster characteristics.   I. Iben, Jr., R. T. Rood.
Bull. American Astron. Soc., Vol. 2, 201 (1970). − Abstr. AAS.

154.011   Photometrische Beobachtungen an RR Lyrae-Sternen. III. Die RR Lyrae-Veränderlichenlücke im Farben-Helligkeitsdiagramm von Omega Centauri (NGC 5139).   E. H. Geyer, B. Szeidl.
Astron. Astrophys., Vol. 4, 40 - 52 (1970).

Mean magnitudes, mean colours and amplitudes for 45 RR Lyrae-variables in the globular cluster ω Cen are presented. These are based on light curves in two colours measured on 78 *B, V* plates taken with the ADH Baker-Schmidt camera of the Boyden Observatory.

154.012   On the evolutionary phase of Cepheids in globular clusters.   M. Schwarzschild, R. Härm.
Astrophys. Journ., Vol. 160, 341 - 344 (1970).

The beginning of the helium shell burning phase has been computed through the first flash cycle for each of nine globular-cluster stars (differing in mass and initial composition). While seven of these stars remained on the red-giant branch throughout the cycle, two stars did not. The latter passed, after each flash, through a loop in the H-R diagram reaching well into the strip occupied by population II Cepheids. It is suggested that the occurrence of these evolutionary loops might be the reason for the occurrence of Cepheids in globular clusters.

154.013   On the incidence of Cepheids in globular clusters.   G. Wallerstein.
Astrophys. Journ., Vol. 160, 345 - 347 (1970).

The available data on Cepheids in globular clusters have been assembled and discussed. The data support the ideas presented by Schwarzschild and Härm because Cepheids are found only in globular clusters with blue horizontal branches. The paucity of Cepheids in type II systems outside the Galaxy is also discussed.

154.014   A bibliography of color-magnitude diagrams for globular clusters.   R. E. White.
Astrophys. Journ., Suppl. Series, Vol. 19, 343 - 366 (1970).

A bibliography of all the published color-magnitude (C-M) diagrams for the galactic globular clusters is presented in tabular form. Certain C-M diagrams are reproduced.

154.015   Observational aspects of black holes in globular

clusters. A. A. Wyller.
Astrophys. Journ., Vol. 160, 443 - 449 (1970).

Gravitationally collapsed cores (black holes) are postulated to exist as remnants of the initial formation process at the centers of globular clusters. It is shown that this hypothesis can be tested by spectroscopic observations and observations of gravitational radiation, although both types of tests strain the limits of present-day techniques.

154.016 Erratum: The RR Lyrae gap in the colour-magnitude diagram of Omega Centauri (NGC 5139). [Astron. Astrophys., Vol. 4, 40 - 52 (1970)].
E. H. Geyer, B. Szeidl.
Astron. Astrophys., Vol. 5, 492 (1970).

154.017 Interpretation of the colour-colour diagram of M92. R. A. Bell.
Monthly Notices, Roy. Astron. Soc., Vol. 149, 179 - 196 (1970).

Sandage & Walker have shown that, at the same $B-V$, the stars in the subgiant branch of the globular cluster M92 differ systematically in $U-B$ from the stars in the asymptotic branch. A number of synthetic stellar spectra have been computed to determine the effect of line absorption upon the colours of the low-metal abundance stars. The models do satisfactorily predict the colours of both the intrinsically brighter subgiant stars and all the asymptotic branch stars.

154.018 On the question of finding the distribution of stars in poor globular clusters.
M. A. Mnatsakanyan.
Dokl. AN Arm. SSR, Vol. 49, No. 1, p. 33 - 37 (1969).
In Russian. — Abstr. in Referativ. Zhurn. 51. Astron., 4.51.761 (1970).

154.019 Studio degli ammassi M53 ed NGC 5053.
A. Blaghikh.
Mem. Soc. Astron. Italiana, Nuova Serie, Vol. 41, 189 - 203 (1970).

The spatial distributions of stars for the clusters M53 and NGC 5053 have been obtained. The diameters derived (62' and 29') are much greater than the previous estimates. The structure of the two clusters has been studied and the spatial

distribution of the stars in both cases has been compared with that of six other clusters. From the comparison of the normalized distributions it results that for all these clusters the slopes of the distribution functions are the same in the central regions, while they become more and more different in the coronal regions.

154.020 On four variable stars in the globular cluster M 2.
B. Kukarkin.
Inform. Bull. Variable Stars (I.A.U. Commission 27), Konkoly Obs., Budapest, No. 422 (1970).

154.021 On the stellar composition of halo-type globular clusters. F. D. Talbert.
Thesis, Univ. Texas, Austin. [Available from Univ. Microfilms, Ann Arbor, Mi.], 145 pp. (1969).

Low-dispersion photoelectric spectral scans of a selection of stars with low abundances of metals coupled with similar data for four halo-type globular cluster (M 3, M 5, M 13, M 92) are used to estimate the stellar content of these clusters. The wavelength range of the measurements is 3350 Å to 6100 Å; spectral purity of the observations is 40 – 60 Å. Cross-correlation of the instrumental profiles (cluster or stellar) provides a simple method of data smoothing and also aids in the determination of spectral features present in the scans. The effect of interstellar absorbing matter is taken into account.

154.022 Variations of periods of variables in Omega Centauri. H. Wilkens.
Asoc. Argentina Astron. Bol., No. 12, (see 012.019), p. 18 - 20 (1968). In Spanish. — Abstract.

Star clusters and associations.
See Abstr. 153.029.

Magnitudes and colors of globular clusters in the Fornax system. See Abstr. 158.029.

The brightest star clusters in galaxies as distance indicators. See Abstr. 158.030.

Intermediate bandpass photometry of galaxies and globular clusters. See Abstr. 158.094.

155 Structure and Evolution of the Galaxy

155.001 Lunar occultation of the galactic center region in the 6-cm formaldehyde line.
F. J. Kerr, A. Sandqvist.
Astrophys. Letters, Vol. 5, 59 - 62 (1970).

A lunar occultation showed that the +40 km/sec feature of the CH_2O profile in the direction of the galactic center region comes from the same cloud that gives rise to the +40 km/sec components of the OH and H I absorption profiles. A more detailed CH_2O absorption profile towards Sgr A, with a velocity resolution of 0.5 km/sec per channel, is also presented.

155.002 Space distribution and kinematics of local early F stars. A. E. Rydgren.
Astron. Journ., Vol. 75, 35 - 38 (1970).

The apparent local concentration of early F stars within several hundred parsecs of the sun appears to be real. It is definitely not due to a statistical fluctuation of randomly distributed stars. The theoretical calculations of other workers show that the typical age of an early F star near the main sequence is about one billion years. The galactic space motions of 96 nearby single F0–F5 stars place a maximum age of about 20 million years on the present local concentration. The reason for the apparent local concentration of early F stars in not found.

155.003 On the question of the aggregations of A type stars investigated by the Monte-Carlo method.
G. Th. Kevanishvili.
Byull. Abastumansk. Astrofiz. Obs. No. 37, p. 117 - 140 (1969). In Russian.

The distribution of about 55.000 A stars of HD Catalogue has been investigated. In a number of cases they turned out to form real condensations. This conclusion is supported by application of the method of random tests (Monte-Carlo-method) and Poisson's formula. The condensations are mostly distributed near the galactic plane. Their angular diameters vary from 2° to 6°; mean dimensions and densities are of the order of those for O and T associations. The condensations of the A stars in Cyg and Lac may be considered as expanding.

155.004 The spiral structure of our Galaxy – II.
B. J. Bok.
Sky Telescope, Vol. 39, 21 - 25 (1970).

155.005 An investigation of the galactic structure in three regions of the Milky Way in Taurus (in the direction of the anti-centre of the Galaxy). N. B. Kalandadze.
Byull. Abastumansk. Astrofiz. Obs. No. 38, p. 3 - 17 (1969). In Russian.

Interstellar absorption and distribution of absorption matter have been studied in three regions of Taurus on the basis of data obtained through two-dimensional classification and three-colour photometry of stars.

155.006 The galactic halo.
J. W. Truran, A. G. W. Cameron.
Nature, Vol. 225, 710 - 711 (1970).

Numerical studies of the chemical evolution of the galaxy are briefly described. A modified stellar birth-rate function, forming only stars of mass $M \geqslant 10 \, M_\odot$, is adopted for the early history of the galaxy. The implosions of stars of mass $M \geqslant 20 \, M_\odot$ are assumed to result in the formation of Schwarzschild singularities. The computed galactic histories which provide the best fits to the various observed abundance criteria also predict that 20 to 50 percent of the present mass of the galaxy may be in the form of black holes. An approximately spherical distribution of these objects throughout the galactic halo would not be in conflict with observations.

155.007 The nature of the far-infrared radiation of the galactic center. J. Lequeux.
Astrophys. Journ., Vol. 159, 459 - 462 (1970).

The strong far-infrared emission discovered at the galactic center by Hoffmann and Frederick is shown to be probably emission by dust particles; the large total mass of these particles suggests that the mass of the interstellar medium at the center of the Galaxy may be larger than indicated by observations of the 21-cm line.

155.008 The background brightness of the milky way near 2500 Å between $l^{II} = 72°$ and 126°.
G. C. Sudbury, M. F. Ingham.
Nature, Vol. 226, 526 - 528 (1970).

We present and discuss here some sounding rocket data derived from four scans across the galactic plane between $l^{II} = 72°$ and 126° at an effective wavelength of 2425 Å. These observations, while agreeing qualitatively with the ground-based observations, show certain details and quantitative aspects which may relate to the distribution of dust and stars and, when compared with other space observations, to the scattering and absorption properties of the interstellar dust in the ultraviolet.

155.009 Far-infrared observations of the galactic center.
H. H. Aumann, F. J. Low.
Astrophys. Journ. (*Letters*), Vol. 159, L159 - L164 (1970).

The center of our Galaxy has been observed between 40 and 350μ. The measured peak flux of $(8 \pm 3) \times 10^{-21} \, Wm^{-2} \, Hz^{-1}$ occurs at $(4.2 \pm .2) \times 10^{12} \, Hz$. If a distance of 10^4 pc is assumed, the total infrared luminosity of the galactic nucleus is $(8 \pm 3) \times 10^7 \, L_\odot$. The far-infrared diameter of the nucleus is less than 3', and its position agrees to within 6' with the position of Sgr A. Size and luminosity considerations strongly favor a nonthermal model of the galactic nucleus which consists of multiple sources.

155.010 The diffusion of Lyman Alpha through the galaxy.
T. F. Adams.
Bull. American Astron. Soc., Vol. 2, 179 - 180 (1970).
Abstr. AAS.

155.011 The distribution and kinematics of supergiant stars. R. M. Humphreys.
Bull. American Astron. Soc., Vol. 2, 200 - 201 (1970).
Abstr. AAS.

155.012 Compton scattering of galactic infrared radiation.
F. M. Ipavich, A. M. Lenchek.
Bull. American Astron. Soc., Vol. 2, 201 (1970). – Abstr. AAS.

155.013 Mechanisms in galactic evolution and spiral structure. R. H. Miller, K. H. Prendergast, W. J. Quirk.
Bull. American Astron. Soc., Vol. 2, 209 (1970). – Abstr. AAS.

155.014 The space distribution of stars at galactic latitude +45°. A. G. D. Philip, L. J. Relyea.
Bull. American Astron. Soc., Vol. 2, 214 (1970). – Abstr. AAS.

155.015 A comparison of the radial velocities of the neutral hydrogen gas and H II regions within 60 degrees of the galactic center.
P. G. Mezger, T. L. Wilson, F. F. Gardner, D. K. Milne.
Astron. Astrophys., Vol. 4, 96 - 100 (1970).

A comparison of the NRAO-MIT and NRAO-CSIRO-MIT H 109α surveys and the Parkes H I 21-cm line survey shows that the kinematics of the H I gas and the H II regions agree rather well. H II regions are predominantly found in regions of high surface brightness of the 21-cm emission of the H I gas which are probably also regions of high space density. Clusters of H II regions appear to be associated with bumps in the maximum velocity curve of the H I gas.

155.016 Fluctuations in brightness of the Milky Way and interstellar clouds. G. Peters.
Astron. Astrophys., Vol. 4, 134 - 143 = Mitt. Landessternw. Heidelberg-Königstuhl No. 159 (1970).

A statistical analysis of the fluctuations in brightness of the Milky Way is presented, using the surface photometry of Pannekoek and Koelbloed. Long-wave variations, which are produced by the large-scale star distribution, are determined and subtracted. The remaining fluctuations, due to the cloud-like structure of the interstellar extinction, are used to derive structural parameters of the interstellar dust distribution. The discussion of the mean square deviation of the brightness shows, that the contribution of the diffuse galactic light to the observed intensity is 20 ... 30% of the direct star light.

155.017 Evidence for a possible expulsion of gas from the galactic nucleus. P. C. van der Kruit.
Astron. Astrophys., Vol. 4, 462 - 481 (1970).

A survey of the high-velocity part of the 21-cm profiles of neutral hydrogen has been made in a region between $l = 350°$ and $l = 10°$ and between $b = -5°$ and $b = +5°$. The observations suggest an ejection of clouds from the nucleus in two roughly opposite directions at a large angle with the plane, which may be identical with that of the secondary ridge of continuum radiation found by Kerr and Sinclair. The total hydrogen mass involved is at least $10^6 \, M_\odot$. Similarities with the motions in the nuclei of Seyfert galaxies indicate that a number of explosive events may have caused the expulsion of these clouds.

155.018 A photometric search for distant OB stars in Norma and an investigation of the Norma cloud.
G. Schnur.
Astron. Astrophys., Vol. 5, 431 - 443 = Mitt. Landessternw. Heidelberg-Königstuhl No. 160 (1970).

About 1500 stars in Norma ($l^{II} \approx 331°$, $b^{II} \approx -2°$) have been measured by means of a photographic UBV photometry. With respect to the interpretation of the observations frequency distributions in the $(U - B, B - V)$-diagram have been calculated for different models of interstellar absorption. In the Norma cloud an increased density of A and F stars at a distance of 0.7 - 1.8 kpc, coinciding with the spiral arm - I, is found. No distant OB stars could be detected in the Norma cloud. In a region near to the galactic equator ($b^{II} = -1°0$) numerous B stars have been identified, which belong to the spiral arm - I.

155.019 Spiral arm structure and high velocity gas at intermediate latitudes. M. Kepner.
Astron. Astrophys., Vol. 5, 444 - 469 (1970).

An intermediate latitude 21-cm survey was made within the region $228° \geqq l \geqq 48°$ and $20° \geqq b \geqq 6°$ in order to investigate the structure of the spiral arms away from the galactic plane and to search for high-velocity features beyond the immediate solar neighbourhood. An approximate gaussian analysis was used to separate the profile contributions resulting from simple extensions of spiral arm structure

and those of other features. This analysis indicated that the spiral arms extend to 1 to 2 kpc above the galactic plane, with possible extensions to more than 3 kpc above the arm density maximum in a few cases. Five features were identified which are not simple continuations of spiral arm structure.

155.020 Infrared spectrum of the galactic centre.
H. Okuda, N. C. Wickramasinghe.
Nature, Vol. 226, 134 - 135 (1970).

The observed infrared spectrum of the extended IR source near the galactic centre may be accounted for by emission from a distribution of dust grains surrounding an intense ultraviolet source situated at the galactic centre.

155.021 Possible influence of the spiral galactic structure on the local distributions of residual stellar velocities.
M. Mayor.
Astron. Astrophys., Vol. 6, 60 - 66 (1970).

The distributions of residual velocities for nearby stars moving in the galactic disk seem to show, independently of their age, a deviation of the vertex for orbits of small eccentricity. We interpret this deviation not in terms of initial conditions but as a local property of a stellar velocity distribution perturbed by a spiral density wave of the type considered by Lin and Shu. The description of the spiral arms by means of a spiral density wave can also qualitatively account for the decrease of the vertex deviation with the dispersion of the velocities. We explain simultaneously the difference found between the calculated and observed asymmetrical drift for the objects with little velocity dispersions.

155.022 Mariner 5 measurements of ultraviolet emission from the galaxy. C. A. Barth.
IAU Symposium No. 36, (see 012.014), p. 334 - 340 (1970).

The Mariner 5 ultraviolet measurements obtained while the spacecraft was in interplanetary flight are interpreted as Lyman-α radiation. The Mariner 5 measurements show a symmetry with respect to the galactic equator which suggests that the major source of the observed emission is the diffuse galactic radiation. Some of the Mariner 5 measurements may be attributed to Lyman-α emission from an H II region since the field of observations scanned the edge of the Gum nebula.

155.023 Survey of spiral structure problems.
J. H. Oort.
IAU Symposium No. 38, (see 012.013), p. 1 - 5 (1970).

155.024 Spiral structure of our Galaxy and of other galaxies.
B. A. Vorontsov-Velyaminov.
IAU Symposium No. 38, (see 012.013), p. 15 - 17 (1970).

155.025 Statistics of spiral patterns and comparison of our Galaxy with other galaxies.
G. de Vaucouleurs.
IAU Symposium No. 38, (see 012.013), p. 18 - 25 (1970).

Apparent relative frequencies of various types of spirals are given for 900 spirals with the best revised Hubble classifications. Mean diameters of inner ring structures vary from 2.1 kpc in ordinary spirals (SA) to 4.4 kpc in barred spirals (SB) with a total range of 10 to 1 within each type. The probable morphological type of our Galaxy is estimated.

155.026 Spiral structure of neutral hydrogen in our Galaxy.
F. J. Kerr.
IAU Symposium No. 38, (see 012.013), p. 95 - 106 (1970).

This paper discusses the evidence on the hydrogen spiral structure which is available from 21-cm observations. The problems involved in deriving a hydrogen map are discussed, and one interpretation of the spiral pattern is presented. Some of the major characteristics of the H I in spiral arms are discussed, and a comparison is made between the kinematics of H I and H II.

155.027 The distribution of H II regions.   P. G. Mezger.
IAU Symposium No. 38, (see 012.013), p. 107 - 121 (1970).

Review paper: Spiral arms and star formation; Star formation and H II regions; Optical observations of H II regions; Radio continuum surveys; Radio recombination line surveys; Spiral structure and radial distribution of H II regions and neutral gas.

155.028 Spiral structure of the Galaxy derived from the Hat Creek survey of neutral hydrogen.   H. Weaver.
IAU Symposium No. 38, (see 012.013), p. 126 - 139 (1970).

The extensive Hat Creek survey of neutral hydrogen combined with southern observations provides the basis for a new discussion of the spiral structure of the Galaxy. The purpose of this investigation is to provide a general picture of the Galaxy. It is found that the pitch of the spiral arms is approximately $12°.5$ and that there are many spurs and interarm features as we observe in external galaxies.

155.029 Matter far from the galactic plane associated with spiral arms.   J. H. Oort.
IAU Symposium No. 38, (see 012.013), p. 142 - 146 (1970).

Neutral hydrogen concentrations were studied in the region between $48°$ and $200°$ longitude (new) and $+6°$ and $+20°$ latitude. Gas associated with individual arms beyond the sun has been found up to distances from 1 to 2 kpc from the galactic plane with an average density between 1 and 2% of that in the arm centres. The gas with the highest negative velocities shows a different behaviour.

155.030 On a possible corrugation of the galactic plane.
C. M. Varsavsky, R. J. Quiroga.
IAU Symposium No. 38, (see 012.013), p. 147 - 150 (1970).

We have studied the rotation curve of the Galaxy at different heights below and above the equator. In the course of this work we noticed that the maximum brightness temperature of hydrogen oscillates around the galactic plane following a fairly sinusoidal pattern.

155.031 A study of spiral structure for $270° \leqslant l^{II} \leqslant 310°$.
S. L. Garzoli.
IAU Symposium No. 38, (see 012.013), p. 151 - 153 (1970).

155.032 A peculiar neutral hydrogen concentration at $l^{II} = 280°, b^{II} = -18°$.
E. Bajaja, F. R. Colomb.
IAU Symposium No. 38, (see 012.013), p. 154 - 156 (1970).

155.033 Local stellar distribution and galactic spiral structure.   S. W. McCuskey.
IAU Symposium No. 38, (see 012.013), p. 189 - 198 (1970).

Aside from the well-known spiral arm tracers such as the OB associations, young galactic clusters, WR stars and possibly the long-period classical cepheids, the more common stars in the neighborhood of the sun within 2 kpc show little or no relationship to the local spiral structure of the Galaxy.

155.034 Remarks on local structure and kinematics.
A. Blaauw.
IAU Symposium No. 38, (see 012.013), p. 199 - 204 (1970).

Attention is drawn to a few aspects of the state of motion of the local population which may become of importance for the study of local spiral structure. Uncertainties in the present knowledge of the local standard of rest are discussed, (a) with regard to the possible outward or inward motion with respect to the galactic centre, and (b) with regard to the component in the direction of circular motion.

155.035 A new interpretation of the galactic structure from H II regions.   G. Courtès, Y. P. Georgelin, Y. M. Georgelin, G. Monnet.
IAU Symposium No. 38, (see 012.013), p. 209 - 212 (1970).

From 6000 optical radial velocities of H II regions a new spiral structure (4 arms of pitch angle $20°$) is found. The radial velocities of the observed H II regions are the same with the velocities of the H I regions. The kinematics of H II regions is similar to that of Cepheids and B stars.

155.036 The distribution of H II regions in the local spiral arm in the direction of Cygnus.
H. R. Dickel, H. J. Wendker, J. H. Bieritz.
IAU Symposium No. 38, (see 012.013), p. 213 - 218 (1970).

The distribution and physical properties of the gaseous nebulae in our local spiral arm should help to reveal the structure within the arm. We will present possible evidence for a symmetry about the axis of the arm and also some preliminary results concerning the orientation and geometry of the local region of the Orion arm as deduced from our study of the nebulae in the Cygnus X complex.

155.037 Reflection nebulae and spiral structure.
R. Racine, S. van den Bergh.
IAU Symposium No. 38, (see 012.013), p. 219 - 221 (1970).

It is shown that stars embedded in reflection nebulosity may be used as spiral-arm tracers.

155.038 The galactic structure and the appearance of the Milky Way.   E. D. Pavlovskaya, A. S. Sharov.
IAU Symposium No. 38, (see 012.013), p. 222 - 224 (1970).

We consider the surface brightness of the Milky Way in detail using different models of the galactic structure. It is assumed that the model of our Galaxy consists of the disk with a radius of 15 kpc, and of several arms in the form of logarithmic spirals.

155.039 The study of the Milky Way integrated spectrum and the spiral structure of the Galaxy.
E. B. Kostjakova.
IAU Symposium No. 38, (see 012.013), p. 225 - 227 (1970).

155.040 Space distribution of interstellar dust in connection with the galactic spiral structure.
T. A. Uranova.
IAU Symposium No. 38, (see 012.013), p. 228 - 231 (1970).

A maximum of dust density along the inner edge of the Cygnus arm is found. It seems that the same happens along the inner edges of the Perseus and the Sagittarius arms.

155.041 A comparison of the density gradients of main sequence stars obtained by two different methods in different galactic latitudes.   W. Becker, R. Fenkart.
IAU Symposium No. 38, (see 012.013), p. 232 - 235 (1970).

The aim of the investigations in the galactic disc is to study stellar clouds and their content of stars of different luminosity, based on the finding that in other galaxies stellar clouds are closely related with spiral structure.

155.042 The galactic distribution of young cepheids.
G. A. Tammann.
IAU Symposium No. 38, (see 012.013), p. 236 - 245 (1970).

The distribution of long-period population I cepheids is studied. The distances of cepheids are determined from a revised period-luminosity-colour relation and from colour excesses, for the determination of which a new, purely photometric method is given. The resulting distribution of cepheids shows a good correlation with the spiral arms, as traced by young clusters.

155.043 A progress report on the Carina spiral feature.

B. J. Bok, A. A. Hine, E. W. Miller.
IAU Symposium No. 38, (see 012.013), p. 246 - 261 (1970).

We have undertaken a study in Carina of the distribution of OB stars, emission nebulae, H I and cosmic dust as a function of galactic longitude. Such a study should enable us to make cross cuts through the spiral feature at various distances and thus determine where the OB stars, emission nebulae, H I and dust are located. This paper is a progress report on our work to date.

155.044 The space distribution of the OB-type stars in Carina.   J. A. Graham.
IAU Symposium No. 38, (see 012.013), p. 262 - 264 (1970).

Distances have been determined for 436 OB-type stars in Carina. A sharp outside edge is found to the OB star distribution with respect to the center of the Galaxy. The layer of OB stars follows the galactic plane to distances of the order of 3 kpc, but at distances greater than 4 kpc it appears to bend away from the plane to negative latitudes by 2° or 3° out to distances of the order of 10 kpc.

155.045 A study of O and B stars in Vela, along the galactic equator.   A. G. Velghe.
IAU Symposium No. 38, (see 012.013), p. 278 - 280 (1970).

155.046 Progress report of the current research on the galactic structure in Vela.
J. Denoyelle.
IAU Symposium No. 38, (see 012.013), p. 281 - 282 (1970).

155.047 Neutral hydrogen in the Sagittarius and Scutum spiral arms.   W. B. Burton, W. W. Shane.
IAU Symposium No. 38, (see 012.013), p. 397 - 414 (1970).

Observations of the neutral hydrogen in the first quadrant of galactic longitude have been analysed. The existence of large-scale streaming motions such as the streaming associated with the Sagittarius arm makes interpretation of the observations in terms of circular galactic rotation unsatisfactory. It is shown that application of the density-wave theory formulated by Lin et al. (1969) leads to a more satisfactory interpretation. Using kinematic models based on this theory the distribution and motion of the neutral hydrogen are studied.

155.048 Deviation of the vertex of the velocity ellipse of young stars and its connection with spiral structure.
R. Woolley.
IAU Symposium No. 38, (see 012.013), p. 423 - 432 (1970).

The present note examines data taken from the velocities of nearby stars, and uses the classification of the late type main sequence stars put forward by Wilson, based on observations of the reversals in the Ca⁺ H and K lines. From this material it is concluded that the vertex deviation is confined to young stars. The paper goes on to show that the deviation of the vertex can be explained by supposing that the stars were formed comparatively recently in a thin strip of the galaxy more or less at right angles to the direction of the galactic centre — in fact, in something like a spiral arm.

155.049 Possible influence of the galactic spiral structure on the local distributions of stellar residual velocities.   M. Mayor.
IAU Symposium No. 38, (see 012.013), p. 433 - 434 (1970).

155.050 On the circulation of gas near the galactic center.
E. A. Spiegel.
IAU Symposium No. 38, (see 012.013), p. 441 - 443 (1970).

A model of a quasi-steady circulation of gas near the galactic center is considered to explain the outward motion of the 3-kpc arm. A hydromagnetic wind from the galactic nucleus reaches the 3-pkc arm, where a shock is formed; then the gas moves out of the plane and eventually returns to the nucleus. The secular behavior of the model is discussed.

155.051 On a corollary to the magnetic dipole theory of the origin of spiral structure.   P. Pişmiş.
IAU Symposium No. 38, (see 012.013), p. 452 - 454 (1970).

Some observable features of our galaxy are explained in terms of a theory of the origin of spiral structure, whose basic feature is a gradually contracting gaseous subsystem with a magnetic dipole of which the axis is very close to the galactic plane.

155.052 A finding list of late-type stars in regions of intermediate galactic latitude.
A. R. Upgren, R. T. Staron.
Astrophys. Journ., Suppl. Series, Vol. 19, 367 - 386 (1970).

This paper continues a survey of four regions of intermediate galactic latitude. A catalog is given of all stars later than spectral class F2; the earlier stars have already been published. The limiting magnitude is about 11.5. Results are presented in the form of star counts. The importance of the giant stars in the catalog is discussed with respect to the problem of the law of galactic force and the inclination to the galactic plane of layers of equal stellar density.

155.053 Lunar occultations of the galactic center region in the 21-cm line of neutral hydrogen.
S. H. Knowles, W. T. Sullivan, III.
Astrophys. Letters, Vol. 6, 21 - 26 (1970).

We have observed five lunar occultations of the galactic center region in the 21-cm neutral hydrogen line. No significant small angular structure is found in either the narrow feature at + 6 km/sec or the general absorption extending from + 15 to - 15 km/sec.

155.054 The space distribution and kinematics of supergiants.   R. M. Humphreys.
Astron. Journ., Vol. 75, 602 - 623 (1970).

The distribution and kinematics of the supergiants of all spectral types are investigated with special emphasis on the correlation of these young stars and the interstellar gas. The stars used for this study are included as a catalogue of supergiants.

155.055 Extension of the Sagittarius-Carina arm from three distant H II regions.
Y. P. Georgelin, Y. M. Georgelin.
Astron. Astrophys., Vol. 7, 133 - 140 (1970).   In French.

A 4°5 × 4°5 region in Carina has been attentively studied in April 1969 at La Silla Observatory ESO with Pérot-Fabry rings. At l^{II} = 290°, radial velocities of 9 H II regions have been obtained. The kinematical distances of these nebulae are spread from 3 to 9 kpc in good agreement with the spectrophotometric distances of exciting stars. This result confirms that at this longitude we are looking tangentially along a spiral arm. With these new results a more complete drawing of the Sagittarius-Carina spiral arm can be made.

155.056 Distribution of diffuse matter in the northern part of the Milky Way.   E. P. Polishchuk.
Astrometriya i Astrofiz., *Kiev*, No. 8, (see 003.024), p. 69 - 72 (1969).   In Russian.

155.057 Zum Problem der Heliumhäufigkeit im Sternsystem.
K.-H. Schmidt.
Sterne, 46. Jahrgang, p. 102 - 106 (1970).

155.058 On the accuracy of determination of the parameters of the law of galactic rotation.   A. Ja. Filin.
Byull. Inst. Astrofiz., *Dushanbe*, No. 50, p. 22 - 29 (1968).
In Russian.

The influence of both random and systematic errors on the parameters of the galactic rotation law is considered. Systematic errors in the interstellar absorption and in the adopted values of R_0 and V_0 are examined.

155.059 Determination of the kinematic parameters of B stars. A. Ja. Filin.
Byull. Inst. Astrofiz., *Dushanbe*, No. 50, p. 30 - 37 (1968). In Russian.

Components of parallactic motion, Oort's constants A and B, parameters of the velocity ellipsoid and other kinematic constants have been determined using both radial and tangential velocities of B0 – B5 stars.

155.060 Structure and dynamics of the galactic system. G. Contopoulos, T. Elvius.
Trans. IAU, Vol. 14A, (see 003.028), 357 - 385 (1970). Report of Commission 33.

155.061 On the upper limit of relativistic stars in the Galaxy. O. H. Guseinov, H. I. Novruzova.
Astron. Tsirk., No. 560, p. 5 - 7 (1970). In Russian.

155.062 On the distribution of single stars and of the main components of multiple systems according to their masses.
O. H. Guseinov, H. I. Novruzova, L. H. Yesojan.
Astron. Tsirk., No. 561, p. 6 - 8 (1970). In Russian.

155.063 Statement of the problem of studying the smoothed structure of the Galaxy and the resulting tasks of investigating the galactovertical column. H. Eelsalu.
Tartu Astron. Obs. Teated, No. 24, 15 pp. (1969).

155.064 The nature of the far infrared radiation of the galactic center. J. Lequeux.
Observations Owens Valley Radio Obs., 1969, No. 7, 11 pp.

A model is proposed for the far-infrared emission source (100 microns) found at galactic centre. Thermal radiation from dust is proposed. – *WMG*

155.065 Optical and radio evidence of large-scale peculiar motions in the Cas–Per arm.
J. J. Rickard.
Thesis, Univ. Maryland, Cantonsville. [Available from Univ. Microfilms, Ann Arbor, Mi.], 79 pp. (1968).

Recent optical data of young open clusters, H II regions, O-associations, interstellar extinction, and interstellar absorption lines together with new high-resolution 21-cm line hydrogen maps of the Cas–Per spiral arm indicate that the interstellar medium in a large section of the arm is moving with peculiar radial velocities of the order of - 20 to - 30 km/sec. A distance estimate based upon the interstellar absorption lines, interstellar extinction and H II regions indicates that the gas, dust, and ionized elements are situated 1.5 to 2.5 kpc from the sun.

155.066 Expansion of the Milky Way and the interpretation of Weber's gravitation signals. P. Jordan.
Zeitschr. Physik, Vol. 233, 84 - 88 (1970). In German.

The expansion of the Galaxy which seemed to be detected by Kerr can probably not be explained as resulting from a rapid diminution of the central mass of the Galaxy according to the hypothesis put forward by Field, Rees, Sciama (1969). It follows as a consequence from the scalar-tensor theory of gravitation. The events in the nucleus of the Galaxy, triggering the Weber signals, must probably be collisions between neutron stars and normal stars.

155.067 Infrared observation of the galactic center. E. E. Becklin.

Thesis, California Inst. Technol., Pasadena. [Available from Univ. Microfilms, Ann Arbor, Mi.], 126 pp. (1969).

Infrared radiation from the nucleus of the Galaxy has been detected at effective wavelengths of 1.65, 2.2, 3.4 and 4.8 μ with angular resolutions from 0.04 to 1.8 arc min. Observations have been made over an area of 1 square degree surrounding the dynamical center of the Galaxy. A comparison of the infrared and radio observations shows that the dominant infrared source and the radio source Sagittarius A have the same coordinates and similar sizes. A comparison is also made between the infrared radiation from the galactic center and that from the nucleus of M 31 which shows agreement in both the apparent structure and infrared luminosity of two nuclei.

155.068 Structure and dynamics of the galactic system. G. Contopoulos.
Contr. Astron. Dep. Univ. Thessaloniki, No. 53, 117 pp. (1970). – Report.

155.069 The Milky Way in Carina–Centaurus. A. Feinstein.
Asoc. Argentina Astron. Bol., No. 12, (see 012.019), p. 14 - 15 (1968). – Abstract.

155.070 Some problems associated with large scale galactic structure. F. Kerr.
Asoc. Argentina Astron. Bol., No. 14 (see 012.020), p. 12 - 13 (1968). – Abstract.

155.071 Studies of galactic arms at negative latitudes. D. Goniadzki, A. Jech.
Asoc. Argentina Astron. Bol., No. 14 (see 012.020), p. 38 (1968). In Spanish. – Abstract.

155.072 Observations of lunar occultations of the galactic center region in the OH and hydrogen lines.
F. Kerr.
Asoc. Argentina Astron. Bol., No. 14 (see 012.020), p. 42 (1968). – Abstract.

155.073 Galactic structure at low latitudes. E. R. Vieyra.
Asoc. Argentina Astron. Bol., No. 14 (see 012.020), p. 51 - 52 (1968). In Spanish. – Abstract.

155.074 Peculiarities in the rotation curve of the galaxy. R. A. Quiroga, C. M. Varsavsky.
Asoc. Argentina Astron. Bol., No. 14 (see 012.020), p. 64 - 65 (1968). In Spanish. – Abstract.

155.075 The high-velocity hydrogen clouds considered as satellites of the galaxy. F. Kerr.
Asoc. Argentina Astron. Bol., No. 14 (see 012.020), p. 67 (1968). – Abstract.

155.076 Interferometric observations of the southern Milky Way. G. J. Carranza, H. Dottori.
Asoc. Argentina Astron. Bol., No. 14 (see 012.020), p. 101 - 102 (1968). In Spanish. – Abstract.

155.077 Structures in low galactic latitudes. D. Goniadzki, A. E. Jech.
Asoc. Argentina Astron. Bol., No. 14 (see 012.020), p. 103 (1968). In Spanish. – Abstract.

155.078 Die Spiralgestalt der Milchstraße. W. Becker.
Bild Wiss., Vol. 6, (No. 10), 922 - 929 (1969).

A catalogue of early-type stars whose spectra have

shown emission lines.   See Abstr. 041.018.

Calcium emission intensities as indicators of stellar age.   See Abstr. 114.077.

Wolf-Rayet stars, ring-type H II regions, and spiral structure.   See Abstr. 114.107.

M supergiants in the Perseus arm.
See Abstr. 114.120.

Ultrashort-period variables and the masses of blue stragglers in the old disk population.
See Abstr. 122.048.

The frequency of supernovae in our Galaxy, estimated from supernova remnants detected at 178 MHz.
See Abstr. 125.011.

Formaldehyde absorption in the galaxy.
See Abstr. 131.004.

A study of interstellar matter in the Cassiopeia-Perseus region. II. Structure of interstellar neutral hydrogen gas in the Perseus arm.   See Abstr. 131.063.

Polarization of southern OB-stars.
See Abstr. 131.093.

Radial velocities and distances of galactic H II regions.   See Abstr. 131.094.

Pulsar distances, spiral structure and the interstellar medium.   See Abstr. 141.144.

Origin of cosmic rays. II. The halo problem. Galactic models.   See Abstr. 143.016.

On cosmic electrons in the galaxy.
See Abstr. 143.020.

Interpretation of large-scale spiral structure.
See Abstr. 151.047.

Nonlinear waves of stellar density and the spiral structure of the Galaxy.   See Abstr. 151.057.

A study of the southern open clusters: Tr 27, Tr 28, NGC 6416 and NGC 6425.   See Abstr. 153.013.

Stellar rings and galactic structure.
See Abstr. 153.018.

Contribution to the study of the distribution of neutral hydrogen in the region $230° \leqslant l^{II} \leqslant 280°$.
See Abstr. 157.017.

A comparison of dynamical models of the Andromeda nebula and the Galaxy.   See Abstr. 158.076.

Les galaxies.   See Abstr. 158.087.

Seeing the Magellanic Clouds and the Milky Way as a physical tripletgalaxy.   See Abstr. 159.006.

Large Magellanic Cloud. List of LMC members and list of galactic stars.   See Abstr. 159.015.

The ultraviolet background (intergalactic gas, the galaxy, and subcosmic rays).   See Abstr. 161.008.

156   Galactic Magnetic Field

156.001   **A measurement of the galactic magnetic field using the pulsar PSR 0833—45.**
R. D. Ekers, J. Lequeux, A. T. Moffet, G. A. Seielstad.
California Inst. Technology, Pasadena, Report 1969—5, 14 pp. (1969).
  The radiation from PSR 0833—45 is strongly linearly polarized. Measurements at 21, 18 and 12 cm show a rotation measure of +33 ± 5 radian/sq m. Combined with the published value of the dispersion measure this yields a mean value of

0.73 ± 0.13 microgauss for the longitudinal component of the galactic magnetic field along the inner face of the Orion arm.

156.002   **Magnetic fields and spiral structure.**
L. Woltjer.
IAU Symposium No. 38, (see 012.013), p. 439 - 440 (1970).

Magnetohydrodynamical models of helical magnetic fields in spiral arms.   See Abstr. 151.049.

157 Galactic Radio Radiation

157.001 A comparison of radio recombination-line results and continuum spectral indices for galactic sources.
T. L. Wilson, W. Altenhoff.
Astrophys. Letters, Vol. 5, 47 - 51 (1970).

From a comparison of the NRAO–MIT H 109α-line survey and the recent Harvard–NRAO continuum surveys, we find that all sources which exhibit radio recombination lines have a thermal continuum spectrum. Most sources with no radio recombination line emission have steep continuum spectra, suggesting they are galactic supernova remnants; however, a small number have a flat spectrum, possibly suggesting that they are objects similar to the Crab nebula. Within the peak flux and angular resolution limits of the surveys, we find that at least one quarter of the sources investigated are non-thermal.

157.002 An 11-cm survey of H II region located in the direction of the galactic anticenter.
E. Churchwell, M. Felli.
Astron. Journ., Vol. 75, 69 - 70 (1970).

Thirty-five of the HII regions listed in Sharpless' catalogue (1959) lying between $120°$ and $230°$ of galactic longitude were observed at a wavelength of 11.1 cm. Of these, only 14 were definitely detected. The peak positions, half-power widths, peak antenna temperatures, and flux densities are given for the detected sources.

157.003 Measurement of 21-cm line calibration regions.
A. A. Penzias, R. W. Wilson, P. J. Encrenaz.
Astron. Journ., Vol. 75, 141 - 143 (1970).

The integrated 21-cm line radiation from four calibration regions was measured. From these measurements we conclude that the temperature scale of the corrected Maryland–Green Bank survey is 2% low.

157.004 The galactic continuum spurs and neutral hydrogen.
E. M. Berkhuijsen, C. G. T. Haslam, C. J. Salter.
Nature, Vol. 225, 364 - 365 (1970).

Low velocity HI with large velocity dispersion has been found situated on the outer gradients of the North Polar Spur in the region $l^{II} = 10°$ to $50°$, $b^{II} = +10°$ to $+50°$ lending new support to a supernova remnant origin of this spur. The mass in the HI shell must be of order $(5 \pm 3) \times 10^4\ M_\odot$. The runaway star ζ Oph has a proper motion which could have taken it close to the spur center some 3×10^6 yr ago.

157.005 Measurements of absolute sky brightness temperatures at 320 and 707 MHz.
J. V. Wall, T. Y. Chu, J. L. Yen.
Australian Journ. Phys., Vol. 23, 45 - 57 (1970).

Measurements of absolute sky brightness temperature have been carried out over limited regions of the sky at 320 and 707 MHz. The results indicate a change in spectrum in this frequency range consistent with addition to the galactic nonthermal radiation of isotropic radiation having a thermal spectrum and a brightness temperature of $3°K$. A power law spectral index of -0.45 ± 0.15 is obtained for the galactic nonthermal emission.

157.006 A 408 MHz survey of the galactic anticentre region.
C. G. T. Haslam, M. J. S. Quigley, C. J. Salter.
Monthly Notices, Roy. Astron. Soc., Vol. 147, 405 - 441 (1970).

The 250-foot aperture Mark I radio telescope at Jodrell Bank has been used to make a 408 MHz continuum survey of the galactic anticentre region. The telescope resolution

was 45 arc minutes. The contours are presented at intervals of 2 °K in beam temperature. The maps are given both in equatorial coordinates and, for the strip $-10° < b^{II} < +10°$, in new galactic coordinates.

157.007 Detection of 166α recombination-line radiation in the direction of the galactic center.
K. W. Riegel, S. D. Kilston.
Astrophys. Journ. (*Letters*), Vol. 159, L155 - L157 (1970).

166α recombination-line emission from the direction of the galactic center, Sgr A, has been detected with the 140-foot telescope of the National Radio Astronomy Observatory. The peak antenna temperature was 0.14°K. If the line is due to hydrogen, the radial velocity is 6.6 km sec^{-1} with respect to the local standard of rest, and the full width at half-maximum power is 28 km sec^{-1}. The emission line may originate in the continuum source or in the interstellar medium between the sun and the galactic center.

157.008 The spectrum of the continuum radio emission from the local spiral arm.
J. K. Alexander, L. W. Brown, T. A. Clark, R. G. Stone, R. R. Weber.
Bull. American Astron. Soc., Vol. 2, 180 (1970). – Abstr. AAS.

157.009 Computer display of 21-cm observations of galactic hydrogen. R. H. Cohen, M. L. Meeks.
Bull. American Astron. Soc., Vol. 2, 188 (1970). – Abstr. AAS.

157.010 The distribution of nonthermal radio emission regions in the Galaxy. R. M. Price.
Bull. American Astron. Soc., Vol. 2, 215 (1970). – Abstr. AAS.

157.011 A survey of H 109 α recombination line emission in galactic H II regions of the northern sky.
E. C. Reifenstein III, T. L. Wilson, B. F. Burke, P. G. Mezger, W. J. Altenhoff.
Astron. Astrophys., Vol. 4, 357 - 377 (1970).

A survey of hydrogen 109 α recombination lines was carried out during 1967 using the 140-foot telescope of the National Radio Astronomy Observatory. One hundred and twenty galactic sources were observed, and hydrogen recombination line emission was detected in a total of 82 of the sources in the survey. Recombination lines corresponding to heavier elements were measured for six of the stronger sources. The helium and hydrogen profiles for M17 and Orion A indicate a number density ratio of helium to hydrogen of 0.08.

157.012 Surveys of the galactic plane at 1.414, 2.695 and 5.000 GHz. W. J. Altenhoff, D. Downes, L. Goad, A. Maxwell, R. Rinehart.
Astron. Astrophys., Suppl. Series, Vol. 1, 319 - 355 (1970).

New surveys of the galactic plane, covering galactic longitudes $l^{II} = 335°$ to $75°$ and latitudes $b^{II} = -4°$ to $+4°$, have been made in the radio continuum at 1.414, 2.695 and 5.000 GHz. The antennas used for the surveys had diameters of 300, 140, and 85 ft, respectively, and the antenna half-power beamwidths in each case were of the order of 11'. The data are presented in the form of contour maps, together with a tabulation of 356 radio sources.

157.013 A low latitude galactic survey from $l^{II} = 6°$ to $26°$ at 2.7 GHz.

W. M. Goss, G. A. Day.
Australian Journ. Phys., Astrophys. Suppl., No. 13, p. 3 - 9 (1970).

The galactic plane has been observed from $l^{II} = 6°$ to 26° with an 8' arc beamwidth at 2.7 GHz. The latitude range covered was ±2°. Positions were determined for 103 sources and flux densities for 95 of these. A contour map for the region is presented. The well-known radio sources associated with M8, W28, M17, and M16 are contained in this survey.

**157.014   A low latitude galactic survey from $l^{II} = 37°$ to 47°
at 2.7 GHz.   G. A. Day, W. G. Warne,
D. J. Cooke.**
Australian Journ. Phys., Astrophys. Suppl., No. 13, p. 11 - 16 (1970).

The results of a survey of the galactic plane at 2.7 GHz from longitudes 37° to 47°, latitudes ±2° are presented as a contour map and a source list giving the positions for 53 radio sources. The angular resolution is 8'.2 arc. A computer-drawn ruled-surface picture of the area is shown. The well-known radio sources associated with W47, W49, and W50 are contained in this section.

**157.015   The distribution of hydrogen in a region in Taurus.
II. High-resolution observations.**
S. L. Garzoli, C. M. Varsavsky.
Astrophys. Journ., Vol. 160, 75 - 82 (1970).

Over 10000 profiles of the 21-cm line of atomic hydrogen were measured in a region of 150 square degrees centered at $\alpha = 4^h 30^m$, $\delta = 27°$. The variation of the number density of hydrogen atoms was compared with measurements of optical absorption made by McCuskey. The general conclusion reached is that the number of hydrogen atoms remains approximately constant with increasing absorption, with one set of optical absorption data giving a decrease of 0.16×10^{21} atoms cm^{-2} mag^{-1} while another set gives an increase of 0.04×10^{21} atoms cm^{-2} mag^{-1}

157.016   A motion picture film of galactic 21-cm line emission.   G. Westerhout.
IAU Symposium No. 38, (see 012.013), p. 122 - 125 (1970).

**157.017   Contribution to the study of the distribution of
neutral hydrogen in the region $230° \leqslant l^{II} \leqslant 280°$.**
D. Goniadzki, A. Jech.
IAU Symposium No. 38, (see 012.013), p. 157 - 163 (1970).

We study the distribution of the local hydrogen and that in the Orion, intermediate and Perseus arms. We find a new structure that starts at $l^{II} = 265°$. We also study the concentrations which lie far below the plane; some of them seem to be related to Lindblad's G arm.

**157.018   Radial velocities of neutral hydrogen in the anti-
center region of the Galaxy.   L. Velden.**
IAU Symposium No. 38, (see 012.013), p. 164 - 168 (1970).

An observational material of 21-cm H I emission-line profiles is investigated by a statistical method to derive the kinematical properties of the interstellar gas in the region of the galactic anticenter. A description of the method used as well as the results obtained, concerning deviations from a circular rotation, are given.

**157.019   Radio observations of some details in the H I local
spiral arm.   N. V. Bystrova, J. V. Gossachinsky,
T. M. Egorova, V. M. Rozanov, N. F. Ryzhkov.**
IAU Symposium No. 38, (see 012.013), p. 169 - 172 (1970).

**157.020   Theoretical 21-cm line profiles: Comparison with
observations.   C. Yuan.**
IAU Symposium No. 38, (see 012.013), p. 391 - 396 (1970).

In order to make a direct comparison with observations

of the 21-cm line of neutral hydrogen, theoretical profiles based on the ideas of the density-wave theory are constructed for a modified Schmidt model of the Galaxy and its theoretical spiral pattern. The comparison has covered galactic longitudes $l^{II} = 30° - 330°$ with 10° intervals in the galactic plane. Good agreement is found in most of the above directions.

**157.021   Low frequency cosmic noise observations of the
constitution of the local system.**
J. K. Alexander, L. W. Brown, T. A. Clark, R. G. Stone.
Astron. Astrophys., Vol. 6, 476 - 480 (1970).

Based on measurements of the low frequency continuum radiation of the Galaxy, estimates have been obtained for the gross distribution of thermal electrons, the synchrotron radiation emissivity, and the flux and spectrum of low energy cosmic ray electrons for the interstellar medium in the local system.

**157.022   On the problem of the existence of the galactic
radio halo.   G. L. Abramjan, E. A. Bene-
diktov, G. G. Getmantsev, V. A. Zinichev.**
Astron. Zhurn. Akad. Nauk SSSR, Vol. 47, 487 - 492 (1970).
In Russian.   English translation in Soviet Astron., AJ, Vol. 14, No. 3.

The experimental data concerning the intensity distribution of cosmic radio emission over the sky, within ultrashort and short wavelengths, may be satisfactorily explained by the simplest quasi-homogeneous models only in the case, when, together with the radiating disk, there will be present in the Galaxy a relatively large "radio halo".

**157.023   On the background radio emission in the interval
of galactic longitudes $l^{II} = 20°.8 - 32°.8$.**
V. I. Ariskin.
Astron. Zhurn. Akad. Nauk SSSR, Vol. 47, 493 - 498 (1970).
In Russian.   English translation in Soviet Astron. AJ, Vol. 14, No. 3.

Observations of the continuous radio emission of the Galaxy carried out with a 22-m radiotelescope of the Physical Institute of the USSR Academy of Sciences in the interval of longitudes $l^{II} = 20°.8 - 32°.8$ at 21.1 cm, allow to ascertain that the background radio emission does not only consist of a population I but of two components: population I and disk population.

**157.024   An extension of the 408 MHz southern Milky Way
survey and assignment of an absolute baselevel.**
R. M. Price.
Australian Journ. Phys., Vol. 23, 227 - 228 (1970). − Short communication.

**157.025   The galactic radio spectrum: Observations at
610 MHz.   T. F. Howell.**
Astrophys. Letters, Vol. 6, 45 - 48 (1970).

The galactic radio radiation at 610 MHz has been observed with an antenna having the same reception pattern as that used for earlier observations at lower frequencies. A comparison of the results confirms the steepening of the galactic spectrum at frequencies above about 200 MHz but suggests that the change of slope is more gradual than was previously thought.

157.026   Galactic radioastronomy.   J. Olmr.
Říše hvězd, Vol. 51, 70 - 72 (1970).   In Czech.

157.027   Radio astronomy.   J. P. Wild.
Trans. IAU, Vol. 14A, (see 003.028), 455 - 479 (1970). − Report of Commission 40. − Solar radio emission, (*S. F. Smerd*); Continuum radiation from the Galaxy, (*G. Westerhout*); Line radiation in the Galaxy, (*R. X. McGee*); Pulsars, (*S. P. Maran*); Extragalactic radio astronomy: Cos-

mology, (*J. Lequeux*); Instruments (*J. W. Findlay*).

157.028 Maryland – Green Bank galactic 21-cm line survey.
         G. Westerhout.
College Park, Maryland, Univ. Maryland Astron. Program,
June 1969. Second Edition. 11 pp. and 1480 maps.
    A survey of 21-cm line profiles in the region $l = 11$ deg to
$l = 235$ deg, $b = +1$ deg to -1 deg is presented in the form of
contour maps giving intensity as a function of right ascension
and velocity. About 77% of the region is covered in 1480
maps. – *JDM*

157.029 Neutral hydrogen in the Carina region.
         S. L. Garzoli.
Asoc. Argentina Astron. Bol., No. 14 (see 012.020), p. 37
(1968).   In Spanish. – Abstract.

157.030 21-cm observations in the region from galactic
         longitude 220° to 270° and latitude between
-10° **and** -20°.   E. Bajaja, R. F. Colomb.

Asoc. Argentina Astron. Bol., No. 14 (see 012.020), p. 100 -
101 (1968).   In Spanish. – Abstract.

    A survey of H 109α recombination line emission
in galactic H II regions of the southern sky.
See Abstr. 131.095.

    A comparison of the radial velocities of the neutral
hydrogen gas and H II regions within 60 degrees of the galactic
center.   See Abstr. 155.015.

    Spiral structure of neutral hydrogen in our
Galaxy.   See Abstr. 155.026.

    Spiral structure of the Galaxy derived from the
Hat Creek survey of neutral hydrogen.
See Abstr. 155.028.

    A study of spiral structure for $270° \leqslant l^{II} \leqslant 310°$.
See Abstr. 155.031.

158 Single and Multiple Galaxies

158.001   Observations of normal galaxies at 5 GHz.
J. B. Whiteoak.
Astrophys. Letters, Vol. 5, 29 - 32 (1970).
   The results are presented of observations at 5 GHz of the positions and flux densities of 81 radio sources located in directions near normal galaxies. Over 80 per cent have positions within 1 arc min of revised optical centres of the galaxies.

158.002   Spectrum and redshift of the optically variable, compact radio galaxy 3C371.
H. Arp, N. Visvanathan.
Astrophys. Letters, Vol. 5, 73 - 74 (1970).
   Only very low contrast emission lines of [O II], [O III] and [Ne III] can be identified in the spectrum of 3C371. Their mean redshift is $z = 0.0501 \pm 0.001$. A few other lines, which appear to be real, cannot be identified. Some other compact galaxies with similar properties are compared with 3C371.

158.003   Compact companions connected to 3C371.
H. Arp.
Astrophys. Letters, Vol. 5, 75 - 79 (1970).
   It is shown that the radio galaxy 3C371 lies in a loose chain of compact galaxies. Luminous extensions connect it to the nearest companions in the chain. The redshift of the cluster is smaller by about a factor two than would be predicted by the standard redshift-apparent magnitude relation for clusters of galaxies.

158.004   Spectrophotométrie du noyau de la galaxie de Seyfert NGC 4051.   Y. Andrillat. S. Souffrin.
Comptes Rendus Acad. Sci. Paris, Sér. B, Vol. 270, 238 - 240 (1970).
   La partie nucléaire de NGC 4051 est constituée, d'une part par un mélange d'étoiles de type A et G contribuant au spectre continu et au spectre en absorption, d'autre part, par un plasma donnant les raies en émission et une proportion du continu très importante dans l'ultraviolet (en particulier du côté ultraviolet de la discontinuité de Balmer).

158.005   A large, asymmetrical light pulse from the N galaxy PKS 0521-36.   O. J. Eggen.
Astrophys. Journ. *(Letters)*, Vol. 159, L95 - L97 (1970).
   The N galaxy associated with PKS 0521-36 showed a rapid, 1-mag increase in brightness early in 1969 and a subsequent slow decline. The colors are very similar to those for other N galaxies and are redder than those for QSSs. If the minimum observed colors and magnitude are assumed to represent the underlying galaxy and are combined with those observed for a QSS in outburst, the result matches the colors of PKS 0521-36 at maximum light.

158.006   Compact nonthermal sources in M87.
G. R. Burbidge.
Astrophys. Journ. *(Letters)*, Vol. 159, L105 - L108 (1970).
   It is shown that the recently discovered compact nonthermal radio source in M87 may give rise to a powerful X-ray source through the Compton effect. It is also suggested that the evidence for small optical structures in the jet may indicate that the optical radiation and some of the X-ray flux comes from electrons and magnetic fields generated in small dense pulsar-like objects which have been ejected from the nucleus of the galaxy.

158.007   On the homogeneity of the spectral-energy distribution among giant E and S0 galaxies and other results.   B. M. Lasker.
Astron. Journ., Vol. 75, 21 - 33 (1970).
   A system of intermediate-bandwidth photometry has been used to observe about 100 galaxies, primarily E's and S0's. For each color index formed, the scatter in colors among the giant E's and S0's is small (0.04–0.11 mag, depending on the specific color index), and at least part of this scatter is real, not observational. Differences among the observed galaxies are discussed in terms of the scatter in their colors. Among the brightest galaxies ($-21 \lesssim M$), the color-magnitude correlation is weak, but significant correlations exist over the larger range in absolute magnitude obtained by including dwarf galaxies ($M \sim -15$). The colors appear to be weakly correlated with σ, the internal velocity dispersion.

158.008   On the application of the normal logarithmic spiral model to galaxies.
R. M. Dzigvashvili, T. M. Borchkhadze.
Byull. Abastumansk. Astrofiz. Obs. No. 37, p. 141 - 146 (1969).   In Russian.
   The characteristic angles μ for spirals of 6 multi-armed galaxies (NGC 1232, 5247, 4303, 4321, 3938, 3184) were determined. The question of application of the normal logarithmic spiral model to the estimation of the number of arms in our Galaxy is considered. On the basis of this model the numbers of the spiral arms for the galaxies mentioned above were determined and compared with the real numbers. It is shown that the normal logarithmic spiral model is undue for such purposes and the number of spiral arms thus obtained is not reliable.

158.009   Die Rotation der Spiralnebel.   W. Fricke.
Bild der Wissenschaft, 7. Jahrgang, No. 2, p. 149 - 157 (1970). — Popular article.

158.010   Beyond the NGC catalogue.   L. J. Robinson.
Sky Telescope, Vol. 39, 100 - 101 (1970).

158.011   The variable compact galaxy Zw 0039.5 + 4003.
F. Zwicky, J. B. Oke, G. Neugebauer, W. L. W. Sargent, A. P. Fairall.
Publ. Astron. Soc. Pacific, Vol. 82, 93 - 98 (1970).
   The optically variable compact galaxy Zw 0039.5 + 4003 has broad emission lines similar to those seen in Seyfert and N-type galaxies. The continuous spectrum shows no evidence of any stellar contribution and is probably largely nonthermal. The object has a redshift $z = 0.1026$ and an absolute visual magnitude of -21.6.

158.012   Physical processes in galaxies.   G. M. Tovmasyan.
Stars, Nebulae, Galaxies, Symposium Byurakan 1968, (see 012.002), p. 209 - 231 (1969).   In Russian.

158.013   Spectral observations of the Markarian galaxies with high dispersion.
D. W. Weedman, Eh. E. Khachikyan.
Stars, Nebulae, Galaxies, Symposium Byurakan 1968, (see 012.002), p. 233 - 235 (1969).   In Russian.

158.014   Polarimetric investigation of extragalactic objects.   V. A. Dombrovskij, V. A. Hagen-Thorn.
Stars, Nebulae, Galaxies, Symposium Byurakan 1968, (see 012.002), p. 237 - 245 (1969).   In Russian.

158.015   On the origin of gas with high radial velocity in the radio galaxy NGC 1275.   V. i. Fronik.
Stars, Nebulae, Galaxies, Symposium Byurakan 1968, (see 012.002), p. 247 - 250 (1969).   In Russian.

158.016 H II–zones in central regions of nine normal galaxies. I. I. Pronik.
Stars, Nebulae, Galaxies, Symposium Byurakan 1968, (see 012.002), p. 251 - 252 (1969). In Russian.

158.017 Remarks on the galaxies NGC 3077 and M 82. A. T. Kalloglyan.
Stars, Nebulae, Galaxies, Symposium Byurakan 1968, (see 012.002), p. 253 - 258 (1969). In Russian.

158.018 Spectrophotometry of the jet in NGC 4486. K. K. Chuvaev, V. I. Pronik.
Stars, Nebulae, Galaxies, Symposium Byurakan 1968, (see 012.002), p. 259 - 265 (1969). In Russian.

158.019 Some characteristics of the nuclei of Seyfert galaxies. Eh. A. Dibaj.
Stars, Nebulae, Galaxies, Symposium Byurakan 1968, (see 012.002), p. 267 - 271 (1969). In Russian.

158.020 Galaxies with ultraviolet continuum. III. B. E. Markarian.
Astrofizika, Vol. 5, 581 - 592 (1969). In Russian. – English translation in Astrophysics, Vol. 5, No. 4.

The third list of galaxies with ultraviolet continuum is presented. Data for 102 objects are given. Twenty objects from this list definitely show emission lines in their spectra, whereas for 48 other objects it is probable. The data of our three lists show that the mean rate of appearance on the sky of galaxies brighter than 17^m with ultraviolet continuum is one galaxy per eight square degrees (adequate to 2 percent).

158.021 On the central condensations in Sa galaxies. K. A. Sahakian.
Astrofizika, Vol. 5, 593 - 602 (1969). In Russian. – English translation in Astrophysics, Vol. 5, No. 4.

The results of the classification of central condensations of 51 Sa galaxies are presented. Observations have shown that about 40% of the Sa galaxies have starlike and semi-stellar nuclei. The histograms of distribution of the central parts according to the five mark classification of the Byurakan Observatory and two color photometric data for the classes 5, 4, and 3 are given. Comparison with earlier results is presented.

158.022 Seyfert galaxies. T. Kwast.
Urania Kraków, Vol. 41, 2 - 7 (1970).
In Polish.

158.023 Galaxies and quasars: Puzzling observations and bizarre theories. R. W. Holcomb.
Science, Vol. 167, 1601 - 1603 (1970).

158.024 Yearly variations of 3C 120. P. D. Usher, B. S. P. Shen, F. W. Wright.
Nature, Vol. 225, 365 - 366 (1970).

The photographic light curve of Seyfert galaxy 3C 120 shows 0.3 to 0.6 magnitude fluctuations with time scale about 360 ± 40 days superimposed on a steady large amplitude decline a brightness from 1934 to 1939. The possibility that these annual variations are due to a selective yearly hour angle effect is investigated, and found to be incapable of explaining the observations. The conclusion is that the observed fluctuations are intrinsic to the galaxy.

158.025 PK 0048-09: a possible radio variable galaxy. M. A. Stull.
Nature, Vol. 225, 832 - 833 (1970).

Observations of sixty radio galaxies north of -20° declination selected from the Parkes Catalogue have been made with the University of Michigan 85-ft radio telescope at a frequency of 8000 Mc/s. For one of these galaxies, PK 0048-09, widely differing values of flux density have been measured on different dates, and it is believed the galaxy is varying. PK 0048-09 is an 18th magnitude galaxy of indeterminate type which is located in a small cluster and appears distinctly brighter on blue sensitive photographs than on red sensitive ones. Its radio spectrum is unusual; its flux density falls at frequencies below 5000 Mc/s. At 8000 Mc/s a 17% decrease in flux density occurred from late 1968 to early 1969. This was followed by an increase of more than 80% over a period of five months. In October, 1969, a decrease in flux density of nearly 20% took place in about ten days.

158.026 Rotation of the Andromeda nebula from a spectroscopic survey of emission regions.
V. C. Rubin, W. K. Ford, Jr.
Astrophys. Journ., Vol. 159, 379 - 403 = Contr. Kitt Peak National Obs. No. 492 (1970).

Spectra of sixty-seven H II regions from 3 to 24 kpc from the nucleus of M31 have been obtained with the DTM image-tube spectrograph at a dispersion of 135 Å mm^{-1}. Radial velocities, principally from Hα, have been determined with an accuracy of ±10 km sec^{-1} for most regions. Rotational velocities have been calculated under the assumption of circular motions only. For the region interior to 3 kpc where no emission regions have been identified, a narrow [N II] λ 6583 emission line is observed. Velocities from this line indicate a rapid rotation in the nucleus, rising to a maximum circular velocity of V = 225 km sec^{-1} at R = 400 pc, and falling to a deep minimum near R = 2 kpc. From the rotation curve for $R \leq$ 24 kpc, a disk model of M31 results. The optical velocities, $R >$ 3 kpc, agree with the 21-cm observations, although the maximum rotational velocity, V = 270 ± 10 km sec^{-1}, is slightly higher than that obtained from 21-cm observations.

158.027 High-velocity gas motions in galactic nuclei. D. W. Weedman.
Astrophys. Journ., Vol. 159, 405 - 413 = Contr. McDonald Obs., Univ. Texas, No. 449 (1970).

Observations at 28 Å mm^{-1} of sixteen galaxies showing nuclear emission lines are used to study the gas motions in the galactic nuclei. The galaxies observed include ten galaxies found by Markarian to have strong emission lines, four nearby spirals with unusually blue nuclei as noted by Tifft, and M51 and NGC 4151. An apparently normal galaxy, NGC 4569 in the Virgo cluster, is found to have nuclear emission lines that are blueshifted by 1300 km sec^{-1} from the radial velocity of the galactic disk, and the forbidden lines are significantly broader than those in the Seyfert galaxy NGC 4151.

158.028 The M87 jet. II. Temperature, ionization, and X-radiation in a secondary-production model.
J. E. Felten, H. C. Arp, C. R. Lynds.
Astrophys. Journ., Vol. 159, 415 - 423 = Contr. Kitt Peak National Obs. No. 520 (1970).

Recent photographs of the M87 jet are presented and discussed. Data on the small sizes of the knots and the lack of spectroscopically observed emission lines are used to examine a theoretical model of the jet in which the optical electrons are continually regenerated by cosmic-ray protons confined to the optical knots. Heating, cooling, and ionization are discussed. If the time scale of the knots is to be as long as 10^5 years, they must contain a large mass of gas ($n \sim$ 400 cm^{-3}, $M \sim 3 \times 10^7 M_\odot$) to maintain themselves against disruption by the cosmic-ray pressure. Observational results to date put some restrictions on the model.

158.029 Magnitudes and colors of globular clusters in the Fornax system. G. de Vaucouleurs, H. D. Ables.
Astrophys. Journ., Vol. 159, 425 - 433 = Contr. McDonald

Obs. Univ. Texas, No. 442 (1970).

Five independent series of photoelectric observations of four globular clusters in the Fornax dwarf elliptical galaxy prove that the $U - B$, $B - V$ colors are normal and match closely those of the bluest globular clusters in the Galaxy.

158.030 The brightest star clusters in galaxies as distance indicators. G. de Vaucouleurs.
Astrophys. Journ., Vol. 159, 435 - 441 = Contr. McDonald Obs. Univ. Texas, No. 443 (1970).

Photometric data on the brightest star clusters in eight galaxies of the local group covering a large interval of absolute magnitudes (–21 to –12) demonstrate the existence of a correlation between the absolute magnitude M (1) of the first-ranked cluster and the total absolute magnitude M (G) of the galaxy. An application to the recent data of Racine and Sandage on the brightest star cluster in M87, the brightest galaxy in the Virgo cluster, leads to a revised value of the Hubble constant $H = 50$ km sec^{-1} Mpc^{-1}, which is less than half the value derived in other recent studies of different distance indicators.

158.031 The spectra and redshifts of thirty Markarian galaxies with ultraviolet continua.
W. L. W. Sargent.
Astrophys. Journ., Vol. 159, 765 - 772 (1970).

Redshifts have been determined from 190 Å mm^{-1} image-tube spectrograms for thirty of the seventy galaxies with ultraviolet continua discovered bv Markarian. These thirty do not constitute a random sample of the Markarian galaxies. Twenty-six of the thirty galaxies have emission-line spectra, and five of these also have absorption lines. The galaxies with ultraviolet continua have at least one-thirtieth the space density of normal galaxies.

158.032 Spectral lines and radial velocities of galaxies in pairs. T. Page.
Astrophys. Journ., Vol. 159, 791 - 797 (1970).

Measures of the spectra of eighty-eight galaxies in forty-five close pairs have been reviewed and corrected. Originally measured for velocity differences used in a determination of the average masses of galaxies, they are now listed in conventional form, together with twenty-six independent measures by others. Evidence is noted for a possible systematic difference in redshift between emission and absorption lines.

158.033 Noncircular gas motions in NGC 253: Evidence for outflow from the center.
M. H. Demoulin, E. M. Burbidge.
Astrophys. Journ., Vol. 159, 799 - 807 (1970).

The velocity field in the ionized gas in the central regions of NGC 253 has been analyzed with the Lick 120-inch telescope. Noncircular motions found earlier by Burbidge et al. have been confirmed. Within 7″– 10″ of the center, strong emission lines give an approach velocity of 120 km sec^{-1} relative to the systemic velocity. The outflow velocity decreases in directions off the minor axis. Emission lines are still strong southeast of the center in a bright fan-shaped area. In these regions, the electron temperature is higher than in spiral-arm regions and turbulent velocities are ~150 km sec^{-1} Various models for the velocity flow are discussed.

158.034 The abundance ratio of helium to hydrogen in extragalactic nebulae. M. Peimbert, H. Spinrad.
Astrophys. Journ., Vol. 159, 809 - 815 (1970).

From photoelectric measures of emission lines the $N(He^+)/N(H^+)$ abundance ratios are obtained for NGC 604, NGC 4449, NGC 5461, NGC 5471, and NGC 7679. Our results are compared with previous determinations of abundance ratios of helium to hydrogen. Upper limits for the helium abundances in M51 and M81 are presented.

158.035 A new determination of the mass of M87.
F. Bertola, M. Capaccioli.
Mem. Soc. Astron. Italiana, Nuova Serie, Vol. 41, 57 - 63 (1970).

The mass of M87 has been computed by means of the virial theorem, using the luminosity profile recently given by de Vaucouleurs (1969). The mass reaches the extremely high value of $2.7 \times 10^{13} M_\odot$ and a mass-to-light ratio of 190 is derived. These values are the largest known for a galaxy.

158.036 A remark on the spectrum of the compact galaxy IV Zw 150. R. Barbon.
Mem. Soc. Astron. Italiana, Nuova Serie, Vol. 41, 129 (1970). Letter.

158.037 Algunos comentarios sobre fotometria de galaxias y el "Atlas de Galaxias Australes" de Jose Luis Sersic. G. de Vaucouleurs.
Revista Astron., Vol. 41, No. 170, p. 27 - 34 (1969).

158.038 Rotation of galaxies.
B. A. Vorontsov-Velyaminov.
Astron. Zhurn. Akad. Nauk SSSR, Vol. 47, 16 - 22 (1970). In Russian. English translation in Soviet Astron. AJ, Vol. 14, No. 1.

From a compilation of existing optical and radio data on the rotation of 170 galaxies or of their nuclear regions the periods and velocities of rotation are discussed statistically.

158.039 Some characteristics of nuclei and integral characteristics of Seyfert galaxies.
A. V. Zasov, E. A. Dibay.
Astron. Zhurn. Akad. Nauk SSSR, Vol. 47, 23 - 26 (1970). In Russian. English translation in Soviet Astron. AJ, Vol. 14, No. 1.

A probable correlation is found between the luminosity of nuclei and the total luminosity and the mass of Seyfert galaxies.

158.040 A high-resolution 5-GHz map of 3C 33.
S. Mitton.
Astrophys. Letters, Vol. 5, 207 - 211 (1970).

The double radio galaxy 3C 33 has recently been mapped at a frequency of 4.995 GHz with the One-Mile radio telescope at Cambridge; the resolving power was 6.5 arc sec (r.a.) by 19 arc sec (dec.). The source consists of two remarkably compact components separated from each other by about 50 times their diameters. The containment of these plasma clouds by the intergalactic medium is discussed.

158.041 Analysis of composite spectra: Independent variables in Wood's twelve-colour photometry.
W. L. Martin, R. G. Bingham.
Observatory, Vol. 90, 13 - 19 (1970).

Twelve-colour photometry of 20 galaxies and two globular clusters is analyzed statistically and it is found that the variation between galaxies may be represented within the limits of observational error by three uncorrelated quantities derived from the observations.

158.042 Rates of stellar collisons in the nucleus of NGC 4151. R. H. Sanders.
Astrophys. Journ., Vol. 159, 1115 - 1118 (1970).

In this note it is assumed that the continuous radiation in the nucleus of NGC 4151 hides a stellar system. Limits on the rate of stellar collisions consistent with the observations of NGC 4151 are calculated.

158.043 Changes in the nuclear spectrum of the Seyfert galaxy NGC 4151. R. Cromwell, R. Weymann.
Astrophys. Journ. (*Letters*), Vol. 159, L147 - L149 (1970).

Evidence is presented that the violet-displaced absorption lines of hydrogen discovered by Anderson and Kraft in the nucleus of NGC 4151 are transient features with a time scale of 1 year. This evidence is used to reassess the rate of mass ejection deduced by Anderson and Kraft.

158.044 Improved photographs of the NGC 1275 phenomenon. R. Lynds.
Astrophys. Journ. (*Letters*), Vol. 159, L151 - L154 = Contr. Kitt Peak National Obs. No. 523 (1970).

New Hα interference-filter photographs of the radio galaxy NGC 1275 are described and illustrated. These photographs show the spectacular extent of the distribution of outlying material in the system and place new limits on the presence of material having velocities differing from those previously known. The character of the distribution of material strongly supports the explosive interpretation of the NGC 1275 phenomenon.

158.045 Observations of infrared galaxies.
D. E. Kleinmann, F. J. Low.
Astrophys. Journ. (*Letters*), Vol. 159, L165 - 172 (1970).

Infrared flux densities have been measured over the last 5 years for a number of galaxies. The continuum out to 25μ, the existence of large variations, and estimates of total infrared luminosities are presented.

158.046 The infrared-galaxy phenomenon. F. J. Low.
Astrophys. Journ. (*Letters*), Vol. 159, L173 - L177 (1970).

An ensemble of identical infrared sources, called irtrons, radiates the quiescent infrared continuum now found to characterize the nuclei of all galaxies. Continuous creation of matter and antimatter within the irtrons releases energies greater than 10^{62} ergs. The observed infrared continuum results from coherent synchrotron decay of electrons and positrons produced by annihilation. The observed infrared luminosities from an evolutionary sequence beginning with QSOs, extending to Seyfert galaxies and exploding galaxies, and ending with large spirals like our own.

158.047 Formation of gas clouds in galactic nuclei.
T. T. Arny.
Monthly Notices, Roy. Astron. Soc., Vol. 148, 63 - 73 (1970).

A model is suggested for the gas in galactic nuclei. The injection of gas from evolving stars and its consequent settling to the inner regions of galaxies is shown to produce a high temperature-low density medium which is strongly thermally unstable. The instability produces clouds whose mass and radius are in rough agreement with observations.

158.048 Masses and mass—luminosity ratio for galaxies.
B. A. Vorontsov-Velyaminov.
Astron. Zhurn. Akad. Nauk SSSR, Vol. 47, 271 - 276 (1970). In Russian. – English translation in Soviet Astron., AJ, Vol. 14, No. 2.

A compilation of 215 determinations of rotation or of dispersion of internal velocities of 141 galaxies and 28 nuclear regions was used. Linear formulae relating masses to luminosities for E–S0 and S0-Irr galaxies were obtained.

158.049 On the problem of the comparison of sizes of galaxies. I. L. Genkin, L. M. Genkina.
Astron. Zhurn. Akad. Nauk SSSR, Vol. 47, 443 - 445 (1970). In Russian. – English translation in Soviet Astron., AJ, Vol. 14, No. 2.

Effective diameters of galaxies and those determined in Holmberg's microphotometric system are compared with micrometric diameters from de Vaucouleurs' Reference Catalogue.

158.050 Luminosity function of elliptical galaxies in extended Virgo cluster region.
G. O. Abell, S. Eastmond.
Bull. American Astron. Soc., Vol. 2, 179 (1970). – Abstr. AAS.

158.051 Emission lines in the nuclear region of M31.
W. K. Ford, Jr., V. C. Rubin.
Bull. American Astron. Soc., Vol. 2, 192 (1970). – Abstr. AAS.

158.052 The radio spectra and brightness distributions of elliptical galaxies. D. S. Heeschen.
Bull. American Astron. Soc., Vol. 2, 197 - 198 (1970). Abstr. AAS.

158.053 A possible radio-variable galaxy.
M. A. Stull.
Bull. American Astron. Soc., Vol. 2, 220 - 221 (1970). Abstr. AAS.

158.054 The extragalactic distance scale.
S. van den Bergh.
Bull. American Astron. Soc., Vol. 2, 223 (1970). – Abstr. AAS.

158.055 The distribution of dwarf galaxies of the Sculptor-type in the Local Supergalaxy.
V. E. Karachentseva.
Vestn. Kiev. Univ., Ser. Astron., No. 11, p. 114 - 119 (1969). In Russian.

Supergalactic coordinates of Sculptor-type dwarf galaxies have been calculated. The statistical analysis of the distribution in longitude and latitude of bright galaxies from A. and G. de Vaucouleurs' catalogue and Sculptor-type dwarfs is carried out.

158.056 N-galaxies as a metagalactic population.
B. V. Komberg, L. M. Ozernoy.
Astrophys. Space Sci., Vol. 6, 450 - 473 (1970). In Russian. – English translation by D. F. Smith, Astrophys. Space Sci., Vol. 7, 31 - 53 (1970).

Optical and radio data about N-galaxies (NG) as known at present are summarised and compared with the corresponding data about Seyfert galaxies (SyG), blue compact galaxies (BCG) and quasistellar objects (QSG and QSS). Information about normal galaxies and Haro and Markaryan objects is also utilized. The luminosity functions of different sources in the optical- and radio-ranges are constructed, based on the observational facts. Along with some general properties of NG and SyG, BCG, QSS and QSG, a number of differences between these objects are noted. Taking into account all these differences, it is concluded that N-galaxies are neither distant SyG nor QSS at small distances and therefore relatively weak, but are the representatives of a different metagalactic population. Arguments about the existence of stars in N-galaxies as well as a central source of nonstellar nature are discussed.

158.057 An electrodynamic model of radio galaxies and quasars. J. H. Piddington.
Monthly Notices, Roy. Astron. Soc., Vol. 148, 131 - 147 (1970).

It is shown by statistical and individual studies that the powerful radio sources exhibit increases in magnetic energy and flux by factors $\sim 10^2 - 10^3$ as sources age $10^3 - 10^5$ years. This requires a previously developed field system extending > 100 kpc along, and twisted around, the rotational axis. Our model comprises a rotating gas cloud from which a galaxy condenses, and a pre-existing field which is amplified until the Rayleigh–Taylor instability causes ejection of magnetic tongues along the axis. In an old galaxy the gas has turned to stars (perhaps burnt out), to a corona and to a small central cloud. This cloud, of mass $\sim 10^9 M_\odot$, radius $\sim 10^{16}$ cm, field $\sim 10^5$ gauss, rotates rapidly to provide the relativistic particles for the radio and optical synchrotron sources. Its thermal plasma provides the QSO line emission.

158.058 A second list of compact and bright-nucleus galaxies.
A. P. Fairall.
Monthly Notes Astron. Soc. Southern Africa, Vol. 29, 48 - 52 (1970).
The list contains positions of compact and bright-nucleus galaxies taken from the Palomar Sky Survey and forms an extension to a region (14^h R.A., +12° Dec.) covered by an earlier list (see Monthly Notes ASSA, Vol. 27, 67 - 69 (1968)).

158.059 The spectra of Markarian galaxies. I.
M. A. Arakelian, E. A. Dibay, V. F. Yesipov.
Astrofizika, Vol. 6, 39 - 47 = Byurakan Astrophys. Obs. No. 53 (1970).   In Russian. – English translation in Astrophysics, Vol. 6, No. 1.
A spectral investigation of thirty three galaxies with strong ultraviolet continuum is carried out with an image tube spectrograph on the 125-cm reflector of the Crimean Station of the Sternberg Institute. Emission lines are found in the spectra of twenty-two objects. The redshifts, photographic absolute magnitudes and equivalent widths of emission lines are determined for twenty objects.

158.060 The Andromeda galaxy M 31. II. Hydrodynamical model. Theory.   J. E. Einasto.
Astrofizika, Vol. 6, 149 - 163 (1970).   In Russian. – English translation in Astrophysics, Vol. 6, No. 1.
The paper considers the problem of constructing a hydrodynamical model of a stellar system on the basis of a mass distribution model. A system of equations is given for the following kinematical functions: the velocity dispersion, the centroid velocity, the ratios of velocity dispersions, the inclination angle α of the major axis of the velocity ellipsoid with respect to the plane of symmetry of the galaxy. For these five unknown functions we have only two hydrodynamical equations. To solve the problem three additional equations are needed. Various methods of completing the system of hydrodynamic equations are described and discussed. The latter is a system of nonlinear differential equations and can be solved by successive approximations.

158.061 Spectroscopic study of the double galaxy NGC 3395 - 3396.   S. D'Odorico.
Astrophys. Journ., Vol. 160, 3 - 9 (1970).
Spectra of NGC 3395 and NGC 3396, a pair of interacting galaxies, show emission lines of H I, He I, [O II], [O III], [N II], and [S II]. In both galaxies the intensity of Hα is greater than that of [N II] λ6583, as in most Sc, irregular, and Seyfert galaxies. The intensity ratios Hα/Hβ = 5.1 in NGC 3395 and Hα/Hβ = 7.5 in NGC 3396 suggest that large amounts of dust may be present in the nuclei of the two galaxies. Within 1.5 kpc of the nucleus, NGC 3395 rotates as a solid body and has an approximate mass of $1.2 \times 10^9 M_\odot$. Velocities in NGC 3396 show no appreciable rotation. The general velocity field is consistent with the hypothesis that NGC 3396 rotates around NGC 3395.

158.062 On the rotation curves of Sb and Sc galaxies.
W. C. Saslaw.
Astrophys. Journ., Vol. 160, 11 - 16 (1970).
The radii, R, and constant angular velocities, ω, of the central regions are correlated in Sb and Sc galaxies. These correlations may differ for the two classes of galaxies. In Sb galaxies, R and ω may also be correlated with the mass of the galaxy. The significance and uncertainties in these results are discussed briefly.

158.063 A spectroscopic survey of compact and peculiar galaxies.   W. L. W. Sargent.
Astrophys. Journ., Vol. 160, 405 - 427 (1970).
Low-dispersion spectrograms have been obtained for 141 objects in Zwicky's first five lists of compact galaxies. Direct photographs were obtained at the 200-inch prime focus of a representative sample of thirty objects. Photoelectric observations were made for forty-four galaxies. Individual descriptions of the 126 galaxies, including their redshifts and absolute magnitudes, are given (the remaining objects were galactic stars and one planetary nebula).

158.064 Physical conditions in the nucleus of M82.
M. Peimbert, H. Spinrad.
Astrophys. Journ., Vol. 160, 429 - 441 (1970).
Narrow-band photoelectric photometry of the nucleus and a bright region northeast of the center of M82 was carried out with the 120-inch telescope at Lick Observatory. The intensity of several emission lines in the nucleus, as well as of the continuum in both regions at different wavelengths, was obtained.

158.065 On the orientation of the magnetic field in M31.
P. D. Noerdlinger.
Astrophys. Journ., Vol. 160, 785 (1970).

158.066 Integrated magnitudes of M33 and M101.
S. Jacobsson.
Astron. Astrophys., Vol. 5, 413 - 415 (1970).
The integrated magnitudes and colours in the UBV-system are found to be $B = 6.^m50$, $B - V = 0.^m65$ and $U - B = 0.^m00$ for M33 and $B = 8.^m43$, $B - V = 0.^m45$ and $U - B = 0.^m08$ for M101.

158.067 Redshifts of companion galaxies.
H. Arp.
Nature, Vol. 225, 1033 - 1035 (1970).
It is shown that smaller galaxies which are known to be companions of nearby, large galaxies, have on the average, significantly larger red shifts than the central galaxy. It is considered that this demonstrates the existence of non velocity red shifts in these companions. Previous observational evidence on the ejection of QSR's, compact radio galaxies and compact companion galaxies predicts this result.

158.068 The hydrogen content and kinematics of Messier 51.
M. S. Roberts, J. L. Warren.
Astron. Astrophys., Vol. 6, 165 - 172 (1970).
The present study is based on observations made with the NRAO 300-foot (91 m) and 140-foot (43 m) telescope in Green Bank, West Virginia. The purpose is two-fold: to attempt to identify H I emission with NGC 5195 and to estimate the parameters of the rotation curve of M51. The available optical and 21-cm data are used as a basis for a general discussion of this system.

158.069 Spectrophotométrie intégrée des galaxies proches dans l'ultraviolet (expérience Persée).
P. Cruvellier, A. Roussin, Y. Valerio.
IAU Symposium No. 36, (see 012.014), p. 130 - 133 (1970).
Preliminary results obtained from a sounding rocket launched in December 1968 show a strong ultraviolet excess in the central part of the Andromeda nebula, as compared to the color of the sky background.

158.070 Spiral structure in external galaxies.
W. W. Morgan.

IAU Symposium No. 38, (see 012.013), p. 9 - 14 (1970).

The need for observing at least four different categories of optical objects for a satisfactory definition of optical spiral structure is emphasized. A sequence of optical form-types for the description of spiral structure is outlined.

158.071 The distribution of dark nebulae in late-type spirals. B. T. Lynds.

IAU Symposium No. 38, (see 012.013), p. 26 - 34 (1970).

Seventeen Sc- galaxies have been studied in order to determine the distribution of dark nebulae within them. Hα plates were used to compare the distribution of obscuring material with the location of H II regions, and it was found that the H II regions are always tangent to or imbedded in regions of high obscuration.

158.072 Angular momenta of late-type spiral galaxies. N. Heidmann.

IAU Symposium No. 38, (see 012.013), p. 35 - 40 (1970).

The angular momenta of galaxies may be evaluated from their photometry. The mass to light ratio in NGC 224 is discussed and the calculation is applied to two late-type spirals.

158.073 Applying the model of a normal logarithmic spiral to galaxies.

R. M. Dzigvashvili, T. M. Borchkhadze.

IAU Symposium No. 38, (see 012.013), p. 41 (1970). – Abstract.

158.074 Density distribution and the radial velocity field in the spiral arms of M 31.

J. Einasto, U. Rümmel.

IAU Symposium No. 38, (see 012.013), p. 42 - 50 (1970).

The density distribution and the radial velocity field in the Andromeda galaxy, M 31, have been studied on the basis of the 21-cm radio-line data from Jodrell Bank and Green Bank. The true density has been obtained from the observed one by solving a two-dimensional integral equation. The mean radial velocities have been derived by solving a two-dimensional non-linear integral equation with the help of hydrogen densities, and a model radial velocity field. The rotational velocity, derived from the radial velocity field, in the central region differs considerably from the velocity curves obtained by earlier authors.

158.075 The rotation curve, mass, light, and velocity distribution of M 31. J. Einasto, U. Rümmel.

IAU Symposium No. 38, (see 012.013), p. 51 - 60 (1970).·

A model for the Andromeda galaxy, M 31, has been derived from the available radio, photometric, and spectroscopic data. The model consists of four components – the nucleus, the bulge, the disc, and the flat component. For all components the following functions have been found: the mass density; the mass-to-light ratio; the velocity dispersions in three perpendicular directions; the deviation angle of the major axis of the velocity ellipsoid from the plane of symmetry; the centroid velocity.

158.076 A comparison of dynamical models of the Andromeda nebula and the Galaxy.

V. C. Rubin, W. K. Ford, Jr.

IAU Symposium No. 38, (see 012.013), p. 61 - 68 (1970).

From new radial velocities of 67 H II regions in M 31, rotational velocities and a mass model of M 31 are derived, and compared with the rotation curve and Schmidt mass model of our Galaxy. It is shown that in M 31 the distribution of H II regions as identified by Baade agrees with the distribution of neutral hydrogen determined from 21-cm observations.

158.077 OB stars on the outermost borders of M 31. F. Börngen, G. Friedrich, G. Lenk, L. Richter,

N. Richter.

IAU Symposium No. 38, (see 012.013), p. 69 - 71 (1970).

The task of the present investigation is to find if there exist OB stars (single or in association) on the outermost borders of this stellar system.

158.078 Observations of radio continuum emission from M 31. G. G. Pooley.

IAU Symposium No. 38, (see 012.013), p. 72 (1970). – Abstract.

158.079 Kinematics of M 33, M 51 and the L. M. C. G. Monnet.

IAU Symposium No. 38, (see 012.013), p. 73 - 78 (1970).

158.080 The stellar velocity field in M 51. S. M. Simkin.

IAU Symposium No. 38, (see 012.013), p. 79 - 82 (1970).

Radial velocities have been measured from the absorption lines on two image tube spectra of M 51. These velocities show large deviations from the 'smoothed' rotation curve for that object. The measurements seem to indicate that both the stars and the gas move in the same way.

158.081 H II regions in NGC 628, NGC 4254 and NGC 5194. K. Chuvaev, I. Pronik.

IAU Symposium No. 38, (see 012.013), p. 83 - 86 (1970).

Multicolour observations of galaxies are being carried out at the prime focus of the 2.6 m Schajn telescope using an image converter and 6 - 9 colour filters. In all three galaxies H II regions are observed from 1 to 8 kpc from the centre, the nearest ones have been found in the H II regions of the central discs.

158.082 Spiral structure and distribution of stellar associations in NGC 6946.

E. Ye. Khachikian, K. A. Sahakian.

IAU Symposium No. 38, (see 012.013), p. 87 - 90 (1970).

The associations of NGC 6946 outline its spiral arms. There is no relation between the colour or magnitude of the associations and their distance from the centre. Their mean absolute magnitude is − 11^m.1 and their mean colour index near zero.

158.083 Barred spiral galaxies. K. C. Freeman.

IAU Symposium No. 38, (see 012.013), p. 351 - 355 (1970).

We point out some properties of barred spiral galaxies which are important for the theory of their formation and spiral structure, and describe some theoretical work on the dynamics of these systems.

158.084 Structure and dynamics of barred spiral galaxies with an asymmetric mass distribution.

G. de Vaucouleurs, K. C. Freeman.

IAU Symposium No. 38, (see 012.013), p. 356 - 362 (1970).

The asymmetric structure and dynamics of late-type barred spirals is analyzed in terms of a model consisting of a small prolate spheroid (the bar) displaced from the center of a large oblate spheroid (the disk).

158.085 Additional evidence for the cataclysmic origin of spiral structure in galaxies.

K. Rudnicki.

IAU Symposium No. 38, (see 012.013), p. 435 - 436 (1970).

The photograph of the galaxy NGC 3486 shows its spiral structure in statu nascendi. The evident explanation is that of a cataclysmic origin of the visible spiral arms.

158.086 Electron-scattering line profiles in nuclei of Seyfert galaxies. R. J. Weymann.

Astrophys. Journ., Vol. 160, 31 - 41 (1970).

The main result of this discussion is that very small, dense, ionized regions seem indicated in those objects like NGC 5548 and NGC 4151 whose Balmer lines have very broad wings. The effects of radiation pressure on material in such a dense region surrounding a small intense source of radiation are significant and might give rise to the observed radial outflow of matter.

158.087 Les galaxies.   P. de la Cotardière.
L'Astronomie, 84e année, p. 189 - 205 (1970).

158.088 Measurements of neutral atomic hydrogen in relatively early type galaxies.
L. Bottinelli, P. Chamaraux, L. Gouguenheim, R. Lauqué.
Astron. Astrophys., Vol. 6, 453 - 459 (1970).

The neutral atomic hydrogen content of 34 galaxies and systemic radial velocities of 44 galaxies with relatively small angular diameter have been measured with the Nançay radio telescope. Determinations of systemic radial velocities have been made for 5 galaxies with previously unknown velocities; their group membership is confirmed except for one of them. Correlations previously found are confirmed and extended to earlier morphological types.

158.089 Die Kerne in Galaxien.   V. A. Ambarzumjan.
Astron. in der Schule, 7. Jahrgang, p. 6 - 13 (1970).

158.090 Verification of radio variability of the galaxy PKS 0048-09.   H. N. Ross.
Nature, Vol. 226, 431 (1970).

As part of a program to investigate about 80 low frequency cutoff sources, PKS 0048-09 was observed at five frequencies over a two-year period. The variability at 8 GHz reported by Stull of Michigan has been confirmed to exist at frequencies both above and below 8 GHz.

158.091 Companion galaxies connected to NGC 772.   H. Arp.
Astrophys. Letters, Vol. 5, 257 - 260 (1970).

Several smaller galaxies in the vicinity of NGC 772 are connected to this disturbed Sb spiral by luminous filaments. The redshift of NGC 772 is 2.430 km/sec. Redshifts of three of the interacting galaxies are 2.450, 20.200, and 19.700 km/sec. It is concluded that the latter two are further examples of small companion galaxies which have redshifts that do not correspond to the standard cosmological-velocity redshifts.

158.092 Erratum: A high resolution 5 GHz map of 3C 33.
        [Astrophys. Letters, Vol. 5, 207 - 211 (1970)].
S. Mitton.
Astrophys. Letters, Vol. 5, 287 (1970). − See Abstr. 158.040.

158.093 Les galaxies.   P. de la Cotardière.
L'Astronomie, 84e année, p. 261 - 268 (1970).

158.094 Intermediate bandpass photometry of galaxies and globular clusters.   J. S. Neff.
Monthly Notices, Roy. Astron. Soc., Vol. 149, 45 - 50 (1970).

Measurements with an intermediate bandpass photometric system are presented for globular clusters and galaxies. The photometric data indicate that the weak line globular clusters show an ultra-violet excess with respect to the strong line clusters if a normal reddening law is assumed. The photometric data are consistent with Morgan's classifications of the galaxies with the exception of the K type galaxies which show a range of near ultra-violet colour.

158.095 The radio spectra and brightness distributions of E and S0 galaxies.   D. S. Heeschen.

Astrophys. Letters, Vol. 6, 49 - 53 (1970).

The radio spectra and brightness distributions of twelve nearby giant E and S0 galaxies are discussed. The galaxies display two distinctly different types of radio characteristics. One group, comprising one-half the observed galaxies, has 'normal' radio spectra, with spectral indices of about -0.9. The second groups shows a wide variety of radio spectral peculiarities and the radio emission originates in a small region in the nucleus of the galaxy.

158.096 Radio observations of E and S0 galaxies.
        D. S. Heeschen.
Astron. Journ., Vol. 75, 523 - 529 (1970).

Observations have been made of 114 of the nearer E, S0, and Sa galaxies, including all those previously known to be radio sources. The observations suggest that there may be a correlation between radio emission and degree of flattening of these galaxies. Radio positions, flux densities at 2-, 6-, and 11-cm wavelengths, and structure at 11-cm wavelength, are given for many of the detected galaxies.

158.097 A neutral hydrogen study of the spiral galaxy NGC 5457.   M. Guélin, L. Weliachew.
Astron. Astrophys., Vol. 7, 141 - 149 (1970).

The neutral atomic hydrogen distribution and dynamic have been investigated in the Sc⁻ galaxy NGC 5457 (M 101) with the Nançay radiotelescope. The resolution was 4' in right ascension, 28' in declination and 59 km/s in velocity. The data were fitted by hydrogen distribution and velocity models.

158.098 Neutral hydrogen in M31—I. The distribution of neutral hydrogen.
R. D. Davies, S. T. Gottesman.
Monthly Notices, Roy. Astron. Soc., Vol. 149, 237 - 261 (1970).

This paper describes an investigation of the neutral hydrogen distribution in M31 based on a survey of the entire object with a frequency resolution of 40 kHz (8.4 km s^{-1}) and a beamwidth of 14' × 18'arc. The integrated neutral hydrogen map shows a neutral hydrogen distribution similar to the Milky Way with a deficiency in the central regions and with elongated spiral features. This distribution has been compared with the distribution of H II regions, OB associations, blue stars and the radio continuum emission. It is found that there is a close correlation between the main neutral hydrogen and H II features. An examination of the data suggests that the rate of star formation is greatest in areas of greatest hydrogen density.

158.099 Neutral hydrogen in M31 − II. The dynamics and kinematics of M31.   S. T. Gottesman, R. D. Davies.
Monthly Notices, Roy. Astron. Soc., Vol. 149, 263 - 290 (1970).

Data from neutral hydrogen maps of M31 taken at 53 velocities have been used to investigate the velocity field over the surface of the galaxy. The rotation curve which is determined in two independent ways shows in each case significant differences between the NF and SP major axes as suggested by previous investigations. Both the velocity and hydrogen distribution information can be used to derive the fundamental parameters of M31.

158.100 Emission-line intensities and radial velocities in the interacting galaxies NGC 4038 - 4039.
V. C. Rubin, W. K. Ford, Jr., S. D'Odorico.
Astrophys. Journ., Vol. 160, 801 - 809 = Contr. Kitt Peak National Obs., No. 529 (1970).

Spectrograms of eighteen emission regions in the peculiar extragalactic system NGC 4038 - 4039 show emission lines of Hβ; [O III] λλ 4959, 5007; He I λ5876; [O I] λ 6300; [N II] λλ 6548, 6583; Hα; and [S II] λλ 6717, 6731. From a discussion of the sizes of the luminous regions, the systemic velocity, the magnitude of the brightest stars, and the magnitude of the 1921 supernova, a distance of 9 ± 3 Mpc is estimated. The line-of-sight velocities show a spread of only 235 km sec^{-1} over all the emission regions. The velocity field exhibits a systematic variation from knot to knot, which can be interpreted as due to two rotating, interacting galaxies.

158.101 **Erratum and addendum: Light curve of the N-type galaxy 3C371.** [Observatory, Vol. 89, 198 - 201 (1969), see 02.158.076].
P. D. Usher, R. D. Cannon, M. V. Penston.
Observatory, Vol. 90, 124 (1970).

158.102 **The log S—log z diagram for radio galaxies and its relation to cosmology.**
F. Hoyle, G. R. Burbidge.
Nature, Vol. 227, 359 - 361 (1970).

An examination of the flux S and redshift z of radio galaxies in the third Cambridge catalogue lends some encouragement to the steady state cosmology.

158.103 **Optical variability of the nucleus of the Seyfert galaxy NGC 4151.**
M. K. Babadzhanjanz, V. A. Hagen-Thorn, V. M. Ljutyi.
Astron. Tsirk., No. 544, p. 6 - 8 (1970). In Russian.

158.104 **On the variability of radio emission of the elliptical galaxy NGC 4278 at centimetre wavelengths.**
A. G. Gorshkov, V. N. Kurilchik, I. G. Moiseev, V. A. Soglasnov.
Astron. Tsirk., No. 545, p. 3 - 4 (1970). In Russian.

158.105 **Galaxies.** G. C. McVittie.
Trans. IAU, Vol. 14A, (see 003.028), 301 - 318 (1970). — Report of Commission 28.

158.106 **The normal elliptical galaxies.** K. C. Freeman.
Proc. Astron. Soc. Australia, Vol. 1, 302 - 303 (1970).

158.107 **Infrared observations of the Seyfert galaxy NGC 1068.** G. V. Khozov.
Astron. Tsirk., No. 557, p. 6 - 7 (1970). In Russian.

158.108 **On the Sculptor-type dwarf galaxies individually associated with normal galaxies.**
V. E. Karachentseva.
Astron. Tsirk., No. 558, p. 4 - 7 (1970). In Russian.

158.109 **Origin of the optical polarization in the nucleus of NGC 1068.** K. Nandy, R. D. Wolstencroft.
Nature, Vol. 225, 621 - 622 (1970).

Measurements of the degree of polarization of the nuclear region of NGC 1068 by Kruszewski and Visvanathan and Oke show that the degree of polarization increases with decreasing diaphragm size, which suggests that the polarization is produced in and close to the nucleus and that the radiation contributed by other parts of the galaxy is unpolarized. In order to find what process or processes are responsible for the optical radiation from the nucleus it is important to determine the intrinsic polarization of the nucleus.

158.110 **Structure and dynamics of barred spiral galaxies, and in particular of the Magellanic type.**
G. de Vaucouleurs, K. C. Freeman.
Univ. Texas, Dep. Astron. McDonald Obs., 170 pp. (1969). [To be published in Vistas in Astronomy, 1970].

158.111 **Blue condensations associated with galaxies.**
A. N. Stockton.

Thesis, Univ. Texas, Austin. [Available from Univ. Microfilms, Ann Arbor, Mi.], 124 pp. (1969).

A number of elliptical and S0 galaxies having nearby blue condensations have been listed by Ambartsumyan and Shachbazyan. Two of these galaxies, NGC 3561 and IC 1182, have visible bridges or jets connecting to the associated blue condensations, and these galaxies have been studied in detail. Spectroscopic observations have also been made of several other blue knots and the galaxies with which they are associated as well as of certain nearby blue galaxies.

158.112 **An optical study of nearby galaxies.**
H. D. Ables.
Thesis, Univ. Texas, Austin. [Available from Univ. Microfilms, Ann Arbor, Mi.], 299 pp. (1969).

Detailed photometric parameters describing the luminosity distribution in six galaxies are derived from photographic and electrographic isophotometry with photoelectric calibrations. Distances to each of the galaxies are derived and the effective and intrinsic optical parameters for each galaxy are compared with the corresponding parameters for other galaxies of similar morphological type.

158.113 **Dvaergsfaeroider i den lokale galaksegruppe.**
H. Stub.
Astron. Tidssk., Vol. 3, 82 - 87 (1970).

158.114 **Spectral properties of galaxies with peculiar nuclei.**
M. G. Pastoriza.
Asoc. Argentina Astron. Bol., No. 12, (see 012.019), p. 11-13 (1968). In Spanish. — Abstract.

158.115 **A search for peculiar galaxies in the Whiteoak atlas.**
J. L. Sérsic, E. Agüero.
Asoc. Argentina Astron. Bol., No. 12, (see 012.019), p. 34 (1968). In Spanish. — Abstract.

158.116 **Interferometry of M 83.**
G. J. Carranza.
Asoc. Argentina Astron. Bol., No. 14 (see 012.020), p. 38 - 40 (1968). In Spanish.

158.117 **Radio emission associated with peculiar galaxies of the Arp catalogue.** R. F. Colomb, C. Varsavsky.
Asoc. Argentina Astron. Bol., No. 14 (see 012.020), p. 72 (1968). In Spanish. — Abstract.

158.118 **Interferometry of NGC 4945.**
G. J. Carranza.
Asoc. Argentina Astron. Bol., No. 14 (see 012.020), p. 90 - 91 (1968). In Spanish. — Abstract.

158.119 **Interferometric observations of southern galaxies.**
G. J. Carranza.
Asoc. Argentina Astron. Bol., No. 14 (see 012.020), p. 91 - 96 (1968). In Spanish.

158.120 **Velocity field in NGC 5128.**
G. Carranza, J. L. Sérsic.
Asoc. Argentina Astron. Bol., No. 14 (see 012.020), p. 102 (1968). In Spanish. — Abstract.

158.121 **Distribution d'hydrogène neutre et dynamique des galaxies.** L. Weliachew.
Thesis, Doct. Sci. Phys., Paris. [Centre Document. C.N.R.S., No. 3831], 77 pp. (1969). — See Bull. Signalét., Vol. 32, Section 120, No. 8984 (1970).

Cosmic sources of infrared radiation.
See Abstr. 114.051.

Cosmic sources of infrared radiation.
See Abstr. 114.074.

New novae in the Andromeda nebula.
See Abstr. 124.007.

Retention of dust grains near galactic nuclei.
See Abstr. 131.001.

The fine-scale structure of Virgo A.
See Abstr. 141.017.

3C 386: Radio galaxy or supernova remnant?
See Abstr. 141.061.

Inverse Compton process in quasars and Seyfert-
ype nuclei.   See Abstr. 141.109.

On a possibility of testing Arp's hypothesis.
See Abstr. 141.114.

Further spectroscopic observations of quasi-stellar
objects and radio galaxies.   See Abstr. 141.125.

Maps of six extra-galactic radio sources at 5 GHz.
See Abstr. 141.164.

Once more about quasars and nuclei of galaxies.
See Abstr. 141.186.

Compact radio source in the nucleus of M 87.
See Abstr. 141.213.

Barred spiral galaxies.
See Abstr. 151.010.

A model for the early evolution of galaxies.
See Abstr. 151.034.

Survey of spiral structure problems.
See Abstr. 155.023.

Spiral structure of our Galaxy and of other galaxies.
See Abstr. 155.024.

Statistics of spiral patterns and comparison of
our Galaxy with other galaxies.   See Abstr. 155.025.

The origin of galaxies.
See Abstr. 162.033.

159 Magellanic Clouds

159.001 A catalogue of stellar associations in the Large Magellanic Cloud. P. B. Lucke, P. W. Hodge.
Astron. Journ., Vol. 75, 171 - 175 (1970).

Stellar associations in the Large Magellanic Cloud are catalogued and described. The number of stars brighter than $V \approx 14.7$, listed for each of the 122 associations, ranges from 2 to 225, and the apparent diameters range from 1' to 20'.

159.002 6-cm recombination line observations of the 30 Doradus nebula.
P. G. Mezger, T. L. Wilson, F. F. Gardner, D. K. Milne.
Astrophys. Letters, Vol. 5, 117 - 121 (1970).

Radio recombination line observations of the thermal component of the 30 Doradus nebula in the Large Magellanic Cloud indicate that the nebula possesses an electron temperature, internal turbulence, and helium abundance that are much higher than average. In addition, the radial velocity from the radio line measurements is somewhat lower than previous estimates made in the optical range. This apparent discrepancy between the radio and optical abundances is not explained; however, the radio results may refer to a dense inner region of the nebula which is not easily observed with optical techniques.

159.003 The extinction law in the Magellanic Clouds.
M. T. Brück. L. C. Lawrence, K. N. Nandy,
A. D. Thackeray, R. Wood.
Nature, Vol. 225, 531 - 532 (1970).

Spectra of reddened early-type stars in the Magellanic Clouds and of similar non-reddened galactic stars were compared using calibrated plates taken at Pretoria. Reddened and less-reddened MC stars were also compared. The differences for 6 pairs were observed in the 4000 Å to 5000 Å region. It is suggested that the extinction law in the MC resembles the Perseus law in the Galaxy.

159.004 Interferometric study of ionized hydrogen in the Large Magellanic Cloud. Y. Georgelin, G. Monnet.
Astrophys. Letters, Vol. 5, 213 - 218 (1970).

First results obtained by Hα Pérot-Fabry interferometry on H II regions in the Large Magellanic Cloud are presented. Evidence is given for a large disk of ionized hydrogen in the center, and its physical parameters are discussed. 240 radial velocities have been measured, and a preliminary rotation curve is given.

159.005 A note on the Magellanic Clouds.
E. E. Mendoza V.
Bol. Obs. Tonantzintla y Tacubaya, Vol. 5 (No. 32), 104 - 106 (1969).

This note shows the intermediate-band photometry of 36 objects that belong to either one or another of the two Magellanic Clouds, obtained with the 36-inch telescope of the Cerro Tololo Inter-American Observatory in November 1968. The results indicate that there exist differences in evolution, atmospheric chemical composition and/or age between the Large Cloud, the Small Cloud and the Galaxy.

159.006 Seeing the Magellanic Clouds and the Milky Way as a physical tripletgalaxy. S. Gaposhkin.
Bull. American Astron. Soc., Vol. 2, 193 - 194 (1970). Abstr. AAS.

159.007 Neutral hydrogen and stars in the wing of the Small Magellanic Cloud. K. C. Turner.
Bull. American Astron. Soc., Vol. 2, 222 (1970). – Abstr. AAS.

159.008 The orbit of the Large Magellanic Cloud.
A. Toomre.
Symposia Mathematica, Vol. 3, 21 - 22 (1970). – Conference paper (see 012.010).

159.009 Note on "A method for determining Magellanic Cloud membership using an objective prism and an absorption filter". C. Fehrenbach.
Monthly Notes Astron. Soc. Southern Africa, Vol. 29, 13 - 16 (1970).

Twenty stars listed as being new members of the LMC by C. J. Butler and M. V. Norris (see Monthly Notes, Astron. Soc. Southern Africa, Vol. 28, 107 - 114 (1969)) are common to our lists. Eighteen of these stars have small radial velocities and are members of the Galaxy. The neodymium chloride method is not suitable for the determination of LMC membership.

159.010 Research on the Magellanic Clouds, 1967 - 1969.
A. D. Thackeray.
Monthly Notes Astron. Soc. Southern Africa, Vol. 29, 19 - 23 (1970).

159.011 UBV magnitude sequences in the Magellanic Clouds.
A. W. J. Cousins.
Monthly Notes Astron. Soc. Southern Africa, Vol. 29, 24 - 36 (1970).

Mean magnitudes and colour indices in the UBV system have been derived from the available observations for stars in the Harvard I and Arp's sequences in the SMC and in the Harvard V and Arp's sequences in the LMC, and for some other stars selected by Wesselink and others. Special attention has been given to the zero points.

159.012 On Butler and Norris's method of determining Large Magellanic Cloud membership. R. Wood.
Monthly Notes Astron. Soc. Southern Africa, Vol. 29, 37 - 39 (1970).

159.013 The magnetic-field structure of the Magellanic Clouds. D. S. Mathewson, V. L. Ford.
Astrophys. Journ. *(Letters)*, Vol. 160, L43 - L46 (1970).

Observations of the polarization of stars in the Magellanic Clouds show the existence of a large-scale, regular magnetic field which is probably associated with the Magellanic System. Strong polarization occurs in the regions rich in Population I components where the magnetic field is twisted with respect to the surrounding magnetic field of the Magellanic System. 30 Doradus has the strongest polarization, and the E-vectors show a remarkable focusing of the magnetic lines of force on this giant emission nebula.

159.014 Polarization measurements and magnetic field structure within the Magellanic Clouds.
T. Schmidt.
Astron. Astrophys., Vol. 6, 294 - 308 (1970).

A photoelectric surface polarimetry of both Magellanic Clouds has been carried out at the Boyden Observatory, South Africa. A discrepancy between polarization angles of single star and surface measurements within the LMC "bar" may be explained by high albedo interstellar dust containing two layers of different star densities and different magnetic field directions. The surface polarization angles in both clouds are predominantly aligned parallel to the LMC-SMC connecting great circle. Therefrom the existence of a general magnetic field is obvious, connecting both clouds and indicating a common origin of the entire "Panmagellanic System".

159.015 **Large Magellanic Cloud. List of LMC members and list of galactic stars.** Charts for recognizing these stars on the maps published by the Smithsonian Institution Astrophysical Observatory.
C. Fehrenbach, M. Duflot, M. Petit.
Astron. Astrophys. Special Suppl. Series, No. 1, 67 + 35 pp. + 37 charts (1970).   In French.
We publish here a list of 495 large radial velocity stars, members of the Large Magellanic Cloud, a list of 1830 galactic stars, and a supplementary list of 110 stars, probable members of the LMC. These stars are drawn on tracings which are superposable to the maps of the Smithsonian Institution Astrophysical Observatory.

159.016 **A method for determining the total reddening and absorption in the Magellanic Clouds.**
P. W. Hodge.
SAO *Cambridge*, Mass., Special Rep. No. 306, 5 + 7 pp. (1969)
Multicolor photometry of galaxies seen behind the Magellanic Clouds is an unexploited method of further examining the total reddening and absorption in them. The method is limited by the intrinsic spread in galaxy colors, but enough galaxies are available for photometry that the gross characteristics of the absorbing layer should be detectable. A list of relatively bright galaxies recognized behind the Large Cloud is assembled in a table. UBV photometry of one of the brightest of these background galaxies, NGC 2150, measured at Cerro Tololo, provides an example.

159.017 **Five-colour photometry of supergiants and the dust-to-gas ratio in the Large Magellanic Cloud.**
A. M. van Genderen.
Astron. Astrophys., Vol. 7, 49 - 58 (1970).
A discussion is presented of photo-electric five-colour observations of about 70 LMC supergiants made in 1966. Slight photometric differences between these supergiants and those in our Galaxy need further confirmation. The dust-to-gas ratio in the LMC appears to be about four times smaller than in the Galaxy.

159.018 **Nova in the Large Magellanic Cloud.**
D. J. MacConnell, A. Gomez.
IAU Circ. No. 2238 (1970).

159.019 **The kinematics of H II regions in the Small Magellanic Cloud.**   M. W. Feast.
Monthly Notices, Roy. Astron. Soc., Vol. 149, 291 - 299 (1970).
Radial velocities, from Hα measurements, are given for 15 SMC H II regions. There is good agreement with the most intense 21-cm peaks in each case and a marked velocity gradient across the SMC. Nine H II regions in the core of the SMC have a low velocity dispersion. The 21-cm profiles have a much higher velocity dispersion in this region and this suggests that the H II regions are confined to the region of highest H I density.

159.020 **The 3323 variable stars in or projected on the Magellanic Clouds down to –0.8 absolute magnitude.**
S. Gaposhkin.
Inform. Bull. Variable Stars (I.A.U. Commission 27), Konkoly Obs., Budapest, No. 420 (1970).

159.021 **A nova in the Large Magellanic Cloud.**
D. J. MacConnell.
Inform. Bull. Variable Stars (I.A.U. Commission 27), Konkoly Obs., Budapest, No. 437 (1970).

159.022 **A deep objective-prism survey for Large Magellanic Cloud members.**   N. Sanduleak.
Contr. Cerro Tololo Inter-American Obs., No. 89, 67 pp. (1970).
The Large Magellanic Cloud has been surveyed on deep objective-prism plates taken with the Curtis Schmidt telescope on Cerro Tololo in Chile. A total of 1272 stars, generally brighter than $m_{pg} \sim 14$, are listed in a catalog as proven or probable LMC members. The stars are identified on the charts in the LMC Atlas recently published by Hodge and Wright.

159.023 **Interferometric observations of the Large Magellanic Cloud.**   G. J. Carranza.
Asoc. Argentina Astron. Bol., No. 14 (see 012.020), p. 101 (1968).   In Spanish. – Abstract.

Absorption interstellaire dans la direction du Grand Nuage de Magellan.   See Abstr. 131.077.

Kinematics of M 33, M 51 and the L. M. C.
See Abstr. 158.079.

160 Clusters of Galaxies

160.001 Significance of the first brightest galaxies in rich clusters. B. A. Peterson.
Astrophys. Journ., Vol. 159, 333 - 340 (1970).

The distributions of absolute magnitudes of the first brightest galaxy in rich clusters of galaxies are computed from a model, and compared with the observed distributions of absolute magnitudes of the first brightest galaxy in clusters of three classes of richness. The model is based on the assumption of a luminosity function for clusters which has a universal form but which allows the masses of clusters to vary. No preference is found for a cluster luminosity function which is or is not cut off at the bright end. A systematic dependence of the average absolute magnitude on richness class is found from the observations, and is predicted by both luminosity functions.

160.002 Structure of the Coma cluster of galaxies. P. J. E. Peebles.
Astron. Journ., Vol. 75, 13 - 20 (1970).

In some cosmologies, a cluster of galaxies is imagined to be a gravitationally bound system which, in analogy with the formation of the Galaxy, originated as a collapsing protocluster. It is shown that a numerical model based on this picture is consistent with the observed features of the Coma Cluster of galaxies. The cluster mass derived from this model agrees with previous values; however, an analysis of the observational uncertainty within the framework of the model shows that the derived mass could be consistent with the estimated total mass provided by the galaxies in the cluster.

160.003 Low-dispersion spectra of galaxies. I. The Coma and Virgo clusters. A. G. D. Philip, N. Sanduleak.
Publ. Astron. Soc. Pacific, Vol. 82, 53 - 68 (1970).

Low-dispersion objective-prism spectra ($\simeq 10,000$ Å/mm) have been obtained of galaxies in the Coma and Virgo clusters. Composite color indices were derived from the spectra which, when combined with magnitudes obtained from the 'Catalogue of Galaxies and Clusters of Galaxies' (Zwicky and Herzog 1963), allowed color-magnitude diagrams for the clusters to be drawn. The differences between the diagrams for different clusters indicate that the method will prove useful in the study of the color distribution within clusters of galaxies.

160.004 Low-dispersion spectra of galaxies. II. Hercules and Abell Nos. 2197 and 2199. A. G. D. Philip.
Publ. Astron. Soc. Pacific, Vol. 82, 69 - 79 (1970).

Low-dispersion spectra ($\simeq 10,000$ Å/mm) have been obtained of the brighter members of the Hercules cluster and Abell Nos. 2197 and 2199. The survey identifies the reddest and bluest galaxies in each cluster. Color-magnitude diagrams are drawn for each cluster.

160.005 Remarks on the youth of groups of galaxies. G. M. Tovmasyan.
Stars, Nebulae, Galaxies, Symposium Byurakan 1968, (see 012.002), p. 279 - 280 (1969). In Russian.

160.006 On the hypothesis that the Coma cluster is stabilized by a massive, ionized intergalactic gas. B. E. Turnrose, H. J. Rood.
Astrophys. Journ., Vol. 159, 773 - 789 (1970).

The possibility that a massive, ionized intergalactic gas stabilizes the Coma cluster is investigated. Analysis of the physical processes in such a gas shows the existence of serious restrictions on the model: Hβ and X-ray data limit the temperature to the range $10^{4\,\circ} - 10^{6\,\circ}$K, substantial heating mechanisms are required to maintain that temperature over the age of the cluster, and some way of preventing collapse of the gas is needed: as long as $T \lesssim 10^{8\,\circ}$K, kinetic pressure cannot balance self-gravitation. Consideration of various possible heating and supportive mechanisms indicates that none are adequate (with the possible exception of effects of a highly turbulent intercluster medium), so that the ionized-gas model appears to be physically implausible.

160.007 Groups and clusters of galaxies. T. P. Snow, Jr.
Astron. Journ., Vol. 75, 237 - 238 (1970).

Descriptions and locations of 34 previously unreported possible groups and clusters of southern galaxies are given. These results are based upon direct inspection of photographic plates taken with the 20-inch double astrograph of the Yale-Columbia Southern Observatory at El Leoncito, Argentina.

160.008 The central part of the cluster of galaxies in Coma Berenices. B. I. Gorbachev.
Astron. Zhurn. Akad. Nauk SSSR, Vol. 47, 224 - 226 (1970). In Russian. English translation in Soviet Astron. AJ, Vol. 14, No. 1.

The surface distributions of different types and ellipticities of galaxies are constructed.

160.009 Unstable groups of galaxies. J. L. Sérsic.
Bull. Astron. Inst. Czechoslovakia, Vol. 21, 92 - 95 (1970).

The observed relationship between mass-luminosity ratios for groups and clusters of galaxies and their population is interpreted on basis of a model developed by the author (1968) in which unstable groups of galaxies are thought to be the result of the fragmentation of a parent galaxy.

160.010 The dissolution of early clusters of galaxies by strong quasi-stellar sources. P. D. Noerdlinger.
Astrophys. Journ. (*Letters*), Vol. 159, L179 - L183 (1970).

All galaxies may once have been in compact clusters with diameters of ~0.4 Mpc, gravitationally bound by much uncondensed gas. In almost all cases, the gas was driven out by the expanding radio halos of QSSs which formed in the clusters, and these fell apart and yielded the field galaxies. The most massive clusters remained bound, but they expanded, and comprise the clusters we see today. The epoch of dissolution and expansion must have run from large redshift ($z \sim 10$) to redshift $\lesssim 2$, with some straggling thereafter.

160.011 Radio mapping of the Hercules cluster of galaxies. F. F. Donivan, Jr., T. D. Carr, G. C. Omer, Jr.
Bull. American Astron. Soc., Vol. 2, 190 - 191 (1970). Abstr. AAS.

160.012 The dissolution of early galaxy clusters by quasars. P. D. Noerdlinger.
Bull. American Astron. Soc., Vol. 2, 211 (1970). – Abstr. AAS.

160.013 The dynamics of clusters of galaxies in universes with non-zero cosmological constant, and the virial theorem mass discrepancy. J. C. Jackson.
Monthly Notices, Roy. Astron. Soc., Vol. 148, 249 - 260 (1970).

The effects on groups of galaxies of a non-zero cosmological constant are considered. A version of the virial theorem which includes a cosmological term is derived.

160.014 **Brightest members of clusters of galaxies.**
B. A. Peterson.
Nature, Vol. 227, 54 - 55 (1970).

The absolute magnitudes of the brightest members of clusters of galaxies have been analysed in the context of the statistical theory of Peebles. Concerning some criticisms by Peach (Nature, Vol. 223, 1140 - 1142, 1969), the author concludes that, at present, the magnitude dispersions predicted by the statistical theory adequately represent the observations.

160.015 **Photometry of intergalactic matter in the Coma cluster.** G. de Vaucouleurs, A. de Vaucouleurs.
Astrophys. Letters, Vol. 5, 219 - 226 (1970).

The diffuse-light emission between galaxies has been measured in the central area of the Coma cluster. Pulse-count photoelectric scans of the brightest cluster member NGC 4889 show a detectable excess over an arc of 30 to 45 arc min centered near the geometric center of the cluster; the peak intergalactic brightness is 26.2 mag sec^{-2}, and the integrated magnitude is 10.0 ± 0.5 or 40 per cent of the integrated luminosity of cluster galaxies as commonly defined.

The observed 'intergalactic' luminosity in the Coma cluster can be substantially accounted for by the overlapping coronas of the two supergiant members NGC 4874 and 4889 and a few others.

160.016 **Unstable groups of galaxies.**
J. L. Sérsic.
Asoc. Argentina Astron. Bol., No. 14 (see 012.020), p. 42 - 47 (1968).

Quasars as protoclusters of galaxies.
See Abstr. 141.112.

Luminosity function of elliptical galaxies in extended Virgo cluster region. See Abstr. 158.050.

The extragalactic distance scale.
See Abstr. 158.054.

A reduction of the mass deficit in clusters of galaxies by means of a negative cosmological constant.
See Abstr. 162.017.

161 Intergalactic Matter

161.001 Photo-ionization of intergalactic hydrogen by quasars. J. Arons, R. McCray.
Astrophys. Letters, Vol. 5, 123 - 128 (1970).

Quasars may photo-ionize intergalactic hydrogen gas of high density ($\rho_0 \approx 10^{-29}$ g cm^{-3}) sufficiently to explain the absence of intergalactic absorption features, provided that: (a) their optical continua extend beyond the Lyman limit, and (b) Schmidt's law for the increase in number density of quasars holds out to redshifts $z \approx 3$. The selection effect of Rees prevents finding quasars beyond $z \approx 2.5$ by standard methods. The temperature of the gas may be well below 10^6 °K.

161.002 On the interaction between intergalactic gas clouds and galaxies of various types.
L. M. Ozernoy.
Astron. Tsirk., No. 542, p. 1 - 3 (1969).   In Russian.

161.003 Accretion of metagalactic gas clouds by galaxies and quasar-like phenomena. L. M. Ozernoy.
Astron. Zhurn. Akad. Nauk SSSR, Vol. 47, 265 - 270 (1970). In Russian. – English translation in Soviet Astron., AJ, Vol. 14, No. 2.

It is shown that the rate of condensation of metagalactic gas into nuclei of galaxies during the "quiet" phase of their evolution must increase along the extended Hubble sequence of galaxies.

161.004 Intergalactic clouds.   J. Silk.
Bull. American Astron. Soc., Vol. 2, 218 (1970).
Abstr. AAS.

161.005 Lukewarm models of the intergalactic medium.
D. A. Mendis, W. I. Axford.
Astrophys. Space Sci., Vol. 7, 160 - 173 (1970).

In this paper we consider the possibility that the intergalactic medium is indeed quite dense at the present epoch, but difficult to observe as a result of being almost fully ionized and relatively cool. It is assumed that the intergalactic medium is in effect an H II region which is maintained by an independently produced flux of ionizing ultraviolet and X-ray photons. We find that it is possible to construct plausible 'lukewarm' models of the intergalactic medium with densities that are significant cosmologically.

161.006 A note on the possibility of existence of a dense intergalactic plasma.   J. Bergeron.
Astron. Astrophys., Vol. 4, 335 - 336 (1970).

Earlier determination of the minimum temperature of a dense, hot, intergalactic gas (Bergeron, 1969) has been deduced from the lack of Lyman α absorption in the emission spectra of the quasi-stellar objects, taking as Lyman α optical depth $\tau \lesssim 0.0625$. If the observations of the emission spectra of the QSO set an upper limit on the Lyman α optical depth of only $\tau \lesssim 0.4$ the hypothesis of a hot dense intergalactic gas ($q_0 = 1$) is not in contradiction with both observations of the isotropic X-ray background around 0.27 keV and the emission spectra of the QSO.

161.007 Search for an intergalactic cloud in Microscopium at 21 cm.   W. G. L. Pöppel.
Astron. Astrophys., Vol. 5, 400 - 406 (1970).

A region in Microscopium, which includes Hoffmeister's proposed intergalactic cloud, was observed in 21-cm. The only hydrogen radiation detected has very low velocity and does not show any correlation with the cloud sketched by Hoffmeister. An upper limit for the total atomic hydrogen mass of the cloud was derived. It is lower than the total masses used tentatively by K. H. Schmidt.

161.008 The ultraviolet background (intergalactic gas, the galaxy, and subcosmic rays).
V. G. Kurt, R. A. Sunyaev.
IAU Symposium No. 36, (see 012.014), p. 341 - 348 (1970).

We discuss observations of the ultraviolet background radiation with lower spectral resolution (~ 100 Å) and the conclusions which can be drawn from the existing observational data concerning (a) the density and temperature of the intergalactic gas ,(b) the integrated ultraviolet radiation of our Galaxy ,(c) the energy density of subcosmic rays in our Galaxy. In addition we discuss indirect methods of determining the flux of background UV radiation in the wavelength range shorter than 912 Å from the distribution of neutral hydrogen in the peripheries of galaxies.

161.009 Erratum: Photo-ionization of intergalactic hydrogen by quasars. [Astrophys. Letters, Vol. 5, 123 - 127 (1970)].   J. Arons, R. McCray.
Astrophys. Letters, Vol. 5, 287 (1970). – See Abstr. 161.001.

161.010 Intergalactic clouds.   J. Silk.
Astrophys. Journ., Vol. 160, 793 - 799 (1970).

The intergalactic medium in clusters of galaxies may be thermally unstable. Intergalactic clouds form, at temperatures in the range 2.5×10^4 to $\sim 10^5$ °K. Observational consequences include an interpretation of the multiple absorption redshifts found in the spectra of some quasi-stellar sources. Recent observations of diffuse X-rays are used to set limits on the amount of uncondensed intracluster gas.

On the hypothesis that the Coma cluster is stabilized by a massive, ionized intergalactic gas.
See Abstr. 160.006.

Photometry of the intergalactic matter in the Coma cluster.   See Abstr. 160.015.

162 Structure and Evolution of the Universe, Cosmology

162.001 Extra-galactic distance scale.
S. van den Bergh.
Nature, Vol. 225, 503 - 505 (1970).
 Nine different methods are used to determine the value of the Hubble constant. A mean value $H = 95^{+15}_{-12}$ km s^{-1} Mpc^{-1} is obtained from all data that are now available.

162.002 The instability of a rotating universe.
J. Silk.
Monthly Notices, Roy. Astron. Soc., Vol. 147, 13 - 19 (1970).
 We show that the Gödel universe, in which there is a uniform rotation relative to the local compass of inertia, is stable to perturbations in the plane of rotation, but unstable to density perturbations along the rotation axis. The qualitative aspects of this result may be applicable to a more realistic universe with expansion provided that there is non-zero rotation and the universe is always matter-dominated.

162.003 On the propagation of light in universes with inhomogeneous mass distribution.
S. Refsdal.
Astrophys. Journ., Vol. 159, 357 - 375 (1970).
 The change in the apparent luminosity and shape of distant light sources due to imhomogeneities in the universe that cause gravitational deflections of the light rays is discussed. The case explored in most detail is a static and flat universe with randomly distributed mass points and with pointlike light sources. Ray-tracing technique by means of a computer is used. It is shown that the results found in the case of static and flat universes in most cases can be applied to expanding and curved universes with fairly good approximation. The validity of the point-source and point deflector approximation is briefly examined, and the changes one has to expect if the approximation is not valid are indicated.

162.004 New limits on the shear and rotation of the universe from the X-ray background. A. M. Wolfe.
Astrophys. Journ. (*Letters*), Vol. 159, L61 - L67 (1970).
 In this letter it is shown that the X-ray background is a sensitive probe of cosmic, large-scale structure; and that the X-ray isotropy leads to new limits on present-day shear and vorticity for approximately half of the sky.

162.005 Champ de gravitation statique d'une masse sphérique plongée dans un univers cosmologique.
J. Eisenstaedt.
Comptes Rendus Acad. Sci. Paris, Sér. A, Vol. 270, 422 - 425 (1970).
 Dans le cadre de la Relativité générale, nous étudierons le champ de gravitation statique créé par une masse sphérique de rayon a, de densité ρ_0 plongée dans un gaz cosmologique de densité ρ_1. Plus précisément nous construirons un modèle qui rassemble les univers statiques d'Einstein et de de Sitter et les solutions intérieures et extérieures de Schwarzschild.

162.006 Possibility of a large evolutionary correction to the magnitude–redshift relation.
B. M. Tinsley.
Astrophys. Space Sci., Vol. 6, 344 - 351 (1970).
 A range of giant elliptical galaxies can be synthesized which are compatible with the available photometry, including Oke and Sandage's (1968) observations of a slow rate of change of color B–V. One such model galaxy is discussed which evolves sufficiently rapidly at short wavelengths to provide an evolutionary correction of several tenths of a magnitude at redshift 0.46.

162.007 Cosmology and relativistic astrophysics.
A. Sapar.
Astron. Geod. Eston. SSR, Tartu, (see 003.004), p. 77 - 91 (1969). In Russian.

162.008 Cosmology: A search for two numbers.
A. R. Sandage.
Phys. Today, Vol. 23, No. 2, p. 34 - 41 (1970).
 Precision measurements of the rate of expansion and the deceleration of the universe may soon provide a major test of cosmological models.

162.009 Neutrinos and arrow of time in cosmology.
P. R. Chaudhuri.
Journ. Phys. A, General Phys., Vol. 3, L5 - L8 (1970).
 The arrow of time in the case of neutrinos is investigated taking into account photon-neutrino weak-coupling theory. It is shown that in both steady-state and Einstein-de Sitter cosmological models, the advanced effect can be avoided.

162.010 The kinetic theory of neutrinos in anisotropic cosmological models.
A. G. Doroshkevich, Y. B. Zeldovich, I. D. Novikov.
Astrofizika, Vol. 5, 539 - 553 (1969). In Russian. – English translation in Astrophysics, Vol. 5, No. 4.
 The situation concerning the question of the role of neutrinos and other weakly interacting particles in anisotropic cosmological models is given. Making use of the kinetic theory and the real properties of neutrinos it has been shown that neutrinos had become free at the stage of anisotropic expansion of the cosmological models when the anisotropy is still large. Later on all processes proceed the same way as described formerly by the authors.

162.011 The cosmological solutions of the Jordan-Dicke field equations and the observed data.
A. V. Manjos.
Astrofizika, Vol. 5, 649 - 653 (1969). In Russian. – English translation in Astrophysics, Vol. 5, No. 4. – Brief information.

162.012 The case for a hierarchical cosmology.
G. de Vaucouleurs.
Science, Vol. 167, 1203 - 1213 (1970).
 Recent observations indicate that hierarchical clustering is a basic factor in cosmology.

162.013 Radiation catastrophe. E. R. Harrison.
Nature, Vol. 225, 245 (1970).
 In the big-bang cosmologies the universe initially consists of almost equal quantities of matter and antimatter. It is shown that the amount of this slight difference does not affect greatly the 3 °K background radiation. Thus, contrary to what has been said, a radiation catastrophe does not occur in a baryon charge symmetric universe.

162.014 Cosmological theory of gravitation.
S. J. Prokhovnik.
Nature, Vol. 225, 359 - 361 (1970).
 A cosmological model, based on a uniformly expanding universe, can be represented as an acceleration field whose properties may provide the basis for the phenomenon of gravitation and for other astronomical phenomena. The gravitational 'constant' emerges as a parametric attribute of the model. The equivalence of gravitational and inertial mass is seen to be due to their common dependence on the field associated with a body.

162.015　**The universe as a hot laboratory for the nuclear and particle physicist.**　Ya. B. Zeldovich.
Comments Astrophys. Space Phys., Vol. 2, 12 - 17 (1970).

Some arguments are presented to the question: How would nulcear and particle physics confirm or disprove the hot-Universe theory?

162.016　**The spectrum of primordial radiation, its distortions and their significance.**
R. A. Sunyaev, Ya. B. Zeldovich.
Comments Astrophys. Space Phys., Vol. 2, 66 - 74 (1970).

After some preliminary remarks the authors describe two principal cases of spectrum distortions of the primordial radiation: (1) energy injection before recombination, at $T > 4000\ °K$, and (2) the same after recombination, at $T < 4000\ °K$. After this some applications of the formulae to cosmological problems are shown.

162.017　**A reduction of the mass deficit in clusters of galaxies by means of a negative cosmological constant.**　W. R. Forman.
Astrophys. Journ., Vol. 159, 719 - 722 (1970).

It is shown that values of the cosmological constant between 0.0 and $-1.0 \times 10^{-34}\ sec^{-2}$ will reduce the mass deficit for clusters of galaxies and will not conflict with observed values of the deceleration parameter q_0 or with derived values of the age of the universe.

162.018　**Can discrete sources produce the cosmic microwave radiation?**　M. G. Smith, R. B. Partridge.
Astrophys. Journ., Vol. 159, 737 - 743 (1970).

After making a few general assumptions about the properties of discrete microwave sources, we calculate the amount of fluctuation the sources will produce in measurements of the temperature of the cosmic microwave background. By comparing these results with the available observations, we show that, if discrete sources produce the microwave background, then either the number of microwave sources per Mpc^3 must exceed the number of galaxies or the deceleration parameter must be less than 0.1.

162.019　**Possible nonequilibrium nature of cosmic-background radiation.**　M. Alexanian.
Astrophys. Journ., Vol. 159, 745 - 752 (1970).

A general result in quantum statistical mechanics allows us to determine, from the observed photon spectrum, whether the cosmic-background radiation is due to a primeval fireball or arises from a super-position of spectra from discrete extragalactic sources. Also, an upper bound to photon spectra is used as a criterion for establishing whether the shorter wavelengths of a photon spectrum can possibly be due to the same entity as that giving rise to the longer wavelengths.

162.020　**The determination of the deceleration parameter and the cosmological constant from the redshift-magnitude relation.**　J. V. Peach.
Astrophys. Journ., Vol. 159, 753 - 763 (1970).

Sandage's photoelectric photometry of brightest cluster galaxies has been used with published photometry by Baum for a redetermination of the deceleration parameter q_0 in cosmological models both with and without a cosmological constant Λ. It is found that the observations are adequate only for setting upper and lower bounds on this constant. Solutions for q_0 have been computed for a number of combinations of the data and are given in the paper.

162.021　**Generation of magnetic fields in the radiation era.**　E. R. Harrison.
Monthly Notices, Roy. Astron. Soc., Vol. 147, 279 - 286 (1970).

It is shown that magnetic fields are generated during the radiation era of the early universe in regions that have rotation. These fields are weak compared with the present intensity of the galactic magnetic field and therefore must be amplified as the Galaxy forms and evolves.

162.022　**Galaxy formation in the early universe.**
E. R. Harrison.
Monthly Notices, Roy. Astron. Soc., Vol. 148, 119 - 130 (1970).

We have proposed that embryonic galaxies exist in the early universe in the form of dense spinning cores. These cores originate from fluctuations in the metric at the threshold of classical cosmology. The present preliminary studies indicate that the 'spinning-core' theory is apparently capable of solving the major basic problems of galaxy formation.

162.023　**Fluid dynamics for cosmology.**
T. Kihara, K. Sakai.
Publ. Astron. Soc. Japan, Vol. 22, 1 - 12 (1970).

The equation of continuity and the equation of motion for fluids in early and later stages of the expanding universe are given in most intuitive forms. On the basis of these equations, the adiabatic invariance of certain integrals is proved. The formulation is effective for explaining the process of the concentration of matter into galaxies.

162.024　**Thermal instability in an expanding medium.**
M.-a. Kondo.
Publ. Astron. Soc. Japan, Vol. 22, 13 - 40 (1970).

Thermal instability in an isotropically expanding (or contracting) medium is studied, including the gravitational and pressure effects, using hydrodynamical and thermal equations which are derived here in simple forms.

162.025　**New limits on the shear and rotation of the universe from the X-ray background.**
A. M. Wolfe.
Bull. American Astron. Soc., Vol. 2, 225 - 226 (1970).
Abstr. AAS.

162.026　**On some solutions of Einstein's equations from the point of view of Petrow's classification.**
N. P. Bondarenko.
Vestn. Kiev. Univ., Ser. Astron., No. 11, p. 10 - 17 (1969).
In Russian.

It is shown that in Petrow's classification Harrison's metric belongs to type I, Gödel's and Petrow's to type D, Friedman's to type 0 (conformal flat space), Pomogaev's to type 0 (flat space).

162.027　**Small-scale fluctuations of relic radiation.**
R. A. Sunyaev, Ya. B. Zeldovich.
Astrophys. Space Sci., Vol. 6, 358 - 376 (1970).
In Russian. English translation by D. F. Smith, Astrophys. Space Sci., Vol. 7, 3 - 19 (1970).

In this article the value of $\delta T/T$ arising from scattering of radiation on moving electrons is calculated; the velocity field is generated by adiabatic or entropy density perturbations. Fluctuations of the relic radiation due to secondary heating of the intergalactic gas are also estimated.

162.028　**The interaction of matter and radiation in the hot model of the universe, II.**
R. A. Sunyaev, Ya. B. Zeldovich.
Astrophys. Space Sci., Vol. 7, 20 - 30 (1970).

Heating of the primaeval plasma prior to the epoch of recombination results in distortions in the Rayleigh-Jeans region of the microwave relic radiation spectrum. The present observational data allow limits to be set to such energy injection from which follow upper limits to (a) the amount of antimatter in the universe; (b) the parameters of primaeval turbulence; and (c) the adiabatic fluctuation spectrum for small masses ($M < 10^{11} M_\odot$).

162.029 **Homogeneous electromagnetic and massive-vector-meson fields in Bianchi cosmologies.**
L. P. Hughston, K. C. Jacobs.
Astrophys. Journ., Vol.160, 147 - 152 (1970).

We investigate the consequences of (a) the source-free Maxwell equations and (b) the equations of massive-vector-meson fields in spatially homogeneous Bianchi cosmologies. We obtain explicit formulae for the electromagnetic field and its stress-energy tensor in the case where the Poynting vector vanishes.

162.030 **Anisotropic, multi-fluid cosmologies with hypersurface orthogonal velocity fields.**
L. P. Hughston, L. C. Shepley.
Astrophys. Journ., Vol. 160, 333 - 336 (1970).

We present new spatially homogeneous cosmological models filled with a perfect fluid and characterized by a fluid velocity orthogonal to homogeneous hypersurfaces of constant curvature. These models include an arbitrary number of fluid components which may be either noninteracting or interacting.

162.031 **Production of helium in the big-bang expansion of a magnetic universe.**
J. J. Matese, R. F. O'Connell.
Astrophys. Journ., Vol. 160, 451 - 458 (1970).

The effects of a magnetic field on the reaction rates for the processes $n + e^+ \rightleftarrows p + \bar{\nu}$, $n + \nu \rightleftarrows p + e^-$, and $n \rightleftarrows p + e^- + \bar{\nu}$ are calculated. The implications for the production of helium in the big-bang expansion of a magnetic universe are discussed. We also consider the influence of a magnetic field on the expansion rate of the universe and find that its effect on the rate of helium production dominates over the effects arising from the changed reaction rates.

162.032 **Nuclear reactions and elementary-particle reactions in a cold universe.** M. Kaufman.
Astrophys. Journ., Vol. 160, 459 - 470 (1970).

The purpose of the present article is (1) to discuss the formation of prestellar helium in a Friedmann universe which was cold enough at nuclear density for matter to be degenerate or partially degenerate and (2) to estimate the changes in temperature which result from elementary-particle decay and nuclear reactions in such a universe.

162.033 **The origin of galaxies.**
M. J. Rees, J. Silk.
Sci. American, Vol. 222, No. 6, p. 26 - 35 (1970).

The size, shape and other properties of the observed galaxies are traced to slight density enhancements in the expanding primordial fireball. Enhancements of certain mass were favored over others.

162.034 **Machian interpretation of general relativity.**
Z. Horák.
Bull. Astron. Inst. Czechoslovakia, Vol. 21, 96 - 109 (1970).

It is shown that the complete field equations with a non zero cosmological term are in full agreement with Mach's principle as well as the original Einstein cosmology which implies that the magnitude of the cosmic potential has a constant value equal to the square of the light velocity. A modification of the ordinary Schwarzschild solution is given for the Einstein quasistatic "superuniverse", consisting of an immense number of oscillating metagalaxies, proposed by the author.

162.035 **Uniform model universes containing matter and blackbody radiation.**

T. L. May, G. C. McVittie.
Monthly Notices, Roy. Astron. Soc., Vol. 148, 407 - 416 (1970).

Models which contain gaseous matter and 'background' blackbody radiation are considered. A general method of solving Einstein's equations for such models is set up. It is applied to the elucidation and extension of previous work by McIntosh and to the discovery of new classes of models. The a priori assumption that the gas pressure is zero is avoided.

162.036 **Fluctuations and galaxy formation in the Einstein universe.** R. Simon.
Astron. Astrophys., Vol. 6, 151 - 154 (1970).

The time evolution of density fluctuations in the Einstein universe are calculated by means of a classical method due to Landau and Lifshitz, taking into account the self gravitation of the medium, as it has been suggested by Saslaw. The spectrum of these density fluctuations shows a very pronounced peak around the Jeans wavelength and a second maximum which develops in the course of time at a larger wavelength which is itself increasing. This behaviour suggests the formation and the clustering of galaxies, as well as a general recession of the various clusters, i.e. a general expansion of the universe, and provides therefore a justification of the Eddington-Lemaître model, on the basis of the theory of fluctuations.

162.037 **On the gravitational pressure in the Friedmann-Lobachevsky space.** I. G. Fikhtengolts.
Dokl. Akad. Nauk SSSR, Ser. Mat. Fiz., Vol. 191, 1028 - 1030 (1970). In Russian.

162.038 **Basic formulae for cosmological models with matter and radiation.** A. Sapar.
Astron. Zhurn. Akad. Nauk SSSR, Vol. 47, 503 - 508 (1970). In Russian. English translation in Soviet Astron.,AJ, Vol. 14, No. 3.

A short review of the results of the author's former papers on the exact solutions both of the evolution of the universe and of the observable dependences for the case of a uniform universe filled with matter and radiation is given. A short, partly critical, review of the papers of other authors on the same topics is added.

162.039 **Model classification for relativistic universes containing matter and radiation.**
A. D. Payne.
Australian Journ. Phys., Vol. 23, 177 - 185 (1970).

Relativistic world models containing both matter and radiation are classified by means of the cosmological constant λ and the density parameter σ_0. The properties of some model parameters are examined in relation to the temperature of the radiation field which is assumed to be Planckian.

162.040 **Lemaître models and the cosmological constant.**
V. Petrosian, E. E. Salpeter.
Comments Astrophys. Space Phys., Vol. 2, 109 - 115 (1970).

The discovery of quasistellar objects with large redshifts has renewed interest in cosmological models. In this note recent views are discussed.

162.041 **Thermodynamics and cosmology.**
E. L. Schucking, E. A. Spiegel.
Comments Astrophys. Space Phys., Vol. 2, 121 - 125 (1970).

162.042 **Vacuum-like state of medium and Friedmann's cosmology.** E. B. Gliner.
Dokl. Akad. Nauk SSSR, Ser. Mat. Fiz., Vol. 192, 771 - 774 (1970). In Russian.

162.043 **Cosmic rays in the expanding universe.**
O. F. Prilutskij, I. L. Rosental'.
Izv. AN SSSR. Ser. fiz., Vol. 33, 1776 - 1786 (1969).
In Russian. – Abstr. in Referativ. Zhurn. 51. Astron.,
4.51.813 (1970).

162.044 **Non-linear development of disturbances in a one-dimensional world.** E. A. Novikov.
Zhurn. ehksperim. i teor. fiz., Vol. 57, 938 - 940 (1969).
In Russian. – Abstr. in Referativ. Zhurn. 51. Astron.,
4.51.821 (1970).

162.045 **Hot model of the universe and the problem of Dirac's mono-field.**
G. V. Domogatskij, I. M. Zheleznykh.
Izv. AN SSSR. Ser. fiz., Vol. 33, 1792 - 1795 (1969).
In Russian. – Abstr. in Referativ. Zhurn. 51. Astron.,
4.51.825 (1970).

162.046 **Zur elliptischen Kosmologie.** R. Zaikov.
Izv. Fiz. in-t s ANEB, Vol. 19, 109 - 154 (1969).
Abstr. in Referativ. Zhurn. 51. Astron., 4.51.827 (1970).

162.047 **Relativistic gas spheres and clusters of point masses with arbitrarily large central redshifts: Can they be stable?** G. S. Bisnovatyi-Kogan, K. S. Thorne.
Astrophys. Journ., Vol. 160, 875 - 885 (1970).
Recently Bisnovatyi-Kogan and Zel'dovich have constructed general-relativistic models of gas spheres and of collisionless star clusters, for which the redshift from the center to infinity is arbitrarily large; and they have speculated that these particular models might be stable against small perturbations. In this paper we prove that the gas spheres are, indeed, stable – at least against radial perturbations. As for the clusters, we show that the most powerful techniques yet devised yield inconclusive results for stability. However, from the behavior of the clusters under scrutiny by those techniques, we believe that probably they are stable.

162.048 **Galaxy formation and the primordial turbulence in the expanding hot universe.**
H. Sato, T. Matsuda, H. Takeda.
Progr. Theor. Phys. Japan, Vol. 43, 1115 - 1117 (1970).
We expand the idea of cosmic turbulence to explain the observed mass of a galaxy (or a cluster of galaxies) from a consideration of the scale of the turbulent eddies.

162.049 **Integral luminosity function for de Sitter's non-empty static world model.** F. M. Gomide.
Mem. Soc. Astron. Italiana, Nuova Serie, Vol. 41, 177 - 181 (1970).
It is considered the possibility of de Sitter's non-empty static cosmological model through the analysis of its integral luminosity function for cosmic radio sources compared with Véron's and Jauncey's statistics for optically identified radio galaxies. The theoretical slope found up to red-shifts of the order of 0.2 is not inconsistent with the ones found on observational grounds by the afore mentioned authors.

162.050 **Evolution of galaxies and secular variation of cosmic rays, magnetic fields and turbulence.**
T. Matsuda.
Progr. Theor. Phys. Japan, Vol. 43, 1491 - 1510 (1970).
Numerical computations of evolution of galaxies are made on the assumption that the galaxies are the homogeneous mixture of gas and stars with constant volume and that the stellar birth rate depends not only on the density but also on the temperature of gas in the galaxies. Evolutional changes of the luminosity, the mass of gas, the temperature, the chemical composition, and the energy densities of cosmic rays, magnetic fields, turbulence, radiation fields and thermal energy are calculated.

162.051 **On the dissipation of primordial turbulence in the expanding universe.**
K. Tomita, H. Narial, H. Sato, T. Matsuda, H. Takeda.
Progr. Theor. Phys. Japan, Vol. 43, 1511 - 1525 (1970).
The decay laws of primordial turbulence and the heating rates by its dissipation are derived in the expanding medium, and it is shown that the matter in the expanding universe cannot be heated and kept at temperatures higher than 10^5 °K which are necessary for the galaxy formation by thermal instability. Moreover the effects of its dissipation on hydrodynamic instability are discussed.

162.052 **Steady state cosmology and quantum electrodynamics.** G. Börner.
Progr. Theor. Phys. Japan, Vol. 43, 1622 - 1623 (1970).
A possible connection between local phenomena of quantum electrodynamics and the large scale structure of the universe has been discussed recently by Hoyle and Narlikar. They develop the Wheeler-Feynman electrodynamics. Now the attempt to formulate the usual quantum electrodynamics in the space-time frame of the steady state theory brings also an interesting result, which is discussed in this paper.

162.053 **Cosmology to-day.** W. H. McCrea.
Revue des Questions Scientifiques, Vol. 141, No. 2, 223 - 241 = Publ. Inst. Astron. Géophys. G. Lemaître, Louvain, Vol. 3, No. 5 (1970). – Inaugural lecture of the Chaire Georges Lemaître, Catholic University of Louvain, 1969 October 28.

162.054 **On the growth of condensations in an expanding universe.** R. W. Michie.
Contr. Kitt Peak National Obs., No. 440, 16 pp. (1969).
This article concerns the growth of small amplitude fluctuations in an evolving Friedmann universe, after the perturbation is inside the particle horizon and before the gas has recombined. The main results are the following: Perturbation masses > 10^{16} $M_\odot$ will not oscillate prior to the cosmological recombination period. Smaller masses do oscillate for some time, but there comes an era when the radiation and matter fields start to slip past each other and the oscillations are damped out. This occurs well before recombination and while the perturbation is still opaque.

162.055 **Turbulence in the hot universe.**
A. D. Chernin.
Nature, Vol. 226, 440 - 441 (1970).
The development of von Weizsäcker's cosmogonical hypothesis has led to work on the combined hydrodynamical motions of the photon gas and the cosmic plasma in the early hot universe, when the radiation density greatly exceeds the density of matter. Here hydrodynamical aspects of the evolution of the cosmic fluid at the matter-dominated stage are considered, assuming cosmological expansion to be described by an open Friedmann model.

162.056 **Space-time metric about a local condensation in a non-empty universe.** Z. Horak.
Phys. Letters, Vol. 30A, No. 2, p. 83 - 84 (1969).

162.057 **Inevitability of a point-singularity in a rotating Newtonian universe.** J. Pachner.
Phys. Letters, Vol. 30A, No. 2, p. 121 (1969).

162.058 **General form of the Einstein equations for a Bianchi type IX universe.** M. P. Ryan, Jr.
Journ. Math. Phys., Vol. 10, (No. 9), 1724 - 1728 (1969).
The Einstein equations for a general Bianchi type IX universe are presented in a form suitable for numerical

solution. As an example, the complete equations for a cosmology with a pure fluid stress tensor are also given.

162.059 **Red shift and epochs in a generalized Friedmann and Lemaitre cosmology.** N. J. Ionescu-Pallas.
Rev. Roumaine Phys., Vol. 14, (No.10), 1105 - 1109 (1969).
  Einstein's gravitional equations are solved under the following circumstances: 1. The metric is conformally flat, 2. The perfect fluid scheme does hold, 3. The cosmical pressure may be neglected, 4. The cosmological constant is related to intrinsic geometry of empty space-time and has a negative value.

162.060 **Discrete isotropies in a class of cosmological models.** B. G. Schmidt.
Commun. Math. Phys., Vol. 15, 329 - 336 (1969).
  It is shown that a certain class of cosmological models admits discrete isotropies. These models are solutions of Einstein's field equations, characterized by: (1) the matter is described as a perfect fluid and, (2) there exists a group of motions simply transitive on three surfaces orthogonal to the fluid flow vector.

162.061 **Relativistic analogs of scalar-tensor cosmologies.** C. B. G. McIntosh.
Journ. Math. Phys., Vol. 11, 250 - 252 (1970).
  Relativistic analogs of Brans-Dicke and Hoyle-Narlikar scalar-tensor cosmologies are given in terms of two- or three-fluid models.

162.062 **Birkhoff gravitational field of Milne's universe.** J. B. Boyling.
Journ. Math. Phys., Vol. 11, 867 - 870 (1970).
  It is shown that Milne's special relativistic cosmology is consistent with Birkhoff's flat-space theory of gravitation and maintains its stability in the presence of gravity. The gravitational correction to the redshift in the light from distant galaxies is calculated.

162.063 **Influence of the cosmological constant in closed-universe models containing matter and radiation.**
A. Agnese, M. La Camera, A. Wataghin.
Nuovo Cimento, Vol. 66B, 202 - 216 (1970).
  The solution of the cosmological equations in isotropic and homogeneous closed-universe models containing simultaneously matter and radiation, having in general the cosmological constant $\Lambda \neq 0$ has been obtained. The limiting cases (matter dominating and radiation dominating) have been studied also numerically.

162.064 **A point mass in a Einstein universe.** C. Leibovitz.
Commun. Math. Phys., Vol. 17, 33 - 34 (1970).
  An exact solution of Einstein's equations is presented for a perfect fluid representing a point mass in a Einstein universe.

162.065 **The hot universe.** J. B. Zeldovics.
Fiz. Szemle, Vol. 19, (No. 12), 353 - 358 (1969).
In Hungarian.
  Consequences of the hot universe theory are discussed and the validity of the theory is examined in the light of experimental results. It is concluded that a great deal of clarification of ideas is needed concerning the theory and for this purpose more experimental work is recommended.

162.066 **The helium problem in cosmology.** G. S. Greenstein.
Thesis, Yale Univ., New Haven, Conn. [Available from Univ. Microfilms, Ann Arbor, Mi.], 142 pp. (1969).
  The author investigates processes leading to the synthesis of He^4 during the early stages of the expansion of a universe away from a 'big bang'. The results depend on the chemical potentials of the neutrinos on the number of neutrino types in excess of those already discovered, and on the present averaged-out density of matter in the universe.

162.067 **Stability of an homogeneous anisotropic relativistic cosmological model.** J. K. Astorga.
Thesis, Univ. Texas, Austin, [Available from Univ. Microfilms, Ann Arbor, Mi.], 104 pp. (1969). – See Phys. Abstr., Vol. 73, No. 45207.

162.068 **Unified field theory, general relativity, gravitation theory, cosmology, Vol. 1.**
Report DDC-TAA-UFT-1, Defence Documentation Center, Alexandria, Va., USA, [Available from CFSTI, Springfield, Va.] 204 pp. (1969). – See Phys. Abstr., Vol. 73, No. 54566 (1970).

162.069 **Critical discussion of accretion theory of Hoyle and Lyttleton.** F. Cernuschi, F. R. Marsicano.
Asoc. Argentina Astron. Bol., No. 14 (see 012.020), p. 72 - 76 (1968). In Spanish.

162.070 **Weltall und Kosmologie.** V. Ginsburg.
Ideen exakt. Wissens (Wiss. Technik Sowjetunion), No. 4.69, p. 223 - 228 (1969).

162.071 **Finite-density nonhomogeneous Newtonian cosmologies.** R. C. Mjolsness.
Phys. Rev., Second Ser., Vol. 187, 1753 - 1761 (1969).
  The model describes cosmologies, or evolving star clusters, in which the usual cosmological principle is modified: Isotropy and homogeneity are approximate symmetries in the interior of the system only, while to present approximation, the Hubble law is still exact. The model agrees with the major observational results of homogeneous isotropic cosmologies, but differs from them in predicting finite (infinite) oscillations of the central density in the gravitationally bound (unbound) case and, in all cases a finite maximum density.

Soviet work suggests Mixmaster singularity at origin.
Phys. Today, Vol. 23, No. 3, p. 59 - 60, 62 (1970).

Cosmology. See Abstr. 003.036.

Gravithermodynamics – III. Phenomenological non-equilibrium theory and finite-time fluctuations.
See Abstr. 022.024.

Thermodynamics of strong interactions at high energy and its consequences for astrophysics.
See Abstr. 022.060.

The origin of magnetic fields.
See Abstr. 061.012.

Gravitational instability: An approximate theory for large density perturbations. See Abstr. 061.016.

On a cosmic background of low-energy neutrinos.
See Abstr. 061.017.

General cosmological solution of the gravitational equations with a singularity in time.
See Abstr. 066.009.

Determination of the radial velocity of a relativistic star. See Abstr. 066.054.

Stellar ages and an extended theory of gravitation. See Abstr. 066.060.

Datations terrestres et cosmiques. See Abstr. 081.006.

Measurement of primordial helium abundance from the star μ Cassiopeiae.   See Abstr. 117.005.

Apparent periodicities in the redshifts of quasi-stellar objects.   See Abstr. 141.054.

Pulsars and the universal X-ray background. See Abstr. 141.057.

An analysis of the redshift–apparent magnitude relation for 147 quasi-stellar objects.   See Abstr. 141.194.

The diffuse cosmic X-ray background radiation. See Abstr. 142.012.

Production of the diffuse background X-ray flux at 44 Å by suprathermal proton bremsstrahlung. See Abstr. 142.016.

Predicted high energy break in the isotropic gamma ray spectrum: A test of cosmological origin. See Abstr. 142.037.

Age of cosmic rays and abundance of antimatter in the Galaxy.   See Abstr. 143.014.

Origin of cosmic rays. I. Introduction. Metagalactic models.   See Abstr. 143.015.

The log S–log z diagram for radio galaxies and its relation to cosmology.   See Abstr. 158.102.

The dynamics of clusters of galaxies in universes with non-zero cosmological constant, and the virial theorem mass discrepancy.   See Abstr. 160.013.

Accretion of metagalactic gas clouds by galaxies and quasar-like phenomena.   See Abstr. 161.003.

Author Index

AARONS, J.
008.010
ABALAKIN, V. K.
007.000
047.001 .020
ABBASOV, G. I.
021.008
ABBEY, S.
094.059
ABDEL-GAWAD, M.
094.123
ABDUSSAMATOV, H. I.
072.018
ABE, O.
124.106
ABELE, M.
031.011
032.008
034.017
ABELE, M. K.
031.032
ABELL, G.
008.014 .075 .113 .114
ABELL, G. O.
158.050
ABELL, P. I.
094.157 .174
ABHYANKAR, K. D.
063.016 .017 .023 .026
.028
091.029
ABLES, H. D.
118.001
158.029 .112
ABLES, J. G.
141.170
ABRAHAM, H. J. M.
045.017
ABRAMENKO, A. N.
034.063 .064
ABRAMI, A.
074.007
075.011
077.039 .053 .054 .055
ABRAMJAN, G. L.
157.022

ABRAMOWICZ, M.
066.019 .047
141.085
ABT, H. A.
112.015
122.051
153.003
ACCARDO, C. A.
079.102
ACKER, A.
041.015
ACKERMAN, M.
055.015
ACKNER, J.
003.046
ACTON, L. W.
008.118
076.030
ADACHI, K.
124.106
ADAMI, L. H.
094.064
ADAMS, A. N.
041.036
ADAMS, J. A. S.
094.071
ADAMS, J. B.
094.146
098.019
ADAMS, L. A.
034.072
ADAMS, T. F.
155.010
ADCOCK, B. S.
010.008
ADE, P.
114.038
ADLER, I.
094.083
AERTS, E.
055.015
AFANAS'EV, V.
103.113
AFANAS'EV, V. L.
103.113

AFANASYEV, V. L.
034.032
AFFRONTI, F.
082.094
AFRICK, S.
126.005
AGADZHANOVA, P. A.
003.023
AGEKIAN, T. A.
117.007
151.009
AGESHIN, P. N.
073.097
AGGSON, V. T. L.
083.065
AGNESE, A.
162.063
AGRAWAL, S. P.
143.025
AGRELL, S. O.
094.081
AGRINIER, B.
141.068 .165
AGUEERO, E.
158.115
AGUILAR, M. L.
114.146 .148
AHLUWALIA, H. S.
073.095
AHNERT, P.
093.006
122.116
AHRENS, L. H.
094.021
AIKIN, A. C.
076.006
AITON, E. J.
004.036
AKASOFU, S.-I.
083.069
084.250
AKHMEDOV, S. B.
077.037
AKHUNDOVA, G. V.
133.010

401

APPARAO, M. V. K.
141.057
APPENZELLER, I.
065.060
131.032
APPLEMAN, D. E.
094.101
APUSHKINSKY, G. P.
077.049
ARAKELIAN, M. A.
141.044 .132 .185
158.059
ARAZOV, G. T.
042.020
ARENDT, P. R.
079.102
ARENSTORF, R. F.
042.077
ARGO, H. V.
076.017
ARGUE, A. N.
103.101
ARIFOV, L. YA.
.066.081
ARISKIN, V. I.
157.023
ARKHIPOVA, V. P.
122.109
133.009
ARMSTRONG, E. B.
082.042
ARMSTRONG, J. C.
084.401
ARMSTRONG, R. J.
083.019
ARMSTRONG, T. P.
073.101 .112
ARNDT, H.
094.277
ARNDT, J.
094.115
ARNETT, W. D.
065.046
125.013
ARNOLD, J. R.
094.078
ARNOLDY, R. L.
084.003
ARNQUIST, W. N.
074.006
ARNY, T.
152.008
ARNY, T. T.
158.047
ARONS, J.
142.021
161.001 .009
ARP, H.
141.031
158.002 .003 .067 .091
ARP, H. C.
158.028
ARRHENIUS, G.
051.009
094.111
ARRIENS, P. A.
094.035
ARSENAULT, J. L.
052.008
ARSENIJEVIC, J.
034.050

ARSENIJEVIC, J.
079.109
122.075
ARTHUR, E.
099.004
ARTIUKHINA, N. M.
152.015
ARTYUKHINA, N. M.
153.004
ASBRIDGE, J. R.
084.269
ASHBROOK, J.
005.004 .007
015.001
032.015
092.017
124.103 .106
ASHPOLE, E.
141.214
ASHWORTH, D. G.
081.029
091.005
ASIMOV, I.
003.058
ASSAF, T.
075.029
ASTAPOVICH, I. S.
104.017
ASTORGA, J. K.
162.067
ASUNMAA, S.
094.111
ATHAY, R. G.
071.029 .047
073.013
080.007
ATKINSON, W. C.
034.031
AUER, L. H.
064.032 .046 .051
AULD, D. R.
084.216
AUMANN, H. H.
155.009
AUSTIN, R. R. D.
121.045
AVE'LALLEMANT, H. G.
094.114
AVRAMCHUK, V. V.
099.043
AVRETT, E. H.
071.032
080.008
AXFORD, W. I.
084.277 .409
093.001
141.064
143.027
161.005
AZIMOV, S. M.
121.014
124.100
BABADZANOV, P. B.
104.054
BABADZHANJANZ, M. K.
158.103
BABADZHANOV, P. B.
003.009
104.045
BABADZHANYANZ, L. K.
042.067

BABADZHANYIANZ, M. K.
141.114
BABAEV, M. B.
103.101
124.103
BABAEV, M. V.
124.104
BABCOCK, H. W.
008.088
BABIJ, B. T.
071.049
BABKOV, V. K.
151.068
BACHOFER, B. T.
054.007
BACKER, D. C.
141.162
BACON, M. E.
022.070
BACSAK, G.
081.037
BAEDECKER, P. A.
094.047
105.107
BAEUMLER, P.
034.010 .037
BAGGALEY, W. J.
083.002
104.020
BAGROV, A. V.
082.088 .089
BAHCALL, J. N.
080.013
141.028
BAIER, F.
141.152
BAILEY, D. K.
084.010 .422
BAILEY, J. C.
094.084
BAILEY, J. M.
107.003
BAIRACHENKO, I. V.
104.027
BAJAJA, E.
141.234
155.032
157.030
BAKER, D.
053.002
BAKER, N. H.
065.030
BAKHAREV, A. M.
082.076
104.049
BAKHRAKH, L. D.
033.026 .027
BALASUBRAHMANYAN, V. K.
078.003
BALDANZA, B.
105.004
BALDWIN, B.
104.006 .008
BALDWIN, R. B.
094.212 .290
BALICK, B.
131.111
BALL, A. W. L.
054.012
BALL, J. A.
131.010 .054 .060 .073

BRUECK, M. T.
034.052
159.003
BRUECKNER, G.
073.059
BRUECKNER, G. E.
076.020
BRUIN, F.
075.029
BRUN, A.
007.000
123.008 .027
BRUNER JR., E. C.
071.031 .042
082.019
BRUNET, J.-P.
131.077
BRUNS, A. V.
076.022
BRUNSTEIN, K.
143.007
BRUWER, J. A.
103.102
BRUZEK, A.
073.069
074.054
BRYAN, W. B.
094.092
BUCAILLE, R.
036.002
BUCCHERI, L.
141.165
BUCCHERI, R.
141.068
BUCHA, V.
084.230
BUCHWALD, V. F.
105.014 .015
BUDDING, E.
072.010
BUDILOV, V. K.
143.054
BUDINE, P. W.
099.040 .057
BUERGER, P. F.
117.010
121.063
BUES, I.
126.010
BUETTNER, K. J. K.
105.076
BUFFONI, L.
054.014
BUHL, D.
073.045
077.029
093.028
094.297
131.044 .055 .056 .065
       .078 .108
BUKATA, R. P.
073.102
BULAVINA, V. I.
032.010
BULLEN, K. E.
091.038
BUNCH, T.
094.116
BUNCH, T. E.
094.086

BUNKER, R.
079.102
BUNNENBERG, E.
094.160
BURBIDGE, E. M.
008.014 .075 .113 .114
065.080
141.084 .125
151.010
158.033
BURBIDGE, G.
142.012
BURBIDGE, G. R.
065.080
114.051 .074
141.084
158.006 .102
BURCKHARDT, J. J.
004.037
BURGHES, D. N.
084.234
BURINSKAS, P. J.
053.010
BURKE, B. F.
033.044
131.075
141.211
157.011
BURKE, J. D.
094.173
BURKHEAD, M. S.
124.103
BURLAGA, L. F.
074.013 .018 .027
078.007
BURLINGAME, A. L.
094.153
BURNASEVA, B. A.
103.110 .113
BURNASOV, V.
113.009
BURNETT, D.
003.051
BURNETT, D. S.
094.030 .182 .183
105.082 .125
BURNETT, J. C.
022.076
BURNS, A. A.
094.186
BURNS, W. R.
033.010
BURROWS, K.
083.005
BURTON, W. B.
155.047
BURTON, W. M.
073.048
BUSCH, H.
122.113
123.024
BUSCOMBE, W.
114.045 .049 .109
BUSECK, P. R.
105.016
BUSELLI, G.
142.050
BUSLAVSKIJ, V. G.
073.018
BUSSE, F. H.
065.014

BUSTATI, N. G.
075.029
BUTCHER, M. E.
033.039
BUTKEVICH, A. V.
003.015
BUTLER, H. E.
012.014
013.010
BUTLER, J. C.
094.107
BUTLER, S. T.
003.091
BUTTMANN, G.
003.088
BYKOV, M. F.
034.058
BYRAM, E. T.
142.010
BYSTROVA, N. V.
131.098
157.019
CABY, M.
022.082
CACCIANI, A.
034.004
CADLE, R. D.
082.011
CAHILL JR., L. J.
084.276
CAHN, J. H.
133.015
CAIN, D. L.
097.047
CALAMAI, G.
065.005
CALAME, O.
094.229 .247
CALDWELL, P. A.
073.081
CALLAWAY, J.
022.039
CALPO, E. V.
047.008 .009 .013
CALVERT, H. R.
004.014
CALVIN, M.
094.153
CAMERON, A. G. W.
065.004 .071
105.087
107.015
155.006
CAMERON, D. J.
010.024
CAMERON, E. N.
094.097
CAMHY-VAL, C.
022.065
CAMICHEL, H.
092.016
CAMPBELL, D. B.
141.052 .065
CAMPBELL, J. W.
113.035
CAMPBELL, M. J.
094.132
CAMPISI CRISTALDI, R.
075.032
CAMPOLATTARO, A.
066.027

GRYGAR, J.
124.103
133.022
GRZEDZIELSKI, S.
074.023
GUBENKO, V. S.
083.036
GUEDUER, N.
103.101
GUELIN, M.
141.149
158.097
GUENTHER, A.
041.007
GUENTHER, O.
051.024
GUENTZEL-LINGNER, U.
002.028
051.012
GUERAULT, G.
094.247
GUERIN, P.
091.037
093.011
100.001
GUERTLER, J.
011.010
GUIDICE, D. A.
074.032
GUINAN, E. F.
101.002
GUINOT, B.
008.098
094.229
GUL'ELMI, A. V.
084.246
GULIELMI, A. V.
084.242 .283
GULKIS, S.
099.053
141.119 .213
GUNN, J. E.
012.018
141.130 .171
GUNTER, S. Z.
097.041
GUPTA, J. C.
084.287
GUREVICH, A. V.
084.048
GUREVICH, L. E.
065.021
GURSKY, H.
091.013
125.007
142.043
GURTOVENKO, E. A.
032.028
071.045
073.071
GUSEINOV, O. H.
064.058
065.093
155.061 .062
GUSEJNOV, R. EH.
073.031 .032
GUSEV, E. B.
103.101
124.103
GUS'KOVA, E.
105.108

GUSSOW, S.
021.011
GUTIERREZ-MORENO, A.
113.053 .063
114.143
121.040
GUTKIN, A. M.
094.007 .320 .333
GUTSCHEWSKI, G.
094.345
GUZMAN, J. S.
075.028
GYLDENKERNE, K.
113.024 .052
121.028 .029
124.109
HAACK, U.
094.074 .279
HAAR, D. TER
107.017
HABER, H.
003.127
HACHENBERG, O.
008.020
033.001
HACK, M.
007.000
HADDOCK, F. T.
077.022 .027
HADLOCK, R.
091.022
HAEGGKVIST, L.
113.025
HAEMEEN-ANTTILA, K. A.
092.016
HAENIG, W.
036.001
HAERENDEL, G.
084.260
HAERM, R.
154.012
HAERTEL, J. C.
008.101
HAEUSSLER, K.
122.113
HAFELE, J. C.
066.074
HAFFNER, H.
008.150
HAFNER, S. S.
094.122
HAGEDORN, R.
022.060
HAGEN-THORN, V. A.
158.014 .103
HAGGERTY, S. E.
094.092
HAINES, E. L.
094.076
HAINES, G. V.
084.271
HAIR, M.
094.074 .279
HAIT, M. H.
094.026
HAJDUK, A.
104.005
HAJKOWICZ, L. A.
084.047
HAKURA, Y.
073.103

HALACY JR., D. S.
003.150
HALL, A. J.
079.102
HALL, D. S.
121.003 .031
HALL, J. E.
079.102
HALL, J. S.
005.002
008.046
091.044
099.026
HALL, N. M.
021.005
HALLAM, K. L.
113.037
HALLBERG, F. C.
034.062
HALLIDAY, I.
065.054
097.055
105.007
HALMOS, F.
046.002
HALPERN, B.
094.159 .160
HAMILTON, N.
122.044
HAMILTON, P. A.
141.170
HAMILTON, P. B.
094.001 .163
HAMON, A.
010.028
HAN, J.
094.153
HANASZ, J.
021.007
033.004
HANBURY BROWN, R.
032.039
HANCOCK, D. A.
105.102
HANEL, R. A.
097.044
HANNAFORD, W.
084.271
HANNEMAN, R. E.
094.076
HANOR, J. S.
094.111
HANSEN, B.
141.216
HANSEN, J. R.
034.073
HANSEN, L.
113.052
HANSEN, R. T.
073.065
HANSEN, S.
073.065
HANSON, W. B.
084.208
HAPKE, B. W.
094.150
HAPPACH, V.
055.001
HARA, H.
032.007

HARA, T.
035.005 .006
HARADA, K.
094.162
HARAMURA, H.
094.091
HARANG, L.
084.038
HARDCASTLE, K.
094.064
HARDIE, R. H.
008.092
HARDING, G. A.
122.084
HARDORP, J.
064.015
102.013
HARDY, D. A.
015.002
HARDY, E.
122.121
HARE, P. E.
094.162
HARGRAVES, R. B.
094.100
HARGREAVES, J. K.
084.009
HARLAN, E. A.
098.030
114.012 .105
134.005
HARMANEC, P.
117.003 .030
HARMER, D. L.
114.001
HARNDEN JR., F. R.
082.008
141.131
142.015
HARO, G.
122.028 .029
153.021 .022
HARPER, R.
065.032
079.102
HARRIES, J. R.
134.016
HARRIS, B. J.
103.101 .102 .115 .118
.128
HARRIS, D. E.
033.054
106.011
141.094
HARRIS, K. K.
083.064
084.203 .257
HARRISON, A. W.
084.007
HARRISON, E. R.
141.001
162.013 .021 .022
HARRISON, J. K.
094.350
HARRISON, M. D.
083.010
HART JR., H. R.
094.076
HARTLE, R. E.
093.007

HARTMANN, G. K.
083.048
HARTMANN, W.
097.042
HARTMANN, W. K.
094.244
HARVEY, G. A.
077.007 .052
HARVEY, J.
071.036
080.004
HARVEY, J. W.
071.006 .013
HARWIT, M.
082.014 .078
106.019
131.025
HARWOOD, D.
103.101 .102 .115
HASEGAWA, A.
084.262
HASEGAWA, H.
143.013
HASEGAWA, I.
103.101
HASHMI, M.
022.048
HASKIN, L. A.
094.040
HASLAM, C. G. T.
157.004 .006
HASSAN, F.
094.022
HAUCK, B.
002.016
113.045
HAUG, A.
083.018 .058
HAUG, H.
100.006
HAUG, U.
113.022 .023
115.005
HAUGE, OE.
071.004
HAUPT, H.
007.000
008.148
047.017
092.006
HAURY, E. W.
084.230
HAVLICEK, K.
034.067
HAYAKAWA, S.
003.053
131.125
143.041
HAYASHI, C.
065.041 .069 .083
HAYAT, G. S.
031.035
HAYES, D. S.
114.005 .086
HAYES, J. M.
094.157 .174
HAYLI, A.
151.060
HAYMES, R. C.
082.008
141.131

HAYMES, R. C.
142.015
HAYS, P. B.
084.019
HAYWARD, R. R.
153.003
HAZARD, C.
141.119 .162
HAZLEHURST, J.
116.002
117.023
HEAP, S. R.
133.012
HEATH, A. W.
100.007
HEDEMAN, E. R.
072.035
073.091
HEDGEPETH, J. M.
033.045
HEESCHEN, D. S.
008.054
158.052 .095 .096
HEGGIE, D. C.
093.020
HEGYI, D.
117.005
HEIDMANN, N.
158.072
HEIKEN, G.
094.145 .206
HEIKKILA, W. J.
084.001
HEILES, C.
131.069 .070 .111
141.052 .065 .129
HEILES, C. E.
131.035
HEINRICH, K. J. F.
094.083
HEINTZ, W. D.
118.011
HEINTZE, J. R. W.
121.013
HEINTZMANN, H.
065.062
066.086
HEISE, J.
065.070
106.012
HEISER, A.
103.101
HEISER, A. M.
082.083 .092
HEISER, E.
122.022
HEITZ, S.
081.040
HELD, S.
022.025
HELDEN, R. VAN
114.022
HELIN, E. F.
098.029
HELLIWELL, R. A.
084.232
HELLYER, B.
098.012
105.111
HELM, T. M.
122.049

HUBE, D. P.
119.010
HUBER, M. C. E.
034.028
073.037
076.012
HUBERMAN, F. P.
064.012
HUDSON, H. S.
142.005 .025
HUEBNER, J. S.
094.101
HUEBNER, W. F.
102.012
HUEBSCHER, J.
124.104
HUEY, J. M.
094.038
HUFFMAN, D. R.
125.003
HUG, K.
094.227
HUGHES, J. A.
008.043
HUGHSTON, L. P.
162.029 .030
HUMMER, D. G.
008.022
064.001 .023
HUMPHREYS, R. M.
152.016
155.011 .054
HUMPHREYS, R. W.
114.120
HUNDHAUSEN, A. J.
074.004 .026
106.003
HUNEKE, J. C.
094.030
HUNG, R. J.
083.012
HUNGER, K.
008.015
064.007
HUNSTEAD, R. W.
141.157
HUNT, G. E.
091.020
HUNT, G. R.
094.235
HUNTEN, D. M.
082.032
093.025
097.006
099.016
HUNTER, C.
131.050 .074
151.039
HUNTER JR., J. H.
141.202
HUNTRESS JR., W. T.
094.133
HUNZIKER, R. R.
052.007
HURLESS, C.
124.103
HURLEY, P. M.
094.034
HURNIK, H.
008.103

HURUHATA, M.
044.027
082.082
HUSS, G. I.
105.027
HUTCHINGS, J. B.
064.024
114.004 .096 .099
124.103
131.090
HUTCHINSON, A. K.
074.047
HUTH, H.
123.032 .036
HYBL, V.
006.000
HYLAND, A. R.
122.053
124.103
HYNDS, R. J.
084.413
HYNEK, J. A.
008.045
IBADINOV, I. KH.
103.134
IBADINOV, KH.
102.017
IBANEZ, J.
094.161
IBEN JR., I.
065.012
154.010
IBRAEV, T. A.
143.064
ICHIMARU, S.
141.080
ICHIMURA, E.
124.103
ICHIMURA, K.
122.062 .098 .102
124.106
IDSO, S. B.
082.033
IERLEY, W. H.
077.032
IHOCHI, H.
094.038
IIJIMA, S.
041.010
IKHSANOV, R. N.
032.020
072.036
ILENCIK, J.
078.012
ILGACH, S. F.
143.058
IL'IN, V. A.
052.009 .025
ILL, M.
054.027
055.007
082.090
IL'YASOV, YU. I.
071.016
IMAI, H.
074.053
IMAI, K.
095.005
IMAZHANOVA, K.
143.057

IMNADZE, M. P.
042.015 .016
IMSHENNIK, V. S.
125.014
INA, T.
032.007
INDZHGIA, R. G.
074.052
INGEMANN-HILBERG, C.
022.006
INGERSOLL, A. P.
097.040
099.031
INGHAM, M. F.
155.008
INGRAM, D. S.
042.008
INNANEN, K. A.
151.025 .032
INOKUTI, Y.
099.060
INOUYE, G. T.
073.021
INOZEMTSEVA, O. I.
143.048 .058
INTRILIGATOR, D. S.
084.213 .273
IONESCU-PALLAS, N. J.
162.059
IOSHPA, B. A.
073.091
IPAVICH, F. M.
155.012
IPSER, J. R.
066.061
151.021
IRELAND, J. G.
131.014
IRIARTE ERRO, B.
113.014 .050 .051
153.005 .023
IRIE, M.
080.015
IRVINE, W. M.
008.003
091.002 .021 .031
IRVING, E.
045.001
ISAACSON, L.
022.072
ISAEV, S. I.
084.043
ISAEVA, R. N.
084.015
ISAMUTDINOV, SH. O.
104.046
ISHCHENKO, I. M.
122.013
ISHIDA, K.
131.126
ISHIKAWA, Y.
094.130
ISHIZUKA, T.
065.026
ISLAM, J. N.
065.025
ISLER, R. C.
022.014
ISLIK ENGIN, S.
114.031

Subject Index